# Stability Loss and Buckling Delamination

Surkay D. Akbarov

# Stability Loss and Buckling Delamination

## Three-Dimensional Linearized Approach for Elastic and Viscoelastic Composites

Springer

Surkay D. Akbarov
Faculty of Mechanics
Yildiz Technical University
Yildiz
Turkey

ISBN 978-3-642-30289-3 ISBN 978-3-642-30290-9 (eBook)
DOI 10.1007/978-3-642-30290-9
Springer Heidelberg New York Dordrecht London

Library of Congress Control Number: 2012944897



Printed on acid-free paper

Springer is part of Springer Science+Business Media (www.springer.com)

*To my father Djabbar Akbarov and my mother Teller Akbarova for their help with my education*

# Preface

This book is the first to focus on the development of the 3D linearized stability loss theory for investigation of internal stability loss in a structure of elastic and viscoelastic composite materials and of stability loss of elements of construction made from these materials within the scope of the initial imperfection approach (criterion). Under the initial imperfection approach we mean that the structure of the composite material or structural element has an initial imperfection and an evolution of this imperfection is studied by employing the 3D geometrically nonlinear field equations of the theory of elasticity and viscoelasticity for anisotropic bodies.

During the last 15 years, my students and I have investigated problems related to the 3D stability loss of cylinders, plates, buckling delamination of plates containing cracks, internal and near-surface stability loss of layered and fibrous elastic and viscoelastic composites, etc. These investigations allowed me to determine how the approximate stability loss theories of beams, plates, and shells can influence the accuracy of the results related to critical time. Moreover, these investigations allowed me to determine the rules of buckling delamination around cracks contained by plates.

This book deals with systematic consideration and analysis of the aforementioned plus related investigations. I assume that the reader is familiar with the theory of elasticity and viscoelasticity for anisotropic bodies, with partial differential equations and integral transformations, and also with the Finite Element Method.

I have designed the book for graduate researchers; for mechanical engineers designing composite materials for automobiles, trucks, flywheels, etc.; for civil engineers contemplating the use of composites in infrastructure; for aerospace engineers studying advanced airframe design; and for biomedical engineers developing lightweight composites for bone replacement and repair.

At the end of each chapter a corresponding reference list is provided.

Surkay D. Akbarov

# Acknowledgments

I must mention several individuals and organizations that were of enormous help in writing this book. First, I am grateful to Prof. Dr. A.N. Guz for his kindly support and encouragement. Also I wish to thank Prof. Dr. N. Yahnioglu for her invaluable help in writing this book.

I also wish to thank my scientific collaborators at Yildiz Technical University (Istanbul, Turkey), the Institute of Mathematics and Mechanics of the National Academy of Science of Azerbaijan (Baku, Azerbaijan) and the Institute of Mechanics of the National Academy of Science of Ukraine (Kiev, Ukraine), for their assistance in the investigations described in this book.

Surkay D. Akbarov

# Contents

# Abbreviations

| | |
|---|---|
| TDLTS | Three-Dimensional Linearized Theory of Stability |
| TSLC | Theoretical Strength Limit in Compression |
| FEM | Finite Element Method |
| 3D | Three-Dimensional |
| 2D | Two-Dimensional |

# Chapter 1
# Introduction

A brief background is given related to general questions on the creation and development of the Three-Dimensional Linearized Theory of Stability (TDLTS) of deformable solid bodies in the historical aspect. It is noted that the intensive study and applications of this theory were caused by investigations of the stability of thick-walled metal structures and thin-walled composite structures with low shear strength; fracture and delamination of composite and pre-stressed materials; stability of rubber-like materials and structural elements; stability of rock workings etc. The sources for development of the TDLTS for elastic and viscoelastic composites within the scope of the initial imperfection criterion are also outlined briefly. A text preview is given and the scope of each chapter of the book is discussed. The list of references contains only works related to the general issues of the TDLTS.

## 1.1 A General Background

As usual, under investigation of the stability loss of structural elements and in trying to simplify the solution of the problems, two- and one-dimensional stability loss theories based on various auxiliary hypotheses, are employed. However, the problems related to the stability of thick-walled metal structures and thin-walled composite structures with low shear strength; stability in the structure of layered and fibrous composites; stability of rubber-like materials and structural elements; stability of rock-workings; and stability loss related to wrinkles in the earth's crust, are only a few examples which are worth considering or which need to be investigated within the scope of the TDLTS, i.e. without reducing them to two or one-dimensional theories with the help of corresponding hypotheses. Some researchers use the notation "General Theory of Stability" instead of the TDLTS (see, for instance, Biezeno and Hencky 1929; Southwell 1913).

S. D. Akbarov, *Stability Loss and Buckling Delamination*,
DOI: 10.1007/978-3-642-30290-9_1, © Springer-Verlag Berlin Heidelberg 2013

The creation of the TDLTS began in the first half of the 20th century and the first attempt in this field was reported in a paper (Southwell 1913) in which the equations of the TDLTS on physical grounds were obtained for small pre-critical homogeneous deformations. Later in (Biezeno and Hencky 1929), again on physical grounds, the equations of the TDLTS were derived for an arbitrary non-homogeneous pre-critical stress–strain state. The equations derived in (Southwell 1913) and (Biezeno and Hencky 1929) in Cartesian and cylindrical co-ordinates, are also described in a book (Biezeno and Grammel 1939). Note that throughout this text, "small pre-critical state" will refer to the pre-critical stress–strain state as determined within the scope of the classical linear theory of elasticity.

Papers (Biot 1934, 1939) were the first in which the equations of the TDLTS are derived from the equations of the non-linear theory of elasticity by the linearization procedure under small pre-critical deformation. The investigations in the aforementioned papers were carried out in Lagrangian co-ordinates, which coincided with Cartesian coordinates in the un-deformed state. In these papers, under linearization it is understood that the magnitude of the perturbations (or increments) of the displacements, strains and stresses are small and their squares and higher order powers can be neglected in the corresponding equations and relations. The main results of the aforementioned and other numerous papers by Biot were summarized in a monograph (Biot 1965). However, in the aforementioned monograph the title "Mechanics of incremental deformation" was used instead of the TDLTS.

In a paper (Neuber 1943), the equations of the TDLTS were constructed in curvilinear co-ordinates with the use of tensor analyses. Moreover, in the aforementioned paper, equations of the TDLTS were also obtained in terms of the displacements for the particular form of a relation between the stresses and strains. At the same time, some methods were proposed for integration of these equations for the case where the pre-critical stress–strain state is homogeneous. It should be noted that the results obtained in the paper (Neuber 1943) for a particular case, i.e. in Cartesian co-ordinates, agree with the corresponding ones obtained in (Biot 1934, 1939, 1965).

In works (Trefftz 1931, 1933), the equations of the TDLTS for small pre-critical deformations were derived by employing the variational principle. Moreover, in these works sufficient conditions of stability were also obtained. In a paper (Kappus 1939), the equations of motion of the elasticity theory under large deformations in Lagrangian co-ordinates, before deformations, coincided with the Cartesian co-ordinates. Moreover, in a paper (Kappus 1939), the equations of the TDLTS were derived under finite (large) pre-critical deformation. These equations were then simplified for a particular form of elastic stress–strain relation and made the transition to the small pre-critical deformations theory.

Equations of the TDLTS for small pre-critical deformations were also detailed in a book (Novozhilov 1953) and these equations were then simplified for small rotation angles.

The most general form of equations for the TDLTS under finite pre-critical deformations for an isotropic elastic solid with an arbitrary form of elastic potential was derived in (Green et al. 1952). In this paper tensor representation was used and a number of particular cases were considered. At the same time, the linearized elasticity theory was named in this paper as the theory of small elastic deformations superimposed on finite elastic deformations.

It should be noted that equations of the TDLTS as well as corresponding three-dimensional linearized equations of motion in terms of stresses and boundary conditions derived in all the works reviewed above are common for all types of stress–strain relations. Consequently, these equations can be specified by substituting into them, the linearized constitutive equations for a particular type of stress–strain relation.

This completes the review of the investigations related to the TDLTS which were carried out in the first half of the 20th century. According to this review, we can conclude that in the first half of the 20th century equations of the TDLTS and corresponding linearized equations of motion were obtained in a rather rigorous way for finite and small pre-critical deformations.

In the second half of the 20th century the development of the TDLTS focused on general issues such as variational principles, the general problem statement, construction of solutions—their existence and uniqueness—and on solving particular groups of problems. The results obtained in these investigations were detailed in monographs (Biot 1965; Guz 1971, 1973, 1977, 1979, 1983, 1999) and others, and outlined in survey papers (Babich and Guz 1983, 2002; Babich et al. 1982, 2001; Bažant 1971; Guz 2001, 2003, 2004, 2009). Equations of the TDLTS and linearized equations of motion were also included in well-known books on continuum mechanics (Green and Adkins 1960, Novozhilov 1953, Truestell 1972; Eringen and Suhubi 1975). It should be noted that the modern level of the equations and relations of the TDLTS were detailed and developed in a monograph (Guz 1999).

It follows from the research in the foregoing and other numerous inquiries that applications of the TDLTS had been made mainly for stability of the elements of construction and for stability in the structure of the composite materials fabricated from time-independent materials. At the same time, these investigations were made using the bifurcation (Euler) approach. It is known that, in general, the Euler approach is not suitable for investigations of stability loss problems of elements of construction fabricated from time-dependent materials, nor stability loss in the structure of these materials under their quasi-static external loading. Therefore in (Guz 1999) to examine these problems, the dynamical investigation method was suggested. Note that under application of this method the sought values are presented with the multiplier $\exp(i\Omega t)$and the critical parameters of the considered problems are determined from the requirement $\mathrm{Im}\Omega = 0$. However, under application of the dynamical investigation method in the framework of the TDLTS other difficulties arise. The main difficulty arises from the complex modulus of elasticity or complex compliance coefficients instead of the ordinary modulus of elasticity or ordinary compliance coefficients respectively. Consequently, finding concrete numerical results becomes significantly more complicated. To the best knowledge of the author, up to the present, no research on stability loss problems on time-dependent materials in the framework of the TDLTS has been carried out with the use of the dynamical investigation method. Therefore, in (Guz 1999) for the investigation of these problems, the critical deformation method (Gerard and Gilbert 1958) was proposed, according to which, it was assumed that the critical deformation of the pure elastic and the corresponding viscoelastic problems are identical. Consequently, using the results of the pure elastic stability problems, the critical time is determined from the corresponding constitutive relations of the considered viscoelastic body. It is evident that the critical deformation method is a very approximate one and can be applied only in the cases where the pre-critical stress–strain state is homogeneous.

The very reliable and frequently used approach for the investigation of the stability loss of the elements of construction, as well as stability loss in the structure of composites fabricated from time dependent materials is based on the study of the evolution of the initial infinitesimal imperfections of the elements of construction and of the structure of composites with flow of time under fixed external compressive forces. Consequently, this approach is based on the stability loss criterion which can be called an analogy and generalization of the well-known stability loss criterion (Hoff 1954) of beams, plates and shells fabricated from time dependent materials for the TDLTS. The following situation made it possible for the author to create such an approach.

Under development of the mechanics of composites with curved structures within the scope of the three-dimensional geometrically non-linear exact equations of the theory of elasticity the boundary form perturbation method is employed. According to this method, sought functions are presented in a power series in a small parameter which characterizes the degree of the initial imperfections or curving of the reinforcing elements (Akbarov and Guz 2000). Substituting these power series into the aforementioned geometrically non-linear equations, and after doing some related mathematical operations, the complete system of equations and boundary and contact conditions are obtained for each approximation. Throughout these procedures the author observed that the equations and boundary and contact conditions derived for the first and subsequent approximations coincide with the corresponding ones of the TDLTS. Based on these observations, the aforementioned approach was presented for the first time by the author of this book and his students in a paper (Akbarov et al. 1997) for investigation of the stability loss in the structure of viscoelastic layered composites. Furthermore, in other papers (Akbarov 1998; Akbarov and Yahnioglu 2001) this approach was developed for the investigation of the stability loss of the plate-type elements of construction fabricated from viscoelastic composites. Since then these approaches have been developed and employed for investigation of a lot of problems related to the stability loss of element of construction, buckling delamination of the composite plates containing cracks, internal stability loss in the structure of layered and unidirectional elastic and viscoelastic composites etc. A review of these investigations carried out during the first seven years of the present century was made in a paper (Akbarov 2007). However, up to the present, there has not been any book related to the systematic consideration and detailing of the results of these investigations. The present book makes an attempt to fill, at least partly, this gap and could be used by graduate researchers; mechanical engineers designing composite materials for automobiles, trucks, flywheels etc. civil engineers contemplating the use of composites in infrastructure; aerospace engineers studying advanced airframe designs; and for biomedical engineers developing lightweight composites for bone replacement and repair.

## 1.2 Text Preview

This book consists of six chapters and two supplements. Chapter 1 is an introduction in which a brief, general background on the TDLTS is given and the reason for the creation of the TDLTS based on the initial imperfection criterion for time-dependent materials is detailed. Moreover, in this chapter a short description of the contents of each chapter of the book is given.

In Chap. 2, within the scope of the initial imperfection criterion, the TDLTS is developed and applied for investigations of the global and local stability loss of circular solid and hollow cylinders fabricated from viscoelastic composite materials. The study is made with the use of 3D geometrically-nonlinear field equations of the theory for viscoelastic orthotropic bodies. By the average-integrating procedure the corresponding approximate global (local) stability loss equations within the scope of the Bernoulli and third order refined beam (of the Kirchhoff-Love and third order shell) theories are derived from the corresponding equations of the TDLTS. The numerical results related to the critical forces and to the critical time obtained within the scope of the TDLTS and approximate approaches are presented and compared.

In Chap. 3, the stability loss problems related to rectangular and circular plates made from elastic and viscoelastic composites are investigated. Analytical, semi-analytical and numerical (2D and 3D FEM) methods are developed and employed for solutions to the corresponding boundary value problems. Each problem is formulated with the use of 3D geometrically nonlinear field equations of the theory of viscoelastic orthotropic bodies. Employing the boundary form perturbation technique the corresponding equations and relations of the TDLTS are derived. By the average-integrating procedure, the corresponding approximate stability loss equations are derived from equations of the TDLTS. The numerical results on the critical forces and on the critical time obtained within the scope of the equations of the TDLTS and approximate stability loss theories are presented and discussed.

Chapter 4 deals with problems related to the buckling delamination of elastic and viscoelastic plates. Both the continuum and piecewise homogeneous body models are employed to describe the plate materials. It is assumed that the plates contain a crack, the edges of which have an initial infinitesimal imperfection. It is also assumed that under the absence of the mentioned imperfections, the crack's edges are parallel to the free-face planes of the plate. With these assumptions, the evolution of the initial infinitesimal imperfections is studied under compression of the plate. This study, as in previous chapters, is performed with the use of the 3D geometrically nonlinear field equations of the theory of viscoelastic orthotropic bodies. Numerical results related to the plane-strain state and axi-symmetric and three-dimensional buckling delamination problems are presented and discussed.

In Chap. 5, the approach developed and employed in the previous chapters is extended for investigation within the scope of the piecewise homogeneous body model of the surface and inner stability loss in the structure of the elastic and viscoelastic layered composites. It is assumed that the composites consist of two alternating layers of which the materials are homogeneous, elastic and viscoelastic. Two types (local and periodic-global) of initial infinitesimal imperfections in the structure of the layered composites are considered. In addition, it is assumed that the initial imperfections of the layers are co-phase. As in the previous chapters, the evolution of the aforementioned initial imperfections is studied within the scope of the 3D geometrically nonlinear field equations of the theory of elasticity and viscoelasticity. Under formulation and solution of the corresponding boundary value problems the mechanics of curved composites developed in works listed in a survey paper (Akbarov and Guz 2004) and detailed in a monograph (Akbarov and Guz 2000) are used. Numerical results on the critical forces and critical time are presented and discussed.

In Chap. 6, the problems related to the internal and surface stability loss of the unidirected fibrous viscoelastic composites are investigated within the scope of the piecewise homogeneous body model with the use of the 3D geometrically nonlinear field equations of the theory of viscoelasticity for isotropic bodies. Concrete numerical results on the internal stability loss of a single, two neighboring and periodically located row of fibers in an infinite viscoelastic matrix are presented and discussed. Moreover, in this chapter the method for investigation of the stability loss of a fiber located near the convex cylindrical surface is presented and employed. Numerical results illustrating the influence of related parameters on the values of the critical forces and critical time obtained for the abovementioned near surface stability loss of a fiber are also presented and discussed.

In Supplement 1 we consider the application of the approach developed in Chap. 4 on the study of the influence of the initial stresses on the Energy Release Rate or Stress Intensity Factor at the crack tips contained by a plate-strip. Moreover, in this supplement the application of the approach developed in Chap. 4 on the study of the influence of the initial tension of a strip with a rectangular hole on the stress concentration caused by additional loading is given.

In Supplement 2 we analyze the numerical results related to the self-balanced stresses which appear as a result of the initial imperfections (periodical curving) of two neighboring fibers, as well as a periodically located row of fibers. The effect of accounting for the geometrical non-linearity and the effect of the interaction between the fibers on these stresses are also discussed. These results are obtained by use of the approach developed in Chap. 6.

# References

Akbarov SD (1998) On the three dimensional stability loss problems of elements of constructions fabricated from the viscoelastic composite materials. Mech Comp Mater 34(6):537–544

Akbarov SD (2007) Three-dimensional instability problems for viscoelastic composite materials and structural members. Int Appl Mech 43 (10):1069–1089

Akbarov SD, Guz AN (2000) Mechanics of curved composites. Kluwer Academic Publishers, Dordrecht

Akbarov SD, Guz AN (2004) Mechanics of curved composites and some related problems for structural members. Mech Advan Mater Struct 11(6):445–515

Akbarov SD, Yahnioğlu N (2001) A method of investigation of the general theory of stability problems of structural elements fabricated from the visco-elastic composite materials. Composites Part B. Engineering 30: 475-482

Akbarov SD, Sisman T, Yahnioğlu N (1997) On the fracture of the unidirectional composites in compression. Int J Eng Sci 35(12/13):1115–1136

Babich IYu, Guz A.N (1983) Stability of bars, plates, and shells of composite materials (three-dimensional formulation): Survey. Int Appl Mech 19(10):835–849

Babich IYu, Guz AN (2002) Stability of composite structural members (three-dimensional formulation). Int Appl Mech 38(9):1048–1075

Babich IYu, Guz AN, Shulga NA (1982) Investigation of the dynamics and stability of composite materials in a 3-dimensional formulation (survey). Int Appl Mech 18(1):1–21

Babich IYu, Guz AN, Chechov VN (2001) The three-dimensional theory stability of fibrous and laminated materials. Int Appl Mech 37(9):1103–1141

Bažant ZP (1971) Correlation study of formulation of incremental deformation and stability of continuous bodies. Trans ASME, Mech Eng Ser A (4):344–358

Biezeno CB, Grammel R (1939) Technische Dynamik. Springer-Verlag, Berlin

Biezeno CB, Hencky H (1929) On the general theory of elastic stability. In: Proceedings Koninklijke Nederlandse Akademie van Wetenschappen, vol 32. Amsterdam, pp 444–456

Biot MA (1934) Sur la stabilité de l'équilibre élastique. Equations de l'élasticité d'un milieu soumis a tention initially. Ann Soc Sci Sect B 54(1):91–109

Biot MA (1939) Nonlinear theory of elasticity and the linearized case for a body under initial stresses. Philos Mag Ser 7, 27:89–115

Biot MA (1965) Mechanics of incremental deformations. New York: Wiley

Eringen AC, Suhubi ES (1975) Elastodynamics, Vol. 1 Finite motions. New York: Academic Press

Gerard F, Gilbert A (1958) A critical strain approach to creep buckling of plates and shells. J Aeroanaut Sci 25(7):429–438

Green AE, Adkins JE (1960) Large elastic deformations and non-linear continuum mechanics. Clarendon Press, Oxford

Green AE, Rivlin RS, Shield RT (1952) General theory of small elastic deformations superimposed on finite elastic deformations. Proceedings of the Royal Society A 211(1104):128–154

Guz AN (1971) Stability of three-dimensional deformable bodies (in Russian). Naukova Dumka, Kiev

Guz AN (1973) Stability of elastic bodies under finite deformations (in Russian). Naukova Dumka, Kiev

Guz AN (1977) Fundamentals of the theory of stability of mine workings (in Russian). Naukova Dumka, Kiev

Guz AN (1979) Stability of elastic bodies under all-round compression (in Russian). Naukova Dumka, Kiev

Guz AN (1983) Mechanics of brittle fracture of materials with initial stresses (in Russian). Naukova Dumka, Kiev

Guz AN (1999) Fundamentals of the Three-Dimensional Theory of Stability of Deformable Bodies. Springer-Verlag, Berlin Heideberg, 1999

Guz AN (2001) Constructing of the three-dimensional theory of stability of deformable bodies. Int Appl Mech 37(1):1–38

Guz AN (2003) Establishing the fundamentals of the theory of stability of mine workings. Int Appl Mech 39(1):20–48

Guz AN (2004) Design models in linearized solid mechanics. Int Appl Mech 40(5):506–516

Guz AN (2009) On the study of nonclassical problems of fracture mechanics and related mechanisms. Int Appl Mech 45(1):3–40

Hoff NJ (1954) Buckling and stability, Journal of the Royal Aeronautical Society, 58(1)

Kappus R (1939) Zur elastizitätstheorie endlicher verschiebungen. ZAMM 19(5):271–285, 19(6):344–361

Neuber H (1943) Die grundgleichungen der elastichen stabilat in allgemeinen koordinaten und ihre integration. ZAMM 23(6):63–82

Novozhilov VV (1953) Foundations of the nonlinear theory of elasticity. Graylock Press,Rochester, New York

Southwell RV (1913) On the general theory of elastic stability. Philosophical Transaction of the Royal Society of London, Serial A, 213:187–244

Trefftz E (1931) Über die ableitung der elastischen gleichgewichts aus der elastizitätstheorie end licher deformation. In: Proceedings of the 3rd international congress for applied mechanics, vol 3. Stockholm, pp 44–50

Trefftz E (1933) Zur theorie der stabilität des elastichen gleichgewichts. ZAMM 12(2):160–165

Truestell C (1972) A first course in rational continuum mechanics. The Johns Hopkins University, Baltimor, Maryland

# Chapter 2
# Stability Loss Problems Related to Solid and Hollow Circular Cylinders

In this chapter the Three-Dimensional Linearized Theory of Stability (TDLTS) is developed and applied to the investigation of global and local stability loss of circular solid and hollow cylinders fabricated from viscoelastic composite materials. The initial imperfection criterion is used for determination of the values of the critical parameters. The study is carried out within the scope of the 3D geometrically non-linear field equations of the theory for viscoelastic orthotropic bodies. The approach is developed and applied for a solution to the corresponding system of non-linear integro-differential equations. By the average-integrating procedure the corresponding approximate global (local) stability loss equations within the scope of the Bernoulli–Euler and the third order refined beam (of the Kirchhoff–Love and the third order shell) theories are derived from the corresponding TDLTS equations. The numerical results related to the critical forces and to the critical times which are obtained within the scope of the TDLTS and approximate approaches are presented and discussed.

The results detailed in the present chapter are based on the investigations carried out in papers (Akbarov and Karakaya 2011, 2012a, b)

## 2.1 Formulation of the Problem Related to the Global Stability Loss

We consider a circular cylinder which has an initial global imperfection in the natural state and determine the position of the points of this cylinder with the Lagrange coordinates in the cylindrical $Or\theta z$ and the Cartesian system of coordinates $Ox_1x_2x_3$ (Fig. 2.1). We differ two cases: in Case 1 (Fig. 2.1a), in Case 2 (Fig. 2.1b) we assume that the cylinder is solid (hollow) one.

The noted initial global imperfection is given through the following equation of the "middle line" of the cylinders

S. D. Akbarov, *Stability Loss and Buckling Delamination*,
DOI: 10.1007/978-3-642-30290-9_2, © Springer-Verlag Berlin Heidelberg 2013

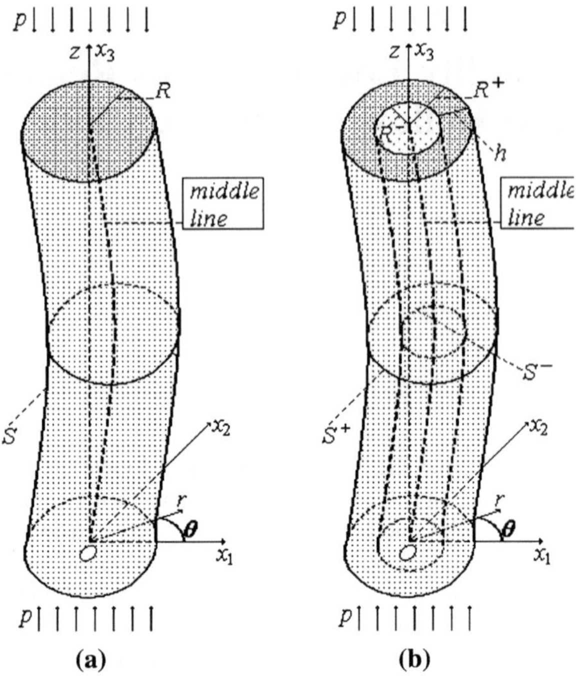


**Fig. 2.1** The geometry of the considered cylinders: **a** solid cylinder, **b** hollow cylinder

$$x_3 = t_3;\ x_1 = L\sin\left(\frac{\pi}{\ell}t_3\right);\ x_2 = 0 \tag{2.1}$$

where $t_3$ is a parameter and $t_3 \in (0, \ell)$, $L$ is the amplitude of the initial imperfection form.

In Case 2 (in Case 1) we assume that the cylinders' cross section is on the plane which is perpendicular to its middle line tangent vector, is a closed circular ring of the constant inner radius $R^-$ and outer radius $R^+$ (a circle of the constant radius $R$). Moreover, we assume that $L \ll l$ and introduce the small parameter

$$\varepsilon = \frac{L}{\ell},\ 0 \leq \varepsilon \ll 1. \tag{2.2}$$

We suppose that the material of the cylinders is viscoelastic transversal isotropic the symmetry axis of which coincides with the $Ox_3$ $(Oz)$ axis. Within the foregoing assumptions, we investigate the evolution of the infinitesimal initial global imperfection of the cylinder with time in the case where the cylinder is loaded by uniformly distributed normal compressing forces with intensity $p$ acting on the ends of the cylinder in the direction of the and $Oz$ axis.

This investigation is made within the scope of the following field equations.

$$\begin{aligned}
&\frac{\partial t_{rr}}{\partial r}+\frac{\partial t_{\theta r}}{r\partial\theta}+\frac{\partial t_{zr}}{\partial z}+\frac{1}{r}(t_{rr}-t_{\theta\theta})=0,\\
&\frac{\partial t_{r\theta}}{\partial r}+\frac{\partial t_{\theta\theta}}{r\partial\theta}+\frac{1}{r}(t_{\theta r}+t_{r\theta})+\frac{\partial t_{z\theta}}{\partial z}=0,\\
&\frac{\partial t_{rz}}{\partial r}+\frac{\partial t_{\theta z}}{r\partial\theta}+\frac{1}{r}t_{rz}+\frac{\partial t_{zz}}{\partial z}=0
\end{aligned} \tag{2.3a}$$

$$\begin{aligned}
t_{rr}&=\sigma_{rr}\left(1+\frac{\partial u_r}{\partial r}\right)+\sigma_{r\theta}\left(\frac{\partial u_r}{r\partial\theta}-\frac{u_\theta}{r}\right)+\sigma_{rz}\frac{\partial u_r}{\partial z},\\
t_{r\theta}&=\sigma_{rr}\frac{\partial u_\theta}{\partial r}+\sigma_{r\theta}\left(1+\frac{\partial u_\theta}{r\partial\theta}+\frac{u_r}{r}\right)+\sigma_{rz}\frac{\partial u_\theta}{\partial z},\\
t_{rz}&=\sigma_{rr}\frac{\partial u_z}{\partial r}+\sigma_{r\theta}\frac{\partial u_z}{r\partial\theta}+\sigma_{rz}\left(1+\frac{\partial u_z}{\partial z}\right),\\
t_{\theta r}&=\sigma_{\theta r}\left(1+\frac{\partial u_r}{\partial r}\right)+\sigma_{\theta\theta}\left(\frac{\partial u_r}{r\partial\theta}-\frac{u_\theta}{r}\right)+\sigma_{\theta z}\frac{\partial u_r}{\partial z},\\
t_{\theta\theta}&=\sigma_{\theta r}\frac{\partial u_\theta}{\partial r}+\sigma_{\theta\theta}\left(1+\frac{\partial u_\theta}{r\partial\theta}+\frac{u_r}{r}\right)+\sigma_{\theta z}\frac{\partial u_\theta}{\partial z},\\
t_{\theta z}&=\sigma_{\theta r}\frac{\partial u_z}{\partial r}+\sigma_{\theta\theta}\frac{\partial u_z}{r\partial\theta}+\sigma_{\theta z}\left(1+\frac{\partial u_z}{\partial z}\right),\\
t_{zr}&=\sigma_{zr}\left(1+\frac{\partial u_r}{\partial r}\right)+\sigma_{z\theta}\left(\frac{\partial u_r}{r\partial\theta}-\frac{u_\theta}{r}\right)+\sigma_{zz}\frac{\partial u_r}{\partial z},\\
t_{z\theta}&=\sigma_{zr}\frac{\partial u_\theta}{\partial r}+\sigma_{z\theta}\left(1+\frac{\partial u_\theta}{r\partial\theta}+\frac{u_r}{r}\right)+\sigma_{zz}\frac{\partial u_\theta}{\partial z},\\
t_{zz}&=\sigma_{zr}\frac{\partial u_z}{\partial r}+\sigma_{z\theta}\frac{\partial u_z}{r\partial\theta}+\sigma_{zz}\left(1+\frac{\partial u_z}{\partial z}\right)
\end{aligned} \tag{2.3b}$$

$$\begin{aligned}
\varepsilon_{rr}&=\frac{\partial u_r}{\partial r}+\frac{1}{2}\left\{\left(\frac{\partial u_r}{\partial r}\right)^2+\left(\frac{\partial u_\theta}{\partial r}\right)^2+\left(\frac{\partial u_z}{\partial r}\right)^2\right\},\\
\varepsilon_{r\theta}&=\frac{1}{2}\left(\frac{\partial u_\theta}{\partial r}+\frac{\partial u_r}{r\partial\theta}-\frac{u_\theta}{r}\right)+\frac{1}{2}\left\{\frac{\partial u_r}{\partial r}\left(\frac{\partial u_r}{r\partial\theta}-\frac{u_\theta}{r}\right)+\frac{\partial u_\theta}{\partial r}\left(\frac{\partial u_\theta}{r\partial\theta}+\frac{u_r}{r}\right)+\frac{\partial u_z}{\partial r}\frac{\partial u_z}{r\partial\theta}\right\}\\
\varepsilon_{rz}&=\frac{1}{2}\left(\frac{\partial u_r}{\partial z}+\frac{\partial u_z}{\partial r}\right)+\frac{1}{2}\left\{\frac{\partial u_r}{\partial r}\frac{\partial u_r}{\partial z}+\frac{\partial u_\theta}{\partial r}\frac{\partial u_\theta}{\partial z}+\frac{\partial u_z}{\partial r}\frac{\partial u_z}{\partial z}\right\}\\
\varepsilon_{\theta\theta}&=\frac{\partial u_\theta}{r\partial\theta}+\frac{u_r}{r}+\frac{1}{2}\left\{\left(\frac{\partial u_r}{r\partial\theta}-\frac{u_\theta}{r}\right)^2+\left(\frac{\partial u_\theta}{r\partial\theta}+\frac{u_r}{r}\right)^2+\frac{1}{r^2}\left(\frac{\partial u_z}{\partial\theta}\right)^2\right\}\\
\varepsilon_{\theta z}&=\frac{1}{2}\left(\frac{\partial u_z}{r\partial\theta}+\frac{\partial u_\theta}{\partial z}\right)+\frac{1}{2}\left\{\frac{\partial u_r}{\partial z}\left(\frac{\partial u_r}{r\partial\theta}-\frac{u_\theta}{r}\right)+\frac{\partial u_\theta}{\partial z}\left(\frac{\partial u_\theta}{r\partial\theta}+\frac{u_r}{r}\right)+\frac{1}{r}\frac{\partial u_z}{\partial\theta}\frac{\partial u_z}{\partial z}\right\},\\
\varepsilon_{zz}&=\frac{\partial u_z}{\partial z}+\frac{1}{2}\left\{\left(\frac{\partial u_r}{\partial z}\right)^2+\left(\frac{\partial u_\theta}{\partial z}\right)^2+\left(\frac{\partial u_z}{\partial z}\right)^2\right\}
\end{aligned} \tag{2.3c}$$

Here the Eq. (2.3a) is equilibrium equation in terms of the physical components $t_{rr}, t_{\theta\theta}, t_{zz}, t_{r\theta}, t_{\theta r}, t_{z\theta}, t_{\theta z}, t_{zr}$ and $t_{rz}$ of the non-symmetric Kirchhoff stress tensor, the Eq. (2.3b) shows the relation between the physical components of non-symmetric Kirchhoff stress tensor and the physical components of the ordinary stress tensor in the cylindrical system of coordinates, the non-linear relation between the physical components $\varepsilon_{rr}, \varepsilon_{\theta\theta}, \varepsilon_{zz}, \varepsilon_{r\theta}, \varepsilon_{\theta r}, \varepsilon_{z\theta}, \varepsilon_{\theta z}, \varepsilon_{zr}$ and $\varepsilon_{rz}$ of the Green strain tensor and the physical components $u_r$, $u_\theta$ and $u_z$ of the displacement vector is illustrated by the Eq. (2.3c).

The constitutive relations for the cylinders material in the cylindrical system of coordinates are given as follows:

$$\begin{aligned}
&\sigma_{rr} = A_{11}^*\varepsilon_{rr} + A_{12}^*\varepsilon_{\theta\theta} + A_{13}^*\varepsilon_{zz};\ \sigma_{\theta\theta} = A_{12}^*\varepsilon_{rr} + A_{11}^*\varepsilon_{\theta\theta} + A_{13}^*\varepsilon_{zz};\\
&\sigma_{zz} = A_{13}^*\varepsilon_{rr} + A_{13}^*\varepsilon_{\theta\theta} + A_{13}^*\varepsilon_{zz};\ \sigma_{r\theta} = \left(A_{11}^* - A_{22}^*\right)\varepsilon_{r\theta};\\
&\sigma_{rz} = 2G^*\varepsilon_{rz};\ \sigma_{\theta z} = 2G^*\varepsilon_{\theta z},
\end{aligned} \tag{2.4}$$

where $A_{ij}^*$ and $G^*$ are following operators

$$\left\{ \begin{matrix} A_{ij}^* \\ G^* \end{matrix} \right\}\phi(t) = \left\{ \begin{matrix} A_{ij0} \\ G_0 \end{matrix} \right\}\phi(t) + \int_0^t \left\{ \begin{matrix} A_{ij1}(t-\tau) \\ G_1(t-\tau) \end{matrix} \right\}\phi(\tau)d\tau \tag{2.5}$$

Here, $A_{ij0}$ and $G_0$ are the instantaneous values of elastic constants, $A_{ij1}(t)$ and $G_1(t)$ are the given functions which determine the hereditary properties of the cylinder material.

In the present section, all further discussions and considerations we will make for the hollow cylinder from which by obvious changing the corresponding ones can be obtained for the solid cylinder.

Assume that on the lateral surfaces $S^\pm$ of the hollow cylinder the following conditions are satisfied

$$\begin{aligned}
&t_{rr}|_{S^\pm}n_r^\pm + t_{\theta r}|_{S^\pm}n_\theta^\pm + t_{zr}|_{S^\pm}n_z^\pm = 0,\ t_{r\theta}|_{S^\pm}n_r^\pm + t_{\theta\theta}|_{S^\pm}n_\theta^\pm + t_{z\theta}|_{S^\pm}n_z^\pm = 0,\\
&t_{rz}|_{S^\pm}n_r^\pm + t_{\theta z}|_{S^\pm}n_\theta^\pm + t_{zz}|_{S^\pm}n_z^\pm = 0
\end{aligned} \tag{2.6}$$

where $n_r^+$, $n_\theta^+$ and $n_z^+(n_r^-$, $n_\theta^-$ and $n_z^-)$ are the components of an unit normal vector to the outer (inner) lateral surface $S^+$ $(S^-)$ of the hollow cylinder.

In the natural state, the upper and lower ends of the cylinder are on the inclined planes with the following unit normal vectors

$$\begin{aligned}
&\mathbf{n}_0 = \frac{-\mathbf{k} - \varepsilon\pi\mathbf{i}}{\sqrt{1+\varepsilon^2\pi^2}} \text{(for the lower end plane)};\\
&\mathbf{n}_\ell = \frac{\mathbf{k} - \varepsilon\pi\mathbf{i}}{\sqrt{1+\varepsilon^2\pi^2}} \text{ (for the upper end plane)},
\end{aligned} \tag{2.7}$$

where $\mathbf{n}_0$ $(\mathbf{n}_\ell)$ is an unit normal vector to the end plane of the cylinder at $z = 0(z = \ell)$, $\mathbf{i}$ and $\mathbf{k}$ ort-unit vectors in the $Ox_3$ and $Ox_1$ axes directions respectively.

Denote the upper (lower) end cross section of the cylinder through $S_\ell$ ($S_0$) and the conditions for the forces on these end cross sections we write as follows:

$$\begin{aligned} t_{rz}|_{S_0} n_{01} + t_{\theta z}|_{S_0} n_{02} + t_{zz}|_{S_0} n_{03} &= p, \\ t_{rz}|_{S_\ell} n_{\ell 1} + t_{\theta z}|_{S_\ell} n_{\ell 2} + t_{zz}|_{S_\ell} n_{\ell 3} &= -p, \end{aligned} \tag{2.8}$$

where $n_{0j}$ $(n_{\ell j})$ is a component of the unit normal vector defined in (2.7). The end conditions for the displacements will be discussed below.

This completes the formulation of the considered problem. It follows from this formulation that the evolution of the infinitesimal initial global imperfection of the cylinder with time for the fixed value of the compressing force $p$ (for the case where the material of the cylinder is viscoelastic) or with initial compressing force $p$ (for the case where the material of the cylinder is pure elastic) will be investigated within the framework of the field equations (2.3), (2.4) and (2.5) and boundary condition (2.6) and (2.8).

## 2.2 Method of Solution for the Global Stability Loss Problem

Now we consider the method of solution of the problem formulated in the previous section. Note that the method employed below can be briefly summarized as follows:

i. By employing the boundary-form perturbation techniques the considered boundary value problem for the non-linear integro-differential Eqs. (2.3)–(2.5) is reduced to the series boundary-value problems for the corresponding system of the linear integro-differential equations;
ii. Owing to both the expressions of the operators (2.5) and the convolution theorem, by using the Laplace transformation with respect to time these series problems for the system of the linear integro-differential equations are reduced to the corresponding series boundary value problems for the linear system of differential equations in the Laplace transformation parameter space;
iii. For each fixed value of the Laplace transformation parameter the linear problems are solved by employing the variable-separation method;
iv. By applying the inverse transformation method (Schapery 1962, 1978) we determine the original of the sought values.

It should be noted that for the case where the material of the cylinder is pure elastic, the operators (2.5) are replaced by mechanical constants and therefore instead of the integro-differential equations we obtain differential equations and the corresponding problems for these equations are also investigated in the framework of the above procedure but without employing the Laplace transformation.

Since, according to the procedure summarized above and to the problem statement, first we derive the equation for the inner and outer lateral surfaces $S^{+}$ and $S^{-}$ of the cylinder. According to the condition of the cylinder's cross section we can conclude that the coordinates of these surfaces must simultaneously satisfy the following equations.

$$\begin{gathered}\varepsilon f'(t_3)(x_{10}^{\pm} - \varepsilon f(t_3)) + x_{30}^{\pm} - t_3 = 0, \\ (x_{20}^{\pm})^2 + (x_{30}^{\pm} - t_3)^2 + (x_{10}^{\pm} - \varepsilon f(t_3))^2 = (R^{\pm})^2, \end{gathered} \tag{2.9}$$

where $f(t_3) = \ell \sin(\pi t_3/\ell)$, $f'(t_3) = \pi \cos(\pi t_3/\ell)$; $x_{10}^{\pm}$, $x_{20}^{\pm}$, $x_{30}^{\pm}$ are coordinates of the surface $S^{\pm}$.

Note that the first equation in (2.9) is an equation of the plane perpendicular to the vector which is the tangent vector to the middle line of the cylinder at the point that corresponds to the fixed value of the parameter $t_3$; but the second equation in (2.9) is an equation of the outer and inner circles with radiuses $R^{+}$ and $R^{-}$ in turn which are counter to the cross section of the hollow cylinder which rises on the foregoing plane. It is also assumed that the thickness $h = R^{+} - R^{-}$ is constant along the cylinder (Fig. 2.1).

Using the relations $x_{10}^{\pm} = r^{\pm}\cos\theta$ and $x_{20}^{\pm} = r^{\pm}\sin\theta$ we obtain the following equation for the surface $S^{\pm}$ in the cylindrical system of coordinates $Or\theta z$:

$$\begin{aligned} r^{\pm} =& r^{\pm}(\theta, t_3, \varepsilon) = \varepsilon f(t_3)\cos\theta \frac{1+\varepsilon^2(f'(t_3))^2}{1+\varepsilon^2(f'(t_3))^2\cos^2\theta} \\ &+ \left\{ \frac{(R^{\pm})^2 - \varepsilon^2(f(t_3))^2\left(1+\varepsilon^2(f'(t_3))^2\right)}{1+\varepsilon^2(f'(t_3))^2\cos^2\theta} \right. \\ &\left. + \varepsilon^2(f(t_3))^2\cos^2\theta \frac{\left(1+\varepsilon^2(f'(t_3))^2\right)}{\left(1+\varepsilon^2(f'(t_3))^2\cos^2\theta\right)^2} \right\}^{\frac{1}{2}} \\ z^{\pm} =& t_3 - \varepsilon f'(t_3)\left(r^{\pm}(\theta, t_3, \varepsilon) - \varepsilon f(t_3)\right),\ f'(t_3) = \frac{df(t_3)}{dt_3}. \end{aligned} \tag{2.10}$$

Using the assumption (2.2) and supposing that $(\varepsilon f'(t_3))^2 \ll 1$, after some mathematical manipulations, we obtain the following equations.

$$\begin{aligned} r^{\pm} &= R^{\pm} + \varepsilon f(t_3)\cos\theta + O(\varepsilon^2),\ z^{\pm} = t_3 - \varepsilon R^{\pm} f'(t_3)\cos\theta + O(\varepsilon^2), \\ n_r^{\pm} &= \pm\left(1 - \varepsilon^2\left(R^{\pm} f''(t_3) - \frac{f(t_3)}{R^{\pm}}\right)^2 + O(\varepsilon^2)\right), \\ n_\theta^{\pm} &= \pm\left(\varepsilon \frac{f(t_3)}{R^{\pm}}\sin\theta + O(\varepsilon^2)\right), \\ n_z^{\pm} &= \pm(-\varepsilon f'(t_3)\cos\theta + O(\varepsilon^2)) \end{aligned} \tag{2.11}$$

where $n_r^{\pm}$, $n_\theta^{\pm}$, $n_z^{\pm}$ are physical components of the unit normal vector to the surface $S^{\pm}$.

Now we write the equation of the planes on which lay the lower and upper inclined ends of the cylinder as follows.

$$\begin{aligned} x_3 &= -\varepsilon\pi x_1 \text{ (for the lower end),} \\ x_3 &= \varepsilon\pi x_1 + \ell \text{ (for the upper end).} \end{aligned} \tag{2.12}$$

According to the Eq. (2.7), we can also present the expression of the components of the normal vectors to these ends as follows

$$n_{01} = n_{\ell 1} = -\varepsilon\pi\left(1 - \frac{1}{2}(\varepsilon\pi)^2 + O(\varepsilon\pi)^4\right),\ n_{03} = -\left(1 - \frac{1}{2}(\varepsilon\pi)^2 + O(\varepsilon\pi)^4\right),$$

$$n_{\ell 3} = \left(1 - \frac{1}{2}(\varepsilon\pi)^2 + O(\varepsilon\pi)^4\right),\ n_{02} = n_{\ell 2} = 0. \tag{2.13}$$

According to the procedures of the boundary perturbation technique, we attempt to solve the considered problem by employing the boundary form perturbation method. For this purpose the unknowns are presented in series form in ε (2.2).

$$\left\{\sigma_{(ij)}; \varepsilon_{(ij)}; u_{(i)}\right\} = \sum_{q=0}^{\infty} \varepsilon^q \left\{\sigma_{(ij)}^{(q)}; \varepsilon_{(ij)}^{(q)}; u_{(i)}^{(q)}\right\}, \tag{2.14}$$

where $(ij) = rr, \theta\theta, zz, r\theta, rz, \theta z$ and $(i) = r, \theta, z$.

Substituting Eq. (2.14) into Eq. (2.3a–2.3c), we obtain set equations for each approximation (2.14). Using Eq. (2.11) we expand the values of each approximation (2.14) in series form in the vicinity of the points $\{r = R^+;\ z_0 = t_3\}$ and $\{r = R^-;\ z_0 = t_3\}$. Substituting these last expressions in the boundary conditions in (2.6) and using the expressions of $n_r^{\pm}$, $n_\theta^{\pm}$ and $n_z^{\pm}$ given in (2.11), after some mathematical transformations we obtain boundary conditions which are satisfied at $\{r = R^+;\ z_0 = t_3\}$ (for outer surface $S^+$) and at $\{r = R^-;\ z_0 = t_3\}$ (for inner surface $S^-$) for each approximation in Eq. (2.14). It is evident that for the zeroth approximation, Eq. (2.3a–2.3c) is valid and condition (2.6) is replaced by the same one satisfied at points $\{r = R^+;\ z_0 = t_3\}$ and $\{r = R^-;\ z_0 = t_3\}$. We assume that the non-linear parts of the strain tensor components are very small and can be ignored with respect to their linear parts. According to this assumption, for the zeroth approximation, we obtain the following system of equations:

$$
\begin{aligned}
&\frac{\partial\sigma_{rr}^{(0)}}{\partial r}+\frac{\partial\sigma_{\theta r}^{(0)}}{r\partial\theta}+\frac{\partial\sigma_{zr}^{(0)}}{\partial z}+\frac{1}{r}\left(\sigma_{rr}^{(0)}-\sigma_{\theta\theta}^{(0)}\right)=0,\\
&\frac{\partial\sigma_{r\theta}^{(0)}}{\partial r}+\frac{\partial\sigma_{\theta\theta}^{(0)}}{r\partial\theta}+\frac{2}{r}\sigma_{\theta r}^{(0)}+\frac{\partial\sigma_{z\theta}^{(0)}}{\partial z}=0,\\
&\frac{\partial\sigma_{rz}^{(0)}}{\partial r}+\frac{\partial\sigma_{\theta z}^{(0)}}{r\partial\theta}+\frac{1}{r}\sigma_{rz}^{(0)}+\frac{\partial\sigma_{zz}^{(0)}}{\partial z}=0,\\
&\varepsilon_{rr}^{(0)}=\frac{\partial u_r^{(0)}}{\partial r},\ \varepsilon_{r\theta}^{(0)}=\frac{1}{2}\left(\frac{\partial u_\theta^{(0)}}{\partial r}+\frac{\partial u_r^{(0)}}{r\partial\theta}-\frac{u_\theta^{(0)}}{r}\right),\ \varepsilon_{rz}^{(0)}=\frac{1}{2}\left(\frac{\partial u_r^{(0)}}{\partial z}+\frac{\partial u_z^{(0)}}{\partial r}\right),\\
&\varepsilon_{\theta\theta}^{(0)}=\frac{\partial u_\theta^{(0)}}{r\partial\theta}+\frac{u_r^{(0)}}{r},\ \varepsilon_{\theta z}^{(0)}=\frac{1}{2}\left(\frac{\partial u_z^{(0)}}{r\partial\theta}+\frac{\partial u_\theta^{(0)}}{\partial z}\right),\ \varepsilon_{zz}^{(0)}=\frac{\partial u_z^{(0)}}{\partial z},
\end{aligned}
\tag{2.15}
$$

and boundary conditions

$$
\sigma_{rr}^{(0)}\Big|_{r=R^{\pm}}=0,\ \sigma_{r\theta}^{(0)}\Big|_{r=R^{\pm}}=0,\ \sigma_{rz}^{(0)}\Big|_{r=R^{\pm}}=0. \tag{2.16}
$$

Moreover, we obtain the following end conditions for the zeroth approximation from (2.7), (2.8) and (2.13).

$$
\sigma_{zz}^{(0)}(r,\theta,0)=\sigma_{zz}^{(0)}(r,\theta,\ell)=-p. \tag{2.17}
$$

Note that the mathematical procedure, according to which the end condition (2.17) is obtained, will be given below.

Taking the last assumption into account, for the subsequent approximations we obtain the following system of equations.

$$
\begin{aligned}
&\frac{\partial\sigma_{rr}^{(q)}}{\partial r}+\frac{\partial\sigma_{\theta r}^{(q)}}{r\partial\theta}+\frac{\partial\sigma_{zr}^{(q)}}{\partial z}+\frac{1}{r}\left(\sigma_{rr}^{(q)}-\sigma_{\theta\theta}^{(q)}\right)+\sigma_{zz}^{(0)}\frac{\partial^2u_r^{(q)}}{\partial z^2}\\
&=\underline{-\frac{\partial S_{rr}^{(q-1)}}{\partial r}-\frac{\partial S_{\theta r}^{(q-1)}}{r\partial\theta}-\frac{\partial S_{zr}^{(q-1)}}{\partial z}-\frac{1}{r}\left(S_{rr}^{(q-1)}-S_{\theta\theta}^{(q-1)}\right)},\\
&\frac{\partial\sigma_{r\theta}^{(q)}}{\partial r}+\frac{\partial\sigma_{\theta\theta}^{(q)}}{r\partial\theta}+\frac{2}{r}\sigma_{\theta r}^{(q)}+\frac{\partial\sigma_{z\theta}^{(q)}}{\partial z}+\sigma_{zz}^{(0)}\frac{\partial^2u_\theta^{(q)}}{\partial z^2}\\
&=\underline{-\frac{\partial S_{r\theta}^{(q-1)}}{\partial r}-\frac{\partial S_{\theta\theta}^{(q-1)}}{r\partial\theta}-\frac{2}{r}\left(S_{\theta r}^{(q-1)}+S_{r\theta}^{(q-1)}\right)-\frac{\partial S_{z\theta}^{(q-1)}}{\partial z}},\\
&\frac{\partial\sigma_{rz}^{(q)}}{\partial r}+\frac{\partial\sigma_{\theta z}^{(q)}}{r\partial\theta}+\frac{1}{r}\sigma_{rz}^{(q)}+\frac{\partial\sigma_{zz}^{(q)}}{\partial z}+\sigma_{zz}^{(0)}\frac{\partial^2u_z^{(q)}}{\partial z^2}\\
&=\underline{-\frac{\partial S_{rz}^{(q-1)}}{\partial r}-\frac{\partial S_{\theta z}^{(q-1)}}{r\partial\theta}-\frac{1}{r}S_{rz}^{(q-1)}-\frac{\partial S_{zz}^{(q-1)}}{\partial z}},
\end{aligned}
$$

$$S_{rr}^{(q-1)} = \sum_{k=1}^{q-1}\left\{\sigma_{rr}^{(k)}\frac{\partial u_r^{(q-k)}}{\partial r} + \sigma_{r\theta}^{(k)}\left(\frac{\partial u_r^{(q-k)}}{r\partial\theta} - \frac{u_\theta^{(q-k)}}{r}\right) + \sigma_{rz}^{(k)}\frac{\partial u_r^{(q-k)}}{\partial z}\right\},$$

$$S_{r\theta}^{(q-1)} = \sum_{k=1}^{q-1}\left\{\sigma_{rr}^{(k)}\frac{\partial u_\theta^{(q-k)}}{\partial r} + \sigma_{r\theta}^{(k)}\left(\frac{\partial u_\theta^{(q-k)}}{r\partial\theta} + \frac{u_r^{(q-k)}}{r}\right) + \sigma_{rz}^{(k)}\frac{\partial u_\theta^{(q-k)}}{\partial z}\right\},$$

$$S_{rz}^{(q-1)} = \sum_{k=1}^{q-1}\left\{\sigma_{rr}^{(k)}\frac{\partial u_z^{(q-k)}}{\partial r} + \sigma_{r\theta}^{(k)}\frac{\partial u_z^{(q-k)}}{r\partial\theta} + \sigma_{rz}^{(k)}\frac{\partial u_z^{(q-k)}}{\partial z}\right\},$$

$$S_{\theta r}^{(q-1)} = \sum_{k=1}^{q-1}\left\{\sigma_{\theta r}^{(k)}\frac{\partial u_r^{(q-k)}}{\partial r} + \sigma_{\theta\theta}^{(k)}\left(\frac{\partial u_r^{(q-k)}}{r\partial\theta} - \frac{u_\theta^{(q-k)}}{r}\right) + \sigma_{\theta z}^{(k)}\frac{\partial u_r^{(q-k)}}{\partial z}\right\},$$

$$S_{\theta\theta}^{(q-1)} = \sum_{k=1}^{q-1}\left\{\sigma_{\theta r}^{(k)}\frac{\partial u_\theta^{(q-k)}}{\partial r} + \sigma_{\theta\theta}^{(k)}\left(\frac{\partial u_\theta^{(q-k)}}{r\partial\theta} + \frac{u_r^{(q-k)}}{r}\right) + \sigma_{\theta z}^{(k)}\frac{\partial u_\theta^{(q-k)}}{\partial z}\right\},$$

$$S_{\theta z}^{(q-1)} = \sum_{k=1}^{q-1}\left\{\sigma_{\theta r}^{(k)}\frac{\partial u_z^{(q-k)}}{\partial r} + \sigma_{\theta\theta}^{(k)}\frac{\partial u_z^{(q-k)}}{r\partial\theta} + \sigma^{(k)}\frac{\partial u_z^{(q-k)}}{\partial z}\right\},$$

$$S_{zz}^{(q-1)} = \sum_{k=1}^{q-1}\left\{\sigma_{rz}^{(k)}\frac{\partial u_z^{(q-k)}}{\partial r} + \sigma_{z\theta}^{(k)}\frac{\partial u_z^{(q-k)}}{r\partial\theta} + \sigma_{zz}^{(k)}\frac{\partial u_z^{(q-k)}}{\partial z}\right\},$$

$$S_{zr}^{(q-1)} = \sum_{k=1}^{q-1}\left\{\sigma_{zr}^{(k)}\frac{\partial u_r^{(q-k)}}{\partial r} + \sigma_{z\theta}^{(k)}\left(\frac{\partial u_r^{(q-k)}}{r\partial\theta} - \frac{u_\theta^{(q-k)}}{r}\right) + \sigma_{zz}^{(k)}\frac{\partial u_z^{(q-k)}}{\partial z}\right\},$$

$$S_{z\theta}^{(q-1)} = \sum_{k=1}^{q-1}\left\{\sigma_{zr}^{(k)}\frac{\partial u_\theta^{(q-k)}}{\partial r} + \sigma_{z\theta}^{(k)}\left(\frac{\partial u_\theta^{(q-k)}}{r\partial\theta} + \frac{u_r^{(q-k)}}{r}\right) + \sigma_{zz}^{(k)}\frac{\partial u_\theta^{(q-k)}}{\partial z}\right\},$$

$$\varepsilon_{rr}^{(q)} = \frac{\partial u_r^{(q)}}{\partial r} + \frac{1}{2}\sum_{k=1}^{q-1}\left\{\left(\frac{\partial u_r^{(k)}}{\partial r}\frac{\partial u_r^{(q-k)}}{\partial r}\right) + \left(\frac{\partial u_\theta^{(k)}}{\partial r}\frac{\partial u_\theta^{(q-k)}}{\partial r}\right) + \left(\frac{\partial u_z^{(k)}}{\partial r}\frac{\partial u_z^{(q-k)}}{\partial r}\right)\right\},$$

$$\varepsilon_{r\theta}^{(q)} = \frac{1}{2}\left(\frac{\partial u_\theta^{(q)}}{\partial r} + \frac{\partial u_r^{(q)}}{r\partial\theta} - \frac{u_\theta^{(q)}}{r}\right) + \frac{1}{2}\sum_{k=1}^{q-1}\left\{\frac{\partial u_r^{(k)}}{\partial r}\left(\frac{\partial u_r^{(q-k)}}{\partial\theta} - u_\theta^{(q-k)}\right),\right.$$

$$\left.\frac{\partial u_\theta^{(k)}}{\partial r}\left(\frac{\partial u_\theta^{(q-k)}}{r\partial\theta} + \frac{u_r^{(q-k)}}{r}\right) + \frac{\partial u_z^{(k)}}{\partial r}\frac{\partial u_z^{(q-k)}}{\partial\theta}\right\},\ \varepsilon_{rz}^{(q)} = \frac{1}{2}\left(\frac{\partial u_r^{(q)}}{\partial z} + \frac{\partial u_z^{(q)}}{\partial r}\right)$$

$$+\frac{1}{2}\sum_{k=1}^{q-1}\left\{\frac{\partial u_r^{(k)}}{\partial r}\frac{\partial u_r^{(q-k)}}{\partial z} + \frac{\partial u_\theta^{(k)}}{\partial r}\frac{\partial u_\theta^{(q-k)}}{\partial z} + \frac{\partial u_z^{(k)}}{\partial r}\frac{\partial u_z^{(q-k)}}{\partial z}\right\},$$

$$\varepsilon_{\theta\theta}^{(q)} = \frac{\partial u_\theta^{(q)}}{r\partial\theta} + \frac{u_r^{(q)}}{r} + \frac{1}{2}\sum_{k=1}^{q-1}\left\{\frac{1}{r^2}\left(\frac{\partial u_r^{(k)}}{\partial\theta} - u_\theta^{(k)}\right)\left(\frac{\partial u_\theta^{(q-k)}}{\partial\theta} - u_\theta^{(q-k)}\right)\right.$$

$$+\underline{\frac{1}{r^2}\left(\frac{\partial u_\theta^{(k)}}{\partial\theta}+u_r^{(k)}\right)\left(\frac{\partial u_\theta^{(q-k)}}{\partial\theta}+u_r^{(q-k)}\right)+\frac{1}{r^2}\left(\frac{\partial u_z^{(k)}}{\partial\theta}\right)\left(\frac{\partial u_z^{(q-k)}}{\partial\theta}\right)\Bigg\}},$$

$$\varepsilon_{\theta z}^{(q)}=\frac{1}{2}\left(\frac{\partial u_z^{(q)}}{r\partial\theta}+\frac{\partial u_\theta^{(q)}}{\partial z}\right)+\underline{\frac{1}{2}\sum_{k=1}^{q-1}\left\{\frac{1}{r}\frac{\partial u_r^{(k)}}{\partial z}\left(\frac{\partial u_r^{(q-k)}}{\partial\theta}-u_\theta^{(q-k)}\right)\right.}$$

$$\underline{\left.+\frac{1}{r}\frac{\partial u_\theta^{(k)}}{\partial z}\left(\frac{\partial u_\theta^{(q-k)}}{\partial\theta}+u_r^{(q-k)}\right)+\frac{1}{r}\frac{\partial u_z^{(k)}}{\partial\theta}\frac{\partial u_z^{(q-k)}}{\partial z}\right\}},$$

$$\varepsilon_{zz}^{(q)}=\frac{\partial u_z^{(q)}}{\partial z}+\underline{\frac{1}{2}\sum_{k=1}^{q-1}\left\{\left(\frac{\partial u_r^{(k)}}{\partial z}\frac{\partial u_r^{(q-k)}}{\partial z}\right)\right.}$$

$$\underline{\left.+\left(\frac{\partial u_\theta^{(k)}}{\partial z}\frac{\partial u_\theta^{(q-k)}}{\partial z}\right)+\left(\frac{\partial u_z^{(k)}}{\partial z}\frac{\partial u_z^{(q-k)}}{\partial z}\right)\right\}} \qquad (2.18)$$

The underlined terms in the Eq. (2.18) are equal to zero for the first approximation. By direct verification it is proven that the Eq. (2.18) without the underlined terms coincide with the corresponding equations of the Three-Dimensional Linearized Theory of Stability (TDLTS) (Guz 1999). Due to linearity, the constitutive relations (2.4) are satisfied by each approximation separately, i.e.

$$\begin{aligned}&\sigma_{rr}^{(q)}=A_{11}^{*}\varepsilon_{rr}^{(q)}+A_{12}^{*}\varepsilon_{\theta\theta}^{(q)}+A_{13}^{*}\varepsilon_{zz}^{(q)};\ \sigma_{\theta\theta}^{(q)}=A_{12}^{*}\varepsilon_{rr}^{(q)}+A_{11}^{*}\varepsilon_{\theta\theta}^{(q)}+A_{13}^{*}\varepsilon_{zz}^{(q)};\\&\sigma_{zz}^{(q)}=A_{13}^{*}\varepsilon_{rr}^{(q)}+A_{13}^{*}\varepsilon_{\theta\theta}^{(q)}+A_{13}^{*}\varepsilon_{zz}^{(q)};\ \sigma_{r\theta}^{(q)}=\left(A_{11}^{*}-A_{22}^{*}\right)\varepsilon_{r\theta}^{(q)};\\&\sigma_{rz}^{(q)}=2G^{*}\varepsilon_{rz}^{(q)};\ \sigma_{\theta z}^{(q)}=2G^{*}\varepsilon_{\theta z}^{(q)}.\end{aligned} \qquad (2.19)$$

Now we write the boundary conditions given on the lateral surface of the cylinder for the first approximation by the physical components of the stress tensor.

$$\begin{aligned}&\sigma_{(ir)}^{(1)}\Big|_{(r=R^{\pm},\,z=t_3)}+f_1\frac{\partial\sigma_{(ir)}^{(0)}}{\partial r}\Bigg|_{(r=R^{\pm},\,z=t_3)}+\varphi_1^{\pm}\frac{\partial\sigma_{(ir)}^{(0)}}{\partial z}\Bigg|_{(r=R^{\pm},\,z=t_3)}\\&+\gamma_\theta^{\pm}\sigma_{(i)\theta}^{(0)}\Big|_{(r=R^{\pm},\,z=t_3)}+\gamma_z^{\pm}\sigma_{(i)z}^{(0)}\Big|_{(r=R^{\pm},\,z=t_3)}=0,\end{aligned} \qquad (2.20)$$

where $(i)=r,\theta,z$. In Eq. (2.20) replacing $(i)$ with $r$, $\theta$ and $z$ we obtain the explicit form of the corresponding boundary conditions in the considered approximation. Moreover, in Eq. (2.20) the following notation is used.

$$
\begin{aligned}
&\gamma_z^{\pm} = \pm(-f'(t_3)\cos(\theta)),\; f_1 = f(t_3)\cos(\theta),\; \varphi_1^{\pm} = -R^{\pm}f'(t_3)\cos(\theta),\\
&\gamma_\theta^{\pm} = \pm\frac{f(t_3)}{R^{\pm}}\sin(\theta),\; f'(t_3) = \frac{df(t_3)}{dt_3},\; f''(t_3) = \frac{d^2f(t_3)}{dt_3^2}.
\end{aligned}
\tag{2.21}
$$

Consider the satisfaction of the end conditions (2.8). To simply the discussion we rewrite these conditions in the Cartesian system of coordinates $Ox_1x_2x_3$.

$$
\sigma_{3n}\left(\delta_n^j + \frac{\partial u_j}{\partial x_n}\right)\bigg|_{S_0} n_{0j} = p;\; \sigma_{3n}\left(\delta_n^j + \frac{\partial u_j}{\partial x_n}\right)\bigg|_{S_\ell} n_{\ell j} = -p \tag{2.22}
$$

where $\delta_n^j$ is a Kronecker symbol, other notation used in (2.22) is conventional.

According to Eqs. (2.12) and (2.13), we can write the following expressions from the conditions (2.22).

$$
\begin{aligned}
&\sigma_{3n}\left(\delta_n^j + \frac{\partial u_j}{\partial x_n}\right)\bigg|_{S_0} n_{0j} = \sigma_{3n}(x_1,x_2,-\varepsilon\pi x_1,t)\\
&\times\left(\delta_n^1 + \frac{\partial u_1(x_1,x_2,-\varepsilon\pi x_1,t)}{\partial x_n}\right)n_{01}\\
&\quad + \sigma_{3n}(x_1,x_2,-\varepsilon\pi x_1,t)\left(\delta_n^3 + \frac{\partial u_1(x_1,x_2,-\varepsilon\pi x_1,t)}{\partial x_n}\right)n_{03} = -p\\
&\sigma_{3n}\left(\delta_n^j + \frac{\partial u_j}{\partial x_n}\right)\bigg|_{S_\ell} n_{\ell j} = \sigma_{3n}(x_1,x_2,\ell+\varepsilon\pi x_1,t)\\
&\times\left(\delta_n^1 + \frac{\partial u_1(x_1,x_2,\ell+\varepsilon\pi x_1,t)}{\partial x_n}\right)n_{\ell 1}\\
&\quad + \sigma_{3n}(x_1,x_2,\ell+\varepsilon\pi x_1,t)\left(\delta_n^3 + \frac{\partial u_3(x_1,x_2,\ell+\varepsilon\pi x_1,t)}{\partial x_n}\right)n_{\ell 3} = p
\end{aligned}
\tag{2.23}
$$

Using the expansions

$$
\begin{aligned}
&\sigma_{in}(x_1,x_2,-\varepsilon\pi x_1,t) = \sum_{q=0}^{\infty}\varepsilon^q\sigma_{in}^{(q)}(x_1,x_2,-\varepsilon\pi x_1,t) = \sigma_{in}^{(0)}(x_1,x_2,0,t)\\
&\quad+\varepsilon\left(\sigma_{in}^{(1)}(x_1,x_2,0,t) + (-\pi x_1)\frac{\partial\sigma_{in}^{(0)}(x_1,x_2,0,t)}{\partial x_3}\right) + o\left((\varepsilon\pi)^2\right);\\
&\frac{\partial u_m(x_1,x_2,-\varepsilon\pi x_1,t)}{\partial x_j} = \sum_{q=0}^{\infty}\varepsilon^q\frac{\partial u_m^{(q)}(x_1,x_2,-\varepsilon\pi x_1,t)}{\partial x_j} = \frac{\partial u_m^{(0)}(x_1,x_2,0,t)}{\partial x_j}\\
&\quad+\varepsilon\left(\frac{\partial u_m^{(1)}(x_1,x_2,0,t)}{\partial x_j} + (-\pi x_1)\frac{\partial^2 u_m^{(0)}(x_1,x_2,0,t)}{\partial x_3\partial x_j}\right) + o\left((\varepsilon\pi)^2\right);
\end{aligned}
$$

$$
\begin{aligned}
\sigma_{in}(x_1,x_2,\ell+\varepsilon\pi x_1,t) &= \sum_{q=0}^{\infty}\varepsilon^q\sigma_{in}^{(q)}\Big(x_1,x_2,\ell+\varepsilon\pi x_1,t\Big) = \sigma_{in}^{(0)}(x_1,x_2,\ell,t) \\
&\quad + \varepsilon\left(\sigma_{in}^{(1)}(x_1,x_2,\ell,t) + (-\pi x_1)\frac{\partial\sigma_{in}^{(0)}(x_1,x_2,\ell,t)}{\partial x_3}\right) \\
&\quad + O\Big((\varepsilon\pi)^2\Big);\ \frac{\partial u_m(x_1,x_2,\ell+\varepsilon\pi x_1,t)}{\partial x_j} \\
&= \sum_{q=0}^{\infty}\varepsilon^q\frac{\partial u_m^{(q)}(x_1,x_2,\ell+\varepsilon\pi x_1,t)}{\partial x_j} = \frac{\partial u_m^{(0)}(x_1,x_2,\ell,t)}{\partial x_j} \\
&\quad + \varepsilon\left(\frac{\partial u_m^{(1)}(x_1,x_2,\ell,t)}{\partial x_j} + (-\pi x_1)\frac{\partial^2 u_m^{(0)}(x_1,x_2,\ell,t)}{\partial x_3\partial x_j}\right) \\
&\quad + O\Big((\varepsilon\pi)^2\Big),
\end{aligned}
\tag{2.24}
$$

we obtain the following expression for the end conditions in (2.22).

$$
\begin{aligned}
&\left\{-\sigma_{3k}^{(0)}\left(\delta_k^3+\frac{\partial u_3^{(0)}}{\partial x_k}\right)+\varepsilon\left[-\pi\sigma_{3k}^{(0)}\left(\delta_k^1+\frac{\partial u_1^{(0)}}{\partial x_k}\right)-\sigma_{3k}^{(0)}\left(\frac{\partial u_3^{(0)}}{\partial x_k}-\pi x_1\frac{\partial^2 u_3^{(0)}}{\partial x_3\partial x_k}\right)\right.\right.\\
&\left.\left.-\left(\sigma_{3k}^{(1)}-\pi x_1\frac{\partial\sigma_{3k}^{(0)}}{\partial x_k}\right)\left(\delta_3^k+\frac{\partial u_3^{(0)}}{\partial x_k}\right)\right]+O(\varepsilon^2)\right\}_{(x_1,x_2,0)}=p,\\
&\left\{\sigma_{3k}^{(0)}\left(\delta_k^3+\frac{\partial u_3^{(0)}}{\partial x_k}\right)+\varepsilon\left[\pi\sigma_{3k}^{(0)}\left(\delta_k^1+\frac{\partial u_1^{(0)}}{\partial x_k}\right)+\sigma_{3k}^{(0)}\left(\frac{\partial u_3^{(0)}}{\partial x_k}+\pi x_1\frac{\partial^2 u_3^{(0)}}{\partial x_3\partial x_k}\right)\right.\right.\\
&\left.\left.+\left(\sigma_{3k}^{(1)}+\pi x_1\frac{\partial\sigma_{3k}^{(0)}}{\partial x_k}\right)\left(\delta_3^k+\frac{\partial u_3^{(0)}}{\partial x_k}\right)\right]+O(\varepsilon^2)\right\}_{(x_1,x_2,\ell)}=-p.
\end{aligned}
\tag{2.25}
$$

In a similar manner, we can write the following expansions for the physical components $u_{(i)}$ of the displacement vector at the ends of the cylinder

$$
\begin{aligned}
&u_{(i)}^{(0)}(x_1,x_2,0,t)+\varepsilon\left(u_{(i)}^{(1)}(x_1,x_2,0,t)\right.\\
&\left.+(-\pi x_1)\frac{\partial u_{(i)}^{(0)}(x_1,x_2,0,t)}{\partial x_3}\right)+O\Big((\varepsilon\pi)^2\Big)=0,\\
&u_{(i)}^{(0)}(x_1,x_2,\ell,t)+\varepsilon\left(u_{(i)}^{(1)}(x_1,x_2,\ell,t)+(-\pi x_1)\frac{\partial u_{(i)}^{(0)}(x_1,x_2,\ell,t)}{\partial x_3}\right)\\
&+O\Big((\varepsilon\pi)^2\Big)=0,\ (i)=r,\theta,z.
\end{aligned}
\tag{2.26}
$$

We assume that the coefficient of $\varepsilon^q$ in the expansion (2.26) for $(i) = r; \theta$ is equal to zero. Consequently, according to this assumption, we obtain the end conditions for the first and subsequent approximations for the displacements $u_r$ and $u_\theta$.

Taking the estimation $\left(\delta_k^3 + \partial u_3^{(0)} \big/ \partial x_k\right) \approx \delta_k^3$, $\left(\delta_k^1 + \partial u_1^{(0)} \big/ \partial x_k\right) \approx \delta_k^1$ and the expansions (2.23)–(2.26) into account we obtain the following end conditions for the stresses for the zeroth and first approximations from the condition (2.22).

For the zeroth approximation:

$$\sigma_{33}^{(0)}(x_1, x_2, 0) = \sigma_{33}^{(0)}(x_1, x_2, \ell) = -p. \tag{2.27}$$

For the first approximation

$$\begin{aligned}
&\pi\sigma_{31}^{(0)}(x_1, x_2, 0, t) + \sigma_{3k}^{(0)}(x_1, x_2, 0, t) \\
&\left(\frac{\partial u_3^{(1)}(x_1, x_2, 0, t)}{k} - \pi x_1 \frac{\partial^2 u_3^{(0)}(x_1, x_2, 0, t)}{\partial x_3 \partial x_k}\right) \\
&+ \left(\sigma_{33}^{(1)}(x_1, x_2, 0, t) - \pi x_1 \frac{\partial \sigma_{33}^{(0)}(x_1, x_2, 0, t)}{\partial x_3}\right) = 0, \\
&\left(\frac{\partial u_3^{(1)}(x_1, x_2, \ell, t)}{\partial x_k} - \pi x_1 \frac{\partial^2 u_3^{(0)}(x_1, x_2, \ell, t)}{\partial x_3 \partial x_k}\right) \\
&+ \left(\sigma_{33}^{(1)}(x_1, x_2, \ell, t) - \pi x_1 \frac{\partial \sigma_{33}^{(0)}(x_1, x_2, \ell, t)}{\partial x_3}\right) = 0.
\end{aligned} \tag{2.28}$$

Thus, rewriting the condition (2.27) in the cylindrical system of coordinates $Or\theta z$ we obtain the condition (2.17).

According to (2.15), (2.16) and (2.17), the values related to the zeroth approximation are determined as follows.

$$\sigma_{zz}^{(0)} = -p, \ \sigma_{(ij)}^{(0)} = 0 \text{ for } (ij) \neq zz. \tag{2.29}$$

It follows from (2.29) that in the zeroth approximation the components of the displacement vector can be presented as follows

$$u_r^{(0)} = a(t)r + a_0, \ u_\theta^{(0)} = b_0, \ u_z^{(0)} = c(t)z + c_0 \tag{2.30}$$

where $a_0$, $b_0$ and $c_0$ are constants, $a(t)$ and $c(t)$ are functions, $t$ is a time. The functions $a(t)$ and $c(t)$ can be easily determined from Eqs. (2.19) and (2.29).

Now we consider determination of the values related to the first approximation. Taking the expression (2.29) into account the following field equations are obtained from Eq. (2.18) for this approximation.

$$\frac{\partial\sigma_{rr}^{(1)}}{\partial r}+\frac{1}{r}\frac{\partial\sigma_{\theta r}^{(1)}}{\partial\theta}+\frac{\partial\sigma_{zr}^{(1)}}{\partial z}+\frac{1}{r}\left(\sigma_{rr}^{(1)}-\sigma_{\theta\theta}^{(1)}\right)+\sigma_{zz}^{(0)}\frac{\partial^2u_r^{(1)}}{\partial z^2}=0,$$
$$\frac{\partial\sigma_{r\theta}^{(1)}}{\partial r}+\frac{1}{r}\frac{\partial\sigma_{\theta\theta}^{(1)}}{\partial\theta}+\frac{\partial\sigma_{z\theta}^{(1)}}{\partial z}+\frac{2}{r}\sigma_{\theta r}^{(1)}+\sigma_{zz}^{(0)}\frac{\partial^2u_\theta^{(1)}}{\partial z^2}=0,$$
$$\frac{\partial\sigma_{rz}^{(1)}}{\partial r}+\frac{1}{r}\frac{\partial\sigma_{\theta z}^{(1)}}{\partial\theta}+\frac{\partial\sigma_{zz}^{(1)}}{\partial z}+\frac{1}{r}\sigma_{rz}^{(1)}+\sigma_{zz}^{(0)}\frac{\partial^2u_z^{(1)}}{\partial z^2}=0, \tag{2.31}$$
$$\varepsilon_{rr}^{(1)}=\frac{\partial u_r^{(1)}}{\partial r},\ \varepsilon_{\theta\theta}^{(1)}=\frac{\partial u_\theta^{(1)}}{r\partial\theta}+\frac{u_r^{(1)}}{r},\ \varepsilon_{r\theta}^{(1)}=\frac{1}{2}\left(\frac{\partial u_r^{(1)}}{r\partial\theta}+\frac{\partial u_\theta^{(1)}}{\partial r}-\frac{u_\theta^{(1)}}{r}\right),$$
$$\varepsilon_{zz}^{(1)}=\frac{\partial u_z^{(1)}}{\partial z},\ \varepsilon_{\theta z}^{(1)}=\frac{1}{2}\left(\frac{\partial u_\theta^{(1)}}{\partial z}+\frac{\partial u_z^{(1)}}{r\partial\theta}\right),\ \varepsilon_{rz}^{(1)}=\frac{1}{2}\left(\frac{\partial u_r^{(1)}}{\partial z}+\frac{\partial u_z^{(1)}}{\partial r}\right).$$

The following conditions on the lateral surface of the cylinder are obtained from (2.20) and (2.29).

$$\sigma_{rr}^{(1)}\left(R^{\pm},\theta,t_3,t\right)=0,\ \sigma_{r\theta}^{(1)}\left(R^{\pm},\theta,t_3,t\right)=0,$$
$$\sigma_{rz}^{(1)}\left(R^{\pm},\theta,t_3,t\right)=\pm\pi\sigma_{zz}^{(0)}\cos(\alpha_1 z)\cos\theta=0,\ \alpha_1=\pi/\ell \tag{2.32}$$

According to Eqs. (2.26) and (2.28), the end conditions for the first approximation can be written as follows.

$$\sigma_{zz}^{(1)}(r,\theta,0,t)+\sigma_{zz}^{(0)}\frac{\partial u_z^{(1)}(r,\theta,0,t)}{\partial z}=0,\ u_r^{(1)}(r,\theta,0,t)=0,\ u_\theta^{(1)}(r,\theta,0,t)=0,$$
$$\sigma_{zz}^{(1)}(r,\theta,\ell,t)+\sigma_{zz}^{(0)}\frac{\partial u_z^{(1)}(r,\theta,\ell,t)}{\partial z}=0,$$
$$u_r^{(1)}(r,\theta,\ell,t)=0,\ u_\theta^{(1)}(r,\theta,\ell,t)=0. \tag{2.33}$$

Thus, the Eq. (2.31), (2.19) and (2.5) and the conditions (2.32) and (2.33) complete the formulation of the problem for determination of the values of the first approximation. For solution to this problem we apply the Laplace transformation

$$\bar{\psi}=\int_0^\infty \psi(t)\,e^{-st}dt \tag{2.34}$$

with parameter $s>0$, to all equations and relations related to the first approximation. After this application to Eq. (2.31), the boundary conditions (2.32) (in which $\sigma_{zz}^{(0)}$ must be replaced with $\sigma_{zz}^{(0)}/s$) and (2.33) are valid for the Laplace transformations of the corresponding sought-for quantities, whilst the constitutive relations (2.19) are transformed to the following ones:

$$\begin{aligned}
&\bar{\sigma}_{rr}^{(1)} = \bar{A}_{11}^{*}\bar{\varepsilon}_{rr}^{(1)} + \bar{A}_{12}^{*}\bar{\varepsilon}_{\theta\theta}^{(1)} + \bar{A}_{13}^{*}\bar{\varepsilon}_{zz}^{(1)},\ \bar{\sigma}_{\theta\theta}^{(1)} = \bar{A}_{12}^{*}\bar{\varepsilon}_{rr}^{(1)} + \bar{A}_{11}^{*}\bar{\varepsilon}_{\theta\theta}^{(1)} + \bar{A}_{13}^{*}\bar{\varepsilon}_{zz}^{(1)},\\
&\bar{\sigma}_{zz}^{(1)} = \bar{A}_{13}^{*}\bar{\varepsilon}_{rr}^{(1)} + \bar{A}_{13}^{*}\bar{\varepsilon}_{\theta\theta}^{(1)} + \bar{A}_{13}^{*}\bar{\varepsilon}_{zz}^{(1)},\ \bar{\sigma}_{r\theta}^{(1)} = \left(\bar{A}_{11}^{*} - \bar{A}_{22}^{*}\right)\bar{\varepsilon}_{r\theta}^{(1)},\\
&\bar{\sigma}_{rz}^{(1)} = 2\bar{G}^{*}\bar{\varepsilon}_{rz}^{(1)};\ \bar{\sigma}_{\theta z}^{(1)} = 2\bar{G}^{*}\bar{\varepsilon}_{\theta z}^{(1)}
\end{aligned} \tag{2.35}$$

where

$$\left\{\begin{matrix}\bar{A}_{ij}^{*}\\ \bar{G}^{*}\end{matrix}\right\}\bar{\phi}(s) = \left\{\begin{matrix}A_{ij0}\\ G_{0}\end{matrix}\right\}\bar{\phi}(s) + \left\{\begin{matrix}\bar{A}_{ij1}(s)\\ G_{1}(s)\end{matrix}\right\}\bar{\phi}(s). \tag{2.36}$$

As has been noted above, Eqs. (2.31)–(2.36) coincide with the corresponding equations of the TDLTS. Therefore, to solve the obtained equation systems, according to the monograph (Guz 1999), in the cylindrical system of coordinates we can use the following representations.

$$\begin{aligned}
&\bar{u}_{r}^{(1)} = \frac{1}{r}\frac{\partial}{\partial\theta}\psi - \frac{\partial^{2}}{\partial r\partial z}\chi,\ \bar{u}_{\theta}^{(1)} = -\frac{1}{r}\psi - \frac{1}{r}\frac{\partial^{2}}{\partial\theta\partial z}\chi,\\
&\bar{u}_{z}^{(1)} = \left(\bar{A}_{13}^{*} + \bar{G}^{*}\right)^{-1}\left(\bar{A}_{11}^{*}\Delta_{1} + \left(\bar{G}^{*} + \sigma_{zz}^{(0)}\right)\frac{\partial^{2}}{\partial z^{2}}\right)\chi,\\
&\Delta_{1} = \frac{\partial^{2}}{\partial r^{2}} + \frac{1}{r}\frac{\partial}{\partial r} + \frac{1}{r}\frac{\partial^{2}}{\partial\theta^{2}}.
\end{aligned} \tag{2.37}$$

The functions $\psi$ and $\chi$ are determined from the equations.

$$\left(\Delta_{1} + \xi_{1}^{2}\frac{\partial^{2}}{\partial z^{2}}\right)\psi = 0,\ \left(\Delta_{1} + \xi_{2}^{2}\frac{\partial^{2}}{\partial z^{2}}\right)\left(\Delta_{1} + \xi_{3}^{2}\frac{\partial^{2}}{\partial z^{2}}\right)\chi = 0, \tag{2.38}$$

where

$$\begin{aligned}
&\xi_{1}^{2} = \frac{2\bar{G}^{*}}{\bar{A}_{11}^{*} - \bar{A}_{12}^{*}},\ \xi_{2,3}^{2} = c \pm \left(c^{2} - \frac{\left(\bar{A}_{33}^{*} + \sigma_{zz}^{(0)}\right)\left(\bar{G}^{*} + \sigma_{zz}^{(0)}\right)}{\bar{A}_{11}^{*}\bar{G}^{*}}\right)^{\frac{1}{2}},\\
&2\bar{A}_{11}^{*}\bar{G}^{*}c = \bar{A}_{11}^{*}\left(\bar{A}_{33}^{*} + \sigma_{zz}^{(0)}\right) + \bar{G}^{*2} - \left(\bar{A}_{13}^{*} + \bar{G}^{*}\right)^{2}.
\end{aligned} \tag{2.39}$$

Taking the expressions of the right sides of the conditions (2.32) and (2.33) we find the solution to the Eq. (2.38) as follows.

$$\begin{aligned}
\psi =& \left(B_{1}I_{1}(\xi_{1}\alpha_{1}r) + B_{4}K_{1}(\xi_{1}\alpha_{1}r)\right)\sin(\alpha_{1}z)\sin\theta,\\
\chi =& \left[B_{2}I_{1}(\xi_{2}\alpha_{1}r) + B_{3}I_{1}(\xi_{3}\alpha_{1}r) + B_{5}K_{1}(\xi_{2}\alpha_{1}r)\right.\\
&\left. + B_{6}K_{1}(\xi_{3}\alpha_{1}r)\right]\cos(\alpha_{1}z)\cos\theta,
\end{aligned} \tag{2.40}$$

where $\alpha_{1} = \pi/\ell$, $I_{1}(x)$ is the first order Bessel function of a purely imaginary argument and $K_{1}(x)$ is the first order Macdonald Function, $B_{1},\ B_{2},\ B_{3},\ B_{4},\ B_{5}$ and $B_{6}$ are unknown constants.

Substituting these solutions into relations (2.37) and (2.35) we obtain the following expressions for Laplace transformation of the south values.

$$\bar{u}_r^{(1)} = \left[B_1 \frac{1}{r} I_1(\xi_2\alpha_1 r) + B_2\xi_2\alpha_1^2 I_1'(\xi_2\alpha_1 r) + B_3\xi_3\alpha_1^2 I_1'(\xi_3\alpha_1 r)\right.$$
$$\left. + B_4\frac{1}{r}K_1(\xi_1\alpha_1 r) + B_5\xi_2\alpha_1^2 K_1'(\xi_2\alpha_1 r) + B_6\xi_3\alpha_1^2 K_1'(\xi_3\alpha_1 r)\right] \sin(\alpha_1 z)\cos\theta,$$
$$\bar{u}_\theta^{(1)} = \left[-B_1\xi_1\alpha_1 I_1'(\xi_1\alpha_1 r) - \frac{\alpha_1}{r}B_2 I_1(\xi_2\alpha_1 r) - \frac{\alpha_1}{r}B_3 I_1(\xi_3\alpha_1 r)\right.$$
$$\left. - B_4\xi_1\alpha_1 K_1'(\xi_1\alpha_1 r) - \tfrac{\alpha_1}{r}B_5 K_1(\xi_2\alpha_1 r) - \tfrac{\alpha}{r}B_6\xi_3\alpha_1^2 K_1(\xi_3\alpha_1 r)\right] \sin(\alpha_1 z)\sin\theta,$$
$$\bar{u}_z^{(1)} = \left[B_2 D_2\alpha_1^2 I_1(\xi_2\alpha_1 r) + B_3 D_2\alpha_1^2 I_1(\xi_3\alpha_1 r)\right.$$
$$\left. + B_5 D_2\alpha_1^2 K_1(\xi_2\alpha_1 r) + B_6 D_3\alpha_1^2 K_1(\xi_3\alpha_1 r)\right] \sin(\alpha_1 z)\cos\theta,$$
$$D_2 = \frac{\bar{A}_{11}^*\xi_2^2 - \bar{G}^* - \sigma_{zz}^{(0)}}{\bar{A}_{13}^* + \bar{G}^*},\quad D_3 = \frac{\bar{A}_{11}^*\xi_3^2 - \bar{G}^* - \sigma_{zz}^{(0)}}{\bar{A}_{13}^* + \bar{G}^*},$$
$$\bar{\sigma}_{rr}^{(1)} = \left\{B_1\left[-(\bar{A}_{11}^* - \bar{A}_{12}^*)\tfrac{1}{r^2}I_1(\xi_1\alpha_1 r) + (\bar{A}_{11}^* - \bar{A}_{12}^*)\tfrac{\xi_1\alpha_1}{r}I_1'(\xi_1\alpha_1 r)\right]\right.$$
$$+B_2\left[\bar{A}_{11}^*\alpha_1^3\xi_2^2 I_1''(\xi_2\alpha_1 r) + \bar{A}_{12}^*\left(\tfrac{-\alpha_1}{r^2}I_1(\xi_2\alpha_1 r) + \tfrac{\xi_2\alpha_1^2}{r}I_1'(\xi_2\alpha_1 r)\right) - \bar{A}_{13}^* D_2\alpha_1^3 I_1(\xi_2\alpha_1 r)\right],$$
$$B_3\left[\bar{A}_{11}^*\alpha_1^3\xi_3^2 I_1''(\xi_3\alpha_1 r) + \bar{A}_{12}^*\left(\frac{-\alpha_1}{r^2}I_1(\xi_3\alpha_1 r) + \tfrac{\xi_3\alpha_1^2}{r}I_1'(\xi_3\alpha_1 r)\right) - \bar{A}_{13}^* D_3\alpha_1^3 I_1(\xi_3\alpha_1 r)\right],$$
$$B_4\left[-(\bar{A}_{11}^* - \bar{A}_{12}^*)\tfrac{1}{r^2}K_1(\xi_1\alpha_1 r) + (\bar{A}_{11}^* - \bar{A}_{12}^*)\tfrac{\xi_1\alpha_1}{r}K_1'(\xi_1\alpha_1 r)\right]$$
$$+B_5\left[\bar{A}_{11}^*\alpha_1^3\xi_2^2 K_1''(\xi_2\alpha_1 r) + \bar{A}_{12}^*\left(\tfrac{-\alpha_1}{r^2}K_1(\xi_2\alpha_1 r)\right.\right.$$
$$\left.\left.+\frac{\xi_2\alpha_1^2}{r}K_1'(\xi_2\alpha_1 r)\right) - \bar{A}_{13}^* D_2\alpha_1^3 K_1(\xi_2\alpha_1 r)\right]$$
$$+B_6\left[\bar{A}_{11}^*\alpha_1^3\xi_3^2 K_1''(\xi_3\alpha_1 r) + \bar{A}_{12}^*\left(\tfrac{-\alpha_1}{r^2}K_1(\xi_3\alpha_1 r) + \tfrac{\xi_3\alpha_1^2}{r}K_1'(\xi_3\alpha_1 r)\right)\right.$$
$$\left.\left. - \bar{A}_{13}^* D_3\alpha_1^3 K_1(\xi_3\alpha_1 r)\right]\right\} \sin(\alpha_1 z)\cos\theta, \bar{\sigma}_{r\theta}^{(1)}$$
$$= \left\{B_1\left[\frac{1}{2}(\bar{A}_{11}^* - \bar{A}_{12}^*)\left(-\xi_1^2\alpha^2 I_1''(\xi_1\alpha_1 r) - \frac{1}{r^2}I_1(\xi_1\alpha_1 r) + \frac{\xi_1\alpha_1}{r}I_1'(\xi_1\alpha_1 r)\right)\right]\right.$$
$$+B_2\left[(\bar{A}_{11}^* - \bar{A}_{12}^*)\left(\tfrac{\alpha_1}{r^2}I_1(\xi_2\alpha_1 r) - \xi_2\tfrac{\alpha_1^2}{r}I_1'(\xi_1\alpha_1 r)\right)\right]$$
$$+B_3\left[(\bar{A}_{11}^* - \bar{A}_{12}^*)\left(\tfrac{\alpha_1}{r^2}I_1(\xi_3\alpha_1 r) - \xi_3\tfrac{\alpha_1^2}{r}I_1'(\xi_3\alpha_1 r)\right)\right]$$
$$+B_4\left[\frac{1}{2}(\bar{A}_{11}^* - \bar{A}_{12}^*)\left(-\xi_1^2\alpha_1^2 K_1''(\xi_1\alpha_1 r) - \frac{1}{r^2}K_1(\xi_1\alpha_1 r) + \frac{\xi_1\alpha_1}{r}K_1'(\xi_1\alpha_1 r)\right)\right]$$
$$+B_5\left[(\bar{A}_{11}^* - \bar{A}_{12}^*)\left(\tfrac{\alpha_1^2}{r^2}K_1(\xi_2\alpha_1 r) - \xi_2\tfrac{\alpha_1^2}{r}K_1'(\xi_2\alpha_1 r)\right)\right]$$
$$\left.+B_6\left[(\bar{A}_{11}^* - \bar{A}_{12}^*)\left(\tfrac{\alpha_1}{r^2}K_1(\xi_3\alpha_1 r) - \xi_3\tfrac{\alpha_1^2}{r}K_1'(\xi_3\alpha_1 r)\right)\right]\right\} \sin(\alpha_1 r)\sin\theta,$$
$$\bar{\sigma}_{rz}^{(1)} = \left\{B_1\bar{G}^*\frac{\alpha_1}{r}I_1(\xi_1\alpha_1 r) + B_2\bar{G}^*\alpha_1^3\xi_2 I_1'(\xi_2\alpha_1 r)(1 + D_2)\right.$$
$$+B_3\bar{G}^*\alpha_1^3\xi_3 I_1'(\xi_3\alpha_1 r)(1 + D_3) + B_4\bar{G}^*\tfrac{\alpha_1}{r}K_1(\xi_1\alpha_1 r)$$
$$\left.+B_5\bar{G}^*\alpha_1^3\xi_2 K_1'(\xi_2\alpha_1 r)(1 + D_2) + B_6\bar{G}^*\alpha_1^3\xi_2 K_1'(\xi_3\alpha_1 r)(1 + D_3)\right\} \cos(\alpha_1 z)\cos\theta,$$

$$\bar{\sigma}_{zz}^{(1)} = \Big\{ B_2 \Big[ \bar{A}_{13}^* \Big( \xi_2^2 \alpha_1^3 I_1''(\xi_2\alpha_1 r) - \frac{\alpha_1}{r^2} I_1(\xi_2\alpha_1 r) + \frac{\alpha_1^2 \xi_2}{r} I_1'(\xi_2\alpha_1 r) \Big)$$
$$- \bar{A}_{33}^* D_2 \alpha_1^3 I_1(\xi_2\alpha_1 r) \Big] + B_3 \Big[ \bar{A}_{13}^* \big( \xi_3^2 \alpha_1^3 I_1'(\xi_3\alpha_1 r)$$
$$- \tfrac{\alpha_1}{r^2} I_1(\xi_3\alpha_1 r) + \tfrac{\alpha_1^2 \xi_2}{r} I_1'(\xi_3\alpha_1 r) \big) - \bar{A}_{33}^* D_3 \alpha_1^3 I_1(\xi_3\alpha_1 r) \Big]$$
$$B_5 \Big[ \bar{A}_{13}^* \Big( \xi_2^2 \alpha_1^3 K_1''(\xi_2\alpha_1 r) - \tfrac{\alpha_1}{r^2} K_1(\xi_2\alpha_1 r) + \tfrac{\alpha_1^2 \xi_2}{r} K_1'(\xi_3\alpha_1 r) \Big)$$
$$- \bar{A}_{33}^* D_2 \alpha_1^3 K_1(\xi_2\alpha_1 r) \big] + B_6 \big[ \bar{A}_{13}^* \big( \xi_3^2 \alpha_1^3 K_1''(\xi_3\alpha_1 r) - \tfrac{\alpha_1}{r^2} K_1(\xi_3\alpha_1 r)$$
$$+ \tfrac{\alpha_1^2 \xi_3}{r} K_1'(\xi_3\alpha_1 r) \big) - \bar{A}_{33}^* D_3 \alpha_1^3 K_1(\xi_3\alpha_1 r) \Big] \Big\} \sin(\alpha_1 z) \cos\theta. \tag{2.41}$$

Here the notation $I_1'(x) = dI_1(x)/dx$, $I_1''(x) = d^2I_1(x)/dx^2$, $K_1'(x) = dK_1(x)/dx$ and $K_1''(x) = d^2K_1(x)/dx^2$ is used. The solution (2.41) to the problem under consideration satisfies automatically the end condition (2.33). Replacing the unknowns $B_1$, $B_2$, $B_3$, $B_4$, $B_5$ and $B_6$ with $\alpha_1^2 B_1$ $(= C_1)$, $\alpha_1^3 B_2$ $(= C_2)$ $\alpha_1^3 B_3$ $(= C_3)$, $\alpha_1^2 B_4$ $(= C_4)$, $\alpha_1^3 B_5$ $(= C_5)$ and $\alpha_1^3 B_6$ $(= C_6)$ respectively, we obtain the following algebraic equation from the boundary condition (2.32) for determination these unknowns.

$$\begin{aligned}
\bar{\sigma}_{rr}^{(1)}(R^+, \theta, t_3, t) =& 0 \Rightarrow C_1 a_{11}(\alpha_1 R^+) + C_2 a_{12}(\alpha_1 R^+) + C_3 a_{13}(\alpha_1 R^+) \\
& + C_4 a_{14}(\alpha_1 R^+) + C_5 a_{15}(\alpha_1 R^+) + C_6 a_{16}(\alpha_1 R^+) = 0, \\
\bar{\sigma}_{r\theta}^{(1)}(R^+, \theta, t_3, t) =& 0 \Rightarrow C_1 a_{21}(\alpha_1 R^+) + C_2 a_{22}(\alpha_1 R^+) + C_3 a_{23}(\alpha_1 R^+) \\
& + C_4 a_{24}(\alpha_1 R^+) + C_5 a_{25}(\alpha_1 R^+) + C_6 a_{26}(\alpha_1 R^+) = 0, \\
\bar{\sigma}_{rz}^{(1)}(R^+, \theta, t_3, t) =& \pi \sigma_{zz}^{(0)} \frac{1}{s} \Rightarrow C_1 a_{31}(\alpha_1 R^+) + C_2 a_{32}(\alpha_1 R^+) + C_3 a_{33}(\alpha_1 R^+) \\
& + C_4 a_{34}(\alpha_1 R^+) + C_5 a_{35}(\alpha_1 R^+) + C_6 a_{36}(\alpha_1 R^+) = \pi \sigma_{zz}^{(0)} \frac{1}{s}, \\
\bar{\sigma}_{rr}^{(1)}(R^-, \theta, t_3, t) =& 0 \Rightarrow C_1 a_{11}(\alpha_1 R^-) + C_2 a_{12}(\alpha_1 R^-) + C_3 a_{13}(\alpha_1 R^-) \\
& + C_4 a_{14}(\alpha_1 R^-) + C_5 a_{15}(\alpha_1 R^-) + C_6 a_{16}(\alpha_1 R^-) = 0, \\
\bar{\sigma}_{r\theta}^{(1)}(R^-, \theta, t_3, t) =& 0 \Rightarrow C_1 a_{21}(\alpha_1 R^-) + C_2 a_{22}(\alpha_1 R^-) + C_3 a_{23}(\alpha_1 R^-) \\
& + C_4 a_{24}(\alpha_1 R^-) + C_5 a_{25}(\alpha_1 R^-) + C_6 a_{26}(\alpha_1 R^-) = 0, \\
\bar{\sigma}_{rz}^{(1)}(R^-, \theta, t_3, t) =& -\pi \sigma_{zz}^{(0)} \frac{1}{s} \Rightarrow C_1 a_{31}(\alpha_1 R^-) + C_2 a_{32}(\alpha_1 R^-) + C_3 a_{33}(\alpha_1 R^-) \\
& + C_4 a_{34}(\alpha_1 R^-) + C_5 a_{35}(\alpha_1 R^-) + C_6 a_{36}(\alpha_1 R^-) = -\pi \sigma_{zz}^{(0)} \frac{1}{s}
\end{aligned} \tag{2.42}$$

Note that expressions for the coefficients $a_{ij}$ $(i;j = 1,2,3,4,5,6)$ can be easily obtained from Eq. (2.41).

Thus, with the foregoing we determine completely the Laplace transformation of the values related to the first approximation. The Laplace transformation of the values of the second and subsequent approximations in (2.14) can also be determined as the values of the first approximation by taking the obvious changes into account. However, as follows from the structure of the Eq. (2.18), for stability loss problems, the consideration of only the zeroth and first approximation is sufficient, because accounting the second and subsequent approximation does not change the values of the critical parameters.

The original of the south values is determined by employing the method (Schapery 1962, 1978), according to which, for instance, the original of the displacement $u_r^{(1)}(r,\theta,t_3,t)$ is determined through the expression

$$u_r^{(1)}(r,\theta,t_3,t) \approx \left( s\bar{u}_r^{(1)}(r,\theta,t_3,t) \right)\Big|_{s=1/(2t)}. \tag{2.43}$$

Now we consider the selection of the stability loss criterion. In the present investigation, the case will be understood under stability loss, where

$$\max_{\substack{t_3\in(0,\ell)\\ r\in(0,R)\\ \theta\in(0,2\pi)}} \left| u_r^{(1)}(r,\theta,t_3,t) \right| \to \infty \text{ as } t \to t_{cr} \text{ (or as } p \to p_{cr} \text{ for the pure elastic case).} \tag{2.44}$$

Thus, the values of the critical time or the values of the critical force are determined from the initial imperfection criterion (2.44).

## 2.3 Approximate Equations for the Stability Loss of the Cylinder-Beam Obtained from Equations of the TDLTS by the Average-Integrating Procedure

### *2.3.1 Bernoulli Beam theory*

For the case under consideration the approximate stability loss equations for the cylinder-beam can be derived from Eqs. (2.31)–(2.36) with the use of the Bernoulli hypothesis, according to which, the displacements of the cylinder are presented as follows.

$$\begin{aligned} u_r^{(1)} &= u_r^{(1)}(\theta,z) = w(z)\cos\theta,\ u_\theta^{(1)} = u_\theta^{(1)}(\theta,z) = -w(z)\sin\theta, \\ u_z^{(1)} &= u_z^{(1)}(r,\theta,z) = -r\frac{dw(z)}{dz}\cos\theta. \end{aligned} \tag{2.45}$$

In writing the expression (2.45), it is assumed that the elongation of the middle line of the cylinder is very small with respect to the term $-rdw(z)/dz\cos\theta$ and is ignored. Thus, according to (2.31) and (2.45), we obtain that

$$\varepsilon_{zz}^{(1)} = -r\cos\theta\frac{d^2w(z)}{dz^2},\ \varepsilon_{(ij)}^{(1)} = 0 \text{ for } (ij) \neq zz,\ \sigma_{rr}^{(1)} = \sigma_{\theta\theta}^{(1)} = \sigma_{r\theta}^{(1)} = 0,$$
$$\sigma_{zz}^{(1)} = E_3^*\varepsilon_{zz}^{(1)} = -r\cos\theta E_3^*\frac{d^2w(z)}{dz^2}, \tag{2.46}$$

where $E_3^*$ is an operator and $E_3^*\phi(t) = E_{30}\phi(t) + \int\limits_0^t E_{31}(t-\tau)\phi(\tau)d\tau$, where $E_{30}$ is the instantaneous value of the modulus of elasticity in the direction of the $Oz$ axis of the cylinder material and $E_{31}(t)$ is a function which determine the relaxation of this material.

Assume that

$$\sigma_{rz}^{(1)} \neq 0,\ \sigma_{\theta z}^{(1)} \neq 0. \tag{2.47}$$

Taking the relations (2.46) and (2.47) into account we obtain the following equation from (2.31).

$$\frac{\partial\sigma_{rz}^{(1)}}{\partial z} + \sigma_{zz}^{(0)}\frac{d^2w}{dz^2}\cos\theta = 0,\ \frac{\partial\sigma_{\theta z}^{(1)}}{\partial z} - \sigma_{zz}^{(0)}\frac{d^2w}{dz^2}\sin\theta = 0,$$
$$\frac{\partial\sigma_{rz}^{(1)}}{\partial r} + \frac{1}{r}\frac{\partial\sigma_{\theta z}^{(1)}}{\partial\theta} + \frac{1}{r}\sigma_{rz}^{(1)} + \frac{\partial\sigma_{zz}^{(1)}}{\partial z} + \sigma_{zz}^{(0)}\frac{d^3w}{dz^3}(-r\cos\theta) = 0. \tag{2.48}$$

Using the expressions of the equations in (2.48) we can write the following presentations.

$$\sigma_{rz}^{(1)} = s_{rz}^{(1)}\cos\theta,\ \sigma_{\theta z}^{(1)} = s_{\theta z}^{(1)}\sin\theta,\ \sigma_{zz}^{(1)} = s_{zz}^{(1)}\cos\theta. \tag{2.49}$$

According to the foregoing presentations, we obtain the following equations from (2.48) and (2.49).

$$\frac{ds_{rz}^{(1)}}{dz} + \sigma_{zz}^{(0)}\frac{d^2w}{dz^2} = 0,\ \frac{ds_{\theta z}^{(1)}}{dz} - \sigma_{zz}^{(0)}\frac{d^2w}{dz^2} = 0. \tag{2.50}$$

Multiplying the last Eq. (2.48) with $(r^2\cos\theta)$ and integrating with respect to $r$ in the interval $[R^+, R^-]$ and with respect to $\theta$ in the interval $[0, 2\pi]$ we can write the following transformations.

$$\int_{R^-}^{R^+}\int_0^{2\pi}\frac{1}{r}\frac{\partial\sigma_{\theta z}^{(1)}}{\partial\theta}r^2\cos\theta\,drd\,\theta=\pi\int_{R^-}^{R^+}s_{rz}^{(1)}rdr;$$

$$\int_{R^-}^{R^+}\int_0^{2\pi}\frac{1}{r}\sigma_{rz}^{(1)}r^2\cos\theta\,drd\,\theta=\pi\int_{R^-}^{R^+}s_{rz}^{(1)}rdr;$$

$$\int_{R^-}^{R^+}\int_0^{2\pi}\frac{1}{r}\frac{\partial\sigma_{rz}^{(1)}}{\partial r}r^2\cos\theta\,drd\,\theta=\pi\int_{R^-}^{R^+}\frac{ds_{rz}^{(1)}}{dr}r^2dr$$

$$=\pi\left[\int_{R^-}^{R^+}\frac{d}{dr}\left(s_{rz}^{(1)}r^2\right)dr-2\int_{R^-}^{R^+}s_{rz}^{(1)}rdr\right]$$

$$=\pi s_{rz}^{(1)}\big|_{r=R^+}\left(R^+\right)^2-\pi s_{rz}^{(1)}\big|_{r=R^-}\left(R^-\right)^2-2\pi\int_{R^-}^{R^+}s_{rz}^{(1)}rdr$$

$$=\pi\sigma_{zz}^{(0)}\left(\left(R^+\right)^2+\left(R^-\right)^2\right)\cos(\alpha_1 z)-2\pi\int_{R^-}^{R^+}s_{rz}^{(1)}rdr,$$

$$\int_{R^-}^{R^+}\int_0^{2\pi}\frac{\partial\sigma_{zz}^{(1)}}{\partial z}r^2\cos\theta\,drd\,\theta=\pi\int_{R^-}^{R^+}E_3^*\frac{d^3w}{dz^3}\left(-r^3\right)dr$$

$$=-\frac{\pi\left(\left(R^+\right)^4-\left(R^-\right)^4\right)}{4}E_3^*\frac{d^3w}{dz^3}=-JE_3^*\frac{d^3w}{dz^3},$$

$$J=\frac{\pi\left(\left(R^+\right)^4-\left(R^-\right)^4\right)}{4} \tag{2.51}$$

Here, the boundary conditions $\sigma_{rz}^{(1)}(R^+,\theta,t_3,t)=2\pi\sigma_{zz}^{(0)}\cos(\alpha_1 z)\cos\theta,\Rightarrow s_{rz}^{(1)}|_{r=R^+}=2\pi\sigma_{zz}^{(0)}\cos(\alpha_1 z)$,

$$\sigma_{rz}^{(1)}(R^-,\theta,t_3,t)=2\pi\sigma_{zz}^{(0)}\cos(\alpha_1 z)\cos\theta,\Rightarrow s_{rz}^{(1)}\big|_{r=R^-}=-2\pi\sigma_{zz}^{(0)}\cos(\alpha_1 z)$$

given in (2.32) are used. Thus, taking the transformation (2.51) and relation (2.50) into account we get the following equation for the displacement $w(z)$ from the last equation in (2.48).

$$\begin{aligned}-\left(E_3^*+\sigma_{zz}^{(0)}\right)J\frac{d^4w}{dz^4}+\pi\left(\left(R^+\right)^2-\left(R^-\right)^2\right)\sigma_{zz}^{(0)}\frac{d^2w}{dz^2}\\=\left(\left(R^+\right)^2-\left(R^-\right)^2\right)\alpha_1\sigma_{zz}^{(0)}\sin(\alpha_1 z).\end{aligned} \tag{2.52}$$

Applying the presentations (2.45) and (2.46) to the end conditions (2.33) and doing the integration over the area of the end cross sections we obtain the following ones for the approximate Bernoulli approach.

$$w|_{z=0;\ell}=0, \quad \left.\frac{d^2w}{dz^2}\right|_{z=0;\ell}=0. \tag{2.53}$$

It is evident that for the real cases the inequality $|\sigma_{zz}^{(0)}|<<|E_{31}^{*}1|$ takes place and taking this inequality into account, the $\sigma_{zz}^{(0)}$ can be ignored in the first term of the Eq. (2.52). After this ignoring with the use of the notation $P=-\pi^2\Big((R^+)^2 -(R^-)^2\Big)\sigma_{zz}^{(0)}$ we obtain the classical stability loss equation of the Bernoulli beam from which the Euler critical force within the scope of the initial imperfection criterion is obtained.

For the solution to the boundary value problem (2.52) and (2.53) we apply the Laplace transformation (2.34) to these equations. Using the convolution theorem for transformation of the term $E_3^* d^4w/dz^4$ we obtain the same equation and end conditions written for $\bar{w}$. For instance, after this transformation we obtain the following equation instead of Eq. (2.52)

$$-\bar{E}_3^* J\frac{d^4\bar{w}}{dz^4}-P\frac{d^2\bar{w}}{dz^2}=-\frac{1}{s\pi}\alpha_1 P\sin(\alpha_1 z). \tag{2.54}$$

According to the end condition (2.53), the solution to the Eq. (2.54) is taken as follows

$$\bar{w}=\bar{A}(s)\sin\alpha_1 z. \tag{2.55}$$

Substituting (2.55) into (2.54) and doing some mathematical calculations we obtain the following expression for the unknown $\bar{A}(s)$.

$$\bar{A}(s)=-\frac{P\ell}{s\pi^2}\left[P-\frac{\pi^2\bar{E}_3^* J}{\ell^2}\right]^{-1}. \tag{2.56}$$

Employing the method (Schapery 1962, 1978), the original of $\bar{A}(s)$ is determined as follows.

$$A(t)=(s\bar{A}(s))|_{s=(1/(2t))}. \tag{2.57}$$

For the cases where the cylinder material is pure elastic we obtain from (2.56) the expression

$$A=-\frac{P\ell}{\pi^2}\left[P-\frac{\pi^2 E_3 J}{\ell^2}\right]^{-1}. \tag{2.58}$$

According to (2.44), the stability loss criterion for the Bernoulli beam can be expressed as

$$\begin{aligned} &A(t) \to \infty \text{ as } t \to t_{cr} \text{ (for the case where the cylinder material is viscoelastic)}, \\ &A \to \infty \text{ as } P \to P_{cr} \text{ (for the case where the cylinder material is pure elastic)} \end{aligned} . \tag{2.59}$$

The following expression is obtained from (2.58) and (2.59) for the critical force which coincide with Euler critical force

$$P_{Eu.cr} = \frac{\pi^2 E_3 J}{\ell^2}. \tag{2.60}$$

## 2.4 The Third Order Refined Beam Theory

For the third order refined beam theory we use the modification of the theory (Kromm 1953, 1955) for the cylinder-beam, according to which, the displacements $u_r^{(1)}$ and $u_\theta^{(1)}$ are presented as in (2.45), but the displacement $u_z^{(1)}$ is presented as

$$\begin{aligned} u_z^{(1)} =& u_z^{(1)}(r,\theta,z) = -r\frac{dw(z)}{dz}\cos\theta \\ &+\frac{1}{4}\left(\left((R^+)^2+(R^-)^2\right)r\cos\theta - \frac{1}{3}r^3\cos^3\theta\right)\varphi(z) \end{aligned}, \tag{2.61}$$

where $\varphi(z)$ is an average shear deformation of the cross section of the cylinder-beam.

Substituting the expressions for $u_r^{(1)}$, $u_\theta^{(1)}$ and $u_z^{(1)}$ into Eq. (2.31) we obtain the expression for strains and stresses given below.

$$\begin{aligned} \varepsilon_{rz}^{(1)} =& \frac{1}{8}((R^+)^2+(R^-)^2)\cos\theta(1-\frac{r^2\cos^2\theta}{(R^+)^2+(R^-)^2})\varphi(z), \\ \varepsilon_{\theta z}^{(1)} =& -\frac{1}{8}((R^+)^2+(R^-)^2)\sin\theta(1-\frac{r^2\cos^2\theta}{(R^+)^2+(R^-)^2})\varphi(z), \\ \varepsilon_{rz}^{(1)} =& -r\cos\theta\frac{d^2w(z)}{dz^2}+\frac{1}{4}(((R^+)^2+(R^-)^2)r\cos\theta-\frac{r^3\cos\theta}{3})\frac{d\varphi(z)}{dz}, \\ \varepsilon_{rr}^{(1)} =& \varepsilon_{\theta\theta}^{(1)} = \varepsilon_{r\theta}^{(1)} = 0,\ \sigma_{rr}^{(1)} = \sigma_{\theta\theta}^{(1)} = \sigma_{r\theta}^{(1)} = 0, \\ \sigma_{rz}^{(1)} =& 2G^*\varepsilon_{rz}^{(1)} = G^*\frac{1}{4}((R^+)^2+(R^-)^2)\cos\theta \\ &\times(1-\frac{r^2}{((R^+)^2+(R^-)^2)}\cos^2\theta)\varphi(z), \\ \sigma_{\theta z}^{(1)} =& 2G^*\varepsilon_{\theta z}^{(1)} = -G^*\frac{1}{4}((R^+)^2+(R^-)^2)\sin\theta\,(1-\frac{r^2}{((R^+)^2+(R^-)^2)}\cos^2\theta)\varphi(z) \\ \sigma_{zz}^{(1)} =& E_3^*\varepsilon_{zz}^{(1)} = E_3^*(-r\cos\theta\frac{d^2w(z)}{dz^2}+\frac{1}{4}((R^+)^2 \\ &+(R^-)^2)r\cos\theta-\frac{r^3\cos\theta}{3})\frac{d\varphi(z)}{dz}. \end{aligned} \tag{2.62}$$

It follows from the expressions given in (2.62) that the Eqs. (2.48) and (2.49) hold also for the considered case. First, we apply the Laplace transformation to the Eqs. (2.48), (2.49) and (2.62) and doing the integration procedure which is similar to that made in the previous subsection we get from Eq. (2.48) the following equation for the function $\bar{w}(z,s)$.

$$E_3^* J\left[-1-\frac{\sigma_{zz}^{(0)}}{E_3^*}-\frac{4}{3}\frac{\Re\sigma_{zz}^{(0)}}{\left((R^+)^2+(R^-)^2\right)G^*}\left(1+\frac{\sigma_{zz}^{(0)}}{E_3^*}\right)\right]\frac{d^4\bar{w}(z,s)}{dz^4}$$

$$+\pi\left((R^+)^2+(R^-)^2\right)\sigma_{zz}^{(0)}\frac{d^2\bar{w}(z,s)}{dz^2}=\frac{\alpha\pi\left((R^+)^2+(R^-)^2\right)}{s}\sigma_{zz}^{(0)}\sin\alpha_1 z, \tag{2.63}$$

where

$$\Re=\left((R^+)^2+(R^-)^2\right)-\frac{(R^+)^6-(R^-)^6}{9\left((R^+)^4-(R^-)^4\right)}. \tag{2.64}$$

Thus, using the presentation (2.55) and the estimation

$$-1-\frac{\sigma_{zz}^{(0)}}{E_3^*}-\frac{4}{3}\frac{\Re\sigma_{zz}^{(0)}}{\left((R^+)^2+(R^-)^2\right)G^*}\left(1+\frac{\sigma_{zz}^{(0)}}{E_3^*}\right)\approx 1-\frac{4}{3}\frac{\Re\sigma_{zz}^{(0)}}{\left((R^+)^2+(R^-)^2\right)G^*}$$

we obtain from (2.63) the following expression for $\bar{A}(s)$

$$\bar{A}(s)=-\frac{P\ell}{s\pi^2}\left[P\left(1+\frac{1}{3}\frac{\bar{E}_3^*\Re}{\bar{G}^*}\left(\frac{\pi}{\ell}\right)^2\right)-\frac{\pi^2\bar{E}_3^*J}{\ell^2}\right]^{-1} \tag{2.65}$$

where, as in the previous case, the notation $P=-\pi^2((R^+)^2+(R^-)^2)\sigma_{zz}^{(0)}$ is used.

By employing the method (2.57) we determine the function $A(t)$. For determination of the values of the critical parameters we can also use the expression (2.59). The following expression can be written for the unknown $A$ from Eq. (2.65) for the case where the material of the cylinder is pure elastic.

$$A=-\frac{P\ell}{s\pi^2}\left[P\left(1+\frac{1}{3}\frac{E_3\Re}{G}\left(\frac{\pi}{\ell}\right)^2\right)-\frac{\pi^2 E_3 J}{\ell^2}\right]^{-1} \tag{2.66}$$

From (2.66) the following expression for the critical force for the case where the cylinder's material is pure elastic one can be written.

$$P=-\frac{\pi^2 E_3 J}{\ell^2}\left(1+\frac{1}{3}\frac{E_3\Re}{G}\left(\frac{\pi}{\ell}\right)^2\right)^{-1} \tag{2.67}$$

## 2.5 Numerical Results and Discussions

We assume that the cylinder is made from viscoelastic unidirectional fibrous composite material and its fibers lie along the $Oz$ axis. In the discussions below, the values related to the matrix and the fibers will be denoted by upper indices (1) and (2), respectively. The material of the fibers is supposed to be pure elastic with Young's modulus $E^{(2)}$, Poisson coefficient $\nu^{(2)}$, shear modulus $\mu^{(2)}$, but the material of the matrix is supposed to be linearly viscoelastic with operators

$$\begin{aligned} E^*(1)\varphi &= E_0^{(1)}\left[\varphi(t) - \omega_0\Pi_\alpha^*(-\omega_0 - \omega_\infty)\varphi\right], \\ \nu^{*(1)}\varphi &= \nu_0^{(1)}\left[\varphi(t) + \frac{1-2\nu_0^{(1)}}{2\nu_0^{(1)}}\omega_0\Pi_\alpha^*(-\omega_0 - \omega_\infty)\varphi\right], \\ \mu^{*(1)}\varphi &= \mu_0^{(1)}\left[\varphi(t) - \frac{3}{2(1+\nu_0^{(1)})}\omega_0\Pi_\alpha^*\left(-\frac{1}{2(1+\nu_0^{(1)})}\omega_0 - \omega_\infty\right)\varphi\right], \end{aligned} \tag{2.68}$$

where $E_0^{(1)}$, $\nu_0^{(1)}$ are the instantaneous values of Young's modulus and the Poisson coefficient, respectively, $\mu_0^{(1)}$ is the instantaneous value of shear modulus, $\alpha$, $\omega_0$ and $\omega_\infty$ are the rheological parameters of the matrix material, $\Pi_\alpha^*$ is the fractional exponential operator (Rabotnov 1977), and this operator is determined as

$$\Pi_\alpha^*(x)\varphi = \int_0^t \Pi_\alpha(x, t-\tau)\varphi(\tau)d\tau, \tag{2.69}$$

where

$$\Pi_\alpha(x,t) = t^\alpha \sum_{n=0}^{\infty} \frac{x^n t^{n(1+\alpha)}}{\Gamma((1+n)(1+\alpha))}, \quad -1<\alpha<0 \tag{2.70}$$

where $\Gamma(x)$is the Gamma function.

Note that the fractional-exponential operators (Rabotnov 1977) allow us to describe with the accuracy required the initial parts of the experimentally and theoretically constructed creep and relaxation graphs. These operators also allow us to determine with very high accuracy, the asymptotic values of these graphs. Operators (2.69) and (2.70) are employed successfully to describe various polymer materials and epoxy-based composites with continuous fibers or layers. The values of the rheological parameters in relation (2.68) were determined for these materials in the monograph (Rabotnov 1977). Moreover, these operators have many simple rules for complicated mathematical transformations, for example, the Laplace transformation, the expression of which is given below.

We introduce the dimensionless rheological parameter $\omega(=\omega_\infty/\omega_0)$ and the dimensionless time $t'(=\omega_0^{1/(1+\alpha)}t)$ and assume that $\nu^{(2)} = \nu_0^{(1)} = 0.3$, $\eta^{(2)} = 0.5$ where $\eta^{(2)}$ is a fiber volume fraction in the composite under consideration.

It is known that, within the scope of the continuum approach this composite can be taken as homogeneous transversal isotropic one the isotropy axis of which coincide with the $Oz$ axis. According to (Christensen 1979) and many others, by replacing the mechanical constants of the constituents of a composite with Laplace transformation of the corresponding operators in the expressions of the effective mechanical properties, we determine the Laplace transform of the effective operators. Therefore, in the Laplace transform of the constitutive relations (2.35) instead of $\bar{A}^*_{ij}$ and $\bar{G}^*$ we write these expressions. For the considered composite material the expressions for $\bar{A}^*_{ij}$ and $\bar{G}^*$ are determined as follows:

$$\bar{A}^*_{33} = \bar{E}^*_3 + 4(\bar{\nu}^*_{31})^2\bar{K}^*_{21},\ \bar{A}^*_{13} = 2\bar{\nu}^*_{31}\bar{K}^*_{21},\ \bar{A}^*_{11} = \bar{\mu}^*_{12} + \bar{K}^*_{21},\ \bar{A}^*_{12} = -\bar{\mu}^*_{12} + \bar{K}^*_{21},$$
$$\bar{G}^* = \bar{\mu}^{(1)}\frac{\bar{\mu}^{(2)}(1+\eta^{(2)}) + \bar{\mu}^{*(1)}(1-\eta^{(2)})}{\bar{\mu}^{(2)}(1-\eta^{(2)}) + \bar{\mu}^*(1)(1+\eta^{(2)})} \tag{2.71}$$

where

$$\bar{K}^*_{12} = \bar{K}^{*(1)} + \frac{1}{3}\bar{\mu}^{*(1)} + \eta^{(2)}\left[\left(\frac{1}{3}\left(\bar{\mu}^{(2)} - \bar{\mu}^{*(1)}\right) + \bar{K}^{(2)} - \bar{K}^{*(1)}\right)^{-1}\right.$$
$$\left. + \frac{(1-\eta^{(2)})}{\bar{K}^{*(1)} + \frac{4}{3}\bar{\mu}^{*(1)}}\right]^{-1},\ \bar{E}^*_3 = \eta^{(2)}\bar{E}^{(2)} + \left(1-\eta^{(2)}\right)\bar{E}^{*(1)}$$
$$+ \frac{4\eta^{(2)}(1-\eta^{(2)})(\bar{\nu}^{(2)} - \bar{\nu}^{*(1)})^2\bar{\mu}^{*(1)}}{(1-\eta^{(2)})\bar{\mu}^{*(1)}(\bar{K}^{(2)} + \bar{\mu}^{(2)}/3)^{-1} + \eta^{(2)}\bar{\mu}^{*(1)}(\bar{K}^{*(1)} + \bar{\mu}^{*(1)}/3)^{-1}},$$
$$\bar{\nu}^*_{31} = \eta^{(2)}\bar{\nu}^{(2)} + \left(1-\eta^{(2)}\right)\bar{\nu}^{*(1)} + 4\eta^{(2)}\left(1-\eta^{(2)}\right)\left(\bar{\nu}^{(2)} - \bar{\nu}^{*(1)}\right)^2$$
$$\times\frac{\left[\bar{\mu}^{*(1)}(\bar{K}^{*(1)} + \bar{\mu}^{*(1)}/3)^{-1} - \bar{\mu}^{*(1)}(\bar{K}^{(2)} + \bar{\mu}^{(2)}/3)^{-1}\right]}{(1-\eta^{(2)})\bar{\mu}^{*(1)}(\bar{K}^{(2)} + \bar{\mu}^{(2)}/3)^{-1} + \eta^{(2)}\bar{\mu}^{*(1)}(\bar{K}^{*(1)} + \bar{\mu}^{*(1)}/3)^{-1}},$$
$$\bar{\mu}^*_{12} = \bar{\mu}^{*(1)}\left\{1 + \eta^{(2)}\left[\frac{\bar{\mu}^{*(1)}}{\bar{\mu}^{(2)} - \bar{\mu}^{*(1)}} + \frac{\bar{K}^{*(1)} + 7\bar{\mu}^{*(1)}/3}{\bar{K}^{*(1)} + 8\bar{\mu}^{*(1)}/3}\right]^{-1}\right\}. \tag{2.72}$$

Here the following notation is used:

$$\bar{K}^{*(1)} = \frac{\bar{E}^{*(1)}}{3(1-2\bar{\nu}^{*(1)})},\ \bar{K}^{(2)} = \frac{\bar{E}^{(2)}}{3(1-2\bar{\nu}^{(2)})},$$
$$\bar{\mu}^{(2)} = \frac{\bar{E}^{(2)}}{3(1+2\bar{\nu}^{(2)})},\ \bar{E}^{(2)} = \frac{1}{s}E^{(2)},\ \bar{\nu}^{(2)} = \frac{1}{s}\nu^{(2)},$$

$$\bar{E}^{*(1)} = E_1^{(0)}\left[\frac{1}{s} - \bar{\Pi}_\alpha(-\omega)\right],\ \bar{v}^{*(1)} = v_1^{(0)}\left[\frac{1}{s} - \frac{1-2v_0^{(1)}}{2v_0^{(1)}}\bar{\Pi}_\alpha(-\omega)\right],$$
$$\bar{\mu}^{*(1)} = \mu_0^{(1)}\left[\frac{1}{s} - \frac{3}{2\left(1+v_0^{(1)}\right)}\bar{\Pi}_\alpha\left(-\frac{3}{2\left(1+v_0^{(1)}\right)} - \omega\right)\right],$$
$$\bar{\Pi}_\alpha(-x) = \frac{1}{s(s^{1+\alpha}+x)}. \tag{2.73}$$

Thus, within the framework of the foregoing preparation we consider the numerical results and first examine the pure elastic stability loss under $t' = 0$ and $t' = \infty$, because in the viscoelstic stability loss of the cylinder the intensity of the external compressed force

$$p = \sigma_{zz}^{(0)} = P\left[\pi\left((R^+)^2 - (R^-)^2\right)\right]^{-1} \tag{2.74}$$

must satisfy the inequality

$$p_{cr.\infty} < p < p_{cr.0} \tag{2.75}$$

where $p_{cr.0}$ ($p_{cr.\infty}$) is the critical force obtained at $t' = 0$ ($t' = \infty$).

We compare the results obtained within the scope of the TDLTS with the corresponding ones obtained within the scope of the Bernoulli beam theory and within the scope of the third order refined beam theory. Through $p'_{3D.c.0}\left(= p_{3D.c.0}/E_0^{(1)}\right)$ and $p'_{3D.c.\infty}\left(= p_{3D.c.\infty}/E_0^{(1)}\right)$ we denote the critical values of the intensity of the dimensionless forces obtained by employing the TDLTS. Here and below the sub-indices 0 and $\infty$ indicate that the critical values of the forces are calculated at $t' = 0$ and at $t' = \infty$ respectively. According to the Eqs. (2.60) and (2.67), the corresponding critical values of the intensity of the compressed forces obtained within the scope of the Bernoulli beam theory (denoted by $p'_{E.c.0}$ and $p'_{E.c.\infty}$) and within the scope of the third order refined beam theory (denoted by $p'_{R.c.0}$ and $p'_{R.c.\infty}$) can be calculated by the use of the following expressions.

$$p'_{E.c.0} = \frac{E_{30}}{4E_0^{(1)}}\left[\left(\frac{\pi R^+}{\ell}\right)^2 + \left(\frac{\pi R^-}{\ell}\right)^2\right],\ p'_{E.c.\infty} = \frac{E_{3\infty}}{4E_0^{(1)}}\left[\left(\frac{\pi R^+}{\ell}\right)^2 + \left(\frac{\pi R^-}{\ell}\right)^2\right],$$
$$p'_{R.c.0} = \frac{E_{30}}{4E_0^{(1)}}\left[\left(\frac{\pi R^+}{\ell}\right)^2 + \left(\frac{\pi R^-}{\ell}\right)^2\right]\left(1 + \frac{1}{3}\frac{E_{30}\Re}{G_0}\left(\frac{\pi}{\ell}\right)^2\right)^{-1},$$
$$p'_{R.c.\infty} = \frac{E_{3\infty}}{4E_0^{(1)}}\left[\left(\frac{\pi R^+}{\ell}\right)^2 + \left(\frac{\pi R^-}{\ell}\right)^2\right]\left(1 + \frac{1}{3}\frac{E_{3\infty}\Re}{G_\infty}\left(\frac{\pi}{\ell}\right)^2\right)^{-1}. \tag{2.76}$$

Substituting $R^+ = R$, $R^- = 0$ into expressions for $J$ (2.51) and $\Re$ (2.64), we obtain from (2.76) the following expressions for the solid cylinder.

**Table 2.1** The values of $p'_{3D.c.0}$, $p'_{R.c.0}$ and $p'_{E.c.0}$ obtained for various values of $E^{(2)}/E_0^{(1)}$ and $\rho$

| $\frac{E^{(2)}}{E_0^{(1)}}$ | $\rho=\frac{\pi R}{\ell}$ | | | | | | | | |
|---|---|---|---|---|---|---|---|---|---|
| | 0.1 | | | 0.2 | | | 0.3 | | |
| | $p'_{3D.c.0}$ | $p'_{R.c.0}$ | $p'_{E.c.0}$ | $p'_{3D.c.0}$ | $p'_{R.c.0}$ | $p'_{E.c.0}$ | $p'_{3D.c.0}$ | $p'_{R.c.0}$ | $p'_{E.c.0}$ |
| 1 | 0.00247 | 0.00248 | 0.0025 | 0.0096 | 0.0097 | 0.010 | 0.0207 | 0.0210 | 0.0224 |
| 5 | 0.00740 | 0.00741 | 0.0075 | 0.0284 | 0.0286 | 0.030 | 0.0603 | 0.0611 | 0.0675 |
| 10 | 0.0134 | 0.01351 | 0.0137 | 0.0510 | 0.0513 | 0.055 | 0.1055 | 0.1066 | 0.1237 |

$$p'_{E.c.0}=\frac{E_{30}}{4E_0^{(1)}}\left(\frac{\pi R}{\ell}\right)^2,\ p'_{R.c.0}=\frac{E_{30}}{4E_0^{(1)}}\left(\frac{\pi R}{\ell}\right)^2\left(1+\frac{8}{27}\frac{E_{30}}{G_0}\right)^{-1},$$
$$p'_{E.c.\infty}=\frac{E_{3\infty}}{4E_0^{(1)}}\left(\frac{\pi R}{\ell}\right)^2,\ p'_{R.c.\infty}=\frac{E_{3\infty}}{4E_0^{(1)}}\left(\frac{\pi R}{\ell}\right)^2\left(1+\frac{8}{27}\frac{E_{3\infty}}{G_0}\right)^{-1}. \tag{2.77}$$

Now we analyze the numerical results obtained within the scope of the foregoing assumptions. The results related to the solid and hollow cylinders will be analyzed separately.

### *2.5.1 Solid Cylinder*

Table 2.1 (Table 2.2) shows the values of $p'_{3D.c.0}$, $p'_{R.c.0}$ and $p'_{E.c.0}$ $(p'_{3D.c.\infty}, p'_{R.c.\infty}$ and $p'_{E.c.\infty})$ calculated for various values of $E^{(2)}/E_0^{(1)}$, $\omega$ and $\rho(=\pi R/\ell)$. Note that the values of $p'_{E.c.0}$, $p'_{R.c.0}$, $p'_{E.c.\infty}$ and $p'_{R.c.\infty}$ are calculated by the use of the expression (2.77). However, the values of $p'_{3D.c.0}$ and $p'_{3D.c.\infty}$ are calculated by the use of the criterion (2.44) and by the use of the solution of the system equation which is obtained from the first three equations of the system equations given in (2.42) by taking the equalities $C_4=0$, $C_5=0$ and $C_6=0$ into account and by replacing the notation $R^+$ with $R$.

It follows from these results that the values of $p'_{R.c.0}$ $(p'_{R.c.\infty})$ are very near to the corresponding values of $p'_{3D.c.0}$ $(p'_{3D.c.\infty})$ and the difference between them is insignificant. However the values of $p'_{E.c.0}$ $(p'_{E.c.\infty})$ are greater than the corresponding values of $p'_{3D.c.0}$ $(p'_{3D.c.\infty})$ and the difference between them becomes more significant with $E^{(2)}/E_0^{(1)}$ and $\rho$.

Consequently, it can be concluded from the foregoing results that for the pure elastic stability loss problems of the considered cylinder, the third order refined beam theory gives results which are sufficiently close to the corresponding ones obtained within the scope of the TDLTS.

**Table 2.2** The values of $p'_{3D.c.\infty}$, $p'_{R.c.\infty}$ and $p'_{E.c.0}$ obtained for various values of $E^{(2)}\big/E_0^{(1)}$, $\rho = \pi R/\ell$ and $\omega$

| $E^{(2)}/E_0^{(1)}$ | $\rho$ = 0.1 | | | 0.2 | | | 0.3 | | |
|---|---|---|---|---|---|---|---|---|---|
| | $p'_{3D.c.\infty}$ | $p'_{R.c.0}$ | $p'_{E.c.0}$ | $p'_{3D.c.\infty}$ | $p'_{R.c.0}$ | $p'_{E.c.0}$ | $p'_{3D.c.\infty}$ | $p'_{R.c.0}$ | $p'_{E.c.0}$ |
| $\omega = 0.5$ | | | | | | | | | |
| 1 | 0.00163 | 0.00163 | 0.00166 | 0.00611 | 0.00614 | 0.0066 | 0.0124 | 0.0125 | 0.015 |
| 5 | 0.00621 | 0.00622 | 0.00666 | 0.02073 | 0.02073 | 0.0266 | 0.0366 | 0.0365 | 0.060 |
| 10 | 0.01138 | 0.01137 | 0.01291 | 0.03368 | 0.03353 | 0.0516 | 0.0530 | 0.0524 | 0.116 |
| $\omega = 1.0$ | | | | | | | | | |
| 1 | 0.00184 | 0.00185 | 0.00187 | 0.00709 | 0.00713 | 0.0075 | 0.0149 | 0.0151 | 0.016 |
| 5 | 0.00664 | 0.00665 | 0.00687 | 0.02411 | 0.02420 | 0.0275 | 0.0471 | 0.0473 | 0.061 |
| 10 | 0.01235 | 0.01236 | 0.01312 | 0.04207 | 0.04212 | 0.0525 | 0.0760 | 0.0759 | 0.118 |
| $\omega = 2.0$ | | | | | | | | | |
| 1 | 0.00206 | 0.00206 | 0.00208 | 0.00796 | 0.0080 | 0.0083 | 0.01698 | 0.0172 | 0.018 |
| 5 | 0.00692 | 0.00693 | 0.00708 | 0.02560 | 0.0261 | 0.0283 | 0.05315 | 0.0536 | 0.063 |
| 10 | 0.01285 | 0.01287 | 0.01333 | 0.04645 | 0.0466 | 0.0533 | 0.09021 | 0.0906 | 0.120 |

Now we consider the case where the cylinder material is viscoelastic and analyze the critical time $t'_{cr}$. To compare the values of $t'_{cr}$ given by the TDLTS with those given by approximate beam theories, we must select the values of $p'(= p\big/E_0^{(1)})$ which fall in the intervals $[p'_{E.c.\infty}, p'_{E.c.0}]$, $[p'_{R.c.\infty}, p'_{R.c.0}]$ and $[p'_{3D.c.\infty}, p'_{3D.c.0}]$ simultaneously. However, as follows from the data given in Tables 2.1 and 2.2, in many cases such common values of $p'(= p\big/E_0^{(1)})$ exist only for the intervals $[p'_{R.c.\infty}, p'_{R.c.0}]$ and $[p'_{3D.c.\infty}, p'_{3D.c.0}]$. For this reason, we will compare only the critical time obtained by employing the TDLTS (denoted by $t'_{3D.cr.}$) with that obtained by employing the third order refined beam theory (denoted by $t'_{Rf.cr.}$).

Thus, consider the data given in Tables 2.3–2.5 which show the values of the $t'_{3D.cr.}$ and $t'_{Rf.cr.}$ obtained for the cases where $\rho = 0.1,\ 0.2$ and $0.3$ respectively for various values of $p'$ and $\alpha$ under $\omega = 0.5,\ E^{(2)}\big/E_0^{(1)} = 5$. Moreover, for a clearer illustration of the results related to the critical time in Fig. 2.2 the graphs of the dependencies among $t'_{3D.cr.}$, $t'_{Rf.cr.}$ and $p'$ are given for various values of $\alpha$ in the case where $\rho = 0.1,\quad \omega = 0,5$.

It follows from the results given in Fig. 2.2 and from other numerical results (which are not given here) that the graphs of the analyzed dependencies constructed for various values of $\alpha$ have a common intersection point. Before (after) this intersection point an increase in the absolute values of $\alpha$ causes an increase (decrease) in the values of the critical time. Figure 2.3 illustrates the parts of the graphs which appear after the aforementioned intersection point and it follows from this investigation that the position of the intersection point does not depend on the values of $\alpha$. According to this statement and according to the expression of

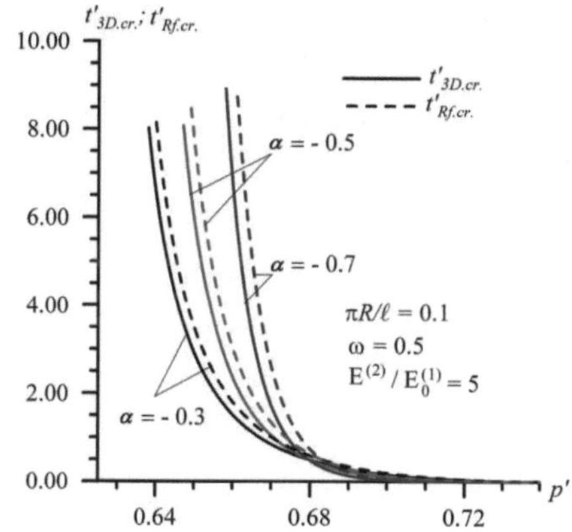


**Fig. 2.2** The graphs of the dependencies among $t'_{3D.cr.}$, $t'_{Rf.cr.}$ and dimensionless intensity of the compressed force $p'$

$\bar{\Pi}_\alpha$ given in (2.73), it can be concluded that the "intersection point" corresponds to the case where the Laplace transformation parameter $s$ is equal to one, i.e. $s = 1$. In other words, according to the method (Schapery 1962, 1978), i.e. according to the expressions (2.43) or (2.57), the "intersection point" corresponds to the case where $t'_{cr} = 0.5$. Consequently, the foregoing conclusion on the influence of the rheological parameter $\alpha$ on the values of $t'_{cr}$ can be rewritten as follows:

If $t'_{cr} > 0.5$ ($t'_{cr} < 0.5$) the increase in the absolute values of $\alpha$ causes an increase (a decrease) in the values of the $t'_{cr}$.

This conclusion holds for all problems considered and studied in the present book. From the viewpoint of the author, in the qualitative sense, the foregoing conclusion follows from the nature of the viscoelastic behavior of deformable solid bodies, but not from either special selection of the operator (2.69) or from the inverse transformation method (Schapery 1962, 1978). For example, it is evident that use of the more accurate inverse transformation method rather than Schapery's method (1962, 1978), does not change significantly the value of $t'_{cr}$ which corresponds to $s = 1$.

Note that the parameter $\alpha$ is one of the rheological parameters of the matrix material and characterizes the mechanical behavior of this viscoelastic material around the initial state of the deformation, i.e. in the near vicinity of $t' = 0$. However, the other dimensionless rheological parameter $\omega$ characterizes the mechanical properties of the matrix material around $t' = \infty$.

As the values of $E_\infty^{(1)}(\nu_\infty^{(1)})$ increase (decrease) with $\omega$, therefore, it can be predicted that the values of $p'_{cr.\infty}$ must increase with $\omega$. This prediction is proved by the data given in Table 2.2. Moreover, the influence of the parameter $\omega$ on the graphs of the dependencies among $t'_{3D.cr.}$, $t'_{Rf.cr.}$ and $p'$ is illustrated by the graphs given in Fig. 2.4.

Analyses of the results given in Tables 2.3, 2.4, 2.5 show that in the case under consideration, i.e. in the case where $E^{(2)} / E_0^{(1)} = 5$, $\omega = 0.5$, $\rho = 0.1$, 0.2 and 0.3 the values of $t'_{Rf.cr.}$ are greater than the corresponding values of $t'_{3D.cr.}$. The

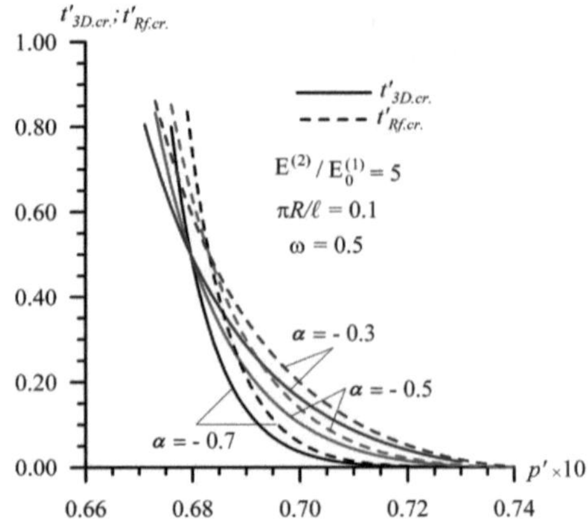


**Fig. 2.3** The parts of the graphs given in the Fig. 2.2 which arise after the intersection pointintersection point

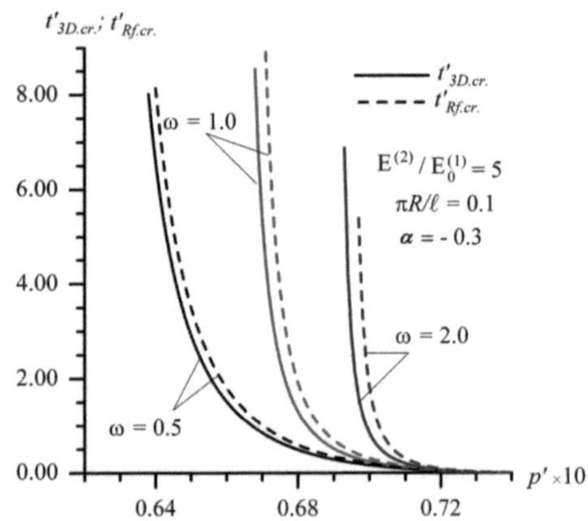


**Fig. 2.4** The graphs of the dependencies among $t'_{3D.cr.}$, $t'_{Rf.cr.}$ and $p'$constructed for various values of rheological parameter $\omega$

difference between $t'_{Rf.cr.}$ and $t'_{3D.cr.}$ depends not only on the rheological parameters $\omega$ and $\alpha$, but also on the values of the intensity of the external compressive forces $p'$ and, as can be predicted, there exist the following relations:

$$\begin{aligned} t'_{3D.cr.} &\to 0 \; as \; p' \to p'_{3D.c.0}; \; t'_{3D.cr.} \to \infty \; as \; p' \to p'_{3D.c.\infty} \\ t'_{Rf.cr.} &\to 0 \; as \; p' \to p'_{R.c.0}; \; t'_{Rf.cr.} \to \infty \; as \; p' \to p'_{R.c.\infty} \end{aligned} \tag{2.78}$$

The results given in Tables2.3, 2.4, 2.5, and many other results (which are not given here), show that the insignificant difference between $p'_{3D.c.0}$ and $p'_{R.c.0}$, as well as the insignificant difference between $p'_{3D.c.\infty}$ and $p'_{R.c.\infty}$, causes the significant difference between the values of $t'_{3D.cr.}$ and $t'_{Rf.cr.}$. In this case, if $p'_{3D.c.0} > p'_{R.c.0}$ (or if $p'_{3D.c.0} < p'_{R.c.0}$) then the values of $t'_{3D.cr.}$ obtained for $p'$ which are close to the $p'_{3D.c.0}$ or to the $p'_{R.c.0}$, are greater (less) than the corresponding values of $t'_{Rf.cr.}$. Moreover, if $p'_{3D.c.\infty} > p'_{R.c.\infty}$ (or if $p'_{3D.c.\infty} < p'_{R.c.\infty}$) then the values of $t'_{3D.cr.}$ obtained for the $p'$ which is close to the $p'_{3D.c.\infty}$ or to the $p'_{R.c.\infty}$, are greater (less) than the corresponding

**Table 2.3** The values of $t'_{3D.cr.}$ (upper number) and $t'_{Rf.cr.}$(lower number) obtained for various values of $p'$ and $\alpha$ under $\rho = 0.1,\ \omega = 0.5$ and $E^{(2)} \big/ E_0^{(1)} = 5$

| $\alpha$ | $p' \times 10^2$ | | | | | | | | |
|---|---|---|---|---|---|---|---|---|---|
| | 0.637 | 0.650 | 0.660 | 0.670 | 0.680 | 0.690 | 0.700 | 0.710 | 0.720 |
| -0.3 | 10.59 / 11.24 | 3.341 / 3.508 | 1.726 / 1.810 | 0.967 / 1.015 | 0.561 / 0.590 | 0.328 / 0.346 | 0.187 / 0.199 | 0.100 / 0.108 | 0.046 / 0.051 |
| -0.5 | 35.92 / 39.06 | 7.143 / 7.648 | 2.835 / 3.030 | 1.259 / 1.347 | 0.588 / 0.631 | 0.277 / 0.299 | 0.187 / 0.199 | 0.052 / 0.058 | 0.018 / 0.020 |
| -0.7 | 620.9 / 713.9 | 42.05 / 47.13 | 9.015 / 10.07 | 2.331 / 2.609 | 0.655 / 0.738 | 0.187 / 0.212 | 0.050 / 0.058 | 0.011 / 0.014 | 0.0019 / 0.0025 |

**Table 2.4** The values of $t'_{3D.cr.}$ (upper number) and $t'_{Rf.cr.}$(lower number) obtained for various values of $p'$ and $\alpha$ under $\rho = 0.2,\ \omega = 0.5$ and $E^{(2)} \big/ E_0^{(1)} = 5$

| $\alpha$ | $p' \times 10^2$ | | | | | | | | |
|---|---|---|---|---|---|---|---|---|---|
| | 2.250 | 2.300 | 2.350 | 2.400 | 2.450 | 2.500 | 2.550 | 2.600 | 2.650 |
| -0.3 | 5.704 / 5.998 | 3.465 / 3.660 | 2.227 / 2.366 | 1.475 / 1.578 | 0.990 / 1.068 | 0.664 / 0.724 | 0.439 / 0.486 | 0.282 / 0.318 | 0.173 / 0.200 |
| -0.5 | 15.10 / 16.20 | 7.517 / 8.117 | 4.049 / 4.408 | 2.275 / 2.501 | 1.301 / 1.447 | 0.743 / 0.840 | 0.417 / 0.480 | 0.225 / 0.266 | 0.113 / 0.139 |
| -0.7 | 146.4 / 165.7 | 45.79 / 52.04 | 16.33 / 18.81 | 6.247 / 7.315 | 2.462 / 2.941 | 0.969 / 1.187 | 0.370 / 0.468 | 0.132 / 0.174 | 0.042 / 0.059 |

**Table 2.5** The values of $t'_{3D.cr.}$ (upper number) and $t'_{Rf.cr.}$(lower number) obtained for various values of $p'$ and $\alpha$ under $\rho = 0.3,\ \omega = 0.5$ and $E^{(2)} \big/ E_0^{(1)} = 5$

| $\alpha$ | $p' \times 10^2$ | | | | | | | | |
|---|---|---|---|---|---|---|---|---|---|
| | 4.100 | 4.200 | 4.300 | 4.400 | 4.500 | 4.600 | 4.700 | 4.800 | 4.900 |
| -0.3 | 8.151 / 8.208 | 5.634 / 5.730 | 4.074 / 4.178 | 3.035 / 3.137 | 2.308 / 2.403 | 1.779 / 1.866 | 1.383 / 1.462 | 1.080 / 1.151 | 0.843 / 0.908 |
| -0.5 | 24.89 / 25.13 | 14.84 / 15.20 | 9.428 / 9.769 | 6.246 / 6.541 | 4.257 / 4.504 | 2.957 / 3.161 | 2.078 / 2.247 | 1.469 / 1.608 | 1.040 / 1.153 |
| -0.7 | 336.9 / 342.4 | 142.3 / 148.0 | 66.80 / 70.87 | 33.62 / 36.31 | 17.74 / 19.50 | 9.670 / 10.81 | 5.373 / 6.119 | 3.015 / 3.504 | 1.695 / 2.013 |

values of $t'_{Rf.cr.}$. The results illustrated in Tables 2.3, 2.4, 2.5 correspond to the case where $p'_{3D.c.0} < p'_{R.c.0}$ and $p'_{3D.c.\infty} < p'_{R.c.\infty}$. Therefore in the all cases considered in these tables the values of $t'_{3D.cr.}$ are less than the corresponding values of $t'_{Rf.cr.}$.

Consider the numerical results related to the other aforementioned cases, namely the case where $p'_{3D.c.0} < p'_{R.c.0}$ and $p'_{3D.c.\infty} > p'_{R.c.\infty}$. The results given in Tables 2.1 and 2.2 show that the values of $p'_{3D.c.0}$, $p'_{R.c.0}$, $p'_{3D.c.\infty}$ and $p'_{R.c.\infty}$ obtained, for example for the case where $\rho = 0.1$, $E^{(2)} \big/ E_0^{(1)} = 10$ and $\omega = 0.5$ satisfy these inequalities. The corresponding values of $t'_{3D.cr.}$ and $t'_{Rf.cr.}$ obtained for the case mentioned are given in Table 2.6. Consequently, according to the foregoing discussion, the following relations must occur:

**Table 2.6** The values of $t'_{3D.cr.}$ (upper number) and $t'_{Rf.cr.}$(lower number) obtained for various values of $p'$ and $\alpha$ under $\rho = 0.1,\ \omega = 0.5$ and $E^{(2)}/E_0^{(1)} = 10$

| $\alpha$ | $p' \times 10^2$ | | | | | | | | |
|---|---|---|---|---|---|---|---|---|---|
| | 1.15 | 1.16 | 1.17 | 1.18 | 1.19 | 1.20 | 1.21 | 1.22 | 1.23 |
| -0.3 | 73.51 / 68.95 | 27.39 / 26.73 | 14.47 / 14.35 | 8.865 / 8.869 | 5.881 / 5.920 | 4.094 / 4.141 | 2.937 / 2.984 | 2.148 / 2.191 | 1.588 / 1.627 |
| -0.5 | 541.1 / 494.7 | 135.8 / 131.3 | 55.61 / 54.96 | 28.00 / 28.02 | 15.76 / 15.91 | 9.493 / 9.648 | 5.964 / 6.098 | 3.849 / 3.958 | 2.522 / 2.608 |

**Table 2.7** The values of $t'_{cdm.cr.}$ (upper number) and $t'_{3D.cr.}$ (lower number) obtained for various values of $p'$ and $\alpha$ under $\rho = 0.1,\ \omega = 0.5$ and $E^{(2)}/E_0^{(1)} = 5$

| $\alpha$ | $p' \times 10^2$ | | | | | | |
|---|---|---|---|---|---|---|---|
| | 0.660 | 0.670 | 0.680 | 0.690 | 0.700 | 0.710 | 0.720 |
| -0.3 | 52.77 / 1.726 | 52.77 / 0.967 | 1.179 / 0.561 | 0.532 / 0.328 | 0.262 / 0.187 | 0.127 / 0.100 | 0.055 / 0.046 |
| -0.5 | 340.2 / 2.835 | 7.577 / 1.259 | 1.662 / 0.588 | 0.545 / 0.277 | 0.202 / 0.126 | 0.074 / 0.052 | 0.023 / 0.018 |
| -0.7 | 26326 / 9.015 | 46.40 / 2.331 | 3.701 / 0.655 | 0.578 / 0.187 | 0.110 / 0.050 | 0.020 / 0.011 | 0.0029 / 0.0019 |

$$
\begin{aligned}
&t'_{3D.cr.} > t'_{Rf.cr.}\ \textit{for } p' \textit{ which is close to the values of } p'_{3D.c.0\ or\ p'R.c.0}\\
&t'_{3D.cr.} < t'_{Rf.cr.}\ \textit{for } p' \textit{ which is close to the values of } p'_{3D.c.\infty\ or\ p'R.c.\infty}
\end{aligned}
\tag{2.79}
$$

The analysis of the results given in Table 2.6 proves the reliability of the relation (2.79).

Compare the values of $t'_{3D.cr.}$ with the corresponding ones calculated by employing the critical deformation method (Gerard and Gilbert 1958). According to this method, it is assumed that the critical deformation of the viscoelastic cylinder is equal to the critical deformation of the corresponding elastic cylinder. Consequently, using this assumption the critical deformation for the pure elastic cylinder is determined within the scope of the TDLTS. Note that in the considered case the critical deformation mentioned corresponds to $p'_{3D.c.0}$. According to this determination, using the relation $p'_{3D.c.0} = p/E_3^*$ the critical time is determined for the selected values of $p$.

The values of the dimensionless critical time (denoted by $t'_{cdm.cr.}$) determined by employing the critical deformation method are given in Table 2.7. These values are calculated for the case where $\rho = 0.1, E^{(2)}/E_0^{(1)} = 5$ and $\omega = 0.5$. At the same time, in this table the corresponding values of $t'_{3D.cr.}$ are also illustrated. Comparison of the $t'_{cdm.cr.}$ with the $t'_{3D.cr.}$ shows that the critical deformation method is not acceptable for determination of the critical time for the stability loss of the cylinder made from viscoelastic composite material.

With this we restrict to consideration of the numerical results related to the solid cylinder.

**Table 2.8** The values of $p'_{3D.c.0}$, $p'_{R.c.0}$ and $p'_{E.c.0}$ obtained for various values of $E^{(2)}/E_0^{(1)}$, $h/R^+$ and $\rho = \pi R/\ell$

| $\frac{h}{R^+}$ | $\frac{E^{(2)}}{E^{(1)}}$ | $\rho$ = 0.2 $p'_{3D.c.0}$ | 0.2 $p'_{R.c.0}$ | 0.2 $p'_{E.c.0}$ | $\rho$ = 0.3 $p'_{3D.c.0}$ | 0.3 $p'_{R.c.0}$ | 0.3 $p'_{E.c.0}$ |
|---|---|---|---|---|---|---|---|
| 0.3 | 1 | 0.0138 | 0.0142 | 0.0148 | 0.028 | 0.030 | 0.033 |
| | 5 | 0.0400 | 0.0417 | 0.0446 | 0.079 | 0.086 | 0.100 |
| | 10 | 0.0697 | 0.0739 | 0.0819 | 0.132 | 0.142 | 0.184 |
| 0.5 | 1 | 0.0117 | 0.0120 | 0.0125 | 0.024 | 0.025 | 0.028 |
| | 5 | 0.0344 | 0.0354 | 0.0374 | 0.070 | 0.074 | 0.108 |
| | 10 | 0.0606 | 0.0630 | 0.0687 | 0.119 | 0.128 | 0.154 |
| 0.8 | 1 | 0.0099 | 0.0100 | 0.0103 | 0.021 | 0.021 | 0.023 |
| | 5 | 0.0294 | 0.0297 | 0.0311 | 0.061 | 0.063 | 0.070 |
| | 10 | 0.0526 | 0.0532 | 0.0571 | 0.107 | 0.110 | 0.128 |

### 2.5.2 Hollow Cylinder

The values of $p'_{3D.c.0}$, $p'_{R.c.0}$ and $p'_{E.c.0}$ are given in Table 2.8 for various values of $E^{(2)}/E_0^{(1)}$, $\rho(=\pi R^+/\ell)$ and $h/R^+$. The corresponding values of $p'_{3D.c.\infty}$, $p'_{R.c.\infty}$ and $p'_{E.c.\infty}$ obtained in case where $E^{(2)}/E_0^{(1)} = 5$, $\omega = 0.5$ are tabulated in Table 2.9. Note that the values of $p'_{E.c.0}$, $p'_{R.c.0}$, $p'_{E.c.\infty}$ and $p'_{R.c.\infty}$ are calculated by the use of the expression (2.76). However, the values of $p'_{3D.c.0}$ and $p'_{3D.c.\infty}$ are calculated by the use of the criterion (2.44) and the solution of the system of Eq. (2.42).

It follows from the foregoing results that for the considered all cases the following relations occur.

$$\begin{gathered}
p'_{3D.c.0} < p'_{R.c.0} < p'_{E.c.0},\; p'_{3D.c.\infty} < p'_{R.c.\infty} < p'_{E.c.\infty}, \\
\left|p'_{E.c.0} - p'_{3D.c.0}\right| > \left|p'_{E.c.0} - p'_{R.c.0}\right|, \\
\left|p'_{E.c.\infty} - p'_{3D.c.\infty}\right| > \left|p'_{E.c.\infty} - p'_{R.c.\infty}\right|.
\end{gathered} \tag{2.80}$$

Moreover, it follows from these results that the values of differences given in (2.80) increase with $E^{(2)}/E_0^{(1)}$ and $\rho$. But, decrease in the values of $h/R^+$ causes to increase of the foregoing dimensionless intensity of the critical forces. If we consider the total dimensionless force

$$P' = \frac{P}{E_0^{(1)}}\left(\frac{\pi}{\ell}\right)^2 = p'\pi\left(\frac{\pi}{\ell}\right)^2\left((R^+)^2 - (R^-)^2\right) \tag{2.81}$$

instead of the $p'$, then decrease in the values of $h/R^+$ must cause to decrease in the critical values of $P'$. This affirmation is proven by data given in Table 2.10 which shows the values of $P'_{3D.c.0}$ for the case where $E^{(2)}/E_0^{(1)} = 5$.

**Table 2.9** The values of $p'_{3D.c.0}$, $p'_{R.c.0}$ and $p'_{E.c.0}$ obtained for various values of $E^{(2)}\big/E_0^{(1)}$, $h/R^+$ and $\rho = \pi R/\ell$ under $\omega = 0.5$

| $\frac{h}{R^+}$ | $\frac{E^{(1)}}{E_0^{(1)}}$ | $\rho$ | | | | | |
|---|---|---|---|---|---|---|---|
| | | 0.2 | | | 0.3 | | |
| | | $p'_{3D.c.0}$ | $p'_{R.c.0}$ | $p'_{E.c.0}$ | $p'_{3D.c.0}$ | $p'_{R.c.0}$ | $p'_{E.c.0}$ |
| 0.3 | 1 | 0.0082 | 0.0087 | 0.0099 | 0.015 | 0.017 | 0.022 |
| | 5 | 0.0236 | 0.0276 | 0.0397 | 0.035 | 0.045 | 0.089 |
| | 10 | 0.0337 | 0.0421 | 0.0769 | 0.045 | 0.060 | 0.173 |
| 0.5 | 1 | 0.0072 | 0.0075 | 0.0083 | 0.013 | 0.015 | 0.018 |
| | 5 | 0.0220 | 0.0244 | 0.0333 | 0.035 | 0.041 | 0.075 |
| | 10 | 0.0331 | 0.0382 | 0.0645 | 0.047 | 0.056 | 0.145 |
| 0.8 | 1 | 0.0062 | 0.0063 | 0.0069 | 0.012 | 0.013 | 0.015 |
| | 5 | 0.0209 | 0.0213 | 0.0277 | 0.036 | 0.037 | 0.062 |
| | 10 | 0.0334 | 0.0343 | 0.0537 | 0.051 | 0.053 | 0.120 |

**Table 2.10** The values of $P'_{3D.c.0} \times 10^2$ obtained in the case where $E^{(2)}\big/E_0^{(1)} = 5$ where $P'_{3D.c.0}$ (2.81) is a dimensionless total critical force

| $\rho = \frac{\pi R^+}{\ell}$ | $h/R^+$ | | | | | | |
|---|---|---|---|---|---|---|---|
| | 0.3 | 0.4 | 0.5 | 0.6 | 0.7 | 0.8 | 0.9 |
| 0.2 | 0.256 | 0.297 | 0.324 | 0.341 | 0.350 | 0.355 | 0.357 |
| 0.3 | 1.152 | 1.354 | 1.494 | 1.587 | 1.647 | 1.682 | 1.701 |

**Table 2.11** The influence of the rheological parameter $\omega$ on the values of $p'_{3D.c.0}$, $p'_{R.c.0}$ and $p'_{E.c.0}$ in the case where $E^{(2)}\big/E_0^{(1)} = 5$ and $h/R^+ = 0.5$

| $\omega$ | $\rho = \pi R^+/\ell$ | | | | | |
|---|---|---|---|---|---|---|
| | 0.2 | | | 0.3 | | |
| | $p'_{3D.c.0}$ | $p'_{R.c.0}$ | $p'_{E.c.0}$ | $p'_{3D.c.0}$ | $p'_{R.c.0}$ | $p'_{E.c.0}$ |
| 0.5 | 0.02208 | 0.02442 | 0.03334 | 0.03524 | 0.04118 | 0.07501 |
| 1 | 0.02756 | 0.02928 | 0.03438 | 0.04991 | 0.05559 | 0.07736 |
| 2 | 0.03067 | 0.03199 | 0.03542 | 0.05927 | 0.06423 | 0.07970 |

The influence of the rheological parameter $\omega$ on the values of the critical forces are illustrated by the data given in Table 2.11 which are obtained for various $\rho$ under $E^{(2)}\big/E_0^{(1)} = 5$ and $h/R^+ = 0.5$.

It follows from the foregoing results that for pure elastic stability loss problems in the cases where the relations $E^{(2)}\big/E_0^{(1)} \le 5$, $h/R^+ \ge 0.5$ and $\rho = 0.2$ satisfy simultaneously the difference between the results obtained in the framework of the TDLTS and in the framework of the third order refined beam theory are not more than 5-6 %.

The comparison of the results given in Tables 2.8 and 2.9 with the results given in Tables 2.1 and 2.2 shows that the values of the critical forces obtained for the

**Table 2.12** The values of $t'_{3D.cr.}$ (upper number) and $t'_{Rf.cr.}$ (lower number) obtained for various values of the rheological parameter $\alpha$ in the case where $h/R^{+} = 0.3$ under $E^{(2)}/E_0^{(1)} = 5$, $\rho = 0.2$ and $\omega = 0.5$

| $\alpha$ | $p' \times 10^2$ | | | | | | |
|---|---|---|---|---|---|---|---|
| | 2.80 | 2.90 | 3.10 | 3.20 | 3.30 | 3.40 | 3.60 |
| −0.3 | 4.088 / 179.1 | 2.688 / 24.79 | 1.277 / 5.306 | 0.895 / 3.153 | 0.625 / 1.988 | 0.430 / 1.289 | 0.182 / 0.546 |
| −0.5 | 9.475 / 1883. | 5.269 / 118.1 | 1.859 / 13.64 | 1.130 / 6.587 | 0.684 / 3.453 | 0.405 / 1.883 | 0.122 / 0.566 |
| −0.7 | 67.35 / $\sim O(10^6)$ | 25.33 / 4516 | 4.463 / 123.7 | 1.948 / 36.74 | 0.843 / 12.52 | 0.351 / 4.561 | 0.047 / 0.614 |

**Table 2.13** The values of $t'_{3D.cr.}$ (upper number) and $t'_{Rf.cr.}$ (lower number) obtained for various values of the rheological parameter $\alpha$ in the case where $h/R^{+} = 0.5$ under $E^{(2)}/E_0^{(1)} = 5$, $\rho = 0.2$ and $\omega = 0.5$

| $\alpha$ | $p' \times 10^2$ | | | | | | |
|---|---|---|---|---|---|---|---|
| | 2.50 | 2.60 | 2.80 | 2.90 | 3.00 | 3.20 | 3.30 |
| -0.3 | 5.348 / 65.03 | 2.969 / 13.23 | 1.092 / 2.827 | 0.673 / 1.579 | 0.405 / 0.907 | 0.117 / 0.280 | 0.046 / 0.135 |
| -0.5 | 13.80 / 455.8 | 6.056 / 49.07 | 1.492 / 5.653 | 0.759 / 2.501 | 0.373 / 1.152 | 0.065 / 0.222 | 0.017 / 0.080 |
| -0.7 | 126.0 / 42853 | 31.94 / 1044 | 3.095 / 28.48 | 1.003 / 7.316 | 0.307 / 2.009 | 0.016 / 0.129 | 0.0019 / 0.0239 |

hollow cylinder approach to the corresponding ones obtained for the solid cylinder with $h/R^{+}$.

Now we consider the case where the cylinder material is viscoelastic and analyze the critical time. As in the previous section, to compare the values of the critical time given by the TDLTS with those given by approximate beam theories we must select the values of $p'(= p/E_0^{(1)})$ which fall in intervals $[p'_{E.c.\infty}, p'_{E.c.0}]$, $[p'_{R.c.\infty}, p'_{R.c.0}]$ and $[p'_{3D.c.\infty}, p'_{3D.c.0}]$ simultaneously. It follows from the data given in Tables 2.8 and 2.9, in many cases such common values of $p'(= p/E_0^{(1)})$ exist only for the intervals $[p'_{R.c.\infty}, p'_{R.c.0}]$ and $[p'_{3D.c.\infty}, p'_{3D.c.0}]$. For this reason, we will compare only the critical time obtained by employing the TDLTS (denoted by $t'_{3D.cr.}$) with that obtained by employing the third order refined beam theory (denoted by $t'_{Rf.cr.}$).

Thus, we consider data given in Tables 2.12, 2.13 and 2.14 which show the values of $t'_{3D.cr.}$(upper number) and $t'_{Rf.cr.}$ (lower number) in the cases where $h/R^{+} = 0.3$, 0.5 and 0.8 respectively under $E^{(2)}/E_0^{(1)} = 5$, $\rho = 0.2$ and $\omega = 0.5$ for various values of rheological parameter $\alpha$.

The foregoing results show that for the cases under considerations the values of $t'_{Rf.cr.}$ are greater than $t'_{3D.cr.}$. In the cases where $p'$ is close to the values of $p'_{3D.c.\infty}$ (or to the values of $p'_{R.c.\infty}$) the difference between $t'_{3D.cr.}$ and $t'_{Rf.cr.}$ becomes more non-negligible. The results given above also show the influence of the rheological parameter $\alpha$ on the values of $t'_{3D.cr.}$ and $t'_{Rf.cr.}$.

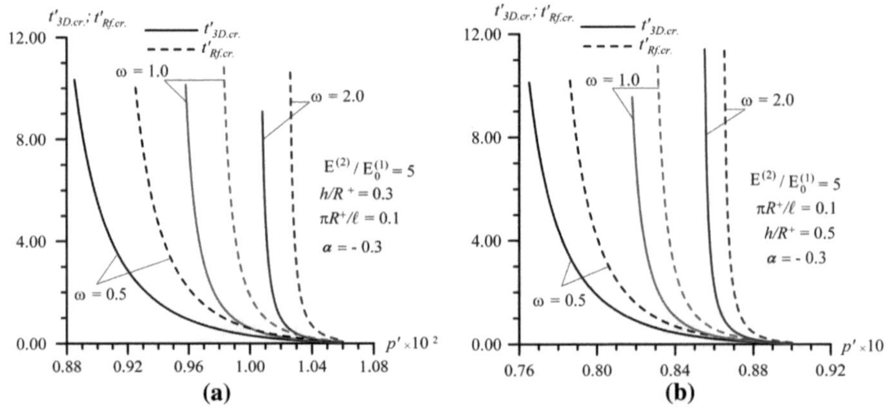


**Fig. 2.7** The graphs of the dependencies among $t'_{3D.cr.}$, $t'_{Rf.cr.}$ and $p'$ constructed for various values of rheological parameter $\omega$ in the case where $h/R^{+} = 0.3$ (**a**) and $h/R^{+} = 0.5$(**b**)

**Table 2.15** The values of $t'_{3D.cr.}$ (upper number) and $t'_{cdm.cr.}$ (lower number) obtained for various values of the rheological parameter $\alpha$ in the case where $\rho = 0.1$ under $E^{(2)} \big/ E_0^{(1)} = 5, h/R^{+} = 0.8$ and $\omega = 0.5$

| $\alpha$ | $p' \times 10^2$ | | | | |
|---|---|---|---|---|---|
| | 0.820 | 0.830 | 0.840 | 0.850 | 0.850 |
| -0.3 | 0.906 / 21.44 | 0.634 / 3.587 | 0.443 / 1.434 | 0.306 / 0.722 | 0.207 / 0.399 |
| -0.5 | 1.150 / 96.45 | 0.698 / 7.891 | 0.4227 / 2.186 | 0.251 / 0.837 | 0.145 / 0.364 |
| -0.7 | 2.006 / 3220. | 0.872 / 49.65 | 0.377 / 5.845 | 0.159 / 1.181 | 0.0641 / 0.2955 |

$$r = r_0(t_3) = R + L\sin(\alpha_1 m t_3); \; \alpha_1 = \pi/\ell \tag{2.82}$$

where $t_3$ is a parameter and $t_3 \in (0, \ell)$, $L$ is the amplitude of the initial imperfection form, and $m$ is the number of the half-length of the period of the initial imperfection.

We assume that the thickness of the cylinder's wall is constant along the cylinder and is equal to $h$. Moreover, we assume that $L \ll l$ and introduce the small parameter:

$$\varepsilon = \frac{L}{\ell}, \; 0 \leq \varepsilon \ll \left(\frac{1}{m\pi}\right). \tag{2.83}$$

We suppose that the material of the cylinder is viscoelastic transversal isotropic, the symmetry axis of which coincides with the $Ox_3$ $(Oz)$ axis. Within the foregoing assumptions, we investigate the evolution of the infinitesimal rotationally symmetric initial imperfection of the cylinder with time for the case where the cylinder is loaded by uniformly distributed normal compressed forces with intensity $p$ acting on the ends of the cylinder in the direction of the $Oz$ axis.

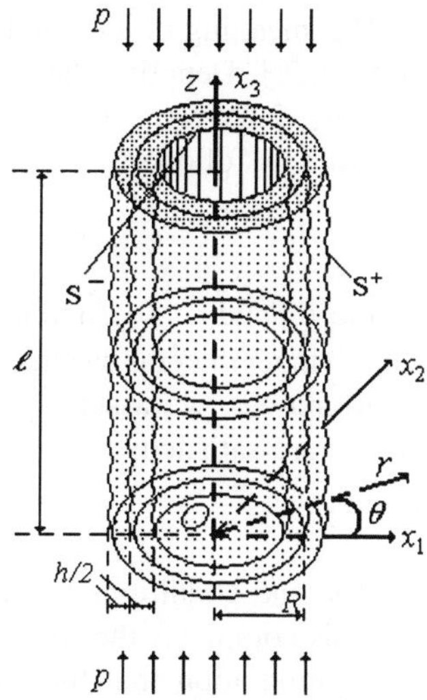


**Fig. 2.8** The geometry of the considered hollow cylinder with rotationally symmetric initial imperfection

The investigation is also made within the scope of the field Eqs. (2.3a), (2.3b), (2.3c), (2.4) and (2.5). For both completeness and clarity of the case under consideration we rewrite these equations for the rotationally symmetric case:

$$\frac{\partial t_{rr}}{\partial r}+\frac{\partial t_{zr}}{\partial z}+\frac{1}{r}(t_{rr}-t_{\theta\theta})=0,\quad \frac{\partial t_{rz}}{\partial r}+\frac{1}{r}t_{rz}+\frac{\partial t_{zz}}{\partial z}=0, \tag{2.84a}$$

$$\begin{aligned} t_{rr} &=\sigma_{rr}\left(1+\frac{\partial u_r}{\partial r}\right)+\sigma_{rz}\frac{\partial u_r}{\partial z},\quad t_{rz}=\sigma_{rr}\frac{\partial u_z}{\partial r}+\sigma_{rz}\left(1+\frac{\partial u_z}{\partial z}\right),\\ t_{\theta\theta} &=\sigma_{\theta r}\frac{\partial u_\theta}{\partial r}+\sigma_{\theta\theta}\left(1+\frac{u_r}{r}\right),\quad t_{zr}=\sigma_{zr}\left(1+\frac{\partial u_r}{\partial r}\right)+\sigma_{zz}\frac{\partial u_r}{\partial z},\\ t_{zz} &=\sigma_{zr}\frac{\partial u_z}{\partial r}+\sigma_{zz}\left(1+\frac{\partial u_z}{\partial z}\right), \end{aligned} \tag{2.84b}$$

$$\begin{aligned} \varepsilon_{rr} &=\frac{\partial u_r}{\partial r}+\frac{1}{2}\left\{\left(\frac{\partial u_r}{\partial r}\right)^2+\left(\frac{\partial u_z}{\partial r}\right)^2\right\},\\ \varepsilon_{rz} &=\frac{1}{2}\left(\frac{\partial u_r}{\partial z}+\frac{\partial u_z}{\partial r}\right)+\frac{1}{2}\left\{\frac{\partial u_r}{\partial r}\frac{\partial u_r}{\partial z}+\frac{\partial u_z}{\partial r}\frac{\partial u_z}{\partial z}\right\}\\ \varepsilon_{\theta\theta} &=\frac{u_r}{r}+\frac{1}{2}\left(\frac{u_r}{r}\right)^2,\quad \varepsilon_{zz}=\frac{\partial u_z}{\partial z}+\frac{1}{2}\left\{\left(\frac{\partial u_r}{\partial z}\right)^2+\left(\frac{\partial u_z}{\partial z}\right)^2\right\}. \end{aligned} \tag{2.84c}$$

$$\begin{aligned} &\sigma_{rr}=A_{11}^*\varepsilon_{rr}+A_{12}^*\varepsilon_{\theta\theta}+A_{13}^*\varepsilon_{zz};\ \sigma_{\theta\theta}=A_{12}^*\varepsilon_{rr}+A_{11}^*\varepsilon_{\theta\theta}+A_{13}^*\varepsilon_{zz};\\ &\sigma_{zz}=A_{13}^*\varepsilon_{rr}+A_{13}^*\varepsilon_{\theta\theta}+A_{33}^*\varepsilon_{zz};\ \sigma_{rz}=2G^*\varepsilon_{rz};\ \sigma_{\theta z}=2G^*\varepsilon_{\theta z} \end{aligned} \tag{2.85}$$

where $A_{ij}^*$ and $G^*$ are the operators determined by expression (2.5).

The meaning of the notation used in Eqs. (2.84a–2.84c) and (2.85) is explained in Sect. 2.1. Denoting the inner and outer surface of the hollow cylinder by $S^-$ and $S^+$ (Fig. 2.8), respectively we may write the free-of-traction conditions on these surfaces as follows:

$$t_{rr}\big|_{S^\pm} n_r^\pm + t_{zr}\big|_{S^\pm} n_z^\pm = 0, \; t_{rz}\big|_{S^\pm} n_r^\pm + t_{zz}\big|_{S^\pm} n_z^\pm = 0 \tag{2.86}$$

where $n_r^+$ and $n_z^+$ ($n_r^-$ and $n_z^-$) are the components of the unit normal vector to the outer (inner) lateral surface $S^+$ ($S^-$) of the hollow cylinder.

Moreover, we assume that at the ends of the cylinder the conditions written below are satisfied:

$$t_{zz}\big|_{z=0} = -p, \;\; t_{zz}\big|_{z=\ell} = -p. \tag{2.87}$$

The end conditions for the displacements will be discussed below.

This completes the formulation of the considered problem and it follows from this formulation that the evolution of the infinitesimal initial rotationally symmetric imperfection of the cylinder with time for the fixed value of the external compressed force $p$ (for the case where the material of the cylinder is viscoelastic) or with initial compressing force $p$ (for the case where the material of the cylinder is pure elastic) will be investigated within the framework of the field Eqs. (2.84) and (2.85) and boundary conditions (2.86) and (2.87).

## 2.7 Method of Solution for the Rotationally Symmetric Problem

We now develop the method summarized and proposed in Sect. 2.2 with respect to the solution to the rotationally symmetric stability loss problem of the hollow cylinder formulated in the previous section.

According to the procedure summarized in Sect. 2.2 and to the problem statement, first we derive the equation for the inner and outer lateral surfaces $S^+$ and $S^-$ of the cylinder shown in Fig. 2.8. Using the condition of the constant thickness of the cylinder's wall we can conclude that the coordinates of these surfaces must simultaneously satisfy the following equations:

$$z^\pm - t_3 = -\frac{dr_0}{dt_3}\left(r^\pm - r_0(t_3)\right), \;\; \left(z^\pm - t_3\right)^2 + \left(r^\pm - r_0(t_3)\right)^2 = \left(\frac{h}{2}\right)^2 \tag{2.88}$$

where $z^\pm$ and $r^\pm$ are the coordinates of the surface $S^\pm$. Note that the first equation in (2.88) is an equation of the straight line (plane) perpendicular to the vector which is the tangent vector to the middle surface of the cylinder at the point that corresponds to the fixed value of the parameter $t_3$, but the second equation in (2.88) shows that the thickness of the cylinder's wall remains constant along the cylinder. We can write the expressions

$$r^{\pm} = r_0 \pm \frac{h}{2}\left(1+\left(\frac{dr_0}{dt_3}\right)^2\right)^{-\frac{1}{2}}, \; z^{\pm} = t_3 \mp \frac{dr_0}{dt_3}\frac{h}{2}\left(1+\left(\frac{dr_0}{dt_3}\right)^2\right)^{-\frac{1}{2}} \tag{2.89}$$

from the Eq. (2.88).

Using the assumption (2.83) and the condition $(dr_0/dt_3)^2 \ll 1$, after some mathematical manipulations, we obtain the following expansions from Eq. (2.89):

$$\begin{aligned}
&r^{\pm} = R \pm \frac{h}{2} + \varepsilon\ell \sin(\alpha_1 m t_3) \mp \frac{h}{4}\varepsilon^2 m^2 \pi^2 \cos^2(\alpha_1 m t_3) + O(\varepsilon^4),\\
&z^{\pm} = t_3 \mp \varepsilon\frac{h}{2}\pi m \cos(\alpha_1 m t_3) \pm \varepsilon^3 \frac{h}{4}\left(m\pi cos^2(\alpha_1 m t_3)\right)^3 + O(\varepsilon^5),\\
&n_r^{\pm} = \pm\frac{dz^{\pm}}{dt_3}(V(t_3))^{-1}, \; n_z^{\pm} = \mp\frac{dr^{\pm}}{dt_3}(V(t_3))^{-1}, \; V(t_3) = \left(\left(\frac{dz^{\pm}}{dt_3}\right)^2 + \left(\frac{dr^{\pm}}{dt_3}\right)^2\right)^{\frac{1}{2}}\\
&\Rightarrow n_r^{\pm} = \pm 1 + O(\varepsilon^2), \; n_z^{\pm} = \mp\varepsilon m\pi \cos(\alpha_1 m t_3) + O(\varepsilon^3),
\end{aligned} \tag{2.90}$$

where $n_r^{\pm}$ and $n_z^{\pm}$ are the physical components of the unit normal vector to the surface $S^{\pm}$ (Fig. 2.8).

As in Sect. 2.2, we attempt to solve the considered problem by employing the boundary form perturbation method. For this purpose, the unknowns are presented in series form in $\varepsilon$ (2.83):

$$\left\{\sigma_{(ij)};\ \varepsilon_{(ij)};\ u_{(i)}\right\} = \sum_{q=0}^{\infty} \varepsilon^q \left\{\sigma_{(ij)}^{(q)};\ \varepsilon_{(ij)}^{(q)};\ u_{(i)}^{(q)}\right\}, \tag{2.91}$$

where $(ij) = rr, \theta\theta, zz, rz$ and $(i) = r, z$.

Substituting Eq. (2.91) into Eq. (2.84a–c), we obtain set equations for each approximation (2.91). Using relation (2.90) we expand the values of each approximation (2.91) in series form in the vicinity of the points $\{r = R + h/2;\ z_0 = t_3\}$ and $\{r = R - h/2;\ z_0 = t_3\}$. Substituting these last expressions in the boundary conditions in (2.86) and using the expressions of $n_r^{\pm}$ and $n_z^{\pm}$ given in (2.90), after some mathematical transformations we obtain boundary conditions which are satisfied at $\{r = R + h/2;\ z = t_3\}$ (for outer surface $S^+$) and $\{r = R - h/2;\ z = t_3\}$(for inner surface $S^-$) for each approximation in Eq. (2.91). It is evident that for the zeroth approximation, equations in (2.84a–2.84c) are valid and condition (2.86) is replaced by the same one satisfied at points $\{r = R + h/2;\ z = t_3\}$ and $\{r = R - h/2;\ z = t_3\}$. We assume that in the zeroth approximation the non-linear parts of the strain tensor components are very small and can be ignored with respect to their linear parts. According to this assumption, for the zeroth approximation, we obtain the following system of equations:

$$\frac{\partial\sigma_{rr}^{(0)}}{\partial r}+\frac{\partial\sigma_{zr}^{(0)}}{\partial z}+\frac{1}{r}\left(\sigma_{rr}^{(0)}-\sigma_{\theta\theta}^{(0)}\right)=0,\ \frac{\partial\sigma_{rz}^{(0)}}{\partial r}+\frac{1}{r}\sigma_{rz}^{(0)}+\frac{\partial\sigma_{zz}^{(0)}}{\partial z}=0,$$
$$\varepsilon_{rr}^{(0)}=\frac{\partial u_r^{(0)}}{\partial r},\ \varepsilon_{rz}^{(0)}=\frac{1}{2}\left(\frac{\partial u_r^{(0)}}{\partial z}+\frac{\partial u_z^{(0)}}{\partial r}\right),\ \varepsilon_{\theta\theta}^{(0)}=\frac{u_r^{(0)}}{r},\ \varepsilon_{zz}^{(0)}=\frac{\partial u_z^{(0)}}{\partial z} \tag{2.92}$$

and boundary conditions:

$$\sigma_{rr}^{(0)}\big|_{r=R\pm h/2}=0,\ \sigma_{rz}^{(0)}\big|_{r=R\pm h/2}=0. \tag{2.93}$$

Moreover, we obtain the following end conditions for the zeroth approximation from (2.87):

$$\sigma_{zz}^{(0)}(r,\theta,0)=\sigma_{zz}^{(0)}(r,\theta,\ell)=-p. \tag{2.94}$$

Taking the assumption $1+\partial u_z^{(0)}/\partial z\approx 1$, $1+u_r^{(0)}\Big/r\approx 1$ and $1+\partial u_z^{(0)}/\partial z\approx 1$into account, for the subsequent approximations we obtain the following system of equations:

$$\frac{\partial\sigma_{rr}^{(q)}}{\partial r}+\frac{\partial\sigma_{zr}^{(q)}}{\partial z}+\frac{1}{r}\left(\sigma_{rr}^{(q)}-\sigma_{\theta\theta}^{(q)}\right)+\sigma_{zz}^{(0)}\frac{\partial^2 u_r^{(q)}}{\partial z^2}$$
$$=\underline{-\frac{\partial S_{rr}^{(q-1)}}{\partial r}-\frac{\partial S_{zr}^{(q-1)}}{\partial z}-\frac{1}{r}\left(S_{rr}^{(q-1)}-S_{\theta\theta}^{(q-1)}\right)},$$
$$\frac{\partial\sigma_{rz}^{(q)}}{\partial r}+\frac{1}{r}\sigma_{rz}^{(q)}+\frac{\partial\sigma_{zz}^{(q)}}{\partial z}+\sigma_{zz}^{(0)}\frac{\partial^2 u_z^{(q)}}{\partial z^2}=\underline{-\frac{\partial S_{rz}^{(q-1)}}{\partial r}-\frac{1}{r}S_{rz}^{(q-1)}-\frac{\partial S_{zz}^{(q-1)}}{\partial z}},$$
$$S_{rr}^{(q-1)}=\sum_{k=1}^{q-1}\left\{\sigma_{rr}^{(k)}\frac{\partial u_r^{(q-k)}}{\partial r}+\sigma_{rz}^{(k)}\frac{\partial u_r^{(q-k)}}{\partial z}\right\},$$
$$S_{rz}^{(q-1)}=\sum_{k=1}^{q-1}\left\{\sigma_{rr}^{(k)}\frac{\partial u_z^{(q-k)}}{\partial r}+\sigma_{rz}^{(k)}\frac{\partial u_z^{(q-k)}}{\partial z}\right\},$$
$$S_{\theta\theta}^{(q-1)}=\sum_{k=1}^{q-1}\left\{\sigma_{\theta\theta}^{(k)}\frac{u_r^{(q-k)}}{r}\right\},\ S_{zz}^{(q-1)}=\sum_{k=1}^{q-1}\left\{\sigma_{rz}^{(k)}\frac{\partial u_z^{(q-k)}}{\partial r}+\sigma_{zz}^{(k)}\frac{\partial u_z^{(q-k)}}{\partial z}\right\},$$
$$S_{zr}^{(q-1)}=\sum_{k=1}^{q-1}\left\{\sigma_{zr}^{(k)}\frac{\partial u_r^{(q-k)}}{\partial r}+\sigma_{zz}^{(k)}\frac{\partial u_z^{(q-k)}}{\partial z}\right\},$$
$$\varepsilon_{rr}^{(q)}=\frac{\partial u_r^{(q)}}{\partial r}+\underline{\frac{1}{2}\sum_{k=1}^{q-1}\left\{\left(\frac{\partial u_r^{(k)}}{\partial r}\frac{\partial u_r^{(q-k)}}{\partial r}\right)+\left(\frac{\partial u_z^{(k)}}{\partial r}\frac{\partial u_z^{(q-k)}}{\partial r}\right)\right\}},$$
$$\varepsilon_{rz}^{(q)}=\frac{1}{2}\left(\frac{\partial u_r^{(q)}}{\partial z}+\frac{\partial u_z^{(q)}}{\partial r}\right)+\underline{\frac{1}{2}\sum_{k=1}^{q-1}\left\{\frac{\partial u_r^{(k)}}{\partial r}\frac{\partial u_r^{(q-k)}}{\partial z}+\frac{\partial u_z^{(k)}}{\partial r}\frac{\partial u_z^{(q-k)}}{\partial z}\right\}},$$

$$\varepsilon_{\theta\theta}^{(q)} = \frac{u_r^{(q)}}{r} + \underline{\frac{1}{2}\sum_{k=1}^{q-1}\left\{\frac{1}{r^2}u_r^{(k)}u_r^{(q-k)}\right\}},$$

$$\varepsilon_{zz}^{(q)} = \frac{\partial u_z^{(q)}}{\partial z} + \underline{\frac{1}{2}\sum_{k=1}^{q-1}\left\{\left(\frac{\partial u_r^{(k)}}{\partial z}\frac{\partial u_r^{(q-k)}}{\partial z}\right) + \left(\frac{\partial u_z^{(k)}}{\partial z}\frac{\partial u_z^{(q-k)}}{\partial z}\right)\right\}}. \quad (2.95)$$

For the problem under consideration the underlined terms in Eq. (2.95) are equal to zero for the first approximation. By direct verification it is proven that Eq. (2.95) without the underlined terms coincides with the corresponding equations of the TDLTS (Guz 1999) related to the rotationally symmetric case. Due to linearity, the constitutive relations in (2.85) are satisfied by each approximation separately, i.e.

$$\begin{aligned}\sigma_{rr}^{(q)} &= A_{11}^*\varepsilon_{rr}^{(q)} + A_{12}^*\varepsilon_{\theta\theta}^{(q)} + A_{13}^*\varepsilon_{zz}^{(q)};\ \sigma_{\theta\theta}^{(q)} = A_{12}^*\varepsilon_{rr}^{(q)} + A_{11}^*\varepsilon_{\theta\theta}^{(q)} + A_{13}^*\varepsilon_{zz}^{(q)};\\ \sigma_{zz}^{(q)} &= A_{13}^*\varepsilon_{rr}^{(q)} + A_{13}^*\varepsilon_{\theta\theta}^{(q)} + A_{33}^*\varepsilon_{zz}^{(q)};\ \sigma_{rz}^{(q)} = 2G^*\varepsilon_{rz}^{(q)}.\end{aligned} \quad (2.96)$$

Now we write the boundary conditions given on the lateral surface of the cylinder for the first approximation by the physical components of the stress tensor:

$$\sigma_{rr}^{(1)}\Big|_{r=R\pm h/2} = 0,\ \sigma_{rz}^{(1)}\Big|_{r=R\pm h/2} = \sigma_{zz}^{(0)} m\pi\cos(\alpha_1 m t_3). \quad (2.97)$$

According to the expressions (2.90) and (2.91), we obtain from Eq. (2.87) the following end conditions for the first approximation:

$$\sigma_{zz}^{(1)}\Big|_{t_3=0;\ell} + \sigma_{zz}^{(0)}\frac{\partial u_z^{(1)}}{\partial z}\Big|_{t_3=0;\ell} = 0. \quad (2.98)$$

Moreover, we assume that:

$$u_r^{(1)}\Big|_{t_3=0;\ell} = 0. \quad (2.99)$$

According to (2.92), (2.93) and (2.94), the values related to the zeroth approximation are determined as follows:

$$\sigma_{zz}^{(0)} = -p,\ \sigma_{(ij)}^{(0)} = 0\, for\ (ij) \neq zz. \quad (2.100)$$

It follows from (2.100) that in the zeroth approximation the components of the displacement vector can be presented as follows:

$$u_r^{(0)} = a(t)r + a_0,\ u_z^{(0)} = c(t)z + c_0 \quad (2.101)$$

where $a_0$ and $c_0$ are constants, $a(t)$ and $c(t)$ are functions and $t$ is a time. The functions $a(t)$ and $c(t)$ can be easily determined from Eqs. (2.92), (2.96) and (2.101).

Now we consider determination of the values related to the first approximation. Taking the expression (2.100) into account the following field equations are obtained from the Eq. (2.95) for this approximation:

$$\frac{\partial \sigma_{rr}^{(1)}}{\partial r}+\frac{\partial \sigma_{zr}^{(1)}}{\partial z}+\frac{1}{r}\left(\sigma_{rr}^{(1)}-\sigma_{\theta\theta}^{(1)}\right)+\sigma_{zz}^{(0)}\frac{\partial^2 u_r^{(1)}}{\partial z^2}=0,$$
$$\frac{\partial \sigma_{rz}^{(1)}}{\partial r}+\frac{\partial \sigma_{zz}^{(1)}}{\partial z}+\frac{1}{r}\sigma_{rz}^{(1)}+\sigma_{zz}^{(0)}\frac{\partial^2 u_z^{(1)}}{\partial z^2}=0, \tag{2.102}$$
$$\varepsilon_{rr}^{(1)}=\frac{\partial u_r^{(1)}}{\partial r},\ \varepsilon_{\theta\theta}^{(1)}=\frac{u_r^{(1)}}{r},\ \varepsilon_{zz}^{(1)}=\frac{\partial u_z^{(1)}}{\partial z},\ \varepsilon_{zr}^{(1)}=\frac{1}{2}\left(\frac{\partial u_z^{(1)}}{\partial r}+\frac{\partial u_r^{(1)}}{\partial z}\right).$$

We attempt to solve these Eqs. (2.96) and (2.102) within the scope of the conditions (2.97), (2.98) and (2.99). For this purpose we apply the Laplace transformation (2.34) to all equations and relations related to the first approximation. After this application, the Eq. (2.96) and the boundary conditions (2.97) (in which $\sigma_{zz}^{(0)}$ must be replaced with $\sigma_{zz}^{(0)}/s$), (2.98) and (2.99) are valid for the Laplace transformation of the corresponding sought-for quantities, whereas the constitutive relations in (2.96) are transformed to the following ones:

$$\bar{\sigma}_{rr}^{(1)}=\bar{A}_{11}^{*}\bar{\varepsilon}_{rr}^{(1)}+\bar{A}_{12}^{*}\bar{\varepsilon}_{\theta\theta}^{(1)}+\bar{A}_{13}^{*}\bar{\varepsilon}_{zz}^{(1)};\ \bar{\sigma}_{\theta\theta}^{(1)}=\bar{A}_{12}^{*}\bar{\varepsilon}_{rr}^{(1)}+\bar{A}_{11}^{*}\bar{\varepsilon}_{\theta\theta}^{(1)}+\bar{A}_{13}^{*}\bar{\varepsilon}_{zz}^{(1)};$$
$$\bar{\sigma}_{zz}^{(1)}=\bar{A}_{13}^{*}\bar{\varepsilon}_{rr}^{(1)}+\bar{A}_{13}^{*}\bar{\varepsilon}_{\theta\theta}^{(1)}+\bar{A}_{33}^{*}\bar{\varepsilon}_{zz}^{(1)};\ \bar{\sigma}_{r\theta}^{(1)}=\left(\bar{A}_{11}^{*}-\bar{A}_{22}^{*}\right)\bar{\varepsilon}_{r\theta}^{(1)}; \tag{2.103}$$
$$\bar{\sigma}_{rz}^{(1)}=2\bar{G}^{*}\bar{\varepsilon}_{rz}^{(1)};\ \bar{\sigma}_{\theta z}^{(1)}=2\bar{G}^{*}\bar{\varepsilon}_{\theta z}^{(1)}$$

where

$$\left\{\begin{matrix}\bar{A}_{ij}^{*}\\ \bar{G}^{*}\end{matrix}\right\}\bar{\phi}(s)=\left\{\begin{matrix}A_{ij0}\\ G_{0}\end{matrix}\right\}\bar{\phi}(s)+\left\{\begin{matrix}\bar{A}_{ij1}\\ \bar{G}_{1}\end{matrix}\right\}\bar{\phi}(s). \tag{2.104}$$

As has been noted above, Eqs. (2.102), (2.103) and (2.104) coincide with the corresponding equations of the TDLTS related to the rotationally symmetric case. Therefore to solve the obtained equation systems, according to the monograph (Guz 1999), in the cylindrical system of coordinates we can use the following representations:

$$\bar{u}_r^{(1)}=-\frac{\partial^2}{\partial r\partial z}\chi,\ \Delta_1=\frac{\partial^2}{\partial r^2}+\frac{1}{r}\frac{\partial}{\partial r},$$
$$\bar{u}_z^{(1)}=\left(\bar{A}_{13}^{*}+\bar{G}^{*}\right)\left(\bar{A}_{11}^{*}\Delta_1+\left(\bar{G}^{*}+\sigma_{zz}^{(0)}\right)\frac{\partial^2}{\partial z^2}\right)\chi. \tag{2.105}$$

The function $\chi$ is determined from the equation:

$$\left(\Delta_1+\xi_2^2\frac{\partial^2}{\partial z^2}\right)\left(\Delta_1+\xi_3^2\frac{\partial^2}{\partial z^2}\right)\chi=0 \tag{2.106}$$

where

$$\xi_{2,3}^2 = c \pm \left( c^2 - \frac{\left(\bar{A}_{33}^* + \sigma_{zz}^{(0)}\right)\left(\bar{G}^* + \sigma_{zz}^{(0)}\right)}{\bar{A}_{11}^*\bar{G}^*} \right)^{\frac{1}{2}},$$
$$2\bar{A}_{11}^*\bar{G}^* c = \bar{A}_{11}^*\left(\bar{A}_{33}^* + \sigma_{zz}^{(0)}\right) + \bar{G}^{*2} - \left(\bar{A}_{13}^* + \bar{G}^*\right)^2. \qquad (2.107)$$

Taking the expressions of the right sides of the conditions (2.97) and (2.98) into account we find the solution to the Eq. (2.106) as follows:

$$\chi = [B_1 I_0(\xi_2\alpha_1 mr) + B_2 K_0(\xi_2\alpha_1 mr) + B_3 I_0(\xi_3\alpha_1 mr) + B_4 K_0(\xi_3\alpha_1 mr)]\cos(\alpha_1 mz) \qquad (2.108)$$

where $I_0(x)$ is the zeroth order Bessel function of a purely imaginary argument, $K_0(x)$ is the zeroth order Macdonald function and $B_1$, $B_2$, $B_3$ and $B_4$ are unknown constants, the values of which also depend on the transformation parameter $s$.

Substituting this solution into relations (2.102), (2.105) and (2.96) we obtain the following expressions for Laplace transformation of the sought functions:

$$\bar{u}_r^{(1)} = \Big[B_1\xi_2(\alpha_1 m)^2 I_0'(\xi_2\alpha_1 mr) + B_3\xi_3(\alpha_1 m)^2 I_0'(\xi_3\alpha_1 mr) + B_2\xi_2(\alpha m)^2 K_0'(\xi_2\alpha_1 mr) + B_4\xi_3(\alpha_1 m)^2 K_0'(\xi_3\alpha_1 mr)\Big]\sin(\alpha_1 mz),$$

$$\bar{u}_z^{(1)} = \Big[B_1 D_2(\alpha_1 m)^2 I_0(\xi_2\alpha_1 mr) + B_3 D_3(\alpha_1 m)^2 I_0(\xi_3\alpha_1 mr) + B_2 D_2(\alpha m)^2 K_0(\xi_2\alpha_1 mr) + B_4 D_3(\alpha_1 m)^2 K_0(\xi_3\alpha_1 mr)\Big]\sin(\alpha_1 mz),$$

$$D_2 = \frac{\bar{A}_{11}^*\xi_2^2 - \bar{G}^* - \sigma_{zz}^{(0)}}{\bar{A}_{13}^* + \bar{G}^*},\quad D_3 = \frac{\bar{A}_{11}^*\xi_3^2 - \bar{G}^* - \sigma_{zz}^{(0)}}{\bar{A}_{13}^* + \bar{G}^*},$$

$$\bar{\sigma}_{rr}^{(1)} = \Big\{B_1\Big[\bar{A}_{11}^*(\alpha_1 m)^3\xi_2^2 I_0''(\xi_2\alpha_1 mr) + \bar{A}_{12}^*\frac{\xi_2(\alpha_1 m)^2}{r} I_0'(\xi_2\alpha_1 mr) + \bar{A}_{13}^*(\alpha_1 m)^3 I_0(\xi_2\alpha_1 mr)\Big] + B_3\Big[\bar{A}_{11}^*(\alpha_1 m)^3\xi_3^2 I_0''(\xi_3\alpha_1 mr) + \bar{A}_{12}^*\frac{\xi_3(\alpha_1 m)^2}{r} I_0'(\xi_2\alpha_1 mr) + \bar{A}_{13}^* D_3(\alpha_1 m)^3 I_0(\xi_3\alpha_1 mr)\Big] + B_2\Big[\bar{A}_{11}^*(\alpha_1 m)^3\xi_2^2 K_0''(\xi_2\alpha_1 mr) + \bar{A}_{12}^*\frac{\xi_2(\alpha_1 m)^2}{r} K_0'(\xi_2\alpha_1 mr) + \bar{A}_{13}^* D_2(\alpha_1 m)^3 K_0(\xi_2\alpha_1 mr)\Big] + B_4\Big[\bar{A}_{11}^*(\alpha_1 m)^3\xi_3^2 K_0''(\xi_3\alpha_1 mr) + \bar{A}_{12}^*\frac{\xi_3(\alpha_1 m)^2}{r} K_0'(\xi_3\alpha_1 mr) + \bar{A}_{13}^* D_3(\alpha_1 m)^3 K_0(\xi_3\alpha_1 mr)\Big]\Big\}\sin(\alpha_1 mz),$$

$$\bar{\sigma}_{rz}^{(1)} = \Big\{B_1\bar{G}^*(\alpha_1 m)^3\xi_2 I_0'(\xi_2\alpha_1 mr)(1 + D_2) + B_3\bar{G}^*(\alpha_1 m)^3\xi_3 I_0'(\xi_3\alpha_1 mr)(1 + D_3) + B_2\bar{G}^*(\alpha_1 m)^3\xi_2 K_0'(\xi_2\alpha_1 mr)(1 + D_2) + B_4\bar{G}^*(\alpha_1 m)^3\xi_2 K_0'(\xi_3\alpha_1 mr)(1 + D_3)\Big\}\cos(\alpha_1 mz),\ \bar{\sigma}_{zz}^{(1)} =$$

$$\Big\{B_1\Big[\bar{A}_{13}^*\Big(\xi_2^2(\alpha_1 m)^3 I_0''(\xi_2\alpha_1 mr) + \frac{(\alpha_1 m)^2\xi_2}{r} I_0'(\xi_2\alpha_1 mr)\Big) - \bar{A}_{33}^* D_2(\alpha_1 m)^3 I_0(\xi_2\alpha_1 mr)\Big]$$

$$
\begin{aligned}
&+B_3\Big[\bar{A}^*_{13}\Big(\xi_3^2(\alpha_1 m)^3 I''_0(\xi_3\alpha_1 mr)+\tfrac{(\alpha_1 m)^2\xi_3}{r}I'_0(\xi_3\alpha_1 mr)\Big)\\
&-\bar{A}^*_{33}D(\alpha m)^3 I_0(\xi_3\alpha_1 mr)\Big]+B_2\Big[\bar{A}^*_{13}\Big(\xi_2^2(\alpha_1 m)^3 K''_0(\xi_2\alpha_1 mr)\\
&+\tfrac{(\alpha_1 m)^2\xi_2}{r}K'_0(\xi_2\alpha_1 mr)\Big)-\bar{A}^*_{33}D_2(\alpha_1 m)^3 K_0(\xi_2\alpha_1 mr)\Big]\\
&+B_4\big[\bar{A}^*_{13}\Big(\xi_3^2(\alpha_1 m)^3 K''_0(\xi_3\alpha_1 mr)\\
&-\tfrac{(\alpha_1 m)^2\xi_3}{r}K'_0(\xi_3\alpha_1 mr)\Big)-\bar{A}^*_{33}D(\alpha_1 m)^3 K_0(\xi_3\alpha_1 mr)\Big]\Big\}\sin(\alpha_1 mz) \qquad (2.109)
\end{aligned}
$$

Here the notation $I'_0(x)=dI_0(x)/dx$, $I''_0(x)=d^2I_0(x)/dx^2$, $K'_0(x)=dK_0(x)/dx$ and $K''_0(x)=d^2K_0(x)/dx^2$ is used. The solution (2.109) to the problem under consideration satisfies automatically the end conditions (2.98) and (2.99). Replacing the unknowns $B_1$, $B_2$, $B_3$ and $B_4$ with $C_1$, $C_2$, $C_3$ and $C_4$, respectively, where $C_i=(\alpha_1 m)^3 B_i$ $(i=1,2,3,4)$, we obtain the following algebraic equation from the boundary condition (2.97) for determination of these unknowns:

$$
\begin{aligned}
\bar{\sigma}^{(1)}_{rr}(R^+,t_3,t)=&0\Rightarrow C_1a_{11}(\alpha_1 mR^+)+C_2a_{12}(\alpha_1 mR^+)\\
&+C_3a_{13}(\alpha_1 mR^+)+C_4a_{14}(\alpha_1 R^+)=0,\\
\bar{\sigma}^{(1)}_{rz}(R^+,t_3,t)=&2\pi m\sigma^{(0)}_{zz}\frac{1}{s}\Rightarrow C_1a_{31}(\alpha_1 mR^+)\\
&+C_3a_{33}(\alpha_1 mR^+)+C_4a_{34}(\alpha_1 R^+)=2\pi m\sigma^{(0)}_{zz}\frac{1}{s},\\
\bar{\sigma}^{(1)}_{rr}(R^-,t_3,t)=&0\Rightarrow C_1a_{11}(\alpha_1 mR^-)+C_2a_{12}(\alpha_1 mR^-)+\\
&C_3a_{13}(\alpha_1 mR^-)+C_4a_{14}(\alpha_1 R^-)=0,\\
\bar{\sigma}^{(1)}_{rz}(R^-,t_3,t)=&2\pi m\sigma^{(0)}_{zz}\frac{1}{s}\Rightarrow C_1a_{31}(\alpha_1 mR^-)\\
&+C_3a_{33}(\alpha_1 mR^-)+C_4a_{34}(\alpha_1 R^-)=2\pi m\sigma^{(0)}_{zz}\frac{1}{s}
\end{aligned}
\qquad (2.110)
$$

where $R^+=R+h/2$, $R^-=R-h/2$. The expressions for the coefficients in Eq. (2.110) can be written easily from the relation (2.109).

Thus, with the foregoing, we determine completely the Laplace transformation of the values related to the first approximation. The Laplace transformation of the values of the second and subsequent approximations in (2.91) can also be determined as the values of the first approximation by taking the obvious changes into account. However, as it has been noted in Sect. 2.2 for stability loss problems, consideration of only the zeroth and the first approximations is sufficient, because, according to the structure of the field equations (2.95) and (2.96), accounting for the second and subsequent approximations does not change the values of the critical parameters.

The original of the sought values is determined by employing the method (Schapery 1962, 1978), according to which, for instance, the original of the displacement $u_r^{(1)}(r,t_3,t)$ is determined through the expression:

$$u_r^{(1)}(r,t_3,t) \approx \left( s\bar{u}_r^{(1)}(r,t_3,s) \right) \Big|_{s=1/(2t)}. \tag{2.111}$$

Now we consider the selection of the stability loss criterion. In the present investigation, as in Sect. 2.2, the case will be understood under stability loss, where:

$$\max_{\substack{t_3(0,\ell) \\ r\in(0,R)}} \left| u_r^{(1)}(r,t_3,t) \right| \to \infty \text{ as } t \to t_{cr.} \; (or \text{ as } p \to p_{cr.} \text{ for the pure elastic case}). \tag{2.112}$$

Thus, the values of the critical time or the values of the critical force are determined from the initial imperfection criterion (2.112).

## 2.8 Approximate Equations of the Stability Loss of the Cylinder-Shell Obtained from Equations of the TDLTS by the Average-Integrating Procedure

### *2.8.1 Kirchhoff–Love Shell Theory*

For the case under consideration, the approximate stability loss equations for the hollow cylinder-shell can be derived from Eqs. (2.97)–(2.103) by using the Kirchhoff–Love hypothesis, according to which, the displacements of the cylinder are presented as follows:

$$u_r^{(1)} = w(z),\; u_z^{(1)} = r' \frac{dw(z)}{dz},\; r' = R - r. \tag{2.113}$$

Underwriting the expression (2.113) is the assumption that the displacement of the middle surface of the cylinder, in the direction of the $Oz$ axis, is very small with respect to the term $-r'dw(z)/dz$ and is ignored. Thus, according to (2.92), (2.96) and (2.113), we obtain:

$$\begin{aligned} \varepsilon_{\theta\theta}^{(1)} &= \frac{w(z)}{r},\; \varepsilon_{zz}^{(1)} = r' \frac{d^2w(z)}{dz^2},\; \varepsilon_{rr}^{(1)} = \varepsilon_{rz}^{(1)} = 0, \\ \sigma_{\theta\theta}^{(1)} &= A_{11}^* \varepsilon_{\theta\theta}^{(1)} + A_{13}^* \varepsilon_{zz}^{(1)},\; \sigma_{zz}^{(1)} = A_{13}^* \varepsilon_{\theta\theta}^{(1)} + A_{33}^* \varepsilon_{zz}^{(1)}. \end{aligned} \tag{2.114}$$

Assume that:

$$\sigma_{rz}^{(1)} \neq 0. \tag{2.115}$$

Taking the relations (2.114) and (2.115) into account we obtain the following equation from (2.102)

$$\begin{aligned} &\frac{\partial \sigma_{rz}^{(1)}}{\partial z} - \frac{1}{r}\sigma_{\theta\theta} + \sigma_{zz}^{(0)}\frac{d^2 w}{dz^2} = 0, \\ &\frac{\partial \sigma_{rz}^{(1)}}{\partial r} + \frac{1}{r}\sigma_{rz}^{(1)} + \frac{\partial \sigma_{zz}^{(1)}}{\partial z} + \sigma_{zz}^{(0)}\frac{d^2 u_z^{(1)}}{dz^2} = 0. \end{aligned} \tag{2.116}$$

We can write the following presentations from Eq. (2.114):

$$\sigma_{\theta\theta}^{(1)} = A_{11}^{*}\frac{w(z)}{r} + A_{13}^{*}r'\frac{d^2 w(z)}{dz^2}, \quad \sigma_{zz}^{(1)} = A_{13}^{*}\frac{w(z)}{r} + A_{33}^{*}r'\frac{d^2 w(z)}{dz^2}. \tag{2.117}$$

Multiplying the first equation in (2.116) by $(1 - r'/R)dr'$ and integrating in the interval $[-h/2, +h/2]$ we obtain the following relation:

$$\int\limits_{-h/2}^{+h/2} \frac{\partial \sigma_{rz}^{(1)}}{\partial z}\left(1 - \frac{r'}{R}\right)dr' = \frac{h}{R^2}A_{11}^{*}w - \sigma_{zz}^{(0)}h\frac{d^2 w}{dz^2}. \tag{2.118}$$

Multiplying the second equation in (2.116) by $(1 - r'/R)r'dr'$, also integrating in the interval $[-h/2, +h/2]$ and taking the relation (2.117) into account, we derive the following equation:

$$\begin{aligned} &\int\limits_{-h/2}^{+h/2} \frac{\partial \sigma_{rz}^{(1)}}{\partial z}\left(1 - \frac{r'}{R}\right)r'dr' + \int\limits_{-h/2}^{+h/2} \frac{1}{R - r'}\sigma_{rz}^{(1)}\left(1 - \frac{r'}{R}\right)r'dr' + \\ &\frac{A_{33}^{*}h^3}{12}\frac{d^3 w}{dz^3} + \frac{\sigma_{zz}^{(0)}h^3}{12}\frac{d^3 w}{dz^3} = 0. \end{aligned} \tag{2.119}$$

We make the following mathematical manipulations:

$$\begin{aligned} &\int\limits_{-h/2}^{+h/2} \frac{\partial \sigma_{rz}^{(1)}}{\partial r}\left(1 - \frac{r'}{R}\right)r'dr' + \int\limits_{-h/2}^{+h/2} \frac{1}{R - r'}\sigma_{rz}^{(1)}\left(1 - \frac{r'}{R}\right)r'dr' \\ &= -\int\limits_{-h/2}^{+h/2} \frac{\partial\left(\sigma_{rz}^{(1)}r'\right)}{\partial r}\left(1 - \frac{r'}{R}\right)r'dr' + \int\limits_{-h/2}^{+h/2} \sigma_{rz}^{(1)}\left(1 - \frac{r'}{R}\right)dr' + \frac{1}{R}\int\limits_{-h/2}^{+h/2} \sigma_{rz}^{(1)}r'dr' \\ &= -\int\limits_{-h/2}^{+h/2} \frac{\partial\left(\sigma_{rz}^{(1)}r'\right)}{\partial r}dr' + \frac{1}{R}\int\limits_{-h/2}^{+h/2} \frac{\partial\left(\sigma_{rz}^{(1)}r'\right)}{\partial r}r'dr' \end{aligned}$$

$$+\int_{-h/2}^{+h/2}\sigma_{rz}^{(1)}\left(1-\frac{r'}{R}\right)dr'+\frac{1}{R}\int_{-h/2}^{+h/2}\sigma_{rz}^{(1)}r'dr'$$
$$=\left(\sigma_{rz}^{(1)}r'\right)\Big|_{-h/2}^{+h/2}+\frac{1}{R}\left(\sigma_{rz}^{(1)}r'^2\right)\Big|_{-h/2}^{+h/2}+\int_{-h/2}^{+h/2}\sigma_{rz}^{(1)}\left(1-\frac{r'}{R}\right)dr'$$
$$=\sigma_{rz}^{(0)}h\left(\frac{h}{2R}-1\right)m\pi\cos(\alpha_1 mz)+\int_{-h/2}^{+h/2}\sigma_{rz}^{(1)}\left(1-\frac{r'}{R}\right)dr' \qquad (2.120)$$

Substituting (2.120) into (2.119), taking the relation (2.118) and $\sigma_{zz}^{(0)}=-p$ into account and assuming that $A_{33}^{*}1-p\approx A_{33}^{*}1$, we obtain the following equation for the stability loss of the cylinder within the scope of the Kirchhoff-Love hypothesis:

$$\frac{h^2}{12}A_{33}^{*}\frac{d^4w}{dz^4}+p\frac{d^2w}{dz^2}+\frac{A_{11}^{*}}{R^2}w=-\alpha_1 mp\left(\frac{h}{2R}-1\right)\sin(\alpha_1 mz). \qquad (2.121)$$

According to (2.113) and (2.114), we obtain from (2.98) and (2.99) the following end conditions for the displacement $w(z)$:

$$w(z)|_{z=0;\ell}=0,\quad \left.\frac{d^2w(z)}{dz^2}\right|_{z=0;\ell}=0. \qquad (2.122)$$

If we assume that the cylinder material is an isotropic one, then it must be written $E^{*}\big/(1-(v^{*})^2)$ instead of $A_{33}^{*}$ and $A_{11}^{*}$. For the case where the material of the cylinder is isotropic and pure elastic, the Eq. (2.121) is transformed to the following one:

$$\frac{T}{h}\frac{d^4w}{dz^4}+p\frac{d^2w}{dz^2}+\frac{E}{R^2(1-v^2)}w=-\alpha_1 mp\left(\frac{h}{2R}-1\right)sin(\alpha_1 mz), \qquad (2.123)$$

where $T=h^3E/(12(1-v^2))$, $E$ and $v$ are the modulus of elasticity and Poisson's coefficient of the cylinder material and $E^{*}$ and $v^{*}$ are the corresponding operators.

If we replace the term $E/(R^2(1-v^2))$ in Eq. (2.123) with $E/R^2$, then the left side of this equation coincides with the well-known classical one given in almost all books related to stability loss of the shells (see, for instance the book (Volmir 1967)). Consequently, obtaining the stability loss equation of the cylindrical shell within the scope of the Kirchhoff–Love hypothesis from the TDLTS Eqs. (2.97)–(2.103) of the circular hollow cylinder by the average-integrating procedure described above gives the foregoing type corrections in the expressions of the coefficients of this equation.

Now we turn to the solution of the Eq. (2.121). This solution is found by employing Laplace transformation (2.34) to this equation. According to the convolution theorem for the Laplace transformation, the Eq. (2.121) is transformed to the following one:

$$\frac{h^2}{12}\bar{A}^*_{33}\frac{d^4\bar{w}}{dz^4}+p\frac{d^2\bar{w}}{dz^2}+\frac{\bar{A}^*_{11}}{R^2}\bar{w}=-\frac{1}{s}\alpha_1 mp\left(\frac{h}{2R}-1\right)sin(\alpha_1 mz). \tag{2.124}$$

Taking the end condition (2.122) and the expression of the right hand-side of the Eq. (2.124), we select the solution to Eq. (2.124) as

$$\bar{w}=\bar{A}(s)\sin(\alpha_1 z). \tag{2.125}$$

Substituting (2.125) into (2.124) and doing some mathematical calculations we obtain the following expression for the unknown $\bar{A}(s)$:

$$\bar{A}(s)=\frac{-\ell p\left(\frac{h}{2R}-1\right)}{m\pi s\left[\frac{A^*_{33}}{12}\left(\frac{h}{R}\right)^2\lambda^4-p\lambda^2+A^*_{11}\right]}, \tag{2.126}$$

where $\lambda=\pi mR/\ell$.

Employing the method (Schapery 1962, 1978), the original of $\bar{A}(s)$ is determined as follows:

$$A(t)=(s\bar{A}(s))|_{s=(1/(2t))}. \tag{2.127}$$

For the cases where the cylinder material is pure elastic we obtain from (2.126) the following expression for determination of the values of the critical force:

$$p_{cr.}=\frac{A_{11}}{\lambda^2}+\frac{A_{33}}{12}\left(\frac{h}{R}\right)\lambda^2. \tag{2.128}$$

It follows from the relation $dp_{cr.}/d(\lambda^2)=0$ that the minimum critical force is obtained under:

$$\lambda\approx\sqrt[4]{12\frac{A_{11}}{A_{33}}}\sqrt{\frac{R}{h}}\Rightarrow m\approx\frac{\ell}{\pi\sqrt{Rh}}\sqrt[4]{12\frac{A_{11}}{A_{33}}}. \tag{2.129}$$

For the case where the material of the cylinder is isotropic, we can write the relations given below from the Eq. (2.129):

$$\lambda\approx\sqrt[4]{12}\sqrt{\frac{R}{h}}\Rightarrow m\approx\frac{\ell}{\pi\sqrt{Rh}}\sqrt[4]{12}. \tag{2.130}$$

By replacing $\sqrt[4]{12}$ with $\sqrt[4]{12(1-\nu^2)}$ we obtain the relations from (2.130) which are given in related papers and books (see, for instance the book (Volmir 1967)).

In the general case, i.e. in the case where the cylinder material is viscoelastic transversal isotropic, the values of the critical parameters, according to (2.112), are defined from the following equation:

$A(t) \to \infty$ *as* $t \to t_{cr.}$ (for the case where the cylinder material is viscoelastic),
$A \to \infty$ *as* $p \to p_{cr.}$ (for the case where the cylinder material is pure elastic).
(2.131)

The values of the parameter $\lambda$ under which the critical time or critical force becomes minimal will be determined by direct verification of the corresponding numerical results.

## 2.9 The Third Order Refined Shell Theory

As for the third order refined shell theory, we use some modification of the theory by (Kromm 1953, 1955) for the cylindrical shell, according to which, the displacement $u_r^{(1)}$ is presented as in (2.113), but the displacement $u_z^{(1)}$ is presented as

$$u_z^{(1)} = r'\frac{dw(z)}{dz} + \left(\frac{h^2}{4}r' - \frac{1}{3}r'^3\right)\varphi(z), \; r' = R - r, \tag{2.132}$$

where $\varphi(z)$ is an average shear deformation of the cross-section of the cylindrical shell.

Substituting the expressions for $u_r^{(1)}$ and $u_z^{(1)}$ into Eqs. (2.101) and (2.96) we obtain the expression for strains and stresses given below:

$$\begin{aligned}
&\varepsilon_{rr}^{(1)} = 0, \; \varepsilon_{\theta\theta}^{(1)} = \frac{w(z)}{r}, \; \varepsilon_{zz}^{(1)} = r'\frac{d^2w(z)}{dz^2} + \left(\frac{h^2}{4}r' - \frac{1}{3}r'^3\right)\frac{d\varphi}{dz}, \\
&\varepsilon_{rz}^{(1)} = -\frac{1}{2}\left(\frac{h^2}{4} - r'^2\right)\varphi(z), \\
&\sigma_{rr}^{(1)} = 0, \; \sigma_{\theta\theta}^{(1)} = A_{11}^*\frac{w}{r} + A_{13}^*\left(r'\frac{d^2w(z)}{dz^2} + + \left(\frac{h^2}{4}r' - \frac{1}{3}r'^3\right)\frac{d\varphi}{dz}\right), \\
&\sigma_{zz}^{(1)} = A_{13}^*\frac{w}{r} + A_{33}^*\left(r'\frac{d^2w(z)}{dz^2} + + \left(\frac{h^2}{4}r' - \frac{1}{3}r'^3\right)\frac{d\varphi}{dz}\right), \\
&\sigma_{rz}^{(1)} = -G^*\left(\frac{h^2}{4} - r'^2\right)\varphi(z).
\end{aligned} \tag{2.133}$$

It follows from the expressions given in (2.133) that the equations in (2.116) hold also for the case considered.

As in the previous subsection, multiplying the first equation in (2.116) by $(1 - r'/R)dr'$, integrating in the interval $[-h/2, +h/2]$ and taking the relation (2.133) into account we obtain the following expressions:

$$\int_{-h/2}^{+h/2} \frac{\partial \sigma_{rz}^{(1)}}{\partial z}\left(1-\frac{r'}{R}\right)dr' = -G^* \frac{h^3}{6}\frac{d\varphi(z)}{dz},$$

$$-\int_{-h/2}^{+h/2} \frac{1}{r}\sigma_{\theta\theta}^{(1)}\left(1-\frac{r'}{R}\right)dr' = -\frac{h}{R^2}A_{11}w(z)$$

$$\int_{-h/2}^{+h/2} \sigma_{zz}^{(1)}\frac{\partial^2 w}{\partial z^2}\left(1-\frac{r'}{R}\right)dr' = -\sigma_{zz}^{(0)} h\frac{d^2 w}{dz^2} \tag{2.134}$$

Thus, after integrating (2.134), the first equation in (2.126) is transformed into the following one:

$$-G^* \frac{h^3}{6}\frac{d\varphi(z)}{dz} - \frac{h}{R^2}A_{11}w(z) + \sigma_{zz}^{(0)} h\frac{d^2 w}{dz^2} = 0. \tag{2.135}$$

Multiplying the second equation in (2.126) by $(1 - r'/R)r'dr'$, also integrating in the interval $[-h/2, +h/2]$ and taking into account the relations

$$\int_{-h/2}^{+h/2} \frac{\partial \sigma_{rz}^{(1)}}{\partial z}\left(1-\frac{r'}{R}\right)r'dr' + \int_{-h/2}^{+h/2} \frac{1}{R-r'}\sigma_{rz}^{(1)}\left(1-\frac{r'}{R}\right)r'dr'$$

$$= \sigma_{zz}^{(0)} h\left(\frac{h}{2R}-1\right)m\pi\cos(\alpha_1 mz) - \frac{h^3}{6}G^*\varphi(z),$$

$$\int_{-h/2}^{+h/2} \frac{\partial \sigma_{zz}^{(1)}}{\partial z}\left(1-\frac{r'}{R}\right)r'dr' = A_{33}^* \frac{h^3}{12}\frac{d^3 w}{dz^3} + A_{33}^* \frac{h^5}{60}\frac{d^2\varphi}{dz^2},$$

$$\int_{-h/2}^{+h/2} \sigma_{zz}^{(1)}\left[r'\frac{d^3 w}{dz^3} + \left(\frac{h^2}{4}r' - \frac{r'^3}{3}\right)\frac{d\varphi}{dz}\right]r'\left(1-\frac{r'}{R}\right)dr'$$

$$= \sigma_{zz}^{(0)}\left[\frac{h^3}{12}\frac{d^3 w}{dz^3} + \frac{h^5}{60}\frac{d^2\varphi}{dz^2}\right] \tag{2.136}$$

the mentioned second equation in (2.126) is transformed into the following one:

$$-\frac{h^3}{6}G^*\varphi + \left(A_{33}^* + \sigma_{zz}^{(0)}\right)\frac{h^3}{12}\frac{d^3 w}{dz^3} + \left(A_{33}^* + \sigma_{zz}^{(0)}\right)\frac{h^5}{60}\frac{d^2\varphi}{dz^2}$$

$$= -h\left(\frac{h}{2R}-1\right)m\pi\sigma_{zz}^{(0)}\cos(\alpha_1 mz). \tag{2.137}$$

Thus, from (2.136) and (2.137) we obtain the following equation:

$$\left[\left(A_{33}^{*}+\sigma_{zz}^{(0)}\right)\frac{h^{2}}{12}+\left(\frac{A_{33}^{*}}{G^{*}}+\frac{\sigma_{zz}^{(0)}}{G^{*}}\right)\frac{\sigma_{zz}^{(0)}h^{2}}{10}\right]\frac{d^{4}w}{dz^{4}}$$
$$+\left[-\sigma_{zz}^{(0)}-\left(\frac{A_{33}^{*}}{G^{*}}+\frac{\sigma_{zz}^{(0)}}{G^{*}}\right)\frac{A_{11}^{*}h^{2}}{10R^{2}}\right]\frac{d^{2}w}{dz^{2}}+\frac{A_{11}^{*}}{R}w$$
$$=\pi\alpha_{1}m^{2}\left(\frac{h}{2R}-1\right)\sigma_{zz}^{(0)}\sin(\alpha_{1}mz). \tag{2.138}$$

The end condition (2.122) holds also for the Eq. (2.138). As in the previous subsection, employing the Laplace transformation (2.34) to the Eq. (2.138), using the presentation (2.125) and the equality $\sigma_{zz}^{(0)}=-p$ we obtain the following expression for the unknown $\bar{A}(s)$:

$$\bar{A}(s)=\frac{-\ell p\left(\frac{h}{2R}-1\right)}{\Theta(s,\lambda,h/R)},$$
$$\Theta(s,\lambda,h/R)=m\pi s\left[A_{11}^{*}-\left[p-\left(\frac{A_{33}^{*}}{G^{*}}-\frac{p}{G^{*}}\right)\frac{A_{11}^{*}}{10}\left(\frac{h}{R}\right)^{2}\right]\lambda^{2}\right.$$
$$\left.+\left[\frac{A_{33}^{*}}{12}-\left(\frac{A_{33}^{*}}{G^{*}}-\frac{p}{G^{*}}\right)\frac{p}{10}\right]\left(\frac{h}{R}\right)^{2}\lambda^{4}\right]. \tag{2.139}$$

For the case where the material of the cylinder is a pure elastic one, within the scope of the assumption $(A_{33}/G-p/G)p/10\approx(A_{33}/G)p/10$, we determine the following expression from the equation for the calculation of the critical values of the external compressed force:

$$p_{cr}=\frac{A_{11}+A_{11}\frac{A_{33}}{10G}\left(\frac{h}{R}\right)^{2}\lambda^{2}+\frac{A_{33}}{12}\lambda^{4}}{\left[1+\frac{A_{11}}{10G}\left(\frac{h}{R}\right)^{2}\right]\lambda^{2}+\frac{A_{33}}{10G}\left(\frac{h}{R}\right)^{2}\lambda^{4}}. \tag{2.140}$$

## 2.10 Numerical Results Related to the Rotationally Symmetric Stability Loss Problem

As in Sect. 2.5, we assume that the cylinder is made from viscoelastic unidirectional fibrous composite materials consisting of a viscoelastic matrix and pure elastic fibers lying along the $Oz$ axis. Moreover, we assume that the notation and expressions (2.68)–(2.73) given in Sect. 2.5 hold also for the case under consideration.

Thus, within the foregoing assumptions we consider the numerical results and, according to the relation (2.75), first examine the pure elastic stability loss under

$t' = 0$ and $t' = \infty$. We compare the results obtained within the scope of the TDLTS with the corresponding ones obtained within the scope of the Kirchhoff–Love shell theory and within the scope of the third order refined shell theory. Through $p'_{3D.c.0}\left(= p_{3D.c.0}\big/E_0^{(1)}\right)$ and $p'_{3D.c.\infty}\left(= p_{3D.c.\infty}\big/E_0^{(1)}\right)$ we denote the critical values of the intensity of the dimensionless forces obtained by employing the TDLTS. Here and below the sub-indices 0 and $\infty$ indicate that the critical values of the forces are calculated at $t' = 0$ and $t' = \infty$ respectively. According to the Eqs. (2.128) and (2.140), the corresponding critical values of the intensity of the compressing forces obtained within the scope of the Kirchhoff–Love shell theory (denoted by $p'_{K.c.0}$ and $p'_{K.c.\infty}$) and within the scope of the third order refined shell theory (denoted by $p'_{R.c.0}$ and $p'_{R.c.\infty}$) can be calculated by the use of the following expressions:

$$
\begin{aligned}
p'_{K.c.0} &= \frac{p_{K.c.0}}{E_0^{(1)}} = \frac{A_{110}}{\lambda^2 E_0^{(1)}} + \frac{A_{330}}{12E_0^{(1)}}\left(\frac{h}{R}\right)^2\lambda^2,\\
p'_{K.c.\infty} &= \frac{p_{K.c.\infty}}{E_0^{(1)}} = \frac{A_{11\infty}}{\lambda^2 E_0^{(1)}} + \frac{A_{33\infty}}{12E_0^{(1)}}\left(\frac{h}{R}\right)^2\lambda^2,\\
p'_{R.c.0} &= \frac{p_{R.c.0}}{E_0^{(1)}} = \frac{A_{110} + A_{110}\frac{A_{330}}{12G_0}\left(\frac{h}{R}\right)^2\lambda^2 + \frac{A_{330}}{12}\lambda^4}{E_0^{(1)}\left[1 + \frac{A_{110}}{10G_0}\left(\frac{h}{R}\right)^2\right]\lambda^2 + \frac{A_{330}}{10G_0}\left(\frac{h}{R}\right)^2\lambda^4},\\
p'_{R.c.\infty} &= \frac{p_{R.c.\infty}}{E_0^{(1)}} = \frac{A_{11\infty} + A_{11\infty}\frac{A_{33\infty}}{12G_\infty}\left(\frac{h}{R}\right)^2\lambda^2 + \frac{A_{33\infty}}{12}\lambda^4}{E_0^{(1)}\left[1 + \frac{A_{11\infty}}{10G_\infty}\left(\frac{h}{R}\right)^2\right]\lambda^2 + \frac{A_{33\infty}}{10G_\infty}\left(\frac{h}{R}\right)^2\lambda^4},
\end{aligned}
\tag{2.141}
$$

where $A_{110} = \bar{A}^*_{11}\big|_{s=\infty}$, $A_{330} = \bar{A}^*_{33}\big|_{s=\infty}$, $A_{11\infty} = \bar{A}^*_{11}\big|_{s=0}$, $A_{33\infty} = \bar{A}^*_{33}\big|_{s=0}$, $G_0 = \bar{G}^*\big|_{s=\infty}$ and $G_\infty = \bar{G}^*\big|_{s=0}$.

Thus, for the problem under consideration we have the dimensionless parameters $h/R$, $\lambda = \pi mR/\ell$, $E^{(2)}\big/E_0^{(1)}$, $\omega$ and $\alpha$. Under numerical investigation, the critical values of the forces are minimized with respect to the parameter $\lambda$. Table 2.16 shows the minimized values of $p'_{3D.c.0}$, $p'_{R.c.0}$ and $p'_{K.c.0}$ for various $E^{(2)}\big/E_0^{(1)}$ and $h/R$. The corresponding values of $p'_{3D.c.\infty}$, $p'_{R.c.\infty}$ and $p'_{K.c.\infty}$ are given in Table 2.17 which are obtained for the case where $\omega = 1.0$. In the foregoing tables the values of the parameter $\lambda$ under which the critical forces have their minimum are also shown.

Note that the parameter $\omega$ characterizes the mechanical behavior of the viscoelastic matrix material in the vicinity of $t' = \infty$. Consequently, an increase in the values of $\omega$ corresponds to an increase (a decrease) in the values of $E_\infty^{(1)}$ $(\nu_\infty^{(1)})$. Therefore the values of $p'_{3D.c.\infty}$, $p'_{R.c.\infty}$ and $p'_{K.c.\infty}$ must be increase with $\omega$. This consideration is confirmed by the data given in Table 2.18 which show the values

**Table 2.16** The values of $p'_{3D.c.0}$, $p'_{R.c.0}$ and $p'_{K.c.0}$ obtained for various values of $E^{(2)}/E_0^{(1)}$ and $h/R$

| $\frac{E^{(2)}}{E_0^{(1)}}$ | $p'_{3D.c.0} \times 10^2$ | $p'_{R.c.0} \times 10^2$ | $p'_{E.c.0} \times 10^2$ |
|---|---|---|---|
| | | $h/R = 0.01$ | |
| 1 | $0.6000(\lambda = 18.2)$ | $0.7678(\lambda = 18.7)$ | $0.7772(\lambda = 18.6)$ |
| 5 | $1.2816(\lambda = 15.8)$ | $1.6471(\lambda = 16.7)$ | $1.6688(\lambda = 16.6)$ |
| 10 | $1.7868(\lambda = 13.9)$ | $2.2674(\lambda = 15.0)$ | $2.3019(\lambda = 14.9)$ |
| | | $h/R = 0.05$ | |
| 1 | $2.8996(\lambda = 8.40)$ | $3.6540(\lambda = 8.60)$ | $3.8860(\lambda = 8.30)$ |
| 5 | $6.2111(\lambda = 7.20)$ | $7.8134(\lambda = 7.70)$ | $8.3442(\lambda = 7.40)$ |
| 10 | $8.6517(\lambda = 6.40)$ | $10.660(\lambda = 6.90)$ | $11.510(\lambda = 6.70)$ |
| | | $h/R = 0.10$ | |
| 1 | $5.5644(\lambda = 6.10)$ | $6.8662(\lambda = 6.30)$ | $7.7721(\lambda = 5.90)$ |
| 5 | $11.958(\lambda = 5.30)$ | $14.618(\lambda = 5.60)$ | $16.691(\lambda = 5.30)$ |
| 10 | $16.634(\lambda = 4.70)$ | $19.765(\lambda = 5.10)$ | $23.019(\lambda = 4.70)$ |
| | | $h/R = 0.20$ | |
| 1 | $10.286(\lambda = 4.60)$ | $12.095(\lambda = 4.70)$ | $15.546(\lambda = 4.20)$ |
| 5 | $22.226(\lambda = 4.00)$ | $25.571(\lambda = 4.20)$ | $33.377(\lambda = 3.70)$ |
| 10 | $30.798(\lambda = 3.60)$ | $33.950(\lambda = 3.90)$ | $46.047(\lambda = 3.30)$ |

**Table 2.17** The values of $p'_{3D.c.\infty}$, $p'_{R.c.\infty}$ and $p'_{K.c.\infty}$ obtained for various values of $E^{(2)}/E_0^{(1)}$ and $h/R$ under $\omega = 0.1$

| $\frac{E^{(2)}}{E_0^{(1)}}$ | $p'_{3D.c.\infty} \times 10^2$ | $p'_{R.c.\infty} \times 10^2$ | $p'_{E.c.\infty} \times 10^2$ |
|---|---|---|---|
| | | $h/R = 0.01$ | |
| 1 | $0.3279(\lambda = 15.8)$ | $0.6277(\lambda = 18.2)$ | $0.6437(\lambda = 18.6)$ |
| 5 | $0.6627(\lambda = 12.1)$ | $1.3315(\lambda = 15.7)$ | $1.3786(\lambda = 15.4)$ |
| 10 | $0.9202(\lambda = 10.4)$ | $1.8103(\lambda = 14.0)$ | $1.8938(\lambda = 13.7)$ |
| | | $h/R = 0.05$ | |
| 1 | $1.5824(\lambda = 7.30)$ | $2.8637(\lambda = 8.50)$ | $3.2106(\lambda = 8.00)$ |
| 5 | $3.1809(\lambda = 5.60)$ | $5.7933(\lambda = 7.50)$ | $6.8929(\lambda = 6.90)$ |
| 10 | $4.3762(\lambda = 4.90)$ | $7.5643(\lambda = 6.90)$ | $9.4694(\lambda = 6.10)$ |
| | | $h/R = 0.10$ | |
| 1 | $3.0307(\lambda = 5.30)$ | $5.1058(\lambda = 6.30)$ | $6.4214(\lambda = 5.70)$ |
| 5 | $6.0469(\lambda = 4.20)$ | $9.7688(\lambda = 5.80)$ | $13.786(\lambda = 4.90)$ |
| 10 | $8.2102(\lambda = 3.70)$ | $12.191(\lambda = 5.40)$ | $18.940(\lambda = 4.30)$ |
| | | $h/R = 0.20$ | |
| 1 | $5.5785(\lambda = 4.00)$ | $8.1647(\lambda = 5.00)$ | $12.842(\lambda = 4.00)$ |
| 5 | $10.914(\lambda = 3.30)$ | $14.270(\lambda = 4.70)$ | $27.582(\lambda = 3.40)$ |
| 10 | $14.368(\lambda = 3.10)$ | $16.706(\lambda = 4.60)$ | $37.885(\lambda = 3.10)$ |

of $p'_{3D.c.\infty}$, $p'_{R.c.\infty}$ and $p'_{K.c.\infty}$ obtained for various $\omega$ under $E^{(2)}/E_0^{(1)} = 5$. It follows from the foregoing numerical results that the relations

$$p'_{3D.c.0} < p'_{R.c.0} < p'_{K.c.0}, \quad p'_{3D.c.\infty} < p'_{R.c.\infty} < p'_{K.c.\infty} \tag{2.142}$$

**Table 2.21** The values of $t'_{3D.cr.}$ (upper number) and $t'_{Rf.cr.}$ (lower number) obtained for various values of $p'$ in the case where $\lambda = 4.0$, $\alpha = -0.3$, $E^{(2)}/E_0^{(1)} = 5$, $h/R = 0.2$ and $\omega = 1.0$

| $p' \times 10^2$ | | | | | | |
|---|---|---|---|---|---|---|
| 20.4 | 20.5 | 20.6 | 20.7 | 20.8 | 20.9 | 21.0 |
| $t'_{3D.cr.}/t'_{cdm.cr.}$ | | | | | | |
| 0.0210<br>118.48 | 0.0190<br>8.4127 | 0.0172<br>3.2494 | 0.0155<br>1.7368 | 0.0138<br>1.0685 | 0.0122<br>0.7095 | 0.0108<br>0.4931 |

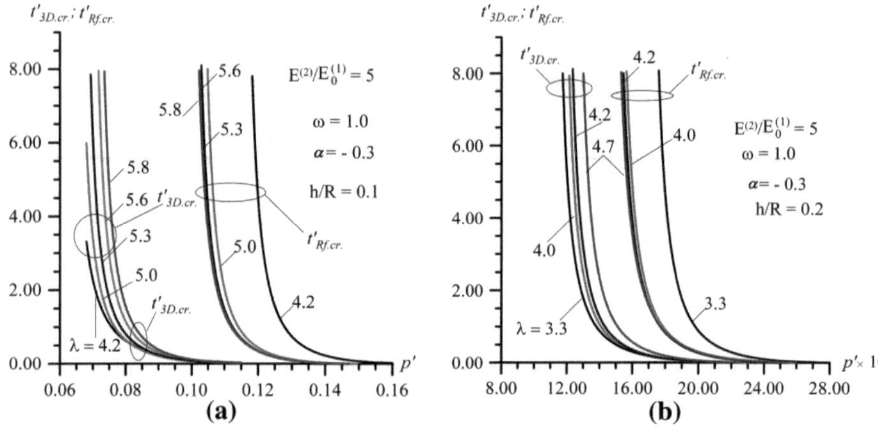


**Fig. 2.9** The graphs of the dependencies among $t'_{3D.cr.}$, $t'_{Rf.cr.}$ and dimensionless intensity of the compressed force $p'$ constructed for various values of the parameter $\lambda$ in the cases where $h/R = 0.1$ (**a**) and $h/R = 0.2$(**b**)

Consider the influence of the rheological parameter $\alpha$ on the values of $t'_{3D.cr.}$. The graphs of the dependencies between $t'_{3D.cr.}$ and $p'$ are given in Fig. 2.10 which are constructed for the cases where $h/R = 0.1$ (Fig. 2.10a) and 0.2 (Fig. 2.10b) under $\omega = 1$ and $\lambda = 4$. It follows from the results given in Fig. 2.10 (and from other numerical results which are not given here) that the graphs of the analyzed dependencies constructed for various $\alpha$ have a common intersection point.

Before (after) this intersection point the increase in the absolute values of $\alpha$ causes an increase (a decrease) in the values of $t'_{3D.cr.}$. Note that the parts of the graphs which appear after the intersection point are illustrated in Fig. 2.11 for the cases where $h/R = 0.1$ (Fig. 2.11a) and $h/R = 0.2$ (Fig. 2.11b) under $\omega = 1$ and $\lambda = 4$. Remind that the explanation of the character of the influence of the parameter $\alpha$ on the values of the critical time is given in Subsection 2.5.1.

Now we consider the question of how the rheological parameter $\omega$ influences the behavior of the graphs related to the dependence between $t'_{3D.cr.}$ and $p'$. The corresponding graphs are given in Fig. 2.12 for various values of the parameter $\omega$ in the case where $E^{(2)}/E_0^{(1)} = 5$, $h/R = 0.2$ and $\alpha = -0.5$. The character of these

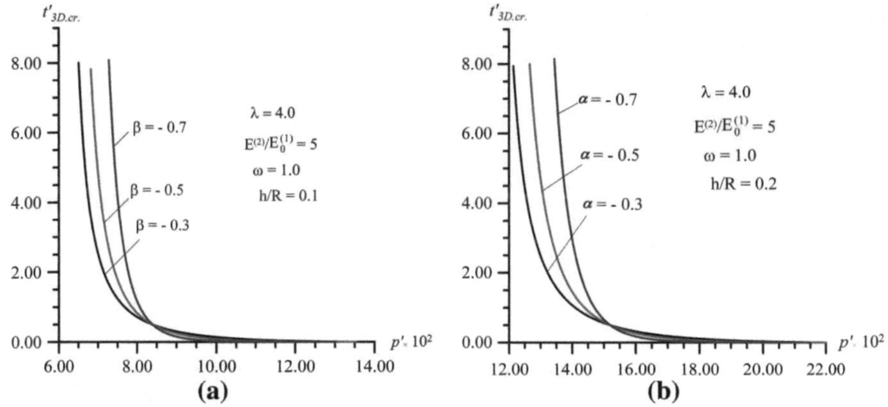


**Fig. 2.10** The influence of the rheological parameter $\alpha$ on the dependencies between $t'_{3D.cr.}$ and compressed force $p'$ in the cases where $h/R = 0.1$(**a**) and $h/R = 0.2$(**b**)

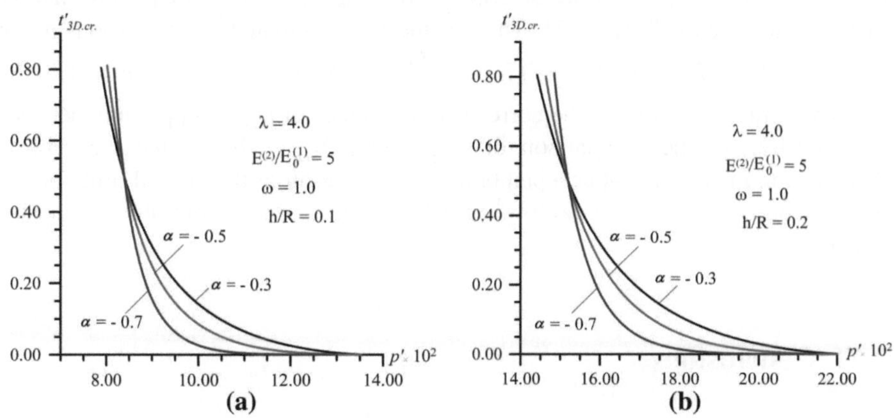


**Fig. 2.11** The parts of the graphs given in Fig. 2.10 which arise after the intersection point in the cases where $h/R = 0.1$(**a**) and $h/R = 0.2$(**b**)

graphs can be explained by the increase in the values of $p'_{3D.c.\infty}$ with $\omega$. Therefore for common $p'$ (with respect to the selected values of $\omega$), the values of $t'_{3D.cr.}$ increase with $\omega$.

As in the previous sections, compare the values of $t'_{3D.cr.}$ with the corresponding ones calculated by employing the critical deformation method (Gerard and Gilbert 1958). According to this method, it is assumed that the critical deformation of the viscoelastic hollow cylinder is equal to the critical deformation of the corresponding elastic cylinder. Using this assumption, the critical deformation of the pure elastic cylinder is determined within the scope of the TDLTS. Note that in the considered case the mentioned critical deformation corresponds to the $p'_{3D.c.0}$.

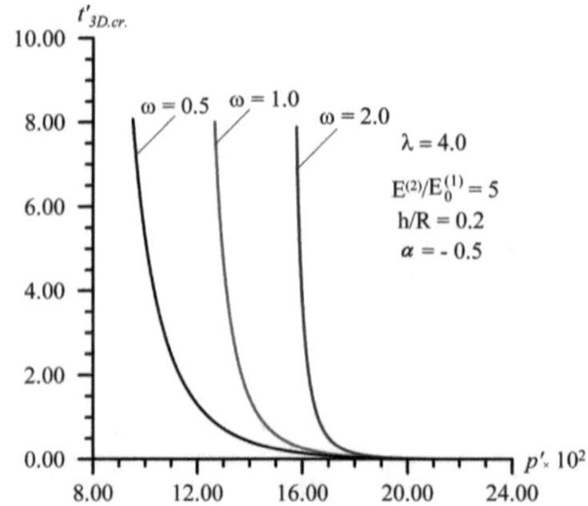


**Fig. 2.12** The graphs of the dependencies between $t'_{3D.cr.}$ and dimensionless intensity of the compressed force $p'$ constructed for various values of the rheological parameter $\omega$

According to this determination, using the relation $p'_{3D.c.0} = p/E_3^*$, the critical time is determined for selected values of $p$. The values of the dimensionless critical time (denoted by $t'_{cdm.cr.}$), which are determined by employing the critical deformation method, are given in Table 2.21. These values are calculated for the various $p'$ for the case where $\lambda = 4.0$, $\alpha = -0.3$, $E^{(2)}\big/E_0^{(1)} = 5$, $h/R = 0.2$ and $\omega = 1.0$. At the same time, in this table the corresponding values of $t'_{3D.cr.}$ (upper number) are also illustrated. The comparison of $t'_{cdm.cr.}$ with $t'_{3D.cr.}$ shows that the critical deformation method is not acceptable for determination of the critical time for the local stability loss of the hollow cylinder made from viscoelastic composite material.

## 2.11 Conclusions

Thus, in the present chapter within the scope of the three-dimensional geometrically non-linear field equations of the theory for viscoelastic orthotropic bodies an approach has been developed for 3D global and local stability loss analyses of solid and hollow cylinders made from viscoelastic composite materials. In studying the global (local) stability loss of the cylinder it is supposed that it has an initial infinitesimal global (local) imperfection and the values of a critical time as well as the values of a critical force are determined from the initial imperfection criterion. According to this criterion, the moment at which the mentioned initial imperfections start to increase and grow indefinitely is taken as a critical time (or critical force). To study the corresponding non-linear boundary value problems the boundary-form disturbance method and small parameter method are employed simultaneously. From the results of these applications, the solution to the aforementioned non-linear problem is reduced to the solution for the corresponding series of linear problems. By direct verification it is proven that the equations and relations related to the first and subsequent approximations coincide with the equations and relations of the well known TDLTS. It is established that for investigation of the stability loss problem it is enough to use only the zeroth and first approximations. By the average-integrating procedure the corresponding approximate global stability loss equations within the scope of the Bernoulli and third order refined beam theories are derived from the equations and relations of the aforementioned first approximations. In addition, by the average-integrating procedure the corresponding approximate local stability loss equations within the scope of the Kirchhoff –Love and third order shell theories are obtained from the equations and relations related to the rotationally symmetric 3D local stability loss of the hollow cylinder. Final numerical results are obtained by employing Laplace transformation and variation separation methods for the case where the cylinder is made from the viscoelastic composite consisting of the viscoelastic matrix and uni-directional elastic fibers lying along the cylinder. First, the numerical results related to the pure elastic stability loss of the cylinder at $t = 0$ and $t = \infty$ are considered. It is established that the critical forces obtained within the scope of the third order refined beam and shell theories are very close to the corresponding results obtained within the scope of the 3D approach. However, the critical time obtained within the scope of the 3D approach can be significantly less than the corresponding values of the critical time obtained within the scope of the third order refined beam and shell theories. The difference between the critical times obtained by employing 3D and third order refined beam and shell theories becomes increasingly non-negligible if the values of the external compressed force is close to the critical compressed force obtained at $t = \infty$. Consequently, for investigation of the global and local stability loss of the cylinders made from viscoelastic composites, in many cases it is necessary to use the TDLTS developed in the present chapter. This approach can be easily extended for investigation of the more complicated stability loss problems, for instance, for non-symmetric local stability loss of the hollow cylinders (shells) made from viscoelastic materials.

## References

Akbarov SD, Karakaya S (2011) 3D analyses of the stability loss of the circular cylinder made from viscoelastic composite material. CMC: Computers, Materials and Continua 22(1):1–38

Akbarov SD, Karakaya S (2012a) 3D analyses of the symmetric local stability loss of the circular hollow cylinder made from viscoelastic composite material. Appl Math Modell 36:4241–4260

Akbarov SD, Karakaya S (2012b) 3D analyses of the global stability loss of the circular hollow cylinder made from viscoelastic composite material. Eur J Mech A/Solids 33:48–66

Chiristensen RM (1979) Mechanics of composite materials. Wiley, New York

Gerard G, Gilbert A (1958) A critical strains approach to creep buckling of plates and shells. J Aeronaut Sci 25:429–438

Guz AN (1999) Fundamentals of the three-dimensional theory of stability of deformable bodies. Springer, Heidelberg

Kromm A (1953) Verallgeneinerte theorie der platen-static. Ingenier-Archiv 21:266–286

Kromm A (1955) Über die ranguer krafter bei gestutzten platen. Z. Angew Math Mech 35(55):231–242
Rabotnov YuN (1977) Elements of hereditary mechanics of solid bodies. Moscow, Nauka (in Russian)
Schapery RA (1962) Approximate method of transform inversion for viscoelastic stress analysis. In:Proceeding 4th US National Congress Appl Mech ASME, 1075–1086
Schapery RA (1978) A viscoelastic behaviour of composite materials, in Composite materials, vol. 1–7. In: Broutman LJ Krock RH (eds.) Mir, Moscow, translated from English), vol. 2. Mechanics of composite materials, 102–195(in Ruıssion, translated from English)
Volmir AS (1967) Stability of deformation systems. Nauko, Moskow

# Chapter 3
# Stability Loss Problems for Viscoelastic Plates

In the present chapter within the scope of the TDLTS an approach is developed for investigation stability loss of plates made of viscoelastic composite materials. Rectangular and circular plates are considered. By average-integrating through the thickness of the plates the stability loss equations within the framework of approximate plate theories are derived from the equations and relations of the TDLTS. For the solution to the corresponding boundary value problems the analytical and numerical methods (semi-analytical and 3D FEM) are developed. Numerical results on the critical forces and critical time obtained by employing various theories are presented and discussed. Concrete investigations are made for the cases where the ends of the plates are clamped or simply supported.

The results detailed in the present chapter are based on investigations carried out in several papers (Akbarov 1998, 2007; Akbarov and Yahnioglu 2001; Akbarov et al. 2001; Kutuk 2009; Kutuk et al. 2003; Selim and Akbarov 2003; Yahnioglu and Akbarov 2002).

## 3.1 Formulation of the Problem and Basic Field Equations

We consider the rectangular plate with an initial infinitesimal imperfection and with constant thickness $2h$. Assume that this plate occupies the region $\Omega$ and with middle plane of that we associate the Lagrangian coordinate system $Ox_1x_2x_3$ which in its natural state coincides with the Cartesian coordinate system. Assume that the region $\Omega$ is restricted by the surfaces $S^+$ and $S^-$ and inclined end sections $M_0$, $M_3$ (Fig. 3.1) under $0 \leq x_1 \leq \ell_1$, $0 \leq x_3 \leq \ell_3$. The middle surface of this plate is given by the equation

$$x_2 = F(x_1, x_3) = \varepsilon f(x_1, x_3), \tag{3.1}$$

where $\varepsilon$ is a small parameter $(0 \leq \varepsilon \ll 1)$ and shows the degree of the initial imperfection of the plate; the function $f(x_1, x_3)$ characterizes the form of this imperfection.

S. D. Akbarov, *Stability Loss and Buckling Delamination*,
DOI: 10.1007/978-3-642-30290-9_3, 

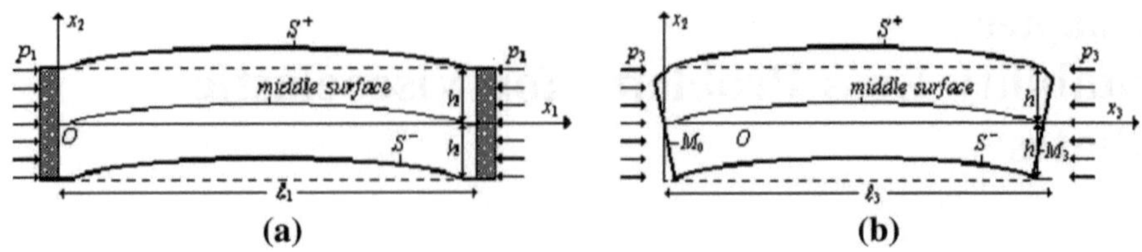


**Fig. 3.1** The geometry of the cross section of the plate at $x_3 = \ell_3/2$ (**a**) and at $x_1 = \ell_1/2$ (**b**); clamped (simply supported) ends at $x_1 = 0$ and $x_1 = \ell_1$ (**a**), (at $x_3 = 0$ and $x_3 = \ell_3$ (**b**))

For generality of the consideration below we will assume that the plate is clamped at $x_1 = 0$ and $\ell_1$ (Fig. 3.1a) and simply supported at $x_3 = 0$ and $\ell_3$ (Fig. 3.1b). According to this assumption, the function $f(x_1, x_3)$ must satisfy the following conditions:

$$\begin{aligned} f(x_1, x_3)|_{x_1=0,\ell_1} = 0, \left.\frac{\partial f(x_1, x_3)}{\partial x_1}\right|_{x_1=0,\,\ell_1} = 0, \\ f(x_1, x_3)|_{x_3=0} = 0, \left.\frac{\partial^2 f(x_1, x_3)}{\partial x_3^2}\right|_{x_3=0,\,\ell_3} = 0 \end{aligned} \tag{3.2}$$

The infinite smallness of the mentioned above initial imperfection requires also the satisfaction of the condition

$$\varepsilon^2\left(\left(\frac{\partial^2 f}{\partial x_1^2}\right)^2 + \left(\frac{\partial^2 f}{\partial x_3^2}\right)\right) \ll 1 \quad \text{for all} \quad x_1 \in (0, \ell_1),\, x_3 \in (0, \ell_3). \tag{3.3}$$

We suppose that the plate material is orthotropic viscoelastic one with the principal axes of elastic symmetry $Ox_1$, $Ox_2$ and $Ox_3$. Attempt to investigate the evolution of the foregoing initial infinitesimal imperfection with flow of a time under compression of the plate along the $Ox_1(Ox_3)$ axis with the uniformly distributed normal forces with intensity $p_1(p_3)$. We make this investigation within the scope of the 3D geometrically non-linear field equations of the theory of viscoelasticity for the anisotropic (orthotropic) material. We write the following governing field equations, which are satisfied within the region $\Omega$.

Equilibrium equation:

$$\frac{\partial}{\partial x_j}\left[\sigma_{jn}\left(\delta_i^n + \frac{\partial u_i}{\partial x_n}\right)\right] = 0, \tag{3.4}$$

Strain-Displacement relations:

$$\varepsilon_{ij} = \frac{1}{2}\left(\frac{\partial u_i}{\partial x_j} + \frac{\partial u_j}{\partial x_i} + \frac{\partial u_n}{\partial x_i}\frac{\partial u_n}{\partial x_j}\right). \tag{3.5}$$

In Eqs. (3.4), (3.5) and below repeated indices $(i;j;n = 1, 2, 3)$ are summed over their ranges and the following notation is used: $\sigma_{ij}$ and $\varepsilon_{ij}$ are components of the stress and strain tensor respectively, $u_i$ is a component of the displacement vector.

The constitutive relations for the plate material are given as follows:

$$\sigma_{11}=A_{11}^{*}\varepsilon_{11}+A_{12}^{*}\varepsilon_{22}+A_{13}^{*}\varepsilon_{33},\ \sigma_{22}=A_{12}^{*}\varepsilon_{11}+A_{22}^{*}\varepsilon_{22}+A_{23}^{*}\varepsilon_{33},$$
$$\sigma_{33}=A_{13}^{*}\varepsilon_{11}+A_{23}^{*}\varepsilon_{22}+A_{33}^{*}\varepsilon_{33},$$
$$\sigma_{23}=2A_{66}^{*}\varepsilon_{13},\ \sigma_{13}=2A_{55}^{*}\varepsilon_{13},\ \sigma_{12}=2A_{66}^{*}\varepsilon_{13}, \tag{3.6}$$

where $A_{11}^{*},\ldots, A_{66}^{*}$ are the following operators

$$A_{ij}^{*}\varphi=A_{ij0}\varphi(t)+\int_{0}^{t}A_{ij1}(t-\tau)\varphi(\tau)d\tau,\ ij=11;\ 22;\ 33;\ 12;\ 13;\ 23;\ 44;\ 55;\ 66. \tag{3.7}$$

We assume that on the upper $S^{+}$ and lower $S^{-}$ surfaces of the plate the tracing-free conditions

$$\left[\sigma_{jn}\left(\delta_{i}^{n}+\frac{\partial u_i}{\partial x_n}\right)\right]\bigg|_{S^{\pm}} n_j^{\pm}=0,\quad i=1,\,2,\,3, \tag{3.8}$$

satisfy, where $n_j^{\pm}$ are the unit ortho-normal components to the surfaces $S^{\pm}$.

Moreover, we assume that the plate is compressed along the $Ox_1$ and $Ox_3$ axes by normal forces with intensity $p_1$ and $p_3$ respectively. However the compression in the direction of $Ox_1$ axis makes through the rigid body which holds the ends of the plate at $x_1=0$ and $x_1=\ell_1$. In this case it is supposed that

$$u_1(0,\,x_2,\,x_3)=-u_1(\ell_1,\,x_2,\,x_3)=U=\text{const},$$
$$u_2(x_1,\,x_2,\,x_3)|_{x_1=0;\,\ell_1}=u_3(x_1,\,x_2,\,x_3)|_{x_1=0;\,\ell_1}=0. \tag{3.9}$$

Note that the relation between $U$ and $p_1$ is determined after solution procedure from the corresponding formulae which will be given below.

The conditions on the inclined edge sections $M_0$, $M_3$ (Fig. 3.1b) are the following ones:

$$u_2(x_1,\,x_2,\,x_3)|_{M_0}=0,\ \left[\sigma_{jn}\left(\delta_{3}^{n}+\frac{\partial u_3}{\partial x_n}\right)\right]\bigg|_{M_0} n_{0j}=p_3,$$
$$u_3(x_1,\,x_2,\,x_3)|_{M_0}=0,\ u_2(x_1,\,x_2,\,x_3)|_{M_3}=0,$$
$$\left[\sigma_{jn}\left(\delta_{3}^{n}+\frac{\partial u_3}{\partial x_n}\right)\right]\bigg|_{M_3} n_{\ell j}=p_3,\ u_1(x_1,\,x_2,\,x_3)|_{M_3}=0. \tag{3.10}$$

where $n_{0j}$ $(n_{\ell j})$ are components of the ortho-normal vector $\mathbf{n}_0(\mathbf{n}_\ell)$ on the end-plane $M_0(M_3)$.

This completes formulation of the problem related to the stability loss of the rectangular plate made from the viscoelastic composite material. Note that this formulation can be easily transformed for the cases where the (3.9) or (3.10) type conditions are given at all ends of the plate. Moreover, note that, according to this formulation, the evolution of the infinitesimal initial imperfection (3.1)–(3.3) of the plate with time for fixed value of the compressed forces $p_1$ and $p_3$ (for the case where the material of the plate is viscoelastic) or with compressed forces $p_1$ and $p_3$ (for the case where the material of the plate is pure elastic) will be investigated within the framework of the field Eqs. (3.4)–(3.7), boundary condition (3.8) and end conditions (3.9) and (3.10).

## 3.2 Approach for the Solution to the Stability Loss Problem for Rectangular Plate

In the principal sense the approach which will be considered below is similar to that which is developed and employed in Chap. 2. Therefore, the steps (i)–(iv) explained in Sect. 2.2 can also be repeated with obvious changing for the problem formulated in the previous section. Since, according to the mentioned steps and problem formulated in the previous section, for employing the boundary-form perturbation technique for solution to the problem under consideration, first we derive the equation of the upper $S^+$ and lower $S^-$ surfaces of the plate. For this purpose we rewrite of the equation of the middle surface of the plate as follows:

$$x_1 = t_1,\ x_3 = t_3,\ x_2 = F(t_1,\ t_3) = \varepsilon f(t_1,\ t_3), \tag{3.11}$$

where $t_1$ and $t_3$ are parameters and $t_1 \in (0,\ \ell_1)$; $t_3 \in (0, \ell_3)$. The slopes of the tangent lines to the surface (3.11) at planes $x_3 = const$ and at planes $x_1 = \text{const}$ can be presented as

$$\frac{\partial x_2}{\partial x_1} = \frac{\partial F(t_1,\ t_3)}{\partial t_1} \text{ and } \frac{\partial x_2}{\partial x_3} = \frac{\partial F(t_1,\ t_3)}{\partial t_3}, \tag{3.12}$$

respectively. Using the expression in (3.12), the equations of the straight lines which are perpendicular to the surface (3.11) and lie on the planes $x_3 = \text{const}$ and $x_1 = \text{const}$ are

$$x_1 - t_1 = -\frac{\partial F(t_1,\ t_3)}{\partial t_1}(x_2 - F(t_1,\ t_3)) \text{ and}$$
$$x_3 - t_3 = -\frac{\partial F(t_1,\ t_3)}{\partial t_3}(x_2 - F(t_1,\ t_3)) \tag{3.13}$$

respectively. As the surfaces $S^+$, $S^-$ and straight lines have common points, and the thickness of the plate is $2h$, therefore the coordinates (denoted by $x_1^\pm,\ x_2^\pm,\ x_3^\pm$) of the surfaces $S^\pm$ must satisfy simultaneously the following three equations.

$$\begin{aligned} x_1^\pm - t_1 &= -\frac{\partial F(t_1,\ t_3)}{\partial t_1}(x_2^\pm - F(t_1,\ t_3)), \\ x_3^\pm - t_3 &= -\frac{\partial F(t_1,\ t_3)}{\partial t_3}(x_2^\pm - F(t_1,\ t_3)), \\ (x_1^\pm - t_1)^2 &+ (x_2^\pm - F(t_1,\ t_3))^2 + (x_3^\pm - t_3)^2 = h^2. \end{aligned} \tag{3.14}$$

Thus, we obtain

$$
\begin{aligned}
x_1^{\pm} &= t_1 \mp \frac{h}{V(t_1, t_3)} \frac{\partial F(t_1, t_3)}{\partial t_1}, \\
x_2^{\pm} &= F(t_1, t_3) \pm \frac{h}{V(t_1, t_3)} \\
x_3^{\pm} &= t_3 \mp \frac{h}{V(t_1, t_3)} \frac{\partial F(t_1, t_3)}{\partial t_3},
\end{aligned}
\tag{3.15}
$$

where

$$
V(t_1, t_3) = \left[1 + \left(\frac{\partial F(t_1, t_3)}{\partial t_1}\right)^2 + \left(\frac{\partial F(t_1, t_3)}{\partial t_3}\right)^2\right]. \tag{3.16}
$$

Through **i**, **j** and **k** we denote the ort-unit vectors in the directions of the $Ox_1$, $Ox_2$ and $Ox_3$ axes respectively and rewrite the Eq. (3.15) in the vector form.

$$
\mathbf{r}^{\pm} = x_1^{\pm}(t_1, t_3)\mathbf{i} + x_2^{\pm}(t_1, t_3)\mathbf{j} + x_3^{\pm}(t_1, t_3)\mathbf{k}. \tag{3.17}
$$

We determine the unit normal vector to the surfaces $S^{\pm}$ from (3.17).

$$
\mathbf{n}^{\pm} = \pm \frac{\mathbf{r}_3^{\pm} \times \mathbf{r}_1^{\pm}}{\left|\mathbf{r}_3^{\pm} \times \mathbf{r}_1^{\pm}\right|} = n_1^{\pm}\mathbf{i} + n_2^{\pm}\mathbf{j} + n_3^{\pm}\mathbf{k}, \tag{3.18}
$$

where

$$
\begin{aligned}
\mathbf{r}_1^{\pm} &= \frac{\partial x_1^{\pm}(t_1, t_3)}{\partial t_1}\mathbf{i} + \frac{\partial x_2^{\pm}(t_1, t_3)}{\partial t_1}\mathbf{j} + \frac{\partial x_3^{\pm}(t_1, t_3)}{\partial t_1}\mathbf{k}, \\
\mathbf{r}_3^{\pm} &= \frac{\partial x_1^{\pm}(t_1, t_3)}{\partial t_3}\mathbf{i} + \frac{\partial x_2^{\pm}(t_1, t_3)}{\partial t_3}\mathbf{j} + \frac{\partial x_3^{\pm}(t_1, t_3)}{\partial t_3}\mathbf{k}, \\
n_i^{\pm} &= \pm \frac{A_i^{\pm}(t_1, t_3)}{A(t_1, t_3)}, \; i = 1, 2, 3.
\end{aligned}
\tag{3.19}
$$

In (3.19) the following notation is used.

$$
\begin{aligned}
A_1^{\pm}(t_1, t_3) &= \frac{\partial x_3^{\pm}(t_1, t_3)}{\partial t_1}\frac{\partial x_2^{\pm}(t_1, t_3)}{\partial t_3} - \frac{\partial x_2^{\pm}(t_1, t_3)}{\partial t_1}\frac{\partial x_3^{\pm}(t_1, t_3)}{\partial t_3}, \\
A_2^{\pm}(t_1, t_3) &= \frac{\partial x_1^{\pm}(t_1, t_3)}{\partial t_1}\frac{\partial x_3^{\pm}(t_1, t_3)}{\partial t_3} - \frac{\partial x_3^{\pm}(t_1, t_3)}{\partial t_1}\frac{\partial x_1^{\pm}(t_1, t_3)}{\partial t_3}, \\
A_3^{\pm}(t_1, t_3) &= \frac{\partial x_2^{\pm}(t_1, t_3)}{\partial t_3}\frac{\partial x_1^{\pm}(t_1, t_3)}{\partial t_1} - \frac{\partial x_1^{\pm}(t_1, t_3)}{\partial t_1}\frac{\partial x_2^{\pm}(t_1, t_3)}{\partial t_3}, \\
A(t_1, t_3) &= \left[\left(A_1^{\pm}(t_1, t_3)\right)^2 + \left(A_2^{\pm}(t_1, t_3)\right)^2 + \left(A_3^{\pm}(t_1, t_3)\right)^2\right].
\end{aligned}
\tag{3.20}
$$

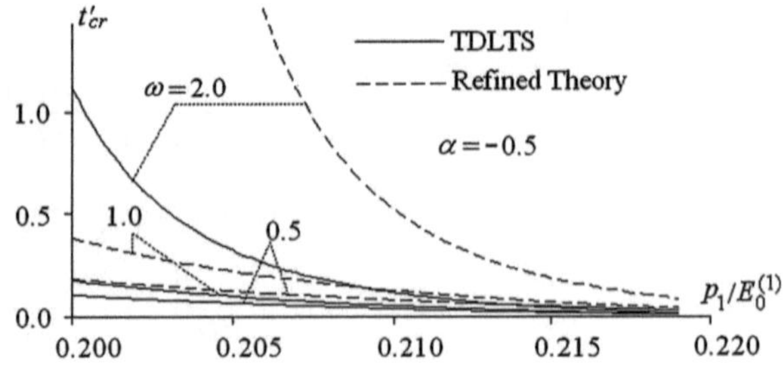


**Fig. 3.2** The influence of the rheological parameter $\omega$ on the dependencies among $t'_{cr.Rf}$, $t'_{cr.3D}$ and $p_1/E_0^{(1)}$ under $\alpha = -0.5$

Using the condition (3.3) we obtain from (3.15), (3.16), (3.19) and (3.20) the following expressions for $x_i^{\pm}$ and $n_i^{\pm}(i = 1, 2, 3)$.

$$x_1^{\pm} = t_1 \mp \varepsilon h \frac{\partial f(t_1, t_3)}{\partial t_1} + O(\varepsilon^3),\ x_2^{\pm} = \pm h + \varepsilon f(t_1, t_3) + O(\varepsilon^2),$$
$$x_3^{\pm} = t_3 \mp \varepsilon h \frac{\partial f(t_1, t_3)}{\partial t_3} + O(\varepsilon^3),\ n_1^{\pm} = \mp \varepsilon \frac{\partial f(t_1, t_3)}{\partial t_1} + O(\varepsilon^3),$$
$$n_2^{\pm} = \pm 1 + O(\varepsilon^2),\ n_3^{\pm} = \mp \varepsilon \frac{\partial f(t_1, t_3)}{\partial t_3} + O(\varepsilon^3). \tag{3.21}$$

Now we consider the series presentation of the equations of the inclined end planes $M_0$, $M_3$ (Fig. 3.2) and the components of the ort-normal vectors $n_{0i}$ and $n_{0\ell}$ to these planes. For this purpose we assume that the function $f(x_1, x_3)$ (3.1) can be presented as

$$f(x_1, x_3) = f_1(x_1) f_3(x_3). \tag{3.22}$$

Taking (3.22) into account and doing corresponding mathematical calculations we obtain:

$$x_3 = -\varepsilon a_0 x_2 \text{ (equation of } M_0 \text{ plane)},$$
$$x_3 = -\varepsilon a_\ell x_2 + \ell_3 \text{ (equation of } M_3 \text{ plane)}$$
$$n_{02} = -\varepsilon a_0 \left(1 - \frac{1}{2}(\varepsilon a_0)^2 + O\left((\varepsilon a_0)^4\right)\right),$$
$$n_{\ell 2} = -\varepsilon a_\ell \left(1 - \frac{1}{2}(\varepsilon a_\ell)^2 + O\left((\varepsilon a_\ell)^4\right)\right),$$
$$n_{03} = -\left(1 - \frac{1}{2}(\varepsilon a_0)^2 + O\left((\varepsilon a_0)^4\right)\right),\ n_{\ell 3} = \left(1 - \frac{1}{2}(\varepsilon a_\ell)^2 + O\left((\varepsilon a_\ell)^4\right)\right),$$
$$n_{01} = n_{\ell 1} = 0, \tag{3.23}$$

where

$$a_0 = \left.\frac{\partial f_3(t_3)}{\partial t_3}\right|_{t_3=0},\ a_\ell = \left.\frac{\partial f_3(t_3)}{\partial t_3}\right|_{t_3=\ell_3}. \tag{3.24}$$

After foregoing preparing procedures, we represent the solution to the problem (3.4)–(3.10) in series form in the small parameter $\varepsilon$:

$$\left\{\sigma_{ij};\ \varepsilon_{ij};\ u_i\right\} = \sum_{q=0}^{\infty} \varepsilon^q \left\{\sigma_{ij}^{(q)};\ \varepsilon_{ij}^{(q)};\ u_i^{(q)}\right\}. \tag{3.25}$$

Doing the procedures explained in Sect. 2.2, i.e. substituting Eq. (3.25) into (3.4)–(3.6) we obtain set equations for each approximation in (3.25). Moreover, using (3.21) and expanding each approximation in (3.25) in series form in the vicinity of the point $(x_1^{\pm} = t_1,\ x_2^{\pm} = \pm h,\ x_3^{\pm} = t_3)$ we obtain from (3.8) the boundary conditions for each approximation on the upper and lower surfaces of the plate. At the same time, expanding each approximation in (3.25) in the vicinity of the points $x_3 = 0$ and $x_3 = \ell_3$ with respect to variable $x_3$ and doing the needed mathematical operations we obtain also the end conditions for each approximation at $x_3 = 0$ and $x_3 = \ell_3$.

Thus, doing the described above mathematical manipulations we obtain the corresponding field equations, boundary and end conditions for each approximation. Here we consider the system equations, boundary and end conditions only for the zeroth and the first approximations. Because, as it has been noted in Sect. 2.2, for investigation of the stability loss of the plate the zeroth and the first approximations are sufficient.

The system of equations, boundary and end conditions for the zeroth approximation are:

$$\frac{\partial}{\partial x_j}\left[\sigma_{jn}^{(0)}\left(\delta_i^n + \frac{\partial u_i^{(0)}}{\partial x_n}\right)\right] = 0,\ \varepsilon_{ij}^{(0)} = \frac{1}{2}\left(\frac{\partial u_i^{(0)}}{\partial x_j} + \frac{\partial u_j^{(0)}}{\partial x_i} + \frac{\partial u_n^{(0)}}{\partial x_i}\frac{\partial u_n^{(0)}}{\partial x_j}\right), \tag{3.26}$$

$$\left.\left[\sigma_{2n}^{(0)}\left(\delta_i^n + \frac{\partial u_i^{(0)}}{\partial x_n}\right)\right]\right|_{x_2=\pm h} = 0, \tag{3.27}$$

$$\begin{gathered} u_1^{(0)}(0, x_2, x_3) = -u_1^{(0)}(\ell_1, x_2, x_3) = U = \text{const}, \\ \left.u_2^{(0)}(x_1, x_2, x_3)\right|_{x_1=0,\,\ell_1} = \left.u_3^{(0)}(x_1, x_2, x_3)\right|_{x_1=0,\,\ell_1} = 0, \end{gathered} \tag{3.28}$$

$$\begin{gathered} \left.u_2^{(0)}(x_1, x_2, x_3)\right|_{x_3=\ell_3} = 0,\ \left.u_1^{(0)}(x_1, x_2, x_3)\right|_{x_3=\ell_3} = 0, \\ \left.\left[\sigma_{3n}^{(0)}\left(\delta_3^n + \frac{\partial u_3^{(0)}}{\partial x_n}\right)\right]\right|_{x_3=0;\ell_3} = -p_3. \end{gathered} \tag{3.29}$$

The system of equations, boundary and end conditions for the first approximation are:

$$\frac{\partial}{\partial x_j} T_{ji}^{(1)} = 0,\ T_{ji}^{(1)} = \left(\delta_i^n + \frac{\partial u_i^{(0)}}{\partial x_i}\right)\sigma_{jn}^{(1)} + \sigma_{jn}^{(0)} \frac{\partial u_i^{(1)}}{\partial x_n},$$

$$\varepsilon_{ij}^{(1)} = \frac{1}{2}\left(\delta_i^n + \frac{\partial u_n^{(0)}}{\partial x_i}\right)\frac{\partial u_n^{(1)}}{\partial x_j} + \frac{1}{2}\left(\delta_j^n + \frac{\partial u_n^{(0)}}{\partial x_j}\right)\frac{\partial u_n^{(1)}}{\partial x_i}, \tag{3.30}$$

$$\begin{aligned} T_{2i}^{(1)}\Big|_{(x_1,\pm h,x_3)} = & \frac{\partial f(x_1,x_3)}{\partial x_1}\left(\delta_i^n + \frac{\partial u_i^{(0)}}{\partial x_n}\right)\sigma_{1n}^{(0)}\Big|_{(x_1,\pm h,x_3)} \\ & + \frac{\partial f(x_1,x_3)}{\partial x_3}\left(\delta_i^n + \frac{\partial u_i^{(0)}}{\partial x_n}\right)\sigma_{3n}^{(0)}\Big|_{(x_1,\pm h,x_3)}, \end{aligned} \tag{3.31}$$

$$\begin{aligned} & u_i^{(1)}\Big|_{x_1=0,\ell_1} = 0\ (i=1,2,3),\ T_{33}^{(1)}\Big|_{x_3=0,\ell_3} = 0, \\ & u_2^{(1)}\Big|_{x_3=0,\ell_3} = 0,\ u_1^{(1)}\Big|_{x_3=0,\ell_3} = 0. \end{aligned} \tag{3.32}$$

Note that the Eqs. (3.26) and (3.30) are satisfied in the region

$$\Omega_0 = \{0<x_1<\ell_1;\ -h<x_2<h;\ 0<x_3<\ell_3\}. \tag{3.33}$$

Owing to the linearity of the mechanical relations considered, the constitutive Eqs. (3.6) and (3.7) will be satisfied for each approximation in (3.25) separately, i.e.

$$\begin{aligned} & \sigma_{11}^{(q)} = A_{11}^*\varepsilon_{11}^{(q)} + A_{12}^*\varepsilon_{22}^{(q)} + A_{13}^*\varepsilon_{33}^{(q)},\ \sigma_{22}^{(q)} = A_{12}^*\varepsilon_{11}^{(q)} + A_{22}^*\varepsilon_{22}^{(q)} + A_{23}^*\varepsilon_{33}^{(q)}, \\ & \sigma_{33}^{(q)} = A_{13}^*\varepsilon_{11}^{(q)} + A_{23}^*\varepsilon_{22}^{(q)} + A_{33}^*\varepsilon_{33}^{(q)},\ \sigma_{23}^{(q)} = 2A_{66}^*\varepsilon_{13}^{(q)}, \\ & \sigma_{13}^{(q)} = 2A_{55}^*\varepsilon_{13}^{(q)},\ \sigma_{12}^{(q)} = 2A_{66}^*\varepsilon_{13}^{(q)}. \end{aligned} \tag{3.34}$$

Thus, the investigation of the growth of the initial insignificant imperfections of the plate with the flow of time under the action of the external static compressive forces is reduced to the solutions of the series of boundary-value problems, such as (3.26)–(3.29), (3.34) (for $q = 0$) and (3.30)–(3.32), (3.34) (for $q = 1$). So that, the values of the zeroth approximation will be determined from the non-linear Eqs. (3.26)–(3.29), and the values of the subsequent approximations will be determined from the linear equations such as (3.30)–(3.32), which are written for the first approximation. The direct verification shows that the linear Eqs. (3.30)–(3.32), (3.34) coincide with the equations of the TDLTS which are presented in (Guz 1999) and elsewhere.

Under concrete investigations we will assume that the material of the plate is a comparatively rigid one and the condition

$$\frac{\partial u_i^{(0)}}{\partial x_j} \ll 1, \quad 1 + \frac{\partial u_i^{(0)}}{\partial x_j} \approx 1, \tag{3.35}$$

satisfies. According to the condition (3.35), the non-linear terms in Eqs. (3.26)–(3.29) related to the zeroth approximation can be neglected with very high accuracy. Consequently, the values of the zeroth approximation will be determined from the following linear boundary-value problem.

$$\frac{\partial}{\partial x_j}\sigma_{ji}^{(0)} = 0,\ \varepsilon_{ij}^{(0)} = \frac{1}{2}\left(\frac{\partial u_i^{(0)}}{\partial x_j} + \frac{\partial u_j^{(0)}}{\partial x_i}\right), \tag{3.36}$$

$$\left.\sigma_{2i}^{(0)}\right|_{x_2=\pm h} = 0,\ i;j = 1,\ 2,\ 3. \tag{3.37}$$

$$\begin{aligned} &u_1^{(0)}(0,\ x_2,\ x_3) = -u_1^{(0)}(\ell_1,\ x_2,\ x_3) = U = \text{const}, \\ &\left.u_2^{(0)}(x_1,\ x_2,\ x_3)\right|_{x_1=0,\ \ell_1} = \left.u_3^{(0)}(x_1,\ x_2,\ x_3)\right|_{x_1=0,\ \ell_1} = 0, \end{aligned} \tag{3.38}$$

$$\begin{aligned} &u_2^{(0)}(x_1,\ x_2,\ 0) = u_2^{(0)}(x_1,\ x_2, \ell_3) = 0, \\ &u_1^{(0)}(x_1,\ x_2,\ 0) = u_1^{(0)}(x_1,\ x_2, \ell_3) = 0,\ \left.\sigma_{33}^{(0)}\right|_{x_3=0;\ell_3} = -p_3. \end{aligned} \tag{3.39}$$

Note that the condition (3.35) simplifies also the system Eqs. (3.30)–(3.32) related to the first approximation and these equations becomes as follows.

$$\frac{\partial}{\partial x_j}\left[\sigma_{ij}^{(1)} + \sigma_{jn}^{(0)}\frac{\partial u_i^{(1)}}{\partial x_n}\right] = 0,\ \varepsilon_{ij}^{(1)} = \frac{1}{2}\left(\frac{\partial u_i^{(1)}}{\partial x_j} + \frac{\partial u_j^{(1)}}{\partial x_i}\right), \tag{3.40}$$

$$\left.\sigma_{i2}^{(1)}\right|_{(x_2=\pm h)} = \frac{\partial f}{\partial x_1}\left.\sigma_{i1}^{(0)}\right|_{(x_2=\pm h)} + \frac{\partial f}{\partial x_3}\left.\sigma_{i3}^{(0)}\right|_{(x_2=\pm h)}, \tag{3.41}$$

$$\begin{aligned} &\left.u_i^{(1)}\right|_{x_1=0,\ \ell_1} = 0,\ i = 1,\ 2,\ 3,\ \left.u_1^{(1)}\right|_{x_3=0,\ell_3} = 0, \\ &\left.u_2^{(1)}\right|_{x_3=0,\ell_3} = 0,\ \left.\left(\sigma_{33}^{(1)} + \sigma_{33}^{(0)}\frac{\partial u_3^{(1)}}{\partial x_3}\right)\right|_{x_3=0,\ \ell_3} = 0. \end{aligned} \tag{3.42}$$

The determination method of values of the zeroth and first approximation will be considered below for the concrete selected end conditions separately. However, for the all considered concrete cases the values of the critical parameters are defined from the following criterion.

$$\max_{x_1,\,x_3\in\Omega_0}\left|u_2^{(1)}\right| \to \infty \text{ as } t \to t_{cr} \text{ (for fixed values of } p_1 \text{ and } p_3\text{) or as } p_1 \to p_{1cr} \tag{3.43}$$

(for fixed value of $p_3$ for the pure elastic case).

## 3.3 Simply Supported Rectangular Plate

### 3.3.1 *Deriving Approximate Equations of the Stability Loss of the Simply Supported Rectangular Plate from Equations of the TDLTS by the Average-Integrating Procedure*

Within the scope of the TDLTS the last three conditions in (3.42) are understood as simply supported end condition with respect to the coordinate $x_3$. Because, as will be shown below, the integration of these conditions and their product with thickness coordinate over plate thickness gives the simply supported end conditions in the classical sense. Moreover, under determination of the values of the zeroth approximation the condition (3.39) is understood as simply supported end condition with respect to the coordinate $x_3$. Consequently, in the present section we assume that not only at the ends $x_3 = 0$ and $x_3 = \ell_3$, but also at the ends $x_1 = 0$ and $x_1 = \ell_1$ the simply supported end conditions occur.

To simplify the consideration we omit the upper index (1) in the quantities related to the first approximation below. Thus, in the case under consideration the end condition (3.39) holds but the condition $u_1^{(0)}(0, x_2, x_3) = -u_1^{(0)}(\ell_1, x_2, x_3) = U = \text{const}$ (3.38) in the zeroth approximation is replaced with the

$$\sigma_{11}^{(0)}\Big|_{x_1=0;\ell_1} = -p_1. \tag{3.44}$$

Consequently, according to this statement, the stresses in the zeroth approximation are determined as follows.

$$\sigma_{11}^{(0)} = -p_1,\ \sigma_{33}^{(0)} = -p_3,\ \sigma_{ij}^{(0)} = 0 \text{ if } ij \neq 11. \tag{3.45}$$

Taking (3.45) into account we obtain from (3.40)–(3.42) the following field equations, boundary and end conditions for determination of the first approximation.

$$\frac{\partial \sigma_{ji}}{\partial x_j} + \sigma_{11}^{(0)}\frac{\partial^2 u_i}{\partial x_1^2} + \sigma_{33}^{(0)}\frac{\partial^2 u_i}{\partial x_3^2} = 0,\ \varepsilon_{ij} = \frac{1}{2}\left(\frac{\partial u_j}{\partial x_i} + \frac{\partial u_i}{\partial x_j}\right), \tag{3.46}$$

$$\sigma_{i2}\big|_{(x_2=\pm h)} = \delta_i^1\frac{\partial f}{\partial x_1}\sigma_{i1}^{(0)} + \delta_i^3\frac{\partial f}{\partial x_3}\sigma_{i3}^{(0)}, \tag{3.47}$$

$$u_2\big|_{x_1=0,\ell_1} = 0,\ \left(\sigma_{11} + \sigma_{11}^{(0)}\frac{\partial u_1}{\partial x_1}\right)\Bigg|_{x_1=0,\ell_1} = 0,\ u_3\big|_{x_1=0,\ell_1} = 0,$$

$$u_2\big|_{x_3=0,\ell_3} = 0,\ \left(\sigma_{33} + \sigma_{33}^{(0)}\frac{\partial u_3}{\partial x_3}\right)\Bigg|_{x_3=0,\ell_3} = 0,\ u_1\big|_{x_3=0,\ell_3} = 0. \tag{3.48}$$

The Eqs. (3.46)–(3.48) and the constitutive relation

$$\sigma_{11} = A_{11}^*\varepsilon_{11} + A_{12}^*\varepsilon_{22} + A_{13}^*\varepsilon_{33},\ \sigma_{22} = A_{12}^*\varepsilon_{11} + A_{22}^*\varepsilon_{22} + A_{23}^*\varepsilon_{33},$$
$$\sigma_{33} = A_{13}^*\varepsilon_{11} + A_{23}^*\varepsilon_{22} + A_{33}^*\varepsilon_{33},\ \sigma_{23} = 2A_{44}^*\varepsilon_{13},$$
$$\sigma_{13} = 2A_{55}^*\varepsilon_{13},\ \sigma_{12} = 2A_{66}^*\varepsilon_{13} \tag{3.49}$$

complete the mathematical formulation of the problem related to the determination of the first approximation, i.e. related to the investigation of the stability loss of the simply supported rectangular plate made from viscoelastic composite material.

Now we derive from (3.46)–(3.49) by average-integrating procedure the stability loss equation for the plate within the scope of the various approximate theories.

#### 3.3.1.1 Kirchhoff-Love Plate Theory

First, we introduce the following notation

$$u(x_1, x_3, t) = u_1(x_1, 0, x_3, t),\ w(x_1, x_3, t) = u_2(x_1, 0, x_3, t),$$
$$v(x_1, x_3, t) = u_3(x_1, 0, x_3, t).$$
$$Q_{11} = \int_{-h}^{+h} \sigma_{11}dx_2,\ Q_{13} = \int_{-h}^{+h} \sigma_{13}dx_2,\ Q_{33} = \int_{-h}^{+h} \sigma_{33}dx_2,\ N_{12} = \int_{-h}^{+h} \sigma_{12}dx_2,$$
$$N_{23} = \int_{-h}^{+h} \sigma_{23}dx_2,\ M_{11} = \int_{-h}^{+h} x_2\sigma_{11}dx_2,\ M_{13} = \int_{-h}^{+h} x_2\sigma_{13}dx_2,$$
$$M_{33} = \int_{-h}^{+h} x_2\sigma_{33}dx_2. \tag{3.50}$$

According to the Kirchhoff-Love plate theory, it is assumed that

$$u_1(x_1, x_2, x_3, t) = u(x_1, x_3, t) - x_2\frac{\partial w(x_1, x_3, t)}{\partial x_1},\ u_2(x_1, x_2, x_3, t) = w(x_1, x_3, t),$$
$$u_3(x_1, x_2, x_3, t) = v(x_1, x_3, t) - x_2\frac{\partial w(x_1, x_3, t)}{\partial x_3}. \tag{3.51}$$

We assume that $u(x_1, x_3, t) \approx 0$, $v(x_1, x_3, t) \approx 0$ and obtain the relation

$$u_1(x_1, x_2, x_3, t) = -x_2\frac{\partial w(x_1, x_3, t)}{\partial x_1},\ u_2(x_1, x_2, x_3, t) = w(x_1, x_3, t),$$
$$u_3(x_1, x_2, x_3, t) = -x_2\frac{\partial w(x_1, x_3, t)}{\partial x_3} \tag{3.52}$$

instead of (3.51). Taking the expressions in (3.52) into account we write the explicit form of the first three equations in (3.46) and integrate that in the interval $[-h, +h]$ with respect to $x_2$.

$$\frac{\partial}{\partial x_1}\int_{-h}^{+h}\sigma_{11}dx_2+\int_{-h}^{+h}\frac{\partial}{\partial x_2}\sigma_{12}dx_2+\frac{\partial}{\partial x_3}\int_{-h}^{+h}\sigma_{33}dx_2$$
$$-\sigma_{11}^{(0)}\frac{\partial^3 w}{\partial x_1^3}\int_{-h}^{+h}x_2dx_2-\sigma_{33}^{(0)}\frac{\partial^3 w}{\partial x_3^2\partial x_1}\int_{-h}^{+h}x_2dx_2=0,$$
$$\frac{\partial}{\partial x_1}\int_{-h}^{h}\sigma_{12}dx_2+\int_{-h}^{h}\frac{\partial}{\partial x_2}\sigma_{22}dx_2+\frac{\partial}{\partial x_3}\int_{-h}^{h}\sigma_{23}dx_2$$
$$-\sigma_{11}^{(0)}2h\frac{\partial^2 w}{\partial x_1^2}-\sigma_{33}^{(0)}2h\frac{\partial^2 w}{\partial x_3^2}=0$$
$$\frac{\partial}{\partial x_1}\int_{-h}^{+h}\sigma_{13}dx_2+\int_{-h}^{+h}\frac{\partial}{\partial x_2}\sigma_{23}dx_2+\frac{\partial}{\partial x_3}\int_{-h}^{+h}\sigma_{33}dx_2.$$
$$-\sigma_{11}^{(0)}\frac{\partial^3 w}{\partial x_1^2\partial x_3}\int_{-h}^{h}x_2dx_2-\sigma_{33}^{(0)}\frac{\partial^3 w}{\partial x_3^3}\int_{-h}^{h}x_2dx_2=0 \quad (3.53)$$

According to the calculation

$$\int_{-h}^{+h}\frac{\partial}{\partial x_2}\sigma_{12}dx_2=\sigma_{12}|_{x_2=h}-\sigma_{12}|_{x_2=-h}=0,$$
$$\int_{-h}^{+h}\frac{\partial}{\partial x_2}\sigma_{23}dx_2=\sigma_{23}|_{x_2=h}-\sigma_{23}|_{x_2=-h}=0,\quad\int_{-h}^{+h}x_2dx_2=0, \quad (3.54)$$

and according to the notation (3.50), we obtain from (3.53) the following equation.

$$\frac{\partial Q_{11}}{\partial x_1}+\frac{\partial Q_{13}}{\partial x_3}=0,\ \frac{\partial N_{12}}{\partial x_1}+\frac{\partial N_{23}}{\partial x_3}+2h\sigma_{11}^{(0)}\frac{\partial^2 w}{\partial x_1^2}+2h\sigma_{33}^{(0)}\frac{\partial^2 w}{\partial x_3^2}=0,$$
$$\frac{\partial Q_{13}}{\partial x_1}+\frac{\partial Q_{33}}{\partial x_3}=0. \quad (3.55)$$

We multiply the first three equations in (3.46) with $x_2$ and integrate them with respect to $x_2$ in the interval $[-h, +h]$

$$\frac{\partial}{\partial x_1}\int_{-h}^{+h} x_2\sigma_{11}dx_2 + \int_{-h}^{+h} x_2\frac{\partial}{\partial x_2}\sigma_{12}dx_2 + \frac{\partial}{\partial x_3}\int_{-h}^{+h} x_2\sigma_{13}dx_2$$

$$-\sigma_{11}^{(0)}\frac{\partial^3 w}{\partial x_1^3}\int_{-h}^{h} x_2^2dx_2 - \sigma_{33}^{(0)}\frac{\partial^3 w}{\partial x_3^2\partial x_1}\int_{-h}^{h} x_2^2dx_2 = 0,$$

$$\frac{\partial}{\partial x_1}\int_{-h}^{h} x_2\sigma_{12}dx_2 + \underbrace{\int_{-h}^{+h} x_2\frac{\partial}{\partial x_2}\sigma_{22}dx_2}_{0} + \frac{\partial}{\partial x_3}\int_{-h}^{h} x_2\sigma_{23}dx_2$$

$$-\sigma_{11}^{(0)}\frac{\partial^2 w}{\partial x_1^2}\underbrace{\int_{-h}^{h} x_2dx_2}_{0} - \sigma_{33}^{(0)}\frac{\partial^2 w}{\partial x_3^2}\underbrace{\int_{-h}^{h} x_2dx_2}_{0} = 0,$$

$$\frac{\partial}{\partial x_1}\int_{-h}^{+h} x_2\sigma_{13}dx_2 + \int_{-h}^{+h} x_2\frac{\partial}{\partial x_2}\sigma_{23}dx_2 + \frac{\partial}{\partial x_3}\int_{-h}^{+h} x_2\sigma_{33}dx_2$$

$$-\sigma_{11}^{(0)}\frac{\partial^3 w}{\partial x_1^2\partial x_3}\int_{-h}^{h} x_2^2dx_2 - \sigma_{33}^{(0)}\frac{\partial^3 w}{\partial x_3^3}\int_{-h}^{h} x_2^2dx_2 = 0. \tag{3.56}$$

Within the scope of the Kirchhoff-Love theory it is assumed that

$$\frac{\partial}{\partial x_1}\int_{-h}^{h} x_2\sigma_{12}dx_2 \approx 0, \quad \frac{\partial}{\partial x_3}\int_{-h}^{h} x_2\sigma_{23}dx_2 \approx 0,$$

consequently, the second equation in (3.56) is dropped. Taking the relations

$$\int_{-h}^{+h} x_2\frac{\partial}{\partial x_2}\sigma_{12}dx_2 = \underbrace{(x_2\sigma_{12})|_{-h}^{+h}}_{0} - \int_{-h}^{+h}\sigma_{12}dx_2 = 2h\sigma_{11}^{(0)}\frac{\partial f}{\partial x_1} - \int_{-h}^{+h}\sigma_{12}dx_2,$$

$$\int_{-h}^{+h} x_2\frac{\partial}{\partial x_2}\sigma_{23}dx_2 = \underbrace{(x_2\sigma_{23})|_{-h}^{+h}}_{0} - \int_{-h}^{+h}\sigma_{23}dx_2 = 2h\sigma_{33}^{(0)}\frac{\partial f}{\partial x_3} - \int_{-h}^{+h}\sigma_{23}dx_2,$$

$$\int_{-h}^{h} x_2^2dx_2 = \frac{2h^3}{3}, \tag{3.57}$$

and notation in (3.50) into account, we obtain from (3.56), the following equation.

$$\frac{\partial M_{11}}{\partial x_1} - N_{12} + \frac{\partial M_{13}}{\partial x_3} - \sigma_{11}^{(0)}\frac{2h^3}{3}\frac{\partial^3 w}{\partial x_1^3} - \sigma_{33}^{(0)}\frac{2h^3}{3}\frac{\partial^3 w}{\partial x_3^2 \partial x_1} = -2h\sigma_{11}^{(0)}\frac{\partial f}{\partial x_1},$$
$$\frac{\partial M_{13}}{\partial x_1} - N_{23} + \frac{\partial M_{33}}{\partial x_3} - \sigma_{11}^{(0)}\frac{2h^3}{3}\frac{\partial^3 w}{\partial x_1^2 \partial x_3} - \sigma_{33}^{(0)}\frac{2h^3}{3}\frac{\partial^3 w}{\partial x_3^3} = -2h\sigma_{33}^{(0)}\frac{\partial f}{\partial x_3}. \tag{3.58}$$

Getting the derivative with respect to $x_1$ and $x_3$ of the first and second equations respectively in (3.58) and summing them we obtain

$$\frac{\partial^2 M_{11}}{\partial x_1^2} - \frac{\partial N_{12}}{\partial x_1} - \frac{\partial N_{23}}{\partial x_3} + \frac{\partial^2 M_{13}}{\partial x_1 \partial x_3} + \frac{\partial^2 M_{33}}{\partial x_3^2} - \sigma_{11}^{(0)}\frac{2h^3}{3}\left(\frac{\partial^4 w}{\partial x_1^4} + \frac{\partial^4 w}{\partial x_1^2 \partial x_3^2}\right)$$
$$- \sigma_{33}^{(0)}\frac{2h^3}{3}\left(\frac{\partial^4 w}{\partial x_3^4} + \frac{\partial^4 w}{\partial x_1^2 \partial x_3^2}\right) = 2h\sigma_{11}^{(0)}\frac{\partial^2 f}{\partial x_1^2} + 2h\sigma_{33}^{(0)}\frac{\partial^2 f}{\partial x_3^2}. \tag{3.59}$$

According to (3.46), (3.49), (3.50) and (3.52) we can write the following relations.

$$\varepsilon_{11} = -x_2\frac{\partial^2 w}{\partial x_1^2}, \ \varepsilon_{12} = \varepsilon_{23} = \varepsilon_{22} = 0,$$
$$\varepsilon_{33} = -x_2\frac{\partial^2 w}{\partial x_3^2}, \ \varepsilon_{13} = -x_2\frac{\partial^2 w}{\partial x_1 \partial x_3}, \tag{3.60}$$

$$\sigma_{11} = A_{11}^*\left(-x_2\frac{\partial^2 w}{\partial x_1^2}\right) + A_{13}^*\left(-x_2\frac{\partial^2 w}{\partial x_3^2}\right), \ \sigma_{13} = 2A_{55}^*\left(-x_2\frac{\partial^2 w}{\partial x_1 \partial x_3}\right),$$
$$\sigma_{33} = A_{13}^*\left(-x_2\frac{\partial^2 w}{\partial x_1^2}\right) + A_{33}^*\left(-x_2\frac{\partial^2 w}{\partial x_3^2}\right), \tag{3.61}$$

$$Q_{11} = 0, \ Q_{33} = 0, \ Q_{13} = 0,$$
$$M_{11} = -A_{11}^*\frac{2h^3}{3}\frac{\partial^2 w}{\partial x_1^2} - A_{13}^*\frac{2h^3}{3}\frac{\partial^2 w}{\partial x_3^2}, \ M_{33} = -A_{13}^*\frac{2h^3}{3}\frac{\partial^2 w}{\partial x_1^2} - A_{33}^*\frac{2h^3}{3}\frac{\partial^2 w}{\partial x_3^2},$$
$$M_{11} = -2A_{55}^*\frac{2h^3}{3}\frac{\partial^2 w}{\partial x_1 \partial x_3}. \tag{3.62}$$

According to (3.61), the first and the third equations in (3.55) satisfy automatically. Substituting the expressions in (3.62) into (3.59) and taking the second equation in (3.55) into account we derive the following stability loss equation for the plate within the scope of the Kirchhoff-Love theory.

$$\frac{2h^3}{3}\left(A_{11}^*+\sigma_{11}^{(0)}\right)\frac{\partial^4 w}{\partial x_1^4}+\frac{2h^3}{3}\left(2A_{13}^*+A_{55}^*+\sigma_{11}^{(0)}+\sigma_{33}^{(0)}\right)$$
$$\times\frac{\partial^4 w}{\partial x_1^2\partial x_3^2}+\frac{2h^3}{3}\left(A_{33}^*+\sigma_{33}^{(0)}\right)\frac{\partial^4 w}{\partial x_3^4} \tag{3.63}$$
$$+2h\sigma_{11}^{(0)}\frac{\partial^2 w}{\partial x_1^2}+2h\sigma_{33}^{(0)}\frac{\partial^2 w}{\partial x_3^2}=-2h\sigma_{11}^{(0)}\frac{\partial^2 f}{\partial x_1^2}-2h\sigma_{33}^{(0)}\frac{\partial^2 f}{\partial x_3^2}.$$

Substituting the expression for $u_1$ in (3.52) into the first end condition in (3.48) and integrating that in the interval $[-h,\ +h]$ with respect to $x_2$, we obtain

$$w|_{x_1=0;\ell_1}=0. \tag{3.64}$$

Moreover, substituting the expressions for $u_1$ in (3.52) and for $\sigma_{11}$ in (3.60) into the second end condition in (3.48) and multiplying that with $x_2$ and integrating in the interval $[-h,\ +h]$ with respect to $x_2$ we derive:

$$\left.\frac{\partial^2 w}{\partial x_1^2}\right|_{x_1=0;\ell_1}=0. \tag{3.65}$$

It should be noted that, under obtaining the Eq. (3.65) it was taken the relation

$$w|_{x_1=0;\ell_1}=0\Rightarrow\left.\frac{\partial^2 w}{\partial x_3^2}\right|_{x_1=0;\ell_1}=0$$

into account.

In the similar manner we obtain from the third and forth conditions in (3.48) the following ones.

$$w|_{x_3=0;\ell_3}=0,\ \left.\frac{\partial^2 w}{\partial x_3^2}\right|_{x_3=0;\ell_3}=0. \tag{3.66}$$

This completes the deriving the stability loss equations and end conditions of the rectangular simply supported plate with initial infinitesimal imperfection within the scope of the Kirchhoff-Love hypotheses from the TDLTS.

#### 3.3.1.2 The Third Order Refined Plate Theory

Consider the derivation of the approximate stability loss equation of the plate within the scope of the third order refined plate theory (Kromm 1953, 1955), from Eqs. (3.46)–(3.49). According to this theory, we assume that

$$
\begin{aligned}
u_1(x_1, x_2, x_3, t) &= -x_2\frac{\partial w(x_1, x_3, t)}{\partial x_1} + \left(h^2 x_2 - \frac{x_2^3}{3}\right)\varphi_1(x_1, x_3, t),\\
u_2(x_1, x_2, x_3, t) &= w(x_1, x_3, t),\\
u_3(x_1, x_2, x_3, t) &= -x_2\frac{\partial w(x_1, x_3, t)}{\partial x_3} + \left(h^2 x_2 - \frac{x_2^3}{3}\right)\varphi_3(x_1, x_3, t).
\end{aligned}
\tag{3.67}
$$

Here $\varphi_1$ and $\varphi_3$ characterize the average shear strains in cross sections of the plate perpendicular to the $Ox_1$ and $Ox_3$ axes respectively.

According to (3.46), (3.49), (3.50) and (3.67), we can write the following relations.

$$
\begin{aligned}
&\varepsilon_{11} = -x_2\frac{\partial^2 w}{\partial x_1^2} + \left(h^2 x_2 - \frac{x_2^3}{3}\right)\frac{\partial\varphi_1}{\partial x_1},\ \varepsilon_{22} = 0,\ \varepsilon_{12} = \frac{1}{2}(h^2 - x_2^2)\varphi_1,\\
&\varepsilon_{23} = \frac{1}{2}(h^2 - x_2^2)\varphi_3,\ \varepsilon_{33} = -x_2\frac{\partial^2 w}{\partial x_3^2} + \left(h^2 x_2 - \frac{x_2^3}{3}\right)\frac{\partial\varphi_3}{\partial x_3},\\
&\varepsilon_{13} = -x_2\frac{\partial^2 w}{\partial x_1\partial x_3} + \frac{1}{2}\left(h^2 x_2 - \frac{x_2^3}{3}\right)\left(\frac{\partial\varphi_1}{\partial x_3} + \frac{\partial\varphi_3}{\partial x_1}\right),\\
&\sigma_{11} = A_{11}^*\left(-x_2\frac{\partial^2 w}{\partial x_1^2} + \left(h^2 x_2 - \frac{x_2^3}{3}\right)\frac{\partial\varphi_1}{\partial x_1}\right) + A_{13}^*\left(-x_2\frac{\partial^2 w}{\partial x_3^2} + \left(h^2 x_2 - \frac{x_2^3}{3}\right)\frac{\partial\varphi_3}{\partial x_3}\right),\\
&\sigma_{33} = A_{13}^*\left(-x_2\frac{\partial^2 w}{\partial x_1^2} + \left(h^2 x_2 - \frac{x_2^3}{3}\right)\frac{\partial\varphi_1}{\partial x_1}\right) + A_{33}^*\left(-x_2\frac{\partial^2 w}{\partial x_3^2} + \left(h^2 x_2 - \frac{x_2^3}{3}\right)\frac{\partial\varphi_3}{\partial x_3}\right)\\
&\sigma_{13} = 2A_{55}^*\left(-x_2\frac{\partial^2 w}{\partial x_1\partial x_3} + \frac{1}{2}\left(h^2 x_2 - \frac{x_2^3}{3}\right)\left(\frac{\partial\varphi_1}{\partial x_3} + \frac{\partial\varphi_3}{\partial x_1}\right)\right),\\
&\sigma_{12} = A_{66}^*(h^2 - x_2^2)\varphi_1,\ \sigma_{23} = A_{44}^*(h^2 - x_2^2)\varphi_3.
\end{aligned}
\tag{3.68}
$$

$$
Q_{11} = 0,\ Q_{33} = 0,\ Q_{13} = 0,\ N_{12} = A_{66}^*\frac{4}{3}h^3\varphi_1,\ N_{23} = A_{44}^*\frac{4}{3}h^3\varphi_3,
\tag{3.69}
$$

$$
\begin{aligned}
M_{11} &= A_{11}^*\left(-\frac{2h^3}{3}\frac{\partial^2 w}{\partial x_1^2} + \frac{8h^5}{15}\frac{\partial\varphi_1}{\partial x_1}\right) + A_{13}^*\left(-\frac{2h^3}{3}\frac{\partial^2 w}{\partial x_3^2} + \frac{8h^5}{15}\frac{\partial\varphi_3}{\partial x_3}\right),\\
M_{33} &= A_{13}^*\left(-\frac{2h^3}{3}\frac{\partial^2 w}{\partial x_1^2} + \frac{8h^5}{15}\frac{\partial\varphi_1}{\partial x_1}\right) + A_{33}^*\left(-\frac{2h^3}{3}\frac{\partial^2 w}{\partial x_3^2} + \frac{8h^5}{15}\frac{\partial\varphi_3}{\partial x_3}\right),\\
M_{13} &= 2A_{55}^*\left(-\frac{2h^3}{3}\frac{\partial^2 w}{\partial x_1\partial x_3} + \frac{4h^5}{15}\left(\frac{\partial\varphi_1}{\partial x_3} + \frac{\partial\varphi_3}{\partial x_1}\right)\right).
\end{aligned}
\tag{3.70}
$$

Substituting (3.67) into the Eq. (3.46) and doing the integration (3.53) we obtain again the Eq. (3.55), but doing the integration (3.56) we obtain the following equation instead of Eq. (3.58).

$$\frac{\partial M_{11}}{\partial x_1} - N_{12} + \frac{\partial M_{13}}{\partial x_3} + \sigma_{11}^{(0)}\left(-\frac{2h^3}{3}\frac{\partial^3 w}{\partial x_1^3} + \frac{8h^5}{15}\frac{\partial^2 \varphi_1}{\partial x_1^2}\right)$$
$$+ \sigma_{33}^{(0)}\left(-\frac{2h^3}{3}\frac{\partial^3 w}{\partial x_3^2 \partial x_1} + \frac{8h^5}{15}\frac{\partial^2 \varphi_1}{\partial x_3^2}\right) = -2h\sigma_{11}^{(0)}\frac{\partial f}{\partial x_1},$$
$$\frac{\partial M_{13}}{\partial x_1} - N_{23} + \frac{\partial M_{33}}{\partial x_3} + \sigma_{11}^{(0)}\left(-\frac{2h^3}{3}\frac{\partial^3 w}{\partial x_1^2 \partial x_3} + \frac{8h^5}{15}\frac{\partial^2 \varphi_3}{\partial x_1^2}\right)$$
$$+ \sigma_{33}^{(0)}\left(-\frac{2h^3}{3}\frac{\partial^3 w}{\partial x_3^3} + \frac{8h^5}{15}\frac{\partial^2 \varphi_3}{\partial x_3^2}\right) = -2h\sigma_{33}^{(0)}\frac{\partial f}{\partial x_3}. \tag{3.71}$$

Substituting the expressions (3.69) into Eq. (3.55) and substituting the expression (3.70) into the Eq. (3.71) we obtain the following system of equations with respect to the unknown functions $\varphi_1$, $\varphi_3$ and $w$.

$$\frac{4h^3}{3}A_{66}^*\frac{\partial \varphi_1}{\partial x_1} + \frac{4h^3}{3}A_{44}^*\frac{\partial \varphi_3}{\partial x_3} + 2h\sigma_{11}^{(0)}\frac{\partial^2 w}{\partial x_1^2} + 2h\sigma_{33}^{(0)}\frac{\partial^2 w}{\partial x_3^2} = 0,$$
$$-\frac{2h^3}{3}\left(A_{11}^* + \sigma_{11}^{(0)}\right)\frac{\partial^3 w}{\partial x_1^3} - \frac{2h^3}{3}\left(A_{13}^* + 2A_{55}^* + \sigma_{33}^{(0)}\right)\frac{\partial^3 w}{\partial x_1 \partial x_3^2}$$
$$+\frac{8h^5}{15}\left(A_{11}^* + \sigma_{11}^{(0)}\right)\frac{\partial^2 \varphi_1}{\partial x_1^2}$$
$$+\frac{8h^5}{15}(A_{13}^* + A_{55}^*)\frac{\partial^2 \varphi_3}{\partial x_1 \partial x_3} - \frac{4h^3}{3}A_{66}^*\varphi_1 + \frac{8h^5}{15}\left(A_{55}^* + \sigma_{33}^{(0)}\right) = -2h\sigma_{11}^{(0)}\frac{\partial f}{\partial x_1},$$
$$-\frac{2h^3}{3}\left(2A_{55}^* + A_{13}^* + \sigma_{11}^{(0)}\right)\frac{\partial^3 w}{\partial x_1^2 \partial x_3} - \frac{2h^3}{3}\left(A_{33}^* + \sigma_{33}^{(0)}\right)\frac{\partial^3 w}{\partial x_3^3}$$
$$+\frac{8h^5}{15}(A_{55}^* + A_{13}^*)\frac{\partial^2 \varphi_1}{\partial x_1 \partial x_3} + \frac{8h^5}{15}\left(A_{55}^* + \sigma_{11}^{(0)}\right)\frac{\partial^2 \varphi_3}{\partial x_1^2}$$
$$+\frac{8h^5}{15}\left(A_{33}^* + \sigma_{33}^{(0)}\right)\frac{\partial^2 \varphi_3}{\partial x_3^2} - \frac{4h^3}{3}A_{44}^*\varphi_3 = -2h\sigma_{33}^{(0)}\frac{\partial f}{\partial x_3}. \tag{3.72}$$

Doing the calculations, which are similar to that made in the previous subsection, we obtain from (3.48) the following end conditions within the scope of the third order refined plate theory.

$$w|_{x_1=0;\ell_1} = 0,\ \left.\frac{\partial^2 w}{\partial x_1^2}\right|_{x_1=0;\ell_1},\ \left.\frac{\partial \varphi_1}{\partial x_1}\right|_{x_1=0;\ell_1} = 0,\ \left.\frac{\partial \varphi_3}{\partial x_1}\right|_{x_1=0;\ell_1} = 0, \tag{3.73a}$$

$$w|_{x_3=0;\ell_3} = 0,\ \left.\frac{\partial^2 w}{\partial x_3^2}\right|_{x_3=0;\ell_3},\ \left.\frac{\partial \varphi_1}{\partial x_3}\right|_{x_3=0;\ell_3} = 0,\ \left.\frac{\partial \varphi_3}{\partial x_3}\right|_{x_3=0;\ell_3} = 0. \tag{3.73b}$$

### 3.3.2 Solution for the Formulated Mathematical Problems for the Simply Supported Plate

Now we attempt to employ the foregoing approaches to the investigation of the stability loss of the rectangular plate the edges of which are simply supported. We consider the solution to the boundary value problems (3.46)–(3.49), (3.63)–(3.66) and (3.72)–(3.73a, b) separately. Note that these problems relate to the study of the stability loss of the simply supported rectangular plate within the scope of the TDLTS, Kirchhoff-Love and the third order refined plate theories respectively.

First, we note that the solution to the foregoing problems depends also on initial imperfection mode. For the case under consideration the function of the initial imperfection mode, i.e. the function $f = f(x_1, x_3)$ satisfies the first, the third and the forth conditions in (3.2), but the second condition in (3.2) is replaced by the

$$\left.\frac{\partial^2 f(x_1, x_3)}{\partial x_1^2}\right|_{x_1=0,\, \ell_1} = 0 \tag{3.74}$$

one.

For further solution procedures it is necessary to give the explicit expression of the initial imperfection mode. Here we consider the case where the function $F(x_1, x_3)$ is given as

$$F = L \sin\left(\frac{\pi x_1}{\ell_1}\right) \sin\left(\frac{\pi x_3}{\ell_3}\right)$$

and assume that $L \ll \ell_1$, according to which, we introduce the small parameter $\varepsilon = L/\ell_1$, and from (3.1) we can conclude that

$$f = \ell_1 \sin\left(\frac{\pi x_1}{\ell_1}\right) \sin\left(\frac{\pi x_3}{\ell_3}\right). \tag{3.75}$$

To solve the problem (3.46)–(3.49) we use principle of correspondence and employ the Laplace transformation (2.34) to all the corresponding equations and relations, after which we obtain the same equations and relations for the Laplace transforms of the sought quantities. Note that under Laplace transformation of the constitutive relations (3.49), (3.7) we use also the convolution theorem. Taking the expression of the initial imperfection mode (3.75), the solution of the Laplace transformation of the problem (3.46)–(3.49) related to the TDLTS of the stability loss of the plate we select as follows.

$$\bar{u}_1 = \bar{\psi}_1(x_2, s) \cos\left(\frac{\pi x_1}{\ell_1}\right) \sin\left(\frac{\pi x_3}{\ell_3}\right), \; \bar{u}_2 = \bar{\psi}_2(x_2, s) \sin\left(\frac{\pi x_1}{\ell_1}\right) \sin\left(\frac{\pi x_3}{\ell_3}\right),$$
$$\bar{u}_3 = \bar{\psi}_3(x_2, s) \sin\left(\frac{\pi x_1}{\ell_1}\right) \cos\left(\frac{\pi x_3}{\ell_3}\right). \tag{3.76}$$

We introduce dimensionless parameter $\gamma = \ell_1/\ell_3$. Substituting the expressions (3.76) into the Laplace transformation of the Eqs. (3.46) and (3.49), we obtain the following equations with respect to the unknown functions $\bar{\psi}_i(x_2, s)$.

$$\begin{aligned}
&\bar{A}^*_{66}\frac{d^2\bar{\psi}_1}{dx_2^2} - (\bar{A}^*_{11} + \gamma^2\bar{A}^*_{55} + \sigma^{(0)}_{11} + \gamma^2\sigma^{(0)}_{33})\bar{\psi}_1 \\
&\quad + (\bar{A}^*_{12} + \bar{A}^*_{66})\frac{d\bar{\psi}_2}{dx_2} - (\bar{A}^*_{13} + \bar{A}^*_{55})\gamma\bar{\psi}_3 = 0, \\
&\bar{A}^*_{22}\frac{d^2\bar{\psi}_2}{dx_2^2} - (\bar{A}^*_{66} + \gamma^2\bar{A}^*_{44} + \sigma^{(0)}_{11} + \gamma^2\sigma^{(0)}_{33})\bar{\psi}_2 \\
&\quad -(\bar{A}^*_{12} + \bar{A}^*_{66})\frac{d\bar{\psi}_1}{dx_2} - (\bar{A}^*_{23} + \bar{A}^*_{44})\gamma\bar{\psi}_3 = 0, \\
&\bar{A}^*_{44}\frac{d^2\bar{\psi}_3}{dx_2^2} - (\bar{A}^*_{55} + \gamma^2\bar{A}^*_{33} + \sigma^{(0)}_{11} + \gamma^2\sigma^{(0)}_{33})\bar{\psi}_3 \\
&\quad +(\bar{A}^*_{44} + \bar{A}^*_{23})\gamma\frac{d\bar{\psi}_2}{dx_2} - (\bar{A}^*_{13} + \bar{A}^*_{55})\gamma\bar{\psi}_1 = 0.
\end{aligned} \tag{3.77}$$

According to the expressions (3.76), the Laplace transformations of the boundary conditions at the ends of the plate, i.e. conditions (3.48) are satisfied automatically. From the Laplace transformations of the boundary conditions on the upper and lower face surfaces of the plate, i.e. from the condition (3.47) we obtain

$$\begin{aligned}
&\left(\bar{A}^*_{22}\frac{d\bar{\psi}_2}{dx_2} - \bar{A}^*_{12}\bar{\psi}_1 - \gamma\bar{A}^*_{23}\bar{\psi}_3\right)\Bigg|_{x_2=\pm h} = 0, \\
&\bar{A}^*_{66}\left(\frac{d\bar{\psi}_1}{dx_2} + \bar{\psi}_3\right)\Bigg|_{x_2=\pm h} = \frac{1}{s}\sigma^{(0)}_{11}\frac{\partial f}{\partial x_1}, \\
&\bar{A}^*_{44}\left(\frac{d\bar{\psi}_3}{dx_2} + \gamma\bar{\psi}_2\right)\Bigg|_{x_2=\pm h} = \frac{1}{s}\sigma^{(0)}_{11}\frac{\partial f}{\partial x_3}.
\end{aligned} \tag{3.78}$$

To solve the problem (3.77)–(3.78) we employ the well known Euler method, according to which, the unknown functions $\bar{\psi}_i(x_2, s)$ are presented as follows.

$$\bar{\psi}_i(x_2, s) = \sum_{j=1}^{6} F_j(s)\, a^{(i)}_j(s)\, e^{k_j x_2}, \tag{3.79}$$

where $a^{(i)}_j$ and $k_j$ are known constants, whose values are found from the Eq. (3.77) and from the corresponding characteristic equation respectively. The unknown constants $F_j(s)$ are determined from the boundary condition (3.78). The inverse of the Laplace transformations is calculated by using the method (Schapery 1962, 1978), according to which

$$\zeta(t) = \left(s\bar{\zeta}(s)\right)\big|_{s=1/(2t)}. \tag{3.80}$$

The values of the critical time or critical forces are determined from the criterion (3.43).

Consider the solution to the problem (3.63)–(3.66) obtained within the scope of the Kirchhoff-Love plate theory. Applying the Laplace transformation (2.34) to the Eqs. (3.63)–(3.66), we arrive at the following representation of the sought-for quantities.

$$\bar{w}(x_1, x_3, s) = \bar{w}_0(s) \sin\left(\frac{\pi x_1}{\ell_1}\right) \sin\left(\frac{\pi x_3}{\ell_3}\right). \tag{3.81}$$

The function (3.81) satisfies automatically the end conditions (3.64), (3.65) and (3.66). The unknown $\bar{w}(s)$ is determined from the Eq. (3.63).

For solution to the problem (3.72)–(3.73a–b) obtained within the scope of the third order refined plate theory we use also the Laplace transformation (2.34) and the functions $\bar{w}(x_1, x_3, s)$, $\bar{\varphi}_1(x_1, x_3, s)$ and $\bar{\varphi}_3(x_1, x_3, s)$ are presented as follows.

$$\begin{aligned}
\bar{w}(x_1, x_3, s) &= \bar{w}_0(s) \sin\left(\frac{\pi x_1}{\ell_1}\right) \sin\left(\frac{\pi x_3}{\ell_3}\right), \\
\bar{\varphi}_1(x_1, x_3, s) &= \bar{\varphi}_{10}(s) \cos\left(\frac{\pi x_1}{\ell_1}\right) \sin\left(\frac{\pi x_3}{\ell_3}\right), \\
\bar{\varphi}_3(x_1, x_3, s) &= \bar{\varphi}_{30}(s) \sin\left(\frac{\pi x_1}{\ell_1}\right) \cos\left(\frac{\pi x_3}{\ell_3}\right).
\end{aligned} \tag{3.82}$$

The presentation (3.82) satisfy automatically the end conditions given in (3.73a, b), and the values of $\bar{w}_0(s)$, $\bar{\varphi}_{10}(s)$ and $\bar{\varphi}_{30}(s)$ are determined from the Eq. (3.72). The original of the sought functions determined within the framework of the foregoing approximate plate theories and the values of the corresponding critical parameters are found through the Eq. (3.80) and criterion (3.48) respectively.

### *3.3.3 Numerical Results and Discussions*

We assume that the plate material consists of the alternating two layers of two isotropic, homogeneous materials and these layers are located in the $Ox_1x_3$ plane. The reinforcing layers are supposed to be pure elastic with mechanical characteristics $E^{(2)}$ (Young's modulus), $\lambda^{(2)}$ and $\mu^{(2)}$ (Lame constants), $\nu^{(2)}$ (Poisson coefficient). The matrix layers material is supposed to be linearly viscoelastic with operators (2.68)–(2.70). Within the framework of the continuum approach the plate material can be considered as homogeneous transversal isotropic viscoelastic one the isotropy axis of which coincide with the $Ox_2$ axis. As has been noted in Chap. 2, the Laplace transformation of the effective operators of the plate material are determined from the expressions of the effective mechanical constants by replacing the mechanical constants related to the matrix and filler materials with the Laplace transformation of the corresponding operators. According to

(Christensen 1979), for the plate material under consideration the Laplace transformation of the effective operators are determined as follows.

$$\bar{A}_{44}^{*} = \bar{A}_{66}^{*} = \frac{\bar{\mu}^{*(1)}\bar{\mu}^{(2)}}{\bar{\mu}^{*(1)}\eta^{(2)} + \bar{\mu}^{(2)}\eta^{(1)}}, \; \bar{A}_{55}^{*} = \bar{\mu}^{*(1)}\eta^{(1)} + \bar{\mu}^{(2)}\eta^{(2)},$$
$$\frac{1}{2}\left(\bar{A}_{11}^{*} - \bar{A}_{13}^{*}\right) = \eta^{(1)}\bar{\mu}^{*(1)} + \eta^{(2)}\bar{\mu}^{(2)},$$
$$\bar{A}_{23}^{*} = \bar{A}_{12}^{*} = \bar{\lambda}^{*(1)}\eta^{(1)} + \bar{\lambda}^{(2)}\eta^{(2)} - \eta^{(1)}\eta^{(2)}\left(\bar{\lambda}^{*(1)} - \bar{\lambda}^{(2)}\right) \times \frac{\left(\bar{\lambda}^{*(1)} + 2\bar{\mu}^{*(1)}\right) - \left(\bar{\lambda}^{(2)} + 2\bar{\mu}^{(2)}\right)}{\left(\bar{\lambda}^{*(1)} + 2\bar{\mu}^{*(1)}\right)\eta^{(2)} + \left(\bar{\lambda}^{(2)} + 2\bar{\mu}^{(2)}\right)\eta^{(1)}},$$
$$\frac{1}{2}\left(\bar{A}_{11}^{*} + \bar{A}_{13}^{*}\right) = \left(\bar{\lambda}^{*(1)} + \bar{\mu}^{*(1)}\right)\eta^{(1)} + \left(\bar{\lambda}^{(2)} + \bar{\mu}^{(2)}\right)\eta^{(2)} - \eta^{(1)}\eta^{(2)}\frac{\left(\bar{\lambda}^{*(1)} - \bar{\lambda}^{(2)}\right)^2}{\left(\bar{\lambda}^{*(1)} + 2\bar{\mu}^{*(1)}\right)\eta^{(2)} + \left(\bar{\lambda}^{(2)} + 2\bar{\mu}^{(2)}\right)\eta^{(1)}},$$
$$\bar{A}_{22}^{*} = \left(\bar{\lambda}^{*(1)} + 2\bar{\mu}^{*(1)}\right)\eta^{(1)} + \left(\bar{\lambda}^{(2)} + 2\bar{\mu}^{(2)}\right)\eta^{(2)} - \eta^{(1)}\eta^{(2)}\frac{\left(\left(\bar{\lambda}^{*(1)} + 2\bar{\mu}^{*(1)}\right) - \left(\bar{\lambda}^{(2)} + 2\bar{\mu}^{(2)}\right)\right)^2}{\left(\bar{\lambda}^{*(1)} + 2\bar{\mu}^{*(1)}\right)\eta^{(2)} + \left(\bar{\lambda}^{(2)} + 2\bar{\mu}^{(2)}\right)\eta^{(1)}}. \quad (3.83)$$

Here $\eta^{(1)} = 1 - \eta^{(2)}$ and $\eta^{(2)}$ is filler layers volume fraction. The meaning of the notation used in (3.83) is given in (2.68). As in Sect. 2.5, we introduce the dimensionless rheological parameter $\omega(= \omega_\infty/\omega_0)$ and the dimensionless time $t'(= \omega_0^{1/(1+\alpha)}t)$ and assume that $\nu^{(2)} = \nu_0^{(1)} = 0.3$, $\eta^{(2)} = 0.5$.

First, we consider pure elastic stability loss of the plate at $t' = 0$ and $t' = \infty$. Introduce the dimensionless parameter $\rho = p_3/p_1$, which characterizes the biaxial character of the external loading of the plate. We consider the cases where $\rho = 0$ and $\rho = 1$ under $\gamma = 0$ and $\gamma = 1$. Remind that the parameter $\gamma = (\ell_1/\ell_3)$ characterizes the wideness of the plate and the case where $\gamma = 0$ corresponds the plate-strip. For fixed values of the parameters $\rho$ and $\gamma$ the pure elastic stability loss of the plate can be determined through the critical values of $p_1$ denoted as $p_{1cr.0}$ (at $t' = 0$) and $p_{1cr.\infty}$(at $t' = \infty$).

Tables 3.1, 3.2 and 3.3 show the values of $p'_{1cr.0}(= p_{1cr.0}/E_0^{(1)})$ and $p'_{1cr.\infty}$ $(= p_{1cr.\infty}/E_0^{(1)})$ for the cases where $\{\gamma = 0,\ \rho = 0\}$, $\{\gamma = 1,\ \rho = 0\}$ and $\{\gamma = 1,\ \rho = 1\}$ respectively under $\omega = 1$. In these tables the abbreviations*K* – *L*, *Rfn* and *3D* stand for the Kirchhoff-Love, the third order refined and the TDLTS theories respectively. Note that these abbreviations will be also used below.

**Table 3.1** Values of $p'_{1cr.0}$ and $p'_{1cr.\infty}$ obtained in the case where $\{\gamma = 0,\ \rho = 0\}$ and $\omega = 1$

| $\frac{E^{(2)}}{E_0^{(1)}}$ | Theories | $h/\ell_1 = 0.05$ | | $h/\ell_1 = 0.075$ | | $h/\ell_1 = 0.10$ | |
|---|---|---|---|---|---|---|---|
| | | $p'_{1cr.0}$ | $p'_{1cr.\infty}$ | $p'_{1cr.0}$ | $p'_{1cr.\infty}$ | $p'_{1cr.0}$ | $p'_{1cr.\infty}$ |
| 1 | $K-L$ | 0.0110 | 0.0099 | 0.0245 | 0.0219 | 0.0429 | 0.0384 |
| | *Rfn* | 0.0107 | 0.0094 | 0.0228 | 0.0198 | 0.0380 | 0.0325 |
| | 3D | 0.0075 | 0.0067 | 0.0180 | 0.0144 | 0.0303 | 0.0241 |
| 5 | $K-L$ | 0.0303 | 0.0294 | 0.0674 | 0.0654 | 0.1182 | 0.1147 |
| | *Rfn* | 0.0287 | 0.0266 | 0.0600 | 0.0529 | 0.0972 | 0.0813 |
| | 3D | 0.0257 | 0.0229 | 0.0542 | 0.0463 | 0.0886 | 0.0724 |
| 10 | $K-L$ | 0.0530 | 0.0522 | 0.1180 | 0.1161 | 0.2069 | 0.2035 |
| | *Rfn* | 0.0486 | 0.0441 | 0.0984 | 0.0828 | 0.1536 | 0.1196 |
| | 3D | 0.0456 | 0.0407 | 0.0929 | 0.0774 | 0.1463 | 0.1136 |
| 20 | $K-L$ | 0.0980 | 0.0972 | 0.2183 | 0.2163 | 0.3826 | 0.3792 |
| | *Rfn* | 0.0846 | 0.0730 | 0.1614 | 0.1247 | 0.2370 | 0.1660 |
| | 3D | 0.0818 | 0.0702 | 0.1571 | 0.1215 | 0.2325 | 0.1637 |

Thus, according to these numerical results, it can be concluded that

$$p^{K-L}_{1cr.\infty} < p^{Rfn}_{1cr.\infty} < p^{3D}_{1cr.\infty} \tag{3.84}$$

for the same fixed values of problem parameters.

Consider numerical results related to the critical time (denoted by $t'_{cr}$). To compare the values of $t'_{cr}$ given by the approximate theories and the TDLTS, we must select the values of $p_1/E_0^{(1)}$ which fall in the intervals $[p^{K-L}_{1cr.\infty}, p^{K-L}_{1cr.0}]$, $[p^{Rfn}_{1cr.\infty}, p^{Rfn}_{1cr.0}]$ and $[p^{3D}_{1cr.\infty}, p^{3D}_{1cr.0}]$ simultaneously. However, as follows from the data given in Tables 3.1, 3.2 and 3.3, such common values of $p_1/E_0^{(1)}$ exist only for the intervals $[p^{Rfn}_{1cr.\infty}, p^{Rfn}_{1cr.0}]$ and $[p^{3D}_{1cr.\infty}, p^{3D}_{1cr.0}]$. For this reason, we compare only the results obtained by using the TDLTS and the third order refined plate theories. These results are given in Table 3.4 from which follow that

$$t'_{cr.Rfn} > t'_{cr.3D} \tag{3.85}$$

where $t'_{cr.Rfn}$ $(t'_{cr.3D})$ denotes the values of the critical time obtained within the scope of the third order refined plate theory (TDLTS).

Let us look at the influence of the rheological parameters $\omega$ and $\alpha$ on the values of $t'_{cr.Rfn}$ and $t'_{cr.3D}$. For this purpose, we consider the graphs depicted in Figs. 3.2 and 3.3, which show the critical times $t'_{cr.Rfn}$ and $t'_{cr.3D}$ as function of $p_1/E_0^{(1)}$ at $h/\ell_1 = 0.1$, $\gamma = 1.0$, $\rho = 1.0$, $\alpha = -0.5$ (Fig. 3.2) and $\omega = 1.0$ (Fig. 3.3) for various values of $\alpha$ (Fig. 3.3) and $\omega$ (Fig. 3.2). It follows from these graphs that $t'_{cr.Rfn}$ and $t'_{cr.3D}$ both increase with $|\alpha|$ and $\omega$. Note that this increase agrees well with the mechanical considerations regarding the nature of viscoelastic materials. Moreover, it is also seen that the difference $\left(t'_{cr.Rfn} - t'_{cr.3D}\right)$ increase with $|\alpha|$ and $\omega$.

**Table 3.2** Values of $p'_{1cr.0}$ and $p'_{1cr.\infty}$ obtained in the case where $\{\gamma = 1,\ \rho = 0\}$ and $\omega = 1$

| $\frac{E^{(2)}}{E_0^{(1)}}$ | Theories | $h/\ell_1 = 0.05$ | | $h/\ell_1 = 0.075$ | | $h/\ell_1 = 0.10$ | |
|---|---|---|---|---|---|---|---|
| | | $p'_{1cr.0}$ | $p'_{1cr.\infty}$ | $p'_{1cr.0}$ | $p'_{1cr.\infty}$ | $p'_{1cr.0}$ | $p'_{1cr.\infty}$ |
| 1 | $K-L$ | 0.0436 | 0.0390 | 0.0961 | 0.0861 | 0.1663 | 0.1488 |
| | *Rfn* | 0.0409 | 0.0357 | 0.0840 | 0.0714 | 0.1334 | 0.1103 |
| | 3D | 0.0336 | 0.0258 | 0.0703 | 0.0533 | 0.1141 | 0.0850 |
| 5 | $K-L$ | 0.1201 | 0.1165 | 0.2648 | 0.2567 | 0.4580 | 0.4445 |
| | *Rfn* | 0.1081 | 0.0962 | 0.2134 | 0.1760 | 0.3249 | 0.2486 |
| | 3D | 0.0974 | 0.0839 | 0.1949 | 0.1576 | 0.3010 | 0.2283 |
| 10 | $K-L$ | 0.2102 | 0.2068 | 0.4636 | 0.4560 | 0.8018 | 0.7887 |
| | *Rfn* | 0.1785 | 0.1522 | 0.3341 | 0.2554 | 0.4824 | 0.3356 |
| | 3D | 0.1683 | 0.1420 | 0.3189 | 0.2434 | 0.4659 | 0.3259 |
| 20 | $K-L$ | 0.3888 | 0.3854 | 0.8573 | 0.8498 | 1.4829 | 1.4699 |
| | *Rfn* | 0.2958 | 0.2327 | 0.5081 | 0.3484 | 0.6802 | 0.4222 |
| | 3D | 0.2877 | 0.2264 | 0.4993 | 0.3447 | 0.6752 | 0.4211 |

**Table 3.3** Values of $p'_{1cr.0}$ and $p'_{1cr.\infty}$ obtained in the case where $\{\gamma = 1,\ \rho = 1\}$ and $\omega = 1$

| $\frac{E^{(2)}}{E_0^{(1)}}$ | Theories | $h/\ell_1 = 0.05$ | | $h/\ell_1 = 0.075$ | | $h/\ell_1 = 0.10$ | |
|---|---|---|---|---|---|---|---|
| | | $p'_{1cr.0}$ | $p'_{1cr.\infty}$ | $p'_{1cr.0}$ | $p'_{1cr.\infty}$ | $p'_{1cr.0}$ | $p'_{1cr.\infty}$ |
| 1 | $K-L$ | 0.0218 | 0.0195 | 0.0481 | 0.0431 | 0.0832 | 0.0744 |
| | *Rfn* | 0.0205 | 0.0179 | 0.0420 | 0.0357 | 0.0667 | 0.0552 |
| | 3D | 0.0168 | 0.0129 | 0.0352 | 0.0267 | 0.0568 | 0.0425 |
| 5 | $K-L$ | 0.0601 | 0.0583 | 0.1324 | 0.1285 | 0.2290 | 0.2223 |
| | *Rfn* | 0.0541 | 0.0481 | 0.1067 | 0.0880 | 0.1625 | 0.1243 |
| | 3D | 0.0487 | 0.0420 | 0.0975 | 0.0788 | 0.1505 | 0.1142 |
| 10 | $K-L$ | 0.1051 | 0.1034 | 0.2318 | 0.2280 | 0.4009 | 0.3944 |
| | *Rfn* | 0.0893 | 0.0761 | 0.1671 | 0.1277 | 0.2412 | 0.1678 |
| | 3D | 0.0842 | 0.0710 | 0.1595 | 0.1217 | 0.2330 | 0.1630 |
| 20 | $K-L$ | 0.1944 | 0.1927 | 0.4287 | 0.4249 | 0.7415 | 0.7350 |
| | *Rfn* | 0.1479 | 0.1164 | 0.2541 | 0.1742 | 0.3401 | 0.2111 |
| | 3D | 0.1439 | 0.1132 | 0.2497 | 0.1724 | 0.3376 | 0.2110 |

It follows from the aforementioned results that

$$\frac{t'_{cr.Rfn}}{t'_{cr.3D}} < \frac{p^{Rfn}_{1cr.0}}{p^{3D}_{1cr.0}}, \quad \frac{t'_{cr.Rfn}}{t'_{cr.3D}} < \frac{p^{Rfn}_{1cr.\infty}}{p^{3D}_{1cr.\infty}}. \tag{3.86}$$

For $p_1 \to p^{3D}_{1cr.0}$ or $p_1 \to p^{3D}_{1cr.\infty}$, inequalities in (3.86) become more stronger, i.e.

$$\frac{t'_{cr.Rfn}}{t'_{cr.3D}} \ll \frac{p^{Rfn}_{1cr.0}}{p^{3D}_{1cr.0}}, \quad \frac{t'_{cr.Rfn}}{t'_{cr.3D}} \ll \frac{p^{Rfn}_{1cr.\infty}}{p^{3D}_{1cr.\infty}}. \tag{3.87}$$

Table 3.4 Values of $t'_{cr}$ obtained for various $\gamma$ and $\rho$ under$\alpha = -0.5$ and $\omega = 1$

| Loading scheme | $\frac{h}{\ell_1} = 0.05$ | $\frac{E^{(2)}}{E_0^{(1)}}$ | $\frac{p_1}{E_0^{(1)}}$ | $t'_{cr.Rfn}$ | $t'_{cr.3D}$ |
|---|---|---|---|---|---|
| $\{\gamma = 0, \rho = 0\}$ | 0.075 | 5 | 0.0536 | 7.9733 | 0.0059 |
| | | 10 | 0.0879 | 1.0008 | 0.1127 |
| | | 20 | 0.1410 | 0.4604 | 0.2491 |
| | 0.10 | 5 | 0.0850 | 1.8487 | 0.0470 |
| | | 10 | 0.1330 | 0.6173 | 0.1823 |
| | | 20 | 0.1993 | 0.3536 | 0.2646 |
| $\{\gamma = 1, \rho = 0\}$ | 0.075 | 5 | 0.1855 | 1.5878 | 0.0606 |
| | | 10 | 0.2872 | 0.5718 | 0.1949 |
| | | 20 | 0.4239 | 0.3378 | 0.2660 |
| | 0.10 | 5 | 0.2748 | 0.8091 | 0.1268 |
| | | 10 | 0.4008 | 0.4031 | 0.2329 |
| | | 20 | 0.5497 | 0.2609 | 0.2512 |
| $\{\gamma = 1, \rho = 1\}$ | 0.075 | 5 | 0.0928 | 1.5560 | 0.0593 |
| | | 10 | 0.1436 | 0.5718 | 0.1949 |
| | | 20 | 0.2120 | 0.3366 | 0.2650 |
| | 0.10 | 5 | 0.1374 | 0.8091 | 0.1268 |
| | | 10 | 0.2000 | 0.4161 | 0.2407 |
| | | 20 | 0.2749 | 0.2603 | 0.2507 |

These relations show that, in many cases, for investigation of the stability loss of the rectangular simply supported plate made of viscoelastic composites, it is necessary to use the approach developed in the present chapter within the scope of the TDLTS.

## 3.4 Rectangular Plate Clamped at Two Opposite Ends and Simply Supported at the Two Other Opposite Ends

### *3.4.1 Solution Procedure of the Equations of the TDLTS. Semi-Analytical FEM Modeling*

First, we consider the case where the ends of the plate at $x_1 = 0$ and $x_1 = \ell_1$ are clamped, but the ends $x_3 = 0$ and $x_3 = \ell_3$ are simply supported, i.e. we consider the problem formulated in Sect. 3.1. As shown in Sect. 3.1, the investigation of the stability loss of the plate under consideration is reduced to the solution to the boundary value problems (3.34), (3.36)–(3.39) (for the zeroth approximation) and (3.34), (3.40)–(3.42) (for the first approximation). As in previous section, we will omit the upper index (1) in the quantities related to the first approximation.

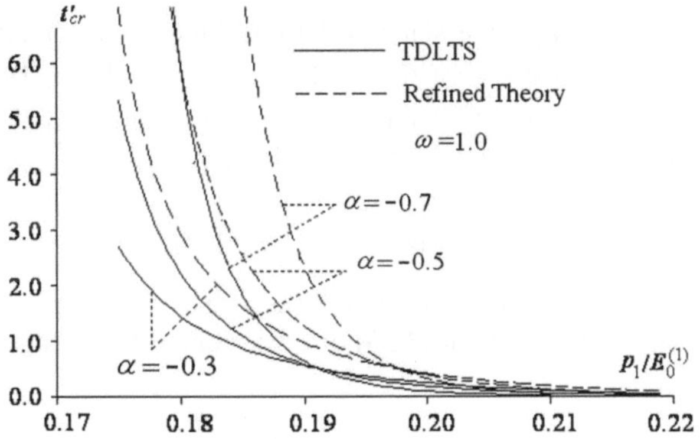


**Fig. 3.3** The influence of the rheological parameter $\alpha$ on the dependencies among $t'_{cr.Rf}$, $t'_{cr.3D}$ and $p_1/E_0^{(1)}$ under $\omega = 1.0$

To solve the problem (3.34), (3.36)–(3.39) we employ Laplace transformation (2.34) and the principle of correspondence. To determine the transformation of the sought values we employ the Finite Element Method (FEM). For this purpose we use the functional

$$\Pi^{(0)} = \frac{1}{2}\int_0^{\ell_1}\left(\int_0^{\ell_3}\left(\int_{-h}^{+h}\bar{\sigma}_{ij}^{(0)}\bar{\varepsilon}_{ij}^{(0)}dx_2\right)dx_3\right)dx_1 - \int_0^{\ell_1}\left(\int_{-h}^{+h}\bar{p}_3\bar{u}_3^{(0)}\Big|_{x_3=0}dx_2\right)dx_1 + \int_0^{\ell_1}\left(\int_{-h}^{+h}\bar{p}_3\bar{u}_3^{(0)}\Big|_{x_3=\ell_3}dx_2\right)dx_1 \quad (3.88)$$

According to the principle of the virtual work, the first equation in (3.30) and boundary conditions (3.37) and (3.39) are obtained from

$$\delta\Pi^{(0)} = \int_0^{\ell_1}\left(\int_0^{\ell_3}\left(\int_{-h}^{+h}\bar{\sigma}_{ij}^{(0)}\delta\bar{\varepsilon}_{ij}^{(0)}dx_2\right)dx_3\right)dx_1 - \int_0^{\ell_1}\left(\int_{-h}^{+h}\bar{p}_3\delta\bar{u}_3^{(0)}\Big|_{x_3=0}dx_2\right)dx_1 + \int_0^{\ell_1}\left(\int_{-h}^{+h}\bar{p}_3\delta\bar{u}_3^{(0)}\Big|_{x_3=\ell_3}dx_2\right)dx_1 = 0, \quad (3.89)$$

where

$$\delta\bar{\varepsilon}_{ij}^{(0)} = \frac{1}{2}\left(\frac{\partial\delta\bar{u}_i^{(0)}}{\partial x_j} + \frac{\partial\delta\bar{u}_j^{(0)}}{\partial x_i}\right). \quad (3.90)$$

Integration by parts applied to Eq. (3.89) yields the first Eq. (3.36) and the boundary conditions (3.37) and (3.39).

Employing the standard Ritz technique, we determine the values of $\bar{\sigma}_{ij}^{(0)}$, $\bar{u}_i^{(0)}$ and with the use of the method (Schapery 1962, 1978), we found the values of $\sigma_{ij}^{(0)}(x_1, x_2, x_3, t)$ which enter the first equation in (3.40). Many numerical results show that for the considered case with very high accuracy we can assume that

$$\sigma_{33}^{(0)} \approx -p_3,\ \sigma_{13}^{(0)} \approx 0,\ \sigma_{23}^{(0)} \approx 0,\ \sigma_{ij}^{(0)} \approx \sigma_{ij}^{(0)}(x_1, x_2, t),\ i;j = 1, 2 \qquad (3.91)$$

After this solution procedure the values of $p_1$ corresponding to the selected possible value of $U$ in (3.38) are determined from the relation.

$$p_1 = \frac{1}{2h}\int_{-h}^{+h} \sigma_{11}^{(0)}\Big|_{x_1=\ell_1/2} dx_2. \qquad (3.92)$$

Consider the solution procedure of the problem related to the first approximation, i.e. of the problem (3.40)–(3.42). It follows from the foregoing discussions and from the Eqs. (3.40) and (3.42) that, as the stresses $\sigma_{ij}^{(0)}$ depend on a time $t$, therefore it is impossible to employ directly the principle of the correspondence to the Eqs. (3.40)–(3.42). Because the Laplace transformation of the term $\sigma_{in}^{(0)}\partial u_j/\partial x_n$ can not be presented as $\bar{\sigma}_{in}^{(0)}\partial \bar{u}_j/\partial x_n$.

For overcome of this difficulty we use the "time freezing" principle, according to which the "own" time is introduced for the first approximation and this "own" time is independent of the "frozen" time of the zeroth approximation. Thus, within the scope of the "time freezing" principle the Laplace transformation of the term $\sigma_{in}^{(0)}\partial u_j/\partial x_n$ can be presented as $\sigma_{in}^{(0)}\partial \bar{u}_j/\partial x_n$, and the principle of the correspondence can be employed for the solution of the problem (3.40)–(3.42).

For the case under consideration, we assume that the initial infinitesimal imperfection mode of the plate is given through the function

$$f = \ell_1 \sin^2\left(\frac{\pi x_1}{\ell_1}\right) \sin\left(\frac{\pi x_3}{\ell_3}\right). \qquad (3.93)$$

According to the boundary and end conditions given in (3.41) and (3.42), we can write the following representations for the Laplace transformation of the first approximation and employ the semi-analytical FEM for determination of the quantities related to this approximation.

$$\bar{u}_1 = \bar{v}_1(x_1, x_2, s)\sin\left(\frac{\pi x_3}{\ell_3}\right),\ \bar{u}_2 = \bar{v}_2(x_1, x_2, s)\sin\left(\frac{\pi x_3}{\ell_3}\right),$$

$$\bar{u}_3 = \bar{v}_3(x_1, x_2, s)\cos\left(\frac{\pi x_3}{\ell_3}\right),$$

$$\bar{\sigma}_{ii} = \bar{S}_{ii}(x_1, x_2, s)\sin\left(\frac{\pi x_3}{\ell_3}\right),\quad i = 1, 2, 3,\ \bar{\sigma}_{12} = \bar{S}_{12}(x_1, x_2, s)\sin\left(\frac{\pi x_3}{\ell_3}\right),$$

$$\bar{\sigma}_{13} = \bar{S}_{13}(x_1, x_2, s)\cos\left(\frac{\pi x_3}{\ell_3}\right),\ \bar{\sigma}_{23} = \bar{S}_{23}(x_1, x_2, s)\cos\left(\frac{\pi x_3}{\ell_3}\right). \tag{3.94}$$

To determine the unknown functions $\bar{v}_i(x_1, x_2, s)$ and $\bar{S}_{ij}(x_1, x_2, s)$ we employ also the FEM with the use of the functional

$$\Pi = \frac{1}{2}\int_0^{\ell_1}\left(\int_{-h}^{+h}\left(\bar{S}_{ij} + \sigma_{in}^{(0)}\frac{\partial \bar{v}_j}{\partial x_n}\right)\frac{\partial \bar{v}_j}{\partial x_i}dx_2\right)dx_1 - \int_0^{\ell_1}\left(\frac{\partial f}{\partial x_1}\sigma_{i1}^{(0)}\Big|_{x_2=+h} + \frac{\partial f}{\partial x_3}\sigma_{i3}^{(0)}\Big|_{x_2=+h}\right)\bar{v}_2\big|_{x_2=+h}dx_1 + \int_0^{\ell_1}\left(\frac{\partial f}{\partial x_1}\sigma_{i1}^{(0)}\Big|_{x_2=-h} + \frac{\partial f}{\partial x_3}\sigma_{i3}^{(0)}\Big|_{x_2=-h}\right)\bar{v}_2\big|_{x_2=-h}dx_1. \tag{3.95}$$

From the relation

$$\delta\Pi = \int_0^{\ell_1}\left(\int_{-h}^{+h}\left(\bar{S}_{ij} + \sigma_{in}^{(0)}\frac{\partial \bar{v}_j}{\partial x_n}\right)\frac{\partial \delta\bar{v}_j}{\partial x_i}dx_2\right)dx_1 - \int_0^{\ell_1}\left(\frac{\partial f}{\partial x_1}\sigma_{i1}^{(0)}\Big|_{x_2=+h} + \frac{\partial f}{\partial x_3}\sigma_{i3}^{(0)}\Big|_{x_2=+h}\right)\delta\bar{v}_2\big|_{x_2=+h}dx_1 + \int_0^{\ell_1}\left(\frac{\partial f}{\partial x_1}\sigma_{i1}^{(0)}\Big|_{x_2=-h} + \frac{\partial f}{\partial x_3}\sigma_{i3}^{(0)}\Big|_{x_2=-h}\right)\delta\bar{v}_2\big|_{x_2=-h}dx_1 = 0, \tag{3.96}$$

the corresponding equations for unknown functions $\bar{v}_i(x_1, x_2, s)$ and $\bar{S}_{ij}(x_1, x_2, s)$ are obtained.

Under FEM modeling of the Eq. (3.96), accounting the symmetry of the problem about $x_1 = \ell_1/2$ all numerical investigations are made in the region $\{0 \le x_1/\ell_1 \le 1/2; -1 \le x_2/h \le 1\}$ which in turn is divided into 80 rectangular Lagrange family quadratic elements, with 369 nodes and 1107 NDOF.

### 3.4.2 Solution Procedure for the Approximate Stability Loss Equations

To obtain the approximate stability loss Eqs. (3.63) and (3.72) by average integrating over the thickness of the plate from the TDLTS Eqs. (3.40) and (3.41) it is necessary to use the assumption

$$\sigma_{11}^{(0)} = -p_1,\ \sigma_{33}^{(0)} = -p_3,\ \sigma_{12}^{(0)} = \sigma_{22}^{(0)} = \sigma_{13}^{(0)} = \sigma_{23}^{(0)} = 0 \tag{3.97}$$

In this case, the end conditions at $x_3 = 0$ and $x_3 = \ell_3$ remain as in (3.66) and (3.73a, b). However, to obtain the average-integrating end conditions at $x_1 = 0$ and $x_1 = \ell_1$ in the form

$$w(0, x_3) = w(\ell_1, x_3) = 0,\ \frac{\partial w(0, x_3)}{\partial x_1} = \frac{\partial w(\ell_1, x_3)}{\partial x_1} = 0, \tag{3.98a}$$

$$\varphi_1(0, x_3) = \varphi_1(\ell_1, x_3) = 0,\ \varphi_3(0, x_3) = \varphi_3(\ell_1, x_3) = 0, \tag{3.98b}$$

we use the conditions $u_i^{(1)}\Big|_{x_1=0;\ell_1} = 0\,(i = 1, 2, 3)$ given in (3.42) and the representations in (3.52) and (3.67). The solution to Laplace transformation of the problem consisting of the Eq. (3.63) and ends conditions (3.66) and (3.98a) we select as follows

$$\bar{w} = \bar{w}_0(s) \sin^2\left(\frac{\pi x_1}{\ell_1}\right) \sin\left(\frac{\pi x_3}{\ell_3}\right), \tag{3.99}$$

However, the Laplace transformation of the problem consisting of the Eq. (3.72) and ends conditions (3.73b) and (3.98a, b) can be selected as

$$\begin{gathered} \bar{w} = \bar{w}_0(s) \sin^2\left(\frac{\pi x_1}{\ell_1}\right) \sin\left(\frac{\pi x_3}{\ell_3}\right),\ \bar{\varphi}_1 = \bar{\varphi}_{10}(s) \sin^2\left(\frac{\pi x_1}{\ell_1}\right) \sin\left(\frac{\pi x_3}{\ell_3}\right), \\ \bar{\varphi}_3 = \bar{\varphi}_{30}(s) \sin\left(2\frac{\pi x_1}{\ell_1}\right) \cos\left(\frac{\pi x_3}{\ell_3}\right). \end{gathered} \tag{3.100}$$

Note that the expressions (3.99) and (3.100) satisfy automatically the end conditions (3.66), (3.98a) and (3.73b), (3.98a, b), respectively. Substituting (3.99) into the Eq. (3.63) and substituting (3.100) into the Eq. (3.72) we obtain the expressions for determination of the quantities $\bar{w}_0(s)$, $\bar{\varphi}_{10}(s)$ and $\bar{\varphi}_{30}(s)$. By employing the method (Schapery 1962, 1978), i.e. the expression (3.80) and the criterion (3.43) we found the values of the critical parameters.

**Table 3.5** The values of $p'_{cr.0}$ (upper number) and $p'_{cr.\infty}$ (lower number) obtained under $\omega = 1.0$, $\gamma = 0$

| Theories | $E^{(2)}/E_0^{(1)}$ | | |
|---|---|---|---|
| | 10 | 20 | 50 |
| $K-L$ | 0.1979<br>0.1939 | 0.3589<br>0.3547 | 0.8378<br>0.8334 |
| *Rfn* | 0.1580<br>0.1313 | 0.2540<br>0.1939 | 0.4403<br>0.2881 |
| 3*D* | 0.1473<br>0.1208 | 0.2450<br>0.1868 | 0.4354<br>0.2865 |

**Table 3.6** The values of $p'_{cr.\infty}$ obtained for various $\omega$ under $E^{(2)}/E_0^{(1)} = 50$, $\gamma = 0$

| Theories | $\omega$ | | |
|---|---|---|---|
| | 1.0 | 2.0 | 5.0 |
| $K-L$ | 0.8334 | 0.8348 | 0.8362 |
| *Rfn* | 0.2881 | 0.3483 | 0.3983 |
| 3*D* | 0.2865 | 0.3459 | 0.3938 |

### *3.4.3 Numerical Results and Discussions*

As in Sect. 3.3.3, we assume that the plate material consists of the alternating two layers of two isotropic, homogeneous materials and these layers are located in the $Ox_1x_3$ plane. The notation and expressions in (3.83) occur also in the present consideration. Assume that$\nu^{(2)} = \nu_0^{(1)} = 0.3$, $2h/\ell_1 = 0.1$, $\eta^{(2)} = 0.5$ and, as in Sect. 3.3.3, introduce dimensionless rheological parameter $\omega = \omega_\infty/\omega_0$ and dimensionless time $t' = t\omega_0^{1/(1+\alpha)}$. First, we consider the case where $\gamma = 0$ and $\rho = 0$ (where $\gamma = \ell_1/\ell_3$ and $\rho = p_3/p_1$), i.e. consider the results related to the stability loss of the clamped plate-strip under $p_3 = 0$.

Table 3.5 shows the values of $p'_{1cr.0}\left(= p_{1cr.0}/E_0^{(1)}\right)$ (upper number) and $p'_{1cr.\infty}\left(= p_{1cr.\infty}/E_0^{(1)}\right)$ (lower number) calculated for various values of $E^{(2)}/E_0^{(1)}$ within the scope of the TDLTS, the third order refined and Kirchhoff-Love stability loss theories. The influence of the rheological parameter $\omega$ on the values of $p'_{cr.\infty}\left(= p_{cr.\infty}/E_0^{(1)}\right)$ is illustrated by the data given in Table 3.6 under $E^{(2)}/E_0^{(1)} = 50$. Note that in Tables 3.5 and 3.6 the abbreviations $K-L$, *Rfn* and 3*D* noted in Sect. 3.3.3 are used.

As before, for consideration of the viscoelastic stability loss we assume that the external compressive force satisfies the relation $p_{1cr.\infty} < p < p_{1cr.0}$. It follows from the data given in Tables 3.5 and 3.6 that in the considered case the values of $p$ satisfied the inequality $p_{1cr.\infty} < p_1 < p_{1cr.0}$ for the TDLTS and third order refined stability loss theories cannot satisfy this inequality for the Kirchhoff-Love theory. Because the values of $p_{1cr.\infty}$ obtained for the Kirchhoff-Love theory are

**Table 3.7** The values of $t'_{cr.3D}$ and $t'_{cr.Rfn}$ with various $\omega$ for $p_1 / E_0^{(1)} = 0.400$, $\alpha = -0.5$, $E^{(2)} / E_0^{(1)} = 50$, $\gamma = 0$

| Theories | $\omega$ | | | | |
|---|---|---|---|---|---|
| | 1.0 | 2.0 | 3.0 | 4.0 | 5.0 |
| *Rfn* | 0.0277 | 0.0475 | 0.0992 | 0.3228 | 8.3692 |
| *3D* | 0.0212 | 0.0338 | 0.0617 | 0.1468 | 0.7000 |

**Table 3.8** The values of $t'_{cr.3D}$ and $t'_{cr.Rfn}$ with various $\alpha$ for $p / E_0^{(1)} = 0.400$, $\omega = 1.0$, $E^{(2)} / E_0^{(1)} = 50$, $\gamma = 0$

| Theories | $\alpha$ | | |
|---|---|---|---|
| | −0.3 | −0.5 | −0.7 |
| *Rfn* | 0.0633 | 0.0277 | 0.0040 |
| *3D* | 0.0524 | 0.0212 | 0.0025 |

**Table 3.9** The values of $p'_{1cr.0} (= p_{1cr.0} / E_0^{(1)})$ and obtained in the framework of the various theories under $\omega = 1.0$, $E^{(2)} / E_0^{(1)} = 50$, $\gamma = 1$

| | Theories | | |
|---|---|---|---|
| | *3D* | *Rfn* | $K - L$ |
| $p_{1cr.0} / E_0^{(1)}$ | 0.5180 | 0.5079 | 1.5184 |
| $p_{1cr.\infty} / E_0^{(1)}$ | 0.3157 | 0.2934 | 1.5129 |

significantly greater than those of $p_{1cr.0}$ obtained for the TDLTS and the third order refined theories. Thus, consider the numerical results related to the critical time. Table 3.7 shows the values of $t'_{cr.3D}$ and $t'_{cr.Rfn}$ obtained for various $\omega$ under $p / E_0^{(1)} = 0.400$, $\alpha = -0.5$ and $E^{(2)} / E_0^{(1)} = 50$.

It follows from these data that as a result of the increasing $\omega$ the difference between $t'_{cr.3D}$ and $t'_{cr.Rfn}$ grows significantly. Similar numerical results obtained with various $\alpha$ are given in Table 3.8.

Analyses of the foregoing numerical results show that for pure elastic stability loss problems the difference between the results obtained in the framework of the TDLTS and in the framework of the third order plate theories as in (Kromm 1953, 1955) is not more than 5–6 %. However, the mentioned difference with respect to the critical time can be more than a few times.

Now we consider numerical results obtained in the case where $\gamma = \ell_1 / \ell_3 = 1$, i.e. consider numerical results related to the square plate. First, we assume that $p_3 = 0$, i.e. assume the plate under consideration is compressed along the $Ox_1$ axis only. Table 3.9 shows the values of $p'_{1cr.0}$ and $p'_{1cr.\infty}$ obtained under $\omega = 1.0$.

**Table 3.10** The values of $t'_{cr.3D}$ and $t'_{cr.Rfn}$ with various $p_1/E_0^{(1)}$ for $\alpha = -0.5$, $\omega = 1.0$, $E^{(2)}/E_0^{(1)} = 50$, $\gamma = 1$

| Theories | $p_1/E_0^{(1)}$ | | | | |
|---|---|---|---|---|---|
| | 0.400 | 0.420 | 0.440 | 0.460 | 0.480 |
| *Rfn* | 0.171 | 0.082 | 0.035 | 0.013 | 0.003 |
| *3D* | 0.132 | 0.068 | 0.034 | 0.015 | 0.005 |

**Table 3.11** The values of $t'_{cr.3D}$ obtained for various $p_1/E_0^{(1)}$ and $p_3/E_0^{(1)}$ under $\omega = 1.0$, $\alpha = -0.5$, $\gamma = 1$

| $p_3/E_0^{(1)}$ | $p_1/E_0^{(1)}$ | | |
|---|---|---|---|
| | 0.400 | 0.420 | 0.440 |
| 0.000 | 0.132 | 0.068 | 0.034 |
| 0.1036 | 0.040 | 0.015 | 0.003 |

The results related to the critical time calculated for various $p_1/E_0^{(1)}$ under $\alpha = -0.5$ and $\omega = 1.0$ are given in Table 3.10.

It follows from Table 3.9 that although in the concrete selected case the values of the critical forces obtained within the scope of the TDLTS are greater than those obtained within the scope of the third order refined plate theory, but the relation between $t'_{cr.3D}$ and $t'_{cr.Rfn}$ depends on the values of $p_1/E_0^{(1)}$. For example, in the case under consideration we can conclude that $t'_{cr.3D} < t'_{cr.Rfn}$ as $p_1 \to p_{1cr.\infty}$ and $t'_{cr.3D} > t'_{cr.Rfn}$ as $p_1 \to p_{1cr.0}$.

We now analyze the other numerical results which are obtained in the framework of the TDLTS. Namely, consider the influence of the existence of the compressing force acting along the $Ox_3$ on the values of critical time $t'_{cr.3D}$. For this purpose we compare the results obtained under $p_3/E_0^{(1)} = 0$ with ones obtained under $p_3/E_0^{(1)} = 0.1036$. Note that under $p_3/E_0^{(1)} = 0.1036$ it is obtained that $p_{1cr.0}/E_0^{(1)} = 0.4663$ and $p_{1cr.\infty}/E_0^{(1)} = 0.2593$.

Table 3.11 shows the values of the $t'_{cr.3D}$ with various $p_1/E_0^{(1)}$ and $p_3/E_0^{(1)}$ under $\omega = 1.0$, $\alpha = -0.5$. It follows from the Table 3.11 that, as a result of the existence of the compression of the plate in the direction of the $Ox_3$ axis the values of $t'_{cr.3D}$ decrease significantly.

The influence of the rheological parameters $\omega$ and $\alpha$ on the values of $t'_{cr.3D}$ is illustrated by data given in Tables 3.12 and 3.13. It follows from the data given in these tables that for the selected values of the problem parameters the values of $t'_{cr.3D}$ increase (decrease) with $\omega$ (with $|\alpha|$).

**Table 3.12** The values of $t'_{cr.3D}$ obtained for various $\omega$, $p_1/E_0^{(1)}$ and $p_3/E_0^{(1)}$ under $\alpha = -0.5$, $\gamma = 1$

| $p_1/E_0^{(1)}$ | $p_3/E_0^{(1)}$ | $\omega$ | | |
|---|---|---|---|---|
| | | 0.5 | 1.0 | 2.0 |
| 0.400 | 0.0000 | 0.083 | 0.132 | 0.560 |
| | 0.1036 | 0.031 | 0.040 | 0.079 |
| 0.420 | 0.0000 | 0.049 | 0.069 | 0.174 |
| | 0.1036 | 0.012 | 0.015 | 0.022 |

**Table 3.13** The values of $t'_{cr.3D}$ obtained for various $\alpha$, $p_1/E_0^{(1)}$ and $p_3/E_0^{(1)}$ under $\omega = 1.0$, $\gamma = 1$

| $p_1/E_0^{(1)}$ | $p_3/E_0^{(1)}$ | $\alpha$ | | |
|---|---|---|---|---|
| | | −0.3 | −0.5 | −0.7 |
| 0.400 | 0.0000 | 0.193 | 0.132 | 0.054 |
| | 0.1036 | 0.083 | 0.040 | 0.007 |
| 0.420 | 0.0000 | 0.121 | 0.068 | 0.018 |
| | 0.1036 | 0.041 | 0.015 | 0.001 |

Note that in the all numerical results the character of the influence of the rheological parameter $\alpha$ on the values of the critical time agrees with the related explanation given in Sect. 2.5.1.

## 3.5 Rectangular Plate Clamped at all Ends

### 3.5.1 *Solution Procedure for the TDLTS Problem. 3D FEM Modeling*

In this case the values of the Laplace transformation of the zeroth approximation are determined from the following boundary value problem

$$\frac{\partial}{\partial x_j}\bar{\sigma}_{ji}^{(0)} = 0,\ \bar{\varepsilon}_{ij}^{(0)} = \frac{1}{2}\left(\frac{\partial \bar{u}_i^{(0)}}{\partial x_j} + \frac{\partial \bar{u}_j^{(0)}}{\partial x_i}\right),\ i;j = 1,\ 2,\ 3, \tag{3.101a}$$

$$\left.\bar{\sigma}_{2i}^{(0)}\right|_{x_2=\pm h} = 0, \tag{3.101b}$$

$$\bar{u}_1^{(0)}(0, x_2, x_3) = -\bar{u}_1^{(0)}(\ell_1, x_2, x_3) = \frac{1}{s}U_1,$$
$$\bar{u}_2^{(0)}(x_1, x_2, x_3)\Big|_{x_1=0,\,\ell_1} = \bar{u}_3^{(0)}(x_1, x_3, x_3)\Big|_{x_1=0,\,\ell_1} = 0,$$
$$\bar{u}_2^{(0)}(x_1, x_2, x_3)\Big|_{x_3=0,\,\ell_3} = \bar{u}_1^{(0)}(x_1, x_3, x_3)\Big|_{x_3=0,\,\ell_3} = 0,$$
$$\bar{u}_3^{(0)}(x_1, x_2, x_3)\Big|_{x_3=0} = -\bar{u}_3^{(0)}(x_1, x_3, x_3)\Big|_{x_3=\ell_3} = \frac{1}{s}U_3. \qquad (3.101c)$$

We obtain also the following boundary value problem for determination of the values of the Laplace transformation of the first approximation.

$$\frac{\partial}{\partial x_j}\left[\bar{\sigma}_{ij} + \sigma_{jn}^{(0)}\frac{\partial \bar{u}_i}{\partial x_n}\right] = 0,\ \bar{\varepsilon}_{ij} = \frac{1}{2}\left(\frac{\partial \bar{u}_i}{\partial x_j} + \frac{\partial \bar{u}_j}{\partial x_i}\right),\ i;j = 1, 2, 3 \qquad (3.102a)$$

$$\bar{\sigma}_{i2}\big|_{(x_2=\pm h)} = \frac{\partial f}{\partial x_1}\bar{\sigma}_{i1}^{(0)}\Big|_{(x_2=\pm h)} + \frac{\partial f}{\partial x_3}\bar{\sigma}_{i3}^{(0)}\Big|_{(x_2=\pm h)}, \qquad (3.102b)$$

$$\bar{u}_i\big|_{x_1=0,\,\ell_1} = 0,\ \bar{u}_i\big|_{x_3=0,\ell_3} = 0,\ i = 1, 2, 3. \qquad (3.102c)$$

For the case under consideration the initial imperfection mode of the plate is described through the function

$$f = \ell_1 \sin^2\left(\frac{\pi x_1}{\ell_1}\right)\sin^2\left(\frac{\pi x_3}{\ell_3}\right). \qquad (3.103)$$

To solve the foregoing problems we employ the 3D FEM. For this purpose we use the functional

$$\Pi^{(0)} = \frac{1}{2}\int_0^{\ell_1}\left(\int_0^{\ell_3}\left(\int_{-h}^{+h}\bar{\sigma}_{ij}^{(0)}\bar{\varepsilon}_{ij}^{(0)}dx_2\right)dx_3\right)dx_1 \qquad (3.104)$$

for the zeroth approximation and the functional

$$\begin{aligned}\Pi = &\frac{1}{2}\int_0^{\ell_3}\left(\int_0^{\ell_1}\left(\int_{-h}^{+h}\left(\bar{\sigma}_{ij} + \sigma_{in}^{(0)}\frac{\partial \bar{u}_j}{\partial x_n}\right)\frac{\partial \bar{u}_j}{\partial x_i}dx_2\right)dx_1\right)dx_3\\ &-\int_0^{\ell_3}\left(\int_0^{\ell_1}\left(\frac{\partial f}{\partial x_1}\sigma_{i1}^{(0)}\Big|_{x_2=+h} + \frac{\partial f}{\partial x_3}\sigma_{i3}^{(0)}\Big|_{x_2=+h}\right)\bar{u}_2\big|_{x_2=+h}dx_1\right)dx_3\\ &+\int_0^{\ell_3}\left(\int_0^{\ell_1}\left(\frac{\partial f}{\partial x_1}\sigma_{i1}^{(0)}\Big|_{x_2=-h} + \frac{\partial f}{\partial x_3}\sigma_{i3}^{(0)}\Big|_{x_2=-h}\right)\bar{u}_2\big|_{x_2=-h}dx_1\right)dx_3\end{aligned} \qquad (3.105)$$

for the first approximation.

Employing the standard Ritz technique, we determine the values of Laplace transformation of the zeroth and first approximations and, as before, with the use of the method (Schapery 1962, 1978) we found the values of the critical parameters from the relation (3.43). Note that under 3D FEM modeling of the foregoing problems, in view of symmetry of the problem about the lines $x_1/\ell_1 = 1/2$ and $x_3/\ell_3 = 1/2$, all numerical investigations are performed for the region $\{0 \le x_1/\ell_1 \le 1/2;\ -1 \le x_2/h \le 1; 0 \le x_3/\ell_3 \le 1/2\}$ which in turn consists of 2304 eight-node rectangular brick elements. In this case, 24, 4 and 24 elements are arranged in the directions of the $Ox_1$, $Ox_2$ and $Ox_3$ axes, respectively.

### *3.5.2 Solution Procedure for the Approximate Stability Loss Problems*

As in the previous section, the approximate stability loss Eqs. (3.63) and (3.72) are obtained from the TDLTS Eqs. (3.40) and (3.41) within the scope of the assumption (3.97). Employing the average-integrating procedure and taking the relation (3.97) into account the following ends conditions are derived from (3.102c) for the Laplace transformation of the sought unknown functions.

For the Kirchhoff-Love plate theory.

$$\bar{w}(0, x_3) = \bar{w}(\ell_1, x_3) = 0, \frac{\partial \bar{w}(0, x_3)}{\partial x_1} = \frac{\partial \bar{w}(\ell_1, x_3)}{\partial x_1} = 0,\ \bar{w}(x_1, 0) = \bar{w}(x_1, \ell_3) = 0,$$
$$\frac{\partial \bar{w}(x_1, 0)}{\partial x_3} = \frac{\partial \bar{w}(x_1, \ell_3)}{\partial x_3} = 0. \tag{3.106}$$

For the third order refined plate theory (Kromm 1953, 1955).

$$\begin{aligned}
&\bar{w}(0, x_3) = \bar{w}(\ell_1, x_3) = 0, \frac{\partial \bar{w}(0, x_3)}{\partial x_1} = \frac{\partial \bar{w}(\ell_1, x_3)}{\partial x_1} = 0,\\
&\bar{\varphi}_1(0, x_3) = \bar{\varphi}_1(\ell_1, x_3) = 0,\ \bar{\varphi}_3(0, x_3) = \bar{\varphi}_3(\ell_1, x_3) = 0,\\
&\bar{w}(x_1, 0) = \bar{w}(x_1, \ell_3) = 0, \frac{\partial \bar{w}(0, \ell_3)}{\partial x_3} = \frac{\partial \bar{w}(x_1, \ell_3)}{\partial x_3} = 0,\\
&\bar{\varphi}_1(x_1, 0) = \bar{\varphi}_1(x_1, \ell_3) = 0,\ \bar{\varphi}_3(x_1, 0) = \bar{\varphi}_3(x_1, \ell_3) = 0.
\end{aligned} \tag{3.107}$$

The solution to the problem (3.63) and (3.106) we select as

$$\bar{w} = \bar{w}_0(s) \sin^2\left(\frac{\pi x_1}{\ell_1}\right) \sin^2\left(\frac{\pi x_3}{\ell_3}\right). \tag{3.108}$$

In the similar manner, the solution to the problem (3.72) and (3.107) is presented as

**Table 3.14** The values of $\delta_{cr.0}$ (upper number) and $\delta_{cr.\infty}$ (lower number) obtained for various $E^{(2)} \big/ E_0^{(1)}$ under $\omega = 1.0$, $\gamma = 1$

| Theories | $E^{(2)} \big/ E_0^{(1)}$ | | |
|---|---|---|---|
| | 1 | 20 | 50 |
| $K-L$ | 0.04326<br>0.03904 | 0.03370<br>0.03220 | 0.03211<br>0.03145 |
| $Rfn$ | 0.03457<br>0.02962 | 0.02320<br>0.01866 | 0.01840<br>0.01486 |
| $3D$ | 0.03207<br>0.02438 | 0.02043<br>0.01381 | 0.01300<br>0.00754 |

$$\bar{w} = \bar{w}_0(s)\sin^2\left(\frac{\pi x_1}{\ell_1}\right)\sin^2\left(\frac{\pi x_3}{\ell_3}\right),\ \bar{\varphi}_1 = \bar{\varphi}_{10}(s)\sin\left(2\frac{\pi x_1}{\ell_1}\right)\sin^2\left(\frac{\pi x_3}{\ell_3}\right),$$
$$\bar{\varphi}_3 = \bar{\varphi}_{30}(s)\sin^2\left(\frac{\pi x_1}{\ell_1}\right)\sin\left(2\frac{\pi x_3}{\ell_3}\right). \tag{3.109}$$

Using the further usual procedure described in the previous sections we determine the values of the critical parameters.

### 3.5.3 Numerical Results and Discussions

Within the scope of the assumptions and notation used in Sect. 3.4.3. we analyze numerical results obtained for the square plate the all ends of which are clamped. Consider the case where $U_1 = U_3$. Remind that $U_1$ and $U_3$ enter the ends conditions (3.101c).

As a critical parameter characterizing the buckling instability, we take the quantity

$$\delta = \frac{2U_1}{\ell_1} = \frac{2U_3}{\ell_3}. \tag{3.110}$$

It is evident that the parameter $\delta$ corresponds to the axial deformation in the zeroth approximation averaged over the whole volume of the plate. For approximate theories the values of $\sigma_{11}^{(0)}$ and $\sigma_{33}^{(0)}$ which enter the Eqs. (3.63) and (3.72) are determined from the relations $\sigma_{11}^{(0)} = (A_{11}^* + A_{13}^*)\delta$ and $\sigma_{33}^{(0)} = (A_{13}^* + A_{33}^*)\delta$.

The critical values of the parameter $\delta$ at $t' = 0$ and $t' = \infty$ we denote through $\delta_{cr.0}$ and $\delta_{cr.\infty}$ respectively. Consider data given in Table 3.14 which shows the values of $\delta_{cr.0}$ (upper number) and $\delta_{cr.\infty}$ (lower number) for various $E^{(2)} \big/ E_0^{(1)}$ under $\omega = 1.0$.

Introducing the symbols $\delta_{cr.0}^{K-L}\left(\delta_{cr.\infty}^{K-L}\right)$, $\delta_{cr.0}^{Rfn}\left(\delta_{cr.\infty}^{Rfn}\right)$ and $\delta_{cr.0}^{3D}\left(\delta_{cr.\infty}^{3D}\right)$ for values of $\delta_{cr.0}(\delta_{cr.\infty})$ obtained by the Kirchhoff-Love, the third order refined and the TDLTS, according to Table 3.14, we can write

**Table 3.15** The values of $t'_{cr.3D}$ and $t'_{cr.Rfn}$ obtained for various $\alpha$ under $\omega = 1.0$, $E^{(2)} \big/ E_0^{(1)} = 20$, $\gamma = 1$

| $\alpha$ | Theories | $\delta$ | | |
|---|---|---|---|---|
| | | 0.0190 | 0.0194 | 0.0198 |
| −0.3 | *Rfn* | 9.6102 | 2.5583 | 1.1356 |
| | 3D | 0.0437 | 0.0023 | 0.0098 |
| −0.5 | *Rfn* | 31.350 | 4.9154 | 1.5766 |
| | 3D | 0.0165 | 0.0068 | 0.0020 |
| −0.7 | *Rfn* | 494.78 | 22.557 | 3.3902 |
| | 3D | 0.0017 | 0.0003 | 0.00005 |

$$\delta^{3D}_{cr.0} < \delta^{Rfn}_{cr.0} < \delta^{K-L}_{cr.0},\ \delta^{3D}_{cr.\infty} < \delta^{Rfn}_{cr.\infty} < \delta^{K-L}_{cr.\infty} \tag{3.111}$$

This concluding is in full agreement discussed in the foregoing sections.

Now, we will consider the numerical results related to the critical time. For this purpose we assume that $\delta_{cr.\infty} < \delta < \delta_{cr.0}$. It follows from Table 3.14 that in the cases where $E^{(2)} \big/ E_0^{(1)} = 1$ and 20, some values of $\delta$ satisfy the assumption $\delta_{cr.\infty} < \delta < \delta_{cr.0}$ for the TDLTS and the third order refined plate theories simultaneously, but there are no such values of $\delta$ in the case $E^{(2)} \big/ E_0^{(1)} = 50$.

Moreover, there are not such values of $\delta$ at which the inequality $\delta_{cr.\infty} < \delta < \delta_{cr.0}$ satisfies for the Kirchhoff-Love, the third order refined and the TDLTS theories simultaneously, since, according to Table 3.14, $\delta^{3D}_{cr.0} < \delta^{K-L}_{cr.\infty}$ and $\delta^{Rfn}_{cr.0} < \delta^{K-L}_{cr.\infty}$. For this reason, we can compare only the results on the critical time obtained by the TDLTS and the third order refined stability loss theories of the plate. These results are given in Table 3.15 for $E^{(2)} \big/ E_0^{(1)} = 20$ and $\omega = 1.0$.

As is seen, the critical time obtained according to the TDLTS can be significantly different from that obtained by the third order refined plate theory. Moreover, the Table 3.15 shows that for structural members made of viscoelastic materials, a small change in the external compressive force can change the critical time considerable. This phenomenon is observed very clearly in the cases where the selected values of the parameter $\delta$ characterizing the external compressive force are close to the critical values of this parameter obtained for $t' = 0$ and $t' = \infty$. It follows from Tables 3.14 and 3.15 that the values of the parameter $\delta$ given in Table 3.15 are very close to the $\delta^{3D}_{cr.0}$ and to the $\delta^{Rfn}_{cr.\infty}$, simultaneously. This is why such a great difference arises between the values of $t'_{cr.3D}$ and $t'_{cr.Rfn}$. Consequently, the third order refined plate theory cannot quantitatively describe the stability loss of the considered type members elements of construction made of viscoelastic composite material. This situation once again confirms the importance of the employing the TDLTS for the problems under considerations.

**Table 3.16** The values of $t'_{cr.3D}$ obtained for various $\alpha$ and $E^{(2)}/E_0^{(1)}$ under $\omega = 1.0$, $\gamma = 1$

| $\frac{E^{(2)}}{E_0^{(1)}}$ | $\delta$ | $\alpha$ | | |
|---|---|---|---|---|
| | | −0.3 | −0.5 | −0.7 |
| 20 | 0.0155 | 1.2389 | 1.7810 | 4.1542 |
| | 0.0172 | 0.2449 | 0.1841 | 0.0946 |
| | 0.0190 | 0.0437 | 0.0165 | 0.0017 |
| 50 | 0.0082 | 4.8634 | 12.081 | 39.143 |
| | 0.0099 | 0.3007 | 0.2546 | 0.1955 |
| | 0.0121 | 0.0217 | 0.0060 | 0.0003 |

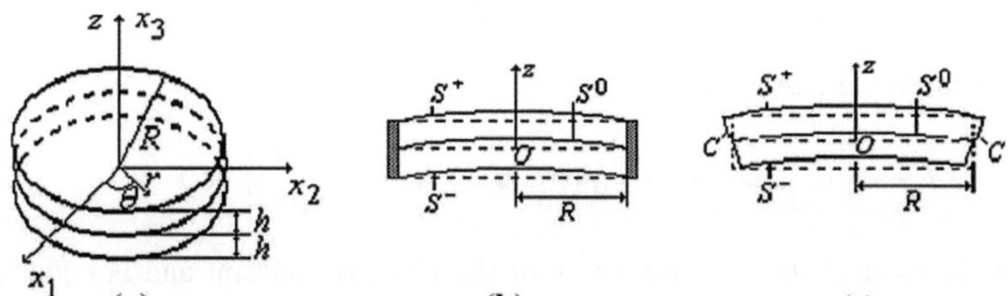


**Fig. 3.4** The geometry of the circular plate (**a**) cross section of the clamped (**b**) and of the simply-supported (**c**) plate under $\theta = \text{const}$

Table 3.16 shows the values of $t'_{cr.3D}$ which are calculated for various values of the rheological parameter $\alpha$ with various $\delta$ and $E^{(2)}/E_0^{(1)}$. It follows from these data that, in the case where $t'_{cr.3D} > 0.5$ $(t'_{cr.3D} < 0.5)$ satisfies the values of $t'_{cr.3D}$ increase (decrease) monotonically as the absolute values of $\alpha$ increase. This result agrees with those discussed in Sect. 2.5.1.

## 3.6 Symmetric Stability Loss of the Circular Plate

### *3.6.1 Formulation of the Problem and Governing Field Equations*

Consider a circular plate which has an insignificant initial imperfection. We associate the Lagrangian cylindrical coordinate system $Or\theta z$ with the middle surface of this plate. We suppose that this plate occupies the region $\Omega$, which is restricted by the surface $S^+$ and $S^-$ under $0 \le r \le R$, $0 \le \theta \le 2\pi$. The middle surface $S_0$ (Fig. 3.4) of the plate is given by the equation

$$z = F(r) = \varepsilon f(r), \tag{3.112}$$

where $\varepsilon$ is a small parameter $(0 \le \varepsilon \ll 1)$ and characterizes the degree of the initial imperfection. For $\varepsilon = 0$ the plate occupies the region $\{0 \le r \le R, 0 \le \theta \le 2\pi, -h \le z \le +h\}$. We also suppose that the function $f(r)$ in (3.112) satisfies the following conditions:

For clamped plate:

$$f(r)|_{r=R} = 0, \quad \left.\frac{df(r)}{dr}\right|_{r=R} = 0 \tag{3.113}$$

For simply supported plate:

$$f(r)|_{r=R} = 0, \quad \left.\frac{d^2 f(r)}{dr^2}\right|_{r=R} = 0. \tag{3.114}$$

Besides these we assume that

$$f(r) > 0, \quad \max_{0 \le r < R} f(r) = f(0), \quad \varepsilon f(r) \ll 1, \quad \varepsilon\left|\frac{df(r)}{dr}\right| \ll 1 \tag{3.115}$$

and in the natural state the thickness of the plate is constant and is equal to $2h$. Using Eq. (3.112) and the constancy condition of the plate thickness we derive the following equation for surfaces $S^{\pm}$ (Fig. 3.4).

$$r^{\pm}(s_1) = s_1 \mp \frac{h}{L(s_1)} \frac{\varepsilon df(s_1)}{ds_1}, \quad z^{\pm}(s_1) = \varepsilon f(s_1) \mp \frac{h}{L(s_1)},$$
$$L(s_1) = \left[1 + \varepsilon^2 \left(\frac{df(s_1)}{ds_1}\right)^2\right]^{\frac{1}{2}}. \tag{3.116}$$

In (3.116) $r^{\pm}(s_1)$ and $z^{\pm}(s_1)$ are the cylindrical coordinates of the points of the surface $S^{\pm}$, $s_1$ is a parameter $(0 \le s_1 \le R)$. Employing the well-known procedure the following expression are also derived for the ortho-normal components of the unit-normal vector to the surfaces $S^{\pm}$.

$$n_r^{\pm} = \pm \frac{-\frac{\partial z^{\pm}(s_1)}{\partial s_1}}{\sqrt{\left(\frac{\partial z^{\pm}(s_1)}{\partial s_1}\right)^2 + \left(\frac{\partial r^{\pm}(s_1)}{\partial s_1}\right)^2}}, \quad n_\theta^{\pm} = 0,$$
$$n_z^{\pm} = \pm \frac{\frac{\partial r^{\pm}(s_1)}{\partial s_1}}{\sqrt{\left(\frac{\partial z^{\pm}(s_1)}{\partial s_1}\right)^2 + \left(\frac{\partial r^{\pm}(s_1)}{\partial s_1}\right)^2}}. \tag{3.117}$$

In the simply supported case, the lateral cylindrical boundary of the plate has also initial imperfection and the equation of this imperfection-surface (denoted by $C$) can be written as follows.

$$r^C = R + \frac{z}{L(s_1)} \frac{\varepsilon df(s_1)}{ds_1}\bigg|_{s_1=R}, \quad -h \leq z \leq +h. \tag{3.118}$$

From (3.118) we obtain the expressions

$$n_r^C = \frac{1}{\sqrt{1+\varepsilon^2\left(\frac{df(s_1)}{L(s_1)ds_1}\Big|_{s_1=R}\right)}}, \quad n_\theta^C = 0,$$
$$n_z^C = \frac{-\varepsilon\frac{df(s_1)}{L(s_1)ds_1}\Big|_{s_1=R}}{\sqrt{1+\varepsilon^2\left(\frac{df(s_1)}{L(s_1)ds_1}\Big|_{s_1=R}\right)}}. \tag{3.119}$$

where $n_r^C$, $n_\theta^C$ and $n_z^C$ are the components of the unit-normal vector to the lateral boundary surface of the plate.

Thus, within the foregoing assumption we investigate the evolution of the initial infinitesimal symmetric imperfection of the plate under compressing inward of that in radial direction. As in the previous sections, we make this investigation within the scope of the geometrically non-linear equations of the theory viscoelasticity for anisotropic materials. For this purpose, we write the corresponding field equations for rotationally symmetric state.

$$\frac{\partial t_{rr}}{\partial r} + \frac{\partial t_{zr}}{\partial z} + \frac{1}{r}(t_{rr} - t_{\theta\theta}) = 0, \quad \frac{\partial t_{rz}}{\partial r} + \frac{\partial t_{zz}}{\partial z} + \frac{1}{r}t_{rz} = 0, \tag{3.120a}$$

$$\sigma_{rr} = A_{11}^*\varepsilon_{rr} + A_{12}^*\varepsilon_{\theta\theta} + A_{13}^*\varepsilon_{zz}, \quad \sigma_{\theta\theta} = A_{12}^*\varepsilon_{rr} + A_{22}^*\varepsilon_{\theta\theta} + A_{23}^*\varepsilon_{zz},$$
$$\sigma_{zz} = A_{13}^*\varepsilon_{rr} + A_{23}^*\varepsilon_{\theta\theta} + A_{33}^*\varepsilon_{zz}, \quad \sigma_{rz} = 2A_{55}^*\varepsilon_{rz} \tag{3.120b}$$

$$\varepsilon_{rr} = \frac{\partial u_r}{\partial r} + \frac{1}{2}\left(\frac{\partial u_r}{\partial r}\right)^2 + \frac{1}{2}\left(\frac{\partial u_r}{\partial z}\right)^2, \quad \varepsilon_{\theta\theta} = \frac{u_r}{r} + \frac{1}{2}\left(\frac{u_r}{r}\right)^2,$$
$$\varepsilon_{rz} = \frac{1}{2}\left(\frac{\partial u_z}{\partial r} + \frac{\partial u_r}{\partial z} + \frac{\partial u_r}{\partial r}\frac{\partial u_r}{\partial z} + \frac{\partial u_z}{\partial r}\frac{\partial u_z}{\partial z}\right),$$
$$\varepsilon_{zz} = \frac{\partial u_z}{\partial z} + \frac{1}{2}\left(\frac{\partial u_z}{\partial z}\right)^2 + \frac{1}{2}\left(\frac{\partial u_r}{\partial z}\right)^2, \tag{3.120c}$$

where

$$A_{ij}^*\varphi = A_{ij0}\varphi(t) + \int_0^t A_{ij1}(t-\tau)\varphi(\tau)d\tau, \quad t_{rr} = \sigma_{rr}\left(1 + \frac{\partial u_r}{\partial r}\right) + \sigma_{rz}\frac{\partial u_r}{\partial z},$$

$$t_{\theta\theta} = \sigma_{\theta\theta}\left(1 + \frac{u_r}{r}\right), \quad t_{rz} = \sigma_{rz}\left(1 + \frac{\partial u_z}{\partial z}\right) + \sigma_{rr}\frac{\partial u_z}{\partial r},$$

$$t_{zr} = \sigma_{zr}\left(1 + \frac{\partial u_r}{\partial r}\right) + \sigma_{zz}\frac{\partial u_r}{\partial z}, \quad t_{zz} = \sigma_{zz}\left(1 + \frac{\partial u_z}{\partial z}\right) + \sigma_{zr}\frac{\partial u_z}{\partial r}. \tag{3.121}$$

The meaning of the notation used in (3.120a–c) and (3.121) is explained in Chap. 2 (Sect. 2.1).

The boundary conditions on the upper and lower surfaces of the plate are

$$n_r^{\pm} t_{rr}\big|_{S^{\pm}} + n_z^{\pm} t_{zr}\big|_{S^{\pm}} = 0,\; n_r^{\pm} t_{rz}\big|_{S^{\pm}} + n_z^{\pm} t_{zz}\big|_{S^{\pm}} = 0. \tag{3.122}$$

We will consider the following three types of conditions on the lateral cylindrical boundary of the plate.

The first type of conditions are taken as in the investigation (Tang 1967 and Babich and Guz 1980), i.e., as

$$\left.\frac{\partial u_z}{\partial r}\right|_{r=R} = 0,\; u_r\big|_{r=R} = -U. \tag{3.123}$$

For the second type conditions, it is assumed that this plate is compressed circumferentially and inwards through a clamp in the boundary $r = R$ by radial normal forces of intensity $p$. In this case it is supposed that

$$u_r\big|_{r=R} = -U,\; u_z\big|_{r=R} = 0. \tag{3.124}$$

The relation between $p$ and $U$ is determined after the solution procedure from the corresponding formulae.

The third type boundary conditions are related to the simply supported plate and can be written as follows

$$u_r\big|_{r=r^C} = 0,\; \left(n_r^C t_{rr} + n_z^C t_{zr}\right)\big|_{r=r^C} = -p. \tag{3.125}$$

Thus the study of the loss of stability of the plate is reduced to the investigations of the evolution of its initial infinitesimal imperfection in the framework of equations (3.120a–c) and (3.121) and boundary conditions (3.122)–(3.125). In this case, as in the previous sections, as a critical time we take the time when this initial infinitesimal imperfection starts to increase and grows indefinitely.

### *3.6.2 Method of Solution*

Using the series presentation (2.14) with respect to the small parameter ε and doing mathematical manipulations similar that made in Chap. 2, we obtain the following boundary-value problems for determination of the zeroth and the first approximations.

The zeroth approximation:

$$\frac{\partial \sigma_{rr}^{(0)}}{\partial r} + \frac{\partial \sigma_{zr}^{(0)}}{\partial z} + \frac{1}{r}\left(\sigma_{rr}^{(0)} - \sigma_{\theta\theta}^{(0)}\right) = 0,\; \frac{\partial \sigma_{rz}^{(0)}}{\partial r} + \frac{\partial \sigma_{zz}^{(0)}}{\partial z} + \frac{1}{r}\sigma_{rz}^{(0)} = 0, \tag{3.126a}$$

$$\sigma_{rr}^{(0)} = A_{11}^{*}\varepsilon_{rr}^{(0)} + A_{12}^{*}\varepsilon_{\theta\theta}^{(0)} + A_{13}^{*}\varepsilon_{zz}^{(0)},\ \sigma_{\theta\theta}^{(0)} = A_{12}^{*}\varepsilon_{rr}^{(0)} + A_{22}^{*}\varepsilon_{\theta\theta}^{(0)} + A_{23}^{*}\varepsilon_{zz}^{(0)}.$$
$$\sigma_{zz}^{(0)} = A_{13}^{*}\varepsilon_{rr}^{(0)} + A_{23}^{*}\varepsilon_{\theta\theta}^{(0)} + A_{33}^{*}\varepsilon_{zz}^{(0)},\ \sigma_{rz}^{(0)} = 2A_{55}^{*}\varepsilon_{rz}^{(0)}, \tag{3.126b}$$

$$\varepsilon_{rr}^{(0)} = \frac{\partial u_r^{(0)}}{\partial r},\ \varepsilon_{\theta\theta}^{(0)} = \frac{u_r^{(0)}}{r},\ \varepsilon_{rz}^{(0)} = \frac{1}{2}\left(\frac{\partial u_z^{(0)}}{\partial r} + \frac{\partial u_r^{(0)}}{\partial z}\right),\ \varepsilon_{zz}^{(0)} = \frac{\partial u_z^{(0)}}{\partial r}, \tag{3.126c}$$

$$\sigma_{rz}^{(0)}\big|_{z=\pm h} = 0,\ \sigma_{zz}^{(0)}\big|_{z=\pm h} = 0, \tag{3.127}$$

$$\left.\frac{\partial u_z^{(0)}}{\partial r}\right|_{r=R} = 0,\ u_r^{(0)}\big|_{r=R} = -U, \tag{3.128}$$

$$u_r^{(0)}\big|_{r=R} = -U,\ u_z^{(0)}\big|_{r=R} = 0, \tag{3.129}$$

$$u_z^{(0)}\big|_{r=R} = 0,\ \sigma_{rr}^{(0)}\big|_{r=R} = -p. \tag{3.130}$$

The first approximation

$$\frac{\partial t_{rr}^{(1)}}{\partial r} + \frac{\partial t_{zr}^{(1)}}{\partial z} + \frac{1}{r}\left(t_{rr}^{(1)} - t_{\theta\theta}^{(1)}\right) = 0,\ \frac{\partial t_{rz}^{(1)}}{\partial r} + \frac{\partial t_{zz}^{(1)}}{\partial z} + \frac{1}{r}t_{rz}^{(1)} = 0, \tag{3.131a}$$

$$\sigma_{rr}^{(1)} = A_{11}^{*}\varepsilon_{rr}^{(1)} + A_{12}^{*}\varepsilon_{\theta\theta}^{(1)} + A_{13}^{*}\varepsilon_{zz}^{(1)},\ \sigma_{\theta\theta}^{(1)} = A_{12}^{*}\varepsilon_{rr}^{(1)} + A_{22}^{*}\varepsilon_{\theta\theta}^{(1)} + A_{23}^{*}\varepsilon_{zz}^{(1)},$$
$$\sigma_{zz}^{(1)} = A_{13}^{*}\varepsilon_{rr}^{(1)} + A_{23}^{*}\varepsilon_{\theta\theta}^{(1)} + A_{33}^{*}\varepsilon_{zz}^{(1)},\ \sigma_{rz}^{(1)} = 2A_{55}^{*}\varepsilon_{rz}^{(1)}, \tag{3.131b}$$

$$\varepsilon_{rr}^{(1)} = \frac{\partial u_r^{(1)}}{\partial r},\ \varepsilon_{\theta\theta}^{(1)} = \frac{u_r^{(1)}}{r},\ \varepsilon_{rz}^{(1)} = \frac{1}{2}\left(\frac{\partial u_z^{(1)}}{\partial r} + \frac{\partial u_r^{(1)}}{\partial z}\right),$$
$$\varepsilon_{zz}^{(1)} = \frac{\partial u_z^{(1)}}{\partial r}, \tag{3.131c}$$

$$t_{zr}^{(1)}\big|_{z=\pm h} = \frac{df(s_1)}{\partial s_1}\sigma_{rr}^{(0)}\big|_{z=\pm h},\ t_{zz}^{(1)}\big|_{z=\pm h} = \frac{df(s_1)}{\partial s_1}\sigma_{zr}^{(0)}\big|_{z=\pm h}, \tag{3.132}$$

$$\left.\frac{\partial u_z^{(1)}}{\partial r}\right|_{r=R} = 0,\ u_r^{(1)}\big|_{r=R} = 0, \tag{3.133}$$

$$u_r^{(1)}\big|_{r=R} = 0,\ u_z^{(1)}\big|_{r=R} = 0, \tag{3.134}$$

$$u_z^{(1)}\big|_{r=R} = 0,\ t_{rr}^{(1)}\big|_{r=R} = \frac{df(s_1)}{ds_1}2\sigma_{rz}^{(0)}\big|_{r=R}, \tag{3.135}$$

where

$$t_{rr}^{(1)} = \sigma_{rr}^{(1)} + \sigma_{rr}^{(0)} \frac{\partial u_r^{(1)}}{\partial r} + \sigma_{rz}^{(0)} \frac{\partial u_r^{(1)}}{\partial z}, \ t_{zz}^{(1)} = \sigma_{zz}^{(1)} + \sigma_{zz}^{(0)} \frac{\partial u_z^{(1)}}{\partial z} + \sigma_{rz}^{(0)} \frac{\partial u_z^{(1)}}{\partial r},$$

$$t_{\theta\theta}^{(1)} = \sigma_{\theta\theta}^{(1)} + \sigma_{\theta\theta}^{(0)} \frac{u_r^{(1)}}{r}, \ t_{zr}^{(1)} = \sigma_{zr}^{(1)} + \sigma_{zr}^{(0)} \frac{\partial u_r^{(1)}}{\partial r} + \sigma_{zz}^{(0)} \frac{\partial u_r^{(1)}}{\partial z},$$

$$t_{rz}^{(1)} = \sigma_{rz}^{(1)} + \sigma_{rr}^{(0)} \frac{\partial u_z^{(1)}}{\partial r} + \sigma_{rz}^{(0)} \frac{\partial u_z^{(1)}}{\partial z}. \tag{3.136}$$

Here the conditions (3.132)–(3.135) are obtained from (3.122)–(3.125) respectively.

Note that the conditions (3.129) and (3.130) for the zeroth approximation and the conditions (3.134) and (3.135) for the first approximation correspond to the clamped and simply-supported plates, respectively. But the condition (3.128) for the zeroth approximation and the condition (3.133) for the first approximation correspond to the clamped plate when no shear forces occur on the cylindrical boundary $r = R$. The condition (3.133) is identical to that in the investigations of (Tang 1967 and Babich and Guz 1980).

#### 3.6.2.1 Approximate Loss of Stability Equations of a Circular Plate with Initial Imperfection

Consider now the approximate loss of stability equations which can be derived from Eqs. (3.131a–c)–(3.136) in the framework of Kirchhoff-Love hypothesis as well as in the framework of the third order refined plate theory (Kromm 1953, 1955). The displacement, of the middle plane of the plate are denoted as $u(r)$ and $w(r)$, i.e.,

$$u(r) = u_r^{(1)}(r, 0), \ w(r) = u_z^{(1)}(r, 0). \tag{3.137}$$

Assume that the inequality $|u(r)| \ll |w(r)|$ is satisfied and $u(r)$ can be neglected.

Thus, according to Kirchhoff-Love hypothesis we can write the following relations:

$$u_r^{(1)}(r) = -z\frac{dw(r)}{dr}, \ u_z^{(1)}(r,z) = w(r). \tag{3.138}$$

Moreover, according to the third order refined plate theory of (Kromm 1953, 1955) we can write the following relations:

$$u_r^{(1)}(r) = -z\frac{dw(r)}{dr} + \frac{1}{4}\left(h^2 z - \frac{z^3}{3}\right)\varphi(r), \ u_z^{(1)}(r,z) = w(r). \tag{3.139}$$

In (3.139) $\varphi(r)$ characterizes the average value of the shear strain in the plate cross section perpendicular to the radius.

Note that within the scope of the approximate theories (3.138) and (3.139), we assume that $\sigma_{rr}^{(0)} = \sigma_{\theta\theta}^{(0)} = -p$, $\sigma_{rz}^{(0)} = \sigma_{zz}^{(0)} = 0$. Taking this into account and integrating the Eqs. (3.131)–(3.136) and their product by $z$, with respect to $z$ from $-h$ to $+h$, after some operations similar to that made in the previous sections, we obtain the following approximate equations and boundary conditions of the stability loss of the circular plate in the framework of the Kirchhoff-Love hypothesis.

$$\frac{2}{3}h^3(A_{11}^* - p)\Delta\Delta w + 2ph\Delta w = 2ph\Delta f, \; \Delta = \frac{d^2}{dr^2} + \frac{d}{rdr}, \tag{3.140}$$

$$w|_{r=R} = 0, \; \left.\frac{dw}{dr}\right|_{r=R} = 0 \; (\text{for clamped plate}) \tag{3.141}$$

$$w|_{r=R} = 0, \; \left.\frac{d^2w}{dr^2}\right|_{r=R} = 0 \; (\text{for simply supported plate}) \tag{3.142}$$

Taking (3.139) into account and performing analogous operations, we obtain the following stability loss equation.

$$\left(\frac{2}{3}h^3 - \frac{4ph^3}{A_{55}^*}\right)(A_{11}^* - p)\Delta\Delta w + 2ph\Delta w = 2ph\Delta f, \tag{3.143}$$

of the third order refined plate theory (Kromm 1953, 1955) and the end conditions (3.141) and (3.142). Note that in the framework of the approximate plate theories the boundary conditions (3.133) and (3.134) coincide and written as (3.141)

#### 3.6.2.2 Solution Procedure for the TDLTS Problems

We consider separately the zeroth and the first approximations.

*The Zeroth Approximation*

The values of this approximation are determined by using the equations in (3.126a–c). According to the problem statement, the solution to these equations for boundary conditions (3.128) and (3.130) can be written as

$$\sigma_{rr}^{(0)} = \sigma_{\theta\theta}^{(0)} = -p, \; \sigma_{rz}^{(0)} = \sigma_{zz}^{(0)} = 0. \tag{3.144}$$

For the boundary condition (3.128) the relation between $p$ and $U$ is determined from the following equation:

$$p = A_{11}^* \frac{U}{R} + A_{13}^* \frac{U}{R}. \tag{3.145}$$

For $p = \text{const}$, $U$ must depend on $t$ (time) and the function $U = U(t)$ can be determined from (3.145). However for the boundary condition (3.129) equations in (3.126a–c) do not have an analytical solution and therefore for this case, we employ the FEM. Note that before employing FEM we apply the Laplace transformation

$$\bar{\varphi}(s) = \int_{0}^{\infty} \varphi(t) e^{-st} dt, \qquad (3.146)$$

with the parameter $s > 0$ to these equations and to the boundary condition (3.129). According to the principle of correspondence, we obtain Eqs. (3.126) and (3.129) for the Laplace transformation of the sought values. The last problem is solved by employing FEM via the principle of virtual work and the original of Laplace transformation is determined with the use of the method (Schapery 1962, 1978). All the algorithms and programs for calculation procedure, in the framework of which all numerical investigations are carried out, have been composed for the case where the region $\{0 \leq r/R \leq 1; -1 \leq z/h \leq +1\}$ is divided into 120 rectangular Lagrange family quadratic elements, with 533 nodes and 1014 NDOF. The traditional Ritz technique is employed for the determination of the Laplace transformation of the sought values.

The numerous numerical results obtained in this way show that in the case where

$$2hp = \int_{-h}^{+h} \sigma_{rr}^{(0)} dz = \text{const}, \qquad (3.147)$$

the stresses $\sigma_{rr}^{(0)}$, $\sigma_{\theta\theta}^{(0)}$, $\sigma_{rz}^{(0)}$ and $\sigma_{zz}^{(0)}$ determined after the solution to the problem (3.126a–c), (3.129) can be taken as independent of the time $t$ with very high accuracy. It should be noted that to satisfy the condition (3.147) the quantities of $U$, which enter (3.129), must be taken as a function of time. This function is determined numerically under calculation procedure from the constitutive (strain-stress) relations.

*The First Approximation*

Taking the above-stated results into account we employ the principle of the correspondence by using the Laplace transformation (3.146) for the determination of the first approximation. So, replacing in (3.131a–c)–(3.135) $\sigma_{(ij)}^{(1)}$, $\varepsilon_{(ij)}^{(1)}$, $u_{(i)}^{(1)}$ and $A_{ij}^{*}$ with $\bar{\sigma}_{(ij)}^{(1)}$, $\bar{\varepsilon}_{(ij)}^{(1)}$, $\bar{u}_{(i)}^{(1)}$ and $\bar{A}_{ij}^{*}$ (where $(ij) = rr, \theta\theta, zz, r\theta, zr, z\theta$, $(i) = r, \theta, z$) respectively, we obtain the corresponding equations and boundary conditions with respect to the Laplace transformations. For the FEM modeling of this problem we introduce the functional

$$\bar{\Pi}\left(\bar{u}_r^{(1)},\bar{u}_z^{(1)}\right)=\frac{1}{2}\int_0^R\int_{-h}^{+h}\left\{t_{rr}^{(1)}\frac{\partial\bar{u}_r^{(1)}}{\partial r}+t_{\theta\theta}^{(1)}\frac{\bar{u}_r^{(1)}}{r}+t_{rz}^{(1)}\frac{\partial\bar{u}_z^{(1)}}{\partial r}+t_{zr}^{(1)}\frac{\partial\bar{u}_r^{(1)}}{\partial z}\right.$$
$$\left.+t_{zz}^{(1)}\frac{\partial\bar{u}_z^{(1)}}{\partial z}\right\}rdrdz-\left[\sigma_{rr}^{(0)}\bar{u}_r^{(1)}+\sigma_{rz}^{(0)}\bar{u}_z^{(1)}\right]\Big|_{z=+h}rdr$$
$$-\int_0^R\frac{df(r)}{dr}\int_0^R\frac{df(r)}{dr}\left[\sigma_{rr}^{(0)}\bar{u}_r^{(1)}+\sigma_{rz}^{(0)}\bar{u}_z^{(1)}\right]\Big|_{z=-h}rdr. \quad (3.148)$$

If we consider the simply supported plate, i.e., in the case where the condition (3.135) satisfies, we add the term

$$-2\int_{-h}^{+h}\left(\frac{df}{dr}\sigma_{rz}^{(0)}\bar{u}_r^{(1)}\right)\Big|_{r=R}dz. \quad (3.149)$$

to the functional (3.148).

From the relation

$$\delta\bar{\Pi}\left(\bar{u}_r^{(1)},\bar{u}_z^{(1)}\right)=0 \quad (3.150)$$

we obtain Eqs. (3.131a), (3.132) and (3.135) rewritten for $\bar{\sigma}_{(ij)}^{(1)}$, $\bar{\varepsilon}_{(ij)}^{(1)}$, $\bar{u}_{(i)}^{(1)}$ and $\bar{A}_{ij}^{*}$. In this way we prove that these equations are the Euler equations of the functional (3.148) and (3.149).

Thus, applying the well-known techniques of the FEM we obtain the corresponding linear algebraic equations for the nodal values of the Laplace transformation of the displacements. Under calculation procedure, according to the method (Schapery 1962, 1978), we assign the corresponding values to the parameters. Based on the results obtained for these values of $s$ and applying the rule (3.80) we determine (numerically) the original values of the sought quantities.

Under numerical investigations the initial imperfection of the plate is selected as

$$z=L\left(1-\left(\frac{r}{R}\right)^2\right)^3=\frac{L}{R}R\left(1-\left(\frac{r}{R}\right)^2\right)^3=\varepsilon R\left(1-\left(\frac{r}{R}\right)^2\right)^3,$$
$$f(r)=R\left(1-\left(\frac{r}{R}\right)^2\right)^3,\ L\ll R,\ \varepsilon=\frac{L}{R}. \quad (3.151)$$

In (3.151) $L$ is the maximum value of the lift of the initial imperfection. Note that the numerous numerical investigations show that the values of the critical parameters do not depend on the initial imperfection form, which satisfies the conditions in (3.115).

As in previous sections, all numerical investigation related to the stability loss of the circular plate are made taking only the zeroth and the first approximations of (2.14), because the subsequent approximations do not change the values of the critical parameters. This statement can be explained by the following consideration. As has been noted in Sect. 2.2 that the homogeneous parts of the equations obtained for the second and subsequent approximations coincide with those obtained for the first order approximation. Consequently, the stiffness matrix **K** constructed for the first and subsequent approximations is the same. According to the selected stability loss criterion, the values of the critical parameters correspond to the case for which $\det\|\mathbf{K}\| = 0$. It is evident that the right hand side of the equations of the second and subsequent order approximations cannot change the values of the critical force or the critical time. Therefore taking the second and the subsequent order approximations into account does not change the values of the critical parameters, which are determined from the criterion

$$\max_{\substack{z\in[-h,+h]\\ r\in[0,R]}} \left|u_z^{(1)}(r,z,t\right| \to \infty \text{ as } t \to t_{cr.} \text{ (for the viscoelastic plate)} \quad \text{or as } p \to p_{cr.} \text{ (for the pure elastic plate)} \tag{3.152}$$

### 3.6.2.3 Solution Procedure for the Approximate Equations of Stability

After applying the Laplace transformation (3.146) to Eqs. (3.140)–(3.143) according to the principle of correspondence, we obtain

$$\Delta\Delta\bar{w} + a(s)\Delta\bar{w} = b(s)\Delta f, \tag{3.153}$$

where the expression of $a(s)$ and $b(s)$ can easily be determined from (3.140) and (3.143) for various theories. Thus from (3.153) we can write

$$\frac{d^2\bar{w}}{d\left(r\sqrt{a(s)}\right)^2} + \frac{1}{r\sqrt{a(s)}}\frac{d\bar{w}}{d\left(r\sqrt{a(s)}\right)} + \bar{w} = b(s)f(r) + c_1(s) + c_2(s)r^{-2}. \tag{3.154}$$

According the condition $|\bar{w}(0,\ s)| < \infty$, we must take $c_2 = 0$. Thus, the solution to the Eq. (3.154) is found as follows.

$$\bar{w}(rs) = D_1(s)J_0\left(r\sqrt{a(s)}\right) + a_6(s)r^6 + a_4(s)r^4 + D_2(s)r^2. \tag{3.155}$$

In (3.155) $J_0(x)$ is the Bessel function of the zeroth order, $a_6(s)$, $a_4(s)$ and $D_2(s)$ are determined via $b(s)$ and $c_1(s)$, but the unknown constants $D_1(s)$ and $c_1(s)$ are determined from the boundary conditions (3.141) (for the clamped plate) or (3.142) (for the simply-supported plate). As before, the original of $\bar{w}\,(rs)$ is determined by the use of the expression (3.80). From the criterion

$$|w(0)| \to \infty \text{ as } t \to t_{cr.} \text{ (for the viscoelastic plate)}$$
$$\text{or as } p \to p_{cr.} \text{ (for the pure elastic plate)} \tag{3.156}$$

the values of the critical parameters are determined.

### 3.6.3 Numerical Results and Discussions

As in Sect. 3.3.3, we assume that the plate is fabricated from the multilayered composite material consisting of the two alternating isotropic layers, and these layers lie in the planes $z = const$. The material of the reinforcing layers is supposed to be elastic with mechanical characteristics $E^{(2)}$ (Young's modulus), $\nu^{(2)}$ (Poisson's coefficient), $\lambda^{(2)}$ and $\mu^{(2)}$ (Lame's constants). The material of the matrix layers is supposed to be linearly viscoelastic with operators

$$\begin{aligned}
E^{*(1)}\varphi(t) &= E_0^{(1)}\left(\varphi(t) - \omega_0 \Pi_\alpha^*(-\omega_0 - \omega_\infty)\varphi\right),\\
\nu^{*(1)}\varphi(t) &= \nu_0^{(1)}\left(\varphi(t) + \frac{1-2\nu_0^{(1)}}{2\nu_0^{(1)}}\omega_0 \Pi_\alpha^*(-\omega_0 - \omega_\infty)\varphi\right),\\
\lambda^{*(1)}\varphi(t) &= \lambda_0^{(1)}\left(\varphi(t) + \frac{\omega_0}{\left(1+\nu_0^{(1)}\right)}\Pi_\alpha^*\left(-\frac{3}{2\left(1+\nu_0^{(1)}\right)}\omega_0 - \omega_\infty\right)\varphi\right),\\
\mu^{*(1)}\varphi(t) &= \mu_0^{(1)}\left(\varphi(t) - \frac{3\omega_0}{2\left(1+\nu_0^{(1)}\right)}\Pi_\alpha^*\left(-\frac{3}{2\left(1+\nu_0^{(1)}\right)}\omega_0 - \omega_\infty\right)\varphi\right),
\end{aligned} \tag{3.157}$$

where $E_0^{(1)}$ and $\nu_0^{(1)}$ are the instantaneous values of the Young's modulus and of the Poisson coefficient respectively, $\lambda_0^{(1)}$ and $\mu_0^{(1)}$ are the instantaneous values of the Lame's constants; $\alpha$, $\omega_0$ and $\omega_\infty$ are the rheological parameters of the matrix material, $\Pi_\alpha^*$ is the fractional-exponential operator (Rabotnov 1977) determined as

$$\Pi_\alpha^*(x)\varphi = \int_0^t \Pi_\alpha(x, t-\tau)\varphi(\tau)d\tau, \tag{3.158}$$

where

$$\Pi_\alpha(x,t) = t^\alpha \sum_{n=0}^{\infty} \frac{x^n t^{n(1+\alpha)}}{\Gamma((1+n)(1+\alpha))}, \quad -1<\alpha<0. \tag{3.159}$$

In (3.159) $\Gamma(x)$ is the Gamma function.

As in Sect. 3.3.3, we introduce the dimensionless rheological parameter $\omega(=\omega_\infty/\omega_0)$ and dimensionless time $t'\left(=t\omega_0^{1/(1+\alpha)}\right)$ and assume that $\nu^{(2)}=\nu_0^{(1)}=0.3$ and $\eta^{(2)}=0.5$ where $\eta^{(2)}$ is a volumetric fraction of the filler layers.

It is known that in the continuum approach this layered composite material is taken as viscoelastic transversal-isotropic material with effective mechanical properties (operators) whose isotropy axis lies on the $Oz$ axis. According to (Chiristensen 1979), by replacing the mechanical constants of components of a composite with the Laplace transformation of corresponding operators in the expressions of the normalized mechanical properties, we determine the Laplace transform of the effective operators. Therefore, while applying the Laplace transformation (3.147) to the constitutive relations (3.128), instead of $\bar{A}^*_{ij}$ we write the mentioned expressions. For the composite material under consideration the expressions of $\bar{A}^*_{ij}$ are determined as follows.

$$
\begin{aligned}
\bar{A}^*_{33} =& \bar{\mu}^{*(1)}\eta^{(1)} + \bar{\mu}^{(2)}\eta^{(2)} + (\bar{\lambda}^{*(1)} + \bar{\mu}^{*(1)})\eta^{(1)} + (\bar{\lambda}^{(2)} + \bar{\mu}^{(2)})\eta^{(2)} \\
& - \eta^{(2)}\eta^{(1)} \frac{\left(\bar{\lambda}^{*(1)} - \bar{\lambda}^{(2)}\right)^2}{\left(\bar{\lambda}^{*(1)} + 2\bar{\mu}^{*(1)}\right)\eta^{(2)} + \left(\bar{\lambda}^{(2)} + 2\bar{\mu}^{(2)}\right)\eta^{(1)}}, \\
\bar{A}^*_{11} =& \bar{A}^*_{22} = (\bar{\lambda}^{*(1)} + \bar{\mu}^{*(1)})\eta^{(1)} + (\bar{\lambda}^{(2)} + \bar{\mu}^{(2)})\eta^{(2)} \\
& - \eta^{(2)}\eta^{(1)} \frac{\left(\left(\bar{\lambda}^{*(1)} + 2\bar{\mu}^{*(1)}\right) - \left(\bar{\lambda}^{(2)} + 2\bar{\mu}^{(2)}\right)\right)^2}{\left(\bar{\lambda}^{*(1)} + 2\bar{\mu}^{*(1)}\right)\eta^{(2)} + \left(\bar{\lambda}^{(2)} + 2\bar{\mu}^{(2)}\right)\eta^{(1)}}, \\
\bar{A}^*_{12} =& \bar{\lambda}^{*(1)}\eta^{(1)} + \bar{\lambda}^{(2)}\eta^{(2)} \\
& - \eta^{(2)}\eta^{(1)} \frac{\left(\left(\bar{\lambda}^{*(1)} + 2\bar{\mu}^{*(1)}\right) - \left(\bar{\lambda}^{(2)} + 2\bar{\mu}^{(2)}\right)\right)^2}{\left(\bar{\lambda}^{*(1)} + 2\bar{\mu}^{*(1)}\right)\eta^{(2)} + \left(\bar{\lambda}^{(2)} + 2\bar{\mu}^{(2)}\right)\eta^{(1)}}, \\
\bar{A}^*_{55} =& \frac{\bar{\mu}^{*(1)}\bar{\mu}^{(2)}}{\bar{\mu}^{*(1)}\eta^{(1)} + \bar{\mu}^{(2)}\eta^{(2)}}, \quad \bar{A}^*_{13} = \bar{A}^*_{23} = \bar{\lambda}^{*(1)}\eta^{(1)} + \bar{\lambda}^{(2)}\eta^{(2)} \\
& - \eta^{(2)}\eta^{(1)}\left(\bar{\lambda}^{*(1)} - \bar{\lambda}^{(2)}\right) \frac{\left(\left(\bar{\lambda}^{*(1)} + 2\bar{\mu}^{*(1)}\right) - \left(\bar{\lambda}^{(2)} + 2\bar{\mu}^{(2)}\right)\right)}{\left(\bar{\lambda}^{*(1)} + 2\bar{\mu}^{*(1)}\right)\eta^{(2)} + \left(\bar{\lambda}^{(2)} + 2\bar{\mu}^{(2)}\right)\eta^{(1)}},
\end{aligned}
\tag{3.160}
$$

**Table 3.17** The values of $p_{cr.0}/E_0^{(1)}$ obtained in the framework of the TDLTS of stability for various $E^{(2)}/E_0^{(1)}$ under $h/R = 0.1$

| $\frac{E^{(2)}}{E_0^{(1)}}$ | $p_{cr.0}/E_0^{(1)}$ | | | |
|---|---|---|---|---|
| | Results for condition (3.133), (Tang 1967, Babich and Guz 1980) | Present FEM results for condition (3.133) | Present FEM results for condition (3.134) | Present FEM results for condition (3.135) |
| 1 | 0.045 | 0.045 | 0.045 | 0.014 |
| 20 | 0.2912 | 0.2941 | 0.2945 | 0.1274 |
| 50 | 0.4357 | 0.4371 | 0.4372 | 0.2418 |
| 100 | 0.5287 | 0.5269 | 0.5296 | 0.3513 |

where

$$\bar{\mu}^{*(1)} = \mu_0^{(1)}\left(\frac{1}{s} - \frac{3\omega_0}{2\left(1+\nu_0^{(1)}\right)}\bar{\Pi}_\alpha^*\left(-\frac{3}{2\left(1+\nu_0^{(1)}\right)}\omega_0 - \omega_\infty\right)\right),$$

$$\bar{\lambda}^{*(1)} = \lambda_0^{(1)}\left(\frac{1}{s} + \frac{\omega_0}{\left(1+\nu_0^{(1)}\right)}\bar{\Pi}_\alpha^*\left(-\frac{3}{2\left(1+\nu_0^{(1)}\right)}\omega_0 - \omega_\infty\right)\right),$$

$$\bar{\Pi}_\alpha^*(x) = \frac{1}{s\left(s^{1+\alpha} - x\right)},\ \bar{\lambda}^{(2)} = \frac{1}{s}\lambda^{(2)},\ \bar{\mu}^{(2)} = \frac{1}{s}\mu^{(2)},\ \eta^{(1)} = 1 - \eta^{(2)}. \tag{3.161}$$

#### 3.6.3.1 Comparison of Some Numerical Results Obtained in the Framework of the TDLTS for the Pure Elastic Plate

Consider the values of $p_{cr.0}/E_0^{(1)}$ which are obtained in (Tang 1967) and in (Babich and Guz 1980) by employing the analytical solution and compare them with those obtained in the present investigation. We also compare these results with the corresponding ones obtained for the clamped (3.134) and simply supported (3.135) plate, which are also obtained in the present investigation.

Tables 3.17 and 3.18 tabulate the values of $p_{cr.0}/E_0^{(1)}$ for various values of the problem parameters and for various end conditions. It follows from these tables that the results of (Tang 1967) and (Babich and Guz 1980) and present results are near each other with very high accuracy. Moreover, these tables indicate that the results obtained in the framework of the conditions (3.133) and (3.134) are also near each other with high accuracy. Further, these results guarantee the trustiness of the approach used and FEM modeling of the considered problems. Taking this

**Table 3.18** The values of $p_{cr.0}\big/E_0^{(1)}$ obtained in the framework of the TDLTS of stability for various $h/R$ under $E^{(2)}\big/E_0^{(1)}=50$

| $\frac{h}{R}$ | $p_{cr.0}\big/E_0^{(1)}$ | | | |
|---|---|---|---|---|
| | Exact results for condition (3.133), (Tang 1967, Babich and Guz 1980) | Present FEM results for condition (3.133) | Present FEM results for condition (3.134–c) | Present FEM results for condition (3.135) |
| 0.10 | 0.4357 | 0.4371 | 0.4372 | 0.2418 |
| 0.15 | 0.5379 | 0.5395 | 0.5395 | 0.3698 |
| 0.20 | 0.5915 | 0.5909 | 0.5931 | 0.4561 |

**Table 3.19** The values of $p_{cr.0}\big/E_0^{(1)}$ (upper) and $p_{cr.\infty}\big/E_0^{(1)}$ (lower) for various $E^{(2)}\big/E_0^{(1)}$ under $h/R=0.1$, $\omega=1$

| $\frac{E^{(2)}}{E_0^{(1)}}$ | Simply-supported plate | | | Clamped plate | | |
|---|---|---|---|---|---|---|
| | Theories | | | Theories | | |
| | 3D | Rfn | $K-L$ | 3D | Rfn | $K-L$ |
| 1 | 0.011<br>0.014 | 0.012<br>0.014 | 0.013<br>0.015 | 0.034<br>0.045 | 0.044<br>0.052 | 0.056<br>0.062 |
| 3 | 0.024<br>0.028 | 0.024<br>0.027 | 0.027<br>0.028 | 0.069<br>0.087 | 0.080<br>0.096 | 0.114<br>0.119 |
| 5 | 0.036<br>0.042 | 0.035<br>0.038 | 0.040<br>0.041 | 0.096<br>0.123 | 0.105<br>0.131 | 0.167<br>0.173 |
| 10 | 0.063<br>0.073 | 0.057<br>0.064 | 0.071<br>0.072 | 0.142<br>0.196 | 0.147<br>0.201 | 0.298<br>0.302 |
| 20 | 0.102<br>0.127 | 0.091<br>0.110 | 0.133<br>0.134 | 0.193<br>0.294 | 0.192<br>0.295 | 0.555<br>0.560 |
| 50 | 0.169<br>0.241 | 0.153<br>0.211 | 0.317<br>0.318 | 0.250<br>0.437 | 0.242<br>0.431 | 1.325<br>1.330 |
| 100 | 0.218<br>0.351 | 0.201<br>0.315 | 0.624<br>0.625 | 0.281<br>0.529 | 0.267<br>0.515 | 2.607<br>2.612 |

situation into account, below we will consider only the numerical results obtained for the condition (3.133) (clamped plate) and (3.135) (simply supported plate).

In Table 3.17 the results related to $E^{(2)}\big/E_0^{(1)}=1$ (i.e., related to the isotropic plate) are also tabulated. Note that in this case the following asymptotic expression had been obtained in (Tang 1967).

$$\frac{p_{cr.0}}{E_0^{(1)}}=\frac{14.68}{12\left(1-\nu_0^2\right)}\left(\frac{2h}{R}\right)^2. \tag{3.162}$$

It follows from (3.162) that for $h/R=0.1$, $p_{cr.0}\big/E_0^{(1)}=0.048$. However, the direct (without the asymptotic expansion) solution of the characteristic equation gives $p_{cr.0}\big/E_0^{(1)}=0.045$ (Table 3.17).

**Table 3.20** The values of $p_{cr.0}\big/E_0^{(1)}$ (upper) and $p_{cr.\infty}\big/E_0^{(1)}$ (lower) for various $h/R$ under $E^{(2)}\big/E_0^{(1)} = 50$, $\omega = 1$

| $\frac{h}{R}$ | Simply-supported plate | | | Clamped plate | | |
|---|---|---|---|---|---|---|
| | Theories | | | Theories | | |
| | 3D | Rfn | $K-L$ | 3D | Rfn | $K-L$ |
| 0.10 | 0.1690<br>0.2418 | 0.1532<br>0.2119 | 0.3173<br>0.3185 | 0.2500<br>0.4372 | 0.2429<br>0.4312 | 1.3251<br>1.3300 |
| 0.15 | 0.2251<br>0.3698 | 0.2089<br>0.3347 | 0.7042<br>0.7068 | 0.2846<br>0.5395 | 0.2692<br>0.5219 | 2.8173<br>2.8276 |
| 0.20 | 0.2567<br>0.4561 | 0.2394<br>0.4203 | 1.2282<br>1.2327 | 0.3023<br>0.5931 | 0.2798<br>0.5636 | 4.6498<br>4.6668 |

#### 3.6.3.2 Comparison of the Results Obtained for the Pure Elastic Plate in the Framework of the TDLTS and Approximate Theories

We consider the pure elastic stability loss problems which take place at $t' = 0$ and $t' = \infty$. In these cases the reduced critical values of the external compressive force determined as $p_{cr.0}\big/E_0^{(1)}$ (for $t' = 0$); $p_{cr.\infty}\big/E_0^{(1)}$ (for $t' = \infty$) with various $E^{(2)}\big/E_0^{(1)}$, $\omega$ and obtained in the framework of the TDLTS, the third order refined plate theory (Kromm 1953, 1955) and Kirchhoff-Love theory are given in Tables 3.19 and 3.20. The values of the problem parameters, which are fixed under calculation of these results, are given in the table headings.

Note that under calculation $p_{cr.\infty}\big/E_0^{(1)}$ the constants

$$E_\infty^{(1)} = E_0^{(1)}\left(1 - \frac{1}{1+\omega}\right),\ \nu_\infty^{(1)} = \nu_0^{(1)}\left(1 + \frac{1-2\nu_0^{(1)}}{2\nu_0^{(1)}(1+\omega)}\right),$$
$$\mu_\infty^{(1)} = \frac{E_\infty^{(1)}}{2\left(1+\nu_\infty^{(1)}\right)},\ \lambda_\infty^{(1)} = \frac{E_\infty^{(1)}\nu_\infty^{(1)}}{\left(1+\nu_\infty^{(1)}\right)\left(1-2\nu_\infty^{(1)}\right)}, \tag{3.163}$$

are taken instead of the operators (3.157).

Tables 3.19 and 3.20 show that the numerical results obtained in the framework of both the TDLTS and of the third order refined plate theory (Kromm 1953, 1955) are significantly different from those obtained in the framework of Kirchhoff-Love theory. Consequently, the results obtained by the TDLTS agree with those obtained by the third order refined theory, but not with those obtained in the framework of Kirchhoff-Love theory. This situation is explained by the following consideration.

According to (3.138), in the framework of the Kirchhoff-Love theory *(i)* the normal to the middle surface of the plate before deformation still remain normal to the same surface after deformation and *(ii)* the length of this normal does not change as a result of the deformation. However, according to (3.139), in the framework of the third order refined plate theory the hypothesis *(i)* is changed by

**Table 3.21** The values of $t'_{cr}$ for the isotropic viscoelastic plate under $\alpha = -0.5$, $\omega = 1$, $h/R = 0.1$

| Theories | Simply-supported plate | | Clamped plate |
|---|---|---|---|
| | $p/E_0^{(1)} = 0.013$ | $p/E_0^{(1)} = 0.0135$ | $p/E_0^{(1)} = 0.045$ |
| $3D$ | 0.172 | 0.047 | 0.000 |
| $Rfn$ | 2.034 | 0.141 | 33.349 |
| $K-L$ | $\infty$ | 198.1 | $\infty$ |

that, according to which, the normal to the middle surface of the plate before the deformation does not remain normal to this surface. The hypothesis *(ii)* holds also for the third order plate theory under consideration. Moreover, in the framework of the Kirchhoff-Love theory the dependence of the $u_r^{(1)}$ on the thickness coordinate $z$ contains only the first order term in $z$, but this dependence in the framework of the third order refined plate theory contains the third order term in $z$. Consequently, the hypotheses of the third order refined plate theory under consideration are nearer the exact statement taken into account within the framework of the TDLTS, than the hypotheses of the theory of Kirchhoff-Love. Therefore, in the considered cases the corresponding results of the TDLTS and of the third order refined plate theories are compared.

### 3.6.3.3 Viscoelastic Plate

As in the previous sections related to the discussions of the numerical results under investigation of the viscoelastic stability loss problems the selected values of the external compressive force $p$ must satisfy the inequality

$$p_{cr.\infty} < p < p_{cr.0}. \tag{3.164}$$

First, taking (3.163) into account we consider the results related to the isotropic viscoelastic plate. These results, i.e., the values of $t'_{cr.}$ under $\alpha = -0.5$ and $\omega = 1$ are given in Table 3.21, from which follows that $p_{cr.0}$ as well as $p_{cr.\infty}$ obtained for the various theories are near each other, but the values of $t'_{cr}$ are significantly different from each other.

Consider the case where the plate material is viscoelastic transversal-isotropic. It follows from the data given in Tables 3.19 and 3.20 that in these cases the values of $p/E_0^{(1)}$ which satisfy the inequality (3.164) simultaneously for the TDLTS and for the third order refined plate theory, but cannot satisfy this inequality for the Kirchhoff-Love theory. Because, the values of $p_{cr.\infty}/E_0^{(1)}$ obtained for Kirchhoff-Love theory are significantly greater than $p_{cr.0}/E_0^{(1)}$ obtained for the TDLTS and for the third order refined plate theory; below we will consider only the results obtained in the framework of both the TDLTS and the third order refined plate theory.

In Table 3.22 for the clamped plate the values of $t'_{cr}$ are given for various $E^{(2)}/E_0^{(1)}$ and $p/E_0^{(1)}$ under $\alpha = -0.5$, $\omega = 1$ and $h/R = 0.1$ In this table the values of the relation

$$\Lambda = \frac{|p - p_{cr.0}|}{|p - p_{cr.\infty}|} \tag{3.165}$$

are also given and they are shown in parentheses. We denote the values of $t'_{cr.}$, $\Lambda$ by $t'_{cr.3D}$, $\Lambda_{3D}$ and $t'_{cr.Rfn}$, $\Lambda_{Rfn}$, the first pair obtained in the framework of the TDLTS and after pair in the framework of the third order refined plate theory respectively. Thus, using this representation and analyzing the data given in Table 3.22 we can conclude that the values of $t'_{cr.3D}\left(t'_{cr.Rfn}\right)$ increase monotonically with increasing $\Lambda_{3D}\left(\Lambda_{Rfn}\right)$, because according to (3.165), the increasing $\Lambda_{3D}\left(\Lambda_{Rfn}\right)$ means that $p \to p_{cr.\infty}$. For fixed problem parameters the following statement occurs: if $\Lambda_{3D} > \Lambda_{Rfn}\left(\Lambda_{3D} < \Lambda_{Rfn}\right)$ then $t'_{cr.3D} > t'_{cr.Rfn}\left(t'_{cr.3D} < t'_{cr.Rfn}\right)$. Moreover, if $\Lambda_{3D} \approx \Lambda_{Rfn}$, then $t'_{cr.3D} \approx t'_{cr.Rfn}$; here "$\approx$" represent the cases which are similar to the cases where $E^2/E_0^{(1)} = 20,\ 50$ (Table 3.22). In the cases where $\Lambda_{3D} \neq \Lambda_{Rfn}$ the values $\left|t'_{cr.3D} - t'_{cr.Rfn}\right|$ grow sharply with $\left|\Lambda_{3D} - \Lambda_{Rfn}\right|$. For example, under $\lambda_\Lambda \geq 0.3$ (where$\lambda_\Lambda = |\Lambda_T - \Lambda_R|/\max\{\Lambda_T, \Lambda_R\}$) we obtain that $\lambda_t \geq 30\,\%$ (where $\lambda_t = 100\,\%\left|t'_{cr.3D} - t'_{cr.Rfn}\right|/\max\left\{t'_{cr.3D}, t'_{cr.Rfn}\right\}$). This character of the numerical results can be explained by the following considerations. For the same selected value of $p/E_0^{(1)}$ and shown in Table 3.22 the value of $\Lambda_{Rfn}$ is significantly greater than that of $\Lambda_{3D}$. Consequently, this value of $p/E_0^{(1)}$ is very close to the $p_{cr.\infty}/E_0^{(1)}$ obtained by the third order refined plate theory than that obtained by the TDLTS. As the same time the values of the critical time increase with $p \to p_{cr.\infty}$, therefore the above-stated difference between the values of $t'_{cr.3D}$ and $t'_{cr.Rfn}$ arises. These results show again that for many cases it is necessary to apply the TDLTS for the investigation of the related stability loss problems. Note that the similar results are also obtained for the simply-supported plate. As an example, in Table 3.23 the values of $t'_{cr.3D}$ and $t'_{cr.Rfn}$ are given as well as for simply supported plate for $E^{(2)}/E_0^{(1)} = 50$, $h/R = 0.2$.

Note that the described type results are also obtained for other values of the rheological parameters $\alpha$ and $\omega$. It follows from the Table 3.24 that in the cases under considerations the values of $t'_{cr}$ increase (decrease) monotonically with increasing $\omega(|\alpha|)$.

**Table 3.22** The values of $t'_{cr}$ and $\Lambda$ (in parentheses) obtained for clamped plate for various $E^{(2)}/E_0^{(1)}$, $p/E_0^{(1)}$ under $\alpha = -0.5$, $\omega = 1$, $h/R = 0.1$

| $E^{(2)}/E_0^{(1)}$ | Theories | $p/E_0^{(1)}$ | | |
|---|---|---|---|---|
| | | 0.082 | 0.084 | 0.086 |
| 3 | 3D | 0.044 (1.154) | 0.013 (0.866) | 0.002 (0.647) |
| | Rfn | 16.29 (7.000) | 2.873 (3.000) | 0.878 (1.666) |
| | | $p/E_0^{(1)}$ | | |
| | | 0.110 | 0.115 | 0.120 |
| 5 | 3D | 0.249(1.076) | 0.052(0.420) | 0.005(0.125) |
| | Rfn | 5.577(4.200) | 0.810(1.600) | 0.174(0.738) |
| | | $p/E_0^{(1)}$ | | |
| | | 0.150 | 0.160 | 0.180 |
| 10 | 3D | 10.29 (5.750) | 1.087 (3.000) | 0.045 (0.421) |
| | Rfn | 92.25 (17.00) | 2.751 (3.154) | 0.110 (0.636) |
| | | $p/E_0^{(1)}$ | | |
| | | 0.200 | 0.230 | 0.260 |
| 20 | 3D | 38.95 (13.42) | 0.633 (1.729) | 0.054 (0.507) |
| | Rfn | 39.03 (11.87) | 0.664 (1.710) | 0.059 (0.514) |
| | | $p/E_0^{(1)}$ | | |
| | | 0.300 | 0.350 | 0.400 |
| 50 | 3D | 1.202 (3.740) | 0.121 (0.870) | 0.009 (0.246) |
| | Rfn | 0.838 (3.258) | 0.091 (0.750) | 0.006 (0.196) |

**Table 3.23** The values of $t'_{cr}$ for various $p/E_0^{(1)}$ under $E^2/E_0^{(1)} = 50$, $\alpha = -0.5$, $\omega = 1$, $h/R = 0.2$

| Theories | Simply-supported plate | | | | Clamped plate | | | | |
|---|---|---|---|---|---|---|---|---|---|
| | $p/E_0^{(1)}$ | | | | | | | | |
| | 0.300 | 0.330 | 0.360 | 0.390 | 0.350 | 0.400 | 0.450 | 0.500 | 0.550 |
| 3D | 2.014 | 0.457 | 0.133 | 0.037 | 3.298 | 0.493 | 0.118 | 0.027 | 0.003 |
| Rfn | 0.640 | 0.161 | 0.040 | 0.006 | 1.146 | 0.229 | 0.055 | 0.010 | 0.0003 |

**Table 3.24** The influence of the rheological parameter

| Theories | Simply-supported plate | | | Clamped plate | | | |
|---|---|---|---|---|---|---|---|
| | $\omega$ | | | | | | |
| | 0.5 | 1.0 | 2.0 | 0.5 | 1.0 | 2.0 | 3.0 |
| *$\omega$ on the values of $t'_{cr}$ under $E^{(2)}\big/E_0^{(1)} = 50,\ \alpha = -0.5,\ h/R = 0.2$* | | | | | | | |
| *3D* | 0.084 | 0.133 | 0.573 | 0.076 | 0.118 | 0.450 | 177.9 |
| *Rfn* | 0.031 | 0.040 | 0.079 | 0.040 | 0.055 | 0.123 | 0.488 |
| Theories | Simply-supported plate | | | Clamped plate | | | |
| | $\alpha$ | | | | | | |
| | −0.3 | −0.5 | −0.7 | −0.3 | −0.5 | −0.7 | |
| *$\alpha$ on the values of $t'_{cr}$ under $E^{(2)}\big/E_0^{(1)} = 50,\ \omega = 1,\ h/R = 0.2$* | | | | | | | |
| *3D* | 0.194 | 0.133 | 0.055 | 0.178 | 0.118 | 0.045 | |
| *Rfn* | 0.083 | 0.040 | 0.007 | 0.103 | 0.055 | 0.012 | |

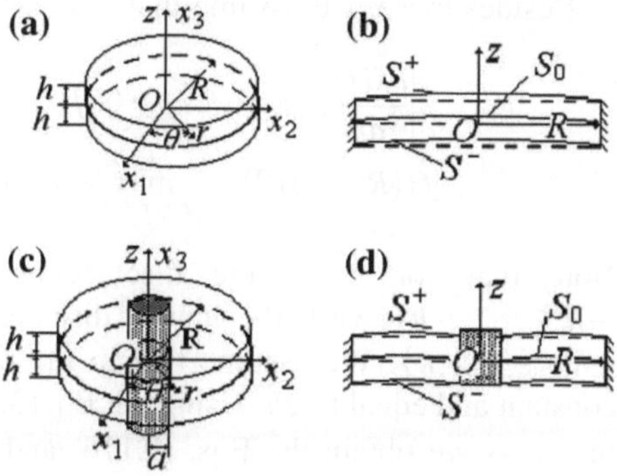


**Fig. 3.5** The geometry of the rotating disc(**a**) and annular disc(**b**); cross section of the rotating disc (**c**) and annular disc (**d**)

## 3.7 Stability Loss of the Rotating Circular and Annular Discs

### *3.7.1 Formulation of the Problem*

In the present section in the framework of the TDLTS, we investigate the rotationally symmetric stability loss problem of the elastic and viscoelastic composite circular and annular discs, which rotate with constant angular velocity $\vartheta$. Assume that circular or annular disc with external radius $R$ has an insignificant rotationally symmetric initial imperfection and with middle surface of the disc we associate a cylindrical system of coordinate $Or\theta z$. The position of points of the disc we determine with the Lagrange coordinates in the mentioned coordinate system. Suppose that the disc occupies the region $\Omega$ (Fig. 3.5) which is restricted by the surfaces $S^+$ and $S^-$ and $0 \le r \le R,\ 0 \le \theta \le 2\pi$ (for the disc Fig. 3.5a) or

$a \le r \le R$, $0 \le \theta \le 2\pi$ (for the annular disc Fig. 3.5c) and equation of the middle surface $S_0$ of the disc (Fig. 3.5b) (or annular disc (Fig. 3.5d) is given by the equation

$$z = F(r) = \varepsilon f(r), \tag{3.166}$$

where, as in the previous section, $\varepsilon$ is a small parameter $(0 \le \varepsilon \ll 1)$ and characterizes the degree of the initial imperfection. We also suppose that the function $f(r)$ must satisfy the following conditions:

For the disc:

$$f(r)|_{r=R} = 0, \left.\frac{df(r)}{dr}\right|_{r=R} = 0. \tag{3.167}$$

For the annular disc:

$$f(r)|_{r=a;R} = 0, \left.\frac{df(r)}{dr}\right|_{r=a;R} = 0. \tag{3.168}$$

Besides these, we assume that

$$\varepsilon\left|\frac{df(r)}{dr}\right| \ll 1, f(r) \ge 0, f(0) = \max_{0 \le r \le R} f(r) \text{ (for the disc)},$$
$$f((R-a)/2) = \max_{a \le r \le R} f(r) \text{ (for the annular disc)}. \tag{3.169}$$

Note that for $\varepsilon = 0$ the disc occupies the region $\{0 \le r \le R; 0 \le \theta \le 2\pi; -h \le z \le +h\}$, while the annular disc occupies the region $\{a \le r \le R; 0 \le \theta \le 2\pi; -h \le z \le +h\}$. We assume that in the natural state the thickness of the disc is constant and equal to $2h$. Using the Eq. (3.166) and constancy condition of the disc thickness we obtain the Eqs. (3.116) and (3.117) for the surfaces $S^{\pm}$ and for the ortho-normal components of the unit-normal vector to these surfaces.

Assume that the disc and annular disc are fabricated from the viscoelastic (hereditary) composite material, which is modeled to be orthotropic with effective mechanical properties and this disc (or annular disc) rotates around the $O_z$ (Fig. 3.5) axis with constant angular velocity $\vartheta$. Moreover, we assume that the field Eqs. (3.120b), (3.120c) and (3.121) are satisfied in the regions occupied by the disc and by the annular disk, but the following ones replace the equilibrium equations in (3.120a)

$$\frac{\partial t_{rr}}{\partial r} + \frac{\partial t_{zr}}{\partial z} + \frac{1}{r}(t_{rr} - t_{\theta\theta}) = \rho\vartheta^2 r, \frac{\partial t_{rz}}{\partial r} + \frac{\partial t_{zz}}{\partial z} + \frac{1}{r}t_{rz} = 0, \tag{3.170}$$

where $\rho$ is a density of the disc's material, $\vartheta$ is a angular velocity.

The boundary conditions on the upper and lower surfaces of the disc are taken as (3.122). We consider the case where the disc and annular disc are clamped on the lateral cylindrical surfaces, i.e.,

$$\begin{aligned} &u_r\big|_{r=R}=0,\ u_z\big|_{r=R}=0\ (\text{for the disc}),\\ &u_r\big|_{r=a;R}=0,\ u_z\big|_{r=a;R}=0\ (\text{for the annular disc}). \end{aligned} \tag{3.171}$$

This completes the mathematical formulation of the problem, according to which, the investigation of the stability loss of the rotationally symmetric disc (or annular disc) is reduced to the solution of the boundary value problem (3.170), (3.120b), (3.120c), (3.121), (3.122) and (3.171). In this case, after replacing the notation $p$ with $\vartheta$ the criterion (3.152) will be used for determination of the values of the critical parameters, i.e. the values of the critical angular velocity $\vartheta$ (denoted by $\vartheta_{cr.}$) or the values critical time.

### 3.7.2 Method of Solution

For solution of the problem formulated above, we employ the procedure developed in the previous section, according to which, for determination the values related to the zeroth approximation we obtain the following equilibrium equation from (3.170).

$$\begin{aligned} &\frac{\partial\sigma_{rr}^{(0)}}{\partial r}+\frac{\partial\sigma_{zr}^{(0)}}{\partial z}+\frac{1}{r}\left(\sigma_{rr}^{(0)}-\sigma_{\theta\theta}^{(0)}\right)=\rho\vartheta^2 r,\\ &\frac{\partial\sigma_{rz}^{(0)}}{\partial r}+\frac{\partial\sigma_{zz}^{(0)}}{\partial z}+\frac{1}{r}\sigma_{rz}^{(0)}=0. \end{aligned} \tag{3.172}$$

The relations (3.126b), (3.126c) and the boundary condition (3.127) hold also for the case under consideration. However, for the zeroth approximation the following conditions, which are satisfied on the lateral cylindrical surface of the disc and annular disc, are obtained from (3.171).

$$u_r^{(0)}\big|_{r=R}=0,\ u_z^{(0)}\big|_{r=R}=0\ (\text{for the disc}), \tag{3.173}$$

$$u_r^{(0)}\big|_{r=a;R}=0,\ u_z^{(0)}\big|_{r=a;R}=0\ (\text{for the annular disc}). \tag{3.174}$$

In the same manner, we derive the field equations in (3.131a–c), (3.136) and boundary condition (3.132) for the values of the first approximation. Moreover, the following conditions are obtained from (3.171) for the displacements of the first approximation.

$$u_r^{(1)}\big|_{r=R}=0,\ u_z^{(1)}\big|_{r=R}=0\ (\text{for the disc}), \tag{3.175}$$

$$u_r^{(1)}\big|_{r=a;R}=0,\ u_z^{(1)}\big|_{r=a;R}=0\ (\text{for the annular disc}). \tag{3.176}$$

We consider briefly the solution procedure of the zeroth and the first approximations separately. Thus, for determination of the zeroth approximation, i.e., for solution of the boundary value problem consisting of Eqs. (3.172), (3.126b),

(3.126c) and the boundary conditions (3.127), (3.173) and (3.174) first we employ the Laplace transformation (3.146). Under this employing we use the principle of correspondence, according to which, we obtain the same Eqs. (3.172), (3.126b), (3.126c) and the boundary conditions (3.127), (3.173) and (3.174) written for the Laplace transformation of the quantities related to the zeroth approximation. For solution of the later problem we apply the FEM through the following functional.l.

$$\Pi^{(0)}\left(\bar{u}_r^{(0)},\bar{u}_z^{(0)}\right)=\frac{1}{2}\int_0^R\left(\int_{-h}^{+h}\left(\bar{\sigma}_{rr}^{(0)}\bar{\varepsilon}_{rr}^{(0)}+\bar{\sigma}_{\theta\theta}^{(0)}\bar{\varepsilon}_{\theta\theta}^{(0)}+\bar{\sigma}_{zz}^{(0)}\bar{\varepsilon}_{zz}^{(0)}\right.\right.$$
$$\left.\left.+2\bar{\sigma}_{rz}^{(0)}\bar{\varepsilon}_{rz}^{(0)}\right)dz\right)dr-\int_0^R\left(\int_{-h}^{+h}\rho\vartheta^2 r u_r^{(0)}dz\right)dr,\ \text{(for the disc);}$$
$$\Pi^{(0)}\left(\bar{u}_r^{(0)},\bar{u}_z^{(0)}\right)=\frac{1}{2}\int_a^R\left(\int_{-h}^{+h}\left(\bar{\sigma}_{rr}^{(0)}\bar{\varepsilon}_{rr}^{(0)}+\bar{\sigma}_{\theta\theta}^{(0)}\bar{\varepsilon}_{\theta\theta}^{(0)}+\bar{\sigma}_{zz}^{(0)}\bar{\varepsilon}_{zz}^{(0)}\right.\right.$$
$$\left.\left.+2\bar{\sigma}_{rz}^{(0)}\bar{\varepsilon}_{rz}^{(0)}\right)dz\right)dr-\int_a^R\left(\int_{-h}^{+h}\rho\vartheta^2 u_r^{(0)}dz\right)dr,\ \text{(for the annular disc).}\tag{3.177}$$

By the use of the traditional Ritz technique we obtain from

$$\delta\Pi^{(0)}(u_r^{(0)},u_z^{(0)})=0.\tag{3.178}$$

the FEM modeling of the foregoing problem for the Laplace transformation of the quantities related to the zeroth approximation. The original of the sought values are obtained through the expression (3.80).

Consider the determination of the values of the first approximation, i.e., consider the solution to the problem (3.131a–c), (3.136), (3.132), (3.175) (for the disc) and (3.176) (for the annular disc). For this purpose, as in the previous section, we assume that the "time freezing" principle discussed in Sect. 3.4.1 is also acceptable for the problem under consideration. Within the scope of this principle we employ the Laplace transformation (3.146) to the equations and relations related to the first approximation. For determination of these transformations we also employ FEM through the functional (3.148). According to the relation (3.150), the traditional Ritz technique is employed for numerical calculations. The original of the sought values are determined by the use of the relation (3.80).

Under the foregoing FEM formulations the region $\{0\leq r/R\leq 1; -1\leq z/h\leq +1;\}$ (for the disc) or $\{a/R\leq r/R\leq 1; -1\leq z/h\leq +1;\}$ (for the annular disc) is divided into 120 rectangular Lagrange family quadratic elements with 533 nodes and 1014 NDOF.

**Table 3.25** The values of the dimensionless critical angular velocity $\kappa_{cr}^{3D}$ and $\kappa_{cr}^{K-L}$ for isotropic rotating disc with various $h/R$

| Theories | $h/R$ | | | | |
|---|---|---|---|---|---|
| | 0.03 | 0.04 | 0.05 | 0.07 | 0.10 |
| $3D$ | 0.545 | 0.915 | 1.337 | 2.237 | 3.484 |
| $K-L$ (Mostaghel and Tadjbakhsh 1973) | 0.594 | 1.057 | 1.652 | 3.237 | 6.607 |

### *3.7.3 Numerical Results and Discussions*

We assume that all assumptions taken in Sect. 3.6.3 and the relations (3.157)–(3.161) occur also for the present case. Under numerical investigations the initial imperfection mode of the disc is selected as

$$\begin{gathered} f(r) = R\left(1-\left(\tfrac{r}{R}\right)^2\right)^2 \text{ (for the disc)}, \\ f(r) = R\left(1-\left(\tfrac{r}{R}\right)^2\right)^2\left(\left(\tfrac{a}{R}\right)^2-\left(\tfrac{r}{R}\right)^2\right)^2 \text{(for the annular disc)}. \end{gathered} \tag{3.179}$$

As the small parameter $\varepsilon$ we take the relation $L/R$, i.e. $\varepsilon = L/R$, assuming that $L \ll R$, where $L$ is the maximum rise of the initial imperfection of the disc. Remind that the values of the critical parameters do not depend on the values of $\varepsilon$ and on the initial imperfection mode which satisfies conditions (3.167)–(3.169). For determination of the values of the critical parameters the criterion (3.152) is employed.

We introduce the dimensionless angular velocity $\kappa = \left(\vartheta^2 \rho 4R^2 \big/ E_0^{(1)}\right)^{1/2}$ and first consider for the numerical results related to the stability loss of the pure elastic disc fabricated from isotropic material, i.e., assume that $E^{(2)} \big/ E_0^{(1)} = 1$, $a/R = 0$ and $t' = 0$. Compare the critical values of $\kappa$ obtained by employing the TDLTS (denoted as $\kappa_{cr}^{3D}$) with those calculated by the approximate analytical expression (denoted as $\kappa_{cr.}^{K-L}$) given in (Mostaghel and Tadjbakhsh 1973). These results for various $h/R$ are given in Table 3.25 from which follows that the difference between the $\kappa_{cr}^{3D}$ and $\kappa_{cr}^{K-L}$ increase with thickness of the disc, i.e., with $h/R$. Consequently, the results of (Mostaghel and Tadjbakhsh 1973) applicable for the thin rotating circular disc and for the comparatively thick ones (i.e., for $h/R \geq 0.05$) are significantly greater than those obtained in the framework of the TDLTS. According to this discussion, we consider below only the results obtained in the scope of the TDLTS.

We analyze the results related to the pure elastic stability loss problems that take place at $t' = 0$ and $t' = \infty$. For these cases the dimensionless critical values of the angular velocity (denoted as $\kappa_{cr.0}^{3D}$ for $t' = 0$ and as $\kappa_{cr.\infty}^{3D}$ for $t' = \infty$) of the circular disc are given in Tables 3.26 and 3.27 where the values of $\kappa_{cr.0}^{3D}\left(\kappa_{cr.\infty}^{3D}\right)$ are shown in the numerator (denominator). Note that the results given in Table 3.26

**Table 3.26** The values of $\kappa_{cr.0}^{3D}$ (numerator) and $\kappa_{cr.\infty}^{3D}$ (denominator) obtained in the case where $h/R = 0.1$, $\omega = 1$

| $a/R$ | $E^{(2)}/E_0^{(1)}$ | | | | |
|---|---|---|---|---|---|
| | 1 | 5 | 10 | 20 | 50 |
| 0.00 | $\frac{3.421}{2.476}$ | $\frac{7.833}{4.879}$ | $\frac{10.23}{5.482}$ | $\frac{11.72}{5.701}$ | $\frac{12.35}{5.828}$ |
| 0.05 | $\frac{5.087}{3.501}$ | $\frac{10.04}{5.322}$ | $\frac{11.39}{5.633}$ | $\frac{12.11}{5.810}$ | $\frac{12.59}{5.927}$ |
| 0.10 | $\frac{5.601}{3.787}$ | $\frac{10.54}{5.733}$ | $\frac{12.28}{6.203}$ | $\frac{13.27}{6.453}$ | $\frac{13.98}{6.621}$ |

**Table 3.27** The values of $\kappa_{cr.0}^{3D}$ (numerator) and $\kappa_{cr.\infty}^{3D}$ (denominator) obtained in the case where $E^{(2)}/E_0^{(1)} = 5$, $\omega = 1$

| $a/R$ | $h/R$ | | | |
|---|---|---|---|---|
| | 0.05 | 0.10 | 0.15 | 0.20 |
| 0.00 | $\frac{3.483}{2.730}$ | $\frac{7.833}{4.879}$ | $\frac{9.763}{5.251}$ | $\frac{10.24}{5.288}$ |
| 0.05 | $\frac{6.040}{4.222}$ | $\frac{10.04}{5.322}$ | $\frac{10.46}{5.776}$ | $\frac{10.54}{5.851}$ |
| 0.10 | $\frac{7.030}{4.701}$ | $\frac{10.54}{5.733}$ | $\frac{11.32}{5.951}$ | $\frac{11.57}{6.015}$ |

have been obtained for various $h/R = 0.1$, $\omega = 1$ with various $E^{(2)}/E_0^{(1)}$ and $a/R$, while the results given in Table 3.27 have been obtained for various $h/R$ and $a/R$ in the case where $E^{(2)}/E_0^{(1)} = 5$, $\omega = 1$. It follows from these results that the values of $\kappa_{cr.0}^{3D}$ and $\kappa_{cr.\infty}^{3D}$ increase monotonically with $h/R$, $E^{(2)}/E_0^{(1)}$ and $a/R$.

With this we restrict the consideration of the numerical results related to the pure elastic stability loss problem.

Consider numerical results related to the stability loss of the viscoelastic disc, i.e., consider numerical results related to the critical time. Under obtaining these results we assume that the dimensionless angular velocity $\kappa$ of the disc must satisfy the following inequality.

$$\kappa_{cr.\infty}^{3D} \leq \kappa \leq \kappa_{cr.0}^{3D}. \tag{3.180}$$

Note that, in Sects. 3.4–3.6 the numerical results related to the critical time were obtained by employing FEM within the scope of the "time freezing" principle. But the "freezing" of the values of the zeroth approximation was made at $t' = 0$. However, in the present sub-section we also analyze the question how the selection of the "freezing" moment effects on the values of the critical time. For this purpose we consider two cases: the values of the zeroth approximation were freezed at moment i) $t' = 0$ and ii) at moment $t' = \infty$. The values of the critical time obtained in the case i (in the case ii) we denote through $t'_{cr.3D.0}$ $\left(t'_{cr.3D.\infty}\right)$.

Tables 3.28 and 3.29 show the values of $t'_{cr.3D.0}$ (upper number) and $t'_{cr.3D.\infty}$ (lower number) obtained for various $\kappa$ and $a/R$ under $E^{(2)}/E_0^{(1)} = 5$ and

**Table 3.28** The values of $t'_{cr.3D.0}$ (upper number) and $t'_{cr.3D.\infty}$ (lower number) obtained in the case where $E^{(2)}/E_0^{(1)} = 5$, $h/R = 0.1$, $\alpha = -0.5$

| $\omega$ | $a/R = 0.00$ | | $a/R = 0.05$ | | $a/R = 0.10$ | |
|---|---|---|---|---|---|---|
| | $\kappa$ | | $\kappa$ | | $\kappa$ | |
| | 7.0 | 7.5 | 8.5 | 9.0 | 8.5 | 9.0 |
| 0.5 | 0.024<br>0.025 | 0.0030<br>0.0032 | 0.028<br>0.029 | 0.011<br>0.011 | 0.047<br>0.047 | 0.022<br>0.022 |
| 1.0 | 0.031<br>0.032 | 0.0032<br>0.0035 | 0.037<br>0.037 | 0.012<br>0.012 | 0.066<br>0.066 | 0.027<br>0.027 |
| 2.0 | 0.055<br>0.057 | 0.0038<br>0.0041 | 0.070<br>0.070 | 0.018<br>0.018 | 0.163<br>0.163 | 0.047<br>0.047 |

**Table 3.29** The values of $t'_{cr.3D.0}$ (upper number) and $t'_{cr.3D.\infty}$ (lower number) obtained in the case where $E^{(2)}/E_0^{(1)} = 5$, $h/R = 0.1$, $\omega = 1$

| $\alpha$ | $a/R = 0.00$ | | $a/R = 0.05$ | | $a/R = 0.10$ | |
|---|---|---|---|---|---|---|
| | $\kappa$ | | $\kappa$ | | $\kappa$ | |
| | 7.0 | 7.5 | 8.5 | 9.0 | 8.5 | 9.0 |
| −0.3 | 0.069<br>0.070 | 0.013<br>0.014 | 0.078<br>0.078 | 0.036<br>0.036 | 0.117<br>0.117 | 0.063<br>0.063 |
| −0.5 | 0.031<br>0.032 | 0.0032<br>0.0035 | 0.037<br>0.037 | 0.012<br>0.012 | 0.066<br>0.066 | 0.027<br>0.027 |
| −0.7 | 0.004<br>0.005 | 0.0001<br>0.0001 | 0.006<br>0.006 | 0.001<br>0.001 | 0.016<br>0.016 | 0.003<br>0.003 |

**Table 3.30** The values of $t'_{cr.3D.0}$ obtained in the case where $h/R = 0.1$, $\alpha = -0.5$, $\omega = 1$, $\kappa = 7.5$ for various $E^{(2)}/E_0^{(1)}$ and $a/R$

| $a/R$ | $E^{(2)}/E_0^{(1)}$ | | | |
|---|---|---|---|---|
| | 5 | 10 | 20 | 50 |
| 0.00 | 0.003 | 0.303 | 0.670 | 0.943 |
| 0.05 | 0.210 | 0.540 | 0.862 | 1.164 |
| 0.10 | 0.513 | 1.738 | 3.598 | 6.090 |

$h/R = 0.1$. Note that the results given in Table 3.28 are obtained in the case where $\alpha = -0.5$ for various $\omega$, while the results given in Table 3.29 are obtained in the case where $\omega = 1$ for various $\alpha$. Moreover, note that under calculation $t'_{cr.3D.\infty}$ for some fixed value of $\omega$ the values of the zeroth approximation are calculated at $t' = \infty$ for the same $\omega$.

Thus, it follows from the foregoing numerical results that for the selected problem parameters the values of $t'_{cr.3D.0}$ are very near the corresponding values of $t'_{cr.3D.\infty}$. Therefore, each of them can be taken as $t'_{cr.3D}$ with very high accuracy. Consequently, according to these results, we can conclude that $t'_{cr.3D}$ increase (decrease) monotonically with $\omega(|\alpha|)$. Moreover according to these results, we can

**Table 3.31** The values of $t'_{cr.3D.0}$ obtained in the case where $E^{(2)}/E_0^{(1)} = 5$, $\alpha = -0.5$, $\omega = 1$, $\kappa = 7.5$ for various$h/R$ and $a/R$

| $a/R$ | $h/R$ | | |
|---|---|---|---|
| | 0.10 | 0.15 | 0.20 |
| 0.00 | 0.003 | 0.163 | 0.211 |
| 0.05 | 0.210 | 0.262 | 0.278 |
| 0.10 | 0.513 | 0.850 | 1.037 |

conclude that the values of $t'_{cr.3D}$ decrease (increase) monotonically with $\kappa(a/R)$.

The influences of $E^{(2)}/E_0^{(1)}$ and $h/R$ on the values of $t'_{cr.3D.0}$ are shown in Tables 3.30 and 3.31, respectively, for $\alpha = -0.5$, $\omega = 1$, $\kappa = 7.5$. Note that under the calculation of the data given in Table 3.30 (Table 3.31) it is supposed that $h/R = 0.1\left(E^{(2)}/E_0^{(1)} = 5\right)$. It follows from these tables that the values of $t'_{cr.3D.0}$ increase monotonically with $E^{(2)}/E_0^{(1)}$ and $h/R$. This situation agrees with the well-known mechanical consideration.

# References

Akbarov SD (1998) On the three dimensional stability loss problems of elements of constructions fabricated from the viscoelastic composite materials. Mech Comp Mater 34(6):537–544

Akbarov SD (2007) Three-dimensional in stability problems for viscoelastic composite materials and structural members. Int Appl Mech 43(10):1069–1089

Akbarov SD, Yanioglu N (2001) A method of investigation of the general theory of stability problems on structural elements fabricated from viscoelastic composite materials composites. Part B: Engineering 32:475–483

Akbarov SD, Yanioglu N, Kutug Z (2001) On the three-dimensional stability loss problem of the viscoelastic composite plate. Int J Eng Sci 39:1443–1457

Babich IYu, Guz AN (1980) Three-dimensional theory of stability of bars, plates and shells. Vitsaya Schola, Kiev (in Russian)

Chiristensen RM (1979) Mechanics of composite materials. Wiley, New York

Guz AN (1999) Fundamentals of the three-dimensional theory of stability of deformable bodies. Springer, Heidelberg

Kromm A (1953) Verallgeneinerte theorie der platen-static. Ingenier-Archiv 21:266–286

Kromm A (1955) Über die ranguer krafter bei gestutzten platen. Z. Angew Math Mech 35(55): 231–242

Kutug Z (2009) On the three-dimensional undulation instability of a rectangular viscoelastic composite plate in biaxial compression. Mech Comp Mater 45(1):65–76

Kutug Z, Yahnioglu N, Akbarov SD (2003) The loss of stability analyses of an elastic and viscoelastic composite circular plate in the framework of the three-dimensional linearized theory. Eur J Mech A/Solids 22:243–256

Mostaghel N, Tadjbakhsh I (1973) Buckling of rotating rods and plates. Int J Mech Sci 15(6):429–434

Rabotnov YuN (1977) Elements of hereditary mechanics of solid bodies. Moscow, Nauka (in Russian)

Schapery RA (1962) Approximate method of transform inversion for viscoelastic stress analysis. In:Proceeding 4th US National Congress Appl Mech ASME :1075–1086

Schapery RA (1978) A viscoelastic behaviour of composite materials, in Composite materials, vol 1–7. In: Broutman LJ Krock RH (eds.) Mir, Moscow, translated from English), vol. 2: Mechanics of composite materials, 102–195 (in Ruıssion, translated from English)

Selim S, Akbarov SD (2003) FEM analyses of the three-dimensional buckling problem for a clamped thick rectangular plate made of viscoelastic composite. Mech Comp Mater 39(6):531–540

Tang S (1967) Elastic stability of thick plates under three-dimensional states of stress. Acta Mech 4:382–397

Yahnioglu N, Akbarov SD (2002) Stability loss analyses of the elastic and viscoelastic composite rotating thick circular plate in the framework of the three-dimensional linearized theory of stability. Int J Mech Sci 44:1225–1244

Robinson [illegible] (1957) [illegible] measurements of solid bodies. [illegible] [illegible]

[illegible] RA (1962) Approximate method of transform inversion for viscoelastic stress analysis. In: Proceedings 4th US National Congress Appl Mech. ASME, [illegible]

[illegible] RA (1975) A [illegible] theory of composite materials [illegible]. In: Broutman LJ, Krock RH (eds) Mechanics of composite materials [illegible]

[illegible] (1983) [illegible] of the threshold [illegible] chopped thick [illegible] plate made of [illegible] composites. Mech Compos Mater [illegible]

[illegible] (197[illegible]) [illegible]

[illegible] (1999) [illegible] circular plate of the framework of the three-dimensional [illegible] theory of [illegible]. Int J Mech Sci 41:1235–1254

# Chapter 4
# Buckling Delamination of Elastic and Viscoelastic Composite Plates with Cracks

The present chapter deals with problems related to buckling delamination of elastic and viscoelastic plates. Both the continuum and the piecewise homogeneous body models are employed to describe the plate materials. It is assumed that the plates contain a crack, the edges of which have an initial infinitesimal imperfection. It is also assumed that under the absence of the mentioned imperfections, the crack's edges are parallel to the free-face planes of the plate. Within these assumptions, the evolution of the initial infinitesimal imperfection is studied under compression of the plate. This study is performed with the use of the three-dimensional geometrically non-linear equations of the theory of viscoelasticity for anisotropic (orthotropic) bodies. Results related to plane-strain state, axi-symmetric and three-dimensional buckling delamination problems are presented and discussed. The initial imperfection criterion is used for determination of the values of the critical parameters. For the solution to the corresponding non-linear boundary value problems the boundary form perturbation technique, Laplace transformation with respect to time and FEM are employed. To describe the viscoelasticity of the plate materials, the fractional-exponential operator (Rabotnov 1977) is used.

This chapter is based on the investigations carried out in many papers (Akbarov and Rzayev 2002a, b, 2003; Akbarov et al. 2007, 2010a, b; Akbarov and Yahnioglu 2010; Rzayev and Akbarov 2002; Rzayev 2002).

## 4.1 Background of Related Problems

One of the most common failure mechanisms in laminated composite materials is a local buckling around the delaminated zone, i.e. the zone in which two adjacent layers are partially debonded at their interface.

Note that this zone may be formed as a consequence of various impact events, poor fabrication processes or fatigue. As is well known, the compressive strength

S. D. Akbarov, *Stability Loss and Buckling Delamination*,
DOI: 10.1007/978-3-642-30290-9_4, 

of the structures made from laminated composite materials may be reduced several times by the presence of this delamination damage, which is modeled as a crack in related research. The region bounded by a crack and the laminate free surface is liable to buckle locally under compressive loads, thereby creating conditions conducive to delamination growth and consequent global failure of the structure. One of the pioneer investigations related to the buckling-failure problem was made in a paper (Kachanov 1976) and numerous studies have so far been carried out in this field. A review of these investigations is given in papers (Kardomateas et al. 1995; Nilsson et al. 1993; Chai et al. 1981; Wang et al. 1995) and others in which it is supposed that there is a crack which is parallel to the free plane and to the direction of the compressed external forces. In this case, the beginning of the delamination growing process is modeled as a buckling of the part of the material which occupies the region between the crack and the free plane and solutions are found in the framework of the approximate stability loss theories of plates or beams. It is evident that the results of these investigations do not apply in the cases where the thickness of this part is equal to or greater than the length of the crack. Moreover, there is a series of works, such as (Hwang and Mao 1999; Short et al. 2001; Arman et al. 2006), investigating the effects of the geometry of the delaminated portion on the buckling force of the laminate. At the same time, in these works the investigations were carried out experimentally and numerically with the use of FEM modeling. For instance, in the paper (Hwang and Mao 1999), the influence of the delaminated zone (crack) on the global buckling critical forces of the plate-strip made from glass-fiber layers was studied. In another paper (Short et al. 2001), the effect of delamination geometry (i.e. the sizes of the rectangular interface crack) on the global buckling critical forces of the layered rectangular plate was investigated experimentally and numerically by employing three-dimensional modeling using the packet programme ABAQUS V.5.8 FEM code. The same problem for the rectangular plate containing the circular cylindrical hole through the thickness of the plate was investigated in a paper by Arman et al. (2006). It was assumed that the circular delamination (crack) is around the circular hole and to obtain the numerical results 3D FEM modeling of the packet programme ANSYS 6.1 was used.

The buckling driven delaminations of compressed films and coatings on substrates were studied experimentally in many papers (Evans and Hutchinson 1995; Gioia and Ortiz 1997; Hutchinson and Suo 1992; Hutchinson et al. 1992; Nilsson and Giannakopoulos 1995; Thouless et al. 1994; Wang and Evans 1998; Moon et al. 2002). In these and other investigations by these authors (for example, Moon et al. 2004 and Hutchinson et al. 2000), the approximate mathematical modeling was also used to describe the experimentally studied problems. It should be noted that in all the foregoing investigations, the fundamental buckling theories based on the three-dimensional nonlinear equations of deformable body mechanics were not proposed or used. One such theory, namely the Three-Dimensional Linearized Theory of Stability (TDLTS) of deformable solid body mechanics for local buckling problems of bodies containing cracks was proposed and employed in the papers (Guz 2000; Guz and Nazarenko 1985a, b; Guz et al. 2004) and others. Note that a detailed description of the field equations and relations of the TDLTS are given in many references, for instance in the monograph (Guz 2008a). A detailed description of some early results was given in other monographs (Guz 2008a, b). The present level of these investigations is detailed in a paper (Bogdanov et al. 2009).

In all the investigations reviewed above it was assumed that the materials of the composites are time-independent. The development of the TDLTS based on the initial imperfection criterion of stability loss for time dependent materials was proposed and employed in works (Akbarov 1998, 2007; Akbarov et al. 1997; Akbarov and Guz 2000; Akbarov and Yahnioglu 2001) and others. Descriptions of related results are detailed in previous chapters. Development and application of the above noted version of the TDLTS on the study of the buckling delamination problems of the elements of construction (such as plate-strip, circular and rectangular plates) fabricated from elastic and viscoelastic materials was undertaken in the works of the author and his students. A review of some of them was considered in the paper (Akbarov 2007).

In these studies, two- and three-dimensional buckling delamination problems were studied for the aforementioned plates containing cracks, the edge faces of which are parallel to their upper and lower face planes. The corresponding boundary value problems were solved by utilizing 2D and 3D FEM modeling and published in several papers (Akbarov and Yahnioglu 2010; Akbarov et al. 2010a, b; Akbarov and Rzayev 2002a, b, 2003; Rzayev and Akbarov 2002; Rzayev 2002). This chapter is based on the investigations carried out in these papers.

## 4.2 Buckling Delamination Problems for Plate-Strips with a Single Crack

### *4.2.1 Formulation of the Problems*

For simplicity of the discussion, first we consider a buckling delamination problem for a plate-strip made from viscoelastic homogeneous material, which contains a single crack and assume that the edges of the crack have an initial imperfection. We will consider two type initial imperfection modes of the crack edges: co-phase (Fig. 4.1a) and anti-phase (Fig. 4.1b), as well as, we will consider two type buckling delamination modes around the crack: co-phase (Fig. 4.2a) and anti-phase (Fig. 4.2b). As will be shown below, the co-phase initial imperfection mode of the crack edges causes the co-phase buckling delamination mode around the crack. But the type of the buckling delamination mode under the anti-phase initial imperfection of the crack's edges depends on the distance of the crack from the free face plane of the plane-strip.

With the plate, we associate Cartesian system of coordinate $Ox_1x_2x_3$ and the position of points of that we determine through Lagrange coordinates. According to the Figs. 4.1 and 4.2, the plate-strip under considerations occupies the region $\{0 \leq x_1 \leq \ell,\ 0 \leq x_2 \leq h,\ -\infty < x_3 < +\infty\}$. As the coordinate axis $Ox_3$ is perpendicular to the figures plane, therefore this axis is not shown in these figures.

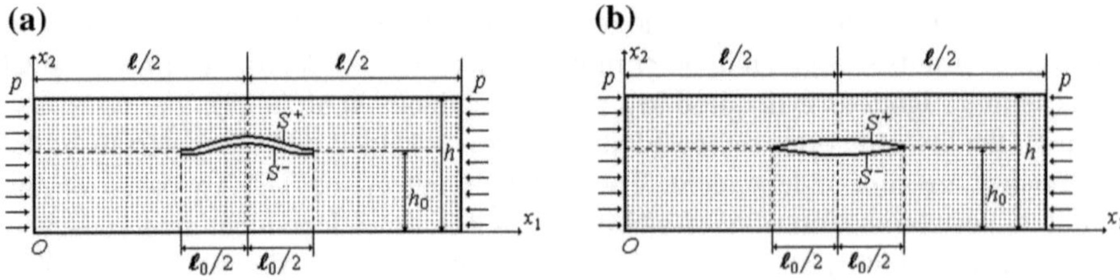


**Fig. 4.1** The geometry of the plate-strip containing a crack and co-phase (**a**) and anti-phase (**b**) crack's edges initial imperfection modes

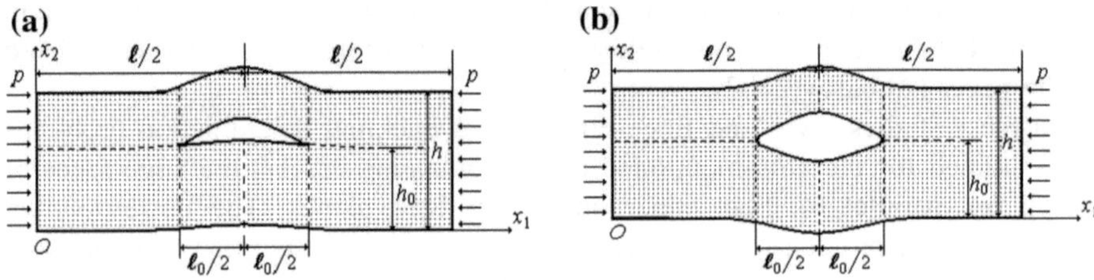


**Fig. 4.2** The geometry of co-phase (**a**) and anti-phase (**b**) buckling delamination modes around the crack

In the selected coordinate system the aforementioned initial imperfection of the crack edge-surfaces, i.e. the equations of the surfaces $S^+$ and $S^-$ (Figs. 4.1 and 4.2) can be described mathematically as follows:

For the co-phase initial imperfection mode (Fig. 4.1a)

$$x_2^+ = h_0 + \varepsilon\ell_0 \sin^2\left(\frac{\pi(x_1^+ - (\ell - \ell_0)/2)}{\ell_0}\right),\ x_1^+ \in \left[\frac{\ell - \ell_0}{2}, \frac{\ell + \ell_0}{2}\right],$$
$$x_2^- = h_0 + \varepsilon\ell_0 \sin^2\left(\frac{\pi(x_1^- - (\ell - \ell_0)/2)}{\ell_0}\right),\ x_1^- \in \left[\frac{\ell - \ell_0}{2}, \frac{\ell + \ell_0}{2}\right]. \tag{4.1}$$

For the anti-phase initial imperfection mode (Fig. 4.1b)

$$x_2^+ = h_0 + \varepsilon\ell_0 \sin^2\left(\frac{\pi(x_1^+ - (\ell - \ell_0)/2)}{\ell_0}\right),\ x_1^+ \in \left[\frac{\ell - \ell_0}{2}, \frac{\ell + \ell_0}{2}\right],$$
$$x_2^- = h_0 - \varepsilon\ell_0 \sin^2\left(\frac{\pi(x_1^- - (\ell - \ell_0)/2)}{\ell_0}\right),\ x_1^- \in \left[\frac{\ell - \ell_0}{2}, \frac{\ell + \ell_0}{2}\right]. \tag{4.2}$$

In Eqs. (4.1) and (4.2) the notation $x_1^+$, $x_2^+(x_1^-, x_2^-)$ shows the coordinates of the point of the surface $S^+(S^-)$. Moreover, in Eqs. (4.1) and (4.2) the notation $\varepsilon$ shows the small parameter and it is assumed that $0 \leq \varepsilon \ll 1$. Note that the small parameter $\varepsilon$ characterizes the degree of the initial infinitesimal imperfections of the crack edges.

Assume that the plate material is orthotropic one with the principal axes of the elastic symmetry $Ox_1$, $Ox_2$, and $Ox_3$.

Thus, within the scope of the foregoing assumptions and with the use of the geometrically non-linear exact equations of the theory of viscoelastic materials for the orthotropic bodies we investigate the evolution of the initial infinitesimal imperfections of the cracks edges with compressing of the plate-strip along the $Ox_1$ axis. Assume that this compressing is caused by the uniformly distributed normal forces with intensity $p$ which act on the ends $x_1 = 0$, $x_1 = \ell$ and the plane-strain state occurs in the $Ox_1x_2$ plane.

According to the foregoing suppositions, the following field equations must be satisfied in the region occupied by the plate-strip.

Equilibrium equations:

$$\frac{\partial}{\partial x_j}\left[\sigma_{jn}\left(\delta_i^n + \frac{\partial u_i}{\partial x_n}\right)\right] = 0,\ i;j;n = 1,2 \tag{4.3}$$

Geometrical relations:

$$\varepsilon_{ij} = \frac{1}{2}\left(\frac{\partial u_i}{\partial x_j} + \frac{\partial u_j}{\partial x_i} + \frac{\partial u_n}{\partial x_i}\frac{\partial u_n}{\partial x_j}\right), \tag{4.4}$$

Constitutive relations:

$$\sigma_{11} = A_{11}^*\varepsilon_{11} + A_{12}^*\varepsilon_{22},\ \sigma_{22} = A_{12}^*\varepsilon_{11} + A_{22}^*\varepsilon_{22},\ \sigma_{12} = 2A_{66}^*\varepsilon_{12}, \tag{4.5}$$

where $A_{11}^*$, $A_{12}^*$, $A_{22}^*$ and $A_{66}^*$ are the following operators:

$$A_{ij}^*\varphi(t) = A_{ij0}\varphi(t) + \int_0^t A_{ij1}(t-\tau)\varphi(\tau)d\tau,\ ij = 11;22;12;66. \tag{4.6}$$

In Eqs. (4.3, 4.4, 4.5) and (4.6) the following notation is used: $\sigma_{jn}$ is a component of the stress tensor, $\varepsilon_{ij}$ is a component Green strain tensor, $u_i$ is a component of the displacement vector, $\delta_i^n$ is a Kronecker symbol.

We assume that on the crack's edge surfaces $S^+$and $S^-$are free of tractions, i.e. the conditions

$$\left[\left(\delta_i^n + \frac{\partial u_i}{\partial x_n}\right)\sigma_{jn}\right]\Bigg|_{S^\pm} n_j^\pm = 0 \tag{4.7}$$

are satisfied. In (4.7) $n_j^\pm$ is a component of the unite normal to the surface $S^\pm$.

Similar conditions are also satisfied on the upper and lower face planes of the plate-strip.

$$\left[\left(\delta_i^n + \frac{\partial u_i}{\partial x_n}\right)\sigma_{2n}\right]\Bigg|_{x_2=0;h} = 0. \tag{4.8}$$

Consider two type conditions, which are satisfied at the ends of the plate-strip. The first type is following one.

$$\left[\left(\delta_i^n + \frac{\partial u_i}{\partial x_n}\right)\sigma_{1n}\right]\Big|_{x_1=0;\ell} = p\delta_i^1,\ u_2|_{x_1=0,\ell}= 0. \tag{4.9}$$

This condition will be called as simply supported end conditions.

The second type end conditions are formulated as follows. It is assumed that the plate-strip is compressed through rigid bodies, that hold the ends, by normal forces of intensity $p$ along the $Ox_1$axis and

$$u_1(x_1,x_2,t)|_{x_1=0} = -u_1(x_1,x_2,t)|_{x_1=\ell} = U(t),\ u_2|_{x_1=0,\ell} = 0. \tag{4.10}$$

The relation between $U(t)$ and $p$ is determined from the equation

$$hp = \int_0^h \sigma_{11}(x_1,x_2,t)|_{x_1=\ell/2}dx_2. \tag{4.11}$$

The end conditions (4.10) will be called as clamped end conditions. This completes the mathematical formulation of the problems related to the buckling delamination of the plate-strip containing a single crack. These problems can be classified as follows.

*Problem 1.* Co-phase buckling-delamination of the simply supported plate-strip, which is reduced to the solution to the boundary-value problem (4.1, 4.3–4.9).

*Problem 2.* Anti-phase buckling-delamination of the simply supported plate-strip, which is reduced to the solution to the boundary-value problem (4.2, 4.3–4.9).

*Problem 3.* Co-phase buckling-delamination of the clamped plate-strip, which is reduced to the solution to the boundary-value problem (4.1, 4.3–4.8, 4.10).

*Problem 4.* Anti-phase buckling-delamination of the clamped plate-strip, which is reduced to the solution to the boundary-value problem (4.2, 4.3–4.8, 4.10).

### 4.2.2 Method of Solution

Now we consider the method of solution of the problem formulated in the previous subsection. Note that the method employed below, which is similar to that made in the previous chapters, can be briefly summarized as follows. By employing the boundary-form perturbation techniques the considered boundary-value problems for the non-linear integro-differential equations (4.3–4.10) are reduced to the series boundary-value problems for the corresponding system of the linear integro-differential equations. Owing to both the expressions of the operators (4.6) and the convolution theorem, by the use of the Laplace transformation with respect to time these series problems are reduced to the corresponding series boundary value problems for the linear system of differential equations in the Laplace transformation parameter space. For each fixed value of this parameter the linear problems are solved by employing FEM and finally, applying the numerical inverse transformation method we determine the sought values. It should be noted that for pure elastic plate-strips the operators in (4.6) are replaced by the corresponding mechanical constants and therefore instead of the integro-differential equations we obtain differential equations and the corresponding problems for these equations are also investigated in the framework of the above procedure but without employing the Laplace transformation.

Since, according to the summarized above procedure and the problems statement, we represent the sought values in the series form in the small parameter $\varepsilon$:

$$\left\{\sigma_{ij};\varepsilon_{ij};u_i\right\}=\sum_{q=0}^{\infty}\varepsilon^q\left\{\sigma_{ij}^{(q)};\varepsilon_{ij}^{(q)};u_i^{(q)}\right\}. \tag{4.12}$$

Substituting the representation (4.12) into Eqs. (4.3–4.10) and comparing equal powers of parameter $\varepsilon$, we obtain a corresponding closed system of equations for each approximation in (4.12). Moreover, using the Eqs. (4.1) and (4.2) and expanding each approximation in series form in the vicinity $x_2^+=h_0+0$ and $x_2^-=h_0-0$ we obtain from (4.7) the boundary conditions for each approximation on the edge surfaces of the crack. For this purpose, we also use following expressions of the components of the unit normal vector to the edge surfaces of the crack:

For co-phase initial imperfection of the crack edges

$$n_1^{\pm}=\pm\varepsilon\pi\sin\left(2\pi\frac{(x_1-(\ell-\ell_0)/2)}{\ell_0}\right)+O(\varepsilon^3),\ n_2^{\pm}=\mp1+O(\varepsilon^2), \tag{4.13}$$

For anti-phase initial imperfection of the crack edges

$$n_1^{\pm}=\varepsilon\pi\sin\left(2\pi\frac{(x_1-(\ell-\ell_0)/2)}{\ell_0}\right)+O(\varepsilon^3),\ n_2^{\pm}=\mp1+O(\varepsilon^2). \tag{4.14}$$

Thus, doing the mathematical manipulations described in the previous chapters and using the assumption (3.35) we obtain the field equations and corresponding boundary conditions for each approximation. Here we consider the zeroth and the first approximations only within the framework of which the investigations are made.

*The Zeroth Approximation*

The following field equations for this approximation are obtained from (4.3–4.6).

$$\frac{\partial\sigma_{ji}^{(0)}}{\partial x_j}=0,\ \varepsilon_{ij}^{(0)}=\frac{1}{2}\left(\frac{\partial u_i^{(0)}}{\partial x_j}+\frac{\partial u_j^{(0)}}{\partial x_i}\right),\ \sigma_{11}^{(0)}=A_{11}^*\varepsilon_{11}^{(0)}+A_{12}^*\varepsilon_{22}^{(0)},$$
$$\sigma_{22}^{(0)}=A_{12}^*\varepsilon_{11}^{(0)}+A_{22}^*\varepsilon_{22}^{(0)},\ \sigma_{12}^{(0)}=2A_{66}^*\varepsilon_{12}^{(0)}. \tag{4.15}$$

The boundary conditions on the upper and lower face planes of the plate-strip and on the crack's edge surfaces for the problems formulated in the previous subsection are following ones.

$$\sigma_{2i}^{(0)}\Big|_{x_2=0;h} = 0,\ \sigma_{2i}^{(0)}\Big|_{x_2=h_0\pm 0} = 0. \tag{4.16}$$

The end conditions are obtained as follows.
For the *Problems 1 and 2*:

$$\sigma_{1i}^{(0)}\Big|_{x_1=0;\ell} = p\delta_i^1,\ u_2^{(0)}\Big|_{x_1=0;\ell} = 0. \tag{4.17}$$

For the *Problems 3 and 4*:

$$u_1^{(0)}\Big|_{x_1=0} = -u_1^{(0)}\Big|_{x_1=\ell} = U(t),\ u_2^{(0)}\Big|_{x_1=0;\ell} = 0. \tag{4.18}$$

Moreover, for the *Problems 3 and 4* the relation

$$hp = \int_0^h \sigma_{11}^{(0)}(x_1,x_2,t)\Big|_{x_1=\ell/2} dx_2 \tag{4.19}$$

is obtained from the Eq. (4.11).
*The first approximation*
The field equations are:

$$\frac{\partial}{\partial x_j}\left[\sigma_{ji}^{(1)} + \sigma_{jn}^{(0)}\frac{\partial u_i^{(1)}}{\partial x_n}\right] = 0,\ \varepsilon_{ij}^{(1)} = \frac{1}{2}\left(\frac{\partial u_i^{(1)}}{\partial x_j} + \frac{\partial u_j^{(1)}}{\partial x_i}\right),$$
$$\sigma_{11}^{(1)} = A_{11}^*\varepsilon_{11}^{(1)} + A_{12}^*\varepsilon_{22}^{(1)},\ \sigma_{22}^{(1)} = A_{12}^*\varepsilon_{11}^{(1)} + A_{22}^*\varepsilon_{22}^{(1)},\ \sigma_{12}^{(1)} = 2A_{66}^*\varepsilon_{12}^{(1)}. \tag{4.20}$$

The boundary conditions on the upper and lower face planes of the plate-strip

$$\sigma_{2i}^{(1)}\Big|_{x_2=0;h} = 0 \tag{4.21}$$

are obtained. The boundary conditions on the ends of the plate-strip are:
For the *Problems 1 and 2*

$$\left[\sigma_{1i}^{(1)} + \sigma_{1n}^{(0)}\frac{\partial u_i^{(1)}}{\partial x_n}\right]\Bigg|_{x_1=0;\ell} = 0,\ u_2^{(1)}\Big|_{x_1=0,\ell} = 0. \tag{4.22}$$

For the *Problems 3 and 4*

$$u_1^{(1)}\Big|_{x_1=0,\ell} = 0,\ u_2^{(1)}\Big|_{x_1=0,\ell} = 0. \tag{4.23}$$

Consider the boundary conditions on the crack's edges. According to (4.13, 4.14, 4.1) and (4.2), we obtain from (4.7):

For the *Problems 1 and 3*

$$\left[\sigma_{2i}^{(1)} + \sigma_{2n}^{(0)} \frac{\partial u_i^{(1)}}{\partial x_n}\right]\Bigg|_{x_2=h_0\pm 0} = \sigma_{1i}^{(0)}\Big|_{x_2=h_0\pm 0} \pi \sin\left(2\pi \frac{(x_1 - (\ell - \ell_0)/2)}{\ell_0}\right)$$
$$\mp \frac{\partial \sigma_{2i}^{(0)}}{\partial x_2}\Bigg|_{x_2=h_0\pm 0} \ell_0 \sin^2\left(\pi \frac{(x_1 - (\ell - \ell_0)/2)}{\ell_0}\right). \tag{4.24}$$

For the *Problems 2 and 4*

$$\left[\sigma_{2i}^{(1)} + \sigma_{2n}^{(0)} \frac{\partial u_i^{(1)}}{\partial x_n}\right]\Bigg|_{x_2=h_0\pm 0} = \mp\, \sigma_{1i}^{(0)}\Big|_{x_2=h_0\pm 0} \pi \sin\left(2\pi \frac{(x_1 - (\ell - \ell_0)/2)}{\ell_0}\right)$$
$$+ \frac{\partial \sigma_{2i}^{(0)}}{\partial x_2}\Bigg|_{x_2=h_0\pm 0} \ell_0 \sin^2\left(\pi \frac{(x_1 - (\ell - \ell_0)/2)}{\ell_0}\right). \tag{4.25}$$

This completes the mathematical formulation of the problems related to the determination of the quantities related to the first approximation. Note that, as in the previous chapters, the buckling delamination problems under consideration are also investigated within the scope of the zeroth and the first approximations only. In this case the values of the critical time or the critical force are determined from the criterion

$$u_2^{(1)}(\ell/2, h_0 + 0, t) \to \infty \text{ as } t \to t_{cr} \text{ (for a viscoelasticplate-strip)}$$
$$\text{or as} p \to p_{cr} \text{ (for a elastic plate-strip)}. \tag{4.26}$$

Now we consider the determination of the values related to the zeroth and the first approximations.

#### 4.2.2.1 Determination of the Zeroth Approximation

For the *Problems 1 and 2* the stresses related to the zeroth approximation are determined by the expressions

$$\sigma_{11}^{(0)} = p,\ \sigma_{12}^{(0)} = 0,\ \sigma_{22}^{(0)} = 0. \tag{4.27}$$

Taking these expressions into account the strains and displacements can be easily found from the constitutive relations in (4.15) and end conditions in (4.17). As follows from the expressions in (4.27), in the zeroth approximation the existence of the crack does not causes any stress concentration around the crack's tips, because the stress $\sigma_{11}^{(0)}(=p)$ is homogeneous and its direction is parallel to the crack's edges.

For solution to the *Problems 3 and 4* it is necessary to involve a numerical method. For this purpose, first, we employ the Laplace transformation

$$\bar{\varphi}(s) = \int_0^\infty \varphi(t) e^{-st} dt \tag{4.28}$$

with $s > 0$, to the Eqs. (4.15, 4.16) and (4.18). For determination of the Laplace transformation of the sought values we use the FEM and for this purpose we introduce the functional

$$\Pi^{(0)}\left(\bar{u}_1^{(0)}, \bar{u}_2^{(0)}\right) = \frac{1}{2} \int_0^\ell \left( \int_0^h \bar{\sigma}_{ij}^{(0)} \bar{\varepsilon}_{ij}^{(0)} dx_2 \right) dx_1 . \tag{4.29}$$

Note that under writing the functional (4.29), the Euler equations of which are the first equations in (4.15) and in (4.16), the existence of the crack does not taken into account. This statement is explained as follows. The numerical results show that the stress distribution obtained within the scope of the variational formulation (4.29) in the near vicinity of the crack can be taken as homogeneous one determined by expressions in (4.27). The non-homogeneous distribution of the stresses arises only in the near vicinity of the clamped ends of the plate-strip. This is why, as in the *Problems 1 and 2*, under determination of the quantities related to the zeroth approximation of the *Problems 3 and 4*, the existence of the crack does not taken into account. At the same time, this simplifies significantly the solution procedure of the problems under consideration.

Thus, taking the foregoing discussion into account, under FEM modeling of the zeroth approximation of the *Problems 3 and 4* we use also the symmetry of the problem about $x_1 = \ell/2$, according to which, only the region $\{0 \le x_1 \le \ell/2, 0 \le x_2 \le h\}$ is considered and this region is divided into 120 quadratic Lagrange-family elements, with 533 nodes and 1026 NDOF. The inverse Laplace transformation of the sought-for values is determined by the method (Schapery 1962, 1978).

The numerous numerical results obtained in this way show, with the condition

$$hp = \int_0^h \sigma_{11}^{(0)}(x_1, x_2, t)\Big|_{x_1=\ell/2} dx_2 = const \tag{4.30}$$

satisfied, that the stresses $\sigma_{ij}^{(0)}$ determined by solving the problem related to the zeroth approximation, can be assumed time-independent with a very high accuracy. It is evident that, in order to ensure the condition (4.30), the function $2U(t)/\ell$ must coincide approximately with the creep function of the plate material.

This completes the consideration of the determination of the values related to the zeroth approximation.

#### 4.2.2.2 Determination of the First Approximation

Taking the time-independent character of the stresses $\sigma_{ij}^{(0)}$ in solving the problems related to the first approximation, we can employ the correspondence principle by using the Laplace transformation (4.28). For determination of the values of these transformations we employ FEM by the use of the following functional.

$$\Pi^{(1)}\left(\bar{u}_1^{(1)},\bar{u}_2^{(1)}\right)=\frac{1}{2}\int_0^{\ell}\left(\int_0^{h}\left[\bar{\sigma}_{ji}^{(1)}+\sigma_{jn}^{(0)}\frac{\partial\bar{u}_i^{(1)}}{\partial x_n}\right]\frac{\partial\bar{u}_i^{(1)}}{\partial x_j}dx_2\right)dx_1$$
$$\underline{-\int_0^{h}\bar{p}\bar{u}_1\big|_{x_1=0}dx_2+\int_0^{h}\bar{p}\bar{u}_1\big|_{x_1=\ell}dx_2+}\int_{(\ell-\ell_0)/2}^{(\ell+\ell_0)/2}\left(\bar{F}_{13i}^{+}\bar{u}_i^{(1)}\right)\Big|_{x_2=h_0+0}dx_1$$
$$+\int_{(\ell-\ell_0)/2}^{(\ell+\ell_0)/2}\left(\bar{F}_{13i}^{-}\bar{u}_i^{(1)}\right)\Big|_{x_2=h_0-0}dx_1,\tag{4.31}$$

where for *Problems 1 and 3*

$$\bar{F}_{13i}^{\pm}=\frac{1}{s}\sigma_{1i}^{(0)}\pi\sin\left(2\pi\frac{(x_1-(\ell-\ell_0)/2)}{\ell_0}\right)$$
$$\mp\frac{1}{s}\frac{\partial\sigma_{2i}^{(0)}}{\partial x_2}\ell_0\sin^2\left(\pi\frac{(x_1-(\ell-\ell_0)/2)}{\ell_0}\right),\tag{4.32}$$

and for *Problems 2 and 4*

$$\bar{F}_{13i}^{\pm}=\mp\frac{1}{s}\sigma_{1i}^{(0)}\pi\sin\left(2\pi\frac{(x_1-(\ell-\ell_0)/2)}{\ell_0}\right)$$
$$+\frac{1}{s}\frac{\partial\sigma_{2i}^{(0)}}{\partial x_2}\ell_0\sin^2\left(\pi\frac{(x_1-(\ell-\ell_0)/2)}{\ell_0}\right).\tag{4.33}$$

Moreover, for *Problems 3 and 4* the underlined terms in (4.31) are removed.

Doing the corresponding mathematical manipulations we obtain the equilibrium equations in (4.20), the boundary condition (4.21), the first two conditions in (4.22), the boundary conditions (4.24) and (4.25) for the listed above problems from the equation

$$\delta\Pi^{(1)}\left(\bar{u}_1^{(1)},\bar{u}_2^{(1)}\right)=\frac{\partial\Pi^{(1)}}{\partial\bar{u}_1^{(1)}}\delta\bar{u}_1^{(1)}+\frac{\partial\Pi^{(1)}}{\partial\bar{u}_2^{(1)}}\delta\bar{u}_2^{(1)}=0.\tag{4.34}$$

As in the previous sub-subsection, under FEM modeling of the Eq. (4.34) we use the symmetry of the problem about $x_1=\ell/2$, according to which, only the region $\{0 \le x_1 \le \ell/2, 0 \le x_2 \le h\}$ is considered and this region, as in the zeroth approximation, is divided into 120 quadratic Lagrange-family elements, with 533 nodes and 1039 NDOF. In this case, the finite elements containing the crack tips are taken as singular types (Benzley 1974; Zienkiewicz and Taylor 1989).

### 4.2.3 Numerical Results and Discussions

As in the previous chapter, we assume that the plate-strip material consists of the alternating two layers of two isotropic, homogeneous materials and these layers are located in the $Ox_1x_3$ plane. The reinforcing layers are supposed to be pure elastic with mechanical characteristics $E^{(2)}$ (Young's modulus), $\lambda^{(2)}$ and $\mu^{(2)}$ (Lame constants), $\nu^{(2)}$ (Poisson coefficient). The matrix layers material is supposed to be linearly viscoelastic with operators (3.68–3.70). Within the framework of the continuum approach the plate-strip material can be considered as homogeneous transversal isotropic viscoelastic one the isotropy axis of which coincide with the $Ox_2$ axis and effective operators of this material are determined through the expressions (3.83). As before, we introduce the dimensionless rheological parameter $\omega(= \omega_\infty/\omega_0)$and the dimensionless time $t'\left(= \omega_0^{1/(1+\alpha)}t\right)$ and assume that $\nu^{(2)} = \nu_0^{(1)} = 0.3$, $\eta^{(2)} = 0.5$ where $\eta^{(2)}$ is filler layers volume fraction.

#### 4.2.3.1 Testing of the Algorithm Used for Obtaining Numerical Results

First, for testing of the algorithm used in the investigations of the foregoing problems we consider the case where the plate-strip material is pure elastic and isotropic and compare the present results with the corresponding ones obtained in a paper (Gladun 2000). Note that in the paper (Gladun 2000) the stability loss of the simply supported plate-strip containing a crack the edges of which are parallel to the plate's free face planes was investigated. It was assumed that the plate-strip is made of elastic isotropic material and Euler (bifurcation) approach was employed. The corresponding eigen-value problem was solved numerically by applying the finite-difference method. Moreover, in the paper (Gladun 2000) it was assumed that $h_0 = h/2$ and concrete numerical results were obtained in the case where $\nu = 0.25$($\nu$ is a Poisson's coefficient of the plate-strip material) and $h/\ell = 0.15$. However in the paper (Gladun 2000) the stability loss mode (buckling delamination mode) does not indicated, as if, the global stability loss of the plate-strip containing the aforementioned crack was considered. The comparison of the numerical results obtained in the present investigations with the corresponding ones obtained in the paper (Gladun 2000) shows that in the results obtained the paper (Gladun) corresponds to the co-phase buckling-delamination mode. The mentioned comparison is illustrated by data given in Table 4.1.

**Table 4.1** Comparison of the values of $p_{cr}/E$ obtained in the present work with the corresponding ones obtained in the paper (Gladun 2000)

| Studies | $\ell_0/\ell$ | | | | | | |
|---|---|---|---|---|---|---|---|
| | 0.9 | 0.8 | 0.7 | 0.6 | 0.5 | 0.3 | 0.0 |
| Present work | 0.0137 | 0.0150 | 0.0168 | 0.0171 | 0.0180 | 0.0180 | 0.0180 |
| Gladun (2000) | 0.0127 | 0.0142 | 0.0158 | 0.0170 | 0.0177 | 0.0182 | 0.0180 |

**Table 4.2** The values of $p_{cr.0}\Big/E_0^{(1)}$ obtained for the co-phase (upper number) and anti-phase (lower number) buckling delamination modes under $h_0/h = 0.5$. Simply supported plate-strip

| $\frac{E^{(2)}}{E_0^{(1)}}$ | $\ell_0/\ell$ | | | | |
|---|---|---|---|---|---|
| | 0.1 | 0.3 | 0.5 | 0.7 | 0.9 |
| 1 | $\frac{0.0180}{0.2195}$ | $\frac{0.0180}{0.0910}$ | $\frac{0.0180}{0.0475}$ | $\frac{0.0168}{0.0286}$ | $\frac{0.0137}{0.0183}$ |
| 5 | $\frac{0.0542}{0.4370}$ | $\frac{0.0542}{0.2130}$ | $\frac{0.0541}{0.1204}$ | $\frac{0.0479}{0.0759}$ | $\frac{0.0381}{0.0474}$ |
| 10 | $\frac{0.0930}{0.5320}$ | $\frac{0.0930}{0.2980}$ | $\frac{0.0914}{0.1820}$ | $\frac{0.0806}{0.1200}$ | $\frac{0.0644}{0.0728}$ |

It follows from this comparison that the numerical results obtained by employing the present algorithm with the use of the FEM are very near to the corresponding ones obtained in the paper (Gladun 2000) which were obtained with the use of the finite-difference method. This statement guaranties the trustiness of the developed algorithm within the scope of the FEM modeling.

#### 4.2.3.2 Buckling Delamination of a Simply Supported Elastic Plate-Strip

Now we consider numerical results obtained within the scope of the foregoing assumption for the buckling delamination problems (*Problems 1 and 2*) for simply supported plate-strip. First we analyze numerical results obtained at $t' = 0$. These results are given in Tables 4.2 and 4.3 which are calculated under $h/\ell = 0.15$ for various values of $E^{(2)}\Big/E_0^{(1)}$ and $\ell_0/\ell$. Note that the results given in Table 4.2 are obtained in the case where $h_0/h = 0.5$, but the results given in Table 4.3 are obtained in the cases where $h_0/h = 0.625$ and 0.750. In other words, the results given in Table 4.2 relate to the case where the crack is on the middle plane of the plate-strip, but the results given in Table 4.3 relate to the case where the crack is near to the upper free plane of the plate-strip rather than the lower free plane of that.

The analyses of the foregoing numerical results allow us to make the following conclusions:

i. The main parameter through which the character of the stability loss or buckling delamination mode can be determined is the parameter

**Table 4.3** The values of $p_{cr.0}\big/E_0^{(1)}$ obtained for the co-phase and anti-phase initial imperfection modes for various $h_0/h$ . The Simply supported plate-strip

| $\frac{E_0^{(2)}}{E_0^{(1)}}$ | $h_0/h$ | | | $\ell_0/\ell$ | | |
|---|---|---|---|---|---|---|
| | | 0.1 | 0.3 | 0.5 | 0.7 | 0.9 |
| 1 | 0.625 | 0.0180 | 0.0180 | 0.0180 | 0.0153 | 0.0105 |
| | 0.750 | 0.0180 | 0.0180 | 0.0153 | 0.0086 | 0.0054 |
| 5 | 0.625 | 0.0542 | 0.0542 | 0.0533 | 0.0423 | 0.0291 |
| | 0.750 | 0.0542 | 0.0542 | 0.0423 | 0.0245 | 0.0155 |
| 10 | 0.625 | 0.0930 | 0.0930 | 0.0898 | 0.0710 | 0.0484 |
| | 0.750 | 0.0930 | 0.0930 | 0.0708 | 0.0421 | 0.0268 |

$$\psi_{h_0} = \frac{h - h_0}{\ell_0}. \tag{4.35}$$

ii. In the case where $\psi_{h_0} \geq h/\ell$ the global stability loss of the plate-strip takes place earlier than the buckling delamination around the crack and the existence of the crack has almost no influence on the values of the critical force.

iii. In the case where $\psi_{h_0} \leq h/\ell$ the buckling delamination of the plate-strip around the crack takes place earlier than the global stability loss and the values of the critical force decrease with a decrease in the values of the parameter $\psi_{h_0}$ (4.35).

iv. Under $\psi_{h_0} \leq h/\ell$ in the case where $h_0/h = 0.5$ (i.e. in the case where the crack (or rather the crack's tips) is on the middle plane of the plate-strip) the co-phase and anti-phase initial imperfection modes of the crack's edges (Fig. 4.1) lead to the co-phase and anti-phase buckling delamination modes (Fig. 4.2) respectively.

v. Under $\psi_{h_0} \leq h/\ell$ in the cases where the crack (or rather the crack's tips) are not on the middle plane of the plate-strip (for example in the cases where $h_0/h =$ 0.625 and 0.750) both the co-phase and anti-phase initial imperfection modes of the crack's edges (Fig. 4.1) lead to only the co-phase buckling delamination mode (Fig. 4.2a).

vi. Under $\psi_{h_0} \leq h/\ell$ the values of the critical forces do not depend on the values of $h/\ell$. This means that the numerical results obtained for the cases under consideration can also be taken as the results for the near surface buckling delamination of the half-plane containing a near-surface crack.

This completes the analyses of the numerical results related to the stability loss of the elastic simply supported plate-strip containing a crack the edges of which have an infinitesimal initial imperfection.

**Table 4.4** The values of $p_{cr.0}/E_0^{(1)}$ obtained for the co-phase (upper number) and anti-phase (lower number) buckling delamination modes under $h_0/h = 0.5$. For the clamped plate-strip

| $\frac{E^{(2)}}{E_0^{(1)}}$ | $\ell_0/\ell$ | | | |
|---|---|---|---|---|
| | 0.1 | 0.3 | 0.5 | 0.7 |
| 1 | 0.0630<br>0.2194 | 0.0635<br>0.0910 | 0.0458<br>0.0474 | 0.0299<br>0.0285 |
| 5 | 0.1638<br>0.4364 | 0.1569<br>0.2126 | 0.1159<br>0.1198 | 0.0801<br>0.0755 |
| 10 | 0.2498<br>0.5320 | 0.2424<br>0.2984 | 0.1842<br>0.1828 | 0.1317<br>0.1212 |

**Table 4.5** The values of $p_{cr.0}/E_0^{(1)}$ obtained for the co-phase and anti-phase initial imperfection modes for various $h_0/h$. For the clamped plate-strip

| $E^{(2)}/E_0^{(1)}$ | $h_0/h$ | $\ell_0/\ell$ | | | |
|---|---|---|---|---|---|
| | | 0.1 | 0.3 | 0.5 | 0.7 |
| 1 | 0.5333 | 0.0630 | 0.0606 | 0.0416 | 0.0259 |
| | 0.625 | 0.0630 | 0.0580 | 0.0305 | 0.0180 |
| | 0.6666 | 0.0630 | 0.0536 | 0.0261 | 0.0149 |
| | 0.750 | 0.0630 | 0.0373 | 0.0162 | 0.0089 |
| 5 | 0.5333 | 0.1638 | 0.1563 | 0.1091 | 0.0702 |
| | 0.625 | 0.1638 | 0.1476 | 0.0819 | 0.0496 |
| | 0.6666 | 0.1638 | 0.1364 | 0.0702 | 0.0416 |
| | 0.750 | 0.1638 | 0.0974 | 0.0447 | 0.0253 |
| 10 | 0.5333 | 0.2498 | 0.2375 | 0.1697 | 0.1139 |
| | 0.625 | 0.2498 | 0.2228 | 0.1310 | 0.0827 |
| | 0.6666 | 0.2498 | 0.2046 | 0.1128 | 0.0694 |
| | 0.750 | 0.2498 | 0.1523 | 0.0746 | 0.0435 |

#### 4.2.3.3 Buckling Delamination of the Clamped Elastic Plate-Strip

Within the scope of the assumptions accepted in the previous sub-subsection consider the numerical results related to the stability loss of the clamped plate-strip containing a crack. In the case where $h_0/h = 0.5$ i.e. for the case where the crack (or rather the crack's tips) is on the middle plane these results are given in Table 4.4. The numerical results obtained for the cases where the crack's location is near to the upper free face plane of the plate-strip are given in Table 4.5. Analyzes of these results show that the conclusion (i.) obtained in the previous sub-subsection also occurs for the considered case. According to this conclusion, the main parameter through which the character of the stability loss or buckling delamination modes is determined, is the parameter $\psi_{h_0}$ (4.35). For clarity, the values of $\psi_{h_0}$ calculated for various values of $h_0/h$ and $\ell_0/\ell$ are given in Table 4.6. However, the conclusions (ii.–vi.) obtained in the previous sub-subsection must be replaced with the following ones for the problem under consideration.

**Table 4.6** The values of the parameter $\psi_{h_0}$ (4.35)

| $h_0/h$ | $\ell_0/\ell$ | | | |
|---|---|---|---|---|
| | 0.1 | 0.3 | 0.5 | 0.7 |
| 0.5333 | 0.7000 | 0.2333 | 0.1400 | 0.0999 |
| 0.6250 | 0.5625 | 0.1875 | 0.1125 | 0.0803 |
| 0.6666 | 0.5000 | 0.1666 | 0.1000 | 0.0714 |
| 0.7500 | 0.3750 | 0.1250 | 0.0750 | 0.0535 |

**Table 4.7** The values of $p_{cr.0}/E_0^{(1)}$ obtained for various $h/\ell$ and $E^{(2)}/E_0^{(1)}$ under $\psi_{h_0} = 0.125$

| $\frac{h}{\ell}$ | $\frac{E_0^{(1)}}{E^{(2)}}$ | $\frac{(h-h_0)/\ell}{\ell_0/\ell}$ | | | |
|---|---|---|---|---|---|
| | | $\frac{0.0500}{0.4000}$ | $\frac{0.04375}{0.35000}$ | $\frac{0.0375}{0.3000}$ | $\frac{0.03125}{0.25000}$ |
| 0.15 | 5 | 0.0952 | 0.0962 | 0.0974 | 0.0993 |
| | 10 | 0.1498 | 0.1508 | 0.1523 | 0.1546 |
| 0.20 | 5 | 0.0971 | 0.0977 | 0.0987 | 0.1002 |
| | 10 | 0.1523 | 0.1530 | 0.1542 | 0.1561 |
| 0.25 | 5 | 0.0978 | 0.0983 | 0.0992 | 0.1007 |
| | 10 | 0.1534 | 0.1540 | 0.1555 | 0.1568 |

i. There exists such a value of $\psi_{h_0}$ (denoted by $\psi_{h_0}^{*}$) before which, (i.e. under $\psi_{h_0} \geq \psi_{h_0}^{*}$), the global stability loss of the clamped plate-strip takes place earlier than the buckling delamination around the crack and the existence of the crack has almost no influence on the values of the critical force. Therefore, as the ends of the plate-strip are clamped, it is found that $\psi_{h_0}^{*} > h/\ell$.
ii. There also exists such a value of $\psi_{h_0}$ (denoted by $\psi_{h_0}^{**}$) after which, (i.e. under $\psi_{h_0} \leq \psi_{h_0}^{**}$), the local buckling delamination around the crack takes place earlier than the global stability loss of the clamped plate-strip and the values of the critical forces do not depend on the values of $h/\ell$.
iii. In the cases where $\psi_{h_0}^{**} \leq \psi_{h_0} \leq \psi_{h_0}^{*}$ the buckling delamination around the crack also takes place earlier than the global stability loss of the clamped plate-strip. However in these cases the values of the critical forces depend on the values of $h/\ell$. According to this statement, the mode of the stability loss which arises in the cases where $\psi_{h_0}^{**} \leq \psi_{h_0} \leq \psi_{h_0}^{*}$ can be called the "mixed" buckling delamination mode.
iv. According to the data given in Tables 4.5 and 4.6, for the cases under consideration, $\psi_{h_0}^{*}$ and $\psi_{h_0}^{**}$ can be approximately taken as 0.2333 and 0.125 respectively. Consequently, in the case where $\psi_{h_0} = 0.125$, the values of $p_{cr.0}/E_0^{(1)}$ must not be dependent on the values of $h/\ell$ and $\ell_0/\ell$. The results given in Table 4.7 prove this conclusion as the difference between the values of $p_{cr.0}/E_0^{(1)}$, obtained for various $h/\ell$ and $\ell_0/\ell$, is not more than 5 %.

**Table 4.8** The values of $p_{cr.0}/E_0^{(1)}$ (upper number) and $p_{cr.\infty}/E_0^{(1)}$ (lower number) obtained for various $E^{(2)}/E_0^{(1)}$, $h_0/h$ (or $\psi_{h_0}$) under $\ell_0/\ell = 0.25$ and $\omega = 1$

| $\frac{E^{(2)}}{E_0^{(1)}}$ | $(h_0/h)/\psi_{h_0}$ | | |
|---|---|---|---|
| | 0.5 / 0.3 | 0.60 / 0.24 | 0.6666 / 0.2000 |
| 1 | 0.0489 / 0.0365 | 0.0356 / 0.0287 | 0.0273 / 0.0206 |
| 10 | 0.1235 / 0.0928 | 0.0932 / 0.0731 | 0.0725 / 0.0584 |
| 20 | 0.2700 / 0.1775 | 0.2242 / 0.1534 | 0.1820 / 0.1316 |
| 50 | 0.3989 / 0.2354 | 0.3480 / 0.2140 | 0.3003 / 0.1933 |

v. Under $\psi_{h_0} < \psi_{h_0}^*$ in the case where the crack tips are on the middle plane of the plate-strip (i.e. in the cases where $h_0/h = 0.5$) the co-phase and anti-phase initial imperfection modes of the crack's edges (Fig. 4.1) lead to the co-phase and anti-phase buckling delamination modes (Fig. 4.2). However, in the cases where the crack's tips are not on the middle plane of the plate-strip and the crack's location is near the free face plane (i.e. in the cases where, for example, $h_0/h = 0.5333, 0.625, 0.6666$ and $0.750$), both the co-phase and anti-phase initial imperfection modes of the crack's edges only lead to the co-phase buckling delamination mode. Note that this conclusion agrees with the (v.)th conclusion made in the previous sub-subsection.

Thus we complete consideration of the results related to the stability loss of the clamped plate strip containing a crack the edges of which have an infinitesimal initial imperfection.

#### 4.2.3.4 Buckling Delamination of the Clamped Viscoelastic Plate-Strip

Consider the case where $\ell_0/\ell = 0.25$, $h/\ell = 0.15$ for which the values of $p_{cr.0}/E_0^{(1)}$ and $p_{cr.\infty}/E_0^{(1)}$ are given in Table 4.8. Note that the numerical results given in this table are obtained for various $E^{(2)}/E_0^{(1)}$, $h_0/h$ and $\psi_{h_0}(= (h - h_0)/\ell_0)$ under $\omega = 1$. As has been noted in previous chapters for realization of the stability loss or buckling delamination of the plate-strip under consideration the magnitude of the intensity of the external compressing force $p$ must satisfy the relation $p_{cr.\infty} < p < p_{cr.0}$. Thus, analyze the results related to the critical time $t'_{cr}$ which are obtained for the values of $p$ satisfied this relation and for the problem parameters $E^{(2)}/E_0^{(1)} = 50$, $\omega = 1.0$ and $\alpha = -0.5$. These results are given in Table 4.9.

It follows from the Table 4.9 that $t'_{cr} \to 0\ (t'_{cr} \to \infty)$ as $p \to p_{cr.0}\ (p \to p_{cr.\infty})$) Moreover, this table shows that for fixed $h/\ell$ and $\ell_0/\ell$ the values of $t'_{cr}$ decrease with decreasing $\psi_{h_0}$. This means that for fixed length $\ell_0$ of the crack $t'_{cr}$ decrease

**Table 4.9** The values of $t'_{cr}$ obtained for various values of $p/E_0^{(1)}$ and $\psi_{h_0}$ under $\ell_0/\ell = 0.25$, $E^{(2)}/E_0^{(1)} = 50$, $\omega = 1$ and $\alpha = -0.5$

| $\psi_{h_0}$ | $p/E_0^{(1)}$ | | | | | | |
|---|---|---|---|---|---|---|---|
| | 0.200 | 0.250 | 0.280 | 0.300 | 0.320 | 0.350 | 0.380 |
| 0.30 | $\infty$ | 17.304 | 1.169 | 0.383 | 0.141 | 0.029 | 0.002 |
| 0.24 | $\infty$ | 1.313 | 0.186 | 0.054 | 0.012 | 0.000 | 0.000 |
| 0.20 | 43.116 | 0.148 | 0.010 | 0.000 | 0.000 | 0.000 | 0.000 |

**Table 4.10** The influence of the rheological parameter $\omega$ on the values $t'_{cr}$ for various values of $p/E_0^{(1)}$ and $\psi_{h_0}$ under $\ell_0/\ell = 0.25$, $E^{(2)}/E_0^{(1)} = 50$ and $\alpha = -0.5$

| $\psi_{h_0}$ | $p/E_0^{(1)}$ | $\omega$ | | |
|---|---|---|---|---|
| | | 0.5 | 1.0 | 2.0 |
| 0.30 | 0.350 | 0.023 | 0.029 | 0.051 |
| | 0.300 | 0.185 | 0.385 | 25.082 |
| 0.24 | 0.320 | 0.010 | 0.012 | 0.018 |
| | 0.300 | 0.041 | 0.054 | 0.130 |
| 0.20 | 0.280 | 0.008 | 0.010 | 0.014 |
| | 0.250 | 0.091 | 0.148 | 0.722 |

**Table 4.11** The influence of the rheological parameter $\alpha$ on the values $t'_{cr}$ for various values of $p/E_0^{(1)}$ and $\psi_{h_0}$ under $\ell_0/\ell = 0.25$, $E^{(2)}/E_0^{(1)} = 50$ and $\omega = 1.0$

| $\psi_{h_0}$ | $p/E_0^{(1)}$ | $\alpha$ | | |
|---|---|---|---|---|
| | | −0.3 | −0.5 | −0.7 |
| 0.30 | 0.350 | 0.066 | 0.029 | 0.004 |
| | 0.300 | 0.414 | 0.385 | 0.321 |
| 0.24 | 0.320 | 0.036 | 0.012 | 0.001 |
| | 0.300 | 0.106 | 0.054 | 0.013 |
| 0.20 | 0.280 | 0.031 | 0.010 | 0.0007 |
| | 0.250 | 0.210 | 0.148 | 0.066 |

with decreasing $(h - h_0)$, i.e. with decreasing the thickness of the region between the crack and upper free face plane of the plate.

Now we consider numerical results illustrated the influence of the rheological parameters $\omega$ and $\alpha$ on the values of $t'_{cr}$. Again we consider the case where $E^{(2)}/E_0^{(1)} = 50$, $\ell_0/\ell = 0.25$ and analyze the data given in Tables 4.10 and 4.11 which show the values of $t'_{cr}$ obtained for various $\omega$ under $\alpha = -0.5$ (Table 4.10) and for various $\alpha$ under $\omega = 1.0$ (Table 4.11). At the same time, the numerical results given in these tables are calculated for various $\psi_{h_0}$.

**Table 4.12** The values of $t'_{cr}$ calculated for various $E^{(2)}/E_0^{(1)}$ and $\eta$

| $\eta$ | $E^{(2)}/E_0^{(1)}$ $(p_{cr.0}/p_{cr.\infty})$ | | |
|---|---|---|---|
| | 1 (0.0162/0.0137) | 5 (0.0447/0.0404) | 10 (0.0746/0.0669) |
| 0.5 | 0.063 | 0.082 | 0.086 |
| 0.9 | 5.169 | 6.467 | 7.051 |

It follows from these tables that in the case under consideration an increase in the values of $\omega$ (in the values of $|\alpha|$) causes to increase (to decrease) in the values of $t'_{cr}$. This conclusion agrees with that made in Sect. 2.5.1.

Now we consider a question how the ratio $E^{(2)}/E_0^{(1)}$ acts on the values of the critical time. For this purpose, we introduce the following expression

$$p = p_{cr.0}(1-\eta) + \eta p_{cr.\infty}, \tag{4.36}$$

where $\eta$ is a parameter $(0 \leq \eta \leq 1)$ through which the value of $p$ can be determined, for example, the value $\eta = 0\,(\eta = 1)$ corresponds the case where $p = p_{cr.0}$ $(p = p_{cr.\infty})$. An increase in the values of the parameter $\eta$ causes to decrease in the values of $p$. Consequently, the value of the parameter $\eta$ shows the "position" of the selected $p$ between the values of $p_{cr.0}$ and $p_{cr.\infty}$. Therefore, in the certain sense, the values of $t'_{cr}$ calculated for various $E^{(2)}/E_0^{(1)}$ under the same value of $\eta$ can be used for estimation of the action of the ratio $E^{(2)}/E_0^{(1)}$ on the values of $t'_{cr}$. Taking this discussion into account consider the results given in Table 4.12 which show the values of $t'_{cr}$ calculated for various $E^{(2)}/E_0^{(1)}$ and $\eta$. Note that these calculations are made in the case where $\ell_0/\ell = 0.5$, $h/\ell = 0.15$, $\omega = 2.0$ and $\psi_{h_0} = 0.075$.

The comparison of the values of the critical time $t'_{cr}$ calculated for various $E^{(2)}/E_0^{(1)}$ under the same value of the parameter $\eta$ shows that an increase in the values of $E^{(2)}/E_0^{(1)}$ causes to increase in the $t'_{cr}$.

## 4.3 Buckling Delamination of the Plate-Strip with Two Parallel Cracks

### 4.3.1 Mathematical Formulation of the Problem

We attempt to investigate the problem related to buckling delamination of the clamped plate-strip containing two parallel cracks with the same length $\ell_0$. The location of the cracks is shown in Fig. 4.3. Assume that the cracks edges have

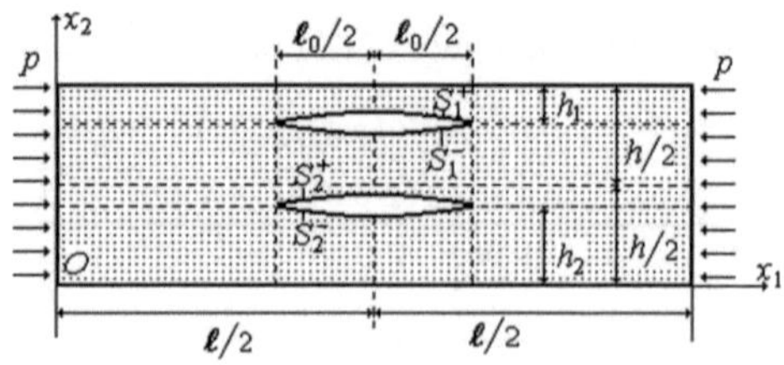


**Fig. 4.3** The geometry of the plate-strip containing two parallel cracks

anti-phase infinitesimal initial imperfections and, as in the previous section, we determine the position of the points of the plate-strip through Lagrange coordinates $x_1$, $x_2$ and $x_3$ in Cartesian coordinate system $Ox_1x_2x_3$. We consider the plane-strain state in the $Ox_1x_2$ plane and investigate the evolution of the noted above initial infinitesimal imperfections of the cracks edges under compressing of that along the $Ox_1$ axis through the rigid bodies, that hold the ends, by the uniformly distributed normal forces with intensity $p$.

In the selected coordinate system the equation of the initial imperfection of the cracks edge-surfaces can be written as follows:

For the surfaces $S_1^+$ and $S_1^-$ (Fig. 4.3) of the upper crack

$$x_{21}^{+} = h - h_1 + \varepsilon\ell_0 \sin^2\left(\tfrac{\pi(x_1^+ - (\ell-\ell_0)/2)}{\ell_0}\right),\ x_1^+ \in \left[\tfrac{\ell-\ell_0}{2}, \tfrac{\ell+\ell_0}{2}\right],$$
$$x_{21}^{-} = h - h_1 - \varepsilon\ell_0 \sin^2\left(\tfrac{\pi(x_1^- - (\ell-\ell_0)/2)}{\ell_0}\right),\ x_1^- \in \left[\tfrac{\ell-\ell_0}{2}, \tfrac{\ell+\ell_0}{2}\right]. \tag{4.37}$$

For the surfaces $S_2^+$ and $S_2^-$ (Fig. 4.3) of the lower crack

$$x_{22}^{+} = h_2 + \varepsilon\ell_0 \sin^2\left(\tfrac{\pi(x_1^+ - (\ell-\ell_0)/2)}{\ell_0}\right),\ x_1^+ \in \left[\tfrac{\ell-\ell_0}{2}, \tfrac{\ell+\ell_0}{2}\right],$$
$$x_{22}^{-} = h_2 - \varepsilon\ell_0 \sin^2\left(\tfrac{\pi(x_1^- - (\ell-\ell_0)/2)}{\ell_0}\right),\ x_1^- \in \left[\tfrac{\ell-\ell_0}{2}, \tfrac{\ell+\ell_0}{2}\right]. \tag{4.38}$$

Assume that the plate material is orthotropic one with the principal axes of the elastic symmetry $Ox_1$, $Ox_2$ and $Ox_3$, and within the scope of the foregoing assumptions we suppose that the field equations (4.3–4.6) satisfy in the region occupied by the plate-strip. The boundary conditions on the cracks edges $S_1^{\pm}$ and $S_2^{\pm}$ are formulated as follows.

$$\left[\left(\delta_i^n + \frac{\partial u_i}{\partial x_n}\right)\sigma_{jn}\right]\Big|_{S_1^{\pm}} n_{j1}^{\pm} = 0,\ \left[\left(\delta_i^n + \frac{\partial u_i}{\partial x_n}\right)\sigma_{jn}\right]\Big|_{S_2^{\pm}} n_{j2}^{\pm} = 0, \tag{4.39}$$

where $n_{j1}^{\pm}$ and $n_{j2}^{\pm}$ are components of the unit normal to the surfaces $S_1^{\pm}$ and $S_2^{\pm}$ respectively. Moreover, we assume that the condition on the upper and lower face planes of the plate (4.8) and the end conditions (4.10) and (4.11) are also satisfied.

Thus, this completes the formulation of the problem.

### 4.3.2 Method of Solution: FEM Modeling

As in the previous section for solution to the formulated above problem we use the series presentation (4.12) with respect to the small parameter $\varepsilon$. Doing the mathematical manipulations which are described in the previous section, and using the presentation

$$n_{11}^{\pm} = n_{12}^{\pm} = \varepsilon\pi \sin\left(2\pi\frac{(x_1 - (\ell - \ell_0)/2)}{\ell_0}\right) + O(\varepsilon^3),\ n_{21}^{\pm} = n_{21}^{\pm} = \mp 1 + O(\varepsilon^2), \tag{4.40}$$

we obtain the field equations in (4.15), boundary condition (4.16) and end conditions (4.18) and (4.19) for determination of the zeroth approximation. Consequently, for the zeroth approximation the existence of the cracks do not taken into account. After employing the Laplace transformation (4.28), for determination of these transformations the FEM modeling is applied through the functional (4.29). The relation between the compressing force $p$ and displacement $U(t)$ of the plate's ends is determined through the Eq. (4.30).

In a similar manner, we obtain the field equations in (4.20), boundary (4.21) and end (4.23) conditions for the first approximation. Moreover, for the first approximation we obtain the following conditions, which are obtained from the Eq. (4.39).

$$\left.\left[\sigma_{2i}^{(1)} + \sigma_{2n}^{(0)}\frac{\partial u_i^{(1)}}{\partial x_n}\right]\right|_{x_2=h-h_1\pm 0} = \mp \left.\sigma_{1i}^{(0)}\right|_{x_2=h-h_1\pm 0} \pi \sin\left(2\pi\frac{(x_1 - (\ell - \ell_0)/2)}{\ell_0}\right)$$
$$+ \left.\frac{\partial \sigma_{2i}^{(0)}}{\partial x_2}\right|_{x_2=h-h_1\pm 0} \ell_0 \sin^2\left(\pi\frac{(x_1 - (\ell - \ell_0)/2)}{\ell_0}\right), \tag{4.41}$$

$$\left.\left[\sigma_{2i}^{(1)} + \sigma_{2n}^{(0)}\frac{\partial u_i^{(1)}}{\partial x_n}\right]\right|_{x_2=h_2\pm 0} = \mp \left.\sigma_{1i}^{(0)}\right|_{x_2=h_2\pm 0} \pi \sin\left(2\pi\frac{(x_1 - (\ell - \ell_0)/2)}{\ell_0}\right)$$
$$+ \left.\frac{\partial \sigma_{2i}^{(0)}}{\partial x_2}\right|_{x_2=h_2\pm 0} \ell_0 \sin^2\left(\pi\frac{(x_1 - (\ell - \ell_0)/2)}{\ell_0}\right). \tag{4.42}$$

Using the correspondence principle, for FEM modeling of the boundary value problem related to the Laplace transformation of the first approximation we introduce the following functional.

$$\Pi^{(1)}\left(\bar{u}_1^{(1)},\bar{u}_2^{(1)}\right)=\frac{1}{2}\int_0^{\ell}\left(\int_0^{h}\left[\bar{\sigma}_{ji}^{(1)}+\sigma_{jn}^{(0)}\frac{\partial\bar{u}_i^{(1)}}{\partial x_n}\right]\frac{\partial\bar{u}_i^{(1)}}{\partial x_j}dx_2\right)dx_1$$
$$+\int_{(\ell-\ell_0)/2}^{(\ell+\ell_0)/2}\left(\bar{F}_{13i}^{+}\bar{u}_i^{(1)}\right)\Big|_{x_2=h-h_1+0}dx_1+\int_{(\ell-\ell_0)/2}^{(\ell+\ell_0)/2}\left(\bar{F}_{13i}^{-}\bar{u}_i^{(1)}\right)\Big|_{x_2=h-h_1-0}dx_1$$
$$+\int_{(\ell-\ell_0)/2}^{(\ell+\ell_0)/2}\left(\bar{F}_{13i}^{+}\bar{u}_i^{(1)}\right)\Big|_{x_2=h_2+0}dx_1+\int_{(\ell-\ell_0)/2}^{(\ell+\ell_0)/2}\left(\bar{F}_{13i}^{-}\bar{u}_i^{(1)}\right)\Big|_{x_2=h_2-0}dx_1, \tag{4.43}$$

where

$$\bar{F}_{13i}^{\pm}=\mp\frac{1}{s}\sigma_{1i}^{(0)}\pi\sin\left(2\pi\frac{(x_1-(\ell-\ell_0)/2)}{\ell_0}\right)$$
$$+\frac{1}{s}\frac{\partial\sigma_{2i}^{(0)}}{\partial x_2}\ell_0\sin^2\left(\pi\frac{(x_1-(\ell-\ell_0)/2)}{\ell_0}\right). \tag{4.44}$$

Using the algorithm which has been employed in the previous section, we obtain numerical results on the critical parameters.

## 4.3.3 Numerical Results and Discussions

Within the scope of the assumptions accepted in Sect. 4.2.2 we consider the numerical results related to the buckling delamination of the pure elastic plate-strip containing two parallel cracks shown in Fig. 4.3. In this case the following calculation procedure is employed. First, from the criterion (4.26) the first critical values of the compressed force (denoted by $p_{Icr.0}$) are calculated. In the next step, it is considered the cases where the values of the compressed force $p$ are greater than the corresponding values of $p_{Icr.0}$ and from the criterion (4.26) the second critical values (denoted by $p_{IIcr.0}$) of the compressed force $p$ are determined.

Now we consider again the values of $p_{cr.0}$ obtained for the *Problem 4* formulated in Sect. 4.2.1, i.e. for the same problem formulated in the previous subsection for the plate-strip, but with a singly crack (Fig. 4.1b). We tabulate the values of $p_{cr.0}/E_0^{(1)}$ which are calculated for various $(h-h_0)/l$, $l_0/l$ and $E^{(2)}/E_0^{(1)}$ in the case where $h/\ell=0.15$. These values of $p_{cr.0}/E_0^{(1)}$ are given in Table 4.13.

We turn to the consideration of the results obtained for the aforementioned $p_{Icr.0}$ and $p_{IIcr.0}$ which are obtained for the plate-strip with two parallel cracks (Fig. 4.3) in the case where $h/\ell=0.20$. Note that the values of $p_{Icr.0}/E_0^{(1)}$ calculated for

**Table 4.13** The values of $p_{cr.0}/E_0^{(1)}$ obtained for *Problem 4* for various $(h-h_0)/\ell$, $\ell_0/\ell$ and $E^{(2)}/E_0^{(1)}$ under $h/\ell = 0.15$

| $\frac{E^{(2)}}{E_0^{(1)}}$ | $\frac{h-h_0}{\ell}$ | $\ell_0/\ell$ | | | | |
|---|---|---|---|---|---|---|
| | | 0.3 | 0.5 | 0.6 | 0.7 | 0.8 |
| 1 | 0.075 | 0.0635 | 0.0458 | 0.0363 | 0.0285 | 0.0230 |
| | 0.070 | 0.0606 | 0.0416 | 0.0326 | 0.0259 | 0.0211 |
| | 0.050 | 0.0536 | 0.0261 | 0.0194 | 0.0149 | 0.0118 |
| 5 | 0.075 | 0.1569 | 0.1159 | 0.0940 | 0.0755 | 0.0622 |
| | 0.070 | 0.1563 | 0.1091 | 0.0869 | 0.0702 | 0.0581 |
| | 0.050 | 0.1364 | 0.0702 | 0.0533 | 0.0416 | 0.0335 |
| 10 | 0.075 | 0.2424 | 0.1828 | 0.1466 | 0.1212 | 0.1034 |
| | 0.070 | 0.2375 | 0.1697 | 0.1381 | 0.1139 | 0.0966 |
| | 0.050 | 0.2046 | 0.1128 | 0.0874 | 0.0694 | 0.0568 |

**Table 4.14** The values of $p_{Icr.0}/E_0^{(1)}$ under $h_1/\ell = 0.05$, $h/\ell = 0.20$ and in the cases where $0.07 \leq h_2/\ell \leq 0.075$

| $\frac{E^{(2)}}{E_0^{(1)}}$ | $\ell_0/\ell$ | | | | |
|---|---|---|---|---|---|
| | 0.3 | 0.5 | 0.6 | 0.7 | 0.8 |
| 1 | 0.0536 | 0.0261 | 0.0194 | 0.0149 | 0.0118 |
| 5 | 0.1364 | 0.0702 | 0.0533 | 0.0416 | 0.0335 |
| 10 | 0.2046 | 0.1128 | 0.0874 | 0.0694 | 0.0568 |

various $\ell_0/\ell$ under $h_1/\ell = 0.05$ are given in Table 4.14. The numerous numerical investigations show that in the cases where $h_1/\ell < h_2/\ell < h/\ell$ the values of $p_{Icr.0}/E_0^{(1)}$ do not depend on the values of $h_2/\ell$. For example, by the direct verification it was established that the results given in Table 4.14 are the same ones for the cases where $0.07 \leq h_2/l \leq 0.075$.

Tables 4.15 and 4.16 show the values of $p_{IIcr.0}/E_0^{(1)}$ obtained for the cases where $h_2/\ell = 0.07$ and 0.075 respectively. The values of $\ell_0/\ell$ and $E^{(2)}/E_0^{(1)}$ are selected as it has been selected under calculation of the results given in Tables 4.13 and 4.14. The numerical results obtained for various $h_1/\ell$ allow us to conclude that for the cases under considerations for $0.00 \leq h_1/\ell \leq 0.07$ the values of $p_{IIcr.0}/E_0^{(1)}$ do not depend on the values of $h_1/\ell$.

From the comparison the values of $p_{Icr.0}/E_0^{(1)}$, $p_{IIcr.0}/E_0^{(1)}$ given in Tables 4.14, 4.15 and 4.16 with the corresponding values of $p_{cr.0}/.E_0^{(1)}$ given in Table 4.13 follows that for the cases under considerations the following relations occur.

**Table 4.15** The values of $p_{IIcr.0}\big/E_0^{(1)}$ under $h_2/\ell = 0.07$, $h/\ell = 0.20$ and in the cases where $0 < h_1/\ell \leq 0.05$

| $\frac{E^{(2)}}{E_0^{(1)}}$ | $\ell_0/\ell$ | | | | |
|---|---|---|---|---|---|
| | 0.3 | 0.5 | 0.6 | 0.7 | 0.8 |
| 1 | 0.0606 | 0.0416 | 0.0326 | 0.0259 | 0.0211 |
| 5 | 0.1563 | 0.1091 | 0.0863 | 0.0702 | 0.0581 |
| 10 | 0.2375 | 0.1697 | 0.1381 | 0.1139 | 0.0966 |

**Table 4.16** The values of $p_{IIcr.0}\big/E_0^{(1)}$ under $h_2/\ell = 0.075$, $h/\ell = 0.20$ and in the cases where $0 < h_1/\ell \leq 0.07$

| $\frac{E^{(2)}}{E_0^{(1)}}$ | $\ell_0/\ell$ | | | | |
|---|---|---|---|---|---|
| | 0.3 | 0.5 | 0.6 | 0.7 | 0.8 |
| 1 | 0.0635 | 0.0458 | 0.0363 | 0.0285 | 0.0230 |
| 5 | 0.1569 | 0.1159 | 0.0940 | 0.0755 | 0.0622 |
| 10 | 0.2424 | 0.1828 | 0.1466 | 0.1212 | 0.1034 |

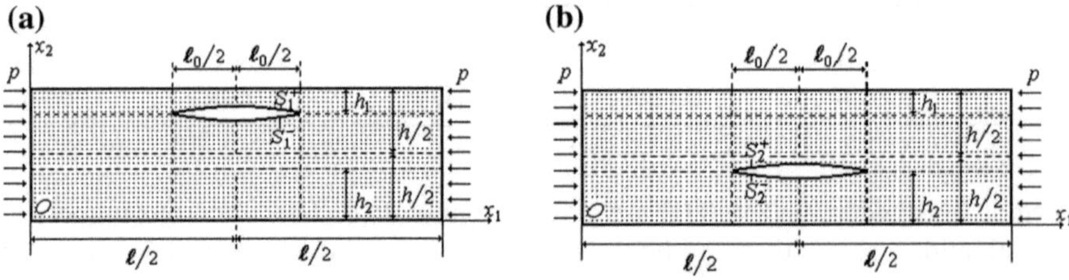


**Fig. 4.4** The geometry of the plate-strip containing only the *upper* (**a**) and *lower* (**b**) crack shown in Fig. 4.3

$$p_{Icr.0}\big/E_0^{(1)} = p_{cr.0}\big/E_0^{(1)} \text{ if } h_1/\ell = (h - h_0)/\ell \text{ and } h_1/\ell < h_2/\ell < h/\ell,$$
$$p_{IIcr.0}\big/E_0^{(1)} = p_{cr.0}\big/E_0^{(1)} \text{ if } h_2/\ell = (h - h_0)/\ell \text{ and } h_1/\ell < h_2/\ell < h/\ell. \tag{4.45}$$

Thus, according to the relation (4.45), it can be conclude that the values of $p_{Icr.0}\big/E_0^{(1)}$ $\left(p_{IIcr.0}\big/E_0^{(1)}\right)$ relate to the buckling delamination of the plate-strip with a single crack shown in Fig. 4.4a (Fig. 4.4b). In the other words, the relation (4.45) means that within the scope of the certain restrictions, such as $h_1/\ell < h_2/\ell < h/\ell$, the solution to the boundary value problem related to the buckling delamination of the plate-strip containing two parallel cracks can be reduced to two separate buckling delamination problems related to the same plate-strip containing a single crack the location geometry of which is show in Fig. 4.4. This conclusion can be

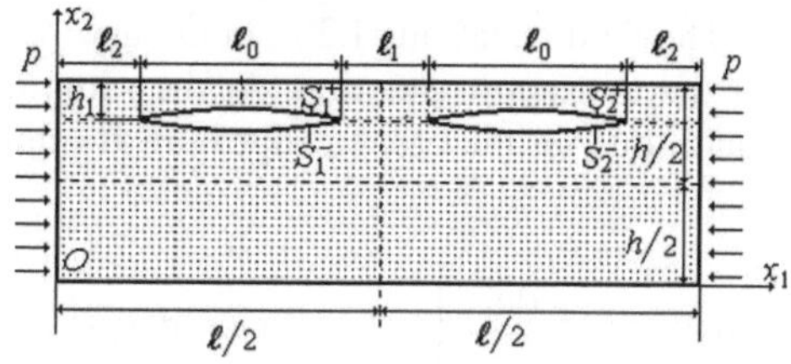

**Fig. 4.5** The geometry of the plate-strip containing two collinear cracks

also extended for the case where the plate-strip contains parallel cracks the number of which is more that two. Moreover, this conclusion can be also extended for the parallel cracks with various lengths.

The foregoing conclusion can simplify significantly the investigation of the buckling delamination problems related to the plate-strip containing multi-cracks which are parallel to each other and to the free face plane of the plate-strip.

## 4.4 Buckling Delamination of the Plate-Strip with Two Collinear Cracks

### *4.4.1 Formulation of the Problem and Solution Method*

Examine the buckling delamination problem for the clamped plate-strip containing two collinear cracks. The geometry of the plate-strip and cracks locations are given in Fig. 4.5.

Suppose that all assumptions accepted in the previous subsections hold also for the case under consideration. In the coordinate system shown in Fig. 4.5 the equations of the edge surfaces $S_1^{\pm}$ and $S_2^{\pm}$ of the left and right cracks are

$$\begin{aligned} x_{21}^{\pm} &= h - h_1 \pm \varepsilon \ell_0 \sin^2\left(\frac{\pi(x_1 - (\ell_2 + \ell_0))}{\ell_0}\right), \; x_1 \in (\ell_2, \ell_2 + \ell_0), \\ x_{22}^{\pm} &= h - h_1 \pm \varepsilon \ell_0 \sin^2\left(\frac{\pi(x_1 - (\ell_2 + \ell_1 + \ell_0))}{\ell_0}\right) \\ & x_1 \in (\ell_2 + \ell_1 + \ell_0, \ell_2 + \ell_1 + 2\ell_0). \end{aligned} \tag{4.46}$$

Suppose that the field equations (4.3–4.6) hold in the region occupied by the plate-strip. The free of traction conditions on the cracks edge are formulated as (4.39). Employing the mathematical approach developed in Sect. 4.2 the field equations (4.15), boundary condition (4.16) and end conditions (4.18) and (4.19) for the zeroth approximation are obtained. This approximation is determined as in the previous section and the existence of the cracks do not taken into account under the mentioned determination.

The field equation (4.20), boundary (4.21) and end (4.23) conditions are derived for the first approximation. At the same time, in the case under consideration the conditions on the cracks edges are formulated as follows.

$$
\begin{aligned}
\left[\sigma_{2i}^{(1)}+\sigma_{2n}^{(0)}\frac{\partial u_i^{(1)}}{\partial x_n}\right]\Bigg|_{x_2=h-h_1\pm 0} &= \mp\,\sigma_{1i}^{(0)}\Big|_{x_2=h-h_1\pm 0}\pi\sin\left(2\pi\frac{(x_1-(\ell_2+\ell_0))}{\ell_0}\right)\\
&\quad+\frac{\partial\sigma_{2i}^{(0)}}{\partial x_2}\Bigg|_{x_2=h-h_1\pm 0}\ell_0\sin^2\left(\pi\frac{(x_1-(\ell_2+\ell_0))}{\ell_0}\right),\\
\left[\sigma_{2i}^{(1)}+\sigma_{2n}^{(0)}\frac{{}^{(1)}_{\;i}}{\partial x_n}\right]\Bigg|_{x_2=h-h_1\pm 0} &= \mp\,\sigma_{1i}^{(0)}\Big|_{x_2=h-h_1\pm 0}\\
&\quad\times\pi\sin\left(2\pi\frac{(x_1-(\ell_2+\ell_1+\ell_0))}{\ell_0}\right)\\
&\quad+\frac{\partial\sigma_{2i}^{(0)}}{2}\Bigg|_{x_2=h-h_1\pm 0}\ell_0\sin^2\left(\pi\frac{(x_1-(\ell_2+\ell_1+\ell_0))}{\ell_0}\right).
\end{aligned}
\tag{4.47}
$$

After employing the Laplace transformation (4.28), the FEM is used to find the values of these transformations for the needed values of the transformation parameter $s>0$. For this purpose the following functional is constructed.

$$
\begin{aligned}
\Pi^{(1)}\left(\bar{u}_1^{(1)},\bar{u}_2^{(1)}\right) &= \frac{1}{2}\int_0^{\ell}\left(\int_0^{h}\left[\bar{\sigma}_{ji}^{(1)}+\sigma_{jn}^{(0)}\frac{\partial\bar{u}_i^{(1)}}{\partial x_n}\right]\frac{\partial\bar{u}_i^{(1)}}{\partial x_j}dx_2\right)dx_1\\
&\quad+\int_{\ell_2}^{\ell_2+\ell_0}\left(\bar{F}_{13i}^{+}\bar{u}_i^{(1)}\right)\Big|_{x_2=h-h_1+0}dx_1+\int_{\ell_2}^{\ell_2+\ell_0}\left(\bar{F}_{13i}^{-}\bar{u}_i^{(1)}\right)\Big|_{x_2=h-h_1-0}dx_1\\
&\quad+\int_{\ell_2+\ell_1+\ell_0}^{\ell_2+\ell_1+2\ell_0}\left(\bar{F}_{23i}^{+}\bar{u}_i^{(1)}\right)\Big|_{x_2=h-h_1+0}dx_1\\
&\quad+\int_{\ell_2+\ell_1+\ell_0}^{\ell_2+\ell_1+2\ell_0}\left(\bar{F}_{23i}^{-}\bar{u}_i^{(1)}\right)\Big|_{x_2=h-h_1-0}dx_1,
\end{aligned}
\tag{4.48}
$$

**Table 4.17** The values of $p_{cr.0}\big/E_0^{(1)}$ obtained for various $E^{(2)}\big/E_0^{(1)}$, $h_1/\ell_0$ and $\ell_1/\ell$. For the plate-strip contained two collinear cracks

| $E^{(2)}\big/E_0^{(1)}$ | $h_1/\ell_0$ | $\ell_1/\ell$ | | | |
|---|---|---|---|---|---|
| | | 0.30 | 0.15 | 0.10 | 0.05 |
| 1 | 0.30 | 0.1105 | 0.1103 | 0.1118 | 0.1181 |
| | 0.24 | 0.0901 | 0.0899 | 0.0901 | 0.0911 |
| | 0.20 | 0.0725 | 0.0724 | 0.0728 | 0.0744 |
| | 0.16 | 0.0546 | 0.0546 | 0.0550 | 0.0560 |
| 10 | 0.30 | 0.3458 | 0.3482 | 0.3547 | 0.3679 |
| | 0.24 | 0.3013 | 0.3023 | 0.3053 | 0.3114 |
| | 0.20 | 0.2581 | 0.2588 | 0.2617 | 0.2686 |
| | 0.16 | 0.2071 | 0.2077 | 0.2100 | 0.2156 |
| 20 | 0.30 | 0.4460 | 0.4488 | 0.4549 | 0.4657 |
| | 0.24 | 0.4013 | 0.4030 | 0.4068 | 0.4137 |
| | 0.20 | 0.3567 | 0.3582 | 0.3622 | 0.3704 |
| | 0.16 | 0.2980 | 0.2993 | 0.3029 | 0.3108 |
| 50 | 0.30 | 0.5628 | 0.5645 | 0.5679 | 0.5739 |
| | 0.24 | 0.5255 | 0.5270 | 0.5299 | 0.5350 |
| | 0.20 | 0.4901 | 0.4919 | 0.4956 | 0.5025 |
| | 0.16 | 0.4329 | 0.4350 | 0.4395 | 0.4481 |

where

$$
\begin{aligned}
\bar{F}_{13i}^{\pm} &= \mp\frac{1}{s}\sigma_{1i}^{(0)}\Big|_{x_2=h-h_1\pm 0}\pi\sin\left(2\pi\frac{(x_1-(\ell_2+\ell_0))}{\ell_0}\right) \\
&\quad +\frac{1}{s}\frac{\partial\sigma_{2i}^{(0)}}{\partial x_2}\Big|_{x_2=h-h_1\pm 0}\ell_0\sin^2\left(\pi\frac{(x_1-(\ell_2+\ell_0))}{\ell_0}\right), \\
\bar{F}_{23i}^{\pm} &= \mp\frac{1}{s}\sigma_{1i}^{(0)}\pi\sin\left(2\pi\frac{(x_1-(\ell_2+\ell_1+\ell_0))}{\ell_0}\right) \\
&\quad +\frac{1}{s}\frac{\partial\sigma_{2i}^{(0)}}{\partial x_2}\ell_0\sin^2\left(\pi\frac{(x_1-(\ell_2+\ell_1+\ell_0))}{\ell_0}\right).
\end{aligned}
\tag{4.49}
$$

Thus, employing the same calculation algorithm proposed in the previous sections we obtain the numerical results on the critical force and critical time.

### 4.4.2 Numerical Results and Discussions

Assume that the assumptions accepted in Sect. 4.2.3 on the plate-strip's material hold also for the case under consideration and consider the case where $\ell_0/\ell = 0.25$ and investigate the influence of the parameters $E^{(2)}\big/E_0^{(1)}$, $h_1/\ell_0$ and $\ell_1/\ell$ on the values of $p_{cr.0}\big/E_0^{(1)}$ which are given in Table 4.17. Note that, according to the numerical results obtained and discussed in the previous sections, results illustrated in Table 4.17 occur with very high accuracy for each value of $h/\ell$ for which the relation $h_1/\ell_0 < h/\ell$ satisfies.

Analyzes of these results show that the values of $p_{cr.0}\big/E_0^{(1)}$ increase insignificantly with decreasing of the distance between the cracks, i.e. with decreasing $\ell_1/\ell$ (Fig. 4.5). Consequently, under investigation the buckling delamination problem for the plate-strip with two collinear cracks the interaction between the cracks can be neglected with very high accuracy. The main geometrical parameter, on which the values of the critical force depend significantly, is the parameter $h_1/\ell_0$. Note that $h_1/\ell_0$ is the parameter $\psi_{h_0}$ (4.35) introduced in Sect. 4.2. Also, it follows from the comparison of the results given in Table 4.17 with the corresponding ones given in Table 4.8 that the values of $p_{cr.0}\big/E_0^{(1)}$ obtained for the case under consideration are significantly greater than corresponding ones obtained for the buckling delamination problem for the clamped plate-strip containing a single crack (Fig. 4.1). Because, in the case under consideration the distance between the left tip of the left crack and left clamped end (or between the right tip and the right clamped end) of the plate-strip is less than that shown in Fig. 4.1, i.e. $\ell_2 < (\ell - \ell_0)/2$.

Now we consider numerical results related to the case where the matrix layer material of the layered composite from which the plate-strip is fabricated, is viscoelastic with operators (3.68–3.70). Introduce the dimensionless rheological parameter $\omega\,(=\omega_\infty/\omega_0)$ and dimensionless time $t'\,(=\omega_0^{1/(1+\alpha)}t)$ and assume that $\nu^{(2)} = \nu_0^{(1)} = 0.3$ (where $\nu^{(2)}$ and $\nu_0^{(1)}$ are Poisson ratios for the filler layer and for the matrix material at $t = 0$ respectively. As for viscoelastic stability loss the intensity of the external compressing force $p$ must satisfy the inequality $p_{cr.\infty} < p < p_{cr.0}$, therefore first we present the values of $p_{cr.\infty}$ which are tabulated in Table 4.18 for various values of $h_1/\ell_0$, $\ell_1/\ell$ and $\omega$ under $E^{(2)}\big/E_0^{(1)} = 50$ and $\ell_0/\ell = 0.25$.

It follows from the results given in Table 4.18 that the character of the influence of the interaction between the cracks under buckling delamination on the values of $p_{cr.\infty}$ is similar to that observed under analyzing the values of $p_{cr.0}$. In other words, the values of $p_{cr.\infty}\big/E_0^{(1)}$ increase monotonically with decreasing the distance between the cracks, i.e. with decreasing $\ell_1/\ell$. This increasing is insignificant and can be neglected practically for the examined cases. Thus, we again see that the interaction between the considered cracks under calculation of the values of the critical forces is negligible when the material of the plate-strip is time independent.

Now we analyze the mentioned above interaction when the material of the plate-strip is viscoelastic. For this purpose we consider the influence of the ratio $\ell_1/\ell$ on the values of $t'_{cr}$ obtained for various values of $h_1/\ell_0$ and $p\big/E_0^{(1)}$. Note that these values of $t'_{cr}$ which are calculated under $\alpha = -0.5$ and $\omega = 2$, are given in Table 4.19. It follows from this table that the values of $t'_{cr}$ increase significantly with decreasing $\ell_1/\ell$ and this increasing becomes more considerable in the cases

**Table 4.18** The values of $p_{cr.\infty}/E_0^{(1)}$ obtained for various $h_1/\ell_0$, $\ell_1/\ell$ and $\omega$ under $E^{(2)}/E_0^{(1)} = 50$ and $\ell_0/\ell = 0.25$

| $\omega$ | $h_1/\ell_0$ | $\ell_1/\ell$ | | | |
|---|---|---|---|---|---|
| | | 0.30 | 0.15 | 0.10 | 0.05 |
| 0.5 | 0.30 | 0.2007 | 0.2009 | 0.2013 | 0.2020 |
| | 0.24 | 0.1949 | 0.1951 | 0.1955 | 0.1960 |
| | 0.20 | 0.1871 | 0.1874 | 0.1879 | 0.1890 |
| | 0.16 | 0.1784 | 0.1787 | 0.1796 | 0.1811 |
| 1.0 | 0.30 | 0.2940 | 0.2944 | 0.2952 | 0.2966 |
| | 0.24 | 0.2826 | 0.2829 | 0.2837 | 0.2850 |
| | 0.20 | 0.2682 | 0.2688 | 0.2699 | 0.2721 |
| | 0.16 | 0.2505 | 0.2514 | 0.2531 | 0.2561 |
| 2.0 | 0.30 | 0.3846 | 0.3854 | 0.3868 | 0.3894 |
| | 0.24 | 0.3662 | 0.3669 | 0.3683 | 0.3707 |
| | 0.20 | 0.3441 | 0.3451 | 0.3471 | 0.3507 |
| | 0.16 | 0.3160 | 0.3173 | 0.3199 | 0.3247 |

**Table 4.19** The influence of the ratio $\ell_1/\ell$ on the values of $t'_{cr}$ obtained for various values of $h_1/\ell_0$ and $p/E_0^{(1)}$ under $\alpha = -0.5$ and $\omega = 2$

| $h_1/\ell_0$ | $\ell_1/\ell$ | $p/E_0^{(1)}$ | | | | | | |
|---|---|---|---|---|---|---|---|---|
| | | 0.550 | 0.520 | 0.500 | 0.450 | 0.400 | 0.390 | 0.388 |
| 0.30 | 0.30 | 0.0002 | 0.004 | 0.015 | 0.163 | 6.269 | 58.90 | 155.3 |
| | 0.15 | 0.0002 | 0.005 | 0.016 | 0.172 | 7.103 | 81.04 | 262.7 |
| | 0.10 | 0.0004 | 0.006 | 0.018 | 0.190 | 9.073 | 178.3 | 1398 |
| | 0.05 | 0.0009 | 0.008 | 0.023 | 0.227 | 14.99 | 6157. | $\infty$ |
| | | 0.480 | 0.460 | 0.440 | 0.400 | 0.380 | 0.360 | 0.350 |
| 0.20 | 0.30 | 0.0000 | 0.002 | 0.013 | 0.140 | 0.522 | 3.824 | 33.33 |
| | 0.15 | 0.0001 | 0.003 | 0.014 | 0.152 | 0.571 | 4.477 | 49.71 |
| | 0.10 | 0.0003 | 0.004 | 0.017 | 0.178 | 0.689 | 6.335 | 153.5 |
| | 0.05 | 0.001 | 0.007 | 0.025 | 0.237 | 0.976 | 13.40 | $\infty$ |
| | | 0.420 | 0.400 | 0.380 | 0.360 | 0.340 | 0.320 | 0.318 |
| 0.16 | 0.30 | 0.0008 | 0.009 | 0.039 | 0.173 | 0.958 | 53.13 | 230.4 |
| | 0.15 | 0.001 | 0.011 | 0.048 | 0.195 | 1.117 | 120.5 | 2093. |
| | 0.10 | 0.002 | 0.015 | 0.061 | 0.250 | 1.574 | $\infty$ | $\infty$ |
| | 0.05 | 0.005 | 0.025 | 0.095 | 0.395 | 3.201 | $\infty$ | $\infty$ |

where the values of $p/E_0^{(1)}$ are near to the corresponding values of $p_{cr.\infty}/E_0^{(1)}$. Consequently, in the cases where the plate-strip's material is viscoelastic one the foregoing interaction between the collinear cracks must be taken into account.

Fig. 4.6 The geometry of the three-layered plate-strip containing two parallel interface cracks

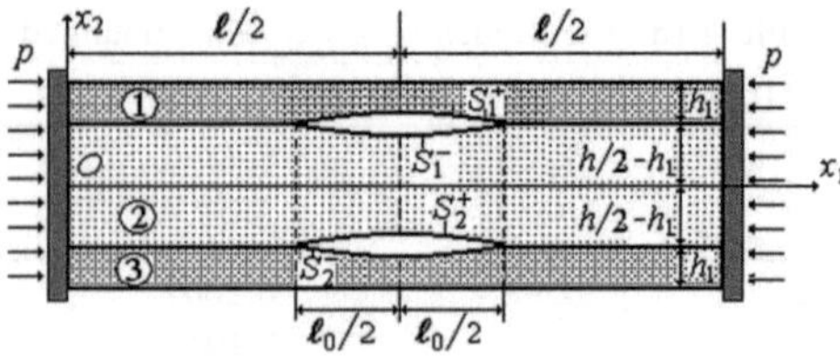

# 4.5 Buckling Delamination of the Three-Layered (Sandwich) Plate-Strip with Two Parallel Interface Cracks

## 4.5.1 Formulation of the Problem

Consider a three-layered (sandwich) plate-strip with a geometry shown in Fig. 4.6. Assume that the materials of the face layers are the same and these materials are isotropic and homogeneous. We determine the position of the points of the layers through the Lagrange coordinates in the Cartesian system of coordinates $Ox_1x_2x_3$. As the direction of the axis $Ox_3$ is perpendicular to the Fig. 4.6 plane, therefore this axis is not shown in this figure. Thus, according to the Fig. 4.6, the region

$$\Omega = \{0 \le x_1 \le \ell, \ -h/2 \le x_2 \le +h/2, \ -\infty < x_3 < +\infty\} \tag{4.50}$$

is occupied by the plate.

Moreover, assume that between the face layers and core layer there exist two cracks the location of which are symmetric with respect to $x_1 = \ell/2$ and with respect to $x_2 = 0$ and the edges of these cracks have initial infinitesimal anti-phase imperfections. Denoting the edge surfaces of the upper (lower) crack through $S_1^{\pm}$ $(S_2^{\pm})$, the equations of these surfaces are given as follows.

For surfaces $S_1^{\pm}$:

$$x_2^{\pm} = \frac{h}{2} - h_1 \pm \varepsilon \ell_0 \sin^2\left(\frac{\pi(x_1 - (\ell - \ell_0)/2)}{\ell_0}\right) \quad \text{for}\, x_1 \in ((\ell - \ell_0)/2, (\ell + \ell_0)/2), \tag{4.51}$$

For surfaces $S_2^{\pm}$:

$$x_2^{\pm} = -\left(\frac{h}{2} - h_1\right) \pm \varepsilon \ell_0 \sin^2\left(\frac{\pi(x_1 - (\ell - \ell_0)/2)}{\ell_0}\right) \quad \text{for}\, x_1 \in ((\ell - \ell_0)/2, (\ell + \ell_0)/2). \tag{4.52}$$

As before, here $\varepsilon$ serves as a measure of the cracks' edges imperfection degree and assume that $\varepsilon \ll 1$. The values related to the upper and lower face layers and core layer of the plate we denote by upper indexes (1), (3) and (2) respectively. Moreover, assume that the three-layered plate-strip is compressed along the $Ox_1$

axis through the rigid bodies that hold the ends and the uniformly distributed normal forces with intensity $p$ act on these rigid bodies.

Thus, within the framework of the piecewise homogeneous body model we investigate the evolution of the cracks edges with the use of the following geometrically non-linear equations of the theory of viscoelasticity in the plane-strain state in the $Ox_1x_2$ plane.

$$\frac{\partial}{\partial x_j}\left[\sigma_{jn}^{(k)}\left(\delta_i^n+\frac{\partial u_i^{(k)}}{\partial x_n}\right)\right]=0,\ \sigma_{ij}^{(k)}=\lambda^{*(k)}\left(\varepsilon_{11}^{(k)}+\varepsilon_{22}^{(k)}\right)\delta_i^j+2\mu^{*(k)}\varepsilon_{ij}^{(k)},$$

$$\varepsilon_{ij}^{(k)}=\frac{1}{2}\left(\frac{\partial u_i^{(k)}}{\partial x_j}+\frac{\partial u_j^{(k)}}{\partial x_i}+\frac{\partial u_n^{(k)}}{\partial x_i}\frac{\partial u_n^{(k)}}{\partial x_j}\right),\ i;j;n=1,2,\ k=1,2,3, \tag{4.53}$$

$$\left\{\begin{array}{c}\lambda^{*(k)}\\ \mu^{*(k)}\end{array}\right\}\varphi(t)=\left\{\begin{array}{c}\lambda_0^{(k)}\\ \mu_0^{(k)}\end{array}\right\}\varphi(t)+\int_0^t\left\{\begin{array}{c}\lambda_1^{(k)}\\ \mu_1^{(k)}\end{array}\right\}(t-\tau)\varphi(\tau)d\tau,$$

where $\lambda_0^{(k)}$ and $\mu_0^{(k)}$ are the instantaneous values of the Lame constants. Other notation used in Eq. (4.53) is conventional.

According to Fig. 4.6, following contact and boundary conditions hold.

$$\left[\sigma_{2n}^{(1)}\left(\delta_i^n+\frac{\partial u_i^{(1)}}{\partial x_n}\right)\right]\Bigg|_{\substack{x_2=h/2\\ x_1\in(0,\ell)}}=0,\ \left[\sigma_{2n}^{(3)}\left(\delta_i^n+\frac{\overset{(3)}{i}}{\partial x_n}\right)\right]\Bigg|_{\substack{x_2=-h/2\\ x_1\in(0,\ell)}}=0$$

$$\left[\sigma_{2n}^{(1)}\left(\delta_i^n+\frac{\partial u_i^{(1)}}{n}\right)\right]\Bigg|_{\substack{x_2=h/2-h_1\\ x_1\in L_1}}=\left[\sigma_{2n}^{(2)}\left(\delta_i^n+\frac{\partial u_i^{(2)}}{n}\right)\right]\Bigg|_{\substack{x_2=h2-h_1\\ x_1\in L_1}},$$

$$\left[\sigma_{2n}^{(2)}\left(\delta_i^n+\frac{\partial u_i^{(2)}}{n}\right)\right]\Bigg|_{\substack{x_2=-(h/2-h_1)\\ x_1\in L_1}}=\left[\sigma_{2n}^{(3)}\left(\delta_i^n+\frac{\partial u_i^{(3)}}{n}\right)\right]\Bigg|_{\substack{x_2=-(h/2-h_1)\\ x_1\in L_1}},$$

$$u_i^{(1)}\Big|_{x_2=h/2-h_1{}_1\in L_1}=u_i^{(2)}\Big|_{x_2=h/2-h_1{}_1\in L_1},\ u_i^{(2)}\Big|_{\substack{x_2=-(h/2-h_1)\\ x_1\in L_1}}=u_i^{(3)}\Big|_{\substack{x_2=-(h/2-h_1)\\ x_1\in L_1}},$$

$$\left[\sigma_{2n}^{(1)}\left(\delta_i^n+\frac{\partial u_i^{(1)}}{n}\right)\right]\Bigg|_{S_1^+}n_{j1}^+=0,\ \left[\sigma_{2n}^{(2)}\left(\delta_i^n+\frac{\partial u_i^{(2)}}{\partial x_n}\right)\right]\Bigg|_{S_1^-}n_{j1}^-=0,$$

$$\left[\sigma_{2n}^{(2)}\left(\delta_i^n+\frac{\overset{(2)}{i}}{\partial x_n}\right)\right]\Bigg|_{S_2^+}n_{j2}^+=0,\ \left[\sigma_{2n}^{(3)}\left(\delta_i^n+\frac{\overset{(3)}{i}}{\partial x_n}\right)\right]\Bigg|_{S_2^-}n_{j2}^-=0, \tag{4.54}$$

where $L_1=((0,(\ell-\ell_0)/2)\cup((\ell+\ell_0)/2,\ell))$.

At the ends of the three-layered plate-strip the conditions

$$
\begin{aligned}
u_1^{(1)}(0,x_2) &= u_1^{(2)}(0,x_2) = u_1^{(3)}(0,x_2) = -u_1^{(1)}(\ell,x_2) \\
&= -u_1^{(2)}(\ell,x_2) = -u_1^{(3)}(\ell,x_2) = U = const. \\
u_2^{(1)}(0,x_2) &= u_2^{(2)}(0,x_2) = u_2^{(3)}(0,x_2) = u_2^{(1)}(\ell,x_2) \\
&= u_2^{(2)}(\ell,x_2) = u_2^{(3)}(\ell,x_2) = 0,
\end{aligned} \tag{4.55}
$$

are satisfied the relation between $U$ and $p$ is obtained from the equation

$$
\begin{aligned}
hp = &\int\limits_{-h/2}^{-h/2+h_1} \sigma_{11}^{(3)}\Big|_{x_1=\ell/2} dx_2 + \int\limits_{-h/2+h_1}^{h/2-h_1} \sigma_{11}^{(2)}\Big|_{x_1=\ell/2} dx_2 \\
&+ \int\limits_{h/2-h_1}^{h/2+h_1} \sigma_{11}^{(1)}\Big|_{x_1=\ell/2} dx_2.
\end{aligned} \tag{4.56}
$$

This completes formulation of the problem on the buckling delamination of the three-layered (sandwich) plate-strip containing two parallel interface cracks.

### *4.5.2 Method of Solution*

As in the previous section, we present the sought values in the series form in a small parameter $\varepsilon$:

$$
\left\{\sigma_{ij}^{(k)}, \varepsilon_{ij}^{(k)}, u_i^{(k)}\right\} = \sum_{q=0}^{\infty} \varepsilon^q \left\{\sigma_{ij}^{(k),q}, \varepsilon_{ij}^{(k),q}, u_i^{(k),q}\right\}. \tag{4.57}
$$

Substituting the series presentation (4.57) into the equations in (4.53) and conditions (4.54–4.56) and doing the mathematical manipulations similar to that made in the previous sections the following set of equations and conditions are obtained for the zeroth and the first approximations.

For the zeroth approximation:

$$
\begin{aligned}
&\frac{\partial \sigma_{ji}^{(k),0}}{\partial x_j} = 0, \ \sigma_{ij}^{(k),0} = \lambda^{*(k)}\left(\varepsilon_{11}^{(k),0} + \varepsilon_{22}^{(k),0}\right)\delta_i^j + 2\mu^{*(k)}\varepsilon_{ij}^{(k),0}, \\
&\varepsilon_{ij}^{(k),0} = \frac{1}{2}\left(\frac{\partial u_i^{(k),0}}{\partial x_j} + \frac{\partial u_j^{(k),0}}{\partial x_i}\right), \ i;j;n = 1,2, \ k = 1,2,3,
\end{aligned} \tag{4.58}
$$

$$\sigma_{2i}^{(1),0}\Big|_{\substack{x_2=h/2\\ x_1\in(0,\ell)}}=0,\ \sigma_{2i}^{(3),0}\Big|_{\substack{x_2=-h/2\\ x_1\in(0,\ell)}}=0,\ \sigma_{2i}^{(1),0}\Big|_{\substack{x_2=h/2-h_1\\ x_1\in L_1}}=\sigma_{2i}^{(2),0}\Big|_{\substack{x_2=h/2-h_1\\ x_1\in L_1}},$$

$$\sigma_{2i}^{(2),0}\Big|_{\substack{x_2=-(h/2-h_1)\\ x_1\in L_1}}=\sigma_{2i}^{(3),0}\Big|_{\substack{x_2=-(h/2-h_1)\\ x_1\in L_1}},$$

$$u_i^{(1),0}\Big|_{\substack{x_2=h/2-h_1\\ x_1\in L_1}}=u_i^{(2),0}\Big|_{\substack{x_2=h/2-h_1\\ x_1\in L_1}},\ u_i^{(2),0}\Big|_{\substack{x_2=-(h/2-h_1)\\ x_1\in L_1}}=u_i^{(3),0}\Big|_{\substack{x_2=-(h/2-h_1)\\ x_1\in L_1}},$$

$$u_1^{(1),0}(0,x_2)=u_1^{(2),0}(0,x_2)=u_1^{(3),0}(0,x_2)=-u_1^{(1),0}(\ell,x_2)$$
$$=-u_1^{(2),0}(\ell,x_2)=-u_1^{(3),0}(\ell,x_2)=U(t)$$
$$u_2^{(1),0}(0,x_2)=u_2^{(2),0}(0,x_2)=u_2^{(3),0}(0,x_2)=u_2^{(1),0}(\ell,x_2)$$
$$=u_2^{(2),0}(\ell,x_2)=u_2^{(3),0}(\ell,x_2)=0.$$

$$hp=\int_{-h/2}^{-h/2+h_1}\sigma_{11}^{(3),0}\Big|_{x_1=\ell/2}dx_2+\int_{-h/2+h_1}^{h/2-h_1}\sigma_{11}^{(2),0}\Big|_{x_1=\ell/2}dx_2$$
$$+\int_{h/2-h_1}^{h/2+h_1}\sigma_{11}^{(1),0}\Big|_{x_1=\ell/2}dx_2. \tag{4.59}$$

For the first approximation:

$$\frac{\partial}{\partial x_j}\left[\sigma_{ji}^{(k),1}+\sigma_{jn}^{(k),0}\frac{\partial u_i^{(k),1}}{\partial x_n}\right]=0,$$
$$\sigma_{ij}^{(k),1}=\lambda^{*(k)}\left(\varepsilon_{11}^{(k),1}+\varepsilon_{22}^{(k),1}\right)\delta_i^j+2\mu^{*(k)}\varepsilon_{ij}^{(k),1}, \tag{4.60}$$
$$\varepsilon_{ij}^{(k),1}=\frac{1}{2}\left(\frac{\partial u_i^{(k),1}}{\partial x_j}+\frac{\partial u_j^{(k),1}}{\partial x_i}\right),\ i;j;n=1,2,\ k=1,2,3.$$

$$
\begin{aligned}
&\sigma_{2i}^{(1),1}\Big|_{\substack{x_2=h/2\\ x_1\in(0,\ell)}}=0,\ \sigma_{2i}^{(3),1}\Big|_{\substack{x_2=-h/2\\ x_1\in(0,\ell)}}=0,\\
&u_1^{(1),1}(0,x_2)=u_1^{(2),1}(0,x_2)=u_1^{(3),1}(0,x_2)=-u_1^{(1),1}(\ell,x_2)\\
&=-u_1^{(2),1}(\ell,x_2)=-u_1^{(3),1}(\ell,x_2)=0\\
&u_2^{(1),1}(0,x_2)=u_2^{(2),1}(0,x_2)=u_2^{(3),1}(0,x_2)=u_2^{(1),1}(\ell,x_2)\\
&=u_2^{(2),1}(\ell,x_2)=u_2^{(3),1}(\ell,x_2)=0.\\
&\left[\sigma_{2i}^{(2),1}+\sigma_{2n}^{(2),0}\frac{\partial u_i^{(2),1}}{\partial x_n}\right]\Bigg|_{\substack{x_2=-(h/2-h_1)\\ x_1\in L_1}}=\left[\sigma_{2i}^{(3),1}+\sigma_{2n}^{(3),0}\frac{\partial u_i^{(3),1}}{\partial x_n}\right]\Bigg|_{\substack{x_2=-(h/2-h_1)\\ x_1\in L_1}}\\
&u_i^{(1),1}\Big|_{x_2=h/2-h_1{}_1\in L_1}=u_i^{(2),1}\Big|_{\substack{x_2=h/2-h_1\\ x_1\in L_1}},\ u_i^{(2),1}\Big|_{\substack{x_2=-(h/2-h_1)\\ x_1\in L_1}}=u_i^{(3),1}\Big|_{\substack{x_2=-(h/2-h_1)\\ x_1\in L_1}},\\
&\left[\sigma_{2i}^{(1),1}+\sigma_{2n}^{(1),0}\frac{{}_i^{(1),1}}{\partial x_n}\right]\Bigg|_{\substack{x_2=h/2-h_1+0\\ x_1\in L_2}}=\sigma_{1i}^{(1),0}\Big|_{\substack{x_2=h/2-h_1+0\\ x_1\in L_2}}\\
&\times\pi\sin\left(2\pi\frac{(x_1-(\ell-\ell_0)/2)}{\ell_0}\right)\\
&+\frac{\partial\sigma_{2i}^{(1),0}}{\partial x_2}\Bigg|_{\substack{x_2=h/2-h_1+0\\ x_1\in L_2}}\ell_0\sin^2\left(\pi\frac{(x_1-(\ell-\ell_0)/2)}{\ell_0}\right),\\
&\left[\sigma_{2i}^{(2),1}+\sigma_{2n}^{(2),0}\frac{{}_i^{(2),1}}{\partial x_n}\right]\Bigg|_{\substack{x_2=h/2-h_1-0\\ x_1\in L_2}}=\sigma_{1i}^{(2),0}\Big|_{\substack{x_2=h/2-h_1-0\\ x_1\in L_2}}\\
&\times\pi\sin\left(2\pi\frac{(x_1-(\ell-\ell_0)/2)}{\ell_0}\right)\\
&+\frac{\partial\sigma_{2i}^{(2),0}}{\partial x_2}\Bigg|_{\substack{x_2=h/2-h_1-0\\ x_1\in L_2}}\ell_0\sin^2\left(\pi\frac{(x_1-(\ell-\ell_0)/2)}{\ell_0}\right),\\
&\left[\sigma_{2i}^{(2),1}+\sigma_{2n}^{(2),0}\frac{{}_i^{(2),1}}{\partial x_n}\right]\Bigg|_{\substack{x_2=-(h/2-h_1)+0\\ x_1\in L_2}}=\sigma_{1i}^{(2),0}\Big|_{\substack{x_2=-(h/2-h_1)+0\\ x_1\in L_2}}\\
&\times\pi\sin\left(2\pi\frac{(x_1-(\ell-\ell_0)/2)}{\ell_0}\right)\\
&+\frac{\partial\sigma_{2i}^{(2),0}}{\partial x_2}\Bigg|_{\substack{x_2=-(h/2-h_1)+0\\ x_1\in L_2}}\ell_0\sin^2\left(\pi\frac{(x_1-(\ell-\ell_0)/2)}{\ell_0}\right),\\
&\left[\sigma_{2i}^{(3),1}+\sigma_{2n}^{(3),0}\frac{{}_i^{(3),1}}{\partial x_n}\right]\Bigg|_{\substack{x_2=-(h/2-h_1)-0\\ x_1\in L_2}}=\sigma_{1i}^{(3),0}\Big|_{\substack{x_2=-(h/2-h_1)-0\\ x_1\in L_2}}\\
&\times\pi\sin\left(2\pi\frac{(x_1-(\ell-\ell_0)/2)}{\ell_0}\right)\\
&+\frac{\partial\sigma_{2i}^{(3),0}}{\partial x_2}\Bigg|_{\substack{x_2=-(h/2-h_1)-0\\ x_1\in L_2}}\ell_0\sin^2\left(\pi\frac{(x_1-(\ell-\ell_0)/2)}{\ell_0}\right),
\end{aligned}
\tag{4.61}
$$

where $L_2 = ((\ell - \ell_0)/2, (\ell + \ell_0)/2)$.

Thus, the determination of the first (the zeroth) approximation is reduced to the solution to the boundary-value problem for the field equations in (4.60) (for the field equations in (4.58)) under satisfaction of the boundary and contact conditions in (4.61) (of the boundary and contact conditions in (4.59)). Employing the Laplace transformation (4.28) to these equations and relations we obtain the same equations and relations for the Laplace transformations of the sought values. It should be noted that under employing Laplace transformation (4.28) into the equations and relations in (4.60) and (4.61) related to the first approximation we use the "time freezing" principle discussed in Sect. 3.4.

Consider the FEM modeling of the foregoing problems in the space of the Laplace transformation parameter. For the FEM modeling of the problem related to the zeroth approximation we use the functional

$$\Pi^{(0)}(\bar{u}_1^{(1),0}, \bar{u}_2^{(1),0}, \bar{u}_1^{(2),0}, \bar{u}_2^{(2),0}, \bar{u}_1^{(3),0}, \bar{u}_2^{(3),0}) = \frac{1}{2}\sum_{k=1}^{3} \iint_{\Omega_k} \bar{\sigma}_{ij}^{(k),0}\bar{\varepsilon}_{ij}^{(k),0} dx_1 dx_2, \tag{4.62}$$

where

$$\begin{aligned} \Omega_1 &= \{0 \leq x_1 \leq \ell,\ h/2 - h_1 \leq x_2 \leq h/2\}, \\ \Omega_2 &= \{0 \leq x_1 \leq \ell,\ -h/2 + h_1 \leq x_2 \leq h/2 - h_1\}, \\ \Omega_3 &= \{0 \leq x_1 \leq \ell,\ -h/2 \leq x_2 \leq -h/2 + h_1\}. \end{aligned} \tag{4.63}$$

Consequently, under the determination of the values of the zeroth approximation we do not take the existence of the cracks into account and employ the usual Ritz technique for obtaining the corresponding equations for the Laplace transformations of the displacements at the nodes of the selected finite elements. Note that these equations are obtained from the relation

$$\begin{aligned} \delta\Pi^{(0)} &= \frac{\partial\Pi^{(0)}}{\partial\bar{u}_1^{(1),0}}\delta\bar{u}_1^{(1),0} + \frac{\partial\Pi^{(0)}}{\partial\bar{u}_2^{(1),0}}\delta\bar{u}_2^{(1),0} + \frac{\partial\Pi^{(0)}}{\partial\bar{u}_1^{(2),0}}\delta\bar{u}_1^{(2),0} \\ &+ \frac{\partial\Pi^{(0)}}{\partial\bar{u}_2^{(2),0}}\delta\bar{u}_2^{(2),0} + \frac{\partial\Pi^{(0)}}{\partial\bar{u}_1^{(3),0}}\delta\bar{u}_1^{(3),0} + \frac{\partial\Pi^{(0)}}{\partial\bar{u}_2^{(3),0}}\delta\bar{u}_2^{(3),0} = 0. \end{aligned} \tag{4.64}$$

By direct verification it is easily obtained that the Laplace transformations of the equilibrium equations in (4.58) and boundary and contact conditions given (4.59) and related to forces are composed the Euler equations of the functional (4.62).

Numerical results show that if the external compressed force $p$ does not depend on the time then the stresses $\sigma_{ij}^{(k),0}$ related to the zeroth approximation can be regarded as time independent with a very high degree of accuracy. Moreover, numerical results show that the end effects on the stress distribution in the three-layered plate-strip arise only in the near vicinity of these ends. For instance, in the cases where $h/\ell = 0.2$, $h_1/\ell = 0.05$, $0.01 \le E_0^{(2)} \big/ E_0^{(1)} \le 50$, $E_0^{(3)} = E_0^{(1)}$ and $\nu_0^{(2)} = \nu_0^{(1)} = 0.3$, $\nu_0^{(3)} = \nu_0^{(1)}$ (where $E_0^{(k)}$ and $\nu_0^{(k)}$ are the instantaneous values of the Young's modulus and Poisson's ratios of the material of the kth layer) the inhomogeneous distribution of the stresses on the interface plane between the layers arise only in the near vicinity of the ends of the plate-strip, i.e. in the regions for which $0 \le x_1/\ell \le 0.1$ and $0.9 \le x_1/\ell \le 1.0$. But in the region for which $0.1 < x_1/\ell < 0.9$ the relations $\sigma_{11}^{(k),0} = const_k$, $\sigma_{12}^{(k),0} = 0$, $\sigma_{22}^{(k),0} = 0$ occur with very high accuracy. Therefore in the cases where the cracks tips are sufficiently far from the plate-strip ends, for example, in the cases where $\ell_0/\ell \le 0.8$, the ignoring of the existence of the cracks is acceptable under determination of the values of the zeroth approximation. Considering the foregoing discussions, we neglect the terms with multiplying $\sigma_{12}^{(k),0}$ and $\partial\sigma_{2i}^{(k),0} \big/ \partial x_2$ in the Laplace transformation of the last four relations related to the traction free conditions on the cracks edge surfaces, i.e. we take the following ones instead of the Laplace transformation of the last four conditions in (4.61).

$$\left[\bar{\sigma}_{2i}^{(1),1} + \sigma_{2n}^{(1),0} \frac{\partial \bar{u}_i^{(1),1}}{\partial x_n}\right]\Bigg|_{\substack{x_2 = h/2 - h_1 + 0 \\ x_1 \in L_2}} = -\frac{1}{s}\delta_i^1 \sigma_{1i}^{(1),0}\Big|_{x_2 = h/2 - h_1 + 0_1 \in L_2} f_0(x_1),$$

$$\left[\bar{\sigma}_{2i}^{(2),1} + \sigma_{2n}^{(2),0} \frac{\partial \bar{u}_i^{(2),1}}{\partial x_n}\right]\Bigg|_{\substack{x_2 = h/2 - h_1 - 0 \\ x_1 \in L_2}} = \frac{1}{s}\delta_i^1 \sigma_{1i}^{(2),0}\Big|_{\substack{x_2 = h/2 - h_1 - 0 \\ x_1 \in L_2}} f_0(x_1),$$

$$\left[\bar{\sigma}_{2i}^{(2),1} + \sigma_{2n}^{(2),0} \frac{\partial \bar{u}_i^{(2),1}}{n}\right]\Bigg|_{\substack{x_2 = -(h/2 - h_1) + 0 \\ x_1 \in L_2}} = -\frac{1}{s}\delta_i^1 \sigma_{1i}^{(2),0}\Big|_{\substack{x_2 = -(h/2 - h_1) + 0 \\ x_1 \in L_2}} f_0(x_1), \quad (4.65)$$

$$\left[\bar{\sigma}_{2i}^{(3),1} + \sigma_{2n}^{(3),0} \frac{\partial \bar{u}_i^{(3),1}}{\partial x_n}\right]\Bigg|_{\substack{x_2 = -(h/2 - h_1) - 0 \\ x_1 \in L_2}} = \frac{1}{s}\delta_i^1 \sigma_{1i}^{(3),0}\Big|_{\substack{x_2 = -(h/2 - h_1) - 0 \\ x_1 \in L_2}} f_0(x_1),$$

$$f_0(x_1) = \pi \sin\left(2\pi \frac{(x_1 - (\ell - \ell_0)/2)}{\ell_0}\right),$$

where $s > 0$ is the Laplace transformation parameter.

After this simplification we introduce the following functional for the FEM modeling of the first approximation.

$$\Pi^{(1)}\left(\bar{u}_1^{(1),1},\bar{u}_2^{(1),1},\bar{u}_1^{(2),1},\bar{u}_2^{(2),1},\bar{u}_1^{(3),1},\bar{u}_2^{(3),1}\right)=\frac{1}{2}\sum_{k=1}^{3}\iint\limits_{\Omega_k''}\left[\bar{t}_{11}^{(k),1}\frac{\partial\bar{u}_1^{(k),1}}{\partial x_1}\right.$$

$$\left.+\bar{t}_{12}^{(k),1}\frac{\partial\bar{u}_2^{(k),1}}{\partial x_1}+\bar{t}_{21}^{(k),1}\frac{\partial\bar{u}_1^{(k),1}}{\partial x_2}+\bar{t}_{22}^{(k),1}\frac{\partial\bar{u}_2^{(k),1}}{\partial x_2}\right]dx_1dx_2$$

$$-\int\limits_{(\ell-\ell_0)/2}^{(\ell+\ell_0)/2}\frac{1}{s}\frac{df_0}{dx_1}\left[\sigma_{11}^{(1),0}\bar{u}_1^{(1),1}\Big|_{x_2=h/2-h+0}-\sigma_{11}^{(2),0}\bar{u}_1^{(2),1}\Big|_{x_2=h/2-h_1-0}\right.$$

$$\left.+\sigma_{11}^{(2),0}\bar{u}_1^{(2),1}\Big|_{x_2=-(h/2-h)+0}-\sigma_{11}^{(3),0}\bar{u}_1^{(3),1}\Big|_{x_2=-(h/2-h_1)-0}\right]dx_1, \tag{4.66}$$

where

$$\begin{aligned}
\Omega_1' &=\{0\le x_1\le\ell,\ h/2-h_1\le x_2\le h/2\}\\
&\quad-\{x_1\in((\ell-\ell_0)/2,(\ell+\ell_0)/2),\ x_2=h/2-h_1+0\},\\
\Omega_2' &=\{0\le x_1\le\ell,\ -h/2+h_1\le x_2\le h/2-h_1\}\\
&\quad-\{x_1\in((\ell-\ell_0)/2,(\ell+\ell_0)/2),\ x_2=h/2-h_1-0\}\\
&\quad-\{x_1\in((\ell-\ell_0)/2,(\ell+\ell_0)/2),\ x_2=-(h/2-h_1)+0\}\\
\Omega_3' &=\{0\le x_1\le\ell,\ -h/2\le x_2\le -h/2+h_1\}\\
&\quad-\{x_1\in((\ell-\ell_0)/2,(\ell+\ell_0)/2),\ x_2=-(h/2-h_1)-0\},\\
\bar{t}_{11}^{(k),1} &=\bar{\sigma}_{11}^{(k),1}+\sigma_{11}^{(k),0}\frac{\partial\bar{u}_1^{(k),1}}{\partial x_1}+\sigma_{12}^{(k),0}\frac{\partial\bar{u}_1^{(k),1}}{\partial x_2}\\
&=\omega_{1111}^{(k)}\frac{\partial\bar{u}_1^{(k),1}}{\partial x_1}+\omega_{1122}^{(k)}\frac{\partial\bar{u}_2^{(k),1}}{\partial x_2}+\omega_{1112}^{(k)}\frac{\partial\bar{u}_1^{(k),1}}{\partial x_2},\\
\bar{t}_{12}^{(k),1} &=\bar{\sigma}_{12}^{(k),1}+\sigma_{11}^{(k),0}\frac{\partial\bar{u}_2^{(k),1}}{\partial x_1}+\sigma_{12}^{(k),0}\frac{\partial\bar{u}_2^{(k),1}}{\partial x_2}\\
&=\omega_{1212}^{(k)}\frac{\partial\bar{u}_1^{(k),1}}{\partial x_2}+\omega_{1222}^{(k)}\frac{\partial\bar{u}_2^{(k),1}}{\partial x_2}+\omega_{1221}^{(k)}\frac{\partial\bar{u}_2^{(k),1}}{\partial x_1},\\
\bar{t}_{21}^{(k),1} &=\bar{\sigma}_{21}^{(k),1}+\sigma_{21}^{(k),0}\frac{\partial\bar{u}_1^{(k),1}}{\partial x_1}+\sigma_{22}^{(k),0}\frac{\partial\bar{u}_1^{(k),1}}{\partial x_2}\\
&=\omega_{2112}^{(k)}\frac{\partial\bar{u}_1^{(k),1}}{\partial x_2}+\omega_{2121}^{(k)}\frac{\partial\bar{u}_2^{(k),1}}{\partial x_1}+\omega_{2111}^{(k)}\frac{\partial\bar{u}_1^{(k),1}}{\partial x_1},\\
\bar{t}_{22}^{(k),1} &=\bar{\sigma}_{22}^{(k),1}+\sigma_{21}^{(k),0}\frac{\partial\bar{u}_2^{(k),1}}{\partial x_1}+\sigma_{22}^{(k),0}\frac{\partial\bar{u}_2^{(k),1}}{\partial x_2}\\
&=\omega_{2211}^{(k)}\frac{\partial\bar{u}_1^{(k),1}}{\partial x_1}+\omega_{2222}^{(k)}\frac{\partial\bar{u}_2^{(k),1}}{\partial x_2}+\omega_{2221}^{(k)}\frac{\partial\bar{u}_2^{(k),1}}{\partial x_1}.
\end{aligned}\tag{4.67}$$

In (4.67) the following notation is used.

$$\begin{aligned}
&\omega_{1111}^{(k)} = \lambda^{(k)} + 2\mu^{(k)} + \sigma_{11}^{(k),0},\ \omega_{1122}^{(k)} = \lambda^{(k)},\ \omega_{1112}^{(k)} = \sigma_{12}^{(k),0},\\
&\omega_{1212}^{(k)} = \mu^{(k)},\ \omega_{1222}^{(k)} = \sigma_{12}^{(k),0},\ \omega_{1221}^{(k)} = \mu^{(k)} + \sigma_{11}^{(k),0},\\
&\omega_{2112}^{(k)} = \mu^{(k)} + \sigma_{22}^{(k),0},\ \omega_{2121}^{(k)} = \mu^{(k)},\ \omega_{2111}^{(k)} = \sigma_{12}^{(k),0},\\
&\omega_{2211}^{(k)} = \lambda^{(k)},\ \omega_{2222}^{(k)} = \lambda^{(k)} + 2\mu^{(k)} + \sigma_{22}^{(k),0},\ \omega_{2221}^{(k)} = \sigma_{12}^{(k),0}.
\end{aligned} \tag{4.68}$$

Thus from the relation

$$\begin{aligned}
\delta\Pi^{(1)} &= \frac{\partial\Pi^{(1)}}{\partial\bar{u}_1^{(1),1}}\delta\bar{u}_1^{(1),1} + \frac{\partial\Pi^{(1)}}{\partial\bar{u}_2^{(1),1}}\delta\bar{u}_2^{(1),1} + \frac{\partial\Pi^{(1)}}{\partial\bar{u}_1^{(2),1}}\delta\bar{u}_1^{(2),1}\\
&+ \frac{\partial\Pi^{(1)}}{\partial\bar{u}_2^{(2),1}}\delta\bar{u}_2^{(2),1} + \frac{\partial\Pi^{(1)}}{\partial\bar{u}_1^{(3),1}}\delta\bar{u}_1^{(3),1} + \frac{\partial\Pi^{(1)}}{\partial\bar{u}_2^{(3),1}}\delta\bar{u}_2^{(3),1} = 0
\end{aligned} \tag{4.69}$$

we obtain the corresponding equations for the displacements of the nodes of the selected finite elements for the first approximation. The values of the critical forces are determined from the criterion

$$\begin{aligned}
&u_2^{(1),1}\Big|_{\substack{x_2=h/2-h_1+0\\ x_1=\ell/2}} \to \infty \text{ as } p \to p_{cr} \text{ (for a pure elastic case)}\\
&\qquad\qquad\text{or as } t \to t_{cr} \text{ (for a viscoelastic case).}
\end{aligned} \tag{4.70}$$

## 4.5.3 Numerical Results and Discussions

First we consider the numerical results related to the case where the materials of the layers of the plate-strip are pure elastic and assume that $E_0^{(1)} = E_0^{(3)}$, $h/\ell = 0.2$, $\nu_0^{(1)} = \nu_0^{(2)} = \nu_0^{(3)} = 0.3$, where $E_0^{(k)}$ and $\nu_0^{(k)}$ are a Young's modulus and Poisson's ratio of the material of the kth layer. Introduce the notation $\tilde{E} = 2h_1E_0^{(1)} + (h-2h_1)E_0^{(2)}$ and $\beta_1 = h_1/\ell_0$. Numerous numerical investigations show that when the relation

$$(h_1/\ell_0)E_0^{(1)} = \beta_1 E_0^{(1)} < \tilde{E}h/\ell \tag{4.71}$$

satisfies the local buckling delamination around the interface cracks takes place earlier than the global stability loss of the three-layered plate-strip. Note that in the case where $E_0^{(1)} = E_0^{(2)}$ this relation becomes as $h_1/\ell_0 < h/\ell$ which has been formulated through the parameter $\psi_{h_0}$ (4.35) in Sect. 4.2.3.

Consider the values of $p_{cr.0}/\tilde{E}$ which are given in Tables 4.20 and 4.21. The results given in Table 4.20 are calculated for various $\ell_0/(2\ell)$ and $E_0^{(1)}/E_0^{(2)}$ under

**Table 4.20** The values of $p_{cr.0}/\tilde{E}$ obtained for various $\ell_0/\ell$ and $E_0^{(1)}/E_0^{(2)}$ under $\beta_1 = 0.005$

| $\ell_0/(2\ell)$ | $E_0^{(1)}/E_0^{(2)}$ | | | | | | |
|---|---|---|---|---|---|---|---|
| | 0.05 | 0.10 | 0.20 | 1.0 | 5.0 | 10.0 | 20.0 |
| 0.3612 | 0.00147 | 0.00161 | 0.00188 | 0.00240 | 0.0133 | 0.0226 | 0.0358 |
| 0.3056 | 0.00201 | 0.00221 | 0.00260 | 0.00550 | 0.0184 | 0.0311 | 0.0493 |
| 0.2500 | 0.00244 | 0.00320 | 0.00376 | 0.00810 | 0.0269 | 0.0456 | 0.0722 |
| 0.1950 | 0.00455 | 0.00505 | 0.00596 | 0.01129 | 0.0430 | 0.0727 | 0.1153 |
| 0.1389 | 0.00807 | 0.00901 | 0.01072 | 0.02340 | 0.0786 | 0.1330 | 0.2107 |

**Table 4.21** The values of $p_{cr.0}/\tilde{E}$ obtained for various $\beta_1$ and $E_0^{(1)}/E_0^{(2)}$ under $\ell_0/(2\ell) = 0.25$

| $\beta_1$ | $E_0^{(1)}/E_0^{(2)}$ | | | | | | |
|---|---|---|---|---|---|---|---|
| | 0.05 | 0.10 | 0.20 | 1.0 | 5.0 | 10.0 | 20.0 |
| 0.005 | 0.00244 | 0.00320 | 0.00376 | 0.0081 | 0.0269 | 0.0456 | 0.0722 |
| 0.0075 | 0.00783 | 0.00848 | 0.00952 | 0.0165 | 0.0471 | 0.0796 | 0.1304 |
| 0.010 | 0.01572 | 0.01686 | 0.01836 | 0.0266 | 0.0608 | 0.0981 | 0.1595 |

**Table 4.22** The values of $p_{cr.0}/\tilde{E}$ obtained for various coatings under $\ell_0/(2\ell) = 0.25$, $h/\ell = 0.2$

| $\beta_1$ | Coatings | | |
|---|---|---|---|
| | *TiN* | *ZrN* | $ZrB_2$ |
| 0.050 | 0.0091 | 0.0096 | 0.0106 |
| 0.075 | 0.0197 | 0.0210 | 0.0241 |
| 0.100 | 0.0335 | 0.0362 | 0.0428 |
| 0.125 | 0.0470 | 0.0506 | 0.0592 |
| 0.150 | 0.0637 | 0.0694 | 0.0829 |

$\beta_1 = 0.005$, but the results given in Table 4.21 are calculated for various $\beta_1$ and $E_0^{(1)}/E_0^{(2)}$ under $\ell_0/(2\ell) = 0.25$. It follows from these tables that for fixed $\ell_0/(2\ell)$ the values of $p_{cr.0}/\tilde{E}$ increase with $\beta_1$ and $E_0^{(1)}/E_0^{(2)}$, as well as, for fixed $\beta_1$ and $E_0^{(1)}/E_0^{(2)}$ the values of $p_{cr.0}/\tilde{E}$ decrease with $\ell_0/(2\ell)$. These results agree well with the mechanical consideration.

The approach developed above can be also applied for the investigation of the buckling delamination (failure) of coatings of the coated materials under the compression of that. For illustration of this application we consider the case where Steel with mechanical constants $E = 200\,GPa$, $\nu = 0.3$ is taken as a basic material (i.e. coated material), but as coatings are taken the materials *TiN* ($E = 251\,GPa$, $\nu = 0.1$), *ZrN* ($E = 397\,GPa$, $\nu = 0.1$) and $ZrB_2$ ($E = 483.5\,GPa$, $\nu = 0.1$) (see Pierson 1996 and Li et al. 2010). Here, the notation $E$ shows the modulus of elasticity, but the notation $\nu$ shows Poisson's ratio of the corresponding material. Table 4.22 illustrate the values of $p_{cr.0}/\tilde{E}$ obtained for various $\beta_1$ for the selected coatings under $\ell_0/(2\ell) = 0.25$, $h/\ell = 0.2$.

It follows from the data given in Table 4.22 that the obtained results are quite real and can by used for estimation of the coatings failure under compression of the coated materials.

Now we consider the numerical results obtained for the case where the material of the coating layers of the plate-strip are viscoelastic with operators

$$E^{*(k)}\varphi = E_0^{(k)}\left[\varphi(t) - \omega_0\Pi_\alpha^*(-\omega_0 - \omega_\infty)\varphi\right],$$
$$\nu^{*(k)}\varphi = \nu_0^{(k)}\left[\varphi(t) + \frac{1 - 2\nu_0^{(k)}}{2\nu_0^{(k)}}\omega_0\Pi_\alpha^*(-\omega_0 - \omega_\infty)\varphi\right], \tag{4.72}$$

where $k = 1, 3$, $E_0^{(1)} = E_0^{(3)}$, $\nu_0^{(1)} = \nu_0^{(3)}$ are the instantaneous values of Young's modulus and the Poisson coefficient, respectively, $\alpha$, $\omega_0$ and $\omega_\infty$ are the rheological parameters of the face layers' materials, $\Pi_\alpha^*$ is the fractional exponential operator (Rabotnov 1977) and this operator is determined as

$$\Pi_\alpha^*(x)\varphi = \int_0^t \Pi_\alpha(x, t - \tau)\varphi(\tau)d\tau, \tag{4.73}$$

where

$$\Pi_\alpha(x, t) = t^\alpha \sum_{n=0}^{\infty} \frac{x^n t^{n(1+\alpha)}}{\Gamma((1 + n)(1 + \alpha))}, \quad -1<\alpha<0, \tag{4.74}$$

where $\Gamma(x)$ is the Gamma function.

As in previous sections, we introduce the dimensionless rheological parameter $\omega = \omega_\infty/\omega_0$ and dimensionless time $t' = t\omega_0^{1/(1+\alpha)}$. As for the stability loss of the viscoelastic plate-strip the external compressed force must satisfies the condition $p_{cr.\infty}<p<p_{cr.0}$, therefore first we consider the results related to the $p_{cr.\infty}/\tilde{E}$.

Note that $E_\infty^{(1)} = E_0^{(1)}(1 - 1/\omega)$ and $\nu_\infty^{(1)} = \nu_0^{(1)}(1 + (1/\omega)(1 - 2\nu_0^{(1)})/2\nu_0^{(1)})$ are taken instead of $E_0^{(1)}$ and $\nu_0^{(1)}$ respectively and the pure elastic buckling delamination problem is solved again under calculation of the values $p_{cr.\infty}/\tilde{E}$. These values are given in Table 4.23 for various $\beta_1$ and $E_0^{(1)}/E_0^{(2)}$ in the cases $\omega = 0.5$ (upper number) and $\omega = 1.0$ (lower number) under $\ell_0/(2\ell) = 0.25$.

Thus, consider the numerical results obtained for $t'_{cr}$ which are tabulated in Tables 4.24 (for various $E_0^{(1)}/E_0^{(2)}$, $\omega$, $p/\tilde{E}$ under $\alpha = -0.5$) and 4.25 (for various $E_0^{(1)}/E_0^{(2)}$, $\alpha$, $p/\tilde{E}$ under $\omega = 1.0$).

**Table 4.23** The values of $p_{cr.\infty}/\tilde{E}$ obtained for various $\beta_1$ and $E_0^{(1)}/E_0^{(2)}$ under $\ell_0/(2\ell) = 0.25$ in the cases where $\omega = 0.5$ (upper number) and $\omega = 1.0$ (lower number)

| $\beta_1$ | $E_0^{(1)}/E_0^{(2)}$ | | | |
|---|---|---|---|---|
| | 1.0 | 5.0 | 10.0 | 20.0 |
| 0.005 | 0.0062<br>0.0067 | 0.0209<br>0.0233 | 0.0319<br>0.0373 | 0.0437<br>0.0542 |
| 0.0075 | 0.0111<br>0.0126 | 0.0367<br>0.0405 | 0.0586<br>0.0655 | 0.0854<br>0.1025 |
| 0.010 | 0.0153<br>0.0184 | 0.0447<br>0.0498 | 0.0718<br>0.0813 | 0.1079<br>0.1273 |

**Table 4.24** The values of $t'_{cr}$ obtained for various $\omega$ under $\alpha = -0.5$

| $E_0^{(1)}/E_0^{(2)}$ | $p/\tilde{E}$ | $\omega$ | | |
|---|---|---|---|---|
| | | 0.5 | 1.0 | 2.0 |
| 1.0 | 0.0080 | 0.0001 | 0.0001 | 0.0001 |
| | 0.0078 | 0.0011 | 0.0012 | 0.0013 |
| | 0.0075 | 0.0051 | 0.0057 | 0.0072 |
| 5.0 | 0.0260 | 0.0010 | 0.0010 | 0.0011 |
| | 0.0250 | 0.0046 | 0.0051 | 0.0064 |
| | 0.0240 | 0.0122 | 0.0143 | 0.0208 |
| 10.0 | 0.0440 | 0.0009 | 0.0009 | 0.0010 |
| | 0.0420 | 0.0055 | 0.0061 | 0.0077 |
| | 0.0400 | 0.0157 | 0.0190 | 0.0293 |

**Table 4.25** The values of $t'_{cr}$ obtained for various $\alpha$ under $\omega = 1.0$

| $E_0^{(1)}/E_0^{(2)}$ | $p/\tilde{E}$ | $\alpha$ | | |
|---|---|---|---|---|
| | | −0.7 | −0.5 | −0.3 |
| 1.0 | 0.0080 | $7 \times 10^{-7}$ | 0.0001 | 0.0015 |
| | 0.0078 | $2 \times 10^{-5}$ | 0.0012 | 0.0069 |
| | 0.0075 | 0.0002 | 0.0057 | 0.0205 |
| 5.0 | 0.0260 | $1 \times 10^{-5}$ | 0.0010 | 0.0061 |
| | 0.0250 | 0.0002 | 0.0051 | 0.0190 |
| | 0.0240 | 0.0013 | 0.0143 | 0.0395 |
| 10.0 | 0.0440 | $1 \times 10^{-5}$ | 0.0009 | 0.0058 |
| | 0.0420 | 0.0003 | 0.0061 | 0.0215 |
| | 0.0400 | 0.0021 | 0.0190 | 0.0483 |

It follows from these tables that the values of $t'_{cr}$ increase with $\omega$. But an increase in the values of $|\alpha|$ causes to decrease of $t'_{cr}$. (Table 4.25)

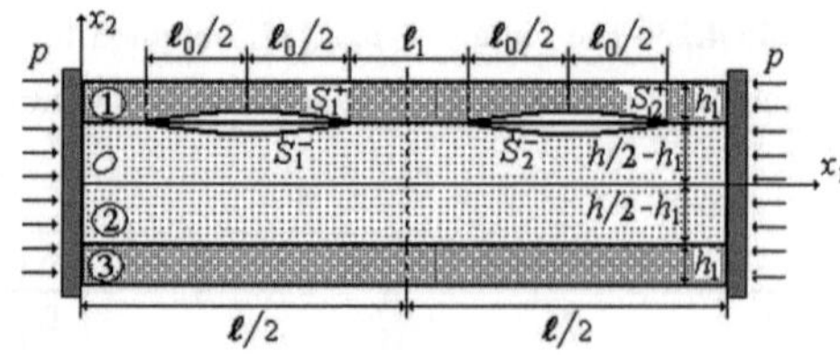

**Fig. 4.7** The geometry of the three-layered plate-strip containing two collinear interface cracks

## 4.6 Buckling Delamination of the Three-Layered (Sandwich) Plate-Strip with Two Collinear Interface Cracks

### *4.6.1 Formulation of the Problem and Method of Solution*

Now we consider again a three-layered sandwich plate-strip, which contains two collinear interface cracks located between the upper and core layers (see Fig. 4.7). Assume that the materials of the face layers are the same ones and the position of points of the layers are determined through the Lagrange coordinates in the Cartesian coordinate system $Ox_1x_2x_3$. The plate-strip occupies the region (4.50) and between the upper face and core layers exist two collinear cracks the location of which are symmetric with respect to $x_1 = \ell/2$ (Fig. 4.7). The edges of these cracks are described through the following equations.

For surfaces $S_1^{\pm}$:

$$x_2^{\pm} = \frac{h}{2} - h_1 \pm \varepsilon\ell_0 \sin^2\left(\frac{\pi(x_1 - (\ell - \ell_1)/2 - \ell_0)}{\ell_0}\right),\quad \text{for}\, x_1 \in ((\ell - \ell_1)/2 - \ell_0, (\ell - \ell_1)/2). \tag{4.75}$$

For surfaces $S_2^{\pm}$:

$$x_2^{\pm} = \frac{h}{2} - h_1 \pm \varepsilon\ell_0 \sin^2\left(\frac{\pi(x_1 - (\ell + \ell_1)/2)}{\ell_0}\right),\quad \text{for}\, x_1 \in ((\ell + \ell_1)/2, (\ell + \ell_1)/2 + \ell_0). \tag{4.76}$$

Within the framework of the assumptions accepted in the previous section, we investigate the evolution of the cracks edges initial imperfections with flow a time

This investigation is made by employing the field equations given in (4.53). However, the contact and boundary conditions (4.54) are rewritten as follows.

$$\left[\sigma_{2n}^{(1)}\left(\delta_i^n+\frac{\partial u_i^{(1)}}{\partial x_n}\right)\right]\Bigg|_{\substack{x_2=h/2\\ x_1\in(0,\ell)}}=0,\ \left[\sigma_{2n}^{(3)}\left(\delta_i^n+\frac{\partial u_i^{(3)}}{\partial x_n}\right)\right]\Bigg|_{\substack{x_2=-h/2\\ x_1\in(0,\ell)}}=0$$

$$\left[\sigma_{2n}^{(1)}\left(\delta_i^n+\frac{\partial u_i^{(1)}}{\partial x_n}\right)\right]\Bigg|_{\substack{x_2=h/2-h_1\\ x_1\in L_1'}}=\left[\sigma_{2n}^{(2)}\left(\delta_i^n+\frac{\partial u_i^{(2)}}{\partial x_n}\right)\right]\Bigg|_{\substack{x_2=h/2-h_1\\ x_1\in L_1'}},$$

$$\left[\sigma_{2n}^{(2)}\left(\delta_i^n+\frac{\partial u_i^{(2)}}{\partial x_n}\right)\right]\Bigg|_{\substack{x_2=-(h/2-h_1)\\ x_1\in(0,\ell)}}=\left[\sigma_{2n}^{(3)}\left(\delta_i^n+\frac{\partial u_i^{(3)}}{\partial x_n}\right)\right]\Bigg|_{\substack{x_2=-(h/2-h_1)\\ x_1\in(0,\ell)}},$$

$$u_i^{(1)}\Big|_{\substack{x_2=h/2-h_1\\ x_1\in L_1'}}=u_i^{(2)}\Big|_{\substack{x_2=h/2-h_1\\ x_1\in L_1'}},\ u_i^{(2)}\Big|_{\substack{x_2=-(h/2-h_1)\\ x_1\in(0,\ell)}}=u_i^{(3)}\Big|_{\substack{2=-(h/2-h_1)\\ x_1\in(0,\ell)}},$$

$$\left[\sigma_{2n}^{(1)}\left(\delta_i^n+\frac{\partial u_i^{(1)}}{\partial x_n}\right)\right]\Bigg|_{S_1^+}n_{j1}^+=0,\ \left[\sigma_{2n}^{(2)}\left(\delta_i^n+\frac{\partial u_i^{(2)}}{\partial x_n}\right)\right]\Bigg|_{S_1^-}n_{j1}^-=0,$$

$$\left[\sigma_{2n}^{(2)}\left(\delta_i^n+\frac{\partial u_i^{(2)}}{\partial x_n}\right)\right]\Bigg|_{S_2^+}n_{j2}^+=0,\ \left[\sigma_{2n}^{(3)}\left(\delta_i^n+\frac{\partial u_i^{(3)}}{\partial x_n}\right)\right]\Bigg|_{S_2^-}n_{j2}^-=0, \qquad (4.77)$$

where
$L_1'=((0,(\ell-\ell_1)/2-\ell_0))\cup((\ell-\ell_1)/2),(\ell+\ell_1)/2))\cup((\ell+\ell_1)/2+\ell_0),\ell))$.

At the ends of the plate-strip, the condition (4.55) satisfies. In addition, the relation (4.56) holds.

Substituting the series presentation (4.57) into the (4.53, 4.77) and (4.55) we obtain the boundary value problem (4.58) and (4.59) for determination of the zeroth approximation. In a similar manner we obtain the field equations given in (4.60) for the first approximation. But the contact and boundary conditions for the first approximation derived from (4.77) and (4.55) for the considered problem can be written as follows.

$$\sigma_{2i}^{(1),1}\Big|_{\substack{x_2=h/2\\ x_1\in(0,\ell)}}=0,\ \sigma_{2i}^{(3),1}\Big|_{\substack{x_2=-h/2\\ x_1\in(0,\ell)}}=0,$$

$$u_1^{(1),1}(0,x_2)=u_1^{(2),1}(0,x_2)=u_1^{(3),1}(0,x_2)=-u_1^{(1),1}(\ell,x_2)$$
$$=-u_1^{(2),1}(\ell,x_2)=-u_1^{(3),1}(\ell,x_2)=0$$
$$u_2^{(1),1}(0,x_2)=u_2^{(2),1}(0,x_2)=u_2^{(3),1}(0,x_2)=u_2^{(1),1}(\ell,x_2)$$
$$=u_2^{(2),1}(\ell,x_2)=u_2^{(3),1}(\ell,x_2)=0.$$

$$\left[\sigma_{2i}^{(1),1}+\sigma_{2n}^{(1),0}\frac{\partial u_i^{(1),1}}{\partial x_n}\right]\Bigg|_{\substack{x_2=h/2-h_1\\ x_1\in L_1'}}=\left[\sigma_{2i}^{(2),1}+\sigma_{2n}^{(2),0}\frac{\partial u_i^{(2),1}}{\partial x_n}\right]\Bigg|_{\substack{x_2=h/2-h_1\\ x_1\in L_1'}},$$

$$\left[\sigma_{2i}^{(2),1}+\sigma_{2n}^{(2),0}\frac{\partial u_i^{(2),1}}{\partial x_n}\right]\Bigg|_{\substack{x_2=-(h/2-h_1)\\ x_1\in(0,\ell)}}=\left[\sigma_{2i}^{(3),1}+\sigma_{2n}^{(3),0}\frac{\partial u_i^{(3),1}}{\partial x_n}\right]\Bigg|_{\substack{x_2=-(h/2-h_1)\\ x_1\in(0,\ell)}},$$

$$u_i^{(1),1}\Big|_{\substack{x_2=h/2-h_1\\ x_1\in L_1'}}=u_i^{(2),1}\Big|_{\substack{x_2=h/2-h_1\\ x_1\in L_1'}},\ u_i^{(2),1}\Big|_{\substack{x_2=-(h/2-h_1)\\ x_1\in(0,\ell}}=u_i^{(3),1}\Big|_{\substack{x_2=-(h/2-h_1)\\ x_1\in(0,\ell)}},$$

$$\left[\sigma_{2i}^{(1),1}+\sigma_{2n}^{(1),0}\frac{\partial u_i^{(1),1}}{\partial x_n}\right]\Bigg|_{\substack{x_2=h/2-h_1+0\\ x_1\in((\ell-\ell_1)/2-\ell_0,(\ell-\ell_1)/2)}}$$
$$=-\sigma_{1i}^{(1),0}\Big|_{\substack{x_2=h/2-h_1+0\\ x_1\in((\ell-\ell_1)/2-\ell_0,(\ell-\ell_1)/2)}}\pi\sin\left(\frac{2\pi(x_1-((\ell-\ell_1)/2)-\ell_0)}{\ell_0}\right)$$
$$+\frac{\partial\sigma_{2i}^{(1),0}}{\partial x_2}\Bigg|_{\substack{x_2=h/2-h_1+0\\ x_1\in((\ell-\ell_1)/2-\ell_0,(\ell-\ell_1)/2)}}\ell_0\sin^2\left(\frac{\pi(x_1-((\ell-\ell_1)/2)-\ell_0)}{\ell_0}\right),$$
$$\left[\sigma_{2i}^{(2),1}+\sigma_{2n}^{(2),0}\frac{\partial u_i^{(2),1}}{\partial x_n}\right]\Bigg|_{\substack{x_2=h/2-h_1-0\\ x_1\in((\ell-\ell_1)/2-\ell_0,(\ell-\ell_1)/2)}}$$
$$=-\sigma_{1i}^{(2),0}\Big|_{\substack{x_2=h/2-h_1-0\\ x_1\in((\ell-\ell_1)/2-\ell_0,(\ell-\ell_1)/2)}}\pi\sin\left(\frac{2\pi(x_1-((\ell-\ell_1)/2)-\ell_0)}{\ell_0}\right)$$
$$+\frac{\partial\sigma_{2i}^{(2),0}}{\partial x_2}\Bigg|_{\substack{x_2=h/2-h_1-0\\ x_1\in((\ell-\ell_1)/2-\ell_0,(\ell-\ell_1)/2)}}\ell_0\sin^2\left(\frac{\pi(x_1-((\ell-\ell_1)/2)-\ell_0)}{\ell_0}\right),$$
$$\left[\sigma_{2i}^{(1),1}+\sigma_{2n}^{(1),0}\frac{\partial u_i^{(1),1}}{\partial x_n}\right]\Bigg|_{\substack{x_2=h/2-h_1+0\\ x_1\in((\ell+\ell_1)/2,(\ell+\ell_1)/2+\ell_0)}}$$
$$=+\sigma_{1i}^{(1),0}\Big|_{\substack{x_2=h/2-h_1+0\\ x_1\in((\ell+\ell_1)/2,(\ell+\ell_1)/2+\ell_0)}}\pi\sin\left(\frac{2\pi(x_1-(\ell+\ell_1)/2)}{\ell_0}\right)$$
$$+\frac{\partial\sigma_{2i}^{(1),0}}{\partial x_2}\Bigg|_{\substack{x_2=h/2-h_1+0\\ x_1\in((\ell+\ell_1)/2,(\ell+\ell_1)/2+\ell_0)}}\ell_0\sin^2\left(\frac{\pi(x_1-(\ell+\ell_1)/2)}{\ell_0}\right),$$
$$\left[\sigma_{2i}^{(2),1}+\sigma_{2n}^{(2),0}\frac{\partial u_i^{(2),1}}{\partial x_n}\right]\Bigg|_{\substack{x_2=h/2-h_1-0\\ x_1\in((\ell+\ell_1)/2,(\ell+\ell_1)/2+\ell_0)}}$$
$$=+\sigma_{1i}^{(2),0}\Big|_{\substack{x_2=h/2-h_1-0\\ x_1\in((\ell+\ell_1)/2,(\ell+\ell_1)/2+\ell_0)}}\pi\sin\left(\frac{2\pi(x_1-(\ell+\ell_1)/2)}{\ell_0}\right)$$
$$+\frac{\partial\sigma_{2i}^{(2),0}}{\partial x_2}\Bigg|_{\substack{x_2=h/2-h_1-0\\ x_1\in((\ell+\ell_1)/2,(\ell+\ell_1)/2+\ell_0)}}\ell_0\sin^2\left(\frac{\pi(x_1-(\ell+\ell_1)/2)}{\ell_0}\right). \tag{4.78}$$

The FEM modeling of the problem related to the zeroth approximation is made as in the previous section, i.e. the functional (4.62) and (4.63) are used and the values of the Laplace transformation of displacements at nodes of the selected finite elements are derived from equation (4.64). Consequently, the discussions made in the previous section on the distribution character of the stresses related to the zeroth approximation hold also for the case under consideration. The following functional is used for the FEM modeling of the problem related to the first approximation.

$$\Pi^{(1)}\left(\bar{u}_1^{(1),1},\bar{u}_2^{(1),1},\bar{u}_1^{(2),1},\bar{u}_2^{(2),1},\bar{u}_1^{(3),1},\bar{u}_2^{(3),1}\right)$$
$$=\frac{1}{2}\sum_{k=1}^{3}\iint\limits_{\Omega_k''}\left[\bar{t}_{11}^{(k),1}\frac{\partial\bar{u}_1^{(k),1}}{\partial x_1}+\bar{t}_{12}^{(k),1}\frac{\partial\bar{u}_2^{(k),1}}{\partial x_1}+\bar{t}_{21}^{(k),1}\frac{\partial\bar{u}_1^{(k),1}}{\partial x_2}+\bar{t}_{22}^{(k),1}\frac{\partial\bar{u}_2^{(k),1}}{\partial x_2}\right]dx_1dx_2$$
$$-\int\limits_{(\ell-\ell_1)/2-\ell_0}^{(\ell-\ell_1)/2}\frac{1}{s}\frac{df_{01}}{dx_1}\left[\sigma_{11}^{(1),0}\bar{u}_1^{(1),1}\Big|_{x_2=h/2-h+0}-\sigma_{11}^{(2),0}\bar{u}_1^{(2),1}\Big|_{x_2=h/2-h_1-0}\right]dx_1$$
$$-\int\limits_{(\ell+\ell_1)/2}^{(\ell+\ell_1)/2+\ell_0}\frac{1}{s}\frac{df_{02}}{dx_1}\left[\sigma_{11}^{(1),0}\bar{u}_1^{(1),1}\Big|_{x_2=h/2-h+0}-\sigma_{11}^{(2),0}\bar{u}_1^{(2),1}\Big|_{x_2=h/2-h_1-0}\right]dx_1, \quad (4.79)$$

where

$$\frac{df_{01}}{dx_1}=\pi\sin\left(2\pi\frac{x_1-(\ell-\ell_1)/2-\ell_0}{\ell_0}\right),$$
$$\frac{df_{02}}{dx_1}=\pi\sin\left(2\pi\frac{x_1-(\ell+\ell_1)/2}{\ell_0}\right),$$
$$\Omega_1''=\{0\le x_1\le\ell,\ h/2-h_1\le x_2\le h/2\}$$
$$-\{x_1\in((\ell-\ell_1)/2-\ell_0,(\ell-\ell_1)/2),\ x_2=h/2-h_1+0\}$$
$$-\{x_1\in((\ell+\ell_1)/2,(\ell+\ell_1)/2+\ell_0),x_2=h/2-h_1+0\},$$
$$\Omega_2''=\{0\le x_1\le\ell,\ -h/2+h_1\le x_2\le h/2-h_1\},$$
$$-\{x_1\in((\ell-\ell_1)/2-\ell_0,(\ell-\ell_1)/2),\ x_2=h/2-h_1-0\}$$
$$-\{x_1\in((\ell+\ell_1)/2,(\ell+\ell_1)/2+\ell_0),x_2=h/2-h_1-0\}$$
$$\Omega_3''=\{0\le x_1\le\ell,\ -h/2\le x_2\le -h/2+h_1\}. \quad (4.80)$$

Note that the notation in (4.67) for $\bar{t}_{ij}^{(k)}$ and the notation (4.68) hold also for the case under consideration. Thus, through the relation (4.69) we obtain the corresponding equations for the Laplace transformations of the displacements of the nodes of the selected finite elements and from the criterion (4.70) (after replacing $x_1=\ell/2$ with $x_1=(\ell-\ell_1-\ell_0)/2$) we determine the values of the critical parameters.

### *4.6.2 Numerical Results and Discussions*

Within the scope of the assumptions and notation used in Sect. 4.5.3 consider numerical results obtained for the values of $p_{cr.0}/\tilde{E}$ and $p_{cr.\infty}/\tilde{E}$. Suppose that $\ell_0/\ell=0.275$ and $E_0^{(1)}/E_0^{(2)}=0.2$. Tables 4.26 and 4.27 show the values of $p_{cr.0}/\tilde{E}$ and $p_{cr.\infty}/\tilde{E}$ respectively calculated for various $\beta_1\ (=h_1/\ell_0)$, $\ell_1/\ell$ and $\omega$. It follows from these tables that interaction between the cracks under buckling

**Table 4.26** The values of $p_{cr.0}/\tilde{E}$ obtained for the case where $\ell_0/\ell = 0.275$ and $E_0^{(1)}/E_0^{(2)} = 0.2$

| $\beta_1$ | $\ell_1/\ell$ | | | |
|---|---|---|---|---|
| | 0.225 | 0.150 | 0.075 | 0.025 |
| 1/11 | 0.02148 | 0.02153 | 0.02168 | 0.02181 |
| 3/22 | 0.04020 | 0.04036 | 0.04095 | 0.04136 |
| 2/11 | 0.06011 | 0.06053 | 0.06160 | 0.06233 |

**Table 4.27** The values of $p_{cr.\infty}/\tilde{E}$ obtained for the case where $\ell_0/\ell = 0.275$ and $E_0^{(1)}/E_0^{(2)} = 0.2$

| $\omega$ | $\beta_1$ | $\ell_1/\ell$ | | | |
|---|---|---|---|---|---|
| | | 0.225 | 0.150 | 0.075 | 0.025 |
| 1.0 | 1/11 | 0.01150 | 0.01150 | 0.01150 | 0.01166 |
| | 3/22 | 0.02166 | 0.02166 | 0.02166 | 0.02216 |
| | 2/11 | 0.03216 | 0.03216 | 0.03250 | 0.03350 |
| 2.0 | 1/11 | 0.01508 | 0.01514 | 0.01538 | 0.01569 |
| | 3/22 | 0.02813 | 0.02836 | 0.02905 | 0.02981 |
| | 2/11 | 0.04174 | 0.04174 | 0.04201 | 0.04328 |

**Table 4.28** The values of $t'_{cr}$ obtained for various $p/\tilde{E}$ under $\omega = 1.0$

| $\beta_1$ | $p/\tilde{E}$ | $\ell_1/\ell$ | | | |
|---|---|---|---|---|---|
| | | 0.225 | 0.150 | 0.075 | 0.025 |
| 1/11 | 0.01544 | 0.365 | 0.382 | 0.471 | 0.673 |
| | 0.01764 | 0.058 | 0.059 | 0.067 | 0.084 |
| | 0.011985 | 0.050 | 0.005 | 0.006 | 0.009 |
| 3/22 | 0.03315 | 0.053 | 0.055 | 0.066 | 0.095 |
| | 0.03536 | 0.017 | 0.018 | 0.022 | 0.034 |
| | 0.03757 | 0.003 | 0.004 | 0.005 | 0.010 |
| 2/11 | 0.04871 | 0.067 | 0.069 | 0.090 | 0.144 |
| | 0.05314 | 0.015 | 0.015 | 0.020 | 0.036 |
| | 0.05757 | 0.001 | 0.001 | 0.002 | 0.007 |

delamination causes to increase insignificantly $p_{cr.0}/\tilde{E}$ and $p_{cr.\infty}/\tilde{E}$, i.e. the values of $p_{cr.0}/\tilde{E}$ and $p_{cr.\infty}/\tilde{E}$ increase insignificantly with decreasing the $\ell_1/\ell$. Consequently, the interaction between the two collinear cracks under their buckling delamination is negligible with very high accuracy.

Results related to the $t'_{cr}$ are given in Tables 4.28 and 4.29 which are obtained for the cases where $\omega = 1.0$ and 2.0 respectively. These results show that the value of $t'_{cr}$ increase sharply with decreasing the distance between the collinear cracks, i.e. with decreasing $\ell_1/\ell$. Accordingly, the mentioned interaction between the cracks is not negligible in the cases where the face layers materials of the sandwich plate- strip are viscoelastic ones.

**Table 4.29** The values of $t'_{cr}$ obtained for various $p/\tilde{E}$ under $\omega = 2.0$

| $\beta_1$ | $p/\tilde{E}$ | $\ell_1/\ell$ | | | |
|---|---|---|---|---|---|
| | | 0.225 | 0.150 | 0.075 | 0.025 |
| 1/11 | 0.01875 | 0.033 | 0.034 | 0.0409 | 0.057 |
| | 0.01588 | 0.686 | 0.736 | 1.01 | 1.90 |
| | 0.01544 | 17.3 | 24.2 | 544 | $\infty$ |
| 3/22 | 0.03866 | 0.0012 | 0.0013 | 0.0022 | 0.0059 |
| | 0.03425 | 0.0565 | 0.0591 | 0.0769 | 0.134 |
| | 0.03094 | 0.657 | 0.724 | 1.2070 | 3.683 |
| 2/11 | 0.05535 | 0.0074 | 0.0077 | 0.0112 | 0.0259 |
| | 0.04871 | 0.167 | 0.177 | 0.271 | 0.680 |
| | 0.04428 | 1.276 | 1.276 | 2.141 | 16.23 |

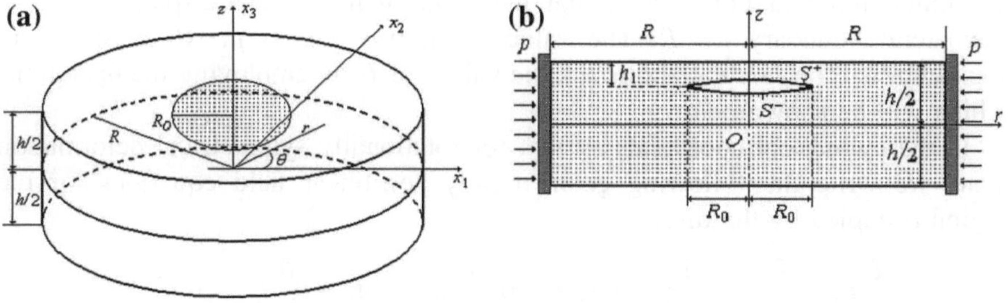


**Fig. 4.8** The geometry of the disc: **a** whole disc with a crack; **b** the geometry of the disc's section under $\theta = \text{const}$

## 4.7 Buckling Delamination of the Elastic and Viscoelstic Composite Circular Plate-Disc with a Penny-Shaped Crack

### *4.7.1 Formulation of the Problem*

Consider a circular plate and the position of points of that we determine through the Lagrange coordinates in the cylindrical system of coordinates $Or\theta z$ associated with the middle plane of the plate. Assume that the plate-disc occupies the region $\Omega = \{0 \le r \le R, 0 \le \theta \le 2\pi, -h/2 \le z \le +h/2\}$ and has a penny shaped crack , which occurs in $\{0 \le r \le R_0, 0 \le \theta \le 2\pi, z = h/2 - h_1\}$. Moreover, we assume that the surfaces (edges) of this crack have infinitesimal rotationally symmetric imperfection. Denote these surfaces as $S^+$ (upper surface) and $S^-$ (lower surface) (see Fig. 4.8) and the equation of those we write as follows:

$$z^+ = (h/2 - h_1) + \varepsilon f(r) \text{ for } S^+ \text{ and } z^- = (h/2 - h_1) - \varepsilon f(r) \text{ for } S^-, \quad (4.81)$$

where, as before, $\varepsilon$ is a small parameter $(0 \le \varepsilon \ll 1)$ and characterizes the degree of the initial imperfection. We also assume that the function $f(r)$ in (4.81) must satisfy the following conditions.

$$f(r) \ge 0,\ |\varepsilon f(r)| \ll 1,\ \varepsilon\left|\frac{df}{dr}\right| \ll 1 \text{ for } r \in (0, R_0),$$
$$|f(0)| = \max|f(r)|,\ \left.\frac{df}{dr}\right|_{r=0,R_0} = 0,\ f|_{r=R_0} = 0. \tag{4.82}$$

Using (4.81) and (4.82) we derive the expressions

$$n_r^{\pm} = \pm\left(-\frac{\partial z^{\pm}(r)}{\partial r}\right) \bigg/ \sqrt{\left(\frac{\partial z^{\pm}(r)}{\partial r}\right)^2 + 1},\ n_\theta^{\pm} = 0,\ n_z^{\pm} = \pm 1 \bigg/ \sqrt{\left(\frac{\partial z^{\pm}(r)}{\partial r}\right)^2 + 1} \tag{4.83}$$

for ortho-normal components $n_r^{\pm}$, $n_\theta^{\pm}$ and $n_z^{\pm}$ of the surfaces $S^{\pm}$.

In the framework of the stated above we suppose that the disc is compressed circumferentially and inwards through the clamp by fixed radial displacement $U$ in the lateral boundary $r = R$. The values of normal forces $p$, which shown in Fig. 4.8b, will be calculated through the values of $U$ by employing the operations which will be shown below.

In the cylindrical coordinate system for rotationally symmetrical deformation state, we write the following geometrically non-linear field equations for the region occupied by the disc.

$$\frac{\partial}{\partial r}t_{rr} + \frac{\partial}{\partial z}t_{zr} + \frac{1}{r}(t_{rr} - t_{\theta\theta}) = 0,\ \frac{\partial}{\partial r}t_{rz} + \frac{1}{r}t_{rz} + \frac{\partial}{\partial z}t_{zz} = 0,$$
$$t_{rr} = \sigma_{rr}\left(1 + \frac{\partial u_r}{\partial r}\right) + \sigma_{rz}\frac{\partial u_r}{\partial z},\ t_{zr} = \sigma_{zr}\left(1 + \frac{\partial u_r}{\partial r}\right) + \sigma_{zz}\frac{\partial u_r}{\partial z},$$
$$t_{\theta\theta} = \sigma_{\theta\theta}\left(1 + \frac{u_r}{r}\right),\ t_{rz} = \sigma_{rz}\left(1 + \frac{\partial u_z}{\partial z}\right) + \sigma_{rr}\frac{\partial u_z}{\partial r},$$
$$t_{zz} = \sigma_{zz}\left(1 + \frac{\partial u_z}{\partial z}\right) + \sigma_{rz}\frac{\partial u_z}{\partial r}, \tag{4.84}$$

$$\sigma_{rr} = A_{11}^*\varepsilon_{rr} + A_{12}^*\varepsilon_{\theta\theta} + A_{13}^*\varepsilon_{zz},\ \sigma_{\theta\theta} = A_{12}^*\varepsilon_{rr} + A_{22}^*\varepsilon_{\theta\theta} + A_{23}^*\varepsilon_{zz},$$
$$\sigma_{zz} = A_{13}^*\varepsilon_{rr} + A_{23}^*\varepsilon_{\theta\theta} + A_{33}^*\varepsilon_{zz},\ \sigma_{rz} = 2A_{55}^*\varepsilon_{rz},$$
$$A_{ij}^*\varphi = A_{ij0}\varphi(t) + \int_0^t A_{ij1}(t-\tau)\varphi(\tau)d\tau,\ ij = 11; 22; 33; 12; 13; 23; 55, \tag{4.85}$$

$$\varepsilon_{rr} = \frac{\partial u_r}{\partial r} + \frac{1}{2}\left(\frac{\partial u_r}{\partial r}\right)^2 + \frac{1}{2}\left(\frac{\partial u_z}{\partial r}\right)^2,\ \varepsilon_{\theta\theta} = \frac{u_r}{r} + \frac{1}{2}\left(\frac{u_r}{r}\right)^2,$$
$$\varepsilon_{rz} = \frac{1}{2}\left(\frac{\partial u_r}{\partial z} + \frac{\partial u_z}{\partial r} + \frac{\partial u_r}{\partial r}\frac{\partial u_r}{\partial z} + \frac{\partial u_z}{\partial r}\frac{\partial u_z}{\partial z}\right),$$
$$\varepsilon_{zz} = \frac{\partial u_z}{\partial r} + \frac{1}{2}\left(\frac{\partial u_z}{\partial z}\right)^2 + \frac{1}{2}\left(\frac{\partial u_r}{\partial z}\right)^2. \tag{4.86}$$

The notation used in the field equations (4.84) and (4.85) is conventional. Similar notation and its explanations have been given in Chap. 2. Using this notation we also write the following conditions which are satisfied, on the upper and lower surfaces of the crack.

$$\left(n_r^{\pm} t_{rr} + n_z^{\pm} t_{zr}\right)\Big|_{\substack{z=(h/2-h_1)\pm(r) \\ 0\le r\le R_0}} = 0,$$
$$\left(n_r^{\pm} t_{rz} + n_z^{\pm} t_{zz}\right)\Big|_{\substack{z=(h/2-h_1)\pm(r) \\ 0\le r\le R_0}} = 0. \tag{4.87}$$

Moreover, on the face planes of the disc the following conditions are satisfied.

$$t_{zr}\big|_{z=\pm h/2} = 0,\ t_{zz}\big|_{z=\pm h/2} = 0\ \text{for}\ 0 \le r \le R. \tag{4.88}$$

On the lateral cylindrical boundary of the disc the conditions

$$u_r\big|_{r=R} = -U,\ u_z\big|_{r=R} = 0,\ \text{for}\ -h/2 \le z \le +h/2 \tag{4.89}$$

are also given.

This completes formulation of the problem on the buckling delamination of the circular plate-disc, which contains a penny shaped crack and made of viscoelastic orthotropic material. The foregoing formulation allows us to investigate the evolution of the initial infinitesimal imperfections of the crack's edges with compressed force (or with $U$) for a pure elastic disc or with flow a time for a viscoelastic disc within the scope of the TDLTS.

### 4.7.2 Method of Solution

Using the series presentation with respect to the small parameter ε, for instance, series presentation (2.14), and doing some mathematical manipulations described in the previous chapters, we obtain the following equations and boundary conditions for the zeroth and the first approximations.

For the zeroth approximation

$$\frac{\partial\sigma_{rr}^{(0)}}{\partial r} + \frac{\partial\sigma_{zr}^{(0)}}{\partial z} + \frac{1}{r}\left(\sigma_{rr}^{(0)} - \sigma_{\theta\theta}^{(0)}\right) = 0, \quad \frac{\partial\sigma_{rz}^{(0)}}{\partial r} + \frac{\partial\sigma_{zz}^{(0)}}{\partial z} + \frac{1}{r}\sigma_{rz}^{(0)} = 0,$$
$$\sigma_{rr}^{(0)} = A_{11}^{*}\varepsilon_{rr}^{(0)} + A_{12}^{*}\varepsilon_{\theta\theta}^{(0)} + A_{13}^{*}\varepsilon_{zz}^{(0)}, \quad \sigma_{\theta\theta}^{(0)} = A_{12}^{*}\varepsilon_{rr}^{(0)} + A_{22}^{*}\varepsilon_{\theta\theta}^{(0)} + A_{23}^{*}\varepsilon_{zz}^{(0)},$$
$$\sigma_{zz}^{(0)} = A_{13}^{*}\varepsilon_{rr}^{(0)} + A_{23}^{*}\varepsilon_{\theta\theta}^{(0)} + A_{33}^{*}\varepsilon_{zz}^{(0)}, \quad \sigma_{rz}^{(0)} = 2A_{55}^{*}\varepsilon_{rz}^{(0)}, \tag{4.90}$$

$$\sigma_{rz}^{(0)}\Big|_{z=\pm h/2} = 0, \quad \sigma_{zz}^{(0)}\big|_{z=\pm h/2} = 0\ \text{for}\ 0 < r < R,$$
$$u_r^{(0)}\Big|_{r=R} = -U, \quad u_z^{(0)}\big|_{r=R} = 0. \tag{4.91}$$

For the first approximation

$$\frac{\partial t_{rr}^{(1)}}{\partial r}+\frac{\partial t_{zr}^{(1)}}{\partial z}+\frac{1}{r}\left(t_{rr}^{(1)}-t_{\theta\theta}^{(1)}\right)=0,\ \frac{\partial t_{rz}^{(1)}}{\partial r}+\frac{\partial t_{zz}^{(1)}}{\partial z}+\frac{1}{r}t_{rz}^{(1)}=0,$$
$$\sigma_{rr}^{(1)}=A_{11}^{*}\varepsilon_{rr}^{(1)}+A_{12}^{*}\varepsilon_{\theta\theta}^{(1)}+A_{13}^{*}\varepsilon_{zz}^{(1)},\ \sigma_{\theta\theta}^{(1)}=A_{12}^{*}\varepsilon_{rr}^{(1)}+A_{22}^{*}\varepsilon_{\theta\theta}^{(1)}+A_{23}^{*}\varepsilon_{zz}^{(1)},$$
$$\sigma_{zz}^{(1)}=A_{13}^{*}\varepsilon_{rr}^{(1)}+A_{23}^{*}\varepsilon_{\theta\theta}^{(1)}+A_{33}^{*}\varepsilon_{zz}^{(1)},\ \sigma_{rz}^{(1)}=2A_{55}^{*}\varepsilon_{rz}^{(1)},$$
$$\varepsilon_{rr}^{(1)}=\frac{\partial u_r^{(1)}}{\partial r},\ \varepsilon_{\theta\theta}^{(1)}=\frac{u_r^{(1)}}{r},\ \varepsilon_{rz}^{(1)}=\frac{1}{2}\left(\frac{\partial u_z^{(1)}}{\partial r}+\frac{\partial u_r^{(1)}}{\partial z}\right),\ \varepsilon_{zz}^{(1)}=\frac{\partial u_z^{(1)}}{\partial r},\tag{4.92}$$

$$t_{zr}^{(1)}\big|_{z=\pm h/2}=0,\ t_{zz}^{(1)}\big|_{z=\pm h/2}=0,\ u_r^{(1)}\big|_{r=R}=0,\ u_z^{(1)}\big|_{r=R}=0,$$
$$t_{zr}^{(1)}\big|_{z=h/2-h_1\pm 0}=\frac{df}{dr}\sigma_{rr}^{(0)}\big|_{z=h/2-h_1\pm 0},\ t_{zz}^{(1)}\big|_{z=h/2-h_1\pm 0}=\frac{df}{dr}\sigma_{zr}^{(0)}\big|_{z=h/2-h_1\pm 0},\tag{4.93}$$
$$\text{for}\ 0<r<R_0,$$

where

$$t_{rr}^{(1)}=\sigma_{rr}^{(1)}+\sigma_{rr}^{(0)}\frac{\partial u_r^{(1)}}{\partial r}+\sigma_{rz}^{(0)}\frac{\partial u_r^{(1)}}{\partial z},\ t_{zz}^{(1)}=\sigma_{zz}^{(1)}+\sigma_{zz}^{(0)}\frac{\partial u_z^{(1)}}{\partial z}+\sigma_{rz}^{(0)}\frac{\partial u_z^{(1)}}{\partial r},$$
$$t_{\theta\theta}^{(1)}=\sigma_{\theta\theta}^{(1)}+\sigma_{\theta\theta}^{(0)}\frac{u_r^{(1)}}{r},\ t_{zr}^{(1)}=\sigma_{zr}^{(1)}+\sigma_{zr}^{(0)}\frac{\partial u_r^{(1)}}{\partial r}+\sigma_{zz}^{(0)}\frac{\partial u_r^{(1)}}{\partial z},$$
$$t_{rz}^{(1)}=\sigma_{rz}^{(1)}+\sigma_{rr}^{(0)}\frac{\partial u_z^{(1)}}{\partial r}+\sigma_{rz}^{(0)}\frac{\partial u_z^{(1)}}{\partial z}.\tag{4.94}$$

As follows from the formulation of the boundary conditions in (4.91), the existence of the crack does not taken into account under determination of the values of the zeroth approximation.

We consider the determination of the zeroth and of the first approximations separately.

#### 4.7.2.1 The Zeroth Approximation

The values of this approximation are determined from the boundary-value problem (4.90) and (4.91) for investigation of which we employ the FEM. Note that before employing FEM we apply the Laplace transformation (4.28) to the Eqs. (4.90) and (4.91). As before, the transformed problem is solved by employing FEM and the original functions of the Laplace transformations are determined with the use of the method (Schapery 1962, 1978). Under FEM modeling the region $\{0\leq r\leq R;-h/2\leq z\leq +h/2\}$ is divided into 120 rectangular Lagrange family quadratic elements with 533 nodes and 1027 NDOF. The traditional Ritz technique is employed for the determination of the Laplace transformation of the sought values and all operations are performed by the displacement based FEM.

Under numerical investigation procedure we give the function $U(t)$ so that the relation

$$hp = \int_{-h/2}^{h/2} \sigma_{rr}^{(0)}\big|_{r=0} dz = const \tag{4.95}$$

is satisfied and from the relation (4.95) we determine the dependence between the values of $p$ and $U(t)$. It should be noted that the function $U(t)$ is constructed under calculation procedure from the constitutive (stress-strain) relation.

The analyzes of the numerical results show that due to the relation (4.95) the stresses $\sigma_{rr}^{(0)}$, $\sigma_{\theta\theta}^{(0)}$, $\sigma_{zz}^{(0)}$ and $\sigma_{zr}^{(0)}$ can be taken as independent of time with very high accuracy. Moreover analyzes of these results show that the stresses $\sigma_{zz}^{(0)}$ and $\sigma_{zr}^{(0)}$ differ from zero in the very near vicinity of the lateral boundary $r = R$ and can be taken as zero outside of this region. Therefore the existence of a crack in the plate does not change the stress distributions from the one without crack in the zeroth approximation.

#### 4.7.2.2 The First Approximation

Taking the time-independent character of the stresses in the zeroth approximation into account for the solution of the Eqs. (4.92–4.94) we can employ the principle of the correspondence by using the Laplace transformation (4.28). So, replacing in (4.92–4.94) $\sigma_{(ij)}^{(1)}$, $\varepsilon_{(ij)}^{(1)}$, $u_{(i)}^{(1)}$ and $A_{ij}^{*}$ (where $(ij) = rr; zz; \theta\theta; rz$, $(i) = r; z$, $ij = 11; 22; 33; 12; 13; 23$) by $\bar{\sigma}_{(ij)}^{(1)}$, $\bar{\varepsilon}_{(ij)}^{(1)}$, $\bar{u}_{(i)}^{(1)}$ and $\bar{A}_{ij}^{*}$ respectively, we obtain the corresponding equations and boundary conditions with respect to the Laplace transformations of the values of the first approximation. For the FEM modeling of this problem we introduce the following functional.

$$\Pi(\bar{u}_r^{(1)}, \bar{u}_z^{(1)}) = \frac{1}{2}\int_0^R \int_{-h/2}^{+h/2} \left\{\left[\bar{t}_{rr}^{(1)}\bar{\varepsilon}_{rr}^{(1)} + \bar{t}_{\theta\theta}^{(1)}\bar{\varepsilon}_{\theta\theta}^{(1)} + \bar{t}_{rz}^{(1)}\bar{\varepsilon}_{rz}^{(1)} + \bar{t}_{zr}^{(1)}\bar{\varepsilon}_{zr}^{(1)}\right]\right\} rdrdz$$
$$- \int_0^{R_0} \frac{df}{dr}\sigma_{rr}^{(0)}\bar{u}_r^{(1)}\big|_{z=h/2-h_1+0} rdr - \int_0^{R_0} \frac{df}{dr}\sigma_{rr}^{(0)}\bar{u}_r^{(1)}\big|_{z=h/2-h_1-0} rdr. \tag{4.96}$$

Note that under writing (4.96) it is taken into accounts that $\sigma_{rz}^{(0)} = 0$ and $\sigma_{zz}^{(0)} = 0$ for $z = h/2 - h_1 \pm 0$, $0 \le r \le R_0$.

From the relation

$$\delta\Pi(\bar{u}_r^{(1)}, \bar{u}_z^{(1)}) = \frac{\partial\Pi}{\partial\bar{u}_r^{(1)}}\delta\bar{u}_r^{(1)} + \frac{\partial\Pi}{\partial\bar{u}_z^{(1)}}\delta\bar{u}_z^{(1)} = 0 \tag{4.97}$$

we obtain the first two equations in (4.92) and boundary conditions for forces in (4.93) rewritten for $\bar{\sigma}_{(ij)}^{(1)}$, $\bar{\varepsilon}_{(ij)}^{(1)}$, $\bar{u}_{(i)}^{(1)}$ and $\bar{A}_{ij}^{*}$. In this way we prove that these equations are the Euler equations of the functional (4.96). Thus, applying the well-known techniques of the FEM to the Eq. (4.97) we obtain the corresponding linear equations for the nodal values of the Laplace transformation of the displacements. The originals of the sought values are obtained by employing of the method (Schapery 1962, 1978). The values of the critical time or the critical force are determined from the criterion

$$\begin{aligned} u_z^{(1)}(0, h/2 - h_1, t) \to \infty \text{ as } t \to t_{cr} \text{ (for a viscoelasticdisc-plate)}, \\ \text{or as } p \to p_{cr} \text{ (for a elastic disc-plate)}. \end{aligned} \tag{4.98}$$

### 4.7.3 Numerical Results and Discussions

The initial imperfection of the crack edges we select as follows:

$$\varepsilon f(r) = \varepsilon R_0 \cos^2\left(\frac{\pi r}{2R_0}\right) = L\cos^2\left(\frac{\pi r}{2R_0}\right),\ L \ll R_0,\ \varepsilon = \frac{L}{R_0}, \tag{4.99}$$

where $L$ is the maximum value of the lift of the initial imperfection. As a result of the analyses of the numerous numerical results it is established that the critical values of the external forces for pure elastic problem or the values of the critical values of the critical time for the viscoelastic problems do not depend on the initial imperfection form which satisfies the conditions (4.82).

We assume that the plate is fabricated from the multilayered composite material consisting of alternating two isotropic layers, and also we assume that these layers lie in the planes $z = const$. The material of the reinforcing layers is supposed to be elastic with mechanical constants $E^{(2)}$ (Young's modulus) and $\nu^{(2)}$ (Poisson coefficient). The material of the matrix layers is supposed to be linearly viscoelastic with operators (3.157–3.159). In the continuum approach this layered composite material is taken as viscoelastic transversally-isotropic material with normalized mechanical properties whose isotropy axis lies on the $Oz$ axis. These normalized mechanical properties (operators) are determined through the expressions (3.160) and (3.161).

We introduce the dimensionless rheological parameter $\omega\,(=\omega_{\infty}/\omega_0)$ and dimensionless time $t'\ (=\omega_0^{1/(1+\alpha)}t)$ and assume that $h/(2R) = 0.15$, $\nu_0^{(1)} = \nu_0^{(2)} = 0.3$, $\eta^{(2)} = 0.5$ (unless otherwise not specified), where $\eta^{(2)}$ is a volumetric fraction of the filler material in the plate. Moreover, we introduce the dimensionless parameter $L_E = (R - R_0)/(2R)$, $H_u = h_1/(2R)$ and $\beta = h_1/R_0$ which

**Table 4.30** The values of $p_{cr.0}\big/E_0^{(1)}$ for various $L_E=(R-R_0)/(2R)$ and $H_u=h_1/(2R)$ under $E^{(2)}\big/E_0^{(1)}=50,\ \beta=h_1/R_0=1/8,\ \eta^{(2)}=0.5$

| $L_E$ | | | | |
|---|---|---|---|---|
| 0.05 | 0.10 | 0.15 | 0.20 | 0.25 |
| | | $H_u$ | | |
| 0.05625 | 0.0500 | 0.04375 | 0.0375 | 0.03125 |
| | | $p_{cr.}\big/E_0^{(1)}$ | | |
| 0.2662 | 0.2571 | 0.2526 | 0.2509 | 0.2510 |

**Table 4.31** The values of $p_{cr.0}\big/(\eta^{(1)}E_0^{(1)}+\eta^{(2)}E^{(2)})$ for various $E^{(2)}\big/E_0^{(1)}$ under $\eta^{(2)}=0.3$, $L_E=0.25,\ \beta=h_1/R_0=1/8,\ H_u=0.03215$

| $E^{(2)}\big/E_0^{(1)}$ | Results of Guz and Nazarenko (1985a) | Present results |
|---|---|---|
| 1 | 0.0167 | 0.0178 |
| 10 | 0.0140 | 0.0154 |
| 25 | 0.0126 | 0.0129 |

characterize the distance of the crack tip from the plate edge, the distance of the crack from the upper plane of the plate and the ratio of the thickness of the layer between the crack and the upper face plane of the plate to the radius of the penny-shaped crack, respectively.

First, we assume that the material of the disc is a pure elastic and investigate the delamination problem for the cases where $t'=0$ and $t'=\infty$. In these cases the reduced critical values of the external compressive force will be denoted as $p_{cr.0}\big/E_0^{(1)}$ (for $t'=0$) and $p_{cr.\infty}\big/E_0^{(1)}$ (for $t'=\infty$). According to the well known mechanical consideration, under keeping $\beta=const$, the values of $p_{cr.0}\big/E_0^{(1)}$ must approach their asymptotic value with an increase in $L_E$ along with a simultaneous decrease in $H_u$. This conclusion is proven by the results given in Table 4.30. Moreover note that this asymptotic value of $p_{cr.0}$ is a related to the buckling delamination problem of the half-space with the penny-shaped near-surface crack. The last problem had been investigated in (Guz and Nazarenko 1985a, b) with employing double integral equation technique and for obtaining the numerical results the Galerkin method had been employed.

Table 4.31 shows the values of $p_{cr.0}\big/(\eta^{(1)}E_0^{(1)}+\eta^{(2)}E^{(2)})$ obtained by applying both present and (Guz and Nazarenko 1985a) approaches in the case where $\eta^{(2)}=0.3,\ \eta^{(1)}=1-\eta^{(2)},\ \beta=1/8,\ L_E=0.25,\ H_u=0.03125$ for various $E_0^{(1)}\big/E^{(2)}$. The comparison of the corresponding results given in Table 4.31 shows that the numerical results obtained in the framework of the present approach are very close to those obtained in (Guz and Nazarenko 1985a). At the same time, the

**Table 4.32** The values of $p_{cr.0}\big/E_0^{(1)}$ (numerator) and $p_{cr.\infty}\big/E_0^{(1)}$ (denominator) for various $L_E$ under $H_u = 0.075$, $\omega = 1.0$, $E^{(2)}\big/E_0^{(1)} = 50$

| $L_E$ | | | |
|---|---|---|---|
| 0.25 | 0.40 | 0.425 | 0.50 |
| $\frac{0.4949}{0.2721}$ | $\frac{0.5455}{0.2899}$ | $\frac{0.6015}{0.3092}$ | $\frac{0.5395}{0.2846}$ |

results given in Tables 4.30 and 4.31 guarantee the validity of the algorithm and programs used in the present investigation.

Consider some numerical results related to $p_{cr.0}\big/E_0^{(1)}$ and $p_{cr.\infty}\big/E_0^{(1)}$ which are obtained for various values of the problem parameters. When calculating $p_{cr.\infty}\big/E_0^{(1)}$ the operators $E^{(1)*}$ and $\nu^{(1)*}$ in (3.157–3.159) are replaced by the constants $E_\infty^{(1)} = E_0^{(1)}(1 - 1/(1+\omega))$ and $\nu_\infty^{(1)} = \nu_0^{(1)}(1 + (1 - 2\nu_0^{(1)})\big/(2\nu_0^{(1)}(1+\omega))$, respectively.

Table 4.32 shows the values of $p_{cr.0}\big/E_0^{(1)}$ (numerator) and $p_{cr.\infty}\big/E_0^{(1)}$ (denominator) for various $L_E$ under $H_u = 0.075$, $\omega = 1$ and $E^{(2)}\big/E_0^{(1)} = 50$. Note that in this table the case where $L_E = 0.50$ corresponds to the stability loss of the whole disk (plate), i.e. the plate without the crack. Under obtaining procedure of the numerical results given in Tables 4.30, 4.31 and 4.32 it was established that in the cases where $\beta = h_1/R_0 \le (h - h_1)/R_0 < h/R$ the values of $p_{cr.0}\big/E_0^{(1)}$ and $p_{cr.\infty}\big/E_0^{(1)}$ depend only on $\beta$ but not depend on $(h - h_1)/R_0$ and on $h/R$. According to these results, we can conclude that the values of $p_{cr.0}\big/E_0^{(1)}$ and $p_{cr.\infty}\big/E_0^{(1)}$ obtained for the case where $\beta_1 < h/R$ is satisfied, correspond to the buckling of the delaminated part of the cracked disc.

Here and below under "delamination part" we mean the part of the disc which occupies the region between the crack and the face plane of the disc, which has a short distance from the crack. Consequently, it follows from the results given in Table 4.32 that in the cases where $L_E \ge 0.4$ the stability loss of the whole disc takes place at an earlier stage of the loading than the buckling of the delaminated part of the disc. Taking this situation into account below we will consider only the cases where $\beta \le (h - h_1)/R_0$.

The graphs given in Fig. 4.9 show the relationships among $p_{cr.0}\big/E_0^{(1)}$ (solid lines), $p_{cr.\infty}\big/E_0^{(1)}$ (dashed lines) and $E^{(2)}\big/E_0^{(1)}$ for the case where $L_E = 0.25$, $\omega = 1.0$. In this figure the graphs indicated by 1, 2, 3 and 4 (for $p_{cr.0}\big/E_0^{(1)}$) and by 1′, 2′, 3′ and 4′ (for $p_{cr.\infty}\big/E_0^{(1)}$) correspond to the cases where $H_u = 0.00, 0.075, 0.06$ and $0.05$ respectively. It follows from these graphs that for all the considered cases the values of $p_{cr.0}\big/E_0^{(1)}$ and $p_{cr.\infty}\big/E_0^{(1)}$ increase monotonically with $E^{(2)}\big/E_0^{(1)}$. Note that the

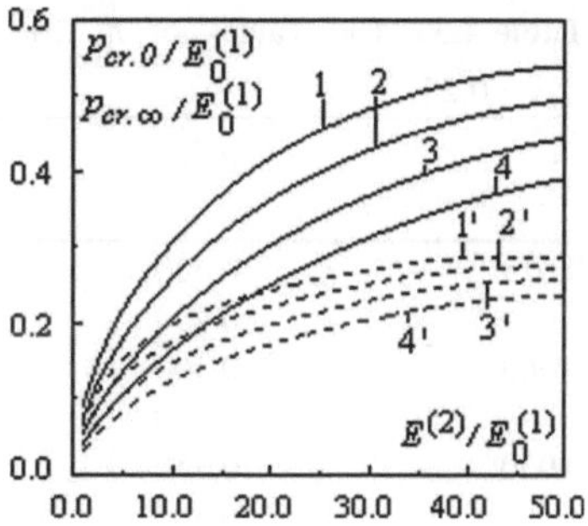


**Fig. 4.9** The graphs of the relationships among $p_{cr.0}\big/E_0^{(1)}$, $p_{cr.\infty}\big/E_0^{(1)}$ and $E^{(2)}\big/E_0^{(1)}$

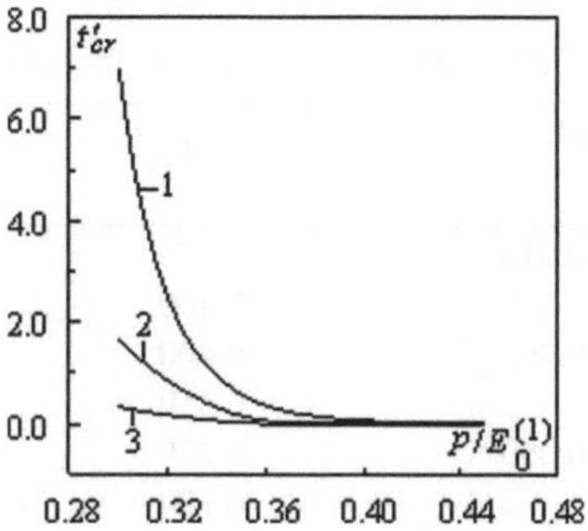


**Fig. 4.10** The graphs of the relationships between $t'_{cr}$ and $p\big/E_0^{(1)}$

case where $H_u = 0.00$ (as the case where $L_E = 0.50$) corresponds to the whole disc without a crack and buckling of the whole disc is studied separately with the use of the approach developed in the previous chapter. In this case the initial insignificant imperfection is given for the whole disc and the development of this imperfection is investigated by the use of the TDLTS.

Now we analyze the numerical results related to the viscoelastic disc. In these cases the selected values of $p\big/E_0^{(1)}$ satisfy the inequality $p_{cr.\infty}\big/E_0^{(1)} < p\big/E_0^{(1)} < p_{cr.0}\big/E_0^{(1)}$. Consider the graphs given in Fig. 4.10 which show the relation between $t'_{cr}$ and $p\big/E_0^{(1)}$ for various $H_u$ under $E^{(2)}\big/E_0^{(1)} = 50$, $\alpha = -0.5$, $\omega = 1.0$ and $L_E = 0.25$. The graphs indicated by 1, 2 and 3 correspond to the cases where $H_u = 0.075$, 0.060 and 0.050 respectively. According to these graphs, we can conclude that $t'_{cr}$ decreases monotonically with increasing $p\big/E_0^{(1)}$ and with decreasing $H_u$.

Tables 4.33 and 4.34 show the influence of the rheological parameters $\omega$ and $\alpha$ respectively on the values of $t'_{cr}$ with various $H_u$ and $p\big/E_0^{(1)}$ under $E^{(2)}\big/E_0^{(1)} = 50$, $\alpha = -0.5$ (for Table 4.33), $\omega = 1.0$ (for Table 4.34) and $L_E = 0.25$. It follows from these tables that for the considered values of $p\big/E_0^{(1)}$ the values of $t'_{cr}$ increase (decrease) monotonically with increasing $\omega$ ($|\alpha|$).

**Table 4.33** The values of $t'_{cr}$ for various $H_u$ and $\omega$ under $E^{(2)}/E_0^{(1)} = 50$, $\alpha = -0.5$ and $L_E = 0.25$

| $H_u$ | $p/E_0^{(1)}$ | $\omega$ | | |
|---|---|---|---|---|
| | | 0.5 | 1.0 | 2.0 |
| 0.075 | 0.450 | 0.008 | 0.009 | 0.012 |
| | 0.400 | 0.055 | 0.080 | 0.222 |
| 0.060 | 0.400 | 0.012 | 0.014 | 0.020 |
| | 0.350 | 0.094 | 0.154 | 0.780 |
| 0.050 | 0.350 | 0.017 | 0.020 | 0.033 |
| | 0.300 | 0.167 | 0.331 | 9.544 |

**Table 4.34** The values of $t'_{cr}$ for various $H_u$ and $\alpha$ under $E^{(2)}/E_0^{(1)} = 50$, $\omega = 1.0$ and $L_E = 0.25$

| $H_u$ | $p/E_0^{(1)}$ | $\alpha$ | | |
|---|---|---|---|---|
| | | −0.3 | −0.5 | −0.7 |
| 0.075 | 0.450 | 0.028 | 0.009 | 0.0006 |
| | 0.400 | 0.135 | 0.080 | 0.023 |
| 0.060 | 0.400 | 0.039 | 0.014 | 0.001 |
| | 0.350 | 0.215 | 0.154 | 0.070 |
| 0.050 | 0.350 | 0.051 | 0.020 | 0.002 |
| | 0.300 | 0.372 | 0.331 | 0.251 |

## 4.8 Buckling Delamination of the Three-Layered (Sandwich) Circular Plate-Disc with Two Parallel Interface Penny-Shaped Cracks

### *4.8.1 Formulation of the Problem*

Consider a circular sandwich plate whose geometry shown in Fig. 4.11 and assume that the materials of the layers are isotropic, homogeneous and linear viscoelastic. Assume also that the materials of the upper and lower layers are the same. We suppose that between the middle and the upper face layer as well as and the lower face layer of the plate there is a penny-shaped crack whose location is shown in Fig. 4.11. Attempt to investigate the stability loss (delamination) around these cracks.

We associate with the middle layer of the plate the cylindrical coordinate system $Or\theta z$ (Fig. 4.11) and the position of the points of the plate we determine through the Lagrange coordinates in this system. Assume that the plate occupies the region $\{0 \le r \le R, 0 \le \theta \le 2\pi, -h/2 \le z \le h/2\}$ and the penny shaped cracks occur in $\{z = \mp(h/2 - h_u), 0 \le r \le R_0\}$. Moreover we assume that in the initial (natural) state the edge-surfaces of the cracks have an insignificant rotationally symmetric imperfection. In Fig. 4.11 the upper (lower) edge of the upper and

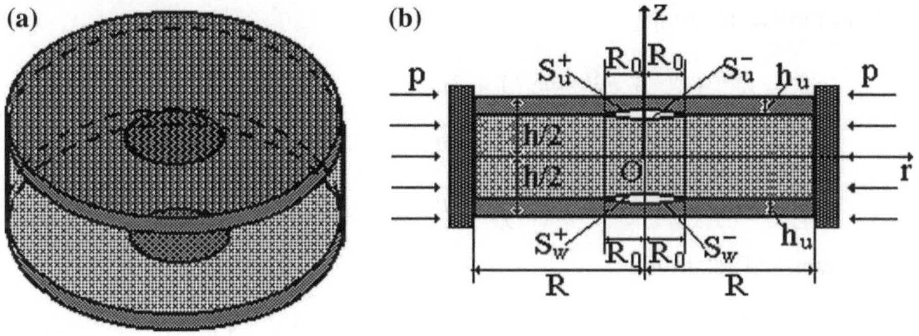


**Fig. 4.11** The geometry of the sandwich plate and interface penny shaped cracks

lower cracks are denoted by $S_u^+(S_u^-)$ and $S_w^+(S_w^-)$ respectively. The equations of these surfaces are given as follows:

for $S_u^\pm$

$$z = \frac{h}{2} - h_u \pm \varepsilon f(r),\ r \in [0, R_0], \tag{4.100}$$

And for $S_w^\pm$

$$z = -\left(\frac{h}{2} - h_u\right) \pm \varepsilon f(r),\ r \in [0, R_0] \tag{4.101}$$

where $\varepsilon$, as in the previous sections, is the infinitesimal parameter and $f(r)$ is a function which characterizes the form of the crack edges imperfection. The parameter $\varepsilon$ and the function $f(r)$ will be specified under consideration of the numerical results.

In the framework of the foregoing assumptions, we suppose that the plate is compressed circumferentially and inwards through the clamp by fixed radial displacement $U$ in the lateral boundary $r = R$. The values of the normal forces with intensity $p$, which is shown in Fig. 4.11b, is calculated through the $U$ after solution procedure through the well-known operations.

We will denote the values related to the upper and lower layers by upper indices (1) and (3) respectively, whereas the values related to the middle layer by (2). Within each layer in the cylindrical coordinate system for rotationally symmetric deformation state we write the following exact, geometrically non-linear field equation for the region occupied by the plate.

The equilibrium equation:

$$\frac{\partial}{\partial r} t_{rr}^{(k)} + \frac{\partial}{\partial z} t_{zr}^{(k)} + \frac{1}{r}\left(t_{rr}^{(k)} - t_{\theta\theta}^{(k)}\right) = 0,\ \frac{\partial}{\partial r} t_{rz}^{(k)} + \frac{\partial}{\partial z} t_{zz}^{(k)} + \frac{1}{r} t_{rz}^{(k)} = 0. \tag{4.102}$$

Constitutive relations:

$$\sigma_{rr}^{(k)} = \lambda^{*(k)}\left(\varepsilon_{rr}^{(k)} + \varepsilon_{\theta\theta}^{(k)} + \varepsilon_{zz}^{(k)}\right) + 2\mu^{*(k)}\varepsilon_{rr}^{(k)},$$
$$\sigma_{\theta\theta}^{(k)} = \lambda^{*(k)}\left(\varepsilon_{rr}^{(k)} + \varepsilon_{\theta\theta}^{(k)} + \varepsilon_{zz}^{(k)}\right) + 2\mu^{*(k)}\varepsilon_{\theta\theta}^{(k)},$$
$$\sigma_{zz}^{(k)} = \lambda^{*(k)}\left(\varepsilon_{rr}^{(k)} + \varepsilon_{\theta\theta}^{(k)} + \varepsilon_{zz}^{(k)}\right) + 2\mu^{*(k)}\varepsilon_{zz}^{(k)},\ \sigma_{rz}^{(k)} = 2\mu^{*(k)}\varepsilon_{rz}^{(k)}. \tag{4.103}$$

Geometrical relations:

$$\varepsilon_{rr}^{(k)} = \frac{\partial u_r^{(k)}}{\partial r} + \frac{1}{2}\left(\frac{\partial u_r^{(k)}}{\partial r}\right)^2 + \frac{1}{2}\left(\frac{\partial u_z^{(k)}}{\partial r}\right)^2,\ \varepsilon_{\theta\theta}^{(k)} = \frac{u_r^{(k)}}{r} + \frac{1}{2}\left(\frac{u_r^{(k)}}{r}\right)^2,$$
$$\varepsilon_{rz}^{(k)} = \frac{1}{2}\left(\frac{\partial u_r^{(k)}}{\partial z} + \frac{\partial u_z^{(k)}}{\partial r} + \frac{\partial u_r^{(k)}}{\partial r}\frac{\partial u_r^{(k)}}{\partial z} + \frac{\partial u_z^{(k)}}{\partial z}\frac{\partial u_z^{(k)}}{\partial z}\right),$$
$$\varepsilon_{zz}^{(k)} = \frac{\partial u_z^{(k)}}{\partial z} + \frac{1}{2}\left(\frac{\partial u_z^{(k)}}{\partial z}\right)^2 + \frac{1}{2}\left(\frac{\partial u_r^{(k)}}{\partial z}\right)^2. \tag{4.104}$$

In (4.102) the following notation is used:

$$t_{rr}^{(k)} = \sigma_{rr}^{(k)}\left(1 + \frac{\partial u_r^{(k)}}{\partial r}\right) + \sigma_{rz}^{(k)}\frac{\partial u_r^{(k)}}{\partial z},\ t_{zr}^{(k)} = \sigma_{rz}^{(k)}\left(1 + \frac{\partial u_r^{(k)}}{\partial r}\right) + \sigma_{zz}^{(k)}\frac{\partial u_r^{(k)}}{\partial z},$$
$$t_{\theta\theta}^{(k)} = \sigma_{\theta\theta}^{(k)}\left(1 + \frac{u_r^{(k)}}{r}\right),\ t_{rz}^{(k)} = \sigma_{rz}^{(k)}\left(1 + \frac{\partial u_z^{(k)}}{\partial z}\right) + \sigma_{rr}^{(k)}\frac{\partial u_z^{(k)}}{\partial r},$$
$$t_{zz}^{(k)} = \sigma_{zz}^{(k)}\left(1 + \frac{\partial uz^{(k)}}{\partial z}\right) + \sigma_{rz}^{(k)}\frac{\partial u_z^{(k)}}{\partial r}. \tag{4.105}$$

Moreover, in (4.103) $\lambda^{*(k)}$ and $\mu^{*(k)}$ are the following operators.

$$\left\{\begin{matrix}\lambda^{*(k)}\\ \mu^{*(k)}\end{matrix}\right\}\varphi(t) = \left\{\begin{matrix}\lambda_0^{(k)}\\ \mu_0^{(k)}\end{matrix}\right\}\varphi(t) + \int_0^t \left\{\begin{matrix}\lambda^{(k)}\\ \mu^{(k)}\end{matrix}\right\}(t-\tau)\varphi(\tau)d\tau. \tag{4.106}$$

Using the notation (4.105) we write the conditions which satisfy on the edge surfaces of the cracks. For this purpose we use also the expressions (4.83) for the components of the unite normal vectors on the surfaces $S_u^{\pm}$ and $S_w^{\pm}$. Thus, these conditions are:

$$t_{rr}^{(1)}\big|_{S_u^+} n_r^+ + t_{zr}^{(1)}\big|_{S_u^+} n_z^+ = 0,\ t_{rz}^{(1)}\big|_{S_u^+} n_r^+ + t_{zz}^{(1)}\big|_{S_u^+} n_z^+ = 0,$$
$$t_{rr}^{(2)}\big|_{S_u^-} n_r^- + t_{zr}^{(2)}\big|_{S_u^-} n_z^- = 0,\ t_{rz}^{(2)}\big|_{S_u^-} n_r^- + t_{zz}^{(2)}\big|_{S_u^+} n_z^- = 0,$$
$$t_{rr}^{(2)}\big|_{S_w^+} n_r^+ + t_{zr}^{(2)}\big|_{S_w^+} n_z^+ = 0,\ t_{rz}^{(2)}\big|_{S_w^+} n_r^+ + t_{zz}^{(2)}\big|_{S_w^+} n_z^+ = 0,$$
$$t_{rr}^{(3)}\big|_{S_w^-} n_r^- + t_{zr}^{(3)}\big|_{S_w^-} n_z^- = 0,\ t_{rz}^{(3)}\big|_{S_w^-} n_r^- + t_{zz}^{(3)}\big|_{S_w^-} n_z^- = 0. \tag{4.107}$$

Note that the conditions (4.107) are satisfied for the region $r \in [0, R_0]$. But for the region $r \in [R_0, R]$ the following complete contact conditions take place between the layers of the plate.

$$t_{zz}^{(1)}\Big|_{\substack{z=(h/2-h_u)\\ r\in[R_0,R]}} = t_{zz}^{(2)}\Big|_{\substack{z=(h/2-h_u)\\ r\in[R_0,R]}},\ t_{zr}^{(1)}\Big|_{\substack{z=(h/2-h_u)\\ r\in[R_0,R]}} = t_{zr}^{(2)}\Big|_{\substack{z=(h/2-h_u)\\ r\in[R_0,R]}},$$
$$u_r^{(1)}\Big|_{\substack{z=(h/2-h_u)\\ r\in[R_0,R]}} = u_r^{(2)}\Big|_{\substack{z=(h/2-h_u)\\ r\in[R_0,R]}},\ u_z^{(1)}\Big|_{\substack{z=(h/2-h_u)\\ r\in[R_0,R]}} = u_z^{(2)}\Big|_{\substack{z=(h/2-h_u)\\ r\in[R_0,R]}},$$
$$t_{zz}^{(2)}\Big|_{\substack{z=-(h/2-h_u)\\ r\in[R_0,R]}} = t_{zz}^{(3)}\Big|_{\substack{z=-(h/2-h_u)\\ r\in[R_0,R]}},\ t_{zr}^{(2)}\Big|_{\substack{z=-(h/2-h_u)\\ r\in[R_0,R]}} = t_{zr}^{(3)}\Big|_{\substack{z=-(h/2-h_u)\\ r\in[R_0,R]}},$$
$$u_r^{(2)}\Big|_{\substack{z=-(h/2-h_u)\\ r\in[R_0,R]}} = u_r^{(3)}\Big|_{\substack{z=-(h/2-h_u)\\ r\in[R_0,R]}},$$
$$u_z^{(2)}\Big|_{\substack{z=-(h/2-h_u)\\ r\in[R_0,R]}} = u_z^{(3)}\Big|_{\substack{z=-(h/2-h_u)\\ r\in[R_0,R]}}. \tag{4.108}$$

Moreover, on the face planes of the covering layers the conditions

$$t_{zz}^{(1)}\Big|_{\substack{z=h/2\\ r\in[0,R]}} = 0,\ t_{zr}^{(1)}\Big|_{\substack{z=h/2\\ r\in[0,R]}} = 0,\ t_{zz}^{(3)}\Big|_{\substack{z=-h/2\\ r\in[0,R]}} = 0,\ t_{zr}^{(3)}\Big|_{\substack{z=-h/2\\ r\in[0,R]}} = 0 \tag{4.109}$$

are satisfied. Finally, we write the following conditions on the lateral surface of the plate.

$$u_r^{(k)}(R,z,t) = -U,\ u_z^{(k)}(R,z,t) = 0,\ k = 1,2,3,\ \text{for}\, z \in [-h/2, h/2]. \tag{4.110}$$

Thus, in the framework of the Eqs. (4.102–4.106), boundary (4.107, 4.109) and (4.110) and contact (4.108) conditions we investigate the evolution of the crack edges' initial infinitesimal imperfections with $U$ for an elastic disk or with time under fixed $U$ for a viscoelastic disk. As in the previous sections, in this case as a critical time we will take a time for which these initial infinitesimal imperfections start to increase indefinitely.

### 4.8.2 Method of Solution

For the solution of the problem formulated above we employ the approach developed in the previous section, according to which, all sought values are presented in the series form in the small parameter $\varepsilon$:

$$\left\{\sigma_{(ij)}^{(k)};\varepsilon_{(ij)}^{(k)};u_{(i)}^{(k)}\right\}=\sum_{q=0}^{\infty}\varepsilon^{q}\left\{\sigma_{(ij)}^{(k),q};\varepsilon_{(ij)}^{(k),q};u_{(i)}^{(k),q}\right\},$$
$$(ij)=rr,\theta\theta,zz,rz,\ (i)=r,z. \tag{4.111}$$

Doing the mathematical manipulations described in the previous sections, we obtain the corresponding equations, contact and boundary conditions for the determination of the zeroth and the first approximations. These equations for the zeroth approximation are:

Equilibrium equations

$$\frac{\partial\sigma_{rr}^{(k),0}}{\partial r}+\frac{\partial\sigma_{rz}^{(k),0}}{\partial z}+\frac{1}{r}\left(\sigma_{rr}^{(k),0}-\sigma_{\theta\theta}^{(k),0}\right)=0,$$
$$\frac{\partial\sigma_{rz}^{(k),0}}{\partial r}+\frac{\partial\sigma_{zz}^{(k),0}}{\partial z}+\frac{1}{r}\sigma_{rz}^{(k),0}=0, \tag{4.112}$$

Constitutive relations

$$\sigma_{rr}^{(k),0}=\lambda^{*(k)}\left(\varepsilon_{rr}^{(k),0}+\varepsilon_{\theta\theta}^{(k),0}+\varepsilon_{zz}^{(k),0}\right)+2\mu^{*(k)}\varepsilon_{rr}^{(k),0},\ \sigma_{rz}^{(k),0}=2\mu^{*(k)}\varepsilon_{rz}^{(k),0}$$
$$\sigma_{\theta\theta}^{(k),0}=\lambda^{*(k)}\left(\varepsilon_{rr}^{(k),0}+\varepsilon_{\theta\theta}^{(k),0}+\varepsilon_{zz}^{(k),0}\right)+2\mu^{*(k)}\varepsilon_{\theta\theta}^{(k),0},$$
$$\sigma_{zz}^{(k),0}=\lambda^{*(k)}\left(\varepsilon_{rr}^{(k),0}+\varepsilon_{\theta\theta}^{(k),0}+\varepsilon_{zz}^{(k),0}\right)+2\mu^{*(k)}\varepsilon_{zz}^{(k),0}, \tag{4.113}$$

Geometrical relations

$$\varepsilon_{rr}^{(k),0}=\frac{\partial u_r^{(k),0}}{\partial r},\ \varepsilon_{\theta\theta}^{(k),0}=\frac{u_r^{(k),0}}{r},$$
$$\varepsilon_{rz}^{(k),0}=\frac{1}{2}\left(\frac{\partial u_r^{(k),0}}{\partial z}+\frac{\partial u_z^{(k),0}}{\partial r}\right),\ \varepsilon_{zz}^{(k),0}=\frac{\partial u_z^{(k),0}}{\partial z}. \tag{4.114}$$

Boundary conditions on the cracks' edges:

$$\sigma_{rz}^{(1),0}\Big|_{\substack{z=(h/2-h_u)+0\\ r\in[0,R_0]}}=0,\ \sigma_{zz}^{(1),0}\Big|_{\substack{z=(h/2-h_u)+0\\ r\in[0,R_0]}}=0,$$
$$\sigma_{rz}^{(2),0}\Big|_{\substack{z=(h/2-h_u)-0\\ r\in[0,R_0]}}=0,\ \sigma_{zz}^{(2),0}\Big|_{\substack{z=(h/2-h_u)-0\\ r\in[0,R_0]}}=0,$$
$$\sigma_{rz}^{(2),0}\Big|_{\substack{z=-(h/2-h_u)+0\\ r\in[0,R_0]}}=0,\ \sigma_{zz}^{(2),0}\Big|_{\substack{z=-(h/2-h_u)-0\\ r\in[0,R_0]}}=0,$$
$$\sigma_{rz}^{(3),0}\Big|_{\substack{z=-(h/2-h_u)-0\\ r\in[0,R_0]}}=0,\ \sigma_{zz}^{(3),0}\Big|_{\substack{z=-(h/2-h_u)-0\\ r\in[0,R_0]}}=0. \tag{4.115}$$

Contact conditions between the layers of the plate-disc:

$$\begin{aligned}
&\sigma_{rz}^{(1),0}\Big|_{\substack{z=(h/2-h_u)\\ r\in[R_0,R]}} = \sigma_{rz}^{(2),0}\Big|_{\substack{z=(h/2-h_u)\\ r\in[R_0,R]}},\ u_r^{(1),0}\Big|_{\substack{z=(h/2-h_u)\\ r\in[R_0,R]}} = u_r^{(2),0}\Big|_{\substack{z=(h/2-h_u)\\ r\in[R_0,R]}},\\
&\sigma_{zz}^{(1),0}\Big|_{\substack{z=(h/2-h_u)\\ r\in[R_0,R]}} = \sigma_{zz}^{(2),0}\Big|_{\substack{z=(h/2-h_u)\\ r\in[R_0,R]}},\ u_z^{(1),0}\Big|_{\substack{z=(h/2-h_u)\\ r\in[R_0,R]}} = u_z^{(2),0}\Big|_{\substack{z=(h/2-h_u)\\ r\in[R_0,R]}},\\
&\sigma_{rz}^{(2),0}\Big|_{\substack{z=-(h/2-h_u)\\ r\in[R_0,R]}} = \sigma_{rz}^{(3),0}\Big|_{\substack{z=-(h/2-h_u)\\ r\in[R_0,R]}},\ u_r^{(2),0}\Big|_{\substack{z=-(h/2-h_u)\\ r\in[R_0,R]}} = u_r^{(3),0}\Big|_{\substack{z=-(h/2-h_u)\\ r\in[R_0,R]}},\\
&\sigma_{zz}^{(2),0}\Big|_{\substack{z=-(h/2-h_u)\\ r\in[R_0,R]}} = \sigma_{zz}^{(3),0}\Big|_{\substack{z=-(h/2-h_u)\\ r\in[R_0,R]}},\\
&u_z^{(2),0}\Big|_{\substack{z=-(h/2-h_u)\\ r\in[R_0,R]}} = u_z^{(3),0}\Big|_{\substack{z=-(h/2-h_u)\\ r\in[R_0,R]}}.
\end{aligned} \tag{4.116}$$

Boundary conditions on the face planes of the plate:

$$\begin{aligned}
&\sigma_{rz}^{(1),0}\Big|_{\substack{z=h/2\\ r\in[0,R]}} = 0,\ \sigma_{zz}^{(1),0}\Big|_{\substack{z=h/2\\ r\in[0,R]}} = 0,\\
&\sigma_{rz}^{(3),0}\Big|_{\substack{z=-h/2\\ r\in[0,R]}} = 0,\ \sigma_{zz}^{(3),0}\Big|_{\substack{z=-h/2\\ r\in[0,R]}} = 0.
\end{aligned} \tag{4.117}$$

Boundary conditions on the lateral face-surface of the plate:

$$u_r^{(k),0}\Big|_{\substack{r=R\\ z\in(-h/2,+h/2)}} = -U,\ u_z^{(k),0}\Big|_{\substack{r=R\\ z\in(-h/2,+h/2)}} = 0,\ k=1,\ 2,\ 3. \tag{4.118}$$

So, the determination of the values of the zeroth approximation is reduced to solution to the boundary value problem (4.112–4.118). Now we write the corresponding equations and conditions for the determination of the values of the first approximation.

The equilibrium equation:

$$\frac{\partial t_{rr}^{(k),1}}{\partial r} + \frac{\partial t_{zr}^{(k),1}}{\partial z} + \frac{1}{r}\left(t_{rr}^{(k),1} - t_{\theta\theta}^{(k),1}\right) = 0,\ \frac{\partial t_{rz}^{(k),1}}{\partial r} + \frac{\partial t_{zz}^{(k),1}}{\partial z} + \frac{1}{r}t_{rz}^{(k),1} = 0. \tag{4.119}$$

Constitutive relations:

$$\begin{aligned}
&\sigma_{rr}^{(k),1} = \lambda^{*(k)}\left(\varepsilon_{rr}^{(k),1} + \varepsilon_{\theta\theta}^{(k),1} + \varepsilon_{zz}^{(k),1}\right) + 2\mu^{*(k)}\varepsilon_{rr}^{(k),1},\ \sigma_{rz}^{(k),1} = 2\mu^{*(k)}\varepsilon_{rz}^{(k),1},\\
&\sigma_{\theta\theta}^{(k),1} = \lambda^{*(k)}\left(\varepsilon_{rr}^{(k),1} + \varepsilon_{\theta\theta}^{(k),1} + \varepsilon_{zz}^{(k),1}\right) + 2\mu^{*(k)}\varepsilon_{\theta\theta}^{(k),1},\\
&\sigma_{zz}^{(k),1} = \lambda^{*(k)}\left(\varepsilon_{rr}^{(k),1} + \varepsilon_{\theta\theta}^{(k),1} + \varepsilon_{zz}^{(k),1}\right) + 2\mu^{*(k)}\varepsilon_{zz}^{(k),1}.
\end{aligned} \tag{4.120}$$

Geometrical relations:

$$\varepsilon_{rr}^{(k),1} = \frac{\partial u_r^{(k),1}}{\partial r}, \; \varepsilon_{\theta\theta}^{(k),1} = \frac{u_r^{(k),1}}{r},$$
$$\varepsilon_{rz}^{(k),1} = \frac{1}{2}\left(\frac{\partial u_r^{(k),1}}{\partial z} + \frac{\partial u_z^{(k),1}}{\partial r}\right), \; \varepsilon_{zz}^{(k),1} = \frac{\partial u_z^{(k),1}}{\partial z}. \tag{4.121}$$

Boundary conditions on the cracks' edges surfaces.

$$t_{zr}^{(1),1}\Big|_{\substack{z=(h/2-h_u)+0\\ r\in[0,R_0]}} = \frac{df(r)}{dr}\sigma_{rr}^{(1),0}\Big|_{\substack{z=(h/2-h_u)+0\\ r\in[0,R_0]}},$$
$$t_{zz}^{(1),1}\Big|_{\substack{z=(h/2-h_u)+0\\ r\in[0,R_0]}} = \frac{df(r)}{dr}\sigma_{rz}^{(1),0}\Big|_{\substack{z=(h/2-h_u)+0\\ r\in[0,R_0]}},$$
$$t_{zr}^{(2),1}\Big|_{\substack{z=(h/2-h_u)-0\\ r\in[0,R_0]}} = -\frac{df(r)}{dr}\sigma_{rr}^{(2),0}\Big|_{\substack{z=(h/2-h_u)-0\\ r\in[0,R_0]}},$$
$$t_{zz}^{(2),1}\Big|_{\substack{z=(h/2-h_u)-0\\ r\in[0,R_0]}} = -\frac{df(r)}{dr}\sigma_{rz}^{(2),0}\Big|_{\substack{z=(h/2-h_u)-0\\ r\in[0,R_0]}},$$
$$t_{zr}^{(2),1}\Big|_{\substack{z=-(h/2-h_u)+0\\ r\in[0,R_0]}} = \frac{df(r)}{dr}\sigma_{rr}^{(2),0}\Big|_{\substack{z=-(h/2-h_u)+0\\ r\in[0,R_0]}},$$
$$t_{zz}^{(2),1}\Big|_{\substack{z=-(h/2-h_u)+0\\ r\in[0,R_0]}} = \frac{df(r)}{dr}\sigma_{rz}^{(2),0}\Big|_{\substack{z=-(h/2-h_u)+0\\ r\in[0,R_0]}},$$
$$t_{zr}^{(3),1}\Big|_{\substack{z=-(h/2-h_u)-0\\ r\in[0,R_0]}} = -\frac{df(r)}{dr}\sigma_{rr}^{(3),0}\Big|_{\substack{z=-(h/2-h_u)-0\\ r\in[0,R_0]}},$$
$$t_{zz}^{(3),1}\Big|_{\substack{z=-(h/2-h_u)-0\\ r\in[0,R_0]}} = -\frac{df(r)}{dr}\sigma_{rz}^{(3),0}\Big|_{\substack{z=-(h/2-h_u)-0\\ r\in[0,R_0]}}. \tag{4.122}$$

Contact conditions between the layers of the plate:

$$t_{zr}^{(1),1}\Big|_{\substack{z=(h/2-h_u)\\ r\in[R_0,R]}} = t_{zr}^{(2),1}\Big|_{\substack{z=(h/2-h_u)\\ r\in[R_0,R]}}, \; u_r^{(1),1}\Big|_{\substack{z=(h/2-h_u)\\ r\in[R_0,R]}} = u_r^{(2),1}\Big|_{\substack{z=(h/2-h_u)\\ r\in[R_0,R]}},$$
$$t_{zz}^{(1),1}\Big|_{\substack{z=(h/2-h_u)\\ r\in[R_0,R]}} = t_{zz}^{(2),1}\Big|_{\substack{z=(h/2-h_u)\\ r\in[R_0,R]}}, \; u_z^{(1),1}\Big|_{\substack{z=(h/2-h_u)\\ r\in[R_0,R]}} = u_z^{(2),1}\Big|_{\substack{z=(h/2-h_u)\\ r\in[R_0,R]}},$$
$$t_{zr}^{(2),1}\Big|_{\substack{z=-(h/2-h_u)\\ r\in[R_0,R]}} = t_{zr}^{(3),1}\Big|_{\substack{z=-(h/2-h_u)\\ r\in[R_0,R]}}, \; u_r^{(2),1}\Big|_{\substack{z=-(h/2-h_u)\\ r\in[R_0,R]}} = u_r^{(3),1}\Big|_{\substack{z=-(h/2-h_u)\\ r\in[R_0,R]}}$$
$$t_{zz}^{(2),1}\Big|_{\substack{z=-(h/2-h_u)\\ r\in[R_0,R]}} = t_{zz}^{(3),1}\Big|_{\substack{z=-(h/2-h_u)\\ r\in[R_0,R]}},$$
$$u_z^{(2),1}\Big|_{\substack{z=-(h/2-h_u)\\ r\in[R_0,R]}} = u_z^{(3),1}\Big|_{\substack{z=-(h/2-h_u)\\ r\in[R_0,R]}}. \tag{4.123}$$

Boundary conditions on the face planes of the plate:

$$\left.t_{zr}^{(1),1}\right|_{\substack{z=h/2\\ r\in(0,R)}}=0,\quad \left.t_{zz}^{(1),1}\right|_{\substack{z=h/2\\ r\in(0,R)}}=0,$$
$$\left.t_{zr}^{(3),1}\right|_{\substack{z=-h/2\\ r\in(0,R)}}=0,\quad \left.t_{zz}^{(3),1}\right|_{\substack{z=-h/2\\ r\in(0,R)}}=0. \tag{4.124}$$

Boundary conditions on the lateral face surface of the plate:

$$\left.u_r^{(k)}\right|_{\substack{r=R\\ z\in(-h/2,h/2)}}=0,\ \left.u_z^{(k)}\right|_{\substack{r=R\\ z\in(-h/2,h/2)}}=0. \tag{4.125}$$

In (4.119, 4.122, 4.123) and (4.124) the following notation is used.

$$t_{rr}^{(k),1}=\sigma_{rr}^{(k),1}+\sigma_{rr}^{(k),0}\frac{\partial u_r^{(k),1}}{\partial r}+\sigma_{rz}^{(k),0}\frac{\partial u_r^{(k),1}}{\partial z},\ t_{\theta\theta}^{(k),1}=\sigma_{\theta\theta}^{(k),1}+\sigma_{\theta\theta}^{(k),0}\frac{u_r^{(k),1}}{r},$$
$$t_{zr}^{(k),1}=\sigma_{zr}^{(k),1}+\sigma_{zr}^{(k),0}\frac{\partial u_r^{(k),1}}{\partial r}+\sigma_{zz}^{(k),0}\frac{\partial u_r^{(k),1}}{\partial z},$$
$$t_{rz}^{(k),1}=\sigma_{rz}^{(k),1}+\sigma_{rr}^{(k),0}\frac{\partial u_z^{(k),1}}{\partial r}+\sigma_{rz}^{(k),0}\frac{\partial u_z^{(k),1}}{\partial z},$$
$$t_{zz}^{(k),1}=\sigma_{zz}^{(k),1}+\sigma_{zr}^{(k),0}\frac{\partial u_z^{(k),1}}{\partial r}+\sigma_{zz}^{(k),0}\frac{\partial u_z^{(k),1}}{\partial z}. \tag{4.126}$$

This completes the formulation of the problem related to the determination of the values of the first approximation. We consider the determination of the zeroth and of the first approximation separately.

#### 4.8.2.1 The Zeroth Approximation

The values of this approximation are determined from the boundary value problem (4.112–4.118) for the investigation of which we employ FEM and, as in the previous sections, the existence of the cracks does not taken into account. Note that before employing FEM we apply the Laplace transformations (4.28) to these equations and conditions. According to the correspondence principle we obtain the Laplace transformation of the Eqs. (4.112–4.118). The transformed problem is solved by employing FEM and the originals of the sought values are determined with the use of the method (Schapery 1962, 1978).

The author and his students have composed all programs in FTN77 for the calculation procedure, in the framework of which all numerical investigations are carried out. In this case taking the problem symmetry with respect to $z=0$ plane into account, we consider only the half-region $\{0\le r\le R,\ 0\le z\le h/2\}$ and this sub-region is divided into 120 Lagrange family quadratic elements with 533 nodes and 1027 NDOF. The traditional Ritz technique is employed for the determination of the Laplace transformation of the sought values and all operations are performed by the displacement based FEM.

**Fig. 4.12** The graphs of the relation between $p_{cr}/E^{(2)}$ and $\log\left(E_0^{(1)}/E^{(2)}\right)$ for various $\beta$

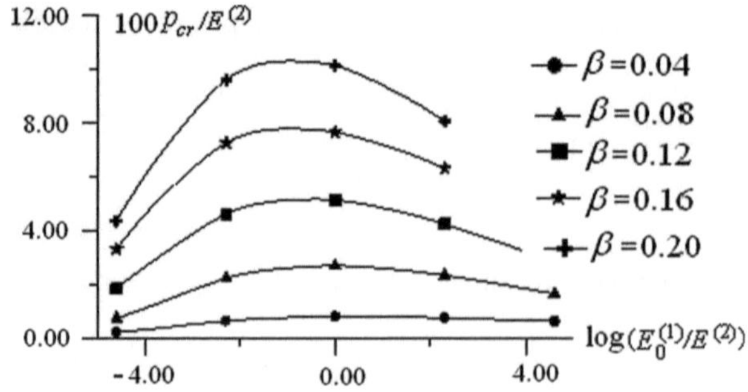


Under the numerical investigation procedure we give the function $U(t)$ so that the relation

$$hp = \int_{-h/2}^{-(h/2-h_u)} \sigma_{rr}^{(3),0}\Big|_{r=0} + \int_{-(h/2-h_u)}^{(h/2-h_u)} \sigma_{rr}^{(2),0}\Big|_{r=0} + \int_{(h/2-h_u)}^{h/2} \sigma_{rr}^{(1),0}\Big|_{r=0} = const \tag{4.127}$$

is satisfied and from the relation (4.127) we determine the values of $p$. It should be noted that the function $U(t)$ is constructed under the calculation procedure from the constitutive (strain-stress) relations.

The analyses of the numerical results show that due to relation (4.127) the stresses $\sigma_{rr}^{(k),0}$, $\sigma_{\theta\theta}^{(k),0}$, $\sigma_{zz}^{(k),0}$ and $\sigma_{rz}^{(k),0}$ can be taken as independent of time with very high accuracy. The analyses of these results show that the stresses $\sigma_{zz}^{(k),0}$ and $\sigma_{rz}^{(k),0}$ differ from zero in the near vicinity of the lateral boundary $r = R$ and can be taken as zero outside this region. Therefore the existence of the cracks in the plate does not change the stress distribution from the one without cracks in the zeroth approximation. The values of this approximation correspond to those obtained for the whole plate without cracks.

#### 4.8.2.2 The First Approximation

Taking the time-independent character of the stresses in the zeroth approximation into account for the solution of the boundary value problem (4.119–4.126) and for the determination of the first approximation, we can employ the principle of the correspondence by using the Laplace transformation (4.28). So, replacing $\sigma_{(ij)}^{(k),1}$, $\varepsilon_{(ij)}^{(k),1}$, $u_{(i)}^{(k),1}$, $\lambda^{*(k)}$ and $\mu^{*(k)}$ in (4.119–4.126) by $\bar{\sigma}_{(ij)}^{(k),1}$, $\bar{\varepsilon}_{(ij)}^{(k),1}$, $\bar{u}_{(i)}^{(k),1}$, $\bar{\lambda}^{*(k)}$ and $\bar{\mu}^{*(k)}$ respectively, we obtain the corresponding equations, boundary and contact conditions with respect to the Laplace transformations of the values of the first approximation. For the FEM modeling of this problem we introduce the following functional:

$$\bar{\Pi}\left(\bar{u}_r^{(1),1},\bar{u}_z^{(1),1},\bar{u}_r^{(2),1},\bar{u}_z^{(2),1},\bar{u}_r^{(3),1},\bar{u}_z^{(3),1}\right)$$
$$=\frac{1}{2}\sum_{k=1}^{3}\left\{\iint\limits_{Q'^{(k)}}\left[\bar{t}_{rr}^{(k),1}\frac{\partial\bar{u}_r^{(k),1}}{\partial r}+\bar{t}_{\theta\theta}^{(k),1}\frac{\bar{u}_r^{(k),1}}{r}+\bar{t}_{rz}^{(k),1}\frac{\partial\bar{u}_z^{(k),1}}{\partial r}\right.\right.$$
$$\left.\left.+\bar{t}_{zr}^{(k),1}\frac{\partial\bar{u}_r^{(k),1}}{\partial z}+\bar{t}_{zz}^{(k),1}\frac{\partial\bar{u}_z^{(k),1}}{\partial z}\right]rdrdz\right\}-\int_0^{R_0}\frac{df(r)}{dr}\sigma_{rr}^{(1),0}\bar{u}_r^{(1),1}\Big|_{z=(h/2-h_u)+0}rdr$$
$$-\int_0^{R_0}\frac{df(r)}{dr}\sigma_{rr}^{(2),0}\bar{u}_r^{(2),1}\Big|_{z=(h/2-h_u)-0}rdr-\int_0^{R_0}\frac{df(r)}{dr}\sigma_{rr}^{(2),0}$$
$$\times\bar{u}_r^{(2),1}\Big|_{z=-(h/2-h_u)+0}rdr-\int_0^{R_0}\frac{df(r)}{dr}\sigma_{rr}^{(3),0}\bar{u}_r^{(3),1}\Big|_{z=-(h/2-h_u)-0}rdr, \tag{4.128}$$

where

$$Q'^{(1)}=\{0\le r\le R,(h/2-h_u)\le z\le h/2\}-\{z=(h/2-h_u)+0,0\le r\le R_0\},$$
$$Q'^{(2)}=\{0\le r\le R,-(h/2-h_u)\le z\le(h/2-h_u)\}$$
$$-\{z=(h/2-h_u)-0,0\le r\le R_0\}-\{z=-(h/2-h_u)+0,0\le r\le R_0\},$$
$$Q'^{(3)}=\{0\le r\le R,-h/2\le z\le-(h/2-h_u)\}$$
$$-\{z=-(h/2-h_u)-0,0\le r\le R_0\}. \tag{4.129}$$

Note that in (4.128) it is taken into account that $\sigma_{rz}^{(k),0}=0$ and $\sigma_{zz}^{(k),0}=0$ for regions $\{z=h/2-h_u\pm 0,0<r<R_0\}$ and $\{z=-(h/2-h_u)\pm 0,\ 0<r<R_0\}$.

From the relation

$$\delta\bar{\Pi}(\bar{u}_r^{(1),1},\bar{u}_z^{(1),1},\bar{u}_r^{(2),1},\bar{u}_z^{(2),1},\bar{u}_r^{(3),1},\bar{u}_z^{(3),1})=0 \tag{4.130}$$

we obtain the equilibrium equations (4.119) and (4.126), contact and boundary conditions in (4.122–4.124) given for stresses which are rewritten for $\bar{\sigma}_{(ij)}^{(k),1}$, $\bar{\varepsilon}_{(ij)}^{(k),1}$, $\bar{u}_{(i)}^{(k),1}$, $\bar{\lambda}^{*(k)}$ and $\bar{\mu}^{*(k)}$. In this way it is proven that the Eqs. (4.199, 4.126, 4.122–4.124) are the Euler equations of the functional (4.128). Thus, applying the well-known techniques of the FEM to the Eq. (4.130) we obtain corresponding system of the linear algebraic equations for the values of the Laplace transformation of the displacements at nodes of the selected finite elements. The originals of the sought values, as before, are determined by employing the method (Schapery 1962, 1978), according to which, as an example, the values of $u_z^{(1),1}(0,(h/2-h_u)+0,t)$ are determined as follows:

$$u_z^{(1),1}(0,(h/2-h_u)+0,t) \approx s\bar{u}_z^{(1),1}(0,(h/2-h_u)+0,s)\big|_{s=1/(2t)}. \tag{4.131}$$

After the foregoing procedures, the values of the critical parameters are determined from the criterion

$$u_z^{(1),1}(0,(h/2-h_u)+0,t) \to \infty \text{ as } t \to t_{cr} \text{ (for fixed } p) \text{ or as } p \to p_{cr} \text{ (for fixed } t).$$

## *4.8.3 Numerical Results and Discussions*

In the numerical investigation the function $f(r)$ which enters the Eqs. (4.100, 4.101) and (4.122) must satisfy the conditions (4.82). The analyses of the numerical results show that the critical values of the external forces or the values of the critical time do not depend on the initial imperfection mode, which satisfies the conditions in (4.82). Considering this situation, as in the previous section, the function $f(r)$ is selected through the expression (4.99).

Thus, we turn to consideration numerical results, which are obtained within the scope of the foregoing assumptions. First, we consider the case where the materials of the face layers and middle layer are elastic and assume that the materials of the face layers are the same. With $E_0^{(1)}$, $\nu_0^{(1)}$ (for face layers) and $E^{(2)}$, $\nu^{(2)}$ (for a core layer) we denote the Young's modulus and Poisson's coefficient. Assume that $\nu_0^{(1)}=\nu^{(2)}=0.3$ and introduce the parameter $\beta=h_u/(2R_0)$ and analyze the influence of the problem parameters on the values of $p_{cr}/E^{(2)}\left(=p_{cr.0}/E^{(2)}\right)$. Note that these analyses will be made below in the framework of the following statement, which is established as a result of various numerical investigations: in the cases where $\beta<h/(2R)$ the values of the $p_{cr}/E^{(2)}$ do not depend on the geometrical parameters of the whole plate, but these values depend only on the values of the geometrical parameter $\beta=h_u/(2R_0)$. Consequently, this situation allows us to conclude that the results obtained for the cases where $\beta<h/(2R)$ relate to the local buckling of the part of the coating layer which is on the crack. Moreover, according to this statement, the interaction between the upper and the lower cracks can be neglected completely under determination of the values of the critical parameters.

Thus, consider graphs given in Fig. 4.12, which show the relations between $p_{cr}/E^{(2)}$ and $\log\left(E_0^{(1)}/E^{(2)}\right)$ under various $\beta$. While constructing these graphs, the values of $E_0^{(1)}/E^{(2)}$ have been changed as follows: $0.01\le E_0^{(1)}/E^{(2)}\le 100$ for $\beta=0.04,\ 0.08$; $0.01\le E_0^{(1)}/E^{(2)}\le 50$ for $\beta=0.12$; $0.01\le E_0^{(1)}/E^{(2)}\le 10$ for $\beta=0.16, 0.20$, because under $E_0^{(1)}/E^{(2)}>50$ (for $\beta=0.12$) and under

**Table 4.35** The values of $p_{cr.0}/E^{(2)}$ (upper numbers) and $p_{cr.\infty}/E^{(2)}$ (lower numbers) obtained for various $\beta$ and $E_0^{(1)}/E^{(2)}$

| $\beta$ | $E_0^{(1)}/E^{(2)}$ | | | |
|---|---|---|---|---|
| | 0.20 | 0.10 | 0.02 | 0.01 |
| 0.12 | 0.0498<br>0.0347 | 0.0460<br>0.0321 | 0.0275<br>0.0193 | 0.0184<br>0.0129 |
| 0.16 | 0.0767<br>0.0531 | 0.0724<br>0.0500 | 0.0476<br>0.0330 | 0.0330<br>0.0229 |
| 0.20 | 0.1016<br>0.0701 | 0.0959<br>0.0660 | 0.0625<br>0.0431 | 0.0433<br>0.0301 |

$E_0^{(1)}/E^{(2)} > 10$ (for $\beta = 0.16, 0.20$) the instability (failure) of the middle layer material arises in an earlier stage of the loading than the local buckling of the coating (face) layer.

The graphs given in Fig. 4.12 show that the dependence between $p_{cr}/E^{(2)}$ and $E_0^{(1)}/E^{(2)}$ has a non-monotonic character for each of the values of $\beta$. Moreover, these graphs indicate that the values of $p_{cr}/E^{(2)}$ increase monotonically with $\beta$.

Now we consider the case where the materials of the face layers are viscoelastic. As before, we assume that the viscoelasticity of these materials are described with the operators

$$E^{*(1)}\varphi(t) = E_0^{(1)}\left(\varphi(t) - \omega_0\Pi_\alpha^*(-\omega_0 - \omega_\infty)\varphi\right),$$
$$v^{*(1)}\varphi(t) = v_0^{(1)}\left(\varphi(t) + \frac{1 - 2v_0^{(1)}}{2v_0^{(1)}}\omega_0\Pi_\alpha^*(-\omega_0 - \omega_\infty)\varphi\right), \tag{4.132}$$

where $E_0^{(1)}$, $v_0^{(1)}$ are the instantaneous values of Young's modulus and of Poisson's ratio at $t = 0$, respectively; $\alpha$, $\omega_0$, $\omega_\infty$ are the rheological parameters, $\Pi_\alpha^*$ is the fractional-exponential operator (Rabotnov 1977) and this operator is determined as

$$\Pi_\alpha^*(\delta)\varphi(t) = \int_0^t \Pi_\alpha(\delta, t - \tau)\varphi(\tau)d\tau,\ \Pi_\alpha(\delta, t) = t^\alpha \sum_{n=0}^{\infty} \frac{\delta^n t^{n(1+\alpha)}}{\Gamma((1+n)(1+\alpha))}, \tag{4.133}$$

where $-1<\alpha<0$ and $\Gamma(x)$ is the Gamma function.

As in the previous sections, we introduce the dimensionless rheological parameter $\omega\,(= \omega_\infty/\omega_0)$ and dimensionless time $t'\ (= \omega_0^{1/(1+\alpha)}t)$. For considering to the real viscoelastic polymer coatings, according to (Rabotnov 1977), we assume that $\omega = 2$, which corresponds to the various epoxy-malein composition. Assume that $E_0^{(1)} < E^{(2)}$ and consider the cases where $t' = 0$ and $t' = \infty$. In these cases we denote the critical values of the external compressive force as $p_{cr.0}/E^{(2)}$ (for $t' = 0$) and $p_{cr.\infty}/E^{(2)}$ (for $t' = \infty$). Table 4.35 shows the values of $p_{cr.0}/E^{(2)}$ (upper number) and $p_{cr.\infty}/E^{(2)}$

**Table 4.36** The values of $t'_{cr}$ obtained for various $\alpha$ and $\eta$ under $\omega = 2$

| $\alpha$ | $\eta$ | | | | |
|---|---|---|---|---|---|
| | 0.25 | 0.50 | 0.75 | 0.90 | 0.95 |
| −0.3 | 0.0220 | 0.1070 | 0.5180 | 2.5 | 7.3 |
| −0.5 | 0.0068 | 0.0600 | 0.5500 | 4.5 | 21.9 |
| −0.7 | 0.0003 | 0.0138 | 0.5400 | 21.4 | 265. |
| −0.8 | 0.0002 | 0.0024 | 0.5900 | 155. | 7398. |

(lower number) for various values of $\beta$ and $E_0^{(1)}/E^{(2)}$. When calculating $p_{cr.\infty}/E^{(2)}$ the operators $E^{*(1)}$ and $v^{*(1)}$ in (4.132) are replaced by the constants $E_\infty^{(1)} = E_0^{(1)}\omega/(1+\omega)$ and $v_\infty^{(1)} = v_0^{(1)}(1+(1-2v_0^{(1)})\omega/(2v_0^{(1)}(1+\omega))$ respectively.

Let us analyze the numerical results related to the viscoelastic coatings (face layers). In these cases selected values of $p/E^{(2)}$ must satisfy the inequality $p_{cr.\infty}/E^{(2)} < p/E^{(2)} < p_{cr.0}/E^{(2)}$ and they can be presented as follows:

$$\frac{p}{p_{cr.0}} = 1 - \eta\frac{\Delta p}{p_{cr.0}}, \tag{4.134}$$

where $\Delta p = p_{cr.0} - p_{cr.\infty}$ and $0 < \eta < 1$.

Note that the case where $\eta = 0$ $(\eta = 1)$ corresponds to the $p = p_{cr.0}$ $(p = p_{cr.\infty})$ and the $p$ closes to the $p_{cr.\infty}$ with the parameter $\eta$. Table 4.36 shows the values of $t'_{cr}$ obtained for various $\alpha$ and $\eta$ under $\omega = 2$, $\beta = 0.12$ and $E_0^{(1)}/E^{(2)} = 0.2$. Numerous numerical investigations show that the values of the critical time $t'_{cr}$ obtained for fixed $\eta$, $\alpha$ and $\omega$, almost, do not depend on the values of $\beta$ and $E_0^{(1)}/E^{(2)}$. However, the values of $p/E^{(2)}$ corresponding to the same $t'_{cr}$ depend significantly on the parameters $\beta$ and $E_0^{(1)}/E^{(2)}$. Moreover, it follows from (4.134) and Table 4.36 that for fixed $p/E^{(2)}$, $\alpha$ and $\omega$ the values of $t'_{cr}$ increase monotonically with $\beta$ and $E_0^{(1)}/E^{(2)}$, in other words the values of $t'_{cr}$ increase monotonically with $\eta$.

## 4.9 Remarks on the FEM Modeling of the Crack's Tips

So far, we passed silently on the question on the FEM modeling of the crack's tips. In the present section, we attempt to analyze this question.

When stress intensity factors (SIFs) are numerically determined by the FEM, special finite elements at crack tips are known to be used in (Benzley 1974; Zienkiewicz and Taylor 1989) and others. These finite elements contain a singularity of the strain-stress field at the crack tip, which equals to the theoretical

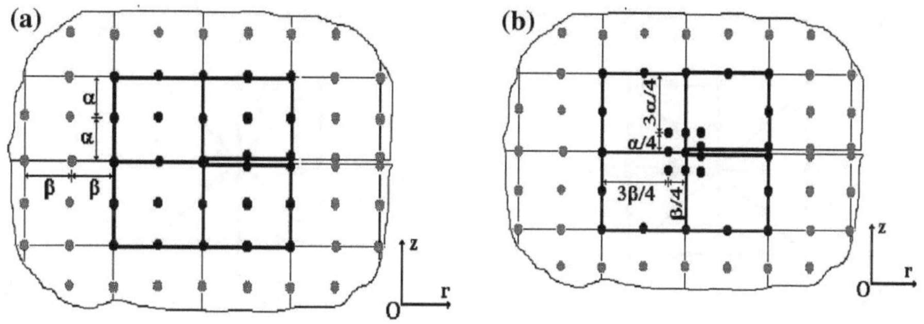


**Fig. 4.13** Rectangular finite elements that contain the crack tip: **a** standard Lagrange-family rectangular finite elements of the second order; **b** singular rectangular elements

singularity. At the same time, these finite elements have various disadvantages such as lack of constant strain and rigid body motion modes, patches testing, etc., which are mentioned in many references. These and other disadvantages of the finite elements sometimes motivate researches to solve the same problems without the use of such elements. For this purpose, various versions of energy release rate methods were employed in (Gosz et al. 1998; Henshell and Shaw 1975; Shih et al. 1986) and others, to determine the SIFs. There are also other problems, such as buckling delamination problems around cracks solved within the framework of the TDLTS in the previous sections of the present chapter, associated with cracks that are numerically solved also using special singular finite elements.

In the present section, we attempt to prove, the using the problems considered in the previous last two sections, that the use of ordinary finite elements instead of the special singular finite elements produces almost the same numerical results for the critical parameters. In other words, we attempt to prove that it is not necessary to use singular finite elements for analysis of the buckling delamination around a crack based on the TDLTS. As examples, we consider the problems considered in the previous last two sections and assume that the materials of the plates are pure elastic. Let us call they as *Problem 1* (Sect. 4.7) and *Problem 2* (Sect. 4.8) and before the analysis of the numerical results, we discuss the selection and location of the finite elements that contain the crack tip. For this purpose, four types of isoparametric finite elements are selected: standart Lagrange-family rectangular finite elements of the second order (Fig. 4.13a); singular rectangular elements (Fig. 4.13b); standard Lagrange-family triangular elements of the second order (Fig. 4.14a); and singular triangular finite elements (Fig. 4.14b).The associated cases are indicated as I, II, III and IV, respectively. Note that the finite elements that do not contain the crack tip are selected to be Lagrange-family rectangular finite elements of the second order in all the cases. By symmetry reasons, we can model just the region $\{0 \le r \le R, -h/2 \le z \le h/2\}$ for *Problem 1* and the region $\{0 \le r \le R, 0 \le z \le h/2\}$ for *Problem 2*.

Let us now return to the analysis of the numerical results obtained by solving *Problems 1* and *2* in the cases I, II, III and IV noted above and associated with the

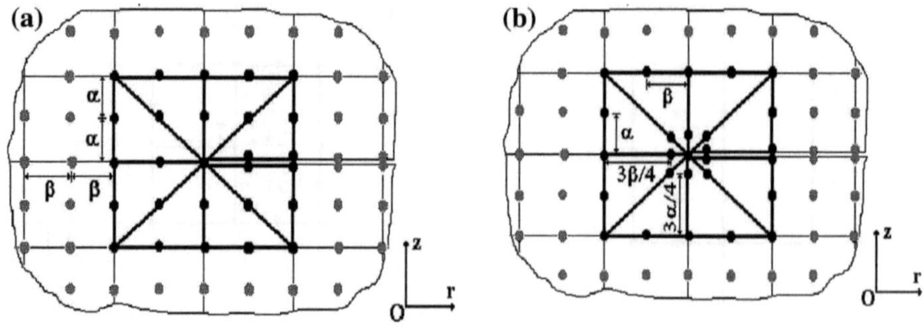


**Fig. 4.14** Triangular finite elements that contain the crack tip: **a** standard Lagrange-family triangular finite elements of the second order; **b** singular triangular elements

finite elements configurations shown in Figs. 4.13a, b, and 4.14a, b, respectively. Note that these analyses will be performed below within the framework of the following two statements established in the previous last two sections:

When $h_1/R_0<h/R$ (in *Problem 1*, Fig. 4.8) and $h_u/R_0<h/R$ (in *Problem 2*, Fig. 4.11), the critical values of the external compressive force $p$ do not depend on the geometrical parameters of the whole plate—these values depend only on parameters $h_1/R_0$ and $h_u/R_0$ in *Problem1* and *Problem 2*, respectively. Consequently, the results obtained in the cases where $h_1/R_0<h/R$ (in *Problem 1*, Fig. 4.8) and $h_u/R_0<h/R$ (in *Problem 2*, Fig. 4.11) relate to local buckling delamination of the plate's part occupied by the sub-region between the crack and the nearest free surface. Moreover, according to this statement, the critical force can be determined neglecting the interaction between the upper and the lower cracks (in *Problem 2*).

The PC programs and algorithms developed and employed are validated by comparing the results with corresponding ones obtained in a paper (Guz and Nazarenko 1985a) where the local buckling of a half-space with a near-surface penny-shaped crack was studied using the double integral equation technique and Galerkin method (to obtain numerical results). Further, the material of the half-space was assumed a multilayer composite consisting of two alternating layers that lie in the planes $z = const$. As in *Problem 1,* in a paper (Guz and Nazarenko 1985a) it was assumed that the near-surface penny-shaped crack is a macro-crack, and the material of the half-space is transversaly isotropic and its isotropy axis coincides with the $Oz$ axis. The numerical results obtained in a paper (Guz and Nazarenko 1985a) for $p_{cr}/(\eta^{(1)}E^{(1)} + \eta^{(2)}E^{(2)})$, $\eta^{(1)} = 1 - \eta^{(2)}$, $\eta^{(2)} = 0.3$ ($\eta^{(2)}$ is a volumetric fraction of the filler material) under $h_1/(2R_0) = 1/16$ and $\nu^{(1)} = \nu^{(2)} = 0.3$ for various $E^{(2)}/E^{(1)}$ are given in the second column of Table 4.37.

Thus, taking the foregoing statements into account, we consider the other numerical results given in Table 4.37 which is obtained by employing FEM for the cases I, II, III and IV. Note that these results have been obtained with the same problem parameters as in a paper (Guz and Nazarenko 1985a). Comparing the results given in Table 4.37 reveals that the use of the singular finite elements

**Table 4.37** The values of $p_{cr}/(\eta^{(1)}E^{(1)} + \eta^{(2)}E^{(2)})$ obtained for Problem 1 under various finite element modeling around the crack tip

| $\frac{E^{(2)}}{E^{(1)}}$ | Guz and Nazarenko (1985a) | FEM results in Case II | FEM results in Case I | FEM results in Case IV | FEM results in Case III |
|---|---|---|---|---|---|
| 1 | 0.0167 | 0.0178 | 0.0174 | 0.0181 | 0.0181 |
| 10 | 0.0140 | 0.0154 | 0.0151 | 0.0157 | 0.0157 |
| 25 | 0.0126 | 0.0129 | 0.0126 | 0.0137 | 0.0137 |

**Table 4.38** The values of $hp_{cr}/(2h_u(E^{(1)} - E^{(2)}) + hE^{(2)})$ obtained for Problem 2 under various finite element modeling around the crack tip

| $h_u/(2R_0)$ | $\frac{E^{(2)}}{E^{(1)}}$ | (20 × 8 FE) Case II | (40 × 12 FE) Case II | (40 × 12 FE) Case I | (40 × 12 FE) Case III | (40 × 12 FE) Case IV |
|---|---|---|---|---|---|---|
| 0.20 | 0.2 | 0.1016 | 0.1013 | 0.1013 | 0.1056 | 0.1056 |
| | 0.1 | 0.0959 | 0.0965 | 0.0955 | 0.1003 | 0.1003 |
| 0.16 | 0.2 | 0.0767 | 0.0769 | 0.0760 | 0.0786 | 0.0786 |
| | 0.1 | 0.0724 | 0.0735 | 0.0722 | 0.0768 | 0.0768 |
| 0.12 | 0.2 | 0.0498 | 0.0511 | 0.0497 | 0.0515 | 0.0515 |
| | 0.1 | 0.0460 | 0.0492 | 0.0477 | 0.0509 | 0.0509 |

changes insignificantly the critical values of compressive external forces in *Problem 1*.

Let us now examine the numerical results obtained for *Problem 2*. The values of $hp_{cr}/(2h_u(E^{(1)} - E^{(2)}) + hE^{(2)})$ for different values of $h_u/(2R_0)$ and $E^{(2)}/E^{(1)}$ are given in Table 4.38. The third and the fourth columns contain the results of the case II for which number of finite elements along $Oz(Or)$ axis is 8(20) and 12(40) respectively. Table 4.38 demonstrates that it is also unnecessary to use the singular finite elements to analyze the buckling-delamination around an interface micro-crack. Taking this result into account, we can significantly simplify the FEM analyses of the similar buckling-delamination problems within the framework of the TDLTS.

## 4.10 Buckling Delamination of a Rectangular Plate Containing a Rectangular Crack

So far, we have investigated two-dimensional problems on the buckling delamination around the cracks. In the present and the next sections we attempt to extend the mentioned investigations for the corresponding three-dimensional problems. Namely, in the present section we consider rectangular plate made from homogeneous viscoelastic anisotropic (orthotropic) material containing a rectangular crack. Three cases are considered. Case 1: it is assumed that the rectangular plate contains a rectangular band crack; Case 2: it is assumed that the rectangular plate contains a rectangular edge-crack; Case 3: it is assumed that the rectangular plate contains a rectangular embedded crack.

## 4.10.1 Formulation of the Problems

Thus, consider a thick rectangular plate which contains a rectangular crack. The Cartesian coordinate system $Ox_1x_2x_3$ is associated with the plate to give Lagrange coordinates of the points of the plate in the natural state. Assume that the plate occupies the region

$$\Omega = \{0 < x_1 < \ell_1;\ 0 < x_2 < h;\ 0 < x_3 < \ell_3\}. \tag{4.135}$$

As noted above, we will distinguish three cases with respect to the form and location of the crack in the plate. In Case 1, we assume that the plate contains the rectangular band-crack (Fig. 4.15a) which occupies the region

$$\Omega_1' = \{(\ell_1 - \ell_{10})/2 < x_1 < (\ell_1 + \ell_{10})/2;\ x_2 = h_A;\ 0 < x_3 < \ell_3\}. \tag{4.136}$$

In Case 2 the edge crack occupies the region (Fig. 4.15b)

$$\Omega_2' = \{(\ell_1 - \ell_{10})/2 < x_1 < (\ell_1 + \ell_{10})/2;\ x_2 = h_A;\ \ell_3 - \ell_{30} < x_3 < \ell_3\}. \tag{4.137}$$

Finally, in Case 3 we assume that the plate contains the rectangular embedded crack at (Fig. 4.15c)

$$\begin{gathered}\Omega_3' = \{(\ell_1 - \ell_{10})/2 < x_1 < (\ell_1 + \ell_{10})/2;\ x_2 = h_A;\\ (\ell_3 - \ell_{30})/2 < x_3 < (\ell_3 + \ell_{30})/2\}.\end{gathered} \tag{4.138}$$

In Eqs. (4.136–4.138) $\ell_{10}(\ell_{30})$ is the length of the crack along the axis $Ox_1(Ox_3)$. It is assumed that edge-surfaces of the crack have an insignificant initial imperfection and this imperfection is symmetric with respect to the $x_1 = \ell_1/2$ and with respect to the $x_2 = h_A$ planes. Moreover in Cases 1 and 3 this imperfection has also symmetry with respect to $x_3 = \ell_3/2$. The equations of the crack's edge-surfaces can be written as follows:

Case 1

$$x_2^{\pm} = h_A + \varepsilon f^{\pm}(x_1),\ \text{for}\ (\ell_1 - \ell_{10})/2 < x_1 < (\ell_1 - \ell_{10})/2,\ 0 < x_3 < \ell_3, \tag{4.139}$$

and the function $f^{\pm}(x_1)$ and its first order derivative are equal to zero at $x_1 = (\ell_1 \pm \ell_{10})/2$.

Case 2

$$\begin{gathered}x_2^{\pm} = h_A + \varepsilon f^{\pm}(x_1, x_3),\ \text{for}\ (\ell_1 - \ell_{10})/2 < x_1 < (\ell_1 - \ell_{10})/2,\\ \ell_3 - \ell_{30} < x_3 < \ell_3,\end{gathered} \tag{4.140}$$

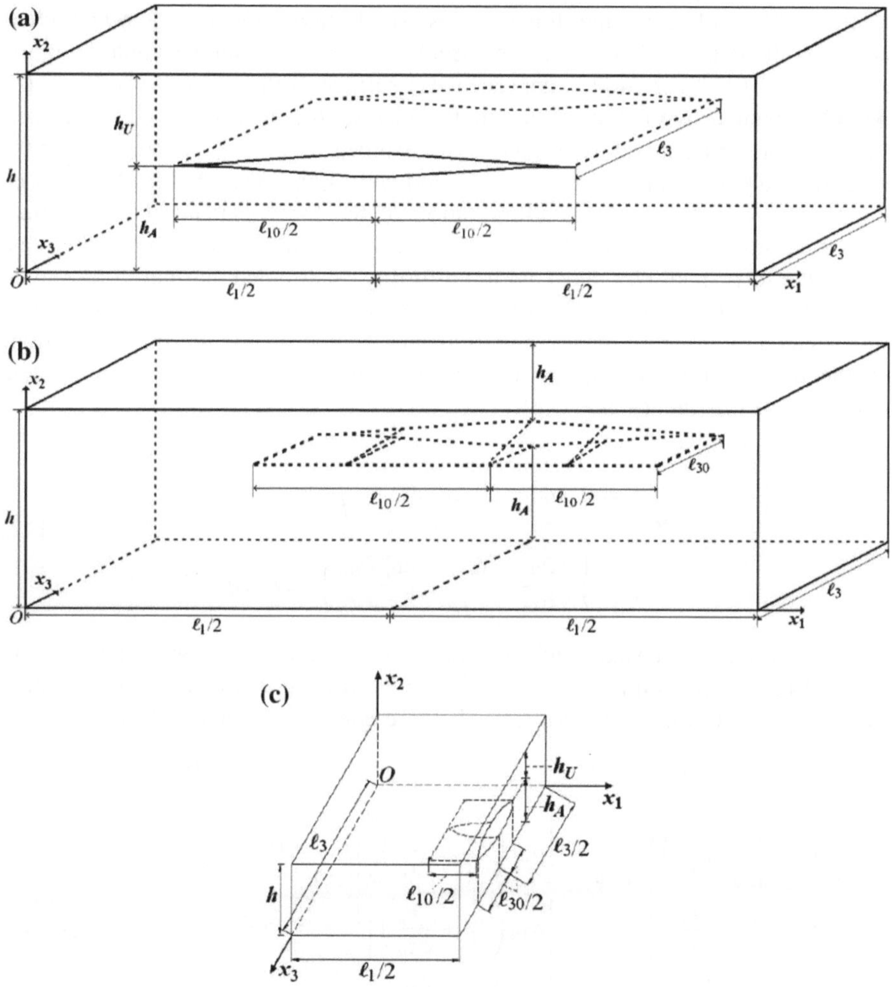


**Fig. 4.15** The geometry of the considered plate with: **a** a band crack (Case 1); **b** an edge crack (Case 2); **c** an embedded crack (Case 3)

and the function $f^{\pm}(x_1, x_3)$ and its first order derivatives with respect to $x_1$ and $x_3$ are equal to zero at $x_1 = (\ell_1 \pm \ell_{10})/2$ and $x_3 = \ell_3 - \ell_{30}$ respectively.

Case 3

$$x_2^{\pm} = h_A + \varepsilon f^{\pm}(x_1, x_3), \text{ for } (\ell_1 - \ell_{10})/2 < x_1 < (\ell_1 - \ell_{10})/2, \quad (\ell_3 - \ell_{30})/2 < x_3 < (\ell_3 + \ell_{30})/2, \tag{4.141}$$

and the function $f^{\pm}(x_1, x_3)$ and its first order derivatives with respect to $x_1$ and $x_3$ are equal to zero at $x_1 = (\ell_1 \pm \ell_{10})/2$ and $x_3 = (\ell_3 \pm \ell_{30})/2$ respectively.

In (4.139–4.141), $\varepsilon$ is the dimensionless small parameter ($\varepsilon \ll 1$) which characterizes the degree of the initial imperfection of the edge-surfaces and the upper symbol "+" ("−") shows the upper (lower) surface of the crack. As in Sect. 4.2, we will consider two modes of he initial imperfection of the crack edge-surfaces. For the first mode, called as co-phase, we assume that $f^{+} = f^{-}$. However, for the second mode we assume that $f^{+} = -f^{-}$ and call as anti-phase mode. Thus, based on the above, we investigate an evolution of the foregoing initial infinitesimal imperfections of the crack edge-surfaces under bi-axial compression of the plate along the $Ox_1$ and $Ox_3$ axes with uniformly distributed normal forces with intensity $p_1$ and $p_3$ respectively. As in the previous sections, this evolution will be investigated by utilizing the three-dimensional geometrically non-linear equations of the theory of viscoelasticity for the anisotropic body. The governing field equations of this theory are

$$\frac{\partial}{\partial x_j}\left[\sigma_{jn}\left(\delta_i^n + \frac{\partial u_i}{\partial x_n}\right)\right] = 0,\ \sigma_{ij} = C_{ijrq0}\varepsilon_{rq}(t) + \int_0^t C_{ijrq}(t-\tau)\varepsilon_{rq}(\tau)d\tau,$$
$$\varepsilon_{rq} = \frac{1}{2}\left(\frac{\partial u_r}{\partial x_q} + \frac{\partial u_q}{\partial x_r} + \frac{\partial u_n}{\partial x_r}\frac{\partial u_n}{\partial x_q}\right),\ i;j;r;q = 1,2,3. \tag{4.142}$$

Throughout, the repeated indices indicate the summation over their ranges. In Eq. (4.142) the conventional notation is used. Now we will consider formulation of the boundary conditions. For Case 1, these can be written as follows:

$$u_2\big|_{x_1=0} = 0,\ u_2\big|_{x_1=\ell_1} = 0,\ u_2\big|_{x_3=0} = 0,\ u_2\big|_{x_3=\ell_3} = 0, \tag{4.143}$$

$$\left[\sigma_{2n}\left(\delta_i^n + \frac{\partial u_i}{\partial x_n}\right)\right]\bigg|_{x_2=0} = \left[\sigma_{2n}\left(\delta_i^n + \frac{\partial u_i}{\partial x_n}\right)\right]\bigg|_{x_2=h} = 0,$$
$$\left[\sigma_{1n}\left(\delta_j^n + \frac{\partial u_j}{\partial x_n}\right)\right]\bigg|_{x_1=0} = \left[\sigma_{1n}\left(\delta_j^n + \frac{\partial u_j}{\partial x_n}\right)\right]\bigg|_{x_1=\ell_1} = p_1\delta_j^1,$$
$$\left[\sigma_{3n}\left(\delta_j^n + \frac{\partial u_j}{\partial x_n}\right)\right]\bigg|_{x_3=0} = \left[\sigma_{3n}\left(\delta_j^n + \frac{\partial u_j}{\partial x_n}\right)\right]\bigg|_{x_3=\ell_3} = p_3\delta_j^3,\ j = 1,3, \tag{4.144}$$

$$\left\{\left[\sigma_{jn}\left(\delta_i^n + \frac{\partial u_i}{\partial x_n}\right)\right]\bigg|_{\substack{x_2^{\pm}=h_A+\varepsilon f^{\pm}(x_1) \\ (\ell_1-\ell_{10})/2<x_1<(\ell_1+\ell_{10})/2 \\ 0<x_3<\ell_3}}\right\} n_j^{\pm} = 0. \tag{4.145}$$

In (4.145) $n_j^{\pm}$ is the orthonormal component of the unit normal vector of the considered surface (i.e. the crack edge-surfaces), $\delta_i^j$ is a Kronecker symbol. The condition (4.144) also holds for Case 3. However for Case 2 the conditions (4.143) and last condition in (4.145) must be taken as follow:

$$u_2|_{x_1=0}=0,\ u_2|_{x_1=\ell_1}=0,\ u_2|_{x_3=0}=0,\ \left[\sigma_{3n}\left(\delta_i^n+\frac{\partial u_i}{\partial x_n}\right)\right]\Big|_{x_3=\ell_3}=0, \tag{4.146}$$

$$\left\{\left[\sigma_{jn}\left(\delta_i^n+\frac{\partial u_i}{\partial x_n}\right)\right]\Big|_{\substack{x_2^{\pm}=h_A+\varepsilon f^{\pm}(x_1,x_3)\\ (\ell_1-\ell_{10})/2<x_1<(\ell_1+\ell_{10})/2\\ \ell_3-\ell_{30}<x_3<\ell_3}}\right\}n_j^{\pm}=0, \tag{4.147}$$

where $i;j=1,2,3$.

Note that not only the condition (4.144) but also the condition (4.143) hold for Case 3. But for Case 3 the condition (4.145) is replaced with the following one.

$$\left\{\left[\sigma_{jn}\left(\delta_i^n+\frac{\partial u_i}{\partial x_n}\right)\right]\Big|_{\substack{x_2^{\pm}=h_A+\varepsilon f^{\pm}(x_1,x_3)\\ (\ell_1-\ell_{10})/2<x_1<(\ell_1+\ell_{10})/2\\ (\ell_3-\ell_{30})/2<x_3<(\ell_3+\ell_{30})/2}}\right\}n_j^{\pm}=0. \tag{4.148}$$

This completes the formulation of the problems which will be investigated in the present section.

### 4.10.2 Solution Method

To simplify the analysis, we consider only the solution procedure for Case 2, from which, as a particular case the solution procedure for Case 1 and Case 3 can be obtained. First, using the equations of crack edge-surfaces given in (4.140), we derive the following expressions for $n_j^{\pm}$ in (4.145):

$$n_j^{\pm}=\frac{\pm\varepsilon\delta_j^1\frac{\partial f^{\pm}(x_1,x_3)}{\partial x_1}\pm\delta_j^2\pm\varepsilon\delta_j^3\frac{\partial f^{\pm}(x_1,x_3)}{\partial x_3}}{\sqrt{1+\varepsilon^2\left(\frac{\partial f^{\pm}(x_1,x_3)}{\partial x_1}\right)^2+\varepsilon^2\left(\frac{\partial f^{\pm}(x_1,x_3)}{\partial x_3}\right)^2}},\ j=1,2,3, \tag{4.149}$$

where $\delta_j^k$ $(j=1,2,3;\ k=1,2,3)$ is the Kronecker symbol.

Assume that $\varepsilon^2\left[\left(\frac{\partial f^{\pm}(x_1,x_3)}{\partial x_1}\right)^2+\left(\frac{\partial f^{\pm}(x_1,x_3)}{\partial x_3}\right)^2\right]\ll 1$. According to this, the expression (4.149) can be represented in series form in terms of the small parameter $\varepsilon$:

$$n_1^{\pm}=\sum_{k=0}^{\infty}\varepsilon^{2k+1}n_{1k}^{\pm}(x_1,x_3),\ n_2^{\pm}=\pm1+\sum_{k=0}^{\infty}\varepsilon^{2k}n_{2k}^{\pm}(x_1,x_3),$$
$$n_3^{\pm}=\sum_{k=0}^{\infty}\varepsilon^{2k+1}n_{3k}^{\pm}(x_1,x_3). \tag{4.150}$$

The explicit expressions of the coefficients $n_{1k}^{\pm}(x_1,x_3)$, $n_{2k}^{\pm}(x_1,x_3)$ and $n_{3k}^{\pm}(x_1,x_3)$ can be easily obtained by employing the well-known power series expansion of the expressions given in (4.149). Note the Eqs. (4.149) and (4.150) are written for Case 2 when substituting $\partial f^{\pm}/\partial x_3 = 0$ and using the equation of the function $f^{\pm}$ in (4.139), we obtain the corresponding expressions for Case 1. Moreover, the expressions (4.149) and (4.150) hold also for Case 3. But in Case 3 the function $f^{\pm}$ is determined through the expression (4.141). As in the previous sections, the sought values are represented in series form in terms of ε as follows:

$$\left\{\sigma_{ij}; \varepsilon_{ij}; u_i\right\} = \sum_{q=0}^{\infty} \varepsilon^q \left\{\sigma_{ij}^{(q)}; \varepsilon_{ij}^{(q)}; u_i^{(q)}\right\}. \tag{4.151}$$

Consider briefly the solution procedure which is similar to that described in the previous sections. According to this procedure, first, we substitute the Eq. (4.151) into Eqs. (4.142, 4.146) and (4.147) and comparing identical powers of ε in equations and boundary conditions to describe of each approximation. Owing to the linearity of the mechanical relations in Eq. (4.142) and the end conditions for displacements in Eq. (4.146), these relations and conditions will be satisfied for each approximation (4.151) separately. The remaining relations obtained from Eqs. (4.142, 4.146) and (4.147) for every $q$th approximation, contain the values of the all previous approximations. At the same time, whilst satisfying the edge-surface condition (4.147), we employ the boundary form perturbation technique, according to which, the values of each approximation in Eq. (4.151) are expanded in series in the vicinity of $(x_1, h_A - 0, x_3)$and $(x_1,\ h_A + 0,\ x_3)$, and using expression (4.150), the corresponding conditions on the crack edge-surfaces are also obtained for the zeroth and the subsequent approximations.

It follows from the well-known mechanical considerations that for comparatively rigid composites under determination of the zeroth approximation we can use the relation $\delta_i^n + \partial u_i^{(0)} \big/ \partial x_n \approx \delta_i^n$, and according to the boundary conditions for the zeroth approximation obtained from Eqs. (4.142, 4.146) and (4.147), the stresses in the zeroth approximation are determined as follows:

$$\sigma_{11}^{(0)} = p_1,\ \sigma_{33}^{(0)} = p_3,\ \sigma_{ij}^{(0)} = 0,\ \text{for}\, ij \neq 11; 33. \tag{4.152}$$

Moreover, according to the relation $\delta_i^n + \partial u_i^{(0)} \big/ \partial x_n \approx \delta_i^n$, and expression (4.152) we obtain the following equilibrium equations and geometrical relations for the first approximation

$$\frac{\partial \sigma_{ji}^{(1)}}{\partial x_j} + \sigma_{11}^{(0)} \frac{\partial^2 u_i^{(1)}}{\partial x_1^2} + \sigma_{33}^{(0)} \frac{\partial^2 u_i^{(1)}}{\partial x_3^2} = 0,\ \varepsilon_{ij}^{(1)} = \frac{1}{2}\left(\frac{\partial u_i^{(1)}}{\partial x_j} + \frac{\partial u_j^{(1)}}{\partial x_i}\right). \tag{4.153}$$

Consider the boundary conditions for the first approximation obtained from the Eq. (4.146) and (4.147). For Case 2 these conditions are

$$u_2^{(1)}\Big|_{x_1=0} = u_2^{(1)}\Big|_{x_1=\ell_1} = 0,\ u_2^{(1)}\Big|_{x_3=0} = 0,\ \sigma_{2i}^{(1)}\Big|_{x_2=0} = \sigma_{2i}^{(1)}\Big|_{x_2=h} = 0,$$

$$\left[\sigma_{i1}^{(1)} + \sigma_{11}^{(0)}\delta_i^1\frac{\partial u_i^{(1)}}{\partial x_1}\right]\Bigg|_{x_1=0} = \left[\sigma_{i1}^{(1)} + \sigma_{11}^{(0)}\delta_i^1\frac{\partial u_i^{(1)}}{\partial x_1}\right]\Bigg|_{x_1=\ell_1} = 0,$$

$$\left[\sigma_{i3}^{(1)} + \sigma_{33}^{(0)}\delta_i^1\frac{\partial u_i^{(1)}}{\partial x_3}\right]\Bigg|_{x_3=0} = \left[\sigma_{i3}^{(1)} + \sigma_{33}^{(0)}\delta_i^1\frac{\partial u_i^{(1)}}{\partial x_3}\right]\Bigg|_{x_3=\ell_3} = 0,\ i = 1,3$$

$$\sigma_{2i}^{(1)}\Bigg|_{\substack{x_2=h_A\pm 0\\ (\ell_1-\ell_{10})/2<x_1<(\ell_1+\ell_{10})/2\\ (\ell_3-\ell_{30})<x_3<\ell_3}} = \mp\sigma_{11}^{(0)}\frac{\partial f^{\pm}(x_1,x_3)}{\partial x_1}\delta_i^1$$

$$\mp\sigma_{33}^{(0)}\frac{\partial f^{\pm}(x_1,x_3)}{\partial x_3}\delta_i^3,\ i = 1,2,3. \tag{4.154}$$

Condition (4.154) also holds for Case 1 and for Case 3, but in Case 1 and Case 3 the additional condition $u_2^{(1)}\Big|_{x_3=0} = 0$ must be added into the Eq. (4.154). Moreover, in Case 1 and Case 3 the last condition in (4.154) must be satisfied for the intervals $x_3 \in (0,\ell_3)$ and $x_3 \in ((\ell_3-\ell_{30})/2,(\ell_3+\ell_{30})/2)$ respectively.

Without limiting generality, let us assume that the material of the considered plate is homogeneous and orthotropic with the principal axes of elastic symmetries $Ox_1$, $Ox_2$ and $Ox_3$. In this case, the constitutive relations of the plate material can be written as follows:

$$\sigma_{ii}^{(q)} = A_{ij}^*\varepsilon_{jj}^{(q)},\ i;j = 1,2,3,\ \sigma_{12}^{(q)}2A_{66}^*\varepsilon_{12}^{(q)},\ \sigma_{13}^{(q)} = 2A_{55}^*\varepsilon_{13}^{(q)},$$
$$\sigma_{32}^{(q)}2A_{44}^*\varepsilon_{32}^{(q)},\ q = 0,1,2,\ldots. \tag{4.155}$$

In Eq. (4.155) $A_{11}^*,\ldots,A_{66}^*$ are the following integral operators

$$A_{ij}^*\varphi(t) = A_{ij0}\varphi(t) + \int_0^t A_{ij}(t-\tau)\varphi(\tau)d\tau. \tag{4.156}$$

The meaning of the notation in Eqs. (4.155) and (4.156) is obvious.

As the zeroth and the first approximations are enough for determination of the values of the critical parameters, we do not consider here the boundary value problems related to the second and subsequent approximations. Thus, we must determine the values of the first approximation for which it is necessary to solve the boundary value problems (4.153–4.156). For this purpose, as in the previous sections, we use the principle of correspondence by using the Laplace transformation (4.28) with the parameter $s > 0$ for the Eqs. (4.153–4.156). So, replacing $\sigma_{ij}^{(1)}$, $\varepsilon_{ij}^{(1)}$, $u_i^{(1)}$ and $A_{ij}^*$ in (4.153–4.156) by $\bar{\sigma}_{ij}^{(1)}$, $\bar{\varepsilon}_{ij}^{(1)}$, $\bar{u}_i^{(1)}$ and $\bar{A}_{ij}^*$, respectively, we obtain the corresponding equations and boundary conditions with respect to the Laplace transformations of the values for the first approximation. For the solution to the problems corresponding to the Laplace transformation of the sought values we employ the FEM.

### 4.10.3 FEM Modeling

In Case 2, the equilibrium equation in (4.153) and the boundary conditions for forces (for stresses) in (4.154) are the Euler equations of the functional (4.157).

$$\Pi\left(\bar{u}_1^{(1)},\bar{u}_2^{(1)},\bar{u}_3^{(1)}\right)=\frac{1}{2}\iiint\limits_{\Omega-\Omega_2}\left[\left(\bar{\sigma}_{11}^{(1)}+\sigma_{11}^{(0)}\frac{\partial\bar{u}_1^{(1)}}{\partial x_1}+\sigma_{33}^{(0)}\frac{\partial\bar{u}_1^{(1)}}{\partial x_3}\right)\frac{\partial\bar{u}_1^{(1)}}{\partial x_1}\right.$$
$$+\bar{\sigma}_{12}^{(1)}\frac{\partial\bar{u}_1^{(1)}}{\partial x_2}+\bar{\sigma}_{13}^{(1)}\frac{\partial\bar{u}_1^{(1)}}{\partial x_3}+\left(\bar{\sigma}_{21}^{(1)}+\sigma_{11}^{(0)}\frac{\partial\bar{u}_2^{(1),1}}{\partial x_1}+\sigma_{33}^{(0)}\frac{\partial\bar{u}_2^{(1)}}{\partial x_3}\right)\frac{\partial\bar{u}_2^{(1)}}{\partial x_1}$$
$$+\bar{\sigma}_{22}^{(1)}\frac{\partial\bar{u}_2^{(1)}}{\partial x_2}+\bar{\sigma}_{13}^{(1)}\frac{\partial\bar{u}_1^{(1)}}{\partial x_3}+\left(\bar{\sigma}_{31}^{(1)}+\sigma_{11}^{(0)}\frac{\partial\bar{u}_3^{(1)}}{\partial x_1}+\sigma_{33}^{(0)}\frac{\partial\bar{u}_3^{(1)}}{\partial x_3}\right)\frac{\partial\bar{u}_3^{(1)}}{\partial x_1}$$
$$\left.+\bar{\sigma}_{32}^{(1)}\frac{\partial\bar{u}_3^{(1)}}{\partial x_2}+\bar{\sigma}_{23}^{(1)}\frac{\partial\bar{u}_2^{(1)}}{\partial x_3}+\bar{\sigma}_{33}^{(1)}\frac{\partial\bar{u}_3^{(1)}}{\partial x_3}\right]dx_1dx_2dx_3$$
$$-\int\limits_{\ell_3-\ell_{30}}^{\ell_3}\int\limits_{(\ell_1-\ell_{10})/2}^{(\ell_1+\ell_{10})/2}\frac{1}{s}\sigma_{11}^{(0)}\frac{\partial f^-}{\partial x_1}\bar{u}_1^{(1)}\bigg|_{x_2=h_A-0}dx_1dx_3$$
$$+\int\limits_{\ell_3-\ell_{30}}^{\ell_3}\int\limits_{(\ell_1-\ell_{10})/2}^{(\ell_1+\ell_{10})/2}\frac{1}{s}\sigma_{33}^{(0)}\frac{\partial f^-}{\partial x_3}\bar{u}_3^{(1)}\bigg|_{x_2=h_A-0}dx_1dx_3$$

$$+\int\limits_{\ell_3-\ell_{30}}^{\ell_3}\int\limits_{(\ell_1-\ell_{10})/2}^{(\ell_1+\ell_{10})/2}\frac{1}{s}\sigma_{11}^{(0)}\frac{\partial f^+}{\partial x_1}\bar{u}_1^{(1)}\bigg|_{x_2=h_A+0}dx_1dx_3$$
$$+\int\limits_{\ell_3-\ell_{30}}^{\ell_3}\int\limits_{(\ell_1-\ell_{10})/2}^{(\ell_1+\ell_{10})/2}\frac{1}{s}\sigma_{33}^{(0)}\frac{\partial f^+}{\partial x_3}\bar{u}_3^{(1)}\bigg|_{x_2=h_A+0}dx_1dx_3. \qquad (4.157)$$

Note that the corresponding functional for Case 1 (Case 3) is obtained from the functional (4.157) by the replacement of the integrating domain $\Omega-\Omega_2$ with $\Omega-\Omega_1$ ($\Omega-\Omega_3$) and by the replacement of the integrating interval $(\ell_3-\ell_{30},\ell_3)$ with the $(0,\ell_3)$ (with $((\ell_3-\ell_{30})/2,\ (\ell_3+\ell_{30})/2)$) in the last four terms in the functional (4.157). Moreover, in Case 1 it must be taken the relation $\partial f^{\pm}/\partial x_3=0$ into account.

By direct verification, it is proven that the equations and boundary conditions with respect to the forces in Cases 1, 2 and 3 are obtained from the equation

$\delta\Pi(\bar{u}_1^{(1)}, \bar{u}_2^{(1)}, \bar{u}_3^{(1)}) = 0$. The virtual work principle in conjunction with the FEM technique is employed. The FEM grid over the solution domain is revised until convergence is achieved. Realization of the corresponding computer programs has been made using the package FTN77.

In the previous section it has been established that under investigation of the buckling delamination around a crack contained in a plate, the numerical results on the critical parameter by the use of the singular type finite elements in the vicinity of the crack tips coincide (with very high accuracy) with those obtained by the use of the ordinary type finite elements in the vicinity of the crack tips. According to this statement, in the present section, the finite elements containing the crack tips (fronts) are also ordinary brick elements. In this way, we simply establish FEM modeling of the problems under consideration.

Thus, by employing the FEM algorithm detailed above we calculate the values of Laplace transformation of the sought values. As before, the originals of the sought functions are determined by the use of the method (Schapery 1962, 1978). The values of the critical parameters are found from the criterion

$$\max_{\substack{0<x_1<\ell_1 \\ 0<x_3<\ell_3}} |u_2^{(1)}|_{x_2=h_A+0}| \to \infty \text{ as } p_1 \to p_{1cr} \text{ (for pure elastic plate} \tag{4.158}$$

under fixed value of the ratio $p_3/p_1$) or as $t \to t_{cr}$ (for viscoelastic plate).

### 4.10.4 Numerical Results and Discussions

For concrete numerical investigations, the function $f^{\pm}(x_1, x_3)$ which characterizes suitable initial imperfection modes of the crack edge-surfaces and enter the Eqs. (4.139–4.141), can be selected as follows:

For Case 1.

The co-phase initial imperfection mode

$$f^{+}(x_1) = f^{-}(x_1) = \ell_{10}\sin^2\left(\pi\left(x_1 - \frac{\ell_1 - \ell_{10}}{2}\right)\Big/\ell_{10}\right), \tag{4.159}$$

and the anti-phase initial imperfection mode

$$f^{\pm}(x_1) = \pm\ell_{10}\sin^2\left(\pi\left(x_1 - \frac{\ell_1 - \ell_{10}}{2}\right)\Big/\ell_{10}\right). \tag{4.160}$$

For Case 2.

The co-phase initial imperfection mode

$$f^{+}(x_1,x_3)=f^{-}(x_1,x_3)=\ell_{10}\sin^2\left(\pi\left(x_1-\frac{\ell_1-\ell_{10}}{2}\right)\Big/\ell_{10}\right)$$
$$\times\sin^2(\pi(x_3-(\ell_3-\ell_{30}))/(2\ell_{30})), \tag{4.161}$$

and the anti-phase initial imperfection mode

$$f^{\pm}(x_1,x_3)=\pm\,\ell_{10}\sin^2\left(\pi\left(x_1-\frac{\ell_1-\ell_{10}}{2}\right)\Big/\ell_{10}\right)$$
$$\times\sin^2(\pi(x_3-(\ell_3-\ell_{30}))/(2\ell_{30})). \tag{4.162}$$

For Case 3.

The co-phase initial imperfection mode

$$f^{+}(x_1,x_3)=f^{-}(x_1,x_3)=\ell_{10}\sin^2\left(\pi\left(x_1-\frac{\ell_1-\ell_{10}}{2}\right)\Big/\ell_{10}\right)$$
$$\times\sin^2\left(\pi\left(x_3-\frac{\ell_3-\ell_{30}}{2}\right)\Big/\ell_{30}\right), \tag{4.163}$$

and the anti-phase initial imperfection mode

$$f^{\pm}(x_1,x_3)=\pm\,\ell_{10}\sin^2\left(\pi\left(x_1-\frac{\ell_1-\ell_{10}}{2}\right)\Big/\ell_{10}\right)$$
$$\times\sin^2\left(\pi\left(x_3-\frac{\ell_3-\ell_{30}}{2}\right)\Big/\ell_{30}\right). \tag{4.164}$$

Figure 4.15 illustrates schematically the anti-phase initial imperfection mode. It is easy to image the co-phase initial imperfection mode. Thus, we turn to the analyses of the numerical results.

#### 4.10.4.1 Numerical Results Obtained in Case 1 (Band Crack)

We assume that $p_3=0$ and consider buckling delamination of the plate under uniaxial compression of that in the direction of the $Ox_1$ axis and use the notation $p$ instead of the notation $p_1$. Moreover, we assume that the material of the plate is pure elastic and orthotropic with principal axes $Ox_1$, $Ox_2$ and $Ox_3$ of the elastic symmetry and take the constants $A_{ij}$ instead of operators $A^{*}_{ij}$ (4.156) in (4.155), where

$$A_{ij}=\frac{(-1)^{i+j}a_{ij}}{\det\|a_{ij}\|},\ a_{11}=\frac{1}{E_1},\ a_{12}=\frac{-\nu_{12}}{E_2},\ a_{13}=\frac{-\nu_{13}}{E_3},\ a_{23}=\frac{-\nu_{23}}{E_3},$$
$$a_{22}=\frac{1}{E_2},\ a_{33}=\frac{1}{E_3},\ a_{ij}=a_{ji}. \tag{4.165}$$

**Table 4.39** The values of $\delta_{cr}$ for the co-phase (upper number) and anti-phase (lower number) buckling modes for various $G_{12}/E_1$, $G_{13}/E_1$ and $G_{23}/E_1$ under $\gamma = \ell_3/\ell_1 = 1$; $E_2/E_1 = 0.5$; $E_3/E_1 = 0.3$ and $\ell_0/\ell_1 = 0.5$

| $G_{12}/E_1$ | $G_{13}/E_1$ | $G_{23}/E_1$ | | |
|---|---|---|---|---|
| | | 0.10 | 0.05 | 0.01 |
| 0.10 | 0.10 | 0.0314<br>0.0461 | 0.0300<br>0.0457 | 0.0253<br>0.0434 |
| | 0.05 | 0.0290<br>0.0457 | 0.0281<br>0.0453 | 0.0245<br>0.0433 |
| | 0.01 | 0.0266<br>0.0453 | 0.0262<br>0.0451 | 0.0242<br>0.0430 |
| 0.05 | 0.10 | 0.0266<br>0.0313 | 0.0255<br>0.0310 | 0.0217<br>0.0297 |
| | 0.05 | 0.0248<br>0.0310 | 0.0241<br>0.0308 | 0.0210<br>0.0296 |
| | 0.01 | 0.0233<br>0.0307 | 0.0230<br>0.0305 | 0.0212<br>0.0295 |

In (4.165) $E_1$, $E_2$ and $E_3$ are the elastic modulus along the symmetry axes $Ox_1$, $Ox_2$ and $Ox_3$, respectively; $\nu_{12}$, $\nu_{13}$ and $\nu_{23}$ are Poisson ratios (see, Lekhnitskii 1981). Note that $\nu_{ij}$ characterizes the shortening (elongation) in the direction of an $i$th axis under tension (compression) along the $j-th$ axis.

We assume that $h_U = h_A = h/2$ (Fig 4.15), $\nu_{12} = \nu_{13} = \nu_{23} = 0.3$ and $h/\ell_1 = 0.15$. The delamination buckling modes at $x_3 = \ell_3/2$ caused by the initial co-phase (4.159) and anti-phase (4.160) imperfections are similar to that depicted in Fig. 4.2. As in the previous sections, these modes are constructed according to the displacement field obtained in the solution procedure for the problem under consideration.

Thus, let us analyze the critical values of the external compressing force, i.e. the values of $\delta_{cr(=|p_{cr}/E_1|)}$, and their dependence on the geometrical and mechanical parameters of the plate. Table 4.39 shows the values of $\delta_{cr}$ for various values of $G_{12}/E_1$, $G_{13}/E_1$ and $G_{23}/E_1$ for the co-phase (numerator) and anti-phase (denominator) buckling delamination modes in the case where $\gamma = \ell_3/\ell_1 = 1$, $E_2/E_1 = 0.5$, $E_3/E_1 = 0.3$ and $\ell_0/\ell_1 = 0.5$. As seen, the values of $\delta_{cr}$ for the anti-phase buckling mode are greater than those obtained for the co-phase mode, which agrees well with the well-known mechanical considerations. In this case, $\delta_{cr}$ decreases with decreasing ratios $G_{12}/E_1$, $G_{13}/E_1$ and $G_{23}/E_1$. This result agrees with the known results on the influence of the shear modulus on the critical forces of instability of beams and plates.

Table 4.40 illustrates the influence of the crack's length, i.e. the influence of the ratio $\ell_{10}/\ell_1$, on the values of $\delta_{cr}$ found for the co-phase (numerator) and anti-phase (denominator) buckling delamination modes for various values of $G_{12}/E_1$ at $G_{13}/E_1 = 0.1$, $E_2/E_1 = 0.5$, $E_3/E_1 = 0.3$, $G_{23}/E_1 = 0.1$ and $\gamma = 1$. As expected, $p_{cr}/E_1$ increases with decreasing crack length. In this table the sign "− "means that the values of $\delta_{cr}$ obtained for co-phase initial imperfection mode coincide with those for the whole plate. Consequently, in such cases, the existence of a crack with an initial co-phase imperfection cannot cause the co-phase buckling delamination mode. Moreover, a comparison of the values of $\delta_{cr}$ found for isotropic and orthotropic plates shows that, in the orthotropy with $G_{ij}/E_1 < 1$, the

**Table 4.40** The influence of the crack length, i.e. of the ratio $\ell_0/\ell_1$, on the values of $\delta_{cr}$ for the co-phase (upper number) and anti-phase (lower number) buckling modes for various $G_{12}/E_1$ under $G_{13}/E_1 = 0.1$, $E_2/E_1 = 0.5$, $E_3/E_1 = 0.3$, $\gamma = 1$ and $G_{23}/E_1 = 0.1$

| $\frac{G_{12}}{E_1}$ | $\ell_0/\ell_1$ | | | | | | | |
|---|---|---|---|---|---|---|---|---|
| | 0.7 | 0.6 | 0.5 | 0.4 | 0.3 | 0.2 | 0.1 | 0.0 |
| 0.10 | 0.0264 / 0.0348 | 0.0287 / 0.0398 | 0.0314 / 0.0461 | 0.0344 / 0.0541 | 0.0377 / 0.0643 | - / 0.0770 | - / 0.0913 | 0.0438 |
| 0.07 | 0.0247 / 0.0300 | 0.0267 / 0.0337 | 0.0291 / 0.0382 | 0.0319 / 0.0436 | 0.0348 / 0.0501 | - / 0.0578 | - / 0.0657 | 0.0403 |
| 0.05 | 0.0227 / 0.0256 | 0.0245 / 0.0283 | 0.0266 / 0.0313 | 0.0290 / 0.0348 | 0.0316 / 0.0389 | 0.0339 / 0.0435 | 0.0427 / 0.0478 | 0.0366 |
| For isotropic material | 0.0336 / 0.0409 | 0.0383 / 0.0481 | 0.0442 / 0.0590 | 0.0508 / 0.0758 | 0.0572 / 0.1032 | 0.0620 / 0.1503 | 0.0646 / 0.2370 | 0.0681 |

**Table 4.41** The influence of $E_3/E_1$ and $E_2/E_1$ on the values of $\delta_{cr}$ for the co-phase (upper number) and anti-phase (lower number) buckling modes under $\gamma = 1$, $G_{12}/E_1 = G_{13}/E_1 = G_{23}/E_1 = 0.1$ and $\ell_0/\ell_1 = 0.5$

| $E_3/E_1$ | $E_2/E_1$ | | | |
|---|---|---|---|---|
| | 0.3 | 0.4 | 0.5 | 0.6 |
| 0.3 | 0.0331 / 0.0495 | 0.0318 / 0.0468 | 0.0314 / 0.0461 | 0.0313 / 0.0459 |
| 0.4 | 0.0296 / 0.0426 | 0.0291 / 0.0415 | 0.0289 / 0.0412 | 0.0289 / 0.0410 |
| 0.5 | 0.0288 / 0.0403 | 0.0284 / 0.0396 | 0.0283 / 0.0393 | 0.0282 / 0.0392 |

values of $\delta_{cr}$ obtained for orthotropic plate are lower than those obtained for the corresponding isotropic plate.

Table 4.41 shows the influence of $E_3/E_1$ and $E_2/E_1$ on the values of $\delta_{cr}$ at $\gamma = 1$, $G_{12}/E_1 = G_{13}/E_1 = G_{23}/E_1 = 0.1$ and $\ell_{10}/\ell_1 = 0.5$. As in the previous tables, the numerator (denominator) corresponds to the co-phase (anti-phase) buckling delamination mode. An analysis of the results confirms that $\delta_{cr}$ decrease with $E_3/E_1$ and $E_2/E_1$.

Consider now the influence of the parameter $\gamma(= \ell_1/\ell_3)$ on $\delta_{cr}$. According to mechanical considerations, at fixed problem parameters, with increasing $\gamma$, $\delta_{cr}$ must approach the corresponding limit value. This value can be taken as $\delta_{cr}$ for the corresponding plane strain state. This conclusion is supported by data given in Table 4.42, which shows values of $p_{cr}/E_1$ for various $\gamma$ and $E_2/E_1$ at $E_3/E_1 = 0.3$, $G_{12}/E_1 = G_{13}/E_1 = G_{23}/E_1 = 0.1$ and $\ell_{10}/\ell_1 = 0.5$. Note that these results confirm the validity of the calculation algorithm developed.

Thus, from the analyses performed above the following conclusions can be drawn for the buckling delamination of the orthotropic plate with a band crack located on the middle plane of the plate:

- The co-phase (anti-phase) initial imperfection mode causes the co-phase (anti-phase) buckling delamination mode;
- The critical values of the compressing force, i.e. the values of $\delta_{cr}(= |p_{cr}/E_1|)$, for the anti-phase buckling delamination mode are greater than those for the co-phase one;

**Table 4.42** The values of $\delta_{cr}$ for the co-phase (upper number) and anti-phase (lower number) buckling modes for various $\gamma$ and $E_2/E_1$ under $E_3/E_1 = 0.3$, $G_{12}/E_1 = G_{13}/E_1 = G_{23}/E_1 = 0.1$ and $\ell_0/\ell_1 = 0.5$

| $\gamma = \ell_3/\ell_1$ | Isotropic material | $E_2/E_1$ | | |
|---|---|---|---|---|
| | | 0.5 | 0.4 | 0.3 |
| 1.0 | 0.0442<br>0.0590 | 0.0314<br>0.0461 | 0.0318<br>0.0468 | 0.0331<br>0.0495 |
| 1.5 | 0.0292<br>0.0539 | 0.0249<br>0.0437 | 0.0251<br>0.0444 | 0.0261<br>0.0469 |
| 2.0 | 0.0243<br>0.0522 | 0.0229<br>0.0429 | 0.0231<br>0.0436 | 0.0240<br>0.0460 |
| 3.0 | 0.0210<br>0.0511 | 0.0216<br>0.0423 | 0.0218<br>0.0430 | 0.0226<br>0.0455 |
| 4.0 | 0.0199<br>0.0507 | 0.0212<br>0.0421 | 0.0214<br>0.0428 | 0.0221<br>0.0452 |
| 5.0 | 0.0194<br>0.0505 | 0.0210<br>0.0420 | 0.0212<br>0.0427 | 0.0219<br>0.0452 |
| 6.0 | 0.0192<br>0.0504 | 0.0209<br>0.0420 | 0.0211<br>0.0427 | 0.0218<br>0.0451 |

- The values of $\delta_{cr}$ increase with a decrease in the crack width, i.e. with decreasing of the parameter $\gamma(= \ell_1/\ell_3)$, because the displacement along the $Ox_2$ axis of the crack's ends at $x_3 = 0$ and $x_3 = \ell_3$ is fixed and is equated to zero;
- The values of $\delta_{cr}$ decrease with decreasing shear modulus of the plate material, i.e., with decreasing values of the ratios $G_{12}/E_1$, $G_{13}/E_1$ and $G_{23}/E_1$;
- The influence of $E_2/E_1$ and $E_3/E_1$ on $\delta_{cr}$ is insignificant;
- The values of $\delta_{cr}$ approach that obtained for the plate-strip under the assumption of the plane-strain state as $\gamma(= \ell_3/\ell_1) \to \infty$.

Now we consider the case where the plate material consists of two alternating layers whose materials are isotropic, homogeneous and these layers are located in the $Ox_1x_3$ plane. The reinforcing layers are supposed to be pure elastic with mechanical characteristics $E^{(2)}$ (Young's modulus) and $\nu^{(2)}$ (Poisson coefficient). The material of matrix layers is supposed to be linearly viscoelastic with the operators

$$E^{(1)} = E_0^{(1)}\left[1 - \omega_0\Pi_\alpha^*(-\omega_0 - \omega_\infty)\right],$$
$$\nu^{(1)} = \nu_0^{(1)}\left[1 + \frac{1 - 2\nu_0^{(1)}}{2\nu_0^{(1)}}\omega_0\Pi_\alpha^*(-\omega_0 - \omega_\infty)\right]. \tag{4.166}$$

where $E_0^{(1)}$ are $\nu_0^{(1)}$ are the instantaneous values of Young's modulus and of Poisson coefficient, respectively; $\alpha$, $\omega_0$ and $\omega_\infty$ are the rheological parameters of the matrix material, $\Pi_\alpha^*$ is the fractional-exponential operator (Rabotnov 1977) and this operator is determined through the expressions (4.73) and (4.74).

As in the previous sections, we introduce the dimensionless rheological parameter $\omega = \omega_\infty/\omega_0$ and the dimensionless time $t' = \omega_0^{1/(1+\alpha)}t$ and assume that $\eta^{(2)} = 0.5$ where $\eta^{(2)}$ is the volume fraction of filler. It is known that in the continuum approach this layered-composite material can be taken as viscoelastic transversally isotropic material with effective mechanical properties whose

**Table 4.43** The values of $\delta_{cr.0}$ (upper number) and $|p_{cr.0}|/E_0^{(1)}$ (lower number) under $\ell_{10}/\ell_1 = 0.5$ for various values of $\gamma$ and $E^{(2)}/E_0^{(1)}$

| $\gamma = \ell_3/\ell_1$ | $E^{(2)}/E_0^{(1)}$ | | | |
|---|---|---|---|---|
| | 1 | 5 | 10 | 50 |
| 1 | 0.0590<br>0.0590 | 0.0499<br>0.1497 | 0.04155<br>0.2285 | 0.01955<br>0.4984 |
| 1.5 | 0.0539<br>0.0539 | 0.0455<br>0.1365 | 0.0377<br>0.2076 | 0.0176<br>0.4479 |
| 2 | 0.0522<br>0.0522 | 0.0440<br>0.1321 | 0.0365<br>0.2007 | 0.0169<br>0.4314 |
| 3 | 0.0511<br>0.0511 | 0.0430<br>0.1291 | 0.0356<br>0.1960 | 0.0165<br>0.4200 |
| 4 | 0.0507<br>0.0507 | 0.0426<br>0.1280 | 0.0353<br>0.1943 | 0.0163<br>0.4160 |
| 5 | 0.0505<br>0.0505 | 0.0425<br>0.1275 | 0.0352<br>0.1935 | 0.0162<br>0.4142 |
| 6 | 0.0504<br>0.0504 | 0.0424<br>0.1273 | 0.0351<br>0.1931 | 0.0162<br>0.4132 |
| $\infty$ | 0.0475<br>0.0475 | 0.0411<br>0.1204 | 0.0346<br>0.1820 | 0.0156<br>0.3989 |

isotropy axis lies on the $Ox_2$ axis. For the case under consideration these normalized mechanical properties are determined through the expression (3.83).

First, we consider the pure elastic stability loss (buckling delamination) which takes place at $t' = 0$ and $t' = \infty$. In these cases the absolute critical values of the averaged strain $\delta = |p|/\tilde{E}$ are calculated for selected values of the problem parameters, where $\tilde{E}$ is an instantaneous value of the effective modulus of elasticity in the direction of the $Ox_1$ axis and $\tilde{E} = E_0^{(1)}(1-\eta^{(2)}) + E^{(2)}\eta^{(2)}$. We assume that $h/\ell_1 = 0.15$, $\nu_0^{(1)} = \nu^{(2)} = 0.3$ and $h_U = h_A = h/2$.

We denote the critical values for $\delta$ attained at $t' = 0$ and at $t' = \infty$ through $\delta_{cr.0}(= |p_{cr.0}|/\tilde{E})$ and $\delta_{cr.\infty}(= |p_{cr.\infty}|/\tilde{E})$ respectively. Table 4.43 shows the values of $\delta_{cr.0}$ (upper number) calculated for various values of $E^{(2)}/E_0^{(1)}$ and $\gamma$ under $\ell_{10}/\ell_1 = 0.5$ in the anti-phase initial imperfection mode (Fig. 4.15a). At the same time, Table 4.43 shows the values of the reduced critical forces $|p_{cr.0}|/E_0^{(1)}$ (lower number). It follows from the data given in this table that the values of $\delta_{cr.0}$ decrease, but the values of $|p_{cr.0}|/E_0^{(1)}$ increase with $E^{(2)}/E_0^{(1)}$ and the order of the critical strains is quite realistic.

According to the well-known mechanical consideration, the numerical results obtained in the case under consideration must approach the corresponding results obtained in Sect. 4.2.3.2 and given in Table 4.2 with $\gamma$, because in the mentioned sub-section the plane-strain state (i.e. the case where $\gamma = \infty$) is considered. The results given in Table 4.43 agree well with this prediction. This statement validates also the reliability of the algorithm and packed programs used in the present investigation. Moreover, it follows from Table 4.43 that the values of $\delta_{cr.0}$ and $|p_{cr.0}|/E_0^{(1)}$ decrease monotonically with $\gamma$. This conclusion also agrees with the mechanical considerations.

**Table 4.44** The values of $\delta_{cr.0}/\delta_{cr.\infty}$ under $\omega = 1,\ h/\ell_1 = 0.15,\ \ell_{10}/\ell_1 = 0.5$ for various values of $\gamma$ and $E^{(2)}\big/E_0^{(1)} = 50$

| $\gamma = \ell_3/\ell_1$ | $\ell_{10}/\ell_1$ | | |
|---|---|---|---|
| | 0.5 | 0.4 | 0.3 |
| 1 | $\frac{0.0196}{0.0115}$ | $\frac{0.0211}{0.0119}$ | $\frac{0.0230}{0.0124}$ |
| 1.5 | $\frac{0.0176}{0.0103}$ | $\frac{0.0197}{0.0110}$ | $\frac{0.0221}{0.0118}$ |
| 2 | $\frac{0.0169}{0.0099}$ | $\frac{0.0192}{0.0108}$ | $\frac{0.0218}{0.0117}$ |
| 3 | $\frac{0.0165}{0.0096}$ | $\frac{0.0188}{0.0106}$ | $\frac{0.0215}{0.0115}$ |
| 4 | $\frac{0.0163}{0.0095}$ | $\frac{0.0187}{0.0105}$ | $\frac{0.0215}{0.0115}$ |
| 5 | $\frac{0.0162}{0.0095}$ | $\frac{0.0187}{0.0105}$ | $\frac{0.0214}{0.0115}$ |
| 6 | $\frac{0.0162}{0.0095}$ | $\frac{0.0186}{0.0105}$ | $\frac{0.0214}{0.0115}$ |

**Table 4.45** The values of $t'_{cr}$ under $E^{(2)}\big/E_0^{(1)} = 1$ and $\alpha = -0.5$ for various values $\omega$

| $\delta$ | $\omega$ | | |
|---|---|---|---|
| | 0.5 | 1 | 2 |
| 0.042 | 0.0089 | 0.0102 | 0.0139 |
| 0.041 | 0.0221 | 0.0276 | 0.0472 |
| 0.040 | 0.0455 | 0.0632 | 0.1523 |

We will analyze below the results related to the critical values of the averaged strain $\delta$, only, because the values of $|p_{cr.0}|\big/E_0^{(1)}$ and $|p_{cr.\infty}|\big/E_0^{(1)}$ can be calculated from the relations $|p_{cr.0}|\big/E_0^{(1)} = \delta_{cr.0}((1-\eta^{(2)}) + E^{(2)}\big/E_0^{(1)}\eta^{(2)})$ and $|p_{cr.\infty}|\big/E_0^{(1)} = \delta_{cr.\infty} \times ((1-\eta^{(2)}) + E^{(2)}\big/E_0^{(1)} \times \eta^{(2)})$ respectively.

Table 4.44 shows the ratio $\delta_{cr.0}/\delta_{cr.\infty}$ obtained for various $\gamma$ and $\ell_{10}/\ell_1$ under $\omega = 1.0$ and $E^{(2)}\big/E_0^{(1)} = 50$ in the anti-phase initial imperfection mode (4.160). These results indicate that the values of the $\delta_{cr.0}$, as well as the values of $\delta_{cr.\infty}$ decrease with $\ell_{10}/\ell_1$. Figures 4.16 and 4.17 show the graphs of the dependencies among $\delta_{cr.0}$, $\delta_{cr.\infty}$ and $\ell_{10}/\ell_1$ for the anti-phase and co-phase initial imperfection modes, respectively. Note that these graphs are constructed for various $E^{(2)}\big/E_0^{(1)}$ under $\gamma = 1$ and $\omega = 1$. It follows from these results that the values of $\delta_{cr.0}$ and $\delta_{cr.\infty}$ attained for the anti-phase initial imperfection mode (4.160) are significantly greater than the corresponding values of $\delta_{cr.0}$ and $\delta_{cr.\infty}$ attained for the co-phase initial imperfection mode (4.159). The difference between the values of critical forces attained for the co-phase and the anti-phase initial imperfection modes decreases with $\ell_{10}/\ell_1$, i.e. with the crack length along the $Ox_1$ axis.

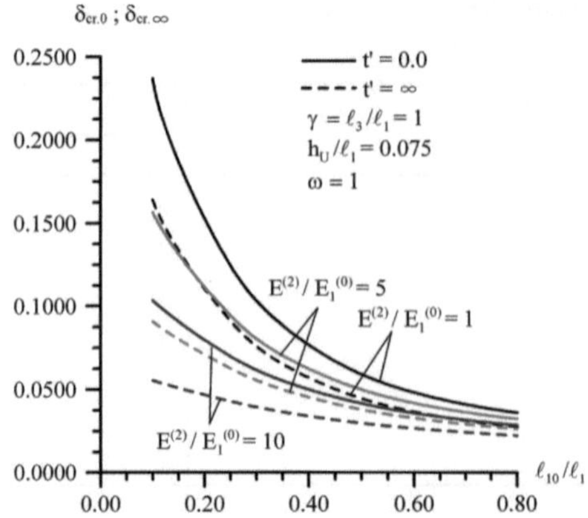


**Fig. 4.16** The graphs of the dependencies among $\delta_{cr.0}$, $\delta_{cr.\infty}$ and $\ell_{10}/\ell_1$ for the anti-phase buckling delamination mode

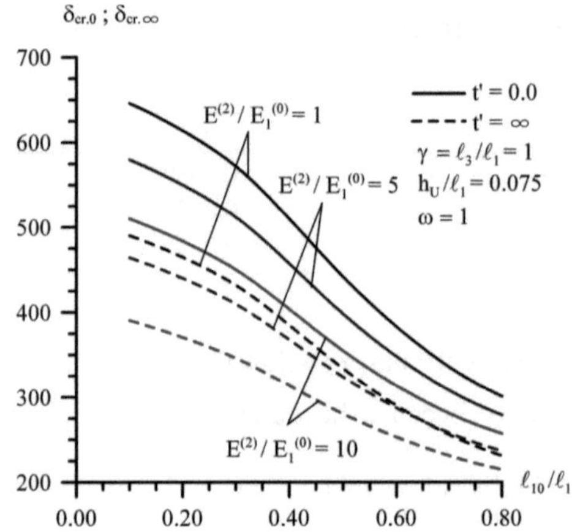


**Fig. 4.17** The graphs of the dependencies among $\delta_{cr.0}$, $\delta_{cr.\infty}$ and $\ell_{10}/\ell_1$ for the co-phase buckling delamination mode

Now we assume that the material of the plate is viscoelastic with the rheological parameters $\alpha = -0.5$, $\omega = 0.5, 1.0$ and $2.0$. Consider the numerical results attained for the critical time $t'_{cr.}$. These results are given in Table 4.45 for the case where$\gamma = 1.0$ and $E^{(2)} \Big/ E_0^{(1)} = 1$ for various $\delta$ under co-phase initial imperfection mode. Note that the external compressive force must satisfy the following relation $|\delta_{cr.\infty}| < \delta < |\delta_{cr.0}|$ so that the viscoelastic buckling delamination of the considered plate will occur.

#### 4.10.4.2 Numerical Results Obtained in Case 2 (Edge Crack)

As in Case 1, we assume that $p_3 = 0$ and use the notation $p$ instead of the notation $p_1$. First, consider the case where the plate material is pure elastic and orthotropic with principal axes $Ox_1$, $Ox_2$ and $Ox_3$ of the elastic symmetry. We take the elastic

**Table 4.46** The influence of the shear modulus $G_{ij}/E_1$ on the values of $\delta_{cr}$ for co-phase (upper number) and anti-phase (lower number) initial imperfection mode under $E_2/E_1 = 0.5$, $E_3/E_1 = 0.3$, $\gamma = 1$, $\ell_{30}/\ell_1 = 0.5$ and $\ell_{10}/\ell_1 = 0.5$

| $\frac{G_{12}}{E_1}$ | $\frac{G_{13}}{E_1}$ | $G_{23}/E_1$ 0.10 | 0.05 | 0.01 |
|---|---|---|---|---|
| 0.10 | 0.10 | $\frac{0.0177}{0.0367}$ | $\frac{0.0174}{0.0363}$ | $\frac{0.0166}{0.0350}$ |
| | 0.05 | $\frac{0.0166}{0.0352}$ | $\frac{0.0164}{0.0348}$ | $\frac{0.0177}{0.0367}$ |
| | 0.01 | $\frac{0.0152}{0.0339}$ | $\frac{0.0151}{0.0330}$ | $\frac{0.0151}{0.0321}$ |
| 0.05 | 0.10 | $\frac{0.0148}{0.0255}$ | $\frac{0.0146}{0.0251}$ | $\frac{0.0139}{0.0243}$ |
| | 0.05 | $\frac{0.0139}{0.0245}$ | $\frac{0.0138}{0.0233}$ | $\frac{0.0135}{0.0226}$ |
| | 0.01 | $\frac{0.0129}{0.0234}$ | $\frac{0.0128}{0.0230}$ | $\frac{0.0128}{0.0199}$ |

constants $A_{ij}$ instead of the operators $A_{ij}^*$ and suppose that these constants are determined through the elastic modulus $E_1$, $E_2$ and $E_3$, Poisson ratios $\nu_{12}$, $\nu_{13}$ and $\nu_{23}$, and shear modulus $G_{12}$, $G_{13}$ and $G_{23}$ by the expressions in (4.165). Assume that $h/\ell_1 = 0.15$, $\nu_{12} = \nu_{13} = \nu_{23} = 0.3$ and $h_U = h_A = h/2$ (Fig. 4.15b). Thus, consider numerical results obtained for $\delta_{cr}(= |p_{cr}/E_1|)$ and analyze the influence of the problem parameters on these results.

Table 4.46 shows the influence of the values of the shear modulus $G_{12}/E_1$, $G_{13}/E_1$ and $G_{23}/E_1$ on the values of $\delta_{cr}$ under $E_2/E_1 = 0.5$, $E_3/E_1 = 0.3$, $\gamma = 1$, $\ell_{30}/\ell_1 = 0.5$ and $\ell_{10}/\ell_1 = 0.5$. It follows from this table that the values of $\delta_{cr}$ decrease with decreasing of the shear modulus of the plate material. However, the influence of the shear modulus $G_{12}/E_1$ on the $\delta_{cr}$ is more significant than the influence of the other shear modulus. The comparison of the results given in Table 4.46 with the corresponding ones given in Table 4.39 shows that the values of $\delta_{cr}$obtained for the edge crack is significantly less than the corresponding one obtained for the band crack. Because in Case 2 the end of the plate at $x_3 = \ell_3$ is free, but in Case 1 the displacement of this end along the plate thickness is stopped.

Tables 4.47 and 4.48 show the influence of the $\ell_{10}/\ell_1$ and $\ell_{30}/\ell_1$ respectively on the $\delta_{cr}$ obtained for various $G_{12}/E_1$ under $G_{13}/E_1 = G_{23}/E_1 = 0.1$, $E_2/E_1 = 0.5$, $E_3/E_1 = 0.3$, $\ell_{30}/\ell_1 = 0.5$ (for Table 4.47) and $\ell_{10}/\ell_1 = 0.5$ (for Table 4.48). As it can be expected, according to these results, it can be concluded that a decrease in the values of the $\ell_{10}/\ell_1$ and $\ell_{30}/\ell_1$ causes to an increase in the values of $\delta_{cr}$. However, under the anti-phase buckling delamination mode the influence of $\ell_{10}/\ell_1$ and $\ell_{30}/\ell_1$ on the values of $\delta_{cr}$ is more considerable than that under co-phase buckling delamination mode of the plate around the end crack.

Consider the influence of the distance of the crack from the upper free face plane, i.e. the influence of the ratio $h_U/\ell_1$ on the values of $\delta_{cr}$.This influence is illustrated by data tabulated in Table 4.49. It follows from this table that the values of $\delta_{cr}$ decrease with decreasing of the $h_U/\ell_1$. Moreover, the results show that under $h_U/\ell_1 = 0.0375$ the values of $\delta_{cr}$ and buckling delamination modes obtained for the both co-phase and anti-phase initial imperfection modes coincide and are the co-phase buckling delamination mode which is similar to that shown in Fig. 4.2.

**Table 4.47** The values of $\delta_{cr}$ obtained for co-phase (upper number) and anti-phase (lower number) initial imperfection mode for various $G_{12}/E_1$ and $\ell_{10}/\ell_1$ under $E_2/E_1 = 0.5$, $E_3/E_1 = 0.3$, $\gamma = 1$, and $\ell_{30}/\ell_1 = 0.5$

| $G_{12}/E_1$ | $\ell_{10}/\ell_1$ | | | | |
|---|---|---|---|---|---|
| | 0.60 | 0.50 | 0.40 | 0.30 | 0.20 |
| 0.10 | 0.0172 / 0.0308 | 0.0177 / 0.0367 | 0.0180 / 0.0447 | 0.0181 / 0.0553 | 0.0181 / 0.0695 |
| 0.07 | 0.0159 / 0.0263 | 0.0163 / 0.0307 | 0.0165 / 0.0364 | 0.0166 / 0.0437 | 0.0167 / 0.0529 |
| 0.05 | 0.0144 / 0.0222 | 0.0148 / 0.0255 | 0.0150 / 0.0295 | 0.0151 / 0.0343 | 0.0151 / 0.0402 |
| Isotropic material | 0.0229 / 0.0479 | 0.0236 / 0.0584 | 0.0240 / 0.0751 | 0.0242 / 0.1025 | 0.0242 / 0.1503 |

**Table 4.48** The values of $\delta_{cr}$ obtained for co-phase (upper number) and anti-phase (lower number) initial imperfection mode for various $G_{12}/E_1$ and $\ell_{30}/\ell_1$ under $E_2/E_1 = 0.5$, $E_3/E_1 = 0.3$, $\gamma = 1$, $\ell_{10}/\ell_1 = 0.5$

| $G_{12}/E_1$ | $\ell_{30}/\ell_1$ | | | | |
|---|---|---|---|---|---|
| | 0.60 | 0.50 | 0.40 | 0.30 | 0.20 |
| 0.10 | 0.0177 / 0.0363 | 0.0177 / 0.0367 | 0.0178 / 0.0378 | 0.0178 / 0.0410 | 0.0179 / 0.0540 |
| 0.07 | 0.0163 / 0.0303 | 0.0163 / 0.0307 | 0.0164 / 0.0318 | 0.0164 / 0.0350 | 0.0165 / 0.0473 |
| 0.05 | 0.0147 / 0.0250 | 0.0148 / 0.0255 | 0.0148 / 0.0266 | 0.0149 / 0.0297 | 0.0149 / 0.0404 |
| Isotropic material | 0.0234 / 0.0557 | 0.0236 / 0.0584 | 0.0237 / 0.0640 | 0.0238 / 0.0777 | 0.0239 / 0.1219 |

**Table 4.49** The influence of $h_U/\ell_1$ on the values of $\delta_{cr}$ obtained for co-phase (upper number) and anti-phase (lower number) initial imperfection mode for various $G_{12}/E_1$ under $E_2/E_1 = 0.5$, $E_3/E_1 = 0.3$, $\gamma = 1$, $\ell_{30}/\ell_1 = 0.5$ and $\ell_{10}/\ell_1 = 0.5$

| $h_U/\ell_1$ | $G_{12}/E_1$ | | |
|---|---|---|---|
| | 0.40 | 0.30 | 0.20 |
| 0.075 | 0.0177 / 0.0367 | 0.0163 / 0.0307 | 0.0148 / 0.0255 |
| 0.050 | 0.0175 / 0.0271 | 0.0161 / 0.0236 | 0.0145 / 0.0204 |
| 0.0375 | 0.0165 / 0.0165 | 0.0151 / 0.0151 | 0.0136 / 0.0136 |

Now, within the scope of the assumptions and notation used in the previous sub-subsection, we analyze the numerical results related to the case where the plate material consists of two alternating layers whose materials are isotropic, homogeneous and these layers are located in the $Ox_1x_3$ plane. The material of matrix layers is supposed to be linearly viscoelastic with the operators (4.166).

We consider the graphs given in Figs. 4.18 and 4.19, which show the dependencies among the $\delta_{cr.0}$, $\delta_{cr.\infty}$ and $\ell_{10}/\ell_1$ for the anti-phase and co-phase initial imperfection modes, respectively. Note that these graphs are constructed for various $E^{(2)} \big/ E_0^{(1)}$ under $h_U/\ell_1 = 0.075$, $\omega = 1$, $\ell_{30}/\ell_1 = 0.5$ and $\gamma = 1$. It follows from the results that, as in Case 1, the values of $\delta_{cr.0}$ and $\delta_{cr.\infty}$ decrease with crack length along the $Ox_1$ axis.

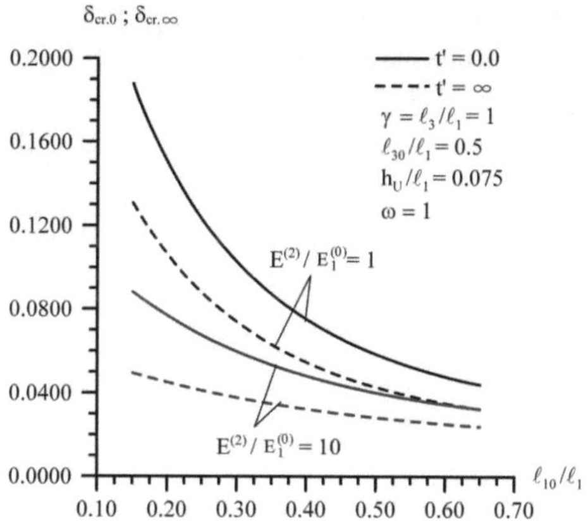


**Fig. 4.18** The graphs of the dependence among $\delta_{cr.0}$, $\delta_{cr.\infty}$ and $\ell_{10}/\ell_1$ for various values of $E^{(2)}/E_0^{(1)}$ in the anti-phase buckling delamination mode for Case 2 under $h_U/\ell_1 = 0.075$, $\omega = 1$, $\ell_{30}/\ell_1 = 0.5$ and $\gamma = 1$

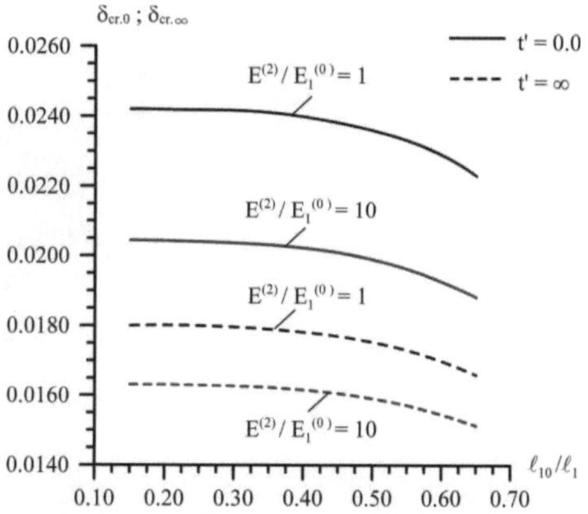


**Fig. 4.19** The graphs of the dependence among $\delta_{cr.0}$, $\delta_{cr.\infty}$ and $\ell_{30}/\ell_1$ for various values of $E^{(2)}/E_0^{(1)}$ in the co-phase buckling delamination mode for Case 2 under $h_U/\ell_1 = 0.075$, $\omega = 1$, $\ell_{10}/\ell_1 = 0.5$ and $\gamma = 1$

Now we handle the question of how the initial imperfection mode can be evaluated with external compressive force $p$ or with duration of a time (for a viscoelastic plate). Note that in the case where $h_A = h_U$ this evaluation does not change the type of the mode, i.e. the buckling delamination mode coincides with the initial imperfection mode. This result occurs for both Case 1 and Case 2. As an example, the buckling delamination mode of the crack's upper surface, i.e. the distribution of the displacement $u_2E_1/(\ell_1 p)$ (where $u_2 = u_2^{(1)}\Big|_{x_2=h_A+0}$), is illustrated in Fig. 4.20 (in this figure the axes $x_1$ and $x_3$ are denoted by $x$ and $z$ respectively).

Figures 4.21 and 4.22 show the influence of the ratio $\ell_{30}/\ell_1$ on the values of $\delta_{cr.0}$ and $\delta_{cr.\infty}$ in the anti-phase and co-phase initial imperfection modes, respectively, under $\omega = 1$ and $\ell_{10}/\ell_1 = 0.5$. As it can be predicted, it follows from the graphs that the values of the critical strains increase monotonously with a decrease in the depth of the crack along the $Ox_3$ axis, i.e. with a decrease in the ratio$\ell_{30}/\ell_1$.

Consider the numerical results attained for the critical time, i.e. for the $t'_{cr}$. As in Case 1, under obtaining these results we assume that the values of the strain $\delta$

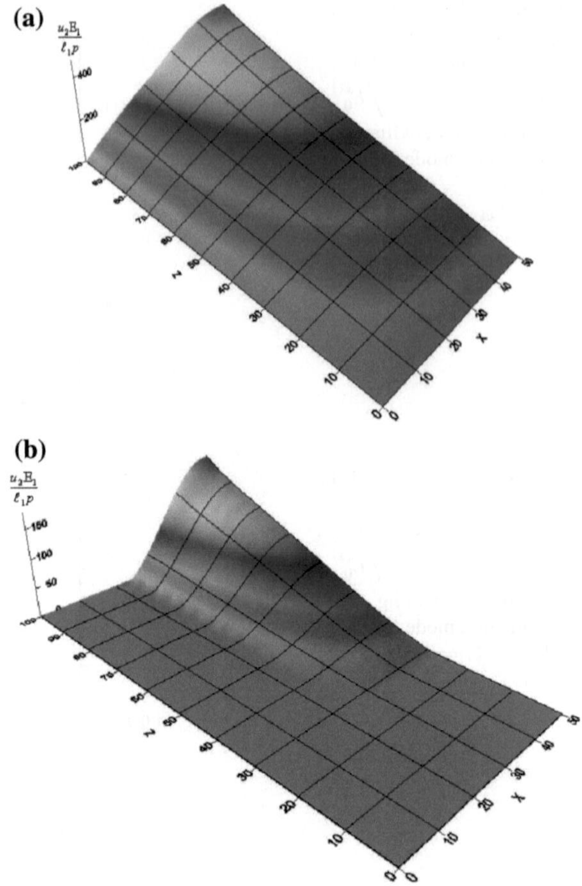


**Fig. 4.20** The buckling delamination mode; **a** co-phase mode, **b** anti-phase mode

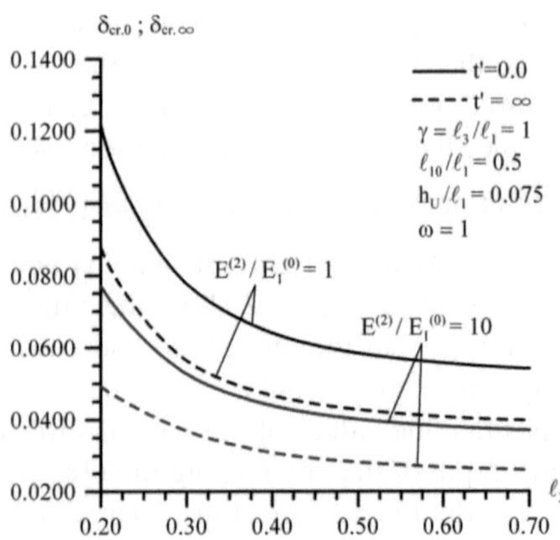


**Fig. 4.21** The graphs of the dependence among $\delta_{cr.0}$, $\delta_{cr.\infty}$ and $\ell_{30}/\ell_1$ for various values of $E^{(2)}\Big/E_0^{(1)}$ in the anti-phase buckling delamination mode for Case 2 under $h_U/\ell_1 = 0.075$, $\omega = 1$, $\ell_{10}\ /\ell_1 = 0.5$ and $\gamma = 1$

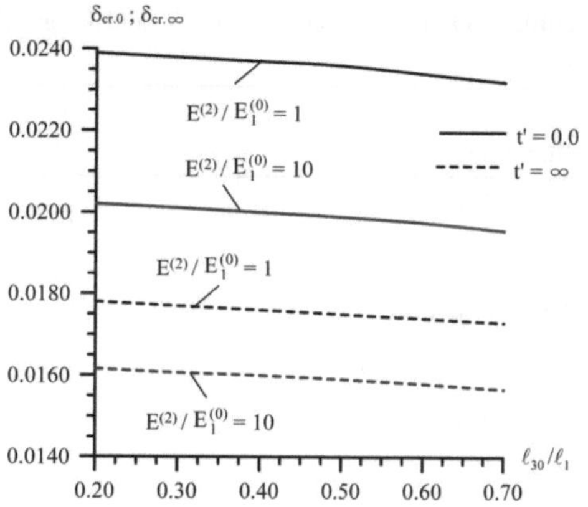


**Fig. 4.22** The graphs of the dependence among $\delta_{cr.0}$, $\delta_{cr.\infty}$ and $\ell_{30}/\ell_1$ for various values of $E^{(2)}\big/E_0^{(1)}$ in the co-phase buckling delamination mode for Case 2 under $h_U/\ell_1 = 0.075,\ \omega = 1,\ \ell_{10}\ /\ell_1 = 0.5$ and $\gamma = 1$

caused by the external compressive force satisfy the inequality-relation $\delta_{cr.\infty} < \delta < \delta_{cr.0}$.

Table 4.50 shows the values of $t'_{cr}$ obtained for various $\ell_{10}/\ell_1$ and $\delta$ under $E^{(2)}\big/E_0^{(1)} = 10,\ \gamma = 1,\ \omega = 1,\ \alpha = -0.5$ and $\ell_{30}/\ell_1 = 0.5$. In the quantitative sense these results agree with the corresponding results obtained in Case 1. The results given in Table 4.51 illustrate the influence of the rheological parameter $\omega$ and $\alpha$ on the values of $t'_{cr}$. It follows from these results that the values of $t'_{cr}$ increase (decrease) with $\omega$ (with absolute values of $\alpha$).

Note that the results discussed above in the qualitative and quantitative sense agree with the results obtained in the previous subsections.

#### 4.10.4.3 Numerical Results Obtained in Case 3 (Embedded Crack)

We consider only the case where the plate material consists of two alternating layers with the properties indicated in the previous sub-subsections. Using the notation and assumptions accepted in these sub-subsections analyze the numerical results related to the values of $\delta_{cr.0}$ and $\delta_{cr.\infty}$. These results obtained for various values of $\ell_{10}/\ell_1$ under $h_A = h_U = h/2,\ \ell_{30}/\ell_1 = 0.5,\ \omega = 1,\ E^{(2)}\big/E_0^{(1)} = 1$ and 10 are given in Table 4.52. The comparison these results with the corresponding ones given in Figs. 4.18 and 4.19 shows that, as it can be expected, the values of $\delta_{cr.0}$ and $\delta_{cr.\infty}$ are greater than corresponding ones obtained for the end crack (Case 2). Moreover, these results show that, as in Cases 1 and 2, the values of $\delta_{cr.0}$ and $\delta_{cr.\infty}$ decrease with $\ell_{10}/\ell_1$. The influence of the crack's length along the $Ox_3$ axis, i.e. the influence of the values of $\ell_{30}/\ell_1$ on the $\delta_{cr.0}$ and $\delta_{cr.\infty}$ is illustrated by data given in Table 4.53. This table shows that the character of the influence of the ratio $\ell_{30}/\ell_1$ on the values of $\delta_{cr.0}$ and $\delta_{cr.\infty}$ is similar to the influence of the ratio $\ell_{10}/\ell_1$ on those.

**Table 4.50** The effects of $\ell_{10}/\ell_1$ and $\delta$ on $t'_{cr}$ for $E^{(2)}/E_0^{(1)} = 10$, $\omega = 1$, $\alpha = 0.5$ and $\ell_{30}/\ell_1 = 0.5$

| $\ell_{10}/\ell_1$ (init. imper. mode) | $\delta$ | $t'_{cr}$ |
|---|---|---|
| 0.6 (Anti-phase) | 0.02909 | 0.4518 |
| | 0.03091 | 0.0916 |
| | 0.03273 | 0.0131 |
| 0.5 (Anti-phase) | 0.03273 | 0.5841 |
| | 0.03636 | 0.0447 |
| | 0.03818 | 0.0078 |
| 0.4 (Anti-phase) | 0.03818 | 0.5752 |
| | 0.04000 | 0.2215 |
| | 0.04182 | 0.0864 |
| 0.6 (Co-phase) | 0.01727 | 0.2977 |
| | 0.01782 | 0.0915 |
| | 0.01818 | 0.0381 |
| 0.5 (Co-phase) | 0.01764 | 0.4281 |
| | 0.01800 | 0.2033 |
| | 0.01891 | 0.0259 |
| 0.4 (Co-phase) | 0.01764 | 0.7770 |
| | 0.01800 | 0.3658 |
| | 0.02000 | 0.0008 |

**Table 4.51** The effects of $\omega$ and $\alpha$ on $t'_{cr}$ in the case where $E^{(2)}/E_0^{(1)} = 10$, $\gamma = 1$, $\ell_{10}/\ell_1 = 0.5$ and $\ell_{30}/\ell_1 = 0.5$

| $\omega$ | $\alpha$ | $t'_{cr.}$ $\delta = 0.03636$ (Anti-phase) | $\delta = 0.01891$ (Co-phase) |
|---|---|---|---|
| 1 | −0.5 | 0.0447 | 0.0259 |
| 2 | | 0.0910 | 0.0436 |
| 3 | | 0.2771 | 0.0878 |
| 1 | −0.3 | 0.0891 | 0.0605 |
| | −0.5 | 0.0447 | 0.0259 |
| | −0.7 | 0.0089 | 0.0036 |

Numerical analyses show that the co-phase initial imperfection mode causes the co-phase buckling delamination mode only. At the same time, if the initial imperfection mode of the crack edges is anti-phase then a decrease in the values of $\ell_{30}/\ell_1$, i.e. a decrease in the crack's length in the $Ox_3$ axis direction can change the character of the buckling delamination mode. Thus, there exists such a value of the ratio $k = \ell_{30}/\ell_{10}$ for which if $\ell_{30}/\ell_{10} \leq k$ then the anti-phase initial imperfection mode causes a more complicated buckling delamination mode than the anti-phase buckling delamination mode. It is established that for the values of the problem

**Table 4.52** The values of $\delta_{cr.0}$ (upper number) and $\delta_{cr.\infty}$ (lower number) obtained for various $\ell_{10}/\ell_1$ and $E^{(2)}\big/E_0^{(1)}$ under $h_A = h_U = h/2,\ \ell_{30}/\ell_1 = 0.5,\ \omega = 1$ and $h/\ell_1 = 0.15$

| $\frac{E^{(2)}}{E_0^{(1)}}$ | Initial imperfection mode | $\ell_{10}/\ell_1$ | | | | | |
|---|---|---|---|---|---|---|---|
| | | 0.15 | 0.20 | 0.30 | 0.40 | 0.50 | 0.60 |
| 1 | Anti-phase | 0.2021<br>0.1411 | 0.1699<br>0.1205 | 0.1342<br>0.0971 | 0.1203<br>0.0882 | 0.1174<br>0.0865 | 0.1167<br>0.0858 |
| | Co-phase | 0.0640<br>0.0484 | 0.0639<br>0.0484 | 0.0636<br>0.0481 | 0.0630<br>0.0477 | 0.0618<br>0.0467 | 0.0597<br>0.0450 |
| 10 | Anti-phase | 0.5196<br>0.2906 | 0.4767<br>0.2784 | 0.4280<br>0.2677 | 0.4135<br>0.2669 | 0.4123<br>0.2617 | 0.4015<br>0.2537 |
| | Co-phase | 0.2801<br>0.2141 | 0.2796<br>0.2138 | 0.2783<br>0.2129 | 0.2756<br>0.2097 | 0.2689<br>0.1987 | 0.2571<br>0.1880 |

**Table 4.53** The values of $\delta_{cr.0}$ (upper number) and $\delta_{cr.\infty}$ (lower number) obtained for various $\ell_{30}/\ell_1$ and $E^{(2)}\big/E_0^{(1)}$ under $h_A = h_U = h/2,\ \ell_{10}/\ell_1 = 0.5,\ \omega = 1$ and $h/\ell_1 = 0.15$

| $\frac{E^{(2)}}{E_0^{(1)}}$ | Initial imperfection mode | $\ell_{30}/\ell_1$ | | | | | | |
|---|---|---|---|---|---|---|---|---|
| | | 0.20 | 0.30 | 0.40 | 0.50 | 0.60 | 0.70 | 0.80 |
| 1 | Anti-phase | 0.2480<br>0.1747 | 0.1852<br>0.1323 | 0.1519<br>0.1099 | 0.1174<br>0.0865 | 0.0934<br>0.0692 | 0.0796<br>0.0590 | 0.0713<br>0.0529 |
| | Co-phase | 0.0639<br>0.0484 | 0.0636<br>0.0481 | 0.0629<br>0.0476 | 0.0618<br>0.0467 | 0.0603<br>0.0455 | 0.0583<br>0.0441 | 0.0561<br>0.0424 |
| 10 | Anti-phase | 0.6600<br>0.4103 | 0.5368<br>0.3114 | 0.4715<br>0.2817 | 0.4123<br>0.2617 | 0.3451<br>0.2356 | 0.2986<br>0.2077 | 0.2682<br>0.1874 |
| | Co-phase | 0.2798<br>0.2139 | 0.2779<br>0.2121 | 0.2745<br>0.2064 | 0.2689<br>0.1987 | 0.2614<br>0.1921 | 0.2529<br>0.1867 | 0.2339<br>0.1818 |

parameters which are used for obtaining the results given in Tables 4.50 and 4.51 the value of $k$ is equal to 0.8.

As an example, we consider buckling delamination mode, i.e. consider the distribution of the dimensionless vertical displacement $u_2 E_0^{(1)}\big/(\ell_1 p)$ of the upper surface of the crack edge for the values of $\delta$ for which $|\delta_{cr.0} - \delta| \sim 10^{-4}$. For the co-phase buckling delamination mode the surface-graph of the mentioned distribution under $\ell_{10}/\ell_1 = 0.6$ and $\ell_{30}/\ell_1 = 0.5$ is given in Fig. 4.23a. However, the surface-graphs of the mentioned distribution for the buckling delamination modes caused by the anti-phase initial imperfection mode in the cases where $\{\ell_{10}/\ell_1 = 0.2,\ \ell_{30}/\ell_1 = 0.5\} \Rightarrow \ell_{30}/\ell_{10} = 2.5$ and $\{\ell_{10}/\ell_1 = 0.7,\ \ell_{30}/\ell_1 = 0.4\}$ $\Rightarrow \ell_{30}/\ell_{10} \approx 0.5714$ are illustrated in Fig. 4.23b and c respectively. Note that all numerical results given in Fig. 4.23 are calculated under $h_A = h_U = h/2$, $E^{(2)}\big/E_0^{(1)} = 10$. Moreover, note that the case shown in Fig. 4.23b is an anti-phase buckling delamination mode causes by the anti-phase initial imperfection mode, but the case shown in Fig. 4.23c is a more complicate buckling delamination mode than the anti-phase buckling delamination one and is also caused by the anti-phase initial imperfection mode.

Table 4.54 shows the values of $\delta_{cr.0}$ (upper numbers) and of $\delta_{cr.\infty}$ (lower numbers) obtained for various values of $h_U/\ell_1$ under $\ell_{10}/\ell_1 = 0.5,\ \ell_{30}/\ell_1 = 0.5$ and $\omega = 1$.

It follows from Table 4.54 that, as in the Cases 1 and 2, a decrease in the values of $h_U/\ell_1$ causes to decrease in the values of critical forces.

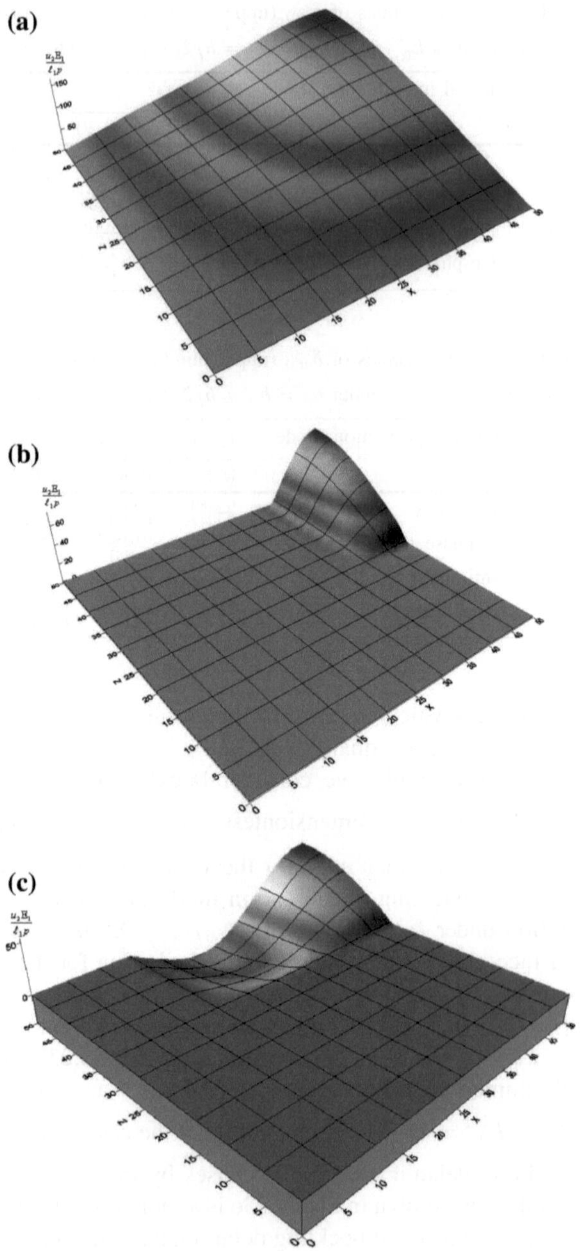


**Fig. 4.23** Buckling delamination modes caused: by the co-phase (under $\ell_{10}/\ell_1 = 0.6$, $\ell_{30}/\ell_1 = 0.5$) (**a**); by the anti-phase (under $\{\ell_{10}/\ell_1 = 0.2, \ell_{30}/\ell_1 = 0.5\} \Rightarrow \ell_{30}/\ell_{10} = 2.5.\ell_{10} = 2.5$) (**b**); also by the anti-phase (under $\{\ell_{10}/\ell_1 = 0.7, \ell_{30} / \ell_1 = 0.4\} \Rightarrow \ell_{30} /\ell_{10} \approx 0.5714$) (**c**) imperfection modes

Now we consider the numerical results related to the case where the plate material is viscoelestic one with effective operators noted in the previous sub-subsections. Assume that $\delta_{cr.\infty} < \delta < \delta_{cr.0}$ and analyze the values of the critical time $t'_{cr}$ given in Table 4.55 which are obtained in the case where $\omega = 1$,

**Table 4.54** The influence of the $h_U/\ell_1$ on the values of the $\delta_{cr.0}$ (upper number) and $\delta_{cr.\infty}$(lower number) under $\ell_{10}/\ell_1 = 0.5$, $\ell_{30}/\ell_1 = 0.5$ and $\omega = 1$

| $\frac{E^{(2)}}{E_0^{(1)}}$ | Initial imperfection mode | $h_U/\ell_1$ | | | | |
|---|---|---|---|---|---|---|
| | | 0.0750 | 0.0625 | 0.0500 | 0.0375 | 0.0250 |
| 1 | Anti-phase | $\frac{0.1174}{0.0865}$ | $\frac{0.0613}{0.0463}$ | $\frac{0.0579}{0.0436}$ | $\frac{0.0436}{0.0329}$ | $\frac{0.0256}{0.0195}$ |
| | Co-phase | $\frac{0.0618}{0.0467}$ | $\frac{0.0613}{0.0463}$ | $\frac{0.0579}{0.0436}$ | $\frac{0.0436}{0.0329}$ | $\frac{0.0256}{0.0195}$ |
| 10 | Anti-phase | $\frac{0.4123}{0.2617}$ | $\frac{0.2649}{0.1939}$ | $\frac{0.2404}{0.1760}$ | $\frac{0.1808}{0.1398}$ | $\frac{0.1058}{0.0874}$ |
| | Co-phase | $\frac{0.2689}{0.1987}$ | $\frac{0.2649}{0.1939}$ | $\frac{0.2404}{0.1760}$ | $\frac{0.1808}{0.1398}$ | $\frac{0.1058}{0.0874}$ |

$h_U = h_A = h/2$, $\ell_{30}/\ell_1 = 0.5$ and $\alpha = -0.5$. This analysis shows that the main parameter on which $t'_{cr}$ depends significantly, is the parameter $\delta$ and the values of $t'_{cr}$ increase (decrease) as $\delta \to \delta_{cr.\infty}$ (as $\delta \to \delta_{cr.0}$).

The influence of the rheological parameters $\omega$ and $\alpha$ on the values of $t'_{cr}$ is given in Tables 4.56 (under $\alpha = -0.5$) and 4.57 (under $\omega = 1$) for $h_U = h_A = h/2$, $\ell_{10}/\ell_1 = 0.5$, $\ell_{30}/\ell_1 = 0.5$ and $E^{(2)}\big/E_0^{(1)} = 10$. It follows from these tables that in the qualitative sense the character of the influence of the rheological parameters $\omega$ and $\alpha$ on the values of $t'_{cr}$ is the same as in Cases 1 and 2.

#### 4.10.4.4 Main Conclusions Based on the Numerical Results

Thus, in the present sub-section numerical results related to the buckling delamination of the rectangular plate with the rectangular band crack (Case 1 shown in Fig. 4.15a), end crack (Case 2 shown in Fig. 4.15b) and embedded crack (Case 3 shown in Fig. 4.15c) under uni-axial compression of the plate along the $Ox_1$ axis. It is assumed that the plate material is homogeneous, orthotropic and viscoelastic one. According to these analyzes, the following main conclusions can be drawn:

- The values of the critical strains (or forces) as well as the values of the critical time decrease with length of the crack along the $Ox_1$ and $Ox_3$ axes (Fig. 4.15);
- The approaching of the crack location to the free surface of the rectangular plate causes the decrease in the values of the critical parameters;
- The results obtained in the Cases 2 and 3 approach the corresponding results obtained in Case 1 with the length of the edge and embedded cracks in the direction of the $Ox_3$ axis;
- With the length of the rectangular band crack along the $Ox_3$ axis, the results approach to those obtained for the corresponding plane-strain state problem studied in Sect. 4.2;
- The buckling delamination mode around the rectangular band crack depends only on the initial imperfection mode, however buckling delamination modes

**Table 4.55** The values of $t'_{cr}$ obtained for various $\delta$, $\ell_{10}/\ell_1$ under $E^{(2)}/E_0^{(1)} = 10$, $\ell_{30}/\ell_1 = 0.5$, $\omega = 1$ and $\alpha = -0.5$

| Initial imperfection mode | $\ell_{10}/\ell_1$ | $\delta$ | $t'_{cr}$ |
|---|---|---|---|
| Anti-phase | 0.6 | 0.30 | 0.9485 |
| | | 0.35 | 0.0562 |
| | | 0.38 | 0.0057 |
| | 0.5 | 0.35 | 0.1063 |
| | | 0.38 | 0.0158 |
| | | 0.40 | 0.0016 |
| | 0.4 | 0.35 | 0.1194 |
| | | 0.38 | 0.0176 |
| | | 0.40 | 0.0020 |
| | 0.3 | 0.35 | 0.1634 |
| | | 0.38 | 0.0330 |
| | | 0.40 | 0.0080 |
| Co-phase | 0.6 | 0.23 | 0.1149 |
| | | 0.24 | 0.0295 |
| | | 0.25 | 0.0035 |
| | 0.5 | 0.23 | 0.4720 |
| | | 0.24 | 0.1453 |
| | | 0.25 | 0.0391 |
| | 0.4 | 0.25 | 0.1026 |
| | | 0.26 | 0.0240 |
| | | 0.27 | 0.0021 |
| | 0.3 | 0.25 | 0.1426 |
| | | 0.26 | 0.0366 |
| | | 0.27 | 0.0050 |

**Table 4.56** The influence of the rheological parameter $\omega$ on the values of $t'_{cr}$ under $\alpha = -0.5$, $h_U = h_A = h/2$, $\ell_{10}/\ell_1 = 0.5$, $\ell_{30}/\ell_1 = 0.5$ and $E^{(2)}/E_0^{(1)} = 10$

| $\omega$ | $t'_{cr}$ | |
|---|---|---|
| | $\delta = 0.38$ (Anti-phase) | $\delta = 0.25$ (Co-phase) |
| 1 | 0.0158 | 0.0391 |
| 2 | 0.0235 | 0.0755 |
| 3 | 0.0384 | 0.2020 |

around the rectangular edge and rectangular embedded cracks depend not only on the initial imperfection modes but also on the ratio of the crack length along the $Ox_1$ and $Ox_3$ axes, i.e. on the $\ell_{30}/\ell_{10}$;

- The numerical results obtained and analyzed in the present subsection can be taken as standard for estimation of the accuracy (in the qualitative and quantitative senses) of the corresponding numerical results obtained within the scope of the approximate plate and bar theories.

**Table 4.57** The influence of the rheological parameter $\alpha$ on the values of $t'_{cr}$ under $\omega = 1$, $h_U = h_A = h/2$, $\ell_{10}/\ell_1 = 0.5$, $\ell_{30}/\ell_1 = 0.5$ and $E^{(2)}/E_0^{(1)} = 10$

| $\alpha$ | $t'_{cr.}$ | |
|---|---|---|
| | $\delta = 0.38$ (Anti-phase) | $\delta = 0.25$ (Co-phase) |
| −0.3 | 0.0425 | 0.0810 |
| −0.5 | 0.0158 | 0.0391 |
| −0.7 | 0.0015 | 0.0071 |

## 4.11 Buckling Delamination of a Sandwich Rectangular Plate with Interface Rectangular Cracks

### *4.11.1 Formulation of the Problem*

Consider a thick sandwich rectangular plate which contains two interface rectangular cracks. The Cartesian coordinate system $Ox_1x_2x_3$ is associated with the plate so as to give Lagrange coordinates of the points of the plate in the natural state. Assume that the plate occupies the region $V = V^{(r_1)} \cup V^{(r_2)} \cup V^{(r_3)}$ where

$$V^{(r_1)} = \{0<x_1<\ell_1;\ 0<x_2<h_F; 0<x_3<\ell_3\},$$
$$V^{(r_2)} = \{0<x_1<\ell_1;\ h_F<x_2<h_F+h_C;\ 0<x_3<\ell_3\},$$
$$V^{(r_3)} = \{0<x_1<\ell_1; h_F+h_C<x_2<h;\ 0<x_3<\ell_3\}.$$

We will consider three cases with respect to the location of the cracks in the plate. In Case 1 we assume that the plate contains two interface rectangular band-cracks (Fig. 4.24a), i.e. upper and lower interface band-cracks at

$$\begin{aligned} \Omega_1 &= \{(\ell_1-\ell_{10})/2<x_1<(\ell_1+\ell_{10})/2;\ x_2=h_F;\ 0<x_3<\ell_3\} \text{ and} \\ \Omega_2 &= \{(\ell_1-\ell_{10})/2<x_1<(\ell_1+\ell_{10})/2; x_2=h_F+h_C;\ 0<x_3<\ell_3\}. \end{aligned} \tag{4.167}$$

In Case 2 the plate contains two rectangular interface edge-cracks (Fig. 4.24b), i.e. upper and lower interface edge-cracks are in the regions

$$\begin{aligned} \Omega_1 &= \{(\ell_1-\ell_{10})/2<x_1<(\ell_1+\ell_{10})/2;\ x_2=h_F;\ 0<x_3<\ell_{30}\} \text{ and} \\ \Omega_2 &= \{(\ell_1-\ell_{10})/2<x_1<(\ell_1+\ell_{10})/2; x_2=h_F+h_C;\ 0<x_3<\ell_{30}\}. \end{aligned} \tag{4.168}$$

Finally, in Case 3 the plate contains two rectangular interface embedded cracks (Fig. 2.24c) in regions

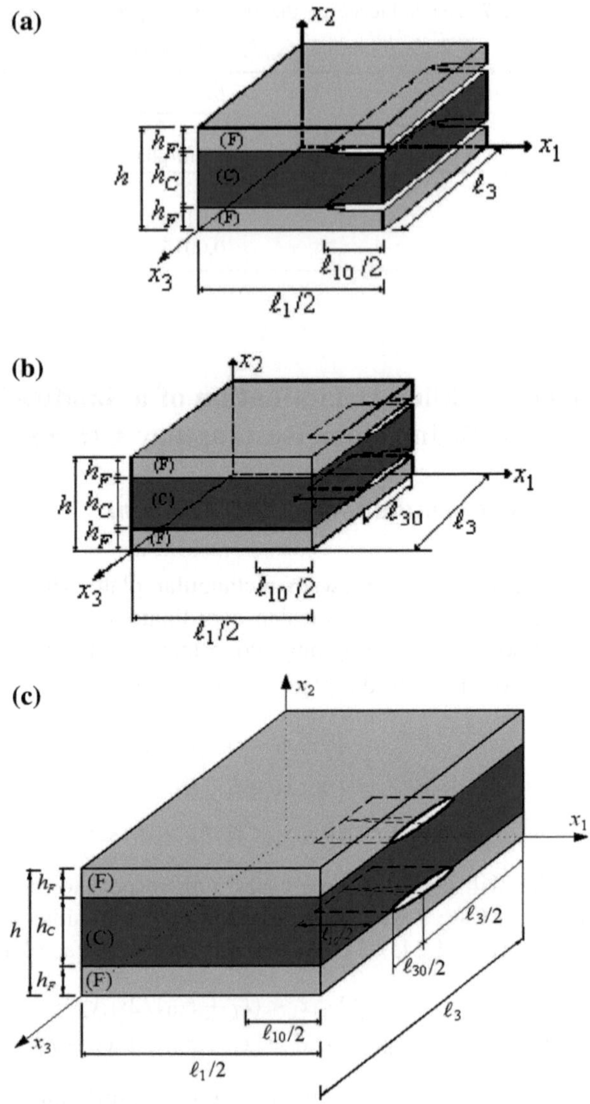


**Fig. 4.24** The geometry of the rectangular sandwich plate and rectangular interface cracks

$$\begin{aligned}\Omega_1 &= \{(\ell_1-\ell_{10})/2 \le x_1 \le (\ell_1+\ell_{10})/2;\\ &\quad x_2 = h_F;\ \ (\ell_3-\ell_{30})/2 \le x_3 \le (\ell_3+\ell_{30})/2\},\\ \Omega_2 &= \{(\ell_1-\ell_{10})/2 \le x_1 \le (\ell_1+\ell_{10})/2;\\ &\quad x_2 = h_F+h_C;\ \ (\ell_3-\ell_{30})/2 \le x_3 \le (\ell_3+\ell_{30})/2\}.\end{aligned} \tag{4.169}$$

In Eqs. (4.167–4.169) $\ell_{10}$ ($\ell_{30}$) is the length of the cracks along the $Ox_1$ ($Ox_3$) axis. In Fig. 4.24 the face layers are indicated by the letter $F$, but the core layer is indicated by the letter $C$. The meaning of the other notation used in Fig. 4.24 and in Eqs. (4.167–4.169) is obvious. Note that in Fig. 4.24 for clarity of the illustration half (i.e. for $0 \le x_1 \le \ell_1$) of the plate is shown and the symmetry of the plate structure with respect to the plane $x_1 = \ell_1/2$ is taken into account.

As in the previous sections, we suppose that edge surfaces of the cracks have an initial infinitesimal imperfection and this imperfection is symmetric with respect to the $x_1 = \ell_1/2$ plane and with respect to the planes $x_2 = h_F$ (for the lower crack) and $x_2 = h_F + h_C$ (for the upper crack). The equations of the edge-surfaces of the upper and the lower cracks can be written as follows:

Case 1

$$\begin{aligned} x_2^{\pm} &= h_F + \varepsilon f^{\pm}(x_1) \text{ (for the lower crack)} \\ \text{and } x_2^{\pm} &= h_F + h_C +^{\pm} (x_1) \text{ (for the upper crack)} \\ \text{for } &(\ell_1 - \ell_{10})/2 < x_1 < (\ell_1 + \ell_{10})/2 \text{ and } 0 < x_3 < \ell_3 \end{aligned} \tag{4.170}$$

and the function $f(x_1)$ and its first order derivatives are equal to zero at $x_1 = (\ell_1 \pm \ell_{10})/2$.

Case 2

$$\begin{aligned} x_2^{\pm} &= h_F + \varepsilon f^{\pm}(x_1, x_3) \text{ (for the lower crack)} \\ \text{and } x_2^{\pm} &= h_F + h_C + \varepsilon f^{\pm}(x_1, x_3) \text{ (for the upper crack)} \\ \text{for } &(\ell_1 - \ell_{10})/2 < x_1 < (\ell_1 + \ell_{10})/2 \text{ and } \quad 0 \le x_3 \le \ell_{30} \end{aligned} \tag{4.171}$$

and the function $f(x_1, x_3)$ and its first order derivatives with respect to $x_1$ and $x_3$ are equal to zero at $x_1 = (\ell_1 \pm \ell_{10})/2$ and $x_3 = \ell_{30}$ respectively.

Case 3

$$\begin{aligned} x_2^{\pm} &= h_F + \varepsilon f^{\pm}(x_1, x_3) \text{ (for the lower crack)} \\ \text{and } x_2^{\pm} &= h_F + h_C + \varepsilon f^{\pm}(x_1, x_3) \text{ (for the upper crack),} \\ \text{for } &(\ell_1 - \ell_{10})/2 \le x_1 \le (\ell_1 + \ell_{10})/2,\ (\ell_3 - \ell_{30})/2 \le x_3 \le (\ell_3 + \ell_{30})/2, \end{aligned} \tag{4.172}$$

and the function $f(x_1, x_3)$ and its first order derivatives with respect to $x_1$ and $x_3$ are equal to zero at $x_1 = (\ell_1 \pm \ell_{10})/2$ and $x_3 = (\ell_3 \pm \ell_{30})/2$, respectively.

In (4.170–4.172), $\varepsilon$ is the dimensionless small parameter ($\varepsilon < < 1$) which characterizes the degree of the initial imperfection of the crack edge-surfaces, $h_F$ ($h_C$) is the thickness of the face-layers (the core layer) and the upper index "+" ("−") represents the upper (lower) edge surface of the considered crack.

Thus, we investigate the evolution of the foregoing initial infinitesimal imperfections of the crack-edge surfaces in Cases 1 and 2 under compression of the plate along the $Ox_1$ axis with uniformly distributed normal forces with intensity $p$, but in Case 3 under bi-axial compression of the plate along the $Ox_1$ and $Ox_3$ axes with uniformly distributed normal forces with intensity $p_1$ and $p_3$ (for the elastic plate) as time elapses at a fixed value of the external compressing force (for the viscoelastic plate). This evolution will be investigated by utilizing the three-dimensional geometrically nonlinear equations of the theory of viscoelasticity in the framework of the piecewise homogeneous body model.

Below, the values related to the core layer and the face layers will be denoted by upper indeces (1) and (2) respectively. At the same time, we will use upper index $r_k$ $(k = 1, 2, 3)$ where $r_1$ and $r_3$ indicate the values related to the lower and the upper layers, respectively, but $r_2$ indicates the values related to the core layer, so that $r_1 = r_3 = 2$, $r_2 = 1$. It is assumed that the face-layers of the sandwich plate are made from the same materials and the plate is symmetric with respect to the middle plane of the core layer.

The governing field equations of the three-dimensional nonlinear theory of viscoelastic bodies is:

Equilibrium equation

$$\frac{\partial}{\partial x_j}\left[\sigma_{jn}^{(r_k)}\left(\delta_i^n + \frac{\partial u_i^{(r_k)}}{\partial x_n}\right)\right] = 0, \tag{4.173}$$
$$i; j; n; k = 1, 2, 3,\ r_1 = r_3 = 2,\ r_2 = 1,$$

Geometrical relation

$$\varepsilon_{ij}^{(r_k)} = \frac{1}{2}\left(\frac{\partial u_i^{(r_k)}}{\partial x_j} + \frac{\partial u_j^{(r_k)}}{\partial x_i} + \frac{\partial u_n^{(r_k)}}{\partial x_i}\frac{\partial u_n^{(r_k)}}{\partial x_j}\right), \tag{4.174}$$

Constitutive relation

$$\sigma_{ij}^{(r_k)} = \lambda^{*(r_k)}\theta^{(r_k)}\delta_i^j + 2\mu^{*(r_k)}\varepsilon_{ij}^{(r_k)},\ \theta^{(r_k)} = \varepsilon_{11}^{(r_k)} + \varepsilon_{22}^{(r_k)} + \varepsilon_{33}^{(r_k)}, \tag{4.175}$$

where $\lambda^{*(r_k)}$ and $\mu^{*(r_k)}$ are the following operators

$$\left\{\begin{matrix}\lambda^{*(r_k)}\\ \mu^{*(r_k)}\end{matrix}\right\}\varphi(t) = \left\{\begin{matrix}\lambda_0^{(r_k)}\\ \mu_0^{(r_k)}\end{matrix}\right\}\varphi(t) + \int_0^t \left\{\begin{matrix}\lambda^{(r_k)}\\ \mu^{(r_k)}\end{matrix}\right\}(t-\tau)\varphi(\tau)d\tau, \tag{4.176}$$

In Eqs. (4.173–4.176) the conventional notation is used. Consider the formulation of the boundary and contact conditions. The boundary conditions for Case 1 can be written as follows.

Boundary conditions at the ends of the plate:

$$u_2^{(r_k)}\Big|_{x_1=0} = 0,\ u_2^{(r_k)}\Big|_{x_1=\ell_1} = 0\, u_2^{(r_k)}\Big|_{x_3=0} = 0,\ u_2^{(r_k)}\Big|_{x_3=\ell_3} = 0, \tag{4.177}$$

$$\left[\sigma_{1n}^{(r_k)}\left(\delta_i^n + \frac{\partial u_i^{(r_k)}}{\partial x_n}\right)\right]\Bigg|_{x_1=0} = \left[\sigma_{1n}^{(r_k)}\left(\delta_i^n + \frac{\partial u_i^{(r_k)}}{\partial x_n}\right)\right]\Bigg|_{x_1=\ell_1} = p_1\delta_i^1, \tag{4.178}$$

$$\left[\sigma_{3n}^{(r_k)}\left(\delta_i^n + \frac{\partial u_i^{(r_k)}}{\partial x_n}\right)\right]\Bigg|_{x_3=0} = \left[\sigma_{3n}^{(r_k)}\left(\delta_i^n + \frac{\partial u_i^{(r_k)}}{\partial x_n}\right)\right]\Bigg|_{x_3=\ell_3} = p_3\delta_3^i, \tag{4.179}$$

where $i = 1, 3$ in conditions (4.178) and (4.179).

Boundary conditions on the free face planes of the plate:

$$\left[\sigma_{2n}^{(r_1)}\left(\delta_k^n + \frac{\partial u_k^{(r_1)}}{\partial x_n}\right)\right]\Bigg|_{x_2=0} = 0, \quad \left[\sigma_{2n}^{(r_3)}\left(\delta_k^n + \frac{\partial u_k^{(r_3)}}{\partial x_n}\right)\right]\Bigg|_{x_2=h} = 0, \tag{4.180}$$

where $k = 1, 2, 3$.

Boundary conditions on the cracks' edge surfaces

$$\begin{gathered}
\left[\sigma_{jn}^{(r_1)}\left(\delta_i^n + \frac{\partial u_i^{(r_1)}}{\partial x_n}\right)\right]\Bigg|_{S_1^-} n_j^- = 0, \quad \left[\sigma_{jn}^{(r_2)}\left(\delta_i^n + \frac{\partial u_i^{(r_2)}}{\partial x_n}\right)\right]\Bigg|_{S_1^+} n_j^+ = 0, \\
\left[\sigma_{jn}^{(r_3)}\left(\delta_i^n + \frac{\partial u_i^{(r_3)}}{\partial x_n}\right)\right]\Bigg|_{S_2^+} n_j^+ = 0, \quad \left[\sigma_{jn}^{(r_2)}\left(\delta_i^n + \frac{\partial u_i^{(r_2)}}{\partial x_n}\right)\right]\Bigg|_{S_2^-} n_j^- = 0,
\end{gathered} \tag{4.181}$$

where

$$\begin{aligned}
S_1^{\pm} &= \{((\ell_1 - \ell_{10})/2 < x_1 < (\ell_1 + \ell_{10})/2), x_2^{\pm} = h_F + \varepsilon f^{\pm}(x_1), 0 < x_3 < \ell_3\}, \\
S_2^{\pm} &= \{((\ell_1 - \ell_{10})/2 < x_1 < (\ell_1 + \ell_{10})/2), x_2^{\pm} = h_F + h_C + \varepsilon f^{\pm}(x_1), 0 < x_3 < \ell_3\}.
\end{aligned} \tag{4.182}$$

Contact conditions between the layers of the plate are:

$$\begin{gathered}
u_i^{(r_1)}\Big|_{\wp_1^-} = u_i^{(r_2)}\Big|_{\wp_1^+}, u_i^{(r_3)}\Big|_{\wp_2^+} = u_i^{(r_2)}\Big|_{\wp_2^-}, \\
\left[\sigma_{2n}^{(r_1)}\left(\delta_i^n + \frac{\partial u_i^{(r_1)}}{\partial x_n}\right)\right]\Bigg|_{\wp_1^-} = \left[\sigma_{2n}^{(r_2)}\left(\delta_i^n + \frac{\partial u_i^{(r_2)}}{\partial x_n}\right)\right]\Bigg|_{\wp_1^+}, \\
\left[\sigma_{2n}^{(r_3)}\left(\delta_i^n + \frac{\partial u_i^{(r_3)}}{\partial x_n}\right)\right]\Bigg|_{\wp_2^+} = \left[\sigma_{2n}^{(r_2)}\left(\delta_i^n + \frac{\partial u_i^{(r_2)}}{\partial x_n}\right)\right]\Bigg|_{\wp_2^-},
\end{gathered} \tag{4.183}$$

where $i = 1, 2, 3$ and

$$\begin{aligned}
\wp_1^{\pm} &= \{x_1 \in ((0, (\ell_1 - \ell_{10})/2) \cup ((\ell_1 + \ell_{10})/2, \ell_1)), \\
&\qquad x_2 = h_F \pm 0, x_3 \in (0, \ell_3)\}, \\
\wp_2^{\pm} &= \{x_1 \in ((0, (\ell_1 - \ell_{10})/2) \cup ((\ell_1 + \ell_{10})/2, \ell_1)), \\
&\qquad x_2 = h_F + h_C \pm 0, x_3 \in (0, \ell_3)\}.
\end{aligned} \tag{4.184}$$

The conditions (4.178–4.184) occur also for Case 2, but the conditions in (4.177) is replaced by the following ones

$$\begin{gathered} u_2^{(r_k)}\Big|_{x_1=0}=0,\ u_2^{(r_k)}\Big|_{x_1=\ell_1}=0,\ u_2^{(r_k)}\Big|_{x_3=\ell_3}=0, \\ \left[\sigma_{3n}^{(r_k)}\left(\delta_i^n+\frac{\partial u_i^{(r_k)}}{\partial x_n}\right)\right]\Big|_{x_3=\ell_3}=0,\ i=1,2,3. \end{gathered} \tag{4.185}$$

At the same time, the conditions in (4.181) satisfy at

$$\begin{aligned} S_1^{\pm}=&\{((\ell_1-\ell_{10})/2<x_1<(\ell_1+\ell_{10})/2), \\ &x_2^{\pm}=h_F+\varepsilon f^{\pm}(x_1,x_3),0<x_3<\ell_{30}\}, \\ S_2^{\pm}=&\{((\ell_1-\ell_{10})/2<x_1<(\ell_1+\ell_{10})/2), \\ &x_2^{\pm}=h_F+h_C+\varepsilon f^{\pm}(x_1,x_3),0<x_3<\ell_{30}\}, \end{aligned} \tag{4.186}$$

and the contact conditions in (4.183) satisfy at

$$\begin{aligned} \wp_1^{\pm}=&\{(\ell_1-\ell_{10})/2_1\le(\ell_1+\ell_{10})/2, \\ &x_2=h_F\pm 0,\ell_{30}<x_3<\ell_3\}, \\ \wp_2^{\pm}=&\{(\ell_1-\ell_{10})/2\le x_1\le(\ell_1+\ell_{10})/2, \\ &x_2=h_F+h_C\pm 0,\ell_{30}<x_3<\ell_3\}. \end{aligned} \tag{4.187}$$

For Case 3 the conditions (4.177–4.180) satisfy completely. But the condition (4.181) satisfies at

$$\begin{aligned} S_1^{\pm}=&\{((\ell_1-\ell_{10})/2\le x_1\le(\ell_1+\ell_{10})/2),x_2^{\pm}=h_F+\varepsilon f^{\pm}(x_1,x_3), \\ &((\ell_3-\ell_{30})/2\le x_3\le(\ell_3+\ell_{30})/2)\}, \\ S_2^{\pm}=&\{((\ell_1-\ell_{10})/2\le x_1\le(\ell_1+\ell_{10})/2),x_2^{\pm}=h_F+h_C+\varepsilon f^{\pm}(x_1,x_3), \\ &((\ell_3-\ell_{30})/2\le x_3\le(\ell_3+\ell_{30})/2)\}, \end{aligned} \tag{4.188}$$

and the contact condition (4.183) satisfies at

$$\begin{aligned} \wp_1^{\pm}=&\{0\le x_1\le\ell_1,\ 0\le x_2\le h_F\pm 0,\ 0\le x_3\le\ell_3\} \\ &-\{(\ell_1-\ell_{10})/2\le x_1(\ell_1+\ell_{10})/2,\ x_2=h_F\pm 0, \\ &(\ell_3-\ell_{30})/2\le x_3\le(\ell_3+\ell_{30})/2\}, \\ \wp_2^{\pm}=&\{0\le x_1\le\ell_1,\ 0\le x_2\le h_F+h_C\pm 0,\ 0\le x_3\le\ell_3\} \\ &-\{(\ell_1-\ell_{10})/2\le x_1(\ell_1+\ell_{10})/2,\ x_2=h_F+h_C\pm 0, \\ &(\ell_3-\ell_{30})/2\le x_3\le(\ell_3+\ell_{30})/2\}. \end{aligned} \tag{4.189}$$

Having thus completed the formulation of the considered problem, now we consider the method of solution.

### 4.11.2 Solution Method

To simplify the analyses, we consider only the solution procedure for Case 3, from which, the solution procedures for Cases 1 and 2 can be obtained as a particular case. As in the previous section, first, using the equations of crack edge surfaces given in (4.172), we derive expressions for $n_j^{\pm}$ in (4.181) in the form (4.149). We derive also the series representation of these expressions as in (4.150). Further, using the series presentation

$$\left\{\sigma_{ij}^{(r_k)};\varepsilon_{ij}^{(r_k)};u_i^{(r_k)}\right\}=\sum_{q=0}^{\infty}\varepsilon^q\left\{\sigma_{ij}^{(r_k),q};\varepsilon_{ij}^{(r_k),q};u_i^{(r_k),q}\right\} \tag{4.190}$$

and doing some mathematical manipulations which are similar to those made in the previous section we obtain the corresponding equations and conditions for the zeroth and the first approximations. In this case, for determination of the values related to the zeroth approximation we use the principle of correspondence by using the Laplace transformation (4.28). So, replacing $\sigma_{ij}^{(r_k),0}$, $\varepsilon_{ij}^{(r_k),0}$, $u_i^{(r_k),0}$, $\lambda^{*(r_k)}$ and $\mu^{*(r_k)}$ in the corresponding equations and relations by $\bar{\sigma}_{ij}^{(r_k),0}$, $\bar{\varepsilon}_{ij}^{(r_k),0}$, $\bar{u}_i^{(r_k),0}$, $\bar{\lambda}^{*(r_k)}$ and $\bar{\mu}^{*(r_k)}$ respectively, we obtain the field equations, boundary and contact conditions for the Laplace transformation of the values of the zeroth approximation. It is evident that, according to the non-homogeneity of the plate material under loading by the bi-axial uniformly distributed normal forces with intensity $p_1$ and $p_3$ in the directions of the $Ox_1$ and $Ox_3$ axes respectively, an inhomogeneous distribution of the stresses and strains appears in the layers in the near vicinity of the ends of the plate. Therefore, these inhomogeneous distributions do not influence on the local buckling delamination of the plate parts around the rectangular cracks and we do not take that into account under determination of the Laplace transformation of the values of the zeroth approximation. Thus, using the relations:

$$\begin{gathered}
2h_F\bar{\sigma}_{11}^{(2),0}(s)+h_C\bar{\sigma}_{11}^{(1),0}(s)=\tfrac{1}{s}p_1h,\ 2h_F\bar{\sigma}_{33}^{(2),0}(s)+h_C\bar{\sigma}_{33}^{(1),0}(s)=\tfrac{1}{s}p_3h,\\
\bar{\varepsilon}_{11}^{(1),0}(s)=\bar{\varepsilon}_{11}^{(2),0}(s),\ \bar{\varepsilon}_{33}^{(1),0}(s)=\bar{\varepsilon}_{33}^{(2),0}(s)\ \Rightarrow\\
\tfrac{1}{\bar{E}^{*(1)}}\bar{\sigma}_{11}^{(1),0}(s)-\tfrac{\bar{\nu}^{*(1)}}{\bar{E}^{*(1)}}\bar{\sigma}_{33}^{(1),0}(s)=\tfrac{1}{\bar{E}^{*(2)}}\bar{\sigma}_{11}^{(2),0}(s)-\tfrac{\bar{\nu}^{*(2)}}{\bar{E}^{*(2)}}\bar{\sigma}_{33}^{(2),0}(s),\\
-\tfrac{\bar{\nu}^{*(1)}}{\bar{E}^{*(1)}}\bar{\sigma}_{11}^{(1),0}(s)+\tfrac{1}{\bar{E}^{*(1)}}\bar{\sigma}_{33}^{(1),0}(s)=-\tfrac{\bar{\nu}^{*(2)}}{\bar{E}^{*(2)}}\bar{\sigma}_{11}^{(2),0}(s)+\tfrac{1}{\bar{E}^{*(2)}}\bar{\sigma}_{33}^{(2),0}(s)
\end{gathered} \tag{4.191}$$

we can write the following expressions for the Laplace transformations of the zeroth approximation:

$$\bar{\sigma}_{11}^{(2),0} = \left(\frac{1}{se_1}(p_1 - \bar{\nu}^{*(1)}p_3) + \frac{e_2}{s(e_1)^2}(p_3 - \bar{\nu}^{*(1)}p_1)\right)\left(1 - \frac{(e_2)^2}{(e_1)^2}\right)^{-1},$$
$$\bar{\sigma}_{11}^{(1),0} = \frac{1}{s}(p_1 - \bar{\sigma}_{11}^{(2),0}2h_F)(h_C)^{-1},\ \bar{\sigma}_{33}^{(2),0} = \frac{1}{s}(p_3 - \bar{\nu}^{(1)}p_1 + e_2\bar{\sigma}_{11}^{(1),0})(e_1)^{-1},$$
$$\bar{\sigma}_{33}^{(1),0} = \frac{1}{s}(p_1 - \bar{\sigma}_{33}^{(2),0}2h_F)(h_C)^{-1},\ \sigma_{ij}^{(r_k),0} = 0 \text{ for } ij \neq 11;33, \tag{4.192}$$

where

$$e_1 = 2h_F + h_c\frac{\bar{E}^{*(1)}}{\bar{E}^{*(2)}},\ e_2 = 2h_F\bar{\nu}^{*(1)} + \bar{\nu}^{*(2)}h_c\frac{\bar{E}^{*(1)}}{\bar{E}^{*(2)}}. \tag{4.193}$$

Selecting a suitable expression for the core function $E^{(r_k)}(t)$ and $\nu^{(r_k)}(t)$ of the integral operators

$$\left\{\begin{matrix} E^{*(r_k)} \\ \nu^{*(r_k)} \end{matrix}\right\}\varphi(t) = \left\{\begin{matrix} E_0^{(r_k)} \\ \nu_0^{(r_k)} \end{matrix}\right\}\varphi(t) + \int_0^t \left\{\begin{matrix} E^{(r_k)} \\ \nu^{(r_k)} \end{matrix}\right\}(t-\tau)\varphi(\tau)d\tau,$$

and employing the inverse Laplace transform we can determine the stresses

$$\sigma_{11}^{(r_k),0} = \sigma_{11}^{(r_k),0}(t),\ \sigma_{33}^{(r_k),0} = \sigma_{33}^{(r_k),0}(t),\ \sigma_{ij}^{(r_k),0} = 0 \text{ for } ij \neq 11;33, \tag{4.194}$$

related to the zeroth approximation.

Now we consider the determination of the first approximation. According to the foregoing assumptions (4.192) and (4.194), we obtain the following equilibrium equations, mechanical and geometrical relations for the first approximation:

$$\frac{\partial\sigma_{ji}^{(r_k),1}}{\partial x_j} + \sigma_{11}^{(r_k),0}(t)\frac{\partial^2 u_i^{(r_k),1}}{\partial x_1^2} + \sigma_{33}^{(r_k),0}(t)\frac{\partial^2 u_i^{(r_k),1}}{\partial x_3^2} = 0,$$
$$\sigma_{ij}^{(r_k),1} = \lambda^{*(r_k)}\theta^{(r_k),1}\delta_i^j + 2\mu^{*(r_k)}\varepsilon_{ij}^{(r_k),1}, \tag{4.195}$$
$$\varepsilon_{ij}^{(r_k),1} = \frac{1}{2}\left(\frac{\partial u_i^{(r_k),1}}{\partial x_j} + \frac{\partial u_j^{(r_k),1}}{\partial x_i}\right).$$

Consider the boundary and contact conditions obtained for the first approximation.

Boundary conditions at the ends of the plate are

$$\begin{gathered}
u_2^{(r_k),1}\Big|_{x_1=0;\ell_1} = 0,\; u_2^{(r_k),1}\Big|_{x_3=0;\ell_3} = 0, \\
\left[\sigma_{11}^{(r_k),1} + \sigma_{11}^{(r_k),0}(t)\frac{\partial u_1^{(r_k),1}}{\partial x_1}\right]\Bigg|_{x_1=0;\ell_1} = 0,\; \sigma_{13}^{(r_k),1}\Big|_{x_1=0;\ell_1} = 0, \\
\sigma_{31}^{(r_k),1}\Big|_{x_3=0;\ell_3} = 0,\; \left[\sigma_{33}^{(r_k),1} + \sigma_{33}^{(r_k),0}(t)\frac{\partial u_3^{(r_k),1}}{\partial x_3}\right]\Bigg|_{x_3=0;\ell_3} = 0.
\end{gathered} \tag{4.196}$$

Boundary conditions on the free face planes of the plate:

$$\begin{gathered}
\sigma_{21}^{(r_1),1}\Big|_{x_2=0} = \sigma_{22}^{(r_1),1}\Big|_{x_2=0} = \sigma_{23}^{(r_1),1}\Big|_{x_2=0} = 0, \\
\sigma_{21}^{(r_3),1}\Big|_{x_2=h} = \sigma_{22}^{(r_3),1}\Big|_{x_2=h} = \sigma_{23}^{(r_3),1}\Big|_{x_2=h} = 0.
\end{gathered} \tag{4.197}$$

Boundary conditions on the cracks' edge surfaces:

$$\begin{gathered}
\sigma_{21}^{(r_1),1}\Big|_{S_1^-} = \sigma_{11}^{(r_1),0}(t)\frac{\partial f^-}{\partial x_1},\; \sigma_{22}^{(r_1),1}\Big|_{S_1^-} = 0,\; \sigma_{23}^{(r_1),1}\Big|_{S_1^-} = \sigma_{33}^{(r_1),0}(t)\frac{\partial f^-}{\partial x_3}, \\
\sigma_{21}^{(r_2),1}\Big|_{S_1^+} = \sigma_{11}^{(r_2),0}(t)\frac{\partial f^+}{\partial x_1},\; \sigma_{22}^{(r_2),1}\Big|_{S_1^+} = 0,\; \sigma_{23}^{(r_2),1}\Big|_{S_1^+} = \sigma_{33}^{(r_2),0}(t)\frac{\partial f^+}{\partial x_3}, \\
\sigma_{21}^{(r_2),1}\Big|_{S_2^-} = \sigma_{11}^{(r_2),0}(t)\frac{\partial f^-}{\partial x_1},\; \sigma_{22}^{(r_2),1}\Big|_{S_2^-} = 0,\; \sigma_{23}^{(r_2),1}\Big|_{S_2^-} = \sigma_{33}^{(r_2),0}(t)\frac{\partial f^-}{\partial x_3}, \\
\sigma_{21}^{(r_3),1}\Big|_{S_2^+} = \sigma_{11}^{(r_3),0}(t)\frac{+}{\partial x_1},\; \sigma_{22}^{(r_3),1}\Big|_{S_2^+} = 0,\; \sigma_{23}^{(r_3),1}\Big|_{S_2^+} = \sigma_{33}^{(r_3),0}(t)\frac{\partial f^+}{\partial x_3}, \\
S_1^{\pm} = \{((\ell_1 - \ell_{10})/2 < x_1 < (\ell_1 + \ell_{10})/2), \\
x_2^{\pm} = h_F \pm 0,\; ((\ell_3 - \ell_{30})/2 \leq x_3 \leq (\ell_3 + \ell_{30})/2)\}, \\
S_2^{\pm} = \{((\ell_1 - \ell_{10})/2 < x_1 < (\ell_1 + \ell_{10})/2), \\
x_2^{\pm} = h_F + h_C \pm 0,\; ((\ell_3 - \ell_{30})/2_3 \leq (\ell_3 + \ell_{30})/2)\}.
\end{gathered} \tag{4.198}$$

Contact conditions between the layers of the plate:

$$\begin{gathered}
u_i^{(r_1),1}\Big|_{\wp_1^-} = u_i^{(r_2),1}\Big|_{\wp_1^+},\; u_i^{(r_3),1}\Big|_{\wp_2^+} = u_i^{(r_2),1}\Big|_{\wp_2^-}, \\
\sigma_{21}^{(r_1),1}\Big|_{\wp_1^-} = \sigma_{21}^{(r_2),1}\Big|_{\wp_1^+},\; \sigma_{22}^{(r_1),1}\Big|_{\wp_1^-} = \sigma_{22}^{(r_2),1}\Big|_{\wp_1^+}, \\
\sigma_{23}^{(r_1),1}\Big|_{\wp_1^-} = \sigma_{23}^{(r_2),1}\Big|_{\wp_1^+},\; \sigma_{21}^{(r_3),1}\Big|_{\wp_2^+} = \sigma_{21}^{(r_2),1}\Big|_{\wp_2^-}, \\
\sigma_{22}^{(r_3),1}\Big|_{\wp_2^+} = \sigma_{22}^{(r_2),1}\Big|_{\wp_2^-},\; \sigma_{23}^{(r_3),1}\Big|_{\wp_2^+} = \sigma_{23}^{(r_2),1}\Big|_{\wp_2^-}.
\end{gathered} \tag{4.199}$$

This completes the formulation of the boundary value problem corresponding to the first approximation.

Now we consider the determination of the values of the first approximation for which it is necessary to solve the problem (4.195–4.199). For this purpose we attempt to use the principle of correspondence by using the Laplace transformation (4.28). As has been noted in the previous sections, under this procedure the following difficulty arises. In the Eq. (4.195) and in the conditions (4.196), $\sigma_{11}^{(r_k),0}(t)$ and $\sigma_{33}^{(r_k),0}(t)$ (as noted above) depend on the time and therefore the Laplace transformation of the terms $\sigma_{11}^{(r_k),0}(t)\partial^2 u_i^{(r_k),1}\Big/\partial x_1^2$ and $\sigma_{33}^{(r_k),0}(t)\partial^2 u_i^{(r_k),1}\Big/\partial x_3^2$ in Eq. (4.195) and the Laplace transformation of the terms $\sigma_{11}^{(r_k),0}(t)\partial u_i^{(r_k),1}\Big/\partial x_1$ and $\sigma_{33}^{(r_k),0}(t)\partial u_i^{(r_k),1}\Big/\partial x_3$ in the conditions (4.196) and (4.198) cannot be written as $\bar{\sigma}_{11}^{(r_k),0}\partial^2\bar{u}_i^{(r_k),1}\Big/\partial x_1^2$, $\bar{\sigma}_{33}^{(r_k),0}\partial^2\bar{u}_i^{(r_k),1}\Big/\partial x_3^2$, $\bar{\sigma}_{11}^{(r_k),0}\partial\bar{u}_i^{(r_k),1}\Big/\partial x_1$ and $\bar{\sigma}_{33}^{(r_k),0}\partial\bar{u}_i^{(r_k),1}\Big/\partial x_3$, respectively. To overcome this difficulty we assume that $\sigma_{11}^{(r_k),0}(t)$ and $\sigma_{33}^{(r_k),0}(t)$ change vary slowly with a time and take the values of $\sigma_{11}^{(r_k),0}(t)$ and $\sigma_{33}^{(r_k),0}(t)$ at some fixed moment $t = t_1$ and consequently, instead of the Laplace transformation of the terms $\bar{\sigma}_{11}^{(r_k),0}\partial^2\bar{u}_i^{(r_k),1}\Big/\partial x_1^2$, $\bar{\sigma}_{33}^{(r_k),0}\partial^2\bar{u}_i^{(r_k),1}\Big/\partial x_3^2$, $\bar{\sigma}_{11}^{(r_k),0}\partial\bar{u}_i^{(r_k),1}\Big/\partial x_1$ and $\bar{\sigma}_{33}^{(r_k),0}\partial\bar{u}_i^{(r_k),1}\Big/\partial x_3$ we write $\sigma_{11}^{(r_k),0}\partial^2\bar{u}_i^{(r_k),1}\Big/\partial x_1^2$, $\sigma_{33}^{(r_k),0}\partial^2\bar{u}_i^{(r_k),1}\Big/\partial x_3^2$, $\sigma_{11}^{(r_k),0}\partial\bar{u}_i^{(r_k),1}\Big/\partial x_1$ and $\sigma_{33}^{(r_k),0}\partial\bar{u}_i^{(r_k),1}\Big/\partial x_3$, respectively. This assumption, also used in Sects. 4.4 and 4.7, allows us to obtain accurate results if the variation of $\sigma_{11}^{(r_k),0}$ with respect to time is insignificant. Thus, taking the foregoing discussions into account and replacing $\sigma_{ij}^{(k),1}$, $\varepsilon_{ij}^{(k),1}$, $u_i^{(k),1}$, $\sigma_{11}^{(r_k),0}(t)$, $\sigma_{33}^{(r_k),0}(t)$, $\lambda^{*(k)}$ and $\mu^{*(k)}$ in (4.195–4.199) with $\bar{\sigma}_{ij}^{(k),1}$, $\bar{\varepsilon}_{ij}^{(k),1}$, $\bar{u}_i^{(k),1}$, $\sigma_{11}^{(r_k),0}(t_1)$, $\sigma_{33}^{(r_k),0}(t_1)$, $\bar{\lambda}^{*(k)}$ and $\bar{\mu}^{*(k)}$ respectively, we obtain the corresponding equations and boundary conditions with respect to the Laplace transformation of the values of the first approximation. For the solution to the problems corresponding to the Laplace transformation of the sought values, we employ the three-dimensional Finite Element Method (3D FEM).

### 4.11.3 FEM Modeling

We construct the following functional for the problems under consideration:

$$\begin{aligned}
\Pi\left(\bar{u}_1^{(r_k),1},\bar{u}_2^{(r_k),1},\bar{u}_3^{(r_k),1}\right) = \sum_{k=1}^{3}\Bigg[\frac{1}{2}\int_{V'^{(r_k)}}\Bigg[\Bigg(\bar{\sigma}_{11}^{(r_k),1} + \sigma_{11}^{(r_k),0}(t_1)\frac{\partial\bar{u}_1^{(r_k),1}}{\partial x_1} \\
+\sigma_{33}^{(r_k),0}(t_1)\frac{\partial\bar{u}_1^{(r_k),1}}{\partial x_3}\Bigg)\frac{\partial\bar{u}_1^{(r_k),1}}{\partial x_1} + \bar{\sigma}_{12}^{(r_k),1}\frac{\partial\bar{u}_1^{(r_k),1}}{\partial x_2} + \bar{\sigma}_{13}^{(r_k),1}\frac{\partial\bar{u}_1^{(r_k),1}}{\partial x_3} \\
+\left(\bar{\sigma}_{21}^{(r_k),1} + \sigma_{11}^{(r_k),0}(t_1)\frac{\partial\bar{u}_2^{(r_k),1}}{\partial x_1} + \sigma_{33}^{(r_k),0}(t_1)\frac{\partial\bar{u}_2^{(r_k),1}}{\partial x_3}\right)\frac{\partial\bar{u}_2^{(r_k),1}}{\partial x_1} \\
+\bar{\sigma}_{22}^{(r_k),1}\frac{\partial\bar{u}_2^{(r_k),1}}{\partial x_2} + \bar{\sigma}_{23}^{(r_k),1}\frac{\partial\bar{u}_2^{(r_k),1}}{\partial x_3} \\
+\left(\bar{\sigma}_{31}^{(r_k),1} + \sigma_{11}^{(r_k),0}(t_1)\frac{\partial\bar{u}_3^{(r_k),1}}{\partial x_1} + \sigma_{33}^{(r_k),0}(t_1)\frac{\partial\bar{u}_3^{(r_k),1}}{\partial x_3}\right)\frac{\partial\bar{u}_3^{(r_k),1}}{\partial x_1} + \\
\bar{\sigma}_{32}^{(r_k),1}\frac{\partial\bar{u}_3^{(r_k),1}}{\partial x_2} + \bar{\sigma}_{33}^{(r_k),1}\frac{\partial\bar{u}_3^{(r_k),1}}{\partial x_3}\Bigg]dx_1dx_2dx_3\Bigg] \\
+\int_{(\ell_3-\ell_{30})/2}^{(\ell_3+\ell_{30})/2}\int_{(\ell_1-\ell_{10})/2}^{(\ell_1+\ell_{10})/2}\frac{1}{s}\,\sigma_{11}^{(r_1),0}(t_1)\frac{\partial f^-}{\partial x_1}\,\bar{u}_1^{(r_1),1}\Big|_{x_2=h_F-0}dx_1dx_3 \\
+\int_{(\ell_3-\ell_{30})/2}^{(\ell_3+\ell_{30})/2}\int_{(\ell_1-\ell_{10})/2}^{(\ell_1+\ell_{10})/2}\frac{1}{s}\,\sigma_{33}^{(r_1),0}(t_1)\frac{\partial f^-}{\partial x_3}\,\bar{u}_3^{(r_1),1}\Big|_{x_2=h_F-0}dx_1dx_3 \\
+\int_{(\ell_3-\ell_{30})/2}^{(\ell_3+\ell_{30})/2}\int_{(\ell_1-\ell_{10})/2}^{(\ell_1+\ell_{10})/2}\frac{1}{s}\sigma_{11}^{(r_2),0}(t_1)\frac{\partial f^+}{\partial x_1}\,\bar{u}_1^{(r_2),1}\Big|_{x_2=h_F+0}dx_1dx_3 \\
+\int_{(\ell_3-\ell_{30})/2}^{(\ell_3+\ell_{30})/2}\int_{(\ell_1-\ell_{10})/2}^{(\ell_1+\ell_{10})/2}\frac{1}{s}\sigma_{33}^{(r_2),0}(t_1)\frac{\partial f^+}{\partial x_3}\,\bar{u}_3^{(r_2),1}\Big|_{x_2=h_F+0}dx_1dx_3 \\
+\int_{(\ell_3-\ell_{30})/2}^{(\ell_3+\ell_{30})/2}\int_{(\ell_1-\ell_{10})/2}^{(\ell_1+\ell_{10})/2}\frac{1}{s}\,\sigma_{11}^{(r_2),0}(t_1)\frac{\partial f^-}{\partial x_1}\,\bar{u}_1^{(r_2),1}\Big|_{x_2=(h_F+hc)-0}dx_1dx_3 \\
+\int_{(\ell_3-\ell_{30})/2}^{(\ell_3+\ell_{30})/2}\int_{(\ell_1-\ell_{10})/2}^{(\ell_1+\ell_{10})/2}\frac{1}{s}\,\sigma_{33}^{(r_2),0}(t_1)\frac{\partial f^-}{\partial x_3}\,\bar{u}_3^{(r_2),1}\Big|_{x_2=(h_F+hc)-0}dx_1dx_3 \\
+\int_{(\ell_3-\ell_{30})/2}^{(\ell_3+\ell_{30})/2}\int_{(\ell_1-\ell_{10})/2}^{(\ell_1+\ell_{10})/2}\frac{1}{s}\,\sigma_{11}^{(r_3),0}(t_1)\frac{\partial f^+}{\partial x_1}\,\bar{u}_1^{(r_3),1}\Big|_{x_2=(h_F+hc)+0}dx_1dx_3 \\
+\int_{(\ell_3-\ell_{30})/2}^{(\ell_3+\ell_{30})/2}\int_{(\ell_1-\ell_{10})/2}^{(\ell_1+\ell_{10})/2}\frac{1}{s}\,\sigma_{33}^{(r_3),0}(t_1)\frac{\partial f^+}{\partial x_3}\,\bar{u}_3^{(r_3),1}\Big|_{x_2=(h_F+hc)+0}dx_1dx_3
\end{aligned} \tag{4.200}$$

where $V'^{(r_1)} = V^{(r_1)} - S_1^-$, $V'^{(r_2)} = V^{(r_2)} - S_1^+ - S_2^-$ and $V'^{(r_3)} = V^{(r_3)} - S_2^+$.

By applying the standard procedure, we obtain the equilibrium equation in (4.195) and all the boundary and contact conditions (4.197–4.199) written for stresses from the relation:

$$\delta\Pi = \sum_{k=1}^{3}\left[\frac{\partial\Pi}{\partial\bar{u}_1^{(r_k)}}\delta\bar{u}_1^{(r_k)} + \frac{\partial\Pi}{\partial\bar{u}_2^{(r_k)}}\delta\bar{u}_2^{(r_k)} + \frac{\partial\Pi}{\partial\bar{u}_3^{(r_k)}}\delta\bar{u}_3^{(r_k)}\right] = 0. \tag{4.201}$$

Thus, after establishing the foregoing functional by usual procedure, the FEM technique is applied to obtain the numerical results. In this case, the domain $V'\left(= V'^{(r_1)} \cup V'^{(r_2)} \cup V'^{(r_3)}\right)$ is divided into a finite number of finite elements in the form of rectangular prism (brick) elements with eight nodes. The number of finite elements is determined from the convergence requirement of the numerical results. According to the results obtained in Sect. 4.8, in the present investigation the finite elements containing the crack tips (fronts) are also ordinary brick elements. In this way, we simply establish FEM modeling of the problems under consideration.

Thus, by employing the FEM algorithm detailed above we calculate the values of the Laplace transformation of the sought values. The values of the inverse Laplace transformation of the sought functions are determined by the use of the method (Schapery 1962, 1978) and from the criterion

$$\max_{\substack{x_1 \in (0,\ell_1) \\ x_3 \in (0,\ell_3)}} u_2^{(r_1)} \Big|_{x_2 = h_F - 0} \to \infty \quad \text{as}\, p_1 \to p_{1cr} \,(\text{for fixed}\, p_3) \tag{4.202}$$

## 4.11.4 Numerical Results and Discussions

The material of the face layers is supposed to be linearly viscoelastic with the operators

$$\begin{aligned} E^{*(2)} &= E_0^{(2)}\left[1 - \omega_0 \Pi_\alpha^*(-\omega_0 - \omega_\infty)\right], \\ \nu^{*(2)} &= \nu_0^{(2)}\left[1 + \frac{1 - 2\nu_0^{(2)}}{2\nu_0^{(2)}} \omega_0 \Pi_\alpha^*(-\omega_0 - \omega_\infty)\right], \end{aligned} \tag{4.203}$$

where $E_0^{(2)}$ and $\nu_0^{(2)}$ are the instantaneous values of Young's modulus and of Poisson coefficient, respectively; $\alpha$, $\omega_0$ and $\omega_\infty$ are the rheological parameters of the covering layers' materials, $\Pi_\alpha^*$ is the fractional-exponential operator (Rabotnov 1977) determined by expression (4.133).

The material of the core layer is supposed to be pure elastic with mechanical characteristics $E^{(1)}$ (Young's modulus) and $\nu^{(1)}$ (Poisson coefficient). We introduce the dimensionless rheological parameter $\omega = \omega_\infty/\omega_0$ and the dimensionless time $t' = \omega_0^{1/(1+\alpha)} t$ and we take $\sigma_{11}^{(r_k),0}(0)$ and $\sigma_{33}^{(r_k),0}(0)$ instead of $\sigma_{11}^{(r_k),0}(t)$ and $\sigma_{33}^{(r_k),0}(t)$ respectively. For concrete numerical investigations, the suitable initial imperfection modes of the crack edge surfaces can be selected as follows.

For Case 1:

$$f^{\pm}(x_1) = \pm \ell_{10} \sin^2\left(\frac{\pi}{\ell_{10}}\left(x_1 - \frac{\ell_1 - \ell_{10}}{2}\right)\right), \tag{4.204}$$

For Case 2:

$$f^{\pm}(x_1,x_3) = \pm\ell_{10}\sin^2\left(\frac{\pi}{\ell_{10}}\left(x_1 - \frac{\ell_1-\ell_{10}}{2}\right)\right)\sin^2\left(\frac{\pi}{2\ell_{30}}(x_3-\ell_{30})\right). \quad (4.205)$$

For Case 3:

$$f^{\pm}(x_1,x_3) = \pm\ell_{10}\sin^2\left(\frac{\pi}{\ell_{10}}\left(x_1 - \frac{\ell_1-\ell_{10}}{2}\right)\right)\sin^2\left(\frac{\pi}{\ell_{30}}\left(x_3 - \frac{\ell_3-\ell_{30}}{2}\right)\right). \quad (4.206)$$

Thus, we turn to the analysis of the numerical results obtained in Cases 1, 2 and 3 separately and assume that $h/\ell_1 = 0.15$, $\nu_0^{(1)} = \nu^{(2)} = 0.3$ and introduce the parameter $\gamma = \ell_3/\ell_1$. Under obtaining all numerical results which will be discussed below, we assume that $t_1 = 0$ in (4.200), i.e. the value of $\sigma_{11}^{(r_k),0}(0)$ will be taken instead of $\sigma_{11}^{(r_k),0}(t_1)$ in (4.200)

#### 4.11.4.1 Numerical Results Obtained in Case 1 (Band Cracks)

In this case we assume that $p_3 = 0$ and use the notation $p$ instead of notation $p_1$ and first, consider pure elastic stability loss buckling delamination which takes place at $t' = 0$ and $t' = \infty$. We introduce a parameter $\delta = ph/(2E_0^{(2)}h_F + h_C E^{(1)})$ and the critical values of $\delta$ attained at $t' = 0$ and at $t' = \infty$ we denote through $\delta_{cr.0}\left(= p_{cr.0}h/(2E_0^{(2)}h_F + h_C E^{(1)})\right)$ and $\delta_{cr.\infty}\left(= p_{cr.\infty}h/(2E_0^{(2)}h_F + h_C E^{(1)})\right)$ respectively.

Table 4.58 shows the values of $\delta_{cr.0}$ calculated for various $E_0^{(2)}/E^{(1)}$, $h_F/\ell_1$ and $\gamma$ under $\ell_{10}/\ell_1 = 0.5$. According to the well-known mechanical consideration, the numerical results attained in the case under consideration must approach the certain asymptote with $\gamma$.

This prediction is proven with the data given in Table 4.58. Moreover, the results obtained in the case under consideration must approach the corresponding results attained in Sect. 4.4 with $\gamma$, because in Sect. 4.4 the plane-strain state (i.e. the case $\gamma = \infty$) is considered.

Table 4.59 shows simultaneously the values of $\delta_{cr.0}$ obtained in the present section (upper number) under $\gamma = 8$ and the results obtained in Sect. 4.4 (lower number). The comparison shows that the results attained in the present section agree quite well with the corresponding ones obtained in Sect. 4.4. This statement also validates the reliability of the algorithm and packed programmes used in the present section.

Moreover, it follows from Tables 4.58 and 4.59 that the values of $\delta_{cr.0}$ decrease with $\gamma$, but increase with $E_0^{(2)}/E^{(1)}$ and $h_F/\ell_1$. A more detailed illustration of the influence of the parameters $E_0^{(2)}/E^{(1)}$ and $h_F/\ell_1$ on the values of the $\delta_{cr.0}$ (upper

**Table 4.58** The values of $\delta_{cr.0}$ attained for various $h'_F$, $e$ and $\gamma$ where $h'_F = h_F/\ell_1$ and $e = E_0^{(2)}/E^{(1)}$ (Case 1)

| $\gamma$ | $h'_F$ | | | | | | | | |
|---|---|---|---|---|---|---|---|---|---|
| | 0.0250 | | | 0.0375 | | | 0.0500 | | |
| | $e$ | | | | | | | | |
| | 1 | 5 | 10 | 1 | 5 | 10 | 1 | 5 | 10 |
| 1 | 0.0120 | 0.0263 | 0.0432 | 0.0217 | 0.0601 | 0.1044 | 0.0336 | 0.1131 | 0.2045 |
| 2 | 0.0111 | 0.0241 | 0.0394 | 0.0198 | 0.0542 | 0.0934 | 0.0304 | 0.1009 | 0.1812 |
| 3 | 0.0109 | 0.0238 | 0.0387 | 0.0195 | 0.0532 | 0.0915 | 0.0298 | 0.0989 | 0.1772 |
| 4 | 0.0109 | 0.0236 | 0.0385 | 0.0193 | 0.0528 | 0.0909 | 0.0296 | 0.0981 | 0.1759 |
| 5 | 0.0108 | 0.0235 | 0.0384 | 0.0193 | 0.0526 | 0.0906 | 0.0295 | 0.0978 | 0.1752 |
| 6 | 0.0108 | 0.0235 | 0.0383 | 0.0193 | 0.0526 | 0.0904 | 0.0295 | 0.0977 | 0.1749 |
| 8 | 0.0108 | 0.0235 | 0.0382 | 0.0192 | 0.0525 | 0.0903 | 0.0294 | 0.0975 | 0.1745 |

**Table 4.59** The values of $\delta_{cr.0}$ obtained in the present section (upper number) and in Sect. 4.4 (lower number) attained under $\ell_{10}/\ell_1 = 0.5$, $h/\ell_1 = 0.15$ and $\gamma = 8$ for various values of $h_F/\ell_1$ and $E_0^{(2)}/E^{(1)}$ (Case 1)

| $h_F/\ell_1$ | $E_0^{(2)}/E^{(1)}$ | | | |
|---|---|---|---|---|
| | 1 | 5 | 10 | 20 |
| 0.0250 | $\frac{0.0108}{0.0081}$ | $\frac{0.0235}{0.0269}$ | $\frac{0.0382}{0.0456}$ | $\frac{0.0656}{0.0722}$ |
| 0.0375 | $\frac{0.0192}{0.0165}$ | $\frac{0.0525}{0.0471}$ | $\frac{0.0903}{0.0796}$ | $\frac{0.1581}{0.1304}$ |

**Table 4.60** The values of $\delta_{cr.0}$ (upper number) and $\delta_{cr.\infty}$ (lower number) obtained under $\gamma = 1$, $\ell_{10}/\ell_1 = 0.5$ and $h_F/\ell_1 = 0.0375$ for various values of $E_0^{(2)}/E^{(1)}$ (Case 1)

| $E_0^{(2)}/E^{(1)}$ | 0.3 | 0.5 | 1 | 2 | 5 | 10 | 20 |
|---|---|---|---|---|---|---|---|
| $\frac{\delta_{cr.0}}{\delta_{cr.\infty}}$ | $\frac{0.0145}{0.0101}$ | $\frac{0.0166}{0.0115}$ | $\frac{0.0217}{0.0153}$ | $\frac{0.0317}{0.0224}$ | $\frac{0.0601}{0.0429}$ | $\frac{0.1044}{0.0749}$ | $\frac{0.1845}{0.1342}$ |

number) and $\delta_{cr.\infty}$ (lower number) is given in Tables 4.60 and 4.61 in the case where $\ell_{10}/\ell_1 = 0.5$. Note that under calculation of the values of the $\delta_{cr.\infty}$ it is assumed that $\omega = 2$.

Figure 4.25 shows the graphs of the dependencies among $\delta_{cr.0}$, $\delta_{cr.\infty}$ and $\ell_{10}/\ell_1$. Note that these graphs are constructed for various $E_0^{(2)}/E^{(1)}$ under $\gamma = 1$, $\omega = 2$. It follows from Fig. 4.25 that, as it can be predicted, the values of $\delta_{cr.0}$ and $\delta_{cr.\infty}$ increase monotonically with a decrease in the values of $\ell_{10}/\ell_1$. Figure 4.26 shows schematically the distribution of $u_2^{(2)}E_0^{(2)}/(\ell_1 p)$ with respect to $x\,(=x_1)$ and $z\,(=\ell_3 - x_3)$ for $x_2 = h_F - 0$, $\ell_3/2 \le x_3 \le \ell_3$, $0 \le x_1 \le \ell_1/2$ in the case where $|\delta - \delta_{cr.0}| = 1 \times 10^{-3}$. In other words Fig. 4.26 shows the buckling delamination mode of the crack's lower edge. Under construction of this buckling delamination

**Table 4.61** The values of $\delta_{cr.0}$ (upper number) and $\delta_{cr.\infty}$ (lower number) obtained under $\gamma = 1$ and $\ell_{10}/\ell_1 = 0.5$ for various values of $E_0^{(2)} \big/ E^{(1)}$ and $h_F/\ell_1$ (Case 1)

| $h_F/\ell_1$ | $E_0^{(2)} \big/ E^{(1)}$ | | | |
|---|---|---|---|---|
| | 1 | 2 | 5 | 10 |
| 0.01250 | 0.0055<br>0.0038 | 0.0063<br>0.0043 | 0.0088<br>0.0061 | 0.0128<br>0.0090 |
| 0.01875 | 0.0083<br>0.0058 | 0.0101<br>0.0071 | 0.0158<br>0.0110 | 0.0247<br>0.0174 |
| 0.02500 | 0.0120<br>0.0084 | 0.0157<br>0.0110 | 0.0263<br>0.0186 | 0.0432<br>0.0308 |
| 0.03125 | 0.0166<br>0.0116 | 0.0228<br>0.0161 | 0.0410<br>0.0291 | 0.0694<br>0.0497 |
| 0.03750 | 0.0217<br>0.0153 | 0.0317<br>0.0224 | 0.0601<br>0.0429 | 0.1044<br>0.0749 |
| 0.04375 | 0.0275<br>0.0193 | 0.0421<br>0.0298 | 0.0841<br>0.0599 | 0.1490<br>0.1070 |
| 0.05000 | 0.0336<br>0.0236 | 0.0543<br>0.0384 | 0.1131<br>0.0805 | 0.2045<br>0.1465 |
| 0.05625 | 0.0400<br>0.0281 | 0.0679<br>0.0479 | 0.1477<br>0.1048 | 0.2725<br>0.1946 |

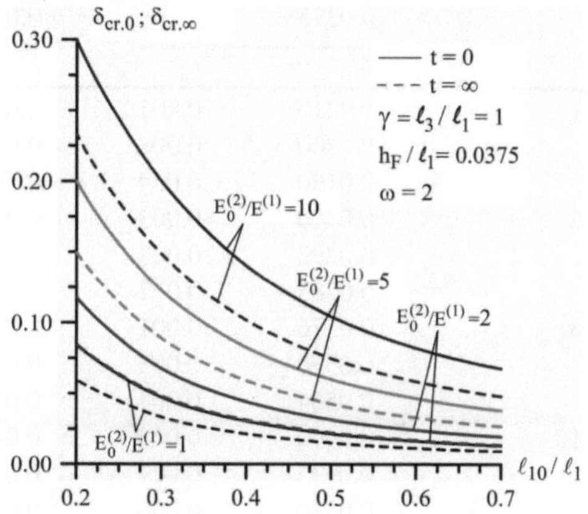


**Fig. 4.25** The graphs of the dependence among $\delta_{cr.0}$, $\delta_{cr.\infty}$ and $\ell_{10}/\ell_1$ constructed for various values of $E_0^{(2)} \big/ E^{(1)}$ (Case 1)

mode the problem symmetry with respect to $x_1 = \ell_1/2$ and $x_3 = \ell_3/2$ is taken into account and it is assumed that $\gamma = 1$, $h_F/\ell_1 = 0.0375$, $\ell_{10}/\ell_1 = 0.5$ and $E_0^{(2)} \big/ E^{(1)} = 10$. It should be noted that, the buckling delamination mode constructed for other values of the problem parameters remain similar to that shown in Fig. 4.26.

Now we assume that the material of the face layers of the sandwich plate is viscoelastic with the rheological parameters $\alpha = -0.5$ and $\omega = 2.0$. Consider the numerical results attained for the critical time $t'_{cr}$. These results are given in Table 4.62 for various $E_0^{(2)} \big/ E^{(1)}$, $h_F/\ell_1$ and $\delta$ for the case where $\gamma = 1.0$ and $\ell_{10}/\ell_1 = 0.5$. Note that the external compressing force, i.e. the values of the parameter $\delta$ must satisfy the relation $\delta_{cr.\infty} < \delta < \delta_{cr.0}$ so that the viscoelastic buckling delamination of the plate considered will occur. Therefore, under obtaining the results illustrated in Table 4.62 the values of $\delta$ are selected according

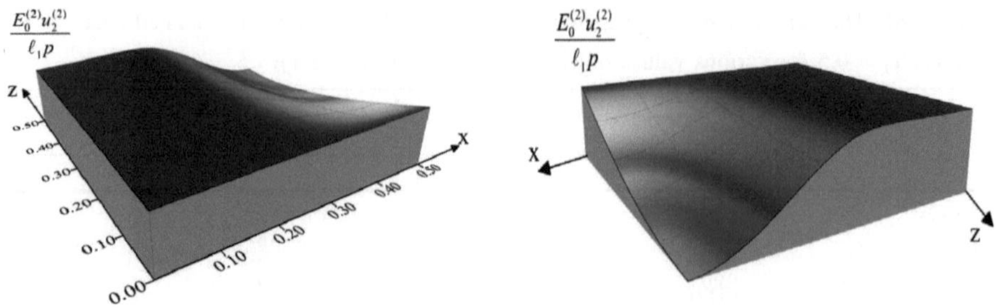


**Fig. 4.26** The buckling delamination mode for the case where $h_F/\ell_1 = 0.0375$, $\ell_{10}/\ell_1 = 0.5$ and $E_0^{(2)} \big/ E^{(1)} = 10$ (Case 1)

**Table 4.62** The values of $t_{cr}$ obtained under $\gamma = 1$, $\ell_{10}/\ell_1 = 0.5$, $\omega = 2$ and $\alpha = -0.5$ for various values of $E_0^{(2)} \big/ E^{(1)}$, $h_F/\ell_1$ and $\delta$(Case 1)

| $E_0^{(2)} \big/ E^{(1)}$ | $h_F/\ell_1$ | | | | | |
|---|---|---|---|---|---|---|
| | 0.0375 | | 0.0250 | | 0.0125 | |
| | $\delta$ | $t_{cr}$ | $\delta$ | $t_{cr}$ | $\delta$ | $t_{cr}$ |
| 1 | 0.0210 | 0.001 | 0.0110 | 0.010 | 0.0052 | 0.003 |
| | 0.0200 | 0.008 | 0.0100 | 0.104 | 0.0050 | 0.010 |
| | 0.0190 | 0.034 | 0.0900 | 1.885 | 0.0048 | 0.029 |
| 2 | 0.0300 | 0.003 | 0.0146 | 0.005 | 0.0058 | 0.007 |
| | 0.0292 | 0.008 | 0.0140 | 0.020 | 0.0052 | 0.093 |
| | 0.0285 | 0.017 | 0.0133 | 0.066 | 0.0046 | 2.261 |
| 5 | 0.0576 | 0.001 | 0.0238 | 0.015 | 0.0083 | 0.003 |
| | 0.0570 | 0.003 | 0.0233 | 0.025 | 0.0079 | 0.011 |
| | 0.0564 | 0.004 | 0.0228 | 0.041 | 0.0076 | 0.033 |
| 10 | 0.0990 | 0.003 | 0.0400 | 0.007 | 0.0120 | 0.004 |
| | 0.0935 | 0.021 | 0.0360 | 0.113 | 0.0115 | 0.016 |
| | 0.0880 | 0.096 | 0.0320 | 5.171 | 0.0110 | 0.048 |

to the relation $\delta_{cr.\infty} < \delta < \delta_{cr.0}$. It follows from Table 4.62 that the values of $t'_{cr}$ decrease with an increase in the values of $\delta$.

Table 4.63 shows the influence of the rheological parameters $\omega$ and $\alpha$ on the values of $t'_{cr}$ for various values of $E_0^{(2)} \big/ E^{(1)}$ in the case where $\delta = 0.036$, $\gamma = 1$, $h_F/\ell_1 = 0.0250$ and $\ell_{10}/\ell_1 = 0.5$. It follows from Table 4.63 that the values of $t'_{cr}$ increase with $\omega$ and with a decrease in the absolute values of the parameter $\alpha$. Note that this result occurs (in the quantitative sense) for the other values of the problem parameters $\delta$, $\gamma$ and $h_F/\ell_1$. Moreover, note that this result agrees with the well-known mechanical consideration, so that the viscoelastic material of the face layers of the plate becomes stiffer with $\omega$, but an increase in the absolute values of the parameter $\alpha$ corresponds to a decrease in the stiffness of the face layer's material at the initial state of loading.

**Table 4.63** The values of $t_{cr}$ obtained under $\gamma = 1$, $\ell_{10}/\ell_1 = 0.5$, $h_F/\ell_1 = 0.0250$ and $\delta = 0.0360$ for various values of $\omega$ and $\alpha$ (Case 1)

| $E_0^{(2)}/E^{(1)}$ | $\delta$ | $\omega$ | $\alpha$ | $t_{cr}$ |
|---|---|---|---|---|
| 2 | 0.0140 | 1 | −0.5 | 0.014 |
| | | 2 | | 0.020 |
| | | 3 | | 0.032 |
| | | 2 | −0.3 | 0.051 |
| | | | −0.5 | 0.020 |
| | | | −0.7 | 0.002 |
| 5 | 0.0233 | 1 | −0.5 | 0.016 |
| | | 2 | | 0.025 |
| | | 3 | | 0.042 |
| | | 2 | −0.3 | 0.059 |
| | | | −0.5 | 0.025 |
| | | | −0.7 | 0.003 |
| 10 | 0.0360 | 1 | −0.5 | 0.052 |
| | | 2 | | 0.113 |
| | | 3 | | 0.414 |
| | | 2 | −0.3 | 0.173 |
| | | | −0.5 | 0.113 |
| | | | −0.7 | 0.042 |

#### 4.11.4.2 Numerical Results Obtained in Case 2 (Edge Cracks)

Within the foregoing assumptions and notation, we analyze the numerical results obtained in Case 2. As in the previous case, first we consider the pure elastic buckling delamination of the rectangular sandwich plate containing two rectangular edge-cracks the edge surfaces of which have an initial imperfection described by the Eq. (4.205). Note that in Case 2 the crack geometry is characterized not only by the initial imperfection mode and the parameters $h_F/\ell_1$ and $\ell_{10}/\ell_1$ but also with the parameter $\ell_{30}/\ell_1$ through which the crack depth along the $Ox_3$ axis is estimated.

Thus we consider the results given in Table 4.64 which shows the values of $\delta_{cr.0}$ (upper number) and $\delta_{cr.\infty}$ (lower number) for various $E_0^{(2)}/E^{(1)}$ under $\gamma = 1$, $\ell_{10}/\ell_1 = 0.5$, $\ell_{30}/\ell_1 = 0.3$ and $h_F/\ell_1 = 0.0375$. Moreover, for the same values of the parameters $\gamma$, $\ell_{10}/\ell_1$ and $\ell_{30}/\ell_1$. Table 4.65 shows the influence of the thickness of the face layer ($h_F/\ell_1$) on the values of $\delta_{cr.0}$ (upper number) and $\delta_{cr.\infty}$ (lower number) for various $E_0^{(2)}/E^{(1)}$.

The comparison of the results given Tables 4.64 and 4.65 with the corresponding results given in Tables 4.60 and 4.61, respectively, show that the values of $\delta_{cr.0}$ and $\delta_{cr.\infty}$ obtained in Case 2 are grater than corresponding ones obtained in Case 1. Moreover, it follows from the results given in Tables 4.64 and 4.65 that in Case 2, as in Case 1, the values of $\delta_{cr.0}$ and $\delta_{cr.\infty}$ increase monotonically with $E_0^{(2)}/E^{(1)}$ and $h_F/\ell_1$.

**Table 4.64** The values of $\delta_{cr.0}$ (upper number) and $\delta_{cr.\infty}$ (lower number) obtained under $\gamma = 1$, $\ell_{10}/\ell_1 = 0.5$, $\ell_{30}/\ell_1 = 0.3$ and $h_F/\ell_1 = 0.0375$ for various values of $E_0^{(2)}/E^{(1)}$ (Case 2)

| $E_0^{(2)}/E^{(1)}$ | 0.3 | 0.5 | 1 | 2 | 5 | 10 | 20 |
|---|---|---|---|---|---|---|---|
| $\delta_{cr.0}/\delta_{cr.\infty}$ | 0.0199 / 0.0134 | 0.0229 / 0.0154 | 0.0300 / 0.0204 | 0.0437 / 0.0299 | 0.0827 / 0.0570 | 0.1428 / 0.0992 | 0.2496 / 0.1768 |

**Table 4.65** The values of $\delta_{cr.0}$ (upper number) and $\delta_{cr.\infty}$ (lower number) obtained under $\gamma = 1$, $\ell_{10}/\ell_1 = 0.5$ and $\ell_{30}/\ell_1 = 0.3$ for various values of $E_0^{(2)}/E^{(1)}$ and $h_F/\ell_1$ (Case 2)

| $h_F/\ell_1$ | $E_0^{(2)}/E^{(1)}$ | | | |
|---|---|---|---|---|
| | 1 | 2 | 5 | 10 |
| 0.01250 | 0.0083 / 0.0055 | 0.0095 / 0.0064 | 0.0133 / 0.0089 | 0.0194 / 0.0131 |
| 0.01875 | 0.0121 / 0.0081 | 0.0148 / 0.0100 | 0.0230 / 0.0156 | 0.0358 / 0.0245 |
| 0.02500 | 0.0171 / 0.0116 | 0.0224 / 0.0152 | 0.0374 / 0.0256 | 0.0611 / 0.0422 |
| 0.03125 | 0.0232 / 0.0157 | 0.0320 / 0.0218 | 0.0572 / 0.0393 | 0.0964 / 0.0668 |
| 0.03750 | 0.0300 / 0.0204 | 0.0437 / 0.0299 | 0.0827 / 0.0570 | 0.1428 / 0.0992 |
| 0.04375 | 0.0375 / 0.0255 | 0.0575 / 0.0393 | 0.1142 / 0.0788 | 0.2015 / 0.1400 |
| 0.05000 | 0.0454 / 0.0309 | 0.0733 / 0.0500 | 0.1523 / 0.1047 | 0.2739 / 0.1899 |
| 0.05625 | 0.0536 / 0.0364 | 0.0910 / 0.0620 | 0.1973 / 0.1352 | 0.3623 / 0.2500 |

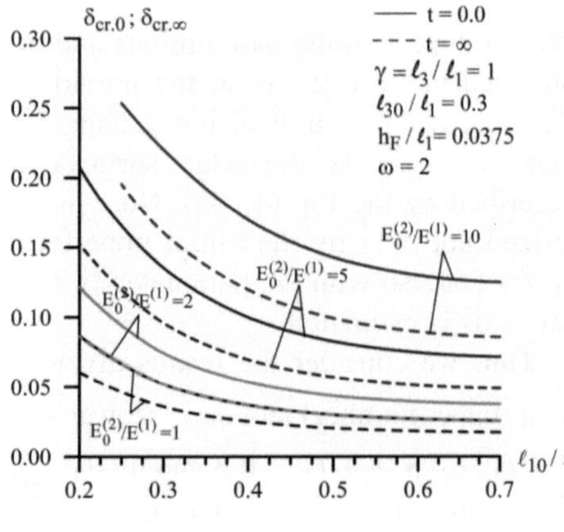


**Fig. 4.27** The graphs of the dependence among $\delta_{cr.0}$, $\delta_{cr.\infty}$ and $\ell_{10}/\ell_1$ constructed for various values of $E_0^{(2)}/E^{(1)}$ (Case 2)

Figure 4.27 shows the graphs of the dependence among $\delta_{cr.0}$, $\delta_{cr.\infty}$ and $\ell_{10}/\ell_1$ for various $E_0^{(2)}/E^{(1)}$ under $\gamma = 1$, $h_F/\ell_1 = 0.0375$, $\ell_{30}/\ell_1 = 0.3$ and $\omega = 2.0$. It follows from these results that the values of the critical strains decrease with the length of the edge crack in the direction of the $Ox_1$ axis. Although this conclusion is illustrated for the selected values of the problem parameters $h_F/\ell_1$ and $\ell_{30}/\ell_1$, but that holds also for the other possible values of these parameters.

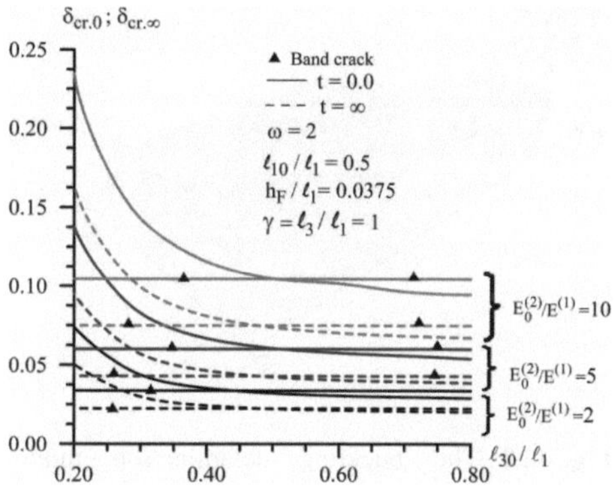


**Fig. 4.28** The graphs of the dependence among $\delta_{cr.0}$, $\delta_{cr.\infty}$ and $\ell_{30}/\ell_1$ constructed for various values of $E_0^{(2)}\big/E^{(1)}$ (Case 2)

Consider the influence of the depth of the edge-crack in the direction of the $Ox_3$ axis on the values of the $\delta_{cr.0}$ and $\delta_{cr.\infty}$. Figure 4.28 illustrates this influence for various values of $E_0^{(2)}\big/E^{(1)}$ in the case where $\ell_{10}/\ell_1 = 0.5$, $h_F/\ell_1 = 0.0375$ and $\gamma = 1$. Note that in these figures the values of $\delta_{cr.0}$ and $\delta_{cr.\infty}$ obtained for the band crack are also indicated. According to the mechanical consideration, there must exist such values of $\ell_{30}/\ell_1$ (denoted by $(\ell_{30}/\ell_1)^*$) before (after) which, i.e. in the cases where $\ell_{30}/\ell_1 < (\ell_{30}/\ell_1)^*$ $(\ell_{30}/\ell_1 > (\ell_{30}/\ell_1)^*)$ the values of $\delta_{cr.0}$ and $\delta_{cr.\infty}$ obtained in Case 2 are greater (less) than the corresponding values of those obtained in Case 1. Because, according to the problem formulation, in Case 1 the plate ends at $x_3 = 0$ and $x_3 = \ell_3$ are simply supported; but in Case 2 the plate end at $x_3 = \ell_3$ is simply supported, but the plate end at $x_3 = 0$ is free.

Now we handle the question of how the buckling delamination mode depends on the crack depth, i.e. on the parameter $\ell_{30}/\ell_1$. Figure 4.29 shows schematically the distribution of $u_2^{(2)}E_0^{(2)}\big/(\ell_1 p)$ with respect to $x$ $(= x_1)$ and $z$ $(= \ell_3 - x_3)$ under $x_2 = h_F - 0$, in other words Fig. 4.29 shows the buckling mode corresponding to the compressing force which is very near the critical force, i.e. for the values of the parameter $\delta$ for which the relation $O(|\delta - \delta_{cr.0}|) = 10^{-4}$ is satisfied. Although the illustrated distributions are constructed for the case where $E_0^{(2)}\big/E^{(1)} = 10$, $h_F/\ell_1 = 0.0375$, $\gamma = 1$ and $\ell_{30}/\ell_1 = 0.3$ (Fig. 4.29), they hold (in a qualitative sense) in the cases where $\ell_{30}/\ell_1 > 0.3$ for the selected values of the problem parameters $E_0^{(2)}\big/E^{(1)}$ and $h_F/\ell_1$. But, in the cases where $\ell_{30}/\ell_{10} < \chi < 0.6$ (or $\ell_{30}/\ell_1 < 0.3$) is satisfied the buckling of the edge crack's surface has a complicated mode. Note that the values of the $\chi$ depend on the problem parameters $h_F/\ell_1$, $\ell_{10}/\ell_1$ and $E_0^{(2)}\big/E^{(1)}$. As an example, we consider the distribution of $u_2^{(2)}E_0^{(2)}\big/(\ell_1 p)$ with respect to $x(= x_1)$ and $z(= \ell_3 - x_3)$ for the same values of the problem parameters under $\ell_{30}/\ell_1 = 0.2$, i.e. under $\ell_{30}/\ell_{10} = 0.5$. Figure 4.30 shows the graphs of this distribution for the cases where $\delta = 0.165$ (Fig. 4.30a), 0.2145 (Fig. 4.30b), 0.2150 (Fig. 4.3c) and 0.2151 (Fig. 4.30d). Consequently,

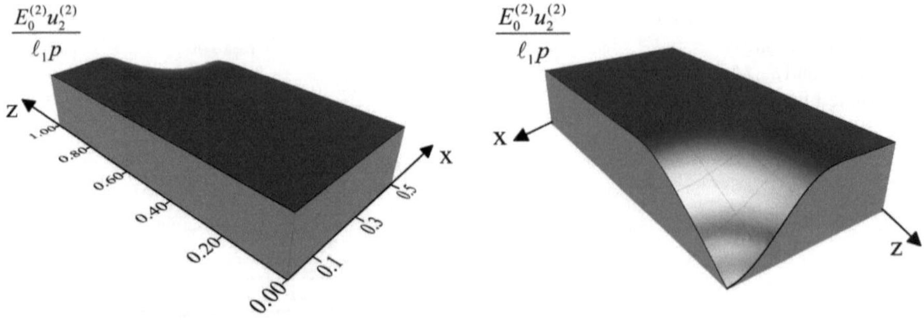


**Fig. 4.29** The buckling delamination mode obtained for the case where $\ell_{30}/\ell_1 = 0.3$ and $E_0^{(2)}/E^{(1)} = 10$ (Case 2)

Fig. 4.30 shows simultaneously the evolution of the initial imperfection mode before the buckling delamination mode. Analyses of the results given in Figs. 4.29 and 4.30, and other results which are not given here show that the buckling mode of the edge rectangular crack depends significantly on the values of the ratio $\ell_{30}/\ell_{10}$.

Thus, there exists such a value of the parameter $\chi$ after which, i.e. for the case where $(\ell_{30}/\ell_{10}) > \chi$ the buckling mode is similar to the initial imperfection mode, but in the cases where $(\ell_{30}/\ell_{10}) < \chi$, the buckling mode has a complicated character, such as shown in Fig. 4.30. As noted above, the values of $\chi$ depend on the values of the problem parameters, especially, on the values of $h_F/\ell_1$. The graphs given in Fig. 4.31 illustrate schematically this dependence in the cases where $E_0^{(2)}/E^{(1)} = 10$ under $\ell_{30}/\ell_1 = 0.2$ and $\ell_{10}/\ell_1 = 0.3$. In Fig. 4.31 the vertical axis shows the values of

$$v = 2u_2^{(2)}\left(\left|\max_{0\le x_1/\ell_1\le 0.5} u_2^{(2)} - \min_{0\le x_1/\ell_1\le 0.5} u_2^{(2)}\right|\right)^{-1} \text{ at } x_2 = h_F - 0,\ x_3 = 0.$$

It follows from the results given in Fig. 4.31 that the appearance of the complicated buckling delamination modes becomes more suitable with decreasing of the parameter $h_F/\ell_1$. Moreover, Fig. 4.32 shows the graphs of the dependence between $\upsilon$ and $x_1/\ell_1$ for various values of $\ell_{30}/\ell_{10}$ under $E_0^{(2)}/E^{(1)} = 10$, $h_F/\ell_1 = 0.0375$, $\ell_{30}/\ell_1 = 0.2$.

It follows from these graphs that, for the considered values of the problem parameters in the case where $\ell_{30}/\ell_{10} = 0.7273$ the buckling delamination mode is similar to the initial imperfection mode. However, for the cases where $(\ell_{30}/\ell_{10}) < \chi = 0.7273$, the buckling delamination mode becomes complicated in the sense shown in Figs. 4.30, 4.31 and 4.32. The results show that the complicated buckling modes which have been observed in the experimental investigations carried out in (Evans and Hutchinson 1995; Gioia and Ortiz 1997; Hutchinson and Suo 1992; Hutchinson et al. 1992; Nilson and Giannakopoulos 1995; Thouless et al. 1994;

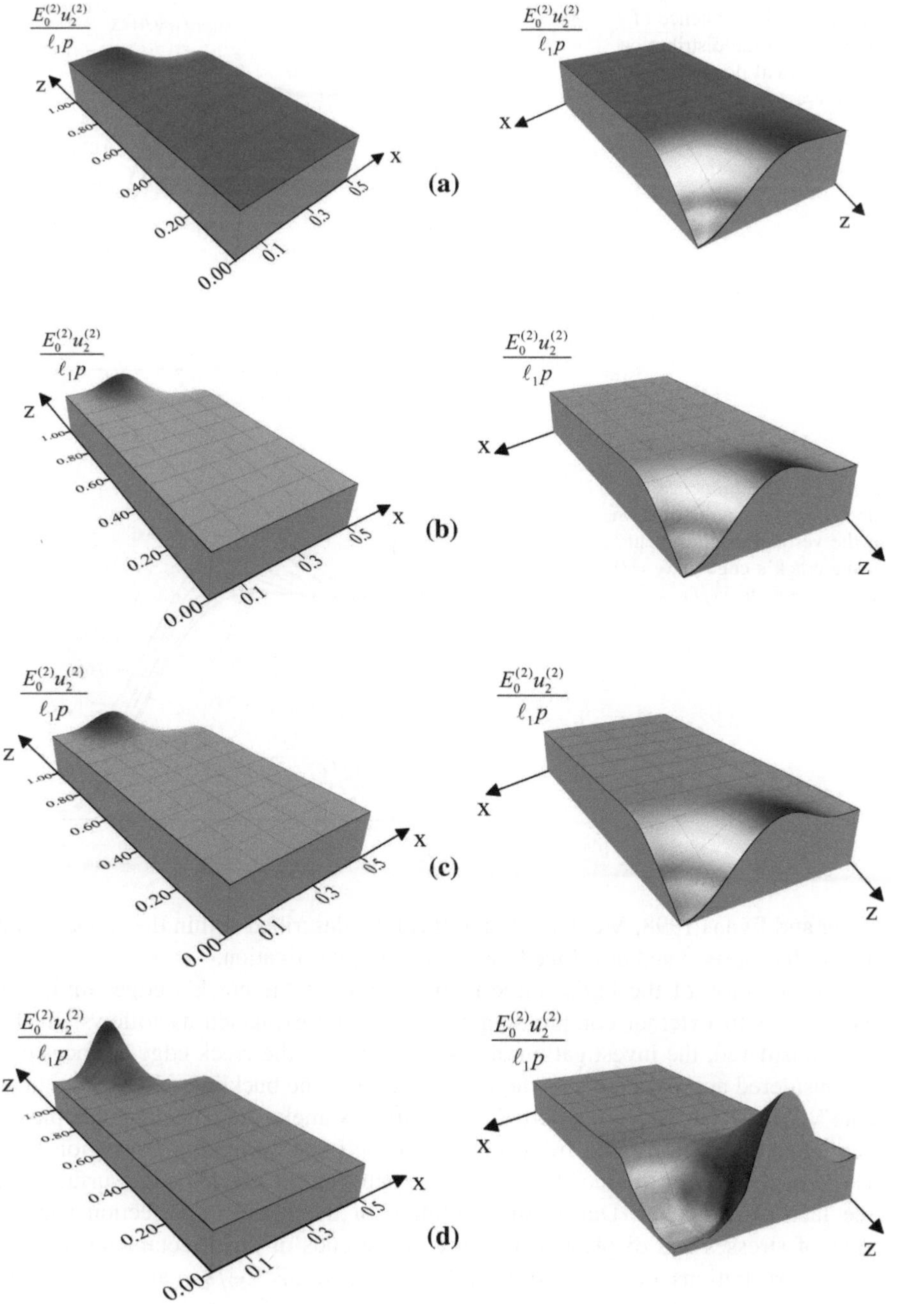


**Fig. 4.30** The evolution of the buckling delamination mode with $\delta$: **a** $\delta = 0.1650$, **b** $\delta = 0.2145$, **c** $\delta = 0.2150$, **d** $\delta = 0.2151$ under $\ell_{30}/\ell_1 = 0.2$, $E_0^{(2)} \big/ E^{(1)} = 10$

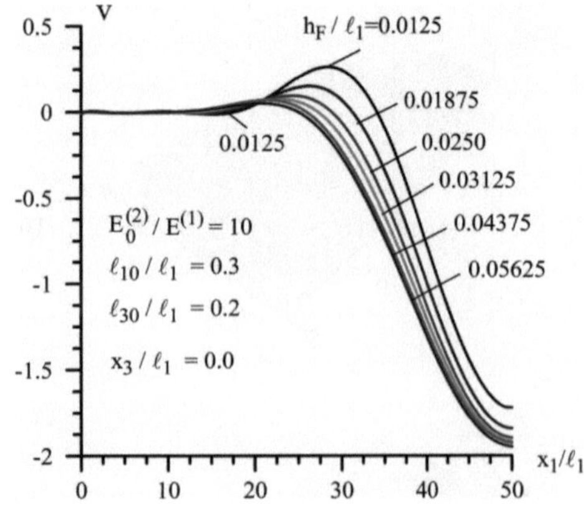


**Fig. 4.31** The influence of the $h_F/\ell_1$ on the distribution of the vertical displacement of the crack's edge at $x_3 = 0$ with respect to $x_1/\ell_1$

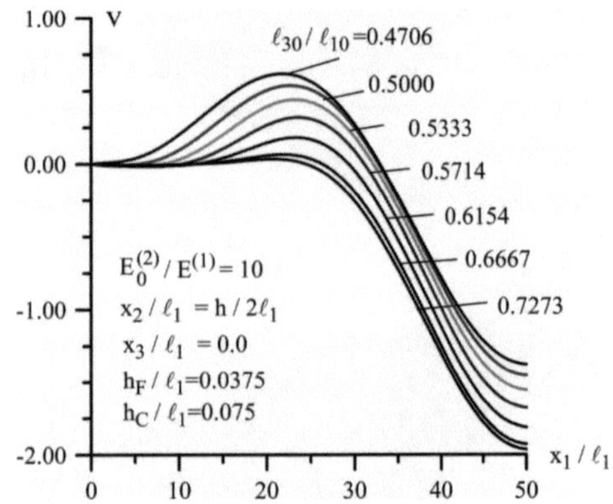


**Fig. 4.32** The influence of the $\ell_{30}/\ell_{10}$ on the distribution of the vertical displacement of the crack's edge at $x_3 = 0$ with respects to $x_1/\ell_1$

Wang and Evans 1998; Moon et al. 2002) can be described within the scope of the approach proposed and developed in the present investigation.

The violation of the initial imperfection mode of the crack's edges under its evolution with external compressing forces can be explained as follows. In the case considered, the investigation on the evolution of the crack edge-surfaces can be considered approximately as the investigation of the buckling of a "rectangular plate", whose edges are supported elastically. Namely, the "rectangular plate" mentioned above is formed from the part of the plate occupying the region consisting of the part of the face layer which is between the crack's edge surface and free face of this layer. During the evolution of the initial imperfection various types of stresses and displacements arise in the ends of this "rectangular plate". These distributions depend significantly on the ratios $\ell_{30}/\ell_{10}$ and $h_F/\ell_1$ and, determine the buckling mode. Moreover, for the clarity of the foregoing discussion the following known fact in the theory of stability of the rectangular plates should be remembered.

**Table 4.66** The values of $t'_{cr}$ attained under $\gamma = 1$, $\ell_{10}/\ell_1 = 0.5$, $\ell_{30}/\ell_1 = 0.3$, $\omega = 2$ and $\alpha = -0.5$ for various values of $E_0^{(2)}/E^{(1)}$, $h_F/\ell_1$ and $\delta$ (Case 2)

| $E_0^{(2)}/E^{(1)}$ | $h_F/\ell_1$ | | | | | |
|---|---|---|---|---|---|---|
| | 0.0375 | | 0.0250 | | 0.0125 | |
| | $\delta$ | $t'_{cr}$ | $\delta$ | $t'_{cr}$ | $\delta$ | $t'_{cr}$ |
| 1 | 0.0280 | 0.004 | 0.0160 | 0.004 | 0.0080 | 0.001 |
| | 0.0270 | 0.012 | 0.0150 | 0.023 | 0.0070 | 0.046 |
| | 0.0260 | 0.029 | 0.0140 | 0.100 | 0.0060 | 1.591 |
| 2 | 0.0420 | 0.001 | 0.0206 | 0.005 | 0.0087 | 0.007 |
| | 0.0412 | 0.002 | 0.0200 | 0.014 | 0.0081 | 0.036 |
| | 0.0405 | 0.005 | 0.0193 | 0.030 | 0.0075 | 0.163 |
| 5 | 0.0780 | 0.002 | 0.0340 | 0.009 | 0.0126 | 0.001 |
| | 0.0720 | 0.029 | 0.0336 | 0.013 | 0.0123 | 0.004 |
| | 0.0660 | 0.197 | 0.0331 | 0.018 | 0.0119 | 0.010 |
| 10 | 0.1375 | 0.001 | 0.0560 | 0.007 | 0.0175 | 0.010 |
| | 0.1320 | 0.006 | 0.0520 | 0.049 | 0.0162 | 0.055 |
| | 0.1265 | 0.020 | 0.0480 | 0.293 | 0.0150 | 0.302 |

Suppose that the stability loss of the rectangular plate without any initial imperfection with the length $\ell_{10}$ and width $\ell_{30}$ is considered within the scope of the Euler (bifurcation) approach. Furthermore, assume that at the ends of this plate the normal and tangential forces act. As it follows from the corresponding theoretical and experimental investigations (see, for example, the monograph (Volmir 1967), the stability loss modes of the rectangular plates depend on the ratio of length to width of that.

A large number of numerical results which are not given here show that under the buckling delamination, the displacement $u_2^{(1)}$ at $x_2 = h_F + 0$, for $(\ell_1 - \ell_{10})/2 < x_1 < (\ell_1 + \ell_{10})/2$, $0 < x_3 < \ell_{30}$, i.e. the vertical displacement of the upper edge surface of the edge crack is significantly less and can be neglected with respect to that considered above, i.e. with respect to the vertical displacement of the lower edge surface of this crack. Therefore, in the cases where $(\ell_{30}/\ell_{10}) < \chi$, i.e. in the cases where the complicated buckling delamination modes appear, some contact areas of the crack's edges may arise in the certain stage of the evolution of the initial imperfection of these edges. It should be noted that this statement is not taken into account in the present investigation. Nevertheless, the results discussed above and relating the cases where $(\ell_{30}/\ell_{10}) < \chi$ give the important information on the evolution stages of the interface crack's edges.

Now, we consider the numerical results attained for the critical time, i.e. for the $t'_{cr}$. As in Case 1, under obtaining these results we assume that the values of the strain caused by the external compressive force satisfy the inequality-relation $\delta_{cr.\infty} < \delta < \delta_{cr.0}$.

Table 4.66 shows the values of $t'_{cr}$ obtained for various $h_F/\ell_1$, $\delta$ and $E_0^{(2)}/E^{(1)}$ under $\omega = 2$, $\alpha = -0.5$, $\gamma = 1$, $\ell_{10}/\ell_1 = 0.5$ and $\ell_{30}/\ell_1 = 0.3$. In the quantitative

**Table 4.67** The values of $t_{cr}$ obtained under $\gamma = 1$, $\ell_{10}/\ell_1 = 0.5$, $\ell_{30}/\ell_1 = 0.3$ and $h_F/\ell_1 = 0.0250$ for various values of $\omega$, $\alpha$ and $E_0^{(2)}/E^{(1)}$ (Case 2)

| $E_0^{(2)}/E^{(1)}$ | $\delta$ | $\omega$ | $\alpha$ | $t_{cr}$ |
|---|---|---|---|---|
| 2 | 0.0200 | 1 | −0.5 | 0.010 |
| | | 2 | | 0.014 |
| | | 3 | | 0.020 |
| | | 2 | −0.3 | 0.038 |
| | | | −0.5 | 0.014 |
| | | | −0.7 | 0.001 |
| 5 | 0.0354 | 1 | −0.5 | 0.009 |
| | | 2 | | 0.013 |
| | | 3 | | 0.018 |
| | | 2 | −0.3 | 0.037 |
| | | | −0.5 | 0.013 |
| | | | −0.7 | 0.001 |
| 10 | 0.0520 | 1 | −0.5 | 0.028 |
| | | 2 | | 0.049 |
| | | 3 | | 0.105 |
| | | 2 | −0.3 | 0.096 |
| | | | −0.5 | 0.049 |
| | | | −0.7 | 0.010 |

sense these results agree with the corresponding results obtained in Case 1. The results given in Table 4.67 illustrate the influence of the rheological parameter $\omega$ and $\alpha$ on the values of $t'_{cr}$. It follows from these results that the values of $t'_{cr}$ increase (decrease) with $\omega$ (with absolute values of $\alpha$).

#### 4.11.4.3 Numerical Results Obtained in Case 3 (Embedded Cracks)

Analyze numerical results related to Case 3, i.e. related to the buckling delamination of the sandwich rectangular plate with embedded interface rectangular cracks the edges of which have initial imperfections determined by the function (4.206). We use the notation and assumptions employed in the previous sub-subsections. However, it should be noted that numerical results related to Case 3 are obtained under bi-axial compression of the plate along the $Ox_1$ and $Ox_3$ axes with uniformly distributed normal forces with intensity $p_1$ and $p_3$, respectively. Therefore, here we introduce the new notation $\delta_1 = \sigma_{11}^{(1),0}/E^{(1)}$ and $\delta_3 = \sigma_{33}^{(1),0}/E^{(1)}$, and assume that $\delta_3 = \xi\delta_1$. Note that for the fixed value of the parameter $\xi$ we can easily determine the corresponding value of ratio $p_3/p_1$ from the relations (4.192) and (4.193). Moreover, note that through the parameter $\xi$ we estimate the influence of the bi-axiality of the external compression of the plate. Consequently, for each fixed $\xi$ the local buckling delamination of the sandwich plate under consideration can be determined through the critical values of the parameter $\delta_1$.

**Table 4.68** The values of $\delta_{1cr.0}$ (upper number) and $\delta_{1cr.\infty}$ (lower number) obtained for various $\xi$, $E_0^{(2)}/E^{(1)}$ and $h_F/\ell_1$ under $\ell_{30}/\ell_1 = \ell_{10}/\ell_1 = 0.5$

| $\xi$ | $E_0^{(2)}/E^{(1)}$ | $h_F/\ell_1$ | | | |
|---|---|---|---|---|---|
| | | 0.01250 | 0.02500 | 0.03750 | 0.05000 |
| 0 | 1 | 0.0111<br>0.0077 | 0.0251<br>0.0176 | 0.0451<br>0.0317 | 0.0684<br>0.0481 |
| | 2 | 0.0220<br>0.0153 | 0.0490<br>0.0346 | 0.0873<br>0.0618 | 0.1322<br>0.0935 |
| | 5 | 0.0535<br>0.0374 | 0.1167<br>0.0829 | 0.2051<br>0.1466 | 0.3103<br>0.2213 |
| | 10 | 0.1036<br>0.0728 | 0.2215<br>0.1586 | 0.3847<br>0.2773 | 0.5828<br>0.4183 |
| 0.5 | 1 | 0.0077<br>0.0054 | 0.0176<br>0.0124 | 0.0319<br>0.0225 | 0.0488<br>0.0344 |
| | 2 | 0.0153<br>0.0107 | 0.0344<br>0.0243 | 0.0618<br>0.0438 | 0.0943<br>0.0668 |
| | 5 | 0.0373<br>0.0261 | 0.0819<br>0.0583 | 0.1451<br>0.1038 | 0.2209<br>0.1578 |
| | 10 | 0.0722<br>0.0508 | 0.1554<br>0.1114 | 0.2716<br>0.1961 | 0.4135<br>0.2980 |
| 1.0 | 1 | 0.0058<br>0.0040 | 0.0133<br>0.0093 | 0.0241<br>0.0170 | 0.0369<br>0.0260 |
| | 2 | 0.0115<br>0.0080 | 0.0260<br>0.0183 | 0.0467<br>0.0331 | 0.0713<br>0.0505 |
| | 5 | 0.0281<br>0.0196 | 0.0618<br>0.0440 | 0.1096<br>0.0784 | 0.1671<br>0.1193 |
| | 10 | 0.0545<br>0.0383 | 0.1173<br>0.0841 | 0.2052<br>0.1482 | 0.3126<br>0.2254 |

First, we consider the pure elastic stability loss buckling delamination which takes place at $t' = 0$ and $t' = \infty$. The critical values of $\delta_1$ obtained at $t' = 0$ and $t' = \infty$ we denote by $\delta_{1cr.0}$ and $\delta_{1cr.\infty}$ respectively. Assume that $h/\ell_1 = 0.15$, $\nu^{(1)} = \nu_0^{(2)} = 0.3$ and $\ell_3/\ell_1 = 1.0$ and consider Table 4.68 which shows the influence of the parameter $\xi$ on the values of $\delta_{1cr.0}$ (upper number) and $\delta_{1cr.\infty}$ (lower number) obtained for various $E_0^{(2)}/E^{(1)}$ and $h_F/\ell_1$ under $\ell_{10}/\ell_1 = 0.5$, $\ell_{30}/\ell_1 = 0.5$ and $\omega = 2$. It follows from this table that, as can be predicted, an increase in the values of $\xi$ causes a decrease in $\delta_{1cr.0}$ and $\delta_{1cr.\infty}$. Moreover, this table shows that the values of $\delta_{1cr.0}$ and $\delta_{1cr.\infty}$ also decrease with decreasing $E_0^{(2)}/E^{(1)}$ and $h_F/\ell_1$, i.e. with a decrease in the face layer's material stiffness and the thickness. These results agree well with the well-known mechanical considerations and give an assurance of the reliability of the developed algorithm and the programs that are used for their calculations.

Consider also the following verification of the validity of the developed and used algorithm and programs. For this purpose we use the results obtained under $p_3 = 0$ which, according to the assumption $\nu_0^{(2)} = \nu^{(1)}$, corresponds to the case where $\xi = 0$ and we compare these results with the corresponding ones obtained for the edge and band cracks. According to the mechanical consideration, for fixed $\ell_{10}/\ell_1$ the values of $\delta_{1cr.0}$ obtained for the noted-above three cases must approach each other with $\ell_{30}/\ell_1$. This consideration is proven with the graphs given in Fig. 4.33 that show the dependence between $\ell_{30}/\ell_1$. and $\ell_{30}/\ell_1$ under $\ell_{10}/\ell_1 = 0.5$, $h_F/\ell_1 = 0.0375$, $E_0^{(2)}/E^{(1)} = 5$ (Fig. 4.33a) and 10 (Fig. 4.33b).

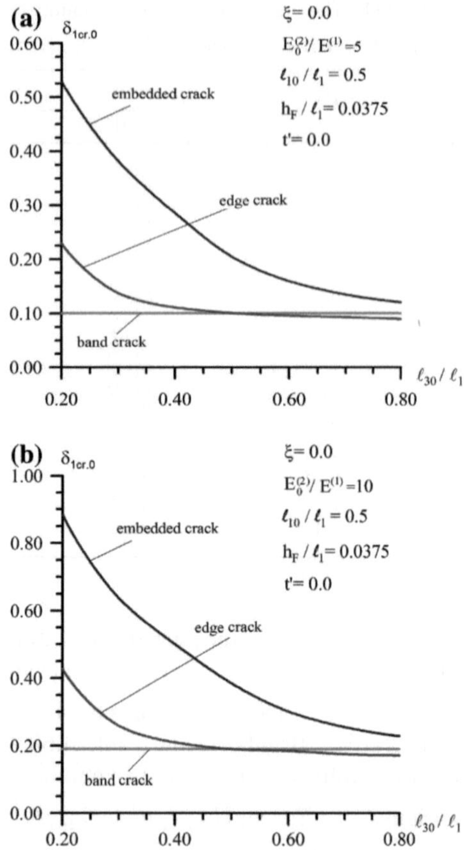


**Fig. 4.33** Comparison of the values of $\delta_{1cr.0}$ and $\delta_{1cr.\infty}$ obtained for the embedded, edge and band cracks

Consider the influence of the cracks' length along the $Ox_1$ and $Ox_3$ axes, i.e. the influence of the parameters $\ell_{10}/\ell_1$ and $\ell_{30}/\ell_1$, on the values of $\delta_{1cr.0}$ and $\delta_{1cr.\infty}$. This influence is illustrated by the graphs given in Fig. 4.34 which are constructed under $\ell_{30}/\ell_1 = 0.5$ and in Fig. 4.35 which are constructed under $\ell_{10}/\ell_1 = 0.5$. Moreover, under construction of these graphs it is assumed that $\omega = 2$, $h_F/\ell_1 = 0.0375$ and $E_0^{(2)}\big/E^{(1)} = 2$ (Figs. 4.34a and 4.35a), 5 (Figs.4.34b and 4.35b) and 10 (Figs. 4.34c and 4.35c). It follows from these results that the values of $\delta_{1cr.0}$ and $\delta_{1cr.\infty}$ decrease with $\ell_{10}/\ell_1$ and $\ell_{30}/\ell_1$, but increase with $E_0^{(2)}\big/E^{(1)}$.

Now we handle the question of how the buckling delamination mode depends on the ratio $\ell_{30}/\ell_{10}$. In this case taking the symmetry of the plate geometry with respect to $x_1 = \ell_1/2$ and with respect to $x_3 = \ell_3/2$ into account, we consider only the parts of the plate for which $\{0 < x_1 < \ell_1/2,\ 0 < x_3 < \ell_3/2,\ 0 < x_2 < h_F + h_C/2\}$.

Figures 4.36 and 4.37 show schematically the distribution of $v = u_2^{(2)}E_0^{(2)}\big/(p_1\ell_1)$ with respect to $x(=x_1)$ and $z(=\ell_3 - x_3)$ under $x_2 = h_F - 0$ in the

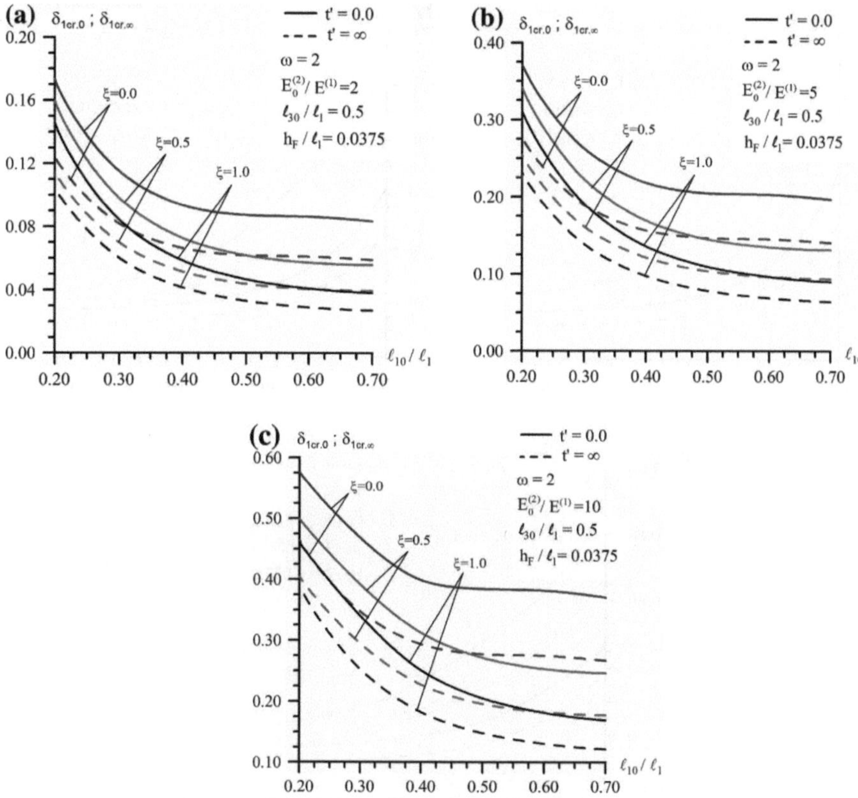


**Fig. 4.34** The graphs of the dependencies among $\delta_{1cr.0}$, $\delta_{1cr.\infty}$ and $\ell_{10}/\ell_1$ under $E_0^{(2)}\Big/E^{(1)} = 2$ (**a**), 5 (**b**) and 10 (**c**), $\omega = 2$, $h_F/\ell_1 = 0.0375$ and $\ell_{30}/\ell_1 = 0.5$

cases where $\ell_{30}/\ell_{10} = 0.6250$, $\ell_{30}/\ell_1 = 0.25$, $\xi = 0.5$ (Fig. 4.36) and $\ell_{30}/\ell_{10} = 0.3571$, $\ell_{30}/\ell_1 = 0.25$, $\xi = 0.5$ (Fig. 4.37) respectively. Under construction of these buckling mode surfaces it is assumed that $\delta_1$ is very near to $\delta_{1cr.0}$, i.e. it is assumed that $O(|\delta_1 - \delta_{1.cr.0}|) = 10^{-4}$ is satisfied.

Although the illustrated mode surfaces are constructed for the case where $E_0^{(2)}\Big/E^{(1)} = 2$ and $h_F/\ell_1 = 0.0375$ they also hold (in a qualitative sense) for the other values of the problem parameters $E_0^{(2)}\Big/E^{(1)}$ and $h_F/\ell_1$. It follows from Fig. 4.36 that in the case where $\ell_{30}/\ell_{10} = 0.6250$, $\ell_{10}/\ell_1 = 0.40$, $\xi = 0.5$ and $\delta_1 = 0.1644$ the buckling delamination mode is similar to the initial imperfection mode of the cracks' edge surfaces, but in the case where $\ell_{30}/\ell_{10} = 0.3571$, $\ell_{10}/\ell_1 = 0.70$, $\xi = 0.5$ and $\delta_1 = 0.1458$ (Fig. 4.37) the buckling delamination mode has a more complicated character. For illustration of the influence of the parameters $\ell_{30}/\ell_{10}$ and $\xi$ on this character we consider the graphs of the

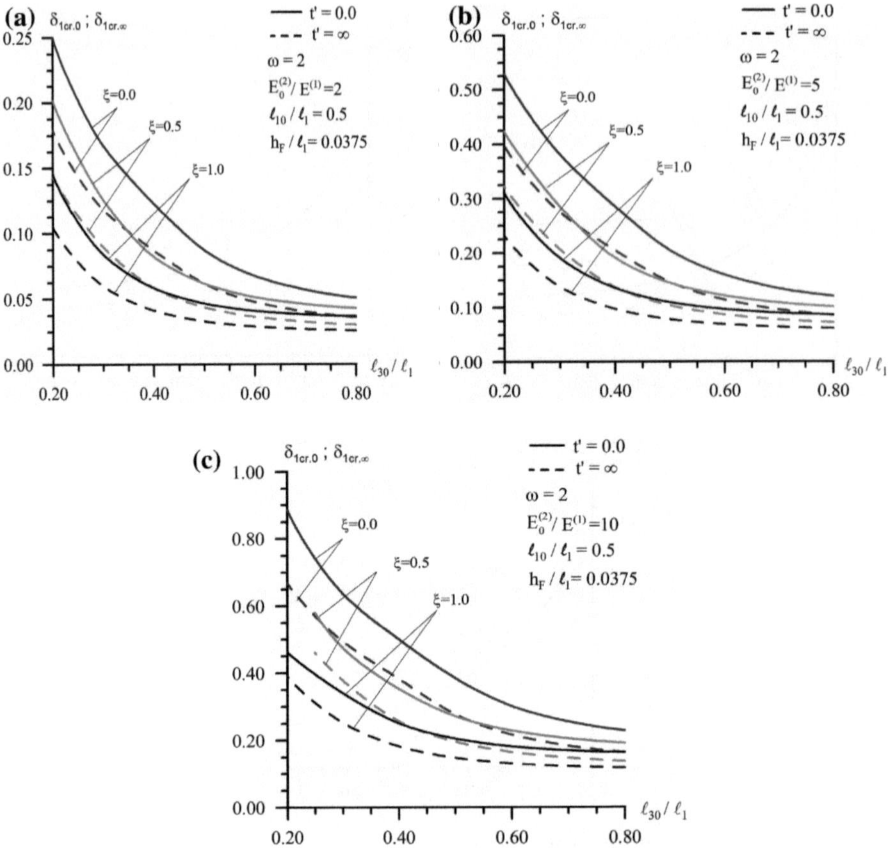


**Fig. 4.35** The graphs of the dependencies among $\delta_{1cr.0}$, $\delta_{1cr.\infty}$ and $\ell_{30}/\ell_1$ under $E_0^{(2)} \big/ E^{(1)} = 2$ (**a**), 5 (**b**) and 10 (**c**) under $\omega = 2$, $h_F/\ell_1 = 0.0375$ and $\ell_{10}/\ell_1 = 0.5$

dependence between the values of $w = 2u_2^{(2)} \left| \max\limits_{0 \le x_1 \le \ell_1/2} \left| u_2^{(2)} \right| - \min\limits_{0 \le x_1 \le \ell_1/2} \left| u_2^{(2)} \right| \right|^{-1}$ at $x_2 = h_F - 0$, $x_3 = \ell_3/2$ and $x_1/\ell_1$ which are constructed under $h_F/\ell_1 = 0.0375$ and $E_0^{(2)} \big/ E^{(1)} = 2$ for various values of these parameters. These graphs are given in Fig. 4.38 and in this figure by the letters a, b and c, the cases where ξ=0.0; 0.5 and 1.0 are indicated respectively. It follows from the graphs that in general, for fixed values of the problem parameters there exists such values of $\chi (= \ell_{30}/\ell_{10})$ (denoted by $\chi*$), according to which, the character of the buckling delamination mode can be determined. Thus, for the case considered in Fig. 4.38, if $\chi > \chi*$, then the buckling delamination mode is similar to the initial imperfection mode, but if $\chi < \chi*$ then the buckling delamination mode is not similar to the initial

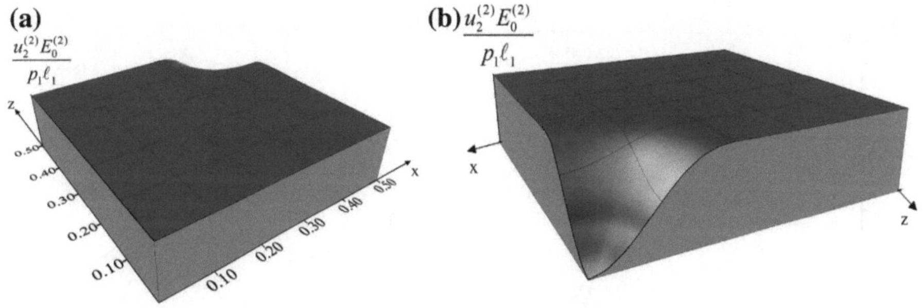


**Fig. 4.36** The buckling delamination mode of the crack's edge surface under $t' = 0$, $E_0^{(2)} \Big/ E^{(1)} = 2$, $h_F/\ell_1 = 0.0375$, $\ell_{30}/\ell_1 = 0.25$, $\ell_{10}/\ell_1 = 0.40$ and $\delta_1 = 0.1644$ : the case where $\ell_{30}/\ell_{10} = 0.6250$ and the buckling delamination mode is similar with the initial imperfection mode of the cracks' edges

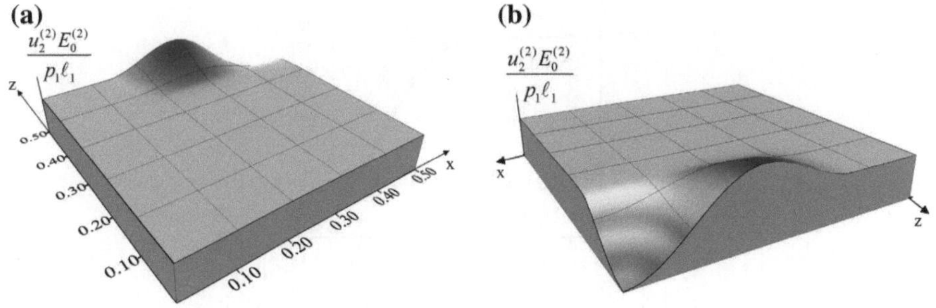


**Fig. 4.37** The buckling delamination mode of the crack's edge surface under $t' = 0$, $E_0^{(2)} \Big/ E^{(1)} = 2$, $h_F/\ell_1 = 0.0375$, $\ell_{30}/\ell_1 = 0.25$, $\ell_{10}/\ell_1 = 0.70$ and $\delta_1 = 0.1458$ : the case where $\ell_{30}/\ell_{10} = 0.3571$ and the buckling delamination mode has a complicate character

imperfection mode and has a complicated character. Note that the value of $\chi*$ decrease with the parameter $\xi$.

Obviously the explanation given in the previous sub-subsection for the rise of the complicated buckling delamination mode around the end cracks can also be used for similar ones with embedded cracks.

Now, we consider the numerical results obtained for the critical time, i.e. for $t'_{cr}$. In obtaining these results, we assume that the values of the strain caused by the external compressing force satisfy the inequality relation $\delta_{1.cr.\infty} < \delta_1 < \delta_{1.cr.0}$. We introduce a parameter:

$$\eta = \frac{\delta_{1cr.0} - \delta_1}{\delta_{1cr.0} - \delta_{1cr.\infty}} \tag{4.207}$$

Note that this parameter characterizes the "distance" of the selected value of $\delta_1$ from the values of $\delta_{1cr.0}$ and of $\delta_{1cr.\infty}$, and $0 \le \eta \le 1$; the values $\eta = 0$ and $\eta = 1$ correspond to the cases where $\delta_1 = \delta_{1cr.0}$ and $\delta_1 = \delta_{1cr.\infty}$, respectively, and for

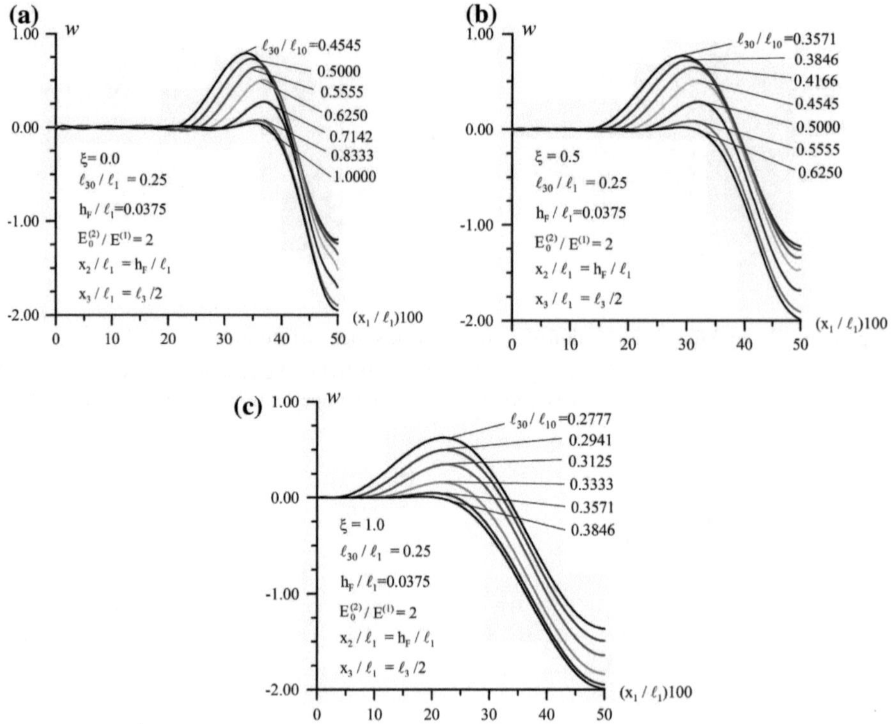


**Fig. 4.38** The distribution of the vertical displacement of the cracks' edge surface with respect to $x_1/\ell_1$ at $x_3 = \ell_3/2$ in the cases where $\xi = 0.0$ (**a**), 0.5 (**b**) and 1 (**c**)

the interval $\delta_{1cr.\infty} < \delta_1 < \delta_{1cr.0}$ the values of $\eta$ increase (decrease) monotonically while $\delta_1 \to \delta_{1cr.0}$ ($\delta_1 \to \delta_{1cr.\infty}$). To use the parameter $\eta$ (4.207) gives us the possibility of determining the influence of the problem parameters (for instance, the parameters $E_0^{(2)} \big/ E^{(1)}$ and $\xi$) on the critical time (i.e. on $t'_{cr}$) obtained for the values of $\delta_1$ which have the same "distance" from $\delta_{1cr.0}$ and $\delta_{1cr.\infty}$. Taking the foregoing discussion into account, consider the results given in Tables 4.69, 4.70 and 4.71 which show the values of $t'_{cr}$ obtained for various $\eta$ (or $\delta_1$) and $E_0^{(2)} \big/ E^{(1)}$ under $h_F/\ell = 0.0375$, $\ell_{30}/\ell_1 = 0.5$, $\ell_{10}/\ell_1 = 0.5$, $\omega = 2$ and $\alpha = -0.5$ in the cases where $\xi = 0.0$, 0.5 and 10.0 respectively. Note that under determination of $t'_{cr}$ the values of $\sigma_{11}^{(r_k)}(0)$ and $\sigma_{33}^{(r_k)}(0)$ are taken instead of $\sigma_{11}^{(r_k)}(t)$ and $\sigma_{33}^{(r_k)}(t)$ respectively, which enter the Eqs. (4.196–4.198).

Thus, it follows from the analyses of the foregoing results that for the fixed value of the parameter $\eta$ the influence of the parameters $E_0^{(2)} \big/ E^{(1)}$ and $\xi$ on the values of $t'_{cr}$ is insignificant. Consequently, the influence of the parameter $\eta$ on the values of $t'_{cr}$ has a dominant character.

**Table 4.69** The values of $t'_{cr}$ obtained for various $\eta$, $E_0^{(2)}/E^{(1)}$ under $\xi = 0$ for the case where $\ell_{30}/\ell_1 = 0.5$, $\ell_{10}/\ell_1 = 0.5$ and $h_F/\ell = 0.0375$

| $E_0^{(2)}/E^{(1)}$ | $\delta_{1cr.0}$ | $\delta_{1cr.\infty}$ | $\delta_1$ | $\eta$ | $t'_{cr}$ |
|---|---|---|---|---|---|
| 1 | 0.0451 | 0.0317 | 0.043882 | 0.09 | 0.0006 |
| | | | 0.041775 | 0.25 | 0.006 |
| | | | 0.040633 | 0.33 | 0.015 |
| | | | 0.038400 | 0.50 | 0.060 |
| | | | 0.036167 | 0.67 | 0.244 |
| | | | 0.035050 | 0.75 | 0.555 |
| | | | 0.033933 | 0.83 | 1.581 |
| | | | 0.033040 | 0.90 | 5.370 |
| | | | 0.032918 | 0.91 | 6.711 |
| | | | 0.032370 | 0.95 | 27.031 |
| 2 | 0.0873 | 0.0618 | 0.084982 | 0.09 | 0.0006 |
| | | | 0.080973 | 0.25 | 0.006 |
| | | | 0.078800 | 0.33 | 0.015 |
| | | | 0.074550 | 0.50 | 0.060 |
| | | | 0.070300 | 0.67 | 0.243 |
| | | | 0.068175 | 0.75 | 0.550 |
| | | | 0.066050 | 0.83 | 1.550 |
| | | | 0.064350 | 0.90 | 5.155 |
| | | | 0.064118 | 0.91 | 6.406 |
| | | | 0.063075 | 0.95 | 24.527 |
| 5 | 0.2051 | 0.1466 | 0.199782 | 0.09 | 0.0006 |
| | | | 0.190585 | 0.25 | 0.006 |
| | | | 0.185600 | 0.33 | 0.015 |
| | | | 0.175850 | 0.50 | 0.060 |
| | | | 0.166100 | 0.67 | 0.242 |
| | | | 0.161225 | 0.75 | 0.547 |
| | | | 0.156350 | 0.83 | 1.529 |
| | | | 0.152450 | 0.90 | 4.999 |
| | | | 0.151918 | 0.91 | 6.185 |
| | | | 0.149525 | 0.95 | 22.765 |

Consider the influence of the rheological parameter $\omega$ on the values of $t'_{cr}$. This influence is tabulated in Table 4.72 under $\alpha = -0.5$, $\xi = 0.0$, $E_0^{(2)}/E^{(1)} = 2$, $h_F/\ell = 0.0375$, $\ell_{30}/\ell_1 = 0.5$ and $\ell_{10}/\ell_1 = 0.5$. Table 4.72 shows that the values of $t'_{cr}$ increase with $\omega$ because the parameter $\omega$ characterizes the dilatational mechanical properties (for instance, the dilatational modulus of elasticity) of the face layers' materials, and an increase in the values of $\omega$ means that the stiffness of these materials increases.

Table 4.73 illustrates the influence of the rheological parameter $\alpha$ on the values of $t'_{cr}$ obtained for the case where $\omega = 2$, $\xi = 0.0$, $E_0^{(2)}/E^{(1)} = 2$, $h_F/\ell = 0.0375$, $\ell_{30}/\ell_1 = 0.5$ and $\ell_{10}/\ell_1 = 0.5$. It follows from Table 4.73 that there exists such a

**Table 4.70** The values of $t'_{cr}$ obtained for various $\eta$, $E_0^{(2)} / E^{(1)}$ under $\xi = 0.5$ for the case where $\ell_{30}/\ell_1 = 0.5$, $\ell_{10}/\ell_1 = 0.5$ and $h_F/\ell = 0.0375$

| $E_0^{(2)} / E^{(1)}$ | $\delta_{1cr.0}$ | $\delta_{1cr.\infty}$ | $\delta_1$ | $\eta$ | $t'_{cr}$ |
|---|---|---|---|---|---|
| 1 | 0.0319 | 0.0225 | 0.031045 | 0.09 | 0.0006 |
| | | | 0.029568 | 0.25 | 0.006 |
| | | | 0.028767 | 0.33 | 0.015 |
| | | | 0.027200 | 0.50 | 0.059 |
| | | | 0.025633 | 0.67 | 0.240 |
| | | | 0.024850 | 0.75 | 0.541 |
| | | | 0.024067 | 0.83 | 1.511 |
| | | | 0.023440 | 0.90 | 4.949 |
| | | | 0.023355 | 0.91 | 6.126 |
| | | | 0.022970 | 0.95 | 22.690 |
| 2 | 0.0618 | 0.0438 | 0.060164 | 0.09 | 0.0006 |
| | | | 0.057334 | 0.25 | 0.006 |
| | | | 0.055800 | 0.33 | 0.015 |
| | | | 0.052800 | 0.50 | 0.060 |
| | | | 0.049800 | 0.67 | 0.241 |
| | | | 0.048300 | 0.75 | 0.546 |
| | | | 0.046800 | 0.83 | 1.533 |
| | | | 0.045600 | 0.90 | 5.057 |
| | | | 0.045436 | 0.91 | 6.274 |
| | | | 0.044700 | 0.95 | 23.559 |
| 5 | 0.1451 | 0.1038 | 0.141345 | 0.09 | 0.0006 |
| | | | 0.134853 | 0.25 | 0.006 |
| | | | 0.131333 | 0.33 | 0.015 |
| | | | 0.124450 | 0.50 | 0.060 |
| | | | 0.117567 | 0.67 | 0.243 |
| | | | 0.114125 | 0.75 | 0.550 |
| | | | 0.110683 | 0.83 | 1.539 |
| | | | 0.107930 | 0.90 | 5.050 |
| | | | 0.107555 | 0.91 | 6.252 |
| | | | 0.105865 | 0.95 | 23.198 |

value of $\delta_1$ before (after) which an increase in the absolute values of the parameter $\alpha$ causes an increase (a decrease) in the values of the critical time $t'_{cr}$. Note that these results in the qualitative sense agree with corresponding ones obtained in the previous two sub-subsections.

#### 4.11.4.4 Conclusions Based on the Numerical Results

In the present section an approach was developed and employed for the study of the buckling delamination of elastic and viscoelastic sandwich rectangular plates containing two parallel interface rectangular cracks. The investigation was made

**Table 4.71** The values of $t'_{cr}$ obtained for various $\eta$, $E_0^{(2)}/E^{(1)}$ under $\xi = 1.0$ for the case where $\ell_{30}/\ell_1 = 0.5$, $\ell_{10}/\ell_1 = 0.5$ and $h_F/\ell = 0.0375$

| $E_0^{(2)}/E^{(1)}$ | $\delta_{1cr.0}$ | $\delta_{1cr.\infty}$ | $\delta_1$ | $\eta$ | $t'_{cr}$ |
|---|---|---|---|---|---|
| 1 | 0.0241 | 0.0170 | 0.023455 | 0.09 | 0.0006 |
| | | | 0.022338 | 0.25 | 0.006 |
| | | | 0.021733 | 0.33 | 0.015 |
| | | | 0.020550 | 0.50 | 0.060 |
| | | | 0.019367 | 0.67 | 0.242 |
| | | | 0.018775 | 0.75 | 0.547 |
| | | | 0.018183 | 0.83 | 1.533 |
| | | | 0.017710 | 0.90 | 5.055 |
| | | | 0.017645 | 0.91 | 6.278 |
| | | | 0.017355 | 0.95 | 23.568 |
| 2 | 0.0467 | 0.0331 | 0.045464 | 0.09 | 0.0006 |
| | | | 0.043326 | 0.25 | 0.006 |
| | | | 0.042167 | 0.33 | 0.015 |
| | | | 0.039900 | 0.50 | 0.060 |
| | | | 0.037633 | 0.67 | 0.242 |
| | | | 0.036500 | 0.75 | 0.549 |
| | | | 0.035367 | 0.83 | 1.541 |
| | | | 0.034460 | 0.90 | 5.102 |
| | | | 0.034336 | 0.91 | 6.336 |
| | | | 0.033780 | 0.95 | 23.967 |
| 5 | 0.1096 | 0.0784 | 0.106764 | 0.09 | 0.0006 |
| | | | 0.101859 | 0.25 | 0.006 |
| | | | 0.099200 | 0.33 | 0.015 |
| | | | 0.094000 | 0.50 | 0.061 |
| | | | 0.088800 | 0.67 | 0.247 |
| | | | 0.086200 | 0.75 | 0.560 |
| | | | 0.083600 | 0.83 | 1.576 |
| | | | 0.081520 | 0.90 | 5.240 |
| | | | 0.081236 | 0.91 | 6.153 |
| | | | 0.079960 | 0.95 | 24.913 |

within the scope of the piecewise homogeneous body model. It was assumed that the edge surfaces of the cracks have initial infinitesimal imperfections and the proposed approach was based on the study of the evolution of these initial imperfections with external compressive forces (with duration of time) for pure elastic (viscoelastic) ones. The noted evolution was determined within the scope of the exact three-dimensional geometrically nonlinear field equations of the theory of elasticity and viscoelasticity. For the solution to the corresponding boundary-value problems, boundary form perturbation techniques, Laplace transform and 3D FEM were employed. The initial imperfection criterion was used as the buckling delamination (stability loss) criterion. For the concrete numerical investigations, three cases were considered. In Case 1 (Case 2) it was assumed that the interface

**Table 4.72** The influence of the rheological parameter $\omega$ on the values of $t'_{cr}$ obtained under which $\alpha = -0.5$, $E_0^{(2)} / E^{(1)} = 2$, $h_F/\ell = 0.0375$, $\ell_{30}/\ell_1 = 0.5$, $\ell_{10}/\ell_1 = 0.5$ and $\xi = 0.0$

| $\delta_1$ | $\omega$ | | |
|---|---|---|---|
| | 0.5 | 1.0 | 2.0 |
| 0.080973 | 0.004 | 0.005 | 0.006 |
| 0.078800 | 0.009 | 0.010 | 0.015 |
| 0.074550 | 0.026 | 0.033 | 0.060 |
| 0.070300 | 0.058 | 0.084 | 0.243 |
| 0.068175 | 0.083 | 0.131 | 0.550 |
| 0.066050 | 0.116 | 0.203 | 1.550 |
| 0.064350 | 0.152 | 0.290 | 5.155 |

**Table 4.73** The influence of the rheological parameter $\alpha$ on the values of $t'_{cr}$ obtained under which $\omega = 2$, $E_0^{(2)} / E^{(1)} = 2$, $h_F/\ell = 0.0375$, $\ell_{30}/\ell_1 = 0.5$, $\ell_{10}/\ell_1 = 0.5$ and $\xi = 0.0$

| $\delta_1$ | $\alpha$ | | |
|---|---|---|---|
| | −0.3 | −0.5 | −0.7 |
| 0.080973 | 0.022 | 0.006 | 0.0003 |
| 0.078800 | 0.040 | 0.015 | 0.001 |
| 0.074550 | 0.110 | 0.060 | 0.014 |
| 0.070300 | 0.298 | 0.243 | 0.150 |
| 0.068175 | 0.535 | 0.550 | 0.587 |
| 0.066050 | 1.112 | 1.550 | 3.296 |
| 0.064350 | 2.646 | 5.155 | 24.419 |

cracks contained by the sandwich plate are rectangular band-cracks (rectangular edge-cracks). But in Case 3 it was assumed that the plate contains embedded cracks. The numerical results on the influence of the problem parameter on the values of the critical strain and critical time as well as on the buckling delamination mode are presented and analyzed. According to these analyses, the following main conclusions can be drawn:

- In Case 1 the mode of buckling delamination around the interface rectangular band-cracks is similar to the initial imperfection mode of these cracks;
- In Cases 2 and 3, the mode of buckling delamination around the interface rectangular edge- and embedded-cracks depends on both the thickness of the face layers of the plate and the values of the ratio $\ell_{30}/\ell_{10}$, where $\ell_{30}$ $(\ell_{10})$ is the length of the crack along the $Ox_3$ $(Ox_1)$ axis;
- The values of the critical strain as well as the values of the critical time decrease with the length and width of the cracks;
- There exists such a length $\ell_{30}$ of the edge cracks after which the critical values of the strain obtained in Case 2 become less than the corresponding ones obtained in Case 1;

- With the length of the rectangular plate along the $Ox_3$ axis, the results obtained for the critical strains in Case 1 approach the results obtained in Sect. 4.4;
- The values of the critical strain and the critical time increase with the stiffness of the material of the face layers;
- In Case3 under unidirectional compression, the values of the critical forces, i.e. the critical values of the parameter $\delta_1$ obtained for the interface rectangular embedded cracks are greater than the corresponding ones obtained for the interface edge and band cracks;
- In Case 3 the bi-axiality of the external compression causes a decrease in the critical values of the parameter $\delta_1$;
- The parameter on which the values of the dimensionless critical time mainly depend, is the parameter $\eta$ (4.207) which characterizes the "distance" of the selected value of the parameter $\delta_1$ from the values of $\delta_{1cr.0}$ and $\delta_{1cr.\infty}$. For a fixed value of the parameter $\eta$ the influence of the problem parameters $E_0^{(2)} / E^{(1)}$ and $\xi$ on the values of the dimensionless critical time is insignificant;
- With the length of the rectangular embedded crack along the $Ox_3$ axis, the results obtained for the critical values of $\delta_1$ approach the corresponding ones obtained in Case 1;
- The critical dimensionless time increases with rheological parameter $\omega$, because an increase in this parameter corresponds to an increase in the dilatational modulus of elasticity of the materials of the face layers;
- There exists such a value of the parameter $\delta_1$ before (after) which an increase in the absolute values of the rheological parameter $\alpha$ causes an increase (a decrease) in the values of the dimensionless critical time;
- The numerical results obtained and analyzed in the present sub-section can be taken as standard for the estimation of the accuracy (in the qualitative and quantitative senses) of the corresponding numerical results obtained within the scope of the approximate plate and bar theories;
- The developed approach gives us the possibility for mathematical modeling and describing of the complicated buckling delamination modes of the stratified layers observed in the corresponding experimental studies.

## References

Akbarov SD (1998) On the three dimensional stability loss problems of elements of constructions fabricated from the viscoelastic composite materials. Mech Comp Mater 34(6):537–544

Akbarov SD (2007) Three-dimensional instability problems for viscoelastic composite materials and structural members. Int Appl Mech 43(10):1069–1089

Akbarov SD, Guz AN (2000) Mechanics of curved composites. Kluwer Academic Publishers, Dordrecht

Akbarov SD, Rzayev OG (2002a) On the buckling of the elastic and viscoelastic composite circular thick plate with a penny-shaped crack. Eur J Mech A Solid 21(2):269–279

Akbarov SD, Rzayev OG (2002b) Delamination of unidirectional viscoelastic composite materials. Mech Comp Mater 38(1):17–24

Akbarov SD, Rzayev OG (2003) On the delamination of a viscoelastic composite circular plate. Int Appl Mech 39(3):368–374

Akbarov SD, Sisman T, Yahnioglu N (1997) On the fracture of the unidirectional composites in compression. Int J Eng Sci 35(12/13):1115–1136

Akbarov SD, Yahnioglu N (2001) The method for investigation of the general theory of stability problems of structural elements fabricated from the viscoelastic composite materials. Compos Part B Eng 32(5):475–482

Akbarov SD, Yahnioglu N, Rzayev OG (2007) On influence of the singular type finite elements to the critical force in studying the buckling of a circular plate with a crack. Int Appl Mech 43(9):120–129

Akbarov SD, Yahnioglu N (2010) Delamination buckling of a rectangular orthotropic composite plate containing a band crack. Mech Comp Mater 46(5):493–504

Akbarov SD, Yahnioglu N, Karatas EE (2010a) Buckling delamination of a rectangular plate containing a rectangular crack and made from elastic and viscoelastic composite materials. Int J Solids Struct 47:3426–3434

Akbarov SD, Yahnioglu N, Tekin A (2010b) 3D FEM analyses of the buckling delamination of a rectangular plate containing interface rectangular cracks and made from elastic and viscoelastic materials. CMES Comput Model Eng Sci 64(2):147–185

Arman Y, Zor M, Aksoy S (2006) Determination of critical delamination diameter of laminated composite plates under buckling loads. Compos Sci Technol 66:2945–2953

Benzley SE (1974) Representation of singularities with isoparametric finite elements. Int J Numer Meth Eng 8:537–545

Bogdanov VL, Guz AN, Nazarenko VM (2009) Fracture of a body with a periodic set of coaxial cracks under forced directed along them: an axisymmetric problem. Inter Appl Mech 45(2):111–124

Chai H, Babcock CD, Knauss WG (1981) One-dimensional modeling of failure in laminated plates by delamination buckling. Int J Solids Struct 17:1069–1083

Evans AG, Hutchinson JW (1995) The thermo mechanical integrity of thin films and multilayers. Acta Metal Mater 43:2507–2530

Gioia F, Ortiz M (1997) Delamination of compresses thin films. Adv Appl Mech 33:120–192

Gladun EYu (2000) Dependence of the critical load on the geometric characteristics of a hinged plate with a crack. Int Appl Mech 39(6):1225–1234

Gosz M, Dolbow L, Moran B (1998) Domain integral formulation for stress intensity factor computation along curved three-dimensional interface cracks. Int J Solids Struct 35:1763–1783

Guz AN (2008a) Fundamentals of the compressive fracture mechanics of composites: fracture in structure of materials, vol 1. Litera, Kiev (in Russian)

Guz AN (2008b) Fundamentals of the compressive fracture mechanics of composites: related mechanics of fracture, vol 2. Litera, Kiev (in Russian)

Guz AN (2000) Description and study of some non-classical problems of fracture mechanics and related mechanisms. Int Appl Mech 36(12):1537–1564

Guz AN, Dysel MS, Nazarenko VM (2004) Fracture and stability of materials and structural members with cracks: approach and results. Int Appl Mech 40(12):1323–1359

Guz AN, Nazarenko VM (1985a) Theory of near-surface delamination of composite materials under compression along the macro-crack. Mech Comp Mater 21(5):826–833

Guz AN, Nazarenko VM (1985b) Symmetric failure of the half-space with penny-shaped crack in compression. Theor Appl Fract Mech 3:233–245

Henshell RD, Shaw KG (1975) Crack tip finite elements are unnecessary. Int J Numer Methods Eng 9:495–507

Hutchinson JW, He MY, Evans AG (2000) The influence of imperfections on the nucleation and propagation of buckling driven delaminations. J Mech Phys Solids 48:709–734

Hutchinson JW, Suo Z (1992) Mixed mode cracking in layered materials. Adv Appl Mech 29:63–191

Hutchinson JW, Thouless MD, Liniger EG (1992) Growth and configurational stability of circular, buckling-driven film delaminations. Acta Metall Mater 40:295–308

Hwang SF, Mao CP (1999) The delamination buckling of single-fibre system and interply hybrid composites. Compos Struct 46:279–287

Kachanov LM (1976) Fracture of composite materials by means a delamination. Mech Polim 5:918–922

Kardomateas GA, Pelgri AA, Malik B (1995) Growth of internal delaminations under cyclic compression in composite plates. J Mech Phys Solids 43(6):847–868

Lekhnitskii SG (1981) Elasticity theory of anisotropic bodies. Mir, Moscow (in Russian)

Li H et al (2010) Cristal structure and elastic properties of ZrB compared with $ZrB_2$ : a first principles study. Comp Mater Sci 49:814–819

Moon MW, Chung JW, Lee KR, Oh KH, Wang R, Evans AG (2002) An experimental study of the influence of imperfections on the buckling of compressed thin films. Acta Materialia 50:1219–1227

Moon MW, Lee KR, Oh KH, Hutchinson JW (2004) Buckle delamination on patterned substrates. Acta Materialia 52:3151–3159

Nilsson KF, Giannakopoulos AE (1995) A finite element analysis of configurational stability and finite growth of buckling driven delamination. J Mech Phys Solids 43:1983–2021

Nilsson KF, Thesken JC, Sindelar P, Giannakopoulos AE, Stoakers B (1993) A theoretical and experimental investigation of buckling induced delamination growth. J Mech Phys Solids 41:749–782

Pierson HO (1996) Handbook of refractory carbides and nitrides. Properties, characteristics, processing and applications. Noyes Publication, Westwood

Rabotnov YuN (1977) Elements of hereditary mechanics of solid bodies. Nauka, Moskow (in Russian)

Rzayev OG, Akbarov SD (2002) Local buckling of the elastic and viscoelastic coating around the penny-shaped interface crack. Int J Eng Sci 40:1435–1451

Rzayev OG (2002) Local buckling around an interfacial crack in a viscoelastic sandwich plate. Mech Comp Mater 38(2):233–242

Schapery RA (1962) Approximate methods of transform inversion for viscoelastic stress analyses. Proceedings of the 4th US national congress on applied mechanics, ASME, pp 1075–1086

Schapery RA (1978) A viscoelastic behaviour of composite materials. In: Broutman LJ, Krock RH (eds) Composite materials, vol 1–7. Mir, Moscow (translated from English) (vol 2: Mechanics of composite materials, pp 102–195 (in Russion, translated from English))

Shih CF, Moran B, Nakamura T (1986) Energy release rate along a three-dimensional crack front in a thermal stressed body. Int J Fracture 30:79–102

Short GJ, Guild FJ, Pavier MJ (2001) The effect of delamination geometry on the compressive failure of composite laminates. Compos Sci Technol 61:2075–2086

Thouless MD, Jensen HM, Liniger EG (1994) Delamination from edge flaws. Proc Royal Soc Lond A447:271–279

Volmir AS (1967) Stability of deformation systems. Nauko, Moskow

Wang JT, Cheng SH, Lin CC (1995) Local buckling of delaminated beams and plates using continuous analysis. J Compos Mater 29:1374–1402

Wang JS, Evans AG (1998) Measurement and analysis of buckling and buckle propagation in compressed oxide layers on superalloy substrates. Acta Mater 46:4993–5505

Zienkiewicz OC, Taylor RL (1989) Basic formulation and linear problems. The finite element method, vol 1, 4th edn. McGraw-Hill, New York



# Chapter 5
# Surface and Internal Stability Loss in the Structure of Elastic and Viscoelastic Layered Composites

In the present chapter, the approach developed and employed in the previous chapters is extended for investigation within the scope of the piecewise homogeneous body model of the surface and inner stability loss in the structure of elastic and viscoelastic layered composites. It is assumed that the composites consist of two alternating layers, the materials of which are homogeneous, elastic and non-aging viscoelastic. Two types (local and periodic-global) of initial infinitesimal imperfections in the structure of the layered composites are considered. In addition, it is assumed that the initial imperfections of the layers are co-phase. As in the previous chapters, the evolution of the above-mentioned initial imperfections is studied using the three-dimensional geometrically non-linear equations and relations of the theory of elasticity and viscoelasticity. The numerical results of the critical forces and critical time are presented and discussed. Under formulation and solution of the corresponding boundary value problems, the mechanics of curved composites developed in works listed in a survey paper (Akbarov and Guz 2004) and detailed in a monograph (Akbarov and Guz 2000) is employed.

Results analyzed in the present chapter are based on papers (Akbarov et al. 1997, 1999; Akbarov and Tekercioglu 2006, 2007; Cilli 2011).

## 5.1 Background of Related Problems

According to experimental results obtained in (Rosen 1965; Rosen and Dow 1972; Greszczuk 1975) and many others described in the monograph (Guz 1990), in many cases surface and internal instability (buckling) of fibrous and layered composite materials can be taken as the initial stage of fracture of these materials under compression along reinforcing components. Two subject areas have emerged in the mechanics of fracture of compressed composite materials due to the buckling of reinforcing elements. The first area includes various approximate

S. D. Akbarov, *Stability Loss and Buckling Delamination*,
DOI: 10.1007/978-3-642-30290-9_5, © Springer-Verlag Berlin Heidelberg 2013

design models (distribution of the compressive load between the filler and the binder, application of one and two-dimensional applied theories to stability analysis, etc.). The studies in (Dow and Gruntfest 1960) and (Rosen 1965) were a few of the first in this area, to which the works reviewed in the paper (Budiansky and Fleck 1994) and the investigations carried out in (Bowden et al. 1998; Huang and Suo 2002; Balint and Hutchinson 2003; Huck et al. 2000) and others may also be assigned.

The second area uses the Three-Dimensional Linearized Theory of Stability (TDLTS) of deformable bodies to study the fracture mechanisms for composite materials. Consistent consideration of these investigations has been made in the monograph (Guz 1990) and a detailed review is given in the paper (Babich et al. 2001). It follows from these references that the investigations regarding this second area have been studied within the scope of the piecewise homogeneous body model as well as within the scope of the continuum approach.

Note that within the scope of the piecewise homogeneous body model the phenomena of internal and surface buckling are defined as follows. It is assumed that the internal as well as the surface stability loss occurs in the structure of the material under certain relations between the stiffness and the geometric parameters of the filler and the binder and does not depend on the dimensions or the shape of the structural element. As applied to various structural elements, the conditions of existence of the internal as well as the surface instability can be represented as $min|p_{cr.}| < |p_{cr.}^{se}|$, $\ell_{cr.} \ll L$, where $min|p_{cr.}|$ and $|p_{cr.}^{se}|$ are the critical loads corresponding to the internal or surface instability and the buckling of the whole structural element in question, respectively. $\ell_{cr.}$ is the half-wavelength of the mode of the internal or surface instability and $L$ is the characteristic (minimum) dimension of the structural element. Thus, the phenomenon of internal or surface instability exists in the case where the dependence $p_{cr.} = p_{cr.}(\chi)$ has a well-defined minimum under $\chi \neq 0$ and the critical value of the external load is determined as $min|p_{cr.}|$ (where $\chi = \pi h/\ell$ for layered materials or $\chi = \pi R/\ell$ for fibrous materials; $h$ is the thickness of the reinforcing layer, $R$ is the radius of the fibre cross-section and $\ell$ is the half-wavelength of the instability mode). The value of the wave-generation parameter $\chi$ corresponding to $min|p_{cr.}|$ is taken as its critical value and is denoted as $\chi_{cr.} = \pi h/\ell_{cr.}$

It should be noted that the foregoing investigations have been made by employing the Euler approach. In these investigations (for example, in the paper (Guz et al. 2000)) the results regarding the viscoelastic materials were obtained in the framework of only the critical deformation method (Gerard and Gilbert 1958) which is very approximate. Consequently, in the above listed investigations in a very strict sense, it was assumed that the mechanical properties of the considered materials are time-independent.

For investigation of the internal and surface stability loss of the time-dependent materials, the approach based on the initial imperfection stability loss criterion may be used within the scope of the three-dimensional geometrically non-linear field equations of the theory of viscoelastic bodies. The period of time during which the initial infinitesimal imperfection of the reinforcing layers or fibres starts to increase and grow indefinitely is taken as the stability loss criterion. Note that this criterion is the analog of the well-known stability loss criterion of beams, plates and shells fabricated from time-dependent materials (Hoff 1954). Attempts in this field were made in works (Akbarov et al. 1997, 1999; Akbarov and Tekercioglu 2006, 2007; Cilli 2011). This chapter is devoted to consideration and discussion of the results of these works.

## 5.2 Stability Loss in the Structure of Elastic and Viscoelastic Layered Composites with Periodical Initial Imperfections

### *5.2.1 Formulation of the Problem on the Determination of the Stress–Strain State in a Layered Composite with an Arbitrary Number of Layers with Initially Infinitesimal Imperfections*

We consider an infinite elastic body, reinforced by an arbitrary number of non-intersecting layers that have infinitesimal initial imperfections. Values related to matrix will be denoted by upper indices (1), to the filler by upper indices (2). Position of points of the $m^{(k)}$th layer we determine by Lagrange coordinates in the Cartesian system of coordinates $O_m^{(k)}x_1^{(m)}x_2^{(k)}x_3^{(k)}$ which are obtained from the system of coordinate $Ox_1x_2x_3$ by parallel transfer along the $Ox_2$ axis. We isolate from the composite the part at $x_{1m}^{(k)} = const$ and $x_{3m}^{(k)} = const$ shown in Fig. 5.1. The reinforcing layers will be assumed to be located in planes $O_m^{(k)}x_1^{(m)}x_3^{(k)}$ and the thickness of each layer will be assumed constant. We suppose that the matrix and filler materials are homogeneous, anisotropic and non-aging viscoelastic.

Now, in the geometrical non-linear statement and within the scope of the piecewise homogeneous body model, we attempt to investigate the stress–strain distribution in the composite considered under action of uniformly distributed normal forces of intensity $p_1$ ($p_3$) in the direction $Ox_1$ ($Ox_3$) axis at infinity. For this purpose for each layer, we write the equilibrium equations, mechanical and geometrical relations as follows:

$$\frac{\partial}{\partial x_{jm}^{(k)}}\left[\sigma_{jn}^{(k)m}\left(\delta_i^n + \frac{\partial u_i^{(k)m}}{\partial x_{nm}^{(k)}}\right)\right] = 0,\ k = 1,2;\ m = 1,2,\ldots,$$

$$\sigma_{ij}^{(k)m} = C_{ijrs0}^{(k)m}\varepsilon_{rs}^{(k)m}(t) + \int_0^t C_{ijrs}^{(k)m}(t-\tau)\varepsilon_{rs}^{(k)m}(\tau)d\tau,$$

$$\varepsilon_{ij}^{(k)m} = \frac{1}{2}\left(\frac{\partial u_i^{(k)m}}{\partial x_{jm}^{(k)}} + \frac{\partial u_j^{(k)m}}{\partial x_{im}^{(k)}} + \frac{\partial u_n^{(k)m}}{\partial x_{jm}^{(k)}}\frac{\partial u_n^{(k)m}}{\partial x_{im}^{(k)}}\right),\ i;j;n;r;s = 1,2,3. \quad (5.1)$$

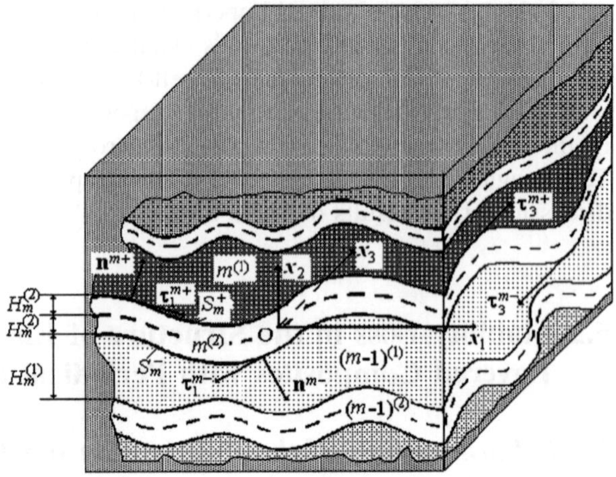


**Fig. 5.1** A part of the composite material at $x_{1m}^{(k)} = const$ and $x_{3m}^{(k)} = const$ with initially imperfections in the structure

In (5.1) conventional notation is used and are not summed with respect to the indices $k$ and $m$.

We assume that there is a complete cohesion between the layers. Denoting the upper and lower surfaces of the $m^{(2)}$th filler by $S_m^+$ and $S_m^-$ respectively, we may write these cohesion conditions as follows:

$$\left[\left(\delta_i^n + \frac{\partial u_i^{(1)m}}{\partial x_{nm}^{(1)}}\right)\sigma_{jn}^{(1)m}\right]\Bigg|_{S_m^+} n_j^{m,\pm} = \left[\left(\delta_i^n + \frac{\partial u_i^{(2)m}}{\partial x_{nm}^{(2)}}\right)\sigma_{jn}^{(2)m}\right]\Bigg|_{S_m^+} n_j^{m,\pm},$$
$$\left[\left(\delta_i^n + \frac{\partial u_i^{(1)m_0}}{\partial x_{nm_0}^{(1)}}\right)\sigma_{jn}^{(1)m_0}\right]\Bigg|_{S_m^-} n_j^{m,-} = \left[\left(\delta_i^n + \frac{\partial u_i^{(2)m}}{\partial x_{nm}^{(2)}}\right)\sigma_{jn}^{(2)m}\right]\Bigg|_{S_m^-} n_j^{m,-},$$
$$u_i^{(1)m}\Big|_{S_m^+} = u_i^{(2)m}\Big|_{S_m^+},\; u_i^{(1)m_0}\Big|_{S_m^-} = u_i^{(2)m}\Big|_{S_m^-}, \tag{5.2}$$

where $m_0 = m - 1$ and $n_j^{m,\pm}$ are the components of the normal to the surfaces $S_m^{\pm}$.

The initial imperfections of the filler layers are given thorough the equations of its middle surface. These equations are presented as follows:

$$x_{2m}^{(2)} = F_m\left(x_{1m}^{(2)}, x_{3m}^{(2)}\right) = \varepsilon f_m\left(x_{1m}^{(2)}, x_{3m}^{(2)}\right), \tag{5.3}$$

where ε is a dimensionless infinitesimal parameter $(0 \leq \varepsilon \ll 1)$ whose geometrical meaning will become apparent in a specific form of the function (5.3).

This gives the general formulation of the problems on the determination of the stress–strain state in the composite under consideration. After the determination of the corresponding initial-imperfection stability loss criterion, which will be detailed below, we find the critical values of the external forces or of the time.

### 5.2.2 Method of Solution

First, we consider some preparatory procedures for the description of the solution method.

#### 5.2.2.1 The Equations of Contact Surfaces

Using the condition of uniform filler layer thickness, and Eq. (5.3) we attempt to obtain the equation for the interface surfaces $S_m^{\pm}$. We introduce the parameters $t_{1m}$ and $t_{3m}$, where $t_{1m}, t_{3m} \in (-\infty, +\infty)$ and rewrite the Eq. (5.3) in the following form

$$x_{1m}^{(2)} = t_{1m},\ x_{3m}^{(2)} = t_{3m},\ x_{2m}^{(2)} = F_m(t_{1m}, t_{3m}) = \varepsilon f_m(t_{1m}, t_{3m}). \tag{5.4}$$

We assume that the function $F_m(t_{1m}, t_{3m})$ and its first order partial derivatives are continuous, and from (5.4) we obtain the expressions

$$\frac{\partial x_{2m}^{(2)}}{\partial x_{1m}^{(2)}} = \frac{\partial F_m(t_{1m}, t_{3m})}{\partial t_{1m}},\ \frac{\partial x_{2m}^{(2)}}{\partial x_{3m}^{(2)}} = \frac{\partial F_m(t_{1m}, t_{3m})}{\partial t_{3m}}, \tag{5.5}$$

for the slopes of the tangent line to the surface (5.4) at $x_{3m}^{(2)} = const$ and at $x_{1m}^{(2)} = const$, respectively. Taking (5.5) into account, we can write the equation of the straight line passing through the point $(t_{1m}, t_{3m}, F_m(t_{1m}, t_{3m}))$ and perpendicular to these tangent lines as follows:

$$\begin{aligned} x_{1m}^{(2)} - t_{1m} &= -\frac{\partial F_m(t_{1m},t_{3m})}{\partial t_{1m}}\left(x_{2m}^{(2)} - F_m(t_{1m}, t_{3m})\right),\\ x_{3m}^{(2)} - t_{3m} &= -\frac{\partial F_m(t_{1m},t_{3m})}{\partial t_{3m}}\left(x_{2m}^{(2)} - F_m(t_{1m}, t_{3m})\right). \end{aligned} \tag{5.6}$$

As the surfaces $S_m^{\pm}$ and the line (5.6) have common points, the coordinates of the intersections with surfaces must satisfy Eq. (5.6). Moreover, taking into account the uniformity of the filler layer thickness, we obtain the following system of equations for the coordinates of the points of the surfaces $S_m^{\pm}$:

$$\begin{aligned} &x_{1m}^{(2)\pm} - t_{1m} = -\frac{\partial F_m(t_{1m},t_{3m})}{\partial t_{1m}}\left(x_{2m}^{(2)\pm} - F_m(t_{1m}, t_{3m})\right),\\ &x_{3m}^{(2)\pm} - t_{3m} = -\frac{\partial F_m(t_{1m},t_{3m})}{\partial t_{3m}}\left(x_{2m}^{(2)\pm} - F_m(t_{1m}, t_{3m})\right),\\ &\left(x_{1m}^{(2)\pm} - t_{1m}\right)^2 + \left(x_{2m}^{(2)\pm} - F_m(t_{1m}, t_{3m})\right)^2 + \left(x_{3m}^{(2)\pm} - t_{3m}\right)^2 = \left(H_m^{(2)}\right)^2. \end{aligned} \tag{5.7}$$

In (5.7) $x_{im}^{(2)\pm}(i = 1, 2, 3)$ are the coordinates of the surfaces $S_m^{\pm}$; $H_m^{(2)}$ is the half-thickness of the $m$th filler layer.

Thus, from (5.7) we obtain the following equation for the surfaces $S_m^{\pm}$:

$$x_{1m}^{(2)\pm} = t_{1m} \mp \frac{H_m^{(2)}}{V(t_{1m}, t_{3m})} \frac{\partial F_m(t_{1m}, t_{3m})}{\partial t_{1m}},$$
$$x_{2m}^{(2)\pm} = F_m(t_{1m}, t_{3m}) \pm \frac{H_m^{(2)}}{V(t_{1m}, t_{3m})},$$
$$x_{3m}^{(2)\pm} = t_{3m} \mp \frac{H_m^{(2)}}{V(t_{1m}, t_{3m})} \frac{\partial F_m(t_{1m}, t_{3m})}{\partial t_{3m}}, \quad (5.8)$$

where

$$V(t_{1m}, t_{3m}) = \left[1 + \left(\frac{\partial F_m(t_{1m}, t_{3m})}{\partial t_{1m}}\right)^2 + \left(\frac{\partial F_m(t_{1m}, t_{3m})}{\partial t_{3m}}\right)^2\right]^{\frac{1}{2}}. \quad (5.9)$$

In all further investigations we will assume that

$$\varepsilon^2 \left[\left(\frac{\partial f_m(t_{1m}, t_{3m})}{\partial t_{1m}}\right)^2 + \left(\frac{\partial f_m(t_{1m}, t_{3m})}{\partial t_{3m}}\right)^2\right] \ll 1. \quad (5.10)$$

Using (5.8) and (5.9) and doing some calculations we obtain the following expressions for the components of the unit normal vector to the surfaces $S_m^{\pm}$:

$$n_i^{m,\pm} = \frac{A_{im}^{\pm}(t_{1m}, t_{3m})}{A_m^{\pm}(t_{1m}, t_{3m})}, \qquad i = 1, 2, 3, \quad (5.11)$$

where

$$A_{1m}^{\pm} = \frac{\partial x_{3m}^{(2)\pm}}{\partial t_{1m}} \frac{\partial x_{2m}^{(2)\pm}}{\partial t_{3m}} - \frac{\partial x_{2m}^{(2)\pm}}{\partial t_{1m}} \frac{\partial x_{3m}^{(2)\pm}}{\partial t_{3m}}, \; A_{2m}^{\pm} = \frac{\partial x_{1m}^{(2)\pm}}{\partial t_{1m}} \frac{\partial x_{3m}^{(2)\pm}}{\partial t_{3m}} - \frac{\partial x_{3m}^{(2)\pm}}{\partial t_{1m}} \frac{\partial x_{1m}^{(2)\pm}}{\partial t_{3m}},$$
$$A_{3m}^{\pm} = \frac{\partial x_{1m}^{(2)\pm}}{\partial t_{1m}} \frac{\partial x_{2m}^{(2)\pm}}{\partial t_{3m}} - \frac{\partial x_{2m}^{(2)\pm}}{\partial t_{1m}} \frac{\partial x_{1m}^{(2)\pm}}{\partial t_{3m}}, \; A_m(t_{1m}, t_{3m})$$
$$= \left[(A_{1m}(t_{1m}, t_{3m}))^2 + (A_{2m}(t_{1m}, t_{3m}))^2 + (A_{3m}(t_{1m}, t_{3m}))^2\right]^{\frac{1}{2}}. \quad (5.12)$$

Taking (5.10) into account we represent the expressions (5.8) and (5.11) as power series in the small parameter $\varepsilon$ as follows:

$$x_{1m}^{(2)^{\pm}} = t_{1m} \mp \varepsilon H_m^{(2)} \tfrac{\partial f_m}{\partial t_{1m}} + O(\varepsilon^3), \; x_{2m}^{(2)^{\pm}} = \pm H_m^{(2)} \pm \varepsilon f_m + O(\varepsilon^2),$$
$$x_{3m}^{(2)^{\pm}} = t_{3m} \mp \varepsilon H_m^{(2)} \tfrac{\partial f_m}{\partial t_{3m}} + O(\varepsilon^3), \; n_1^{m,\pm} = -\varepsilon \tfrac{\partial f_m}{\partial t_{1m}} + O(\varepsilon^3), \quad (5.13)$$
$$n_2^{m,\pm} = 1 + O(\varepsilon^2), \; n_3^{m,\pm} = -\varepsilon \tfrac{\partial f_m}{\partial t_{3m}} + O(\varepsilon^3).$$

#### 5.2.2.2 The Employing the Boundary form Perturbation Technique

First, we represent the sought values of the $m^{(k)}$th layer as power series in the parameter $\varepsilon$ as follows:

$$\left\{\sigma_{ij}^{(k)m};\varepsilon_{ij}^{(k)m};u_i^{(k)m}\right\}=\sum_{q=1}^{\infty}\varepsilon^q\left\{\sigma_{ij}^{(k)m,q};\varepsilon_{ij}^{(k)m,q};u_i^{(k)m,q}\right\}. \tag{5.14}$$

Substituting the series expression (5.14) into the field Eq. (5.1) and grouping by equal powers of $\varepsilon$ we obtain a closed system of field equations for each approximation. Owing to the linearity of the constitutive relations in (5.1) they will be satisfied separately for each approximation in (5.14). The remaining relations obtained from field equations in (5.1) for the $q$th approximation contain the values of all previous approximations. As we will use only the zeroth and the first approximations below, therefore here we write the field equations for the first approximation only.

$$\frac{\partial}{\partial x_{jm}^{(k)}}T_{ji}^{(k)m,1}=0,$$
$$\varepsilon_{ij}^{(k)m,1}=\frac{1}{2}\left(\delta_i^n+\frac{\partial u_n^{(k)m,0}}{\partial x_{im}^{(k)}}\right)\frac{\partial u_n^{(k)m,1}}{\partial x_{jm}^{(k)}}+\frac{1}{2}\left(\delta_i^n+\frac{\partial u_n^{(k)m,0}}{\partial x_{jm}^{(k)}}\right)\frac{\partial u_n^{(k)m,1}}{\partial x_{im}^{(k)}}, \tag{5.15}$$

where

$$T_{ji}^{(k)m,1}=\left(\delta_i^n+\frac{\partial u_i^{(k)m,0}}{\partial x_{nm}^{(k)}}\right)\sigma_{jn}^{(k)m,1}+\sigma_{jn}^{(k)m,0}\frac{\partial u_i^{(k)m,0}}{\partial x_{nm}^{(k)}}. \tag{5.16}$$

Note that the field equations in (5.1) hold completely for the zeroth approximation.

Now we consider obtaining the contact conditions for each approximation by employing boundary form perturbation technique. For this purpose we rewrite the equation of the surfaces $S_m^+$ and $S_m^-$ in the system of coordinates $O_m^{(1)}x_{1m}^{(1)}x_{2m}^{(1)}x_{3m}^{(1)}$ and $O_{m_1}^{(1)}x_{1m_1}^{(1)}x_{2m_1}^{(1)}x_{3m_1}^{(1)}$, respectively. Taking into account the relations $x_{1m}^{(1)}=x_{1m_1}^{(1)}=x_{1m}^{(2)}$, $x_{3m}^{(1)}=x_{3m_1}^{(1)}=x_{3m}^{(2)}$, $x_{2m}^{(2)}=x_{2m}^{(1)}-H_m^{(1)}-H_m^{(2)}$, $x_{2m}^{(2)}=x_{2m_1}^{(1)}+H_m^{(1)}+H_m^{(2)}$ in Eq. (5.13) we obtain:

the equation of the surface $S_m^+$ in the system of coordinates $O_m^{(1)}x_{1m}^{(1)}x_{2m}^{(1)}x_{3m}^{(1)}$:

$$x_{1m}^{(1)+}=t_{1m}+\varepsilon H_m^{(2)}\frac{\partial f_m}{\partial t_{1m}}+O(\varepsilon^3),$$
$$x_{2m}^{(1)^+}=-H_m^{(1)}\pm\varepsilon f_m+O(\varepsilon^2),$$
$$x_{3m}^{(1)^+}=t_{3m}+\varepsilon H_m^{(2)}\frac{\partial f_m}{\partial t_{3m}}+O(\varepsilon^3); \tag{5.17}$$

the equation of the surface $S_m^-$ in the system of coordinates $O_{m_1}^{(1)}x_{1m_1}^{(1)}x_{2m_1}^{(1)}x_{3m_1}^{(1)}$:

$$\begin{aligned} x_{1m_1}^{(1)+} &= t_{1m} - \varepsilon H_m^{(2)} \frac{\partial f_m}{\partial t_{1m}} + O(\varepsilon^3), \\ x_{2m_1}^{(1)^+} &= -H_{m_1}^{(1)} + \varepsilon f_m + O(\varepsilon^2), \\ x_{3m_1}^{(1)^+} &= t_{3m} - \varepsilon H_m^{(2)} \frac{\partial f_m}{\partial t_{3m}} + O(\varepsilon^3). \end{aligned} \tag{5.18}$$

Note that the expressions for the components $n_i^{m,+}$ and $n_i^{m,-}$ in the system of coordinates $O_m^{(1)}x_{1m}^{(1)}x_{2m}^{(1)}x_{3m}^{(1)}$ and $O_{m_1}^{(1)}x_{1m_1}^{(1)}x_{2m_1}^{(1)}x_{3m_1}^{(1)}$ respectively, remain as in (5.13). Thus, taking the Eqs. (5.13), (5.17), (5.18) and (5.14) into account we obtain from (5.2):

$$\begin{aligned} &\left[\left(\delta_i^n + \sum_{q=0}^{\infty} \varepsilon^q \frac{\partial u_i^{(1)m,q}}{\partial x_{nm}^{(1)}}\right) \sum_{r=0}^{\infty} \sigma_{jn}^{(1)m,r}\right](x_{1m}^{(1)+}, x_{2m}^{(1)+}, x_{3m}^{(1)+}) n_j^{m,+} \\ &= \left[\left(\delta_i^n + \sum_{q=0}^{\infty} \varepsilon^q \frac{\partial u_i^{(2)m,q}}{\partial x_{nm}^{(1)}}\right) \sum_{r=0}^{\infty} \sigma_{jn}^{(2)m,r}\right](x_{1m}^{(2)+}, x_{2m}^{(2)+}, x_{3m}^{(2)+}) n_j^{m,+}, \\ &\left[\left(\delta_i^n + \sum_{q=0}^{\infty} \varepsilon^q \frac{\partial u_i^{(1)m_1,q}}{\partial x_{nm}^{(1)}}\right) \sum_{r=0}^{\infty} \sigma_{jn}^{(1)m_1,r}\right](x_{1m_1}^{(1)-}, x_{2m_1}^{(1)-}, x_{3m_1}^{(1)-}) n_j^{m,-} \\ &= \left[\left(\delta_i^n + \sum_{q=0}^{\infty} \varepsilon^q \frac{\partial u_i^{(2)m,q}}{\partial x_{nm}^{(1)}}\right) \sum_{r=0}^{\infty} \sigma_{jn}^{(2)m,r}\right](x_{1m}^{(2)-}, x_{2m}^{(2)-}, x_{3m}^{(2)-}) n_j^{m,-}, \\ &\sum_{q=0}^{\infty} \varepsilon^q u_i^{(1)m,q}(x_{1m}^{(1)+}, x_{2m}^{(1)+}, x_{3m}^{(1)+}) = \sum_{q=0}^{\infty} \varepsilon^q u_i^{(2)m,q}(x_{1m}^{(2)+}, x_{2m}^{(2)+}, x_{3m}^{(2)+}) \\ &\sum_{q=0}^{\infty} \varepsilon^q u_i^{(1)m_1,q}(x_{1m_1}^{(1)-}, x_{2m_1}^{(1)-}, x_{3m_1}^{(1)-}) = \sum_{q=0}^{\infty} \varepsilon^q u_i^{(2)m,q}(x_{1m}^{(2)-}, x_{2m}^{(2)-}, x_{3m}^{(2)-}). \end{aligned} \tag{5.19}$$

Expanding the functions $\sigma_{jn}^{(1)m,r}(x_{1m}^{(1)+}, x_{2m}^{(1)+}, x_{3m}^{(1)+})$, $u_i^{(1)m,q}(x_{1m}^{(1)+}, x_{2m}^{(1)+}, x_{3m}^{(1)+})$ in Taylor's series in the vicinity of $(t_{1m}, -H_m^{(1)}, t_{3m})$, the functions $\sigma_{jn}^{(2)m,r}(x_{1m}^{(2)\pm}, x_{2m}^{(2)\pm}, x_{3m}^{(2)\pm})$, $u_i^{(2)m,q}(x_{1m}^{(2)\pm}, x_{2m}^{(2)\pm}, x_{3m}^{(2)\pm})$ in the vicinity $(t_{1m}, \pm H_m^{(2)}, t_{3m})$, and the functions $\sigma_{jn}^{(1)m_1,r}(x_{1m_1}^{(1)-}, x_{2m_1}^{(1)-}, x_{3m_1}^{(1)-})$, $u_i^{(1)m_1,q}(x_{1m_1}^{(1)-}, x_{2m_1}^{(1)-}, x_{3m_1}^{(1)-})$ in the vicinity $(t_{1m}, +H_{m_1}^{(1)}, t_{3m})$, and grouping by identical powers of $\varepsilon$ we obtain the contact conditions for each approximation. Here we consider the detail of these conditions for the zeroth and the first approximations only. Thus, for the zeroth approximation the contact conditions are

$$\left[\left(\delta_i^n + \frac{\partial u_i^{(1)m,0}}{\partial x_{nm}^{(1)}}\right)\sigma_{jn}^{(1)m,0}\right]\Bigg|_{(t_{1m},-H_m^{(1)},t_{3m})}$$
$$= \left[\left(\delta_i^n + \frac{\partial u_i^{(2)m,0}}{\partial x_{nm}^{(2)}}\right)\sigma_{jn}^{(2)m,0}\right]\Bigg|_{(t_{1m},+H_m^{(2)},t_{3m})},$$
$$\left[\left(\delta_i^n + \frac{\partial u_i^{(1)m_0,0}}{\partial x_{nm_0}^{(1)}}\right)\sigma_{jn}^{(1)m_0,0}\right]\Bigg|_{(t_{1m},+H_{m0}^{(1)},t_{3m})}$$
$$= \left[\left(\delta_i^n + \frac{\partial u_i^{(2)m,0}}{\partial x_{nm}^{(2)}}\right)\sigma_{jn}^{(2)m,0}\right]\Bigg|_{(t_{1m},-H_m^{(2)},t_{3m})},$$
$$u_i^{(1)m,0}\Big|_{(t_{1m},-H_m^{(1)},t_{3m})} = u_i^{(2)m,0}\Big|_{(t_{1m},+H_m^{(2)},t_{3m})},$$
$$u_i^{(1)m_0,0}\Big|_{(t_{1m},+H_{m0}^{(1)},t_{3m})} = u_i^{(2)m,0}\Big|_{(t_{1m},-H_m^{(2)},t_{3m})}. \tag{5.20}$$

For the first approximation we can write the following contact relations from (5.19).

$$T_{i2}^{(1)m,1}\Big|_{(t_{1m},-H_m^{(1)},t_{3m})} - T_{i2}^{(2)m,1}(t_{1m},+H_m^{(2)},t_{3m})$$
$$= \Phi_{1i}(\sigma_{ij}^{(1)m,0}, u_i^{(1)m,0}, \sigma_{ij}^{(2)m,0}, u_i^{(2)m,0}, f_m)$$
$$T_{i2}^{(1)m_0,1}\Big|_{(t_{1m},+H_{m0}^{(1)},t_{3m})} - T_{i2}^{(2)m,1}(t_{1m},-H_m^{(2)},t_{3m})$$
$$= \Phi_{2i}(\sigma_{ij}^{(1)m_0,0}, u_i^{(1)m_0,0}, \sigma_{ij}^{(2)m,0}, u_i^{(2)m,0}, f_m),$$
$$u_i^{(1)m,1}\Big|_{(t_{1m},-H_m^{(1)},t_{3m})} - u_i^{(2)m,1}(t_{1m},+H_m^{(2)},t_{3m}) = \varphi_{1i}(u_i^{(1)m,0}, u_i^{(2)m,0}, f_m)$$
$$u_i^{(1)m_0,1}\Big|_{(t_{1m},+H_{m0}^{(1)},t_{3m})} - u_i^{(2)m,1}(t_{1m},-H_m^{(2)},t_{3m}) = \varphi_{2i}(u_i^{(1)m_0,0}, u_i^{(2)m,0}, f_m). \tag{5.21}$$

In Eq. (5.21) the $\Phi_{1i}$, $\Phi_{2i}$, $\varphi_{1i}$ and $\varphi_{2i}$ are known functions defined in the solution procedure; we omit details. Further we will assume that $\partial u_i^{(k)m,0} / \partial x_{nm}^{(k)} \ll 1$ and $(\delta_i^n + \partial u_i^{(k)m,0} / \partial x_{nm}^{(k)}) \approx \delta_i^n$ in Eqs. (5.15), (5.20) and (5.21). Therefore, in the zeroth approximation, the non-linear terms in equations related to this approximation can be neglected with a sufficiently high accuracy.

We employ the foregoing solution procedure for investigation of the plane-strain state in a laminated composite material with an infinite number of alternating layers in the direction of the $Ox_2$ axis. Consider two cases:

Case 1. Assume that all filler layers have initial infinitesimal imperfections with the co-phase periodically plane-curved form.

Case 2. Assume that a single filler layer has only initial infinitesimal imperfections with the periodically plane-curved form.

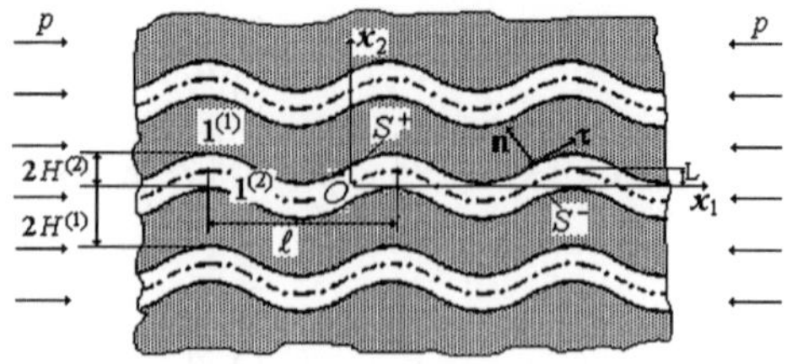


**Fig. 5.2** The structure of the composite material with co-phase periodically plane-curved layers

### 5.2.2.3 Plane-Strain State in Case 1 (Co-Phase Imperfections of Layers)

We assume that the layers are periodically curved in the direction of the $Ox_1$ axis, and the plane-strain takes place in the $Ox_1x_2$ plane under the action, at infinity, of uniformly distributed normal forces with intensity $p$ in the direction of the $Ox_1$ axis. The structure of this material is shown in Fig. 5.2.

The periodicity of the composite structure in the direction of the $Ox_2$ axis has period $2(H^{(1)} + H^{(2)})$, where $2H^{(1)}$ is a thickness of the matrix layer and $2H^{(2)}$ is a thickness of the filler layer. Among the layers we single out two of them, i.e. $1^{(1)}$, $1^{(2)}$ (Fig. 5.2), and discuss them below. The Eq. (5.3) for the middle surface of the $1^{(2)}$th layer we take as follows:

$$x_2^{(2)} = F(x_1^{(2)}) = \varepsilon f(x_1^{(2)}) = L \sin\left(\frac{2\pi x_1^{(2)}}{\ell}\right) = \varepsilon\ell \sin\left(\frac{2\pi x_1^{(2)}}{\ell}\right), \tag{5.22}$$

where $L$ is the curving (initial imperfection) rise, $\ell$ is the wavelength of the initial imperfection mode of the middle surface of the $1^{(2)}$th layer. With the assumption $L \ll \ell$, the value of the small parameter is selected as $\varepsilon = L/\ell \ll 1$.

We will omit the index $m$ which enters the expression given in the previous sub-subsection. We assume that the materials of the layers are homogeneous, isotropic and linear non-aging viscoelastic ones and the constitutive relations for each approximation are given as follows:

$$\sigma_{ij}^{(k),q} = \lambda^{(k)*}\theta^{(k),q}\delta_i^j + 2\mu^{(k)*}\varepsilon_{ij}^{(k),q}, \tag{5.23}$$

where

$$\left\{ \begin{array}{c} \lambda^{(k)*} \\ \mu^{(k)*} \end{array} \right\} g(t) = \left\{ \begin{array}{c} \lambda_0^{(k)} \\ \mu_0^{(k)} \end{array} \right\} g(t) + \int_0^t \left\{ \begin{array}{c} \lambda^{(k)} \\ \mu^{(k)} \end{array} \right\} (t-\tau) g(\tau) d\tau. \tag{5.24}$$

Here $\lambda_0^{(k)}$, $\mu_0^{(k)}$ are time independent Lamé's constants; $\lambda^{(k)}(t)$ and $\mu^{(k)}(t)$ are the kernels of the bounded operators $\lambda^{(k)*}$ and $\mu^{(k)*}$, respectively.

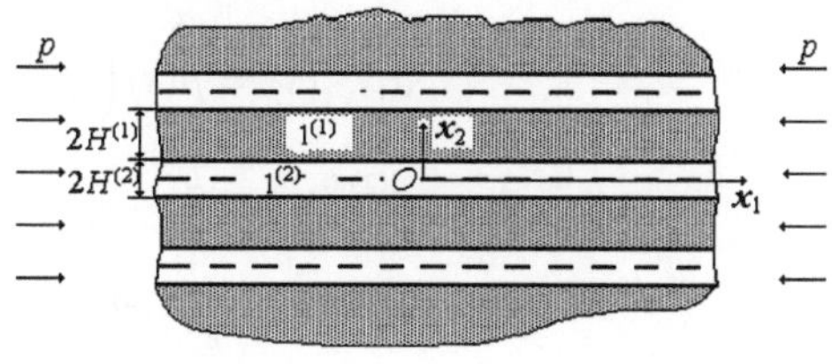


**Fig. 5.3** An ideal layout of the layers in the structure of the composite material

Taking into account the periodicity of the material structure in the direction of the $Ox_2$ axis and the assumption $\partial u_i^{(k)m,0} \big/ \partial x_{nm}^{(k)} \ll 1$ and $(\delta_i^n + \partial u_i^{(k)m,0} \big/ \partial x_{nm}^{(k)}) \approx \delta_i^n$ we can write the following field equations

$$\frac{\partial}{\partial x_j^{(k)}} \sigma_{ji}^{(k),0} = 0, \; \varepsilon_{ij}^{(k),0} = \frac{1}{2}\left(\frac{\partial u_i^{(k),0}}{\partial x_j^{(k)}} + \frac{\partial u_j^{(k),0}}{\partial x_i^{(k)}}\right) \tag{5.25}$$

and contact condition for the zeroth approximation

$$\sigma_{i2}^{(1),0}\Big|_{(t_1,\mp H^{(1)})} = \sigma_{i2}^{(2),0}\Big|_{(t_1,\pm H^{(2)})}, \; u_i^{(1),0}\Big|_{(t_1,\mp H^{(1)})} = u_i^{(2),0}\Big|_{(t_1,\pm H^{(2)})}. \tag{5.26}$$

Thus, the zeroth approximation of the problem considered corresponds to the determination of the stress-strain state in the material with an ideal layout of the layers in the structure of the composite material (Fig. 5.3) within the scope of the classical linear theory of viscoelasticity.

For solution to the problem (5.23)–(5.26) related to the zeroth approximation we employ the correspondence principle with the use of the Laplace transformation

$$\bar{\psi}(s) = \int_0^\infty \psi(t) e^{-st} dt \tag{5.27}$$

with real parameter $s > 0$.

Thus, after applying (5.27) to the Eqs. (5.23)–(5.24) and using the convolution theorem for the Laplace transformation of the constitutive relations (5.24), according to the statement of the problem, we can write the following relations for Laplace transformation of the zeroth approximation.

$$\begin{gathered} \bar{\sigma}_{12}^{(k),0} = \bar{\sigma}_{22}^{(k),0} = 0, \quad \bar{\sigma}_{11}^{(1),0}\eta^{(1)} + \bar{\sigma}_{11}^{(2),0}\eta^{(2)} = \bar{p}, \\ \eta^{(k)} = H^{(k)} \big/ (H^{(1)} + H^{(2)}), \; k = 1, 2. \end{gathered} \tag{5.28}$$

Using the Laplace transformation of the constitutive relations (5.24) and relation (5.28), we find the Laplace transformation of the zeroth approximation as follows:

$$\begin{aligned}
\bar{\sigma}_{11}^{(1),0}(s) &= \bar{p}(s)\left(\eta^{(1)} + \frac{\bar{E}^{(2)*}(s)\left(1-(\bar{\nu}^{(1)*}(s))^2\right)}{\bar{E}^{(1)*}(s)\left(1-(\bar{\nu}^{(2)*}(s))^2\right)}\right)^{-1}, \\
\bar{\sigma}_{11}^{(2),0}(s) &= \frac{\bar{E}^{(2)*}(s)\left(1-(\bar{\nu}^{(1)*}(s))^2\right)}{\bar{E}^{(1)*}(s)\left(1-(\bar{\nu}^{(2)*}(s))^2\right)}\bar{\sigma}_{11}^{(1),0}(s), \\
\bar{u}_1^{(k),0} &= \frac{1-(\bar{\nu}^{(k)*}(s))^2}{\bar{E}^{(k)*}(s)}\bar{\sigma}_{11}^{(k),0}(s)x_1, \qquad k=1,2, \\
\bar{u}_2^{(k),0} &= -\frac{\bar{\nu}^{(k)*}(s)(1+\bar{\nu}^{(k)*}(s))^2}{\bar{E}^{(k)*}(s)}\bar{\sigma}_{11}^{(k),0}(s)x_2^{(k)} + C^{(k)}(s),
\end{aligned} \tag{5.29}$$

where

$$\bar{\nu}^{(k)*}(s) = \frac{\bar{\lambda}^{(k)*}(s)}{2(\bar{\lambda}^{(k)*}(s)+\bar{\mu}^{(k)*}(s))}, \; \bar{E}^{(k)*}(s) = \frac{\bar{\mu}^{(k)*}(s)(3\bar{\lambda}^{(k)*}(s)+\bar{\mu}^{(k)*}(s))}{\bar{\lambda}^{(k)*}(s)+\bar{\mu}^{(k)*}(s)},$$
$$\bar{\lambda}^{(k)*}(s) = \lambda_0^{(k)} + \bar{\lambda}^{(k)}(s), \; \bar{\mu}^{(k)*}(s) = \mu_0^{(k)} + \bar{\mu}^{(k)}(s). \tag{5.30}$$

Consider the determination of the originals of the values of the zeroth approximation. For this purpose, we use also the method (Schapery 1962, 1978). Analyze in more detail this method with respect to determination of the inverse of the Laplace transformation given through the expressions in (5.29) and (5.30). First we calculate the values of $\bar{\sigma}_{ij}^{(k),0}(s)$ and $\bar{u}_i^{(k),0}(s)$ at $s = s_n$ $(n = 1, 2, ..N)$. After this calculation we construct the graphs among $s\bar{\sigma}_{ij}^{(k),0}(s)$, $s\bar{u}_i^{(k),0}(s)$ and $\log s$. If the curvatures of these graphs are sufficiently small, in the sense described in (Schapery 1962, 1978), then we find the inverse transformation by using the approximate relationships

$$\sigma_{ij}^{(k),0}(t) \approx s\bar{\sigma}_{ij}^{(k),0}(s)\Big|_{s=1/(2t)}, \; u_i^{(k),0}(t) \approx s\bar{u}_i^{(k),0}(s)\Big|_{s=1/(2t)}. \tag{5.31}$$

If the smallness condition is not satisfied then the relations (5.31) are not applicable; we proceed as follows. First by using limiting $\lambda^{(k)*}$ and $\mu^{(k)*}$ as $t \to \infty$, instead of these operators, we determine the protracted (long-time) stresses (we denote them by $\sigma_{ijpr}^{(k),0}$, $u_{ipr}^{(k),0}$). The stress tensor and displacement vector components are represented as

$$\sigma_{ij}^{(k),0} = \delta\sigma_{ij}^{(k),0} + \sigma_{ijpr}^{(k),0}, \; u_i^{(k),0} = \delta u_i^{(k),0} + u_{ipr}^{(k),0}. \tag{5.32}$$

To determine $\delta\sigma_{ij}^{(k),0}$ and $\delta u_i^{(k),0}$ we apply the collocation method described in (Schapery 1962, 1978), based on the values $\bar{\sigma}_{ij}^{(k),0}(s), \bar{u}_i^{(k),0}(s)$ at $s = s_q$. As an example, we consider the determination of the values $\delta\sigma_{ij}^{(k),0}$; we assume that it can be approximated by the finite Dirichlet series

$$\delta\sigma_{ij}^{(k),0} \cong \sum_{i=1}^{N} S_i e^{-\lambda_i t}, \tag{5.33}$$

where $\lambda_i$ are chosen as real constants, and $S_i$ are real unknowns. For determination of these unknowns, we obtain the system of equations

$$\sum_{i=1}^{N}\left(1+\frac{\lambda_j}{\lambda_i}\right)S_j = s\delta\bar{\sigma}_{ij}^{(k),0}(s)\Big|_{s=\lambda_i}, \tag{5.34}$$

Which are obtained from the minimization of the square deviation

$$E = \int_0^{\infty}\left(\delta\sigma_{nm}^{(k),0} - \sum_{i=1}^{N}\left(1+\frac{\lambda_j}{\lambda_i}\right)^{-1} S_i\right)^2 dt \tag{5.35}$$

with respect to each $S_i$. In (5.34), the notation $\delta\bar{\sigma}_{nm}^{(k),0}(s)$ is

$$\delta\bar{\sigma}_{nm}^{(k),0}(s) = \bar{\sigma}_{nm}^{(k),0}(s) - \sigma_{nmpr}^{(k),0}(s). \tag{5.36}$$

In (Schapery 1962, 1978) it has been shown how the values for the parameter $\lambda_i$ can be selected from the graphs of $s\bar{\sigma}_{nm}^{(k),0}(s)$ against $\log s$. We can often restrict ourselves to equally spaced $\lambda_i$ and we can improve accuracy by increasing $N$ in (5.33). The accuracy of the numerical results depends on the character of the function $p(t)$ describing the external forces with respect to time $t$. If $p(t)$ is a slowly changing continuous or constant function, method (Schapery 1962, 1978) gives results with very high accuracy.

Thus in the foregoing manner we determine the stresses $\sigma_{11}^{(k),0} = \sigma_{11}^{(k),0}(t)$ which enter the equations and relations of the first approximation. According to (5.29) we obtain from (5.15), (5.16), (5.21) and (5.23) the following equations and relations for the first approximation.

The equilibrium equations

$$\frac{\partial\sigma_{11}^{(k),1}}{\partial x_1} + \sigma_{11}^{(k),0}(t)\frac{\partial^2 u_1^{(k),1}}{\partial x_1^2} + \frac{\partial\sigma_{12}^{(k),1}}{\partial x_2} = 0,$$
$$\frac{\partial\sigma_{12}^{(k),1}}{\partial x_1} + \sigma_{11}^{(k),0}(t)\frac{\partial^2 u_2^{(k),1}}{\partial x_1^2} + \frac{\partial\sigma_{22}^{(k),1}}{\partial x_2} = 0; \tag{5.37}$$

The geometrical relations

$$\varepsilon_{ij}^{(k),1} = \frac{1}{2}\left(\frac{\partial u_i^{(k),1}}{\partial x_j^{(k)}} + \frac{\partial u_j^{(k),1}}{\partial x_i^{(k)}}\right), \quad i;j = 1,2. \tag{5.38}$$

The constitutive relations remain as in (5.23) and (5.24). Moreover, from (5.21) we obtain the following contact conditions.

$$\sigma_{i2}^{(1),1}\Big|_{(t_1;\pm H^{(1)})} - \sigma_{i2}^{(2),1}\Big|_{(t_1;\mp H^{(2)})} = \frac{df(x_1)}{dx_1}\Big|_{(t_1)} (\sigma_{11}^{(1),0} - \sigma_{11}^{(2),0})\delta_i^1,$$
$$u_i^{(1),1}\Big|_{(t_1;\pm H^{(1)})} - u_i^{(2),1}\Big|_{(t_1;\mp H^{(2)})} = \ell \sin(\alpha_1 t_1)\left(\frac{\partial u_2^{(1),0}}{\partial x_2^{(1)}} - \frac{\partial u_2^{(2),0}}{\partial x_2^{(2)}}\right)\delta_i^2. \tag{5.39}$$

Thus the determination of the first approximation is reduced to the solution to the quasi-static problem, given by Eqs. (5.37)–(5.39), (5.23) and (5.24). For further discussions we assume that the material of the filler layers is pure elastic with elastic constant $E^{(2)}$ (modulus of elasticity), $\nu^{(2)}$ (Poisson coefficient) or $\lambda^{(2)}$, $\mu^{(2)}$ (Lamé's constants), but the material of the matrix layers is linearly viscoelastic with operators

$$E^{(1)*} = E_0^{(1)}(1 - \omega_0 \Pi_\alpha^*(-\omega_0 - \omega_\infty)),\ \nu^{(1)*} = \nu_0^{(1)}\left(1 + \frac{1-2\nu_0^{(1)}}{2\nu_0^{(1)}}\omega_0 \Pi_\alpha^*(-\omega_0 - \omega_\infty)\right),$$
$$\lambda^{(1)*} = \lambda_0^{(1)}\left(1 + \frac{1-2\nu_0^{(1)}}{2\nu_0^{(1)}(1+\nu_0^{(1)})}\omega_0 \Pi_\alpha^*\left(-\frac{3}{2(1+\nu_0^{(1)})}\omega_0 - \omega_\infty\right)\right),$$
$$\mu^{(1)*} = \mu_0^{(1)}\left(1 - \frac{3}{2\nu_0^{(1)}(1+\nu_0^{(1)})}\omega_0 \Pi_\alpha^*\left(-\frac{3}{2(1+\nu_0^{(1)})}\omega_0 - \omega_\infty\right)\right), \tag{5.40}$$

where $E_0^{(1)}$ and $\nu_0^{(1)}$ are the instantaneous values of Young's modulus and Poisson coefficient, $\lambda_0^{(1)}$ and $\mu_0^{(1)}$ are the instantaneous values of Lamé's constants; $\alpha$, $\omega_0$ and $\omega_\infty$ are the rheological parameters, $\Pi_\alpha^*$ is the fractional-exponential operator (Rabotnov 1977) and this operator is determined as

$$\Pi_\alpha^*(\beta)\varphi(t) = \int_0^t \Pi_\alpha(\beta, t-\tau)\varphi(\tau)d\tau \tag{5.41}$$

where

$$\Pi_\alpha(\beta, t) = t^\alpha \sum_{n=0}^{\infty} \frac{\beta^n t^{n(1+\alpha)}}{\Gamma((n+1)(1+\alpha))}. \tag{5.42}$$

In (5.42) $\Gamma(x)$ is the Gamma function.

For simplicity we introduce the dimensionless rheological parameter $\omega = \omega_\infty/\omega_0$. As follows from Eq. (5.37) the coefficient of in this equation is a function of time $t'\left(= \omega_0^{1/(1+\beta)}t\right)$; causes difficulties: we cannot employ the principle of correspondence. We propose the following approach.

At first we investigate the behavior the functions

$$\varphi_1(t') = \frac{\sigma_{11}^{(1),0}}{E_0^{(1)}},\ \varphi_2(t') = \frac{\sigma_{11}^{(2),0}}{E^{(2)}} \tag{5.43}$$

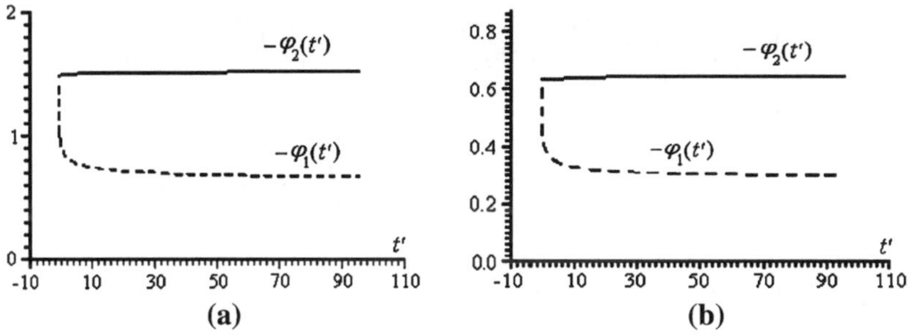


**Fig. 5.4** Graphs of the functions among $\varphi_1(t')$, $\varphi_2(t')$ and $t'$ under $\eta^{(2)} = 0.2$ (**a**) and 0.5 (**b**)

which are the coefficient of equations in (5.37). As an example, consider the case where $\omega = 0.5$, $\alpha = -0.7$, $E^{(2)}\big/E_0^{(1)} = 100$, $p\big/E_0^{(1)} = 0.3$.

Figure 5.4 shows typical graphs of functions (5.43) with $\eta^{(2)} = 0.2$ and 0.5. These graphs show that the function $\varphi_2(t')$ may be taken to be constants with respect to $t'$. However the function $\varphi_1(t')$ depends significantly on $t'$ at an early stage of the compression. Therefore, in Eq. (5.37) we replace the function $\varphi_2(t')$ by its value at $t' = 0$ or $t' = \infty$. Now we apply Laplace transformation (5.27) to Eqs. (5.37)–(5.39), (5.23) and (5.24). The Laplace transformations of the values related to the filler layer are obtained as Eqs. (5.37)–(5.39), (5.23) and (5.24) where $k = 2$ and $\sigma_{ij}^{(k),1}$, $u_i^{(k),1}$ have been replaced by their Laplace transformations $\bar{\sigma}_{ij}^{(k),1}$, $\bar{u}_i^{(k),1}$. The convolution theorem for the Laplace transformation gives the following equations from (5.37) and (5.23) with $k = 1$:

$$\begin{aligned}
&\frac{\partial\bar{\sigma}_{11}^{(1),1}}{\partial x_1} + E_0^{(1)}\int_0^\infty \varphi_1(t)\frac{\partial^2 u_1^{(1),1}}{\partial x_1^2}e^{-st}dt + \frac{\partial\bar{\sigma}_{12}^{(1),1}}{\partial x_2^{(1)}} = 0,\\
&\frac{\partial\bar{\sigma}_{12}^{(1),1}}{\partial x_1} + E_0^{(1)}\int_0^\infty \varphi_1(t)\frac{\partial^2 u_2^{(1),1}}{\partial x_1^2}e^{-st}dt + \frac{\partial\bar{\sigma}_{22}^{(1),1}}{\partial x_2^{(1)}} = 0,\\
&\bar{\sigma}_{ij}^{(1),1} = \bar{\lambda}^{(1)*}\left(\frac{\partial\bar{u}_1^{(1),1}}{\partial x_1} + \frac{\partial\bar{u}_2^{(1),1}}{\partial x_2^{(1)}}\right)\delta_i^j + 2\bar{\mu}^{(1)*}\bar{\varepsilon}_{ij}^{(1),1},\ ij = 11; 12; 22.
\end{aligned} \tag{5.44}$$

Now we consider the integral terms in the first two equations of (5.44), i.e. consider the integrals

$$\int_0^\infty \varphi_1(t)\frac{\partial^2 u_j^{(1),1}}{\partial x_1^2}e^{-st}dt; \qquad j = 1, 2. \tag{5.45}$$

Since

$$\varphi_1(t) \le 0 \text{ and } |\varphi_1(\infty)| \le |\varphi_1(t)| \le |\varphi_1(0)| \text{ for each } t \in (0, +\infty), \tag{5.46}$$

we can write the following relations for each fixed $s > 0$:

$$\left| \varphi_1(\infty) \int_0^{+\infty} \frac{\partial^2 u_i^{(1),1}}{\partial x_1^2} e^{-st} dt \right| \leq \left| \int_0^{+\infty} \varphi_1(t) \frac{\partial^2 u_i^{(1),1}}{\partial x_1^2} e^{-st} dt \right|$$
$$\leq \left| \varphi_1(0) \int_0^{+\infty} \frac{\partial^2 u_i^{(1),1}}{\partial x_1^2} e^{-st} dt \right|. \tag{5.47}$$

So that, if $s > 0$,

$$\int_0^{+\infty} \varphi_1(t) \frac{\partial^2 u_i^{(1),1}}{\partial x_1^2} e^{-st} dt \cong \varphi_1(t_*) \int_0^{+\infty} \frac{\partial^2 u_i^{(1),1}}{\partial x_1^2} e^{-st} dt = \varphi_1(t_*) \frac{\partial^2 \bar{u}_i^{(1),1}}{\partial x_1^2}, \tag{5.48}$$

where $t_*$ is a fixed value of $t$.

Thus, we obtain from (5.48) and (5.44) the equilibrium equation in the displacement terms.

$$c_1^{(k)} \frac{\partial^2 \bar{u}_1^{(k),1}}{\partial x_1^2} + c_2^{(k)} \frac{\partial^2 \bar{u}_2^{(k),1}}{\partial x_1 \partial x_2^{(k)}} + c_3^{(k)} \frac{\partial^2 \bar{u}_1^{(k),1}}{\partial (x_2^{(k)})^2} = 0,$$
$$c_4^{(k)} \frac{\partial^2 \bar{u}_2^{(k),1}}{\partial x_1^2} + c_5^{(k)} \frac{\partial^2 \bar{u}_1^{(k),1}}{\partial x_1 \partial x_2^{(k)}} + c_6^{(k)} \frac{\partial^2 \bar{u}_2^{(k),1}}{\partial (x_2^{(k)})^2} = 0, \tag{5.49}$$

where

$$c_1^{(k)} = \frac{1 - \bar{v}^{(k)}}{1 - 2\bar{v}^{(k)}} + \frac{\sigma_{11}^{(k),0}(t_*)}{E_0^{(k)}}(1 + \bar{v}^{(k)}),\ c_2^{(k)} = \frac{1}{2(1 - 2\bar{v}^{(k)})},\ c_3^{(k)} = \frac{1}{2},$$
$$c_4^{(k)} = \frac{1}{2} + \frac{\sigma_{11}^{(k),0}(t_*)}{E_0^{(k)}}(1 + \bar{v}^{(k)}),\ c_5^{(k)} = c_2^{(k)},\ c_6^{(k)} = \frac{1 - \bar{v}^{(k)}}{1 - 2\bar{v}^{(k)}}. \tag{5.50}$$

According to (5.22) and (5.39), we represent the displacements in the first approximation as follows:

$$\bar{u}_1^{(k),1} = \bar{f}_1^{(k),1}(x_2^{(k)}) \cos\left(\frac{2\pi}{\ell} x_1\right),\ \bar{u}_2^{(k),1} = \bar{f}_2^{(k),1}(x_2^{(k)}) \sin\left(\frac{2\pi}{\ell} x_1\right). \tag{5.51}$$

Substituting (5.51) in (5.49) and doing some operations we derive the following equation for the function $\bar{f}_2^{(k),1}(x_2^{(k)})$:

$$d_1^{(k)} \frac{d^4 \bar{f}_2^{(k),1}}{d(x_2^{(k)})^4} + d_2^{(k)} \frac{d^2 \bar{f}_2^{(k),1}}{d(x_2^{(k)})^2} + d_3^{(k)} \bar{f}_2^{(k),1} = 0, \tag{5.52}$$

where

$$d_1^{(k)} = \frac{c_3^{(k)} c_6^{(k)}}{c_5^{(k)}},\ d_2^{(k)} = c_2^{(k)} - \frac{c_1^{(k)} c_6^{(k)}}{c_5^{(k)}} - \frac{c_3^{(k)} c_4^{(k)}}{c_5^{(k)}},\ d_3^{(k)} = \frac{c_1^{(k)} c_4^{(k)}}{c_5^{(k)}}. \tag{5.53}$$

We select the particular solution to the Eq. (5.52) as

$$\bar{f}_2^{(k),1} = D_1^{(k)} \exp(\lambda^{(k)} \alpha_1 x_2^{(k)}),\ \alpha_1 = 2\pi/\ell. \tag{5.54}$$

Substituting (5.54) in (5.52) we obtain the following characteristic equation:

$$(\lambda^{(k)})^4 + a_1^{(k)} (\lambda^{(k)})^2 + a_2^{(k)} = 0, \tag{5.55}$$

where $a_1^{(k)} = d_2^{(k)} \big/ d_1^{(k)}$, $a_2^{(k)} = d_3^{(k)} \big/ d_1^{(k)}$.

According to the sign of the discriminant $D = (a_1^{(k)})^2 \big/ 4 - a_2^k$, we can determine the solution to the Eq. (5.55) and explicit form of the function $\bar{f}_2^{(k),1}$. As an instance, we assume that $D = (a_1^{(k)})^2 \big/ 4 - a_2^k < 0$, then we can represent the roots of the characteristic Eq. (5.55) as follows:

$$\lambda_1^{(k)} = \alpha_1^{(k)} + i\beta_1^{(k)},\ \lambda_2^{(k)} = \alpha_1^{(k)} - i\beta_1^{(k)},\ \lambda_3^{(k)} = -\lambda_1^{(k)},\ \lambda_4^{(k)} = -\lambda_2^{(k)}, \tag{5.56}$$

where

$$\begin{aligned}
\alpha_1^{(k)} &= \left( \left( (\alpha^{(k)})^2 + (\beta^{(k)})^2 \right)^{\frac{1}{4}} \left( \tfrac{1}{2} + \tfrac{\alpha^{(k)}}{2} \left( (\alpha^{(k)})^2 + (\beta^{(k)})^2 \right)^{-\frac{1}{2}} \right) \right)^{\frac{1}{2}}, \\
\beta_1^{(k)} &= \left( \left( (\alpha^{(k)})^2 + (\beta^{(k)})^2 \right)^{\frac{1}{4}} \left( \tfrac{1}{2} - \tfrac{\alpha^{(k)}}{2} \left( (\alpha^{(k)})^2 + (\beta^{(k)})^2 \right)^{-\frac{1}{2}} \right) \right)^{\frac{1}{2}}, \\
\alpha^{(k)} &= -\tfrac{a_1^{(k)}}{2},\ \beta^{(k)} = \tfrac{1}{2}\sqrt{4a_2^{(k)} - (a_1^{(k)})^2}.
\end{aligned} \tag{5.57}$$

Thus, the function $\bar{f}_2^{(k),1}(x_2^{(k)})$ is determined as follows:

$$\begin{aligned}
\bar{f}_2^{(k),1}(x_2^{(k)}) = & D_1^{(k)} \cos(\beta_1^{(k)} \alpha_1 x_2^{(k)}) \cosh(\alpha_1^{(k)} \alpha_1 x_2^{(k)}) \\
& + D_2^{(k)} \cos(\beta_1^{(k)} \alpha_1 x_2^{(k)}) \sinh(\alpha_1^{(k)} \alpha_1 x_2^{(k)}) \\
& + D_3^{(k)} \sin(\beta_1^{(k)} \alpha_1 x_2^{(k)}) \cosh(\alpha_1^{(k)} \alpha_1 x_2^{(k)}) \\
& + D_4^{(k)} \sin(\beta_1^{(k)} \alpha_1 x_2^{(k)}) \sinh(\alpha_1^{(k)} \alpha_1 x_2^{(k)}).
\end{aligned} \tag{5.58}$$

Taking the expressions of the contact relations (5.39) into account we conclude that the function $\bar{f}_2^{(k),1}(x_2^{(k)})$ must be an even function, i.e. $D_2^{(k)} = D_3^{(k)} = 0$ and

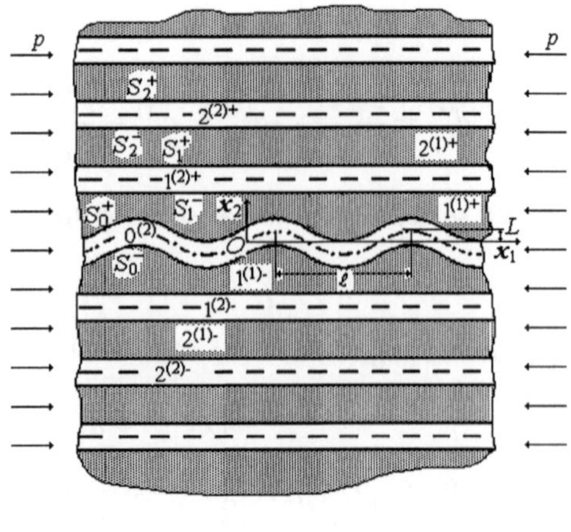


**Fig. 5.5** Structure of the composite material with a single periodically curved layer

$$\bar{f}_2^{(k),1}(x_2^{(k)}) = D_1^{(k)} \cos(\beta_1^{(k)} \alpha_1 x_2^{(k)}) \cosh(\alpha_1^{(k)} \alpha_1 x_2^{(k)}) + D_4^{(k)} \sin(\beta_1^{(k)} \alpha_1 x_2^{(k)}) \sinh(\alpha_1^{(k)} \alpha_1 x_2^{(k)}). \quad (5.59)$$

In a similar manner we can determine the function $\bar{f}_2^{(k),1}(x_2^{(k)})$ in the cases where the discriminant $D = (a_1^{(k)})^2 / 4 - a_2^k \geq 0$.

Using the usual procedure we determine the unknown constants $D_1^{(k)}$ and $D_4^{(k)}$ from the contact condition (5.39) and in this way we also find the Laplace transformations of the values of the first approximation. To find the original of these values we employ the method (Schapery 1962, 1978).

#### 5.2.2.4 Plane-Strain State in Case 2 (Imperfection of a Single Layer)

Consider a composite consisting of alternating layers of two materials and assume that only one filler layer, i.e. the filler layer denoted by $0^{(2)}$ has an initial periodical curving (Fig. 5.5). All other layers are ideal. Suppose that the curving of the filler layer $0^{(2)}$ depends only $x_1$, and the uniformly distributed normal compressing forces are applied at infinity with the intensity $p$ in the direction of the $Ox_1$ axis. We investigate the plane-strain state in this composite within the scope of the zeroth and first approximations.

Because of the symmetry of the problem, the Laplace transformation of the zeroth approximation may be written as follows:

$$\bar{\sigma}_{ij}^{(k)m,0} = \bar{\sigma}_{ij}^{(k),0}, \; \bar{u}_1^{(k)m,0} = \bar{u}_1^{(k),0}$$
$$\bar{u}_2^{(k)m,0} = -\frac{\bar{\nu}^{(k)}(1+\bar{\nu}^{(k)})}{E_0^{(k)}} \bar{\sigma}_{11}^{(k)m,0} x_{2m}^{(k)} + C^{(k)m}, \; C^{(k)m} = const, \quad (5.60)$$

where $\bar{\sigma}_{ij}^{(k),0}$, $\bar{u}_i^{(k),0}$ are given by formulae (5.29).

For the first approximation we obtain the following contact conditions on the surface $S_0^+$ (Fig. 5.5):

$$\bar{\sigma}_{i2}^{(1)1,1}\Big|_{(t_1,-H^{(1)})} - \bar{\sigma}_{i2}^{(2)0,1}\Big|_{(t_1,+H^{(2)})} = 2\pi\cos(\alpha_1 t_1)(\bar{\sigma}_{11}^{(1),0} - \bar{\sigma}_{11}^{(2),0})\delta_i^1,$$
$$\bar{u}_i^{(1)1,1}\Big|_{(t_1,-H^{(1)})} - \bar{u}_i^{(2)0,1}\Big|_{(t_1,+H^{(2)})} = \ell\sin(\alpha_1 t_1)\left(\frac{\partial u_2^{(1),0}}{\partial x_{21}^{(1)}} - \frac{\partial u_2^{(2),0}}{\partial x_{20}^{(2)}}\right)\delta_i^2, \tag{5.61}$$

where $\alpha_1 = 2\pi/\ell$ and $\delta_i^j$ is a Kronecker symbol. The contact conditions between the uncurved filler and matrix layers are obtained as

$$\bar{\sigma}_{i2}^{(1)m,1}\Big|_{(t_1,+H^{(1)})} - \bar{\sigma}_{i2}^{(2)m,1}\Big|_{(t_1,-H^{(2)})} = 0,$$
$$\bar{u}_i^{(1)m,1}\Big|_{(t_1,+H^{(1)})} - \bar{u}_i^{(2)m,1}\Big|_{(t_1,-H^{(2)})} = 0,$$
$$\bar{\sigma}_{i2}^{(2)m,1}\Big|_{(t_1,+H^{(2)})} - \bar{\sigma}_{i2}^{(1)m,1}\Big|_{(t_1,-H^{(1)})} = 0,$$
$$\bar{u}_i^{(2)m,1}\Big|_{(t_1,+H^{(2)})} - \bar{u}_i^{(1)m,1}\Big|_{(t_1,-H^{(1)})} = 0. \tag{5.62}$$

The following limiting condition holds:

$$\bar{\bar{\sigma}}_{ij}^{(k)m,1} = \left(\bar{\sigma}_{ij}^{(k)m,1} - \bar{\sigma}_{ij}^{(k)m,0}\right) \to 0 \text{ as } m \to \infty. \tag{5.63}$$

Using (5.63), we assume that after a certain layer with the number $m*$ all components of the tensor $\bar{\bar{\sigma}}_{ij}^{(k)m,1}$ are zero; an infinite number of layers is transformed into a finite number and Eq. (5.63) is transformed into a finite number and Eq. (5.63) is changed to

$$\bar{\bar{\sigma}}_{ij}^{(k)m,1} = \left(\bar{\sigma}_{ij}^{(k)m,1} - \bar{\sigma}_{ij}^{(k)m,0}\right)\Big|_{x_{2m*}^{(k)}=H^{(k)}} = 0. \tag{5.64}$$

It is evident that the Eqs. (5.49) and (5.52) hold also for Case 2. Therefore, in the case where the discriminant of the characteristic equation (5.55) $D = (a_1^{(k)})^2/4 - a_2^k < 0$, the solution to Eq. (5.52) is found as follows:

$$\bar{f}_2^{(2)0,1}(x_{20}^{(2)}) = A_1^{(2)0}\cos(\beta_1^{(2)}\alpha_1 x_{20}^{(2)})\cosh(\alpha_1^{(2)}\alpha_1 x_{20}^{(2)}) + A_4^{(2)0}\sin(\beta_1^{(2)}\alpha_1 x_{20}^{(2)})\sinh(\alpha_1^{(2)}\alpha_1 x_{20}^{(2)}),$$
$$\bar{f}_2^{(k)m,1}(x_{2m}^{(k)}) = A_1^{(k)m}\cos(\beta_1^{(k)}\alpha_1 x_{2m}^{(k)})\cosh(\alpha_1^{(k)}\alpha_1 x_{2m}^{(k)}) + A_2^{(k)m}\cos(\beta_1^{(k)}\alpha_1 x_{2m}^{(k)})\sinh(\alpha_1^{(k)}\alpha_1 x_{2m}^{(k)}) + A_3^{(k)m}\sin(\beta_1^{(k)}\alpha_1 x_{2m}^{(k)})\cosh(\alpha_1^{(k)}\alpha_1 x_{2m}^{(k)}) + A_4^{(k)m}\sin(\beta_1^{(k)}\alpha_1 x_{2m}^{(k)})\sinh(\alpha_1^{(k)}\alpha_1 x_{2m}^{(k)}). \tag{5.65}$$

From the contact conditions (5.61) and (5.62) we obtain system of linear algebraic equations for the unknowns $A_i^{(k)m}$ $(i = 1, 2, 3, 4)$. After this, by usual procedure we determine the Laplace transformations of all sought values related to Case 2.

### 5.2.3 Numerical Results and Discussions

First, we assume that the materials of both the filler (with Young's modulus $E^{(2)}$ and Poisson coefficient $\nu^{(2)}$) and matrix layers are pure elastic and analyze numerical results related to the critical values of compressive force $p$. Note that these critical values are determined by applying the initial imperfection criterion. In this case, the results for the first approximation on the stress-strain state obtained in the previous sub-section are used.

Introduce the parameter

$$\delta = \frac{10^2 p(1 - (\nu^{(1)})^2)}{E^{(1)}\left(\eta^{(1)} + \eta^{(2)} E^{(2)} (1 - (\nu^{(1)})^2)(1 - (\nu^{(2)})^2)^{-1}\right)} = 10^2 \varepsilon_{11}^{(1),0} = 10^2 \varepsilon_{11}^{(2),0} \tag{5.66}$$

and investigate the dependence between

$$A_{u_2} = \max_{0 \le x_1 \le \ell} \left| u_2^{(2),1}(x_1, 0) \right| / \ell \tag{5.67}$$

and the parameter $\delta$. Introduce also the parameter $\chi = 2\pi H^{(2)} / \ell$.

Through the parameters $A_{u_2}$ and $\delta$ we investigate the evolution of the initial imperfection (curving) of the filler layers and a stability loss criterion we formulate as follows:

$$\max_{0 \le x_1 \le \ell} \left| u_2^{(2),1}(x_1, 0) \right| / \ell \to \infty \text{ as } p \to p_{cr}. \tag{5.68}$$

Consider numerical results related to Case 1 (Fig. 5.2) and Case 2 (Fig. 5.5) separately.

#### 5.2.3.1 Numerical Results Related to Case1 (Co-Phase Curving of Layers)

First, we note that the corresponding stability loss problem within the scope of the Euler approach (i.e. within the scope of the bifurcation approach) with the use of the TDLTS (see Guz 1999) is investigated in the following manner. It was assumed that the ideal layout of the filler layers exists in the structure of the composite as shown in Fig. 5.3 and this composite is compressed at infinity with uniformly distributed normal forces with intensity $p$. For determination of the critical values of this force under co-phase stability loss of the filler layers the expressions of the displacements

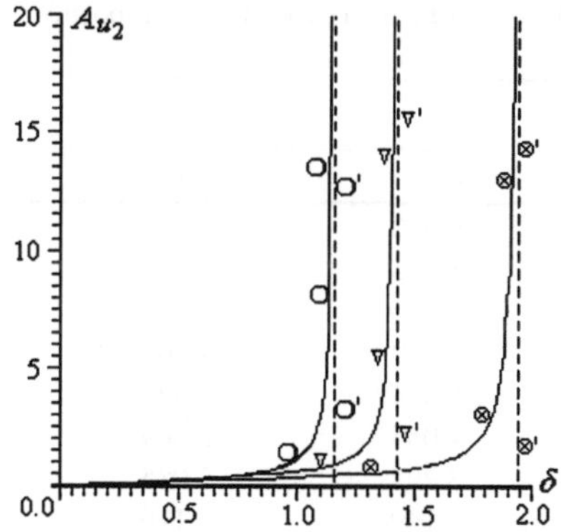


**Fig. 5.6** The relation between $A_{u_2}$ and $\delta$ the case where $\chi = 2\pi H^{(2)}/\ell = 0.1$, $\eta^{(2)} = 0.5$, $\nu^{(1)} = \nu^{(2)} = 0.3$ with various $E^{(2)}/E^{(1)}$ : $\otimes$ $- E^{(2)}/E^{(1)} = 100$, $\nabla$ $- E^{(2)}/E^{(1)} = 150$, O $- E^{(2)}/E^{(1)} = 200$; $\otimes'$, O$'$, $\nabla'$. the critical values of $\delta$ obtained in the framework of Euler approach for $E^{(2)}/E^{(1)} = 100$, 150 and 200 respectively

are also selected as (5.51) and the function $f_2^{(k),1}(x_2^{(k)})$ is selected as in the Eq. (5.59). Furthermore, for the unknown constants $D_1^{(k)}$ and $D_4^{(k)}$ $(k = 1, 2)$ the homogeneous algebraic equations are obtained from the contact conditions which coincide with (5.39) under $f = 0$ and $df/dx_1 = 0$. Consequently, the critical values of $p$ or $\delta$ (5.66) determined by criterion (5.68) must be coincide with the corresponding critical values of that determined by Euler criterion. This prediction is proven with the numerical results illustrated in Fig. 5.6 which shows the dependence between the $A_{u_2}$ (5.67) and the parameter $\delta$ for fixed problem parameters and various $E^{(2)}/E^{(1)}$; the critical values of $\delta$ obtained from the criterion (5.68) coincide with those obtained from the Euler approach.

Other results on the critical values of $\delta$ (denoted by $\delta_{cr}$) are given in Table 5.1. These results show that the critical values of the compressive force $p$ depend significantly on the parameters $E^{(2)}/E^{(1)}$, $\eta^{(2)}$ and $\chi (= 2\pi H^{(2)}/\ell)$ (where $\ell$ is the length of the period of the initial imperfection mode); only the parameter $\chi$ characterizes the initial imperfection mode of the filler layers. Such a parameter is also used under investigation of the corresponding stability loss problem within the scope of the Euler approach for which $\ell$ in expression of $\chi$ is length of the period of the stability loss mode of the layers and the critical value of $\chi$ is determined from the relation between $\delta$ and $\chi$. This relation is constructed from the solution to the corresponding characteristic equation. If this relation is non-monotonic and has a minimum for $\chi \neq 0$, then it is assumed that the internal or structural stability loss occurs and the values of $p$ and $\chi$ corresponding to this minimum are taken as their critical values. However, within the scope of the initial imperfection approach the value of the parameter $\chi$ is given in aforetime as a structural parameter of composites. This raises the possibility for optimal selection of the parameter with respect to stability loss in the structure of the composites.

**Table 5.1** The values of $\delta_{cr.}$ obtained for the composite shown in Fig. 5.2. under $\nu^{(1)} = \nu^{(2)} = 0.3$

| $\frac{E^{(2)}}{E^{(1)}}$ | $\chi$ | $\eta^{(2)}$ | | | |
|---|---|---|---|---|---|
| | | 0.02 | 0.10 | 0.20 | 0.50 |
| 100 | 0.05 | 9.620 | 3.678 | 2.205 | 1.468 |
| | 0.10 | 5.845 | 3.698 | 2.410 | 1.703 |
| | 0.20 | 4.276 | 3.965 | 3.208 | 2.606 |
| | 0.30 | 4.849 | 4.807 | 4.457 | 3.990 |
| 150 | 0.05 | 6.694 | 2.521 | 1.509 | 1.008 |
| | 0.10 | 4.064 | 2.609 | 1.727 | 1.247 |
| | 0.20 | 3.291 | 3.085 | 2.572 | 2.165 |
| | 0.30 | 4.149 | 4.119 | 3.885 | 3.570 |
| 200 | 0.05 | 5.136 | 1.928 | 1.156 | 0.778 |
| | 0.10 | 3.151 | 2.052 | 1.381 | 1.019 |
| | 0.20 | 2.791 | 2.638 | 2.251 | 1.943 |
| | 0.30 | 3.794 | 3.773 | 3.597 | 3.359 |

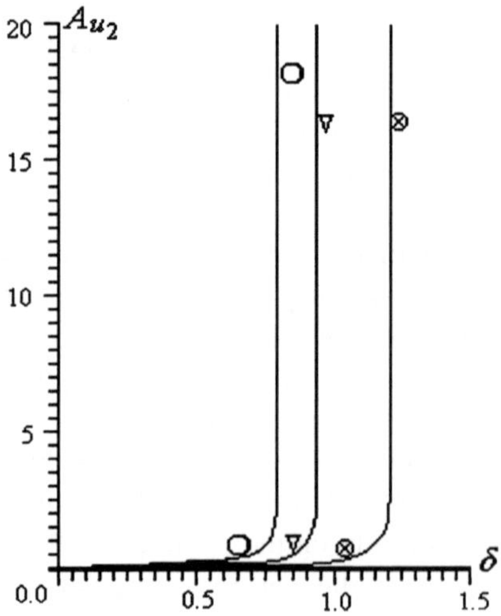


**Fig. 5.7** The relation between $A_{u_2}$ and $\delta$ in the case where $\chi = 0.1$, $\eta^{(2)} = 0.5$, $\nu^{(1)} = \nu^{(2)} = 0.3$ with various $E^{(2)}/E^{(1)}$: $\otimes$ – $E^{(2)}/E^{(1)} = 100$, $\nabla$ – $E^{(2)}/E^{(1)} = 150$, $\bigcirc$ – $E^{(2)}/E^{(1)} = 200$

#### 5.2.3.2 Numerical Results Related to Case 2 (Imperfection of a Single Layer)

Consider the dependence between

$$A_{u_2} = \max_{0 \le x_1 \le \ell} \left| u_2^{(2)0,1}(x_1, 0) \right| \Big/ \ell \tag{5.69}$$

and the parameter $\delta$ (5.66). The graphs of this dependence are given in Fig. 5.7.

According to the results given in this figure and according to criterion

**Table 5.2** The values of $\delta_{cr.}$ obtained for the composite shown in Fig. 5.5 under $\nu^{(1)} = \nu^{(2)} = 0.3$

| $\frac{E^{(2)}}{E^{(1)}}$ | $\chi$ | $\eta^{(2)}$ | | | |
|---|---|---|---|---|---|
| | | 0.02 | 0.10 | 0.20 | 0.50 |
| 100 | 0.05 | 9.619 | 2.504 | 1.443 | 0.774 |
| | 0.10 | 5.802 | 2.671 | 1.749 | 1.189 |
| | 0.20 | 4.201 | 3.005 | 2.668 | 2.180 |
| | 0.30 | 4.846 | 4.742 | 3.981 | 3.621 |
| 150 | 0.05 | 6.690 | 1.716 | 1.004 | 0.578 |
| | 0.10 | 4.002 | 1.888 | 1.299 | 0.916 |
| | 0.20 | 3.201 | 2.461 | 2.216 | 1.880 |
| | 0.30 | 4.145 | 4.100 | 3.575 | 3.326 |
| 200 | 0.05 | 5.111 | 1.306 | 0.780 | 0.479 |
| | 0.10 | 3.146 | 1.480 | 1.048 | 0.769 |
| | 0.20 | 2.780 | 2.133 | 1.985 | 1.729 |
| | 0.30 | 3.791 | 3.788 | 3.345 | 3.178 |

**Table 5.3** The values of $\delta_{cr.}$ obtained in various m* under $E^{(2)}/E^{(1)} = 200$

| $\eta^{(2)}$ | Number of equations (layers) | $\chi$ | | |
|---|---|---|---|---|
| | | 0.05 | 0.10 | 0.20 |
| 0.02 | 10 (3) | 5.111 | 3.106 | 2.740 |
| | 14 (4) | 5.111 | 3.106 | 2.740 |
| | 54 (13) | 5.111 | 3.146 | 2.780 |
| 0.10 | 10 (3) | 1.262 | 1.393 | 2.002 |
| | 14 (4) | 1.306 | 1.480 | 2.133 |
| | 54 (13) | 1.306 | 1.480 | 2.133 |
| 0.20 | 10 (3) | 0.758 | 1.003 | 1.895 |
| | 14 (4) | 0.780 | 1.048 | 1.985 |
| | 54 (13) | 0.780 | 1.048 | 1.985 |
| 0.50 | 10 (3) | 0.470 | 0.769 | 1.711 |
| | 14 (4) | 0.479 | 0.778 | 1.729 |
| | 54 (13) | 0.479 | 0.769 | 1.729 |

$$\max_{0 \leq x_1 \leq \ell} \left| u_2^{(2)0,1}(x_1, 0) \right| / \ell \to \infty \text{ as } p \to p_{cr}, \tag{5.70}$$

we determine the critical values of the external compressive force $p$ or parameter $\delta$. Table 5.2 shows these critical values of $\delta$ calculated for various $E^{(2)}/E^{(1)}$, $\eta^{(2)}$ and $\chi$. It follows from the comparison of the values $\delta_{cr}$ shown in Fig. 5.7 and in Table 5.2 with corresponding ones shown in Fig. 5.6 and in Table 5.1 that the critical values obtained for the composite with initial imperfection illustrated in Fig. 5.2 are greater than the corresponding ones obtained for the composite with initial imperfection illustrated in Fig. 5.5.

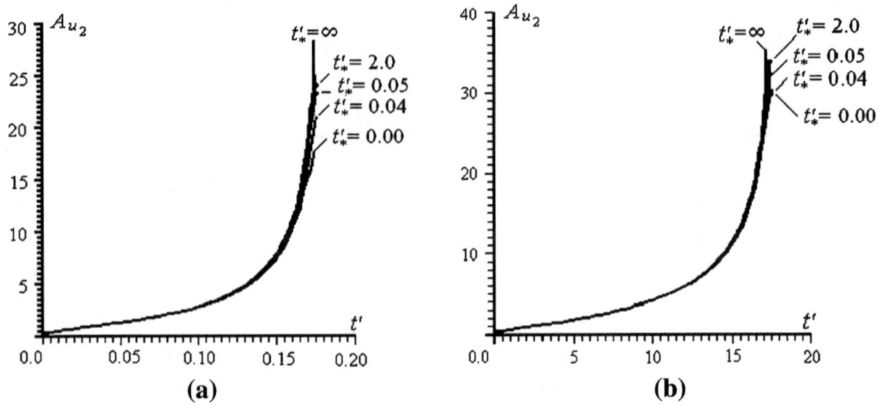


**Fig. 5.8** Relation between $A_{u_2}$ and $t'$ under $\eta^{(2)} = 0.2$, $p\big/E_0^{(1)} = 0.3$ (**a**);$p\big/E_0^{(1)} = 0.2$ (b)

Table 5.3 shows the critical values of $\delta$ obtained for various $m*$ and confirm the effectiveness of the solution method.

#### 5.2.3.3 Numerical Results Related to Viscoelastic Composite Materials

Within the scope of the approach and assumptions developed in Sect. 5.2.2.4 we consider numerical results related to the critical time obtained in Case 1. For this purpose as a stability loss criterion we take

$$\max_{0\le x_1\le \ell}\left|u_2^{(2),1}(x_1,0,t')\right|\Big/\ell \to \infty \text{ as } t' \to t'_{cr}. \tag{5.71}$$

We consider the selection of the values of $t'_*$ in Eqs. (5.48)–(5.50). Physical intuition and Eq. (5.47) show that the exact value of the critical time $t'_{cr}$ must lies between those obtained for $t'_* = 0$ and $t'_* = \infty$, i.e. $t'_{cr.\infty} \le t'_{cr} \le t'_{cr.0}$; $t'_{cr.0}$, $t'_{cr.\infty}$ are the critical times for $t'_* = 0$, $t'_* = \infty$ respectively, and find $t'_{cr.0}$ and $t'_{cr.\infty}$. For this purpose we analyze the change of

$$A_{u_2} = \max_{0\le x_1\le \ell}\left|u_2^{(2),1}(x_1,0,t')\right|\Big/\ell \tag{5.72}$$

with respect to $t'$ under $\chi = 0.1$, $\omega = 0.5$, $\alpha = -0.7$ and $E^{(2)}\big/E_0^{(1)} = 100$.

Consider the graphs given in Figs. 5.8 and 5.9 which show that the values of $A_{u_2}$obtained for various $t'_*$ are bounded by those obtained for $t'_* = \infty$, $t'_* = 0$. With an increase in filler concentration, and decrease in $p\big/E_0^{(1)}$, the values of $t'_{cr}$ grow.

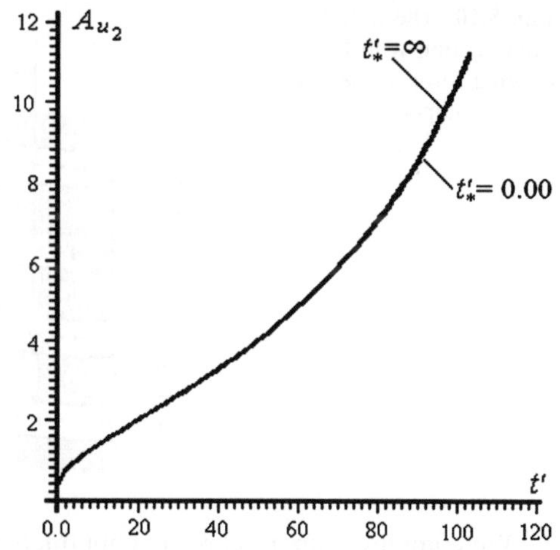


**Fig. 5.9** Relation between $A_{u_2}$ and $t'$under $\eta^{(2)} = 0.5,\ p/E_0^{(1)} = 0.3$

We now compare the results with those obtained by the critical deformation method (Gerard and Gilbert 1958). Assume that there are uniformly distributed normal forces with intensity $p$ acting at infinity in the direction of the $Ox_1$ axis, filler layer is purely elastic, and matrix layer viscoelastic with operators given in Eqs. (5.40)–(5.42). The critical deformation gives $p^e_{cr}$, the elastic critical value from the relation

$$\varepsilon^{(1),0}_{11.cr} = \frac{p^e_{cr}}{E_0^{(1)}}\left(1-\left(\nu_0^{(1)}\right)^2\right)\left(\eta^{(1)} + \eta^{(2)}\frac{\left(1-\left(\nu_0^{(1)}\right)^2\right)E^{(2)}}{\left(1-\left(\nu^{(2)}\right)^2\right)E_0^{(1)}}\right)^{-1}. \tag{5.73}$$

The corresponding viscoelastic strain is

$$\varepsilon^{(1),0}_{11.v.e.} = \frac{p_{v.e.}}{E_0^{(1)}}\left(1-\left(\nu^{(1)*}\right)^2\right)\left(\eta^{(1)} + \eta^{(2)}\frac{\left(1-\left(\nu^{(1)*}\right)^2\right)E^{(2)}}{\left(1-\left(\nu^{(2)}\right)^2\right)E^{(1)*}}\right)^{-1}, \tag{5.74}$$

where $p_{v.e.}$ is the external normal compressive force in the viscoelastic case. We obtain the following equation for the determination of the critical time by equating Eqs. (5.73) and (5.74):

$$\frac{p_{v.e.}}{p^e_{cr}} = \frac{\left(1-\left(\nu^{(1)*}\right)^2\right)E^{(1)*}\left(E_0^{(1)}\right)^{-1}}{\left(1-\left(\nu_0^{(1)}\right)^2\right)} \times \frac{\left(\eta^{(1)} + \eta^{(2)}\left(1-\left(\nu_0^{(1)}\right)^2\right)\left(1-\left(\nu^{(2)}\right)^2\right)^{-1}E^{(2)}\left(E_0^{(1)}\right)^{-1}\right)}{\left(\eta^{(1)} + \eta^{(2)}\left(1-\left(\nu^{(1)*}\right)^2\right)\left(1-\left(\nu^{(2)}\right)^2\right)^{-1}E^{(2)}\left(E^{(1)*}\right)^{-1}\right)}, \tag{5.75}$$

where $E^{(1)*}$ and $\nu^{(1)*}$ are given by (5.40)–(5.42).

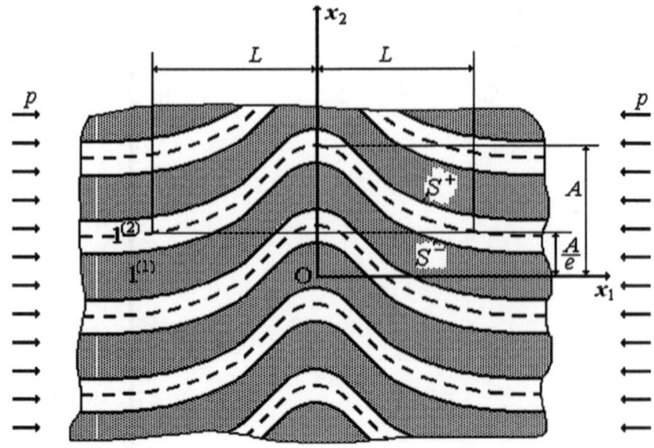


**Fig. 5.10** The initial imperfection with local curving form in the structure of the composite material

We compare numerical results obtained from Eq. (5.75) for the critical time with those obtained in our approach. Assume that $E^{(2)}/E_0^{(1)} = 100$, $\nu^{(2)} = \nu_0^{(1)} = 0.3$, $\eta^{(2)} = 0.2$, $\chi = 0.1$, $\alpha = -0.7$ and $\omega = 0.5$. We find $p_{cr}^e = 0.55E_0^{(1)}$ from Eq. (5.66) and from the numerical results given in Table 5.1. Consider the following two cases: $p_{v.e.} = 0.3E_0^{(1)}$ (case 1); $p_{v.e.} = 0.2E_0^{(1)}$ (case 2). Equation (5.75) gives the critical time: $t'_{cr} = 0.06$ (for case 1); $t'_{cr} = 1.00$ (for case 2). Figure 5.8a (case 1) gives $t'_{cr} = 0.17$, while Fig. 5.8b (case 2) gives $t'_{cr} = 17$. Thus, the critical deformation method (Gerard and Gilbert 1958) underestimates the critical time.

## 5.3 Stability Loss in the Structure of the Elastic and Viscoelastic Layered Composites with Local Initial Imperfections

### *5.3.1 Formulation of the Problem and Method of Solution*

Within the scope of all the equations, relations and suppositions detailed in Sects. 5.2.1–5.2.3 we consider the evolution of the local initial infinitesimal imperfection in the structure of the layered viscoelastic composite material with time. Assume that this composite consists of alternating two layers of two materials and the local initial infinitesimal imperfection is modeled as infinitesimal local co-phase curving of the layers, as shown in Fig. 5.10.

As usual we single out two layers, i.e. $1^{(1)}$, $1^{(2)}$; we assume that the layer $1^{(2)}$ is elastic, the layer $1^{(1)}$ is linearly viscoelastic with operators (5.40)–(5.42). The initial slight local curving form of the layer $1^{(2)}$ we select as follows:

$$x_2^{(2)} = A\exp\left\{-\left(\left(\frac{x_1}{L}\right)^2\right)^{\gamma}\right\}\cos\left(n\frac{x_1}{L}\right) = \varepsilon L\exp\left\{-\left(\left(\frac{x_1}{L}\right)^2\right)^{\gamma}\right\}\cos\left(n\frac{x_1}{L}\right). \tag{5.76}$$

The geometrical meaning of the parameters $\gamma$, $n$ and $L$ is evident. We assume that $A \ll L$, and introduce a small parameter $\varepsilon = A/L$.

Thus we investigate the plane-strain state in the foregoing composite material by employing the approach described in the previous section. In this case, the zeroth approximation is determined as in Sect. 5.2.2.4 and for determination of the first approximation we obtain the Eqs. (5.23), (5.37) and (5.38) and contact condition (5.39). However, in the present case the function $f(x_1)$ in this condition, according to Eq. (5.76), is defined as

$$f(x_1) = L\exp\left\{-\left(\left(\frac{x_1}{L}\right)^2\right)^{\gamma}\right\}\cos\left(n\frac{x_1}{L}\right). \tag{5.77}$$

Taking the Eq. (5.77) in the contact condition (5.39) into account we employ the exponential Fourier transformation

$$\varphi_F(\eta) = \int_0^{\infty}\varphi(x_1)e^{-(i\eta x_1)}dx_1,\ i = \sqrt{-1} \tag{5.78}$$

to Eqs. (5.23), (5.37), (5.38) and contact condition (5.39). After this employing, we obtain the following equation and contact condition for Fourier transformation of the Laplace transformation of the sought values of the first approximation:

Equilibrium equation

$$\eta^2 c_1^{(k)}\bar{u}_{1F}^{(k),1} - i\eta c_2^{(k)}\frac{d\bar{u}_{2F}^{(k),1}}{dx_2^{(k)}} + c_3^{(k)}\frac{d^2\bar{u}_{1F}^{(k),1}}{d(x_2^{(k)})^2} = 0,$$
$$\eta^2 c_4^{(k)}\bar{u}_{2F}^{(k),1} - i\eta c_5^{(k)}\frac{d\bar{u}_{1F}^{(k),1}}{dx_2^{(k)}} + c_6^{(k)}\frac{d^2\bar{u}_{2F}^{(k),1}}{d(x_2^{(k)})^2} = 0, \tag{5.79}$$

Geometrical relation

$$\bar{\varepsilon}_{11F}^{(k),1} = -i\eta\bar{u}_{1F}^{(k),1},\ \bar{\varepsilon}_{22F}^{(k),1} = \frac{d\bar{u}_{2F}^{(k),1}}{dx_2^{(k)}},\ \bar{\varepsilon}_{12F}^{(k),1} = \frac{1}{2}\left(\frac{d\bar{u}_{1F}^{(k),1}}{dx_2^{(k)}} - i\eta\bar{u}_{2F}^{(k),1}\right), \tag{5.80}$$

Contact condition

$$\bar{\sigma}_{i2F}^{(1),1}\Big|_{(t_1;\pm H^{(1)})} - \bar{\sigma}_{i2F}^{(2),1}\Big|_{(t_1;\mp H^{(2)})} = -i\eta f_F(\bar{\sigma}_{11}^{(1),0} - \bar{\sigma}_{11}^{(2),0})\delta_i^1,$$
$$\bar{u}_{iF}^{(1),1}\Big|_{(x_2^{(1)}=\pm H^{(1)})} - \bar{u}_{iF}^{(2),1}\Big|_{(x_2^{(1)}=\mp H^{(2)})} = f_F\left(\frac{\partial\bar{u}_2^{(1),0}}{\partial x_2^{(1)}} - \frac{\partial\bar{u}_2^{(2),0}}{\partial x_2^{(2)}}\right)\delta_i^2. \tag{5.81}$$

From (5.79) we obtain the following equation for $\bar{u}_{2F}^{(k),1}$:

$$d_1^{(k)}\frac{d^4\bar{u}_{2F}^{(k),1}}{d(\eta x_2^{(k)})^4}+d_2^{(k)}\frac{d^2\bar{u}_{2F}^{(k),1}}{d(\eta x_2^{(k)})^2}+d_3^{(k)}\bar{u}_{2F}^{(k),1}=0, \tag{5.82}$$

where $d_1^{(k)}$, $d_2^{(k)}$ and $d_3^{(k)}$ are determined through the expression (5.53). Using the particular solution in the form (5.54) we obtain the characteristic Eq. (5.55). Thus according to the sign of the discriminant $D^{(k)}=(a_1^{(k)})^2/4-a_2^{(k)}$, we can determine the general solution of the Eq. (5.82). For example, in the case where $D^{(k)}=(a_1^{(k)})^2/4-a_2^{(k)}<0$, we obtain the roots (5.56) from the characteristic Eq. (5.55) and the mentioned general solution are determined as (5.58) with replacing $\alpha_1$with $\eta$. Moreover, using the fact that the function $\bar{u}_{2F}^{(k),1}$ must be even function with respect to $x_2^{(k)}$we determine this function $\bar{u}_{2F}^{(k),1}$ as follows:

$$\begin{aligned}\bar{u}_{2F}^{(k),1}(\eta x_2^{(k)}) =&D_1^{(k)}\cos(\beta_1^{(k)}\eta x_2^{(k)})\cosh(\alpha_1^{(k)}\eta x_2^{(k)})\\&+D_4^{(k)}\sin(\beta_1^{(k)}\eta x_2^{(k)})\sinh(\alpha_1^{(k)}\eta x_2^{(k)}).\end{aligned} \tag{5.83}$$

Using the inverse Fourier transformation

$$\bar{u}_2^{(k),1}(x_1,x_2^{(k)})=\frac{1}{2\pi}\int_{-\infty}^{+\infty}\bar{u}_{2F}^{(k),1}(\eta x_2^{(k)})e^{i\eta x_1}d\eta \tag{5.84}$$

we determine the Laplace transformation of the $\bar{u}_2^{(k),1}(x_1,x_2^{(k)})$. Finally, employing the method (5.31) (Schapery 1962, 1978) we find the displacement $u_2^{(k),1}(x_1,x_2^{(k)})$ and other sought values. After foregoing solution procedure, we find the values of the critical force (for purely elastic composite) and of the critical time (for viscoelatic composite) from the criterion

$$\begin{aligned}u_2^{(2),1}(0,0)\big/L\to\infty \,\text{as}\, p\to p_{cr}\ &\text{(for purely elastic case)}\\ \text{or as}\ t'\to t'_{cr}\ &\text{(for viscoelastic case)}.\end{aligned} \tag{5.85}$$

### *5.3.2 Numerical Results and Discussions*

For explanation of the numerical results which will be discussed below, first, we consider the conception on the failure of the layered composites in compression within the scope of the continuum approach developed in (Biot 1965, Guz 1969, 1990, 2008a, b). According to this conception, composite materials with structure shown in Fig. 5.3, i.e. composite material with the ideal layout of the layers, are modeled as a structurally homogeneous orthotropic material with normalized mechanical characteristics. After this procedure, the type change of the equations of TDLTS written for the above homogeneous orthotropic infinite body is analyzed under the action at "infinity" uniformly distributed compressive forces with intensity $p$. For the case under consideration these equations are

$$(A_{11}+p)\frac{\partial^2 u_1}{\partial x_1^2}+G_{12}\frac{\partial^2 u_1}{\partial x_2^2}+(A_{12}+G_{12})\frac{\partial^2 u_2}{\partial x_1 \partial x_2}=0,$$
$$(G_{12}+p)\frac{\partial^2 u_2}{\partial x_1^2}+A_{22}\frac{\partial^2 u_2}{\partial x_2^2}+(A_{12}+G_{12})\frac{\partial^2 u_1}{\partial x_1 \partial x_2}=0, \tag{5.86}$$

where

$$\begin{aligned}
G_{12} &= \frac{\mu^{(1)}\mu^{(2)}}{\mu^{(1)}\eta^{(2)}+\mu^{(2)}\eta^{(1)}},\\
A_{11} &= \mu^{(1)}\eta^{(1)}+\mu^{(2)}\eta^{(2)}+(\lambda^{(1)}+\mu^{(1)})\eta^{(1)}+(\lambda^{(2)}+\mu^{(2)})\eta^{(2)}\\
&\quad -\eta^{(1)}\eta^{(2)}\frac{(\lambda^{(1)}-\lambda^{(2)})}{(\lambda^{(1)}+2\mu^{(1)})\eta^{(2)}+(\lambda^{(2)}+2\mu^{(2)})\eta^{(1)}},\\
A_{12} &= \lambda^{(1)}\eta^{(1)}+\lambda^{(2)}\eta^{(2)}-\eta^{(1)}\eta^{(2)}(\lambda^{(1)}-\lambda^{(2)})\\
&\quad \times\frac{(\lambda^{(1)}+2\mu^{(1)})-(\lambda^{(2)}+2\mu^{(2)})}{(\lambda^{(1)}+2\mu^{(1)})\eta^{(2)}+(\lambda^{(2)}+2\mu^{(2)})\eta^{(1)}},\\
A_{22} &= (\lambda^{(1)}+2\mu^{(1)})\eta^{(1)}+(\lambda^{(2)}+2\mu^{(2)})\eta^{(2)}\\
&\quad -\eta^{(1)}\eta^{(2)}\frac{\left((\lambda^{(1)}+2\mu^{(1)})-(\lambda^{(2)}+2\mu^{(2)})\right)^2}{(\lambda^{(1)}+2\mu^{(1)})\eta^{(2)}+(\lambda^{(2)}+2\mu^{(2)})\eta^{(1)}},
\end{aligned} \tag{5.87}$$

where $\lambda^{(k)}$ and $\mu^{(k)}$ are Lamé's constants of the $k$th layer's material, $\eta^{(k)}=H^{(k)}/(H^{(1)}+H^{(2)})$.

The type of equations in (5.86) is analyzed. For this purpose the roots of the equation

$$\begin{gathered}
((A_{11}+p)\beta_1^2+G_{12}\beta_2^2)((G_{12}+p)\beta_1^2+A_{22}\beta_2^2)-(A_{12}+G_{12})^2\beta_1^2\beta_2^2=0,\\
|\beta_1|^2+|\beta_2|^2=1
\end{gathered} \tag{5.88}$$

are determined with respect to unknowns $\beta_1$ and $\beta_2$. If these roots are complex numbers, then it is said that the type of the system of equations in (5.87) is elliptic. If at least one of these roots is real number, then it is said that the ellipticity of the system of equations in (5.87) is violated and this system has "hyperbolic type" solution. Namely, the value of the external compressive force $p$ corresponding to this case is taken as Theoretical Strength Limit in Compression (TSLC) of the considered composite material. In the case where $\beta_2=0$, $|\beta_1|=1$, it follows from the Eq. (5.88) that $(A_{11}+p)(G_{12}+p)=0$. As, in real cases the relation $A_{11}\gg G_{12}$ takes place, therefore from the equation $(A_{11}+p)(G_{12}+p)=0$ we determine the value of the TSLC (denote it as $p_{cr}^T$) for the material under consideration as

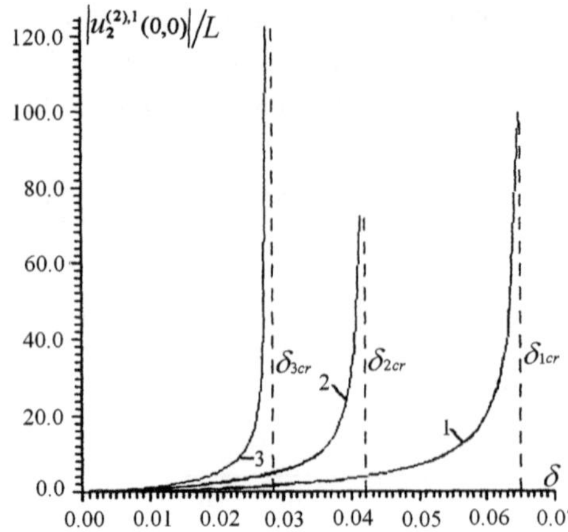


**Fig. 5.11** The dependence between $u_2^{(2),1}(0,0)/L$ and $\delta$ constructed for the cases where $E^{(2)}/E^{(1)}=50$, $\eta^{(2)}=0.5$ and $\chi=0.1$

$$p_{cr}^T = -G_{12}. \tag{5.89}$$

In the mechanical sense, TSLC shows the theoretical limit of the external compressive force for which the load carrying capacity in the macro-volume of the composite material exhausts completely. Therefore the determination of the values of the TSLC has a considerable theoretical and practical significance.

Now we consider the numerical results obtained from the criterion (5.85). For this purpose, as in the previous section, we introduce the parameter

$$\delta = \frac{10^2 p(1-(\nu^{(1)})^2)}{E^{(1)}\left(\eta^{(1)}+\eta^{(2)}E^{(2)}(1-(\nu^{(1)})^2)(1-(\nu^{(2)})^2)^{-1}\right)} \tag{5.90}$$

and investigate the graphs of the dependence between $u_2^{(2),1}(0,0)/L$ and $\delta$. Note that these graphs are given in Fig. 5.11 for the cases where $E^{(2)}/E^{(1)}=50$. Similar graphs are also obtained for the cases where $E^{(2)}/E^{(1)}=100$ and 200. The curves indicated by numbers 1, 2 and 3 show the critical values of the parameter $\delta$, obtained according to the criterion (5.85) and indicated by $\delta_{1cr}$, $\delta_{2cr}$ and $\delta_{3cr}$ in the figures correspond the cases where $\eta^{(2)}=0.1, 0.2 and 0.5$ respectively. Moreover, note that these graphs are constructed under $\chi=H^{(2)}/L=0.1$. Numerical results obtained for various values of the parameter $\chi\,(=H^{(2)}/L)$ show that the change of this parameter does not influence on the values of $\delta_{cr}$. As an example, we consider the graphs given in Fig. 5.12 which show the dependence between $u_2^{(2),1}(0,0)/L$ and $\delta$ constructed for various values of the parameter $\chi\,(=H^{(2)}/L)$ under $E^{(2)}/E^{(1)}=100$ and $\eta^{(2)}=0.5$. In this figure the curves indicated by numbers 1, 2, 3, 4, 5, 6, 7, and 8 correspond the values of $\chi=0.05, 0.1, 0.2, 0.3, 0.5, 0.7, 0.8$ and 1.0 respectively. In a similar manner we

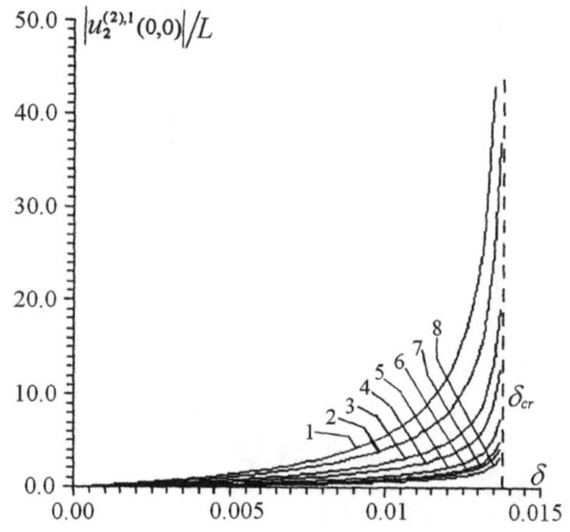


**Fig. 5.12** The dependence between $u_2^{(2),1}(0,0)/L$ and $\delta$ constructed for various values of the parameter $\chi$ under $E^{(2)}/E^{(1)}$ =100 and $\eta^{(2)} = 0.5$

establish that the parameters $\gamma$ and $n$ which enter the Eq. (5.77) and characterize the local curving form of the reinforcing layers also do not influence on the critical values of the parameter $\delta$ (5.90). Consequently the values of $\delta_{cr}$ depend only on the mechanical constants and volumetric fraction of the layers in the composite under consideration. Substituting these values of $\delta_{cr}$ into expression

$$p_{cr} = \frac{\delta_{cr} E^{(1)} \left( \eta^{(1)} + \eta^{(2)} E^{(2)} (1 - (\nu^{(1)})^2)(1 - (\nu^{(2)})^2)^{-1} \right)}{10^2 (1 - (\nu^{(1)})^2)} \tag{5.91}$$

we determine the critical values of the compressive force, i.e. $p_{cr}$ from the criterion (5.85). The comparison of these values of $p_{cr}$ with the corresponding values of $p_{cr}^T$ (5.89) shows that

$$p_{cr} = p_{cr}^T. \tag{5.92}$$

So, by direct verification we prove that the critical values of the compressive force determined for the stability loss in the structure of the composite with the initial infinitesimal local imperfection in the structure coincide with the corresponding values of TSLC determined from the continuum approach.

The conclusion (5.92) allows us to use the approach developed in the present section for determination TSLC for the viscoelastic composites. In this case under TSLC for the viscoelastic composite we will understand the values of $t'_{cr}$ determined from the criterion (5.85). Thus, we consider the numerical results related to TSLC of the viscoelastic composites. These results are obtained for the cases where $t'_* = 0$ and $t'_* = \infty$. Remind that the $t'_*$ enters the expressions (5.48) and (5.50). The critical values of $t'$ related to the cases where $t'_* = 0$ and $t'_* = \infty$we denote as $t'_{cr.0}$ and $t'_{cr.\infty}$ respectively. Since $t'_{cr.\infty} \leq t'_{cr} \leq t'_{cr.0}$, our analysis is reduced to the determination of $t'_{cr.0}$ and $t'_{cr.\infty}$.

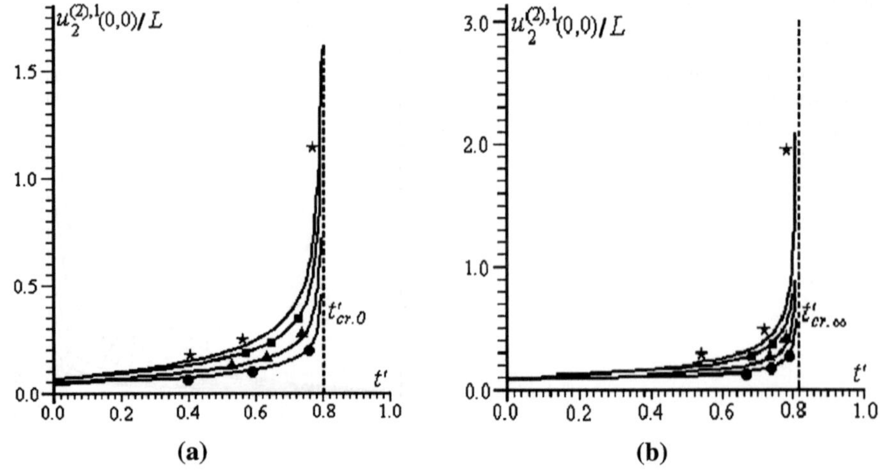


**Fig. 5.13** The graphs of the dependencies between $u_2^{(2),1}(0,0)\big/L$ and $t'$ under $t'_* = 0$ (a) and $t'_* = \infty$ (b), $E^{(2)}\big/E_0^{(1)} = 100$, $p\big/E_0^{(1)} = 0.25$, $\omega = 0.5$, $\alpha = -0.5$; $* - \chi = 0.1$, $n = 0$, $\gamma = 1, 2, 3$; ■ $- \chi = 0.1$, $\gamma = 1$, $n = 0$; ▲ $- \chi = 0.2$, $\gamma = 1$, $n = 0$; • $- \chi = 0.3$, $\gamma = 1$, $n = 0$

We assume that $G_{12.\infty} \le p \le G_{12.0}$ where $G_{12.0}\,(G_{12.\infty})$ is determined from the expression of $G_{12}$ in (5.80) with the change $\mu^{(1)}$ and $\nu^{(1)}$ with $\mu_0^{(1)}$ and $\nu_0^{(1)}$ ($\mu_\infty^{(1)}$ and $\nu_\infty^{(1)}$) respectively. Moreover, assume that $\nu_0^{(1)} = \nu^{(2)} = 0.3$.

To illustrate the independence of the values of $t'_{cr.0}$ and $t'_{cr.\infty}$ on the initial imperfection mode (i.e. on the parameters $\gamma$ and $n$ in Equation (5.76)) and on the structural parameter $\chi\,(= H^{(2)}/L)$ we consider the graphs given in Fig. 5.13. Note that these graphs show the dependence of $u_2^{(2),1}(0,0)\big/L$ on $t'$ for $t'_* = 0$ (Fig. 5.13a) and $t'_* = \infty$ (Fig. 5.13b) in the case where $E^{(2)}\big/E_0^{(1)} = 100$, $n = 0$, $\eta^{(2)} = 0.2$, $p\big/E_0^{(1)} = 0.5$, $\omega = 0.5$ and $\alpha = -0.5$. Note that the graphs which are constructed at $\gamma = 1, 2$ and 3 fit well with very high accuracy, therefore in Fig. 5.13 these graphs coincide. According to the graphs and according to other similar results which are not given here, we find that the values of $t'_{cr.0}$ and $t'_{cr.\infty}$ do not depend on the local initial imperfection mode and on the structural parameter $\chi$.

Table 5.4 shows $t'_{cr.0}$ and $t'_{cr.\infty}$ obtained for various $p\big/E_0^{(1)}$; increasing $p\big/E_0^{(1)}$, the values $t'_{cr.0}$ and $t'_{cr.\infty}$ monotonically decrease, and $p\big/E_0^{(1)}$ approaches the $G_{12.0}\big/E_0^{(1)}$; as it has been noted above, $G_{12.0}$ is the TSLC for the corresponding pure elastic material.

Consider the influence of $\eta^{(2)}$, $E^{(2)}\big/E_0^{(1)}$ and rheological parameters $\omega$ and $\alpha$ on the values of $t'_{cr.0}$ and $t'_{cr.\infty}$. Table 5.5 shows $t'_{cr.0}$ and $t'_{cr.\infty}$ for various

**Table 5.4** The values of $t'_{cr.0}$ and $t'_{cr.\infty}$ for various $p/E_0^{(1)}$ under $\eta^{(2)} = 0.2$; $E^{(2)}/E_0^{(1)} = 100$, $\omega = 0.5$, $\alpha = -0.3$

| $p/E_0^{(1)}$ | $t'_{cr.0}$ | $t'_{cr.\infty}$ |
|---|---|---|
| 0.36 | 0.1150 | 0.1150 |
| 0.38 | 0.0839 | 0.0839 |
| 0.40 | 0.0599 | 0.0590 |
| 0.42 | 0.0350 | 0.0350 |
| 0.44 | 0.0209 | 0.0205 |
| 0.47 | 0.0054 | 0.0054 |

**Table 5.5** The values of $t'_{cr.0}$ and $t'_{cr.\infty}$ for various $E^{(2)}/E_0^{(1)}$ under $\eta^{(2)} = 0.2$, $p/E_0^{(1)} = 0.28$, $\omega = 0.5$, $\alpha = -0.3$

| $E^{(2)}/E_0^{(1)}$ | $t'_{cr.0}$ | $t'_{cr.\infty}$ |
|---|---|---|
| 50 | 0.4900 | 0.4799 |
| 100 | 0.4699 | 0.4650 |
| 150 | 0.4499 | 0.4499 |
| 200 | 0.4000 | 0.4000 |

**Table 5.6** The values of $t'_{cr.0}$ and $t'_{cr.\infty}$ for various $\omega$ under $\eta^{(2)} = 0.2$, $E^{(2)}/E_0^{(1)} = 100$; $p/E_0^{(1)} = 0.42$; $\alpha = -0.3$

| $\omega$ | $t'_{cr.0}$ | $t'_{cr.\infty}$ |
|---|---|---|
| 0.5 | 0.0350 | 0.0350 |
| 1.0 | 0.0390 | 0.0390 |
| 2.0 | 0.0544 | 0.0549 |
| 3.0 | 0.0649 | 0.0649 |
| 5.0 | 0.1890 | 0.1890 |
| 6.0 | 0.6490 | 0.6460 |

$E^{(2)}/E_0^{(1)}$; for increasing $E^{(2)}/E_0^{(1)}$ the TSLC (i.e. $t'_{cr.0}$ and $t'_{cr.\infty}$) decreases.

The influence of the rheological parameter $\omega$ on $t'_{cr.0}$ and $t'_{cr.\infty}$ is given in Table 5.6. It follows from these numerical results as $\omega$ increases the values of $t'_{cr.0}$ and $t'_{cr.\infty}$ increase monotonically.

Table 5.7 shows $t'_{cr.0}$ and $t'_{cr.\infty}$ for various $\alpha$ and $\eta^{(2)}$; as $|\alpha|$ and $\eta^{(2)}$ increase $t'_{cr.0}$ and $t'_{cr.\infty}$ increase monotonically.

The numerical results given in Tables 5.4–5.7 show that $t'_{cr.0}$ and $t'_{cr.\infty}$ close; either may be used for $t'_{cr}$.

**Table 5.7** The values of $t'_{cr.0}$ and $t'_{cr.\infty}$ for various $\alpha$ and $\eta^{(2)}$ under $E^{(2)}\big/E_0^{(1)} = 100$, $p\big/E_0^{(1)} = 0.32$ and $\omega = 0.5$

| $\alpha$ | $\eta^{(2)}$ | $t'_{cr.0}$ | $t'_{cr.\infty}$ |
|---|---|---|---|
| −0.3 | 0.1 | 0.100 | 0.100 |
| | 0.2 | 0.209 | 0.209 |
| | 0.5 | 2.500 | 2.500 |
| −0.5 | 0.1 | 0.075 | 0.075 |
| | 0.2 | 0.170 | 0.170 |
| | 0.5 | 4.900 | 4.900 |
| −0.7 | 0.1 | 0.023 | 0.022 |
| | 0.2 | 0.088 | 0.086 |
| | 0.5 | 22.00 | 22.00 |

## 5.4 The Influence of the Inclination of the Local Initial Imperfections of the Reinforcing Layers on the Values of the Critical Parameters

### *5.4.1 Formulation of the Problem and Solution Method*

As in the previous section, we consider also a composite material consisting of the alternating layers of two materials and these layers are located in planes which are parallel to the plane $Ox_1x_3$ and each reinforcing layer has an infinitesimal initial imperfection (curving). However, we assume that this local imperfection regions of the reinforcing layers move with respect to each other in the direction of the $Ox_1$ axis (Fig. 5.14). This moving length is $\ell_0 = 2(H^{(1)} + H^{(2)})tg\beta$, where $2H^{(2)}$ $(2H^{(1)})$ is the thickness of the reinforcing (matrix) layer and an angle $\beta$ is the inclination angle shown in Fig. 5.14.

It is evident that the case where $\beta = 0$ coincides with that considered in the previous section. The material of the filler layers is assumed to be only elastic with mechanical characteristics of $E^{(2)}$ (Young's modulus) and $\nu^{(2)}$ (Poisson coefficient), and material of the matrix layers is assumed to be linearly viscoelastic with operators (5.40)–(5.42). The values related to filler (matrix) layers will be denoted by the upper index (2) (by the upper index (1)).

Taking the character of the composite structure shown in Fig. 5.14 into account, among the layers considered we single out three of them, i.e. $1^{(1)}$, $1^{(2)}$ and $2^{(2)}$, and discuss them below. We associate the corresponding Lagrangian coordinates $O_m^{(k)}x_{1m}^{(k)}x_{2m}^{(k)}x_{3m}^{(k)}$ with the middle surface of the $k$th layer, where $x_1 = x_{11}^{(1)} = x_{11}^{(2)} = x_{12}^{(2)}$, $x_3 = x_{31}^{(1)} = x_{31}^{(2)} = x_{32}^{(2)}$, although we consider the plane strain state in the $Ox_1x_2$ plane. Note that in the natural state these coordinates coincide with corresponding Cartesian coordinates which are obtained from $Ox_1x_2x_3$ (Fig. 5.14) by parallel transfer along the $Ox_2$ axis.

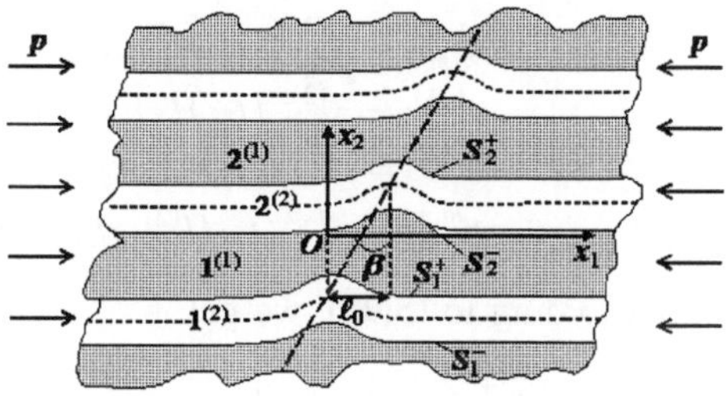

**Fig. 5.14** The structure of the considered composite material with an inclined local insignificant imperfection

We investigate the evolution of the inclined initial imperfections of the filler layers in the structure of the composite with the external compressive force with intensity $p$ (for a purely elastic composite) or with the time $t$ under fixed $p$ (for a viscoelastic composite). As before, we make this investigation within the scope of the piecewise homogeneous body model within the following field equations.

$$\begin{gathered}
\frac{\partial}{\partial x_{jm}^{(k)}}\left[\sigma_{jn}^{(k)m}\left(\delta_i^n+\frac{\partial u_i^{(k)m}}{\partial x_{nm}^{(k)}}\right)\right]=0,\ k=1,2;\ m=1,2,\ldots, \\
\sigma_{ij}^{(k)m}=\lambda^{(k)*}\theta^{(k)m}+2\mu^{(k)*}\varepsilon_{ij}^{(k)m},\ \theta^{(k)m}=\varepsilon_{11}^{(k)m}+\varepsilon_{22}^{(k)m} \\
\varepsilon_{ij}^{(k)m}=\frac{1}{2}\left(\frac{\partial u_i^{(k)m}}{\partial x_{jm}^{(k)}}+\frac{\partial u_j^{(k)m}}{\partial x_{im}^{(k)}}+\frac{\partial u_n^{(k)m}}{\partial x_{jm}^{(k)}}\frac{\partial u_n^{(k)m}}{\partial x_{im}^{(k)}}\right),\ i;j;n;r;s=1,2,
\end{gathered} \tag{5.93}$$

where $\lambda^{(2)*}=\lambda^{(2)}$, $\mu^{(2)*}=\mu^{(2)}$ are Lamé's constants for the filler layers materials, but $\lambda^{(1)*}$ and $\mu^{(1)*}$ are the operators determined by the expressions (5.40)–(5.42).

We assume that, between the filler and matrix layers there is complete cohesion. We denote the upper surface of the $2^{(2)}$ ($1^{(2)}$) layer by $S_2^+$ ($S_1^+$) and lower surface of it by $S_2^-$ ($S_1^-$). The initial insignificant local imperfections (curvings) of the selected filler layers $2^{(2)}$ and $1^{(2)}$ are given through the middle surface of these layers as

$$x_{22}^{(2)}=\varepsilon f_2(x_1)\ \text{and}\ x_{21}^{(2)}=\varepsilon f_1(x_1). \tag{5.94}$$

Here $\varepsilon$ is the dimensionless small parameter $(0\leq\varepsilon\ll 1)$ the geometrical meaning of which is described by the specifically prescribed form of the functions $f_2(x_1)$ and $f_1(x_1)$. According to Fig. 5.14, we can write that

$$f_1(x_1)=f_2(x_1+\ell_0)\ \text{or}\ f_2(x_1)=f_1(x_1-\ell_0). \tag{5.95}$$

It is also supposed that the functions $f_2(x_1)$ and $f_1(x_1)$, and their first order derivatives are continuous and

$$|f_1(x_1)|\to 0,\ |f_2(x_1)|\to 0,\ \left|\frac{df_1(x_1)}{dx_1}\right|\to 0,\ \left|\frac{df_2(x_1)}{dx_1}\right|\to 0\ \text{as}\ |x_1|\to\infty. \tag{5.96}$$

Taking the above into account and according to the expression (5.13), we can write the following equations for the surfaces $S_m^{\pm}$ $(m=1,2)$ and the orthonormal components to these surfaces.

$$\begin{aligned} x_{1m}^{(2)\pm} &= x_{1m}^{(2)\pm}(t_1, H^{(2)}, \varepsilon f_m(t_1)) = t_1 \mp \varepsilon H^{(2)} \tfrac{\partial f_m}{\partial t_1} + O(\varepsilon^3), \\ x_{2m}^{(2)\pm} &= x_{2m}^{(2)\pm}(t_1, H^{(2)}, \varepsilon f_m(t_1)) = \pm H^{(2)} \pm \varepsilon f_m + O(\varepsilon^2), \\ n_1^{m,\pm} &= n_1^{m,\pm}(t_1, H^{(2)}, \varepsilon f_m(t_1)) = -\varepsilon \tfrac{\partial f_m}{\partial t_1} + O(\varepsilon^3), \\ n_2^{m,\pm} &= n_2^{m,\pm}(t_1, H^{(2)}, \varepsilon f_m(t_1)) = 1 + O(\varepsilon^2). \end{aligned} \tag{5.97}$$

According to (5.95), we get

$$\begin{aligned} x_{i2}^{(2)\pm}(t_1 - \ell_0, H^{(2)}, \varepsilon f_m(t_1 - \ell_0)) &= x_{i1}^{(2)\pm}(t_1, H^{(2)}, \varepsilon f_m(t_1)), \\ n_i^{2,\pm}(t_1 - \ell_0, H^{(2)}, \varepsilon f_m(t_1 - \ell_0)) &= n_i^{1,\pm}(t_1, H^{(2)}, \varepsilon f_m(t_1)). \end{aligned} \tag{5.98}$$

Using Eqs. (5.97) and (5.98) we write the contact conditions among the layers $1^{(1)}$, $1^{(2)}$ and $2^{(2)}$ as follows:

$$\begin{aligned} &\left\{\sigma_{jn}^{(1)1}\left(\delta_i^n + \frac{\partial u_i^{(1)1}}{\partial x_{n1}^{(k)}}\right)\left(x_{11}^{(1)-}(t_1, \varepsilon f_1(t_1)), x_{21}^{(1)-}(t_1, \varepsilon f_1(t_1))\right)\right\} n_j^{1,+}(t_1, \varepsilon f_1(t_1)) \\ &= \left\{\sigma_{jn}^{(2)1}\left(\delta_i^n + \frac{\partial u_i^{(2)1}}{\partial x_{n1}^{(k)}}\right)\left(x_{11}^{(2)+}(t_1, \varepsilon f_1(t_1)), x_{21}^{(2)+}(t_1, \varepsilon f_1(t_1))\right)\right\} n_j^{1,+}(t_1, \varepsilon f_1(t_1)), \\ &u_i^{(1)1}(x_{11}^{(1)-}(t_1, \varepsilon f_1(t_1)), x_{21}^{(1)-}(t_1, \varepsilon f_1(t_1))) \\ &= u_i^{(2)1}(x_{11}^{(2)+}(t_1, \varepsilon f_1(t_1)), x_{21}^{(2)+}(t_1, \varepsilon f_1(t_1))), \\ &\left\{\sigma_{jn}^{(1)1}\left(\delta_i^n + \frac{\partial u_i^{(1)1}}{\partial x_{n1}^{(k)}}\right)\left(x_{11}^{(1)+}(t_1, f_2(t_1)), x_{21}^{(1)+}(t_1, f_2(t_1))\right)\right\} n_j^{2,-}(t_1, f_2(t_1)) \\ &= \left\{\sigma_{jn}^{(2)2}\left(\delta_i^n + \frac{\partial u_i^{(2)2}}{\partial x_{n2}^{(k)}}\right)\left(x_{12}^{(2)-}(t_1, f_2(t_1)), x_{22}^{(2)-}(t_1, f_2(t_1))\right)\right\} n_j^{2,-}(t_1, f_2(t_1)) \\ &u_i^{(1)1}(x_{11}^{(1)+}(t_1, f_2(t_1)), x_{21}^{(1)+}(t_1, f_2(t_1))) \\ &= u_i^{(2)2}(x_{12}^{(2)-}(t_1, f_2(t_1)), x_{22}^{(2)-}(t_1, f_2(t_1))), \\ &\left\{\sigma_{jn}^{(2)1}\left(\delta_i^n + \frac{\partial u_i^{(2)1}}{\partial x_{n1}^{(k)}}\right)\left(x_{11}^{(2)-}(t_1, \varepsilon f_1(t_1)), x_{21}^{(2)-}(t_1, \varepsilon f_1(t_1))\right)\right\} n_j^{1,-}(t_1, \varepsilon f_1(t_1)) \\ &= \left\{\sigma_{jn}^{(2)2}\left(\delta_i^n + \frac{\partial u_i^{(2)2}}{\partial x_{n2}^{(k)}}\right)\left(x_{12}^{(2)+}(t_{1*}, f_2(t_{1*})), x_{22}^{(2)+}(t_{1*}, f_2(t_{1*}))\right)\right\} n_j^{2,+}(t_{1*}, f_2(t_{1*})) \\ &u_i^{(2)1}(x_{11}^{(2)-}(t_1, \varepsilon f_1(t_1)), x_{21}^{(2)-}(t_1, \varepsilon f_1(t_1))) \\ &= u_i^{(2)2}(x_{12}^{(2)+}(t_{1*}, f_2(t_{1*})), x_{22}^{(2)+}(t_{1*}, f_2(t_{1*}))), \end{aligned} \tag{5.99}$$

where

$$t_{1*} = t_1 - \ell_0, \tag{5.100}$$

$$\begin{aligned} x_{21}^{(1)-} &= x_{21}^{(2)+} - H^{(1)} - H^{(2)},\ x_{11}^{(1)-} = x_{11}^{(2)+}, \\ x_{21}^{(1)+} &= x_{22}^{(2)-} + H^{(1)} + H^{(2)},\ x_{11}^{(1)+} = x_{12}^{(2)-}. \end{aligned} \tag{5.101}$$

Like to the previous section, we select the functions $f_2(x_1)$ and $f_1(x_1)$ as follows:

$$f_2(x_1) = L\exp\left\{-\left(\left(\frac{x_1-\ell_0}{L}\right)^2\right)^\gamma\right\}\cos\left(n\frac{x_1-\ell_0}{L}\right),$$
$$f_1(x_1) = L\exp\left\{-\left(\left(\frac{x_1}{L}\right)^2\right)^\gamma\right\}\cos\left(n\frac{x_1}{L}\right). \quad (5.102)$$

This completes the formulation of the problem. Now we consider the solution method of that. Values characterizing the stress-deformation state of the selected layers we present in the series form (5.14) in the parameter $\varepsilon$. Doing some mathematical operations which are similar to those made in the previous sections we obtain the complete system of field equations and contact conditions for each approximation. We consider the zeroth and the first approximations only because the second and subsequent approximations do not influence on the values of the critical parameters. For this purpose we employ the Laplace transformation (5.27) and the Laplace transformation of the zeroth approximation is determined as in the previous section, i.e. through the expressions (5.28)–(5.30).

For the first approximation we obtain the field equations in the form (5.37), (5.38), (5.23) and (5.24). However, for the case under consideration the upper index $(k)$ in these equations must be replaced with the upper index $(k)m$. The following contact conditions are obtained from (5.99)–(5.101) for the first approximation.

$$\sigma_{i2}^{(1)1,1}(t_1,-H^{(1)}) - \sigma_{i2}^{(2)1,1}(t_1,H^{(2)}) = \tfrac{df_1(t_1)}{dt_1}(\sigma_{11}^{(1)1,0} - \sigma_{11}^{(2)1,0})\delta_i^1,$$
$$u_i^{(1)1,1}(t_1,-H^{(1)}) - u_i^{(2)1,1}(t_1,H^{(2)}) = f_1(t_1)\left(\frac{\partial u_2^{(1)1,0}}{\partial x_{21}^{(1)}} - \frac{\partial u_2^{(2)1,0}}{\partial x_{21}^{(2)}}\right)\delta_i^2,$$
$$\sigma_{i2}^{(1)1,1}(t_1,+H^{(1)}) - \sigma_{i2}^{(2)2,1}(t_1,-H^{(2)}) = \tfrac{df_2(t_1)}{dt_1}(\sigma_{11}^{(1)1,0} - \sigma_{11}^{(2)2,0})\delta_i^1,$$
$$u_i^{(1)1,1}(t_1,+H^{(1)}) - u_i^{(2)1,1}(t_1,-H^{(2)}) = f_2(t_1)\left(\frac{\partial u_2^{(1)1,0}}{\partial x_{21}^{(1)}} - \frac{\partial u_2^{(2)2,0}}{\partial x_{21}^{(2)}}\right)\delta_i^2,$$
$$\sigma_{i2}^{(2)1,1}(t_1,-H^{(2)}) = \sigma_{i2}^{(2)2,1}(t_1-\ell_0,+H^{(2)}),$$
$$u_i^{(2)1,1}(t_1,-H^{(2)}) = u_i^{(2)2,1}(t_1-\ell_0,+H^{(2)}). \quad (5.103)$$

For the solution to the problem (5.37), (5.38), (5.23), (5.24) and (5.103) we employ also the Laplace transformation (5.27) and exponential Fourier transformation (5.78). Doing the same operations (with obvious changing) which are made in the previous section we determine also the values of the first approximation. The values of the critical parameters we find from the criterion

$$u_2^{(2)1,1}(0,0)\big/L \to \infty \text{ as } p \to p_{cr} \text{ (for purely elastic case)}$$
$$\text{or as } t' \to t'_{cr} \text{ (for a viscoelastic case)}. \quad (5.104)$$

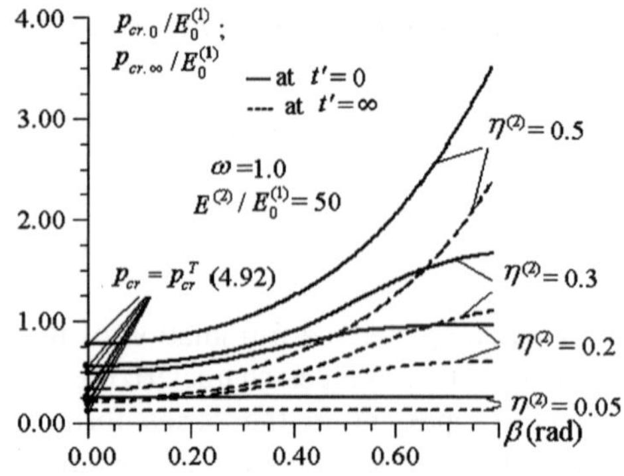


**Fig. 5.15** The dependence among $p_{cr.0}\big/E_0^{(1)}, p_{cr.\infty}\big/E_0^{(1)}$ and $\beta$ for various $\eta^{(2)}$

### *5.4.2 Results and Discussions*

We assume that $\nu^{(2)} = \nu_0^{(1)} = 0.3$ and $E^{(2)}\big/E_0^{(1)} = 50$, and use the notation on the dimensionless rheological parameter and dimensionless time introduced in the previous sections. Investigate the influence of the inclination angle $\beta$ on the values of $p_{cr.0}\big/E_0^{(1)}$, $p_{cr.\infty}\big/E_0^{(1)}$ and $t'_{cr}$. Note that under calculation $t'_{cr}$ we assume that $t'_* = 0$ in equations (5.48)–(5.50). Moreover, note that under calculation $p_{cr.0}\big/E_0^{(1)}$ ($p_{cr.\infty}\big/E_0^{(1)}$) the constants $E_0^{(1)}$ and $\nu_0^{(1)}$ ($E_\infty^{(1)}$ and $\nu_\infty^{(1)}$) are taken instead of the operators $E^{(1)*}$ and $\nu^{(1)*}$.

First, we consider the results which illustrate the influence of the angle $\beta$ on the values of $p_{cr.0}\big/E_0^{(1)}$ and $p_{cr.\infty}\big/E_0^{(1)}$. As it was established in the previous section these critical values are not depend on the initial imperfection mode, i.e. on the parameters $\gamma$ and $n$ in the Eq. (5.102). At the same time, the values of $p_{cr.0}\big/E_0^{(1)}$ and $p_{cr.\infty}\big/E_0^{(1)}$ are not depend also on the parameter $\chi\,(= H^{(2)}/L)$, $2H^{(2)}$ is the thickness of the filler layer and $L$ is a geometric parameter which also enters the Eq. (5.102). Therefore in obtaining the numerical results discussed below we assume that $\gamma = 1$, $n = 0$ and $\chi = 0.1$.

Figure 5.15 shows the graphs of the dependencies among $p_{cr.0}\big/E_0^{(1)}$, $p_{cr.\infty}\big/E_0^{(1)}$ and $\beta$ for various $\eta^{(2)}$ in the cases where $0 \le \beta \le 0.8$ ($0° \le \beta \le 45°$) under $\omega = 1.0$. The influence of the values $\omega$ on the foregoing dependencies is illustrated by the graphs given in Fig. 5.16.

It follows from Figs. 5.15 and 5.16 that in the case where $\beta = 0$ the values of $p_{cr.0}\big/E_0^{(1)}$ ($p_{cr.\infty}\big/E_0^{(1)}$) coincide with the corresponding values of the TSLC determined in the previous section by expression (5.92), i.e. coincide with $G_{12.0}\big/E_0^{(1)}$ ($G_{12.\infty}\big/E_0^{(1)}$) where $G_{12.0}$ ($G_{12.\infty}$) is determined from the expression of

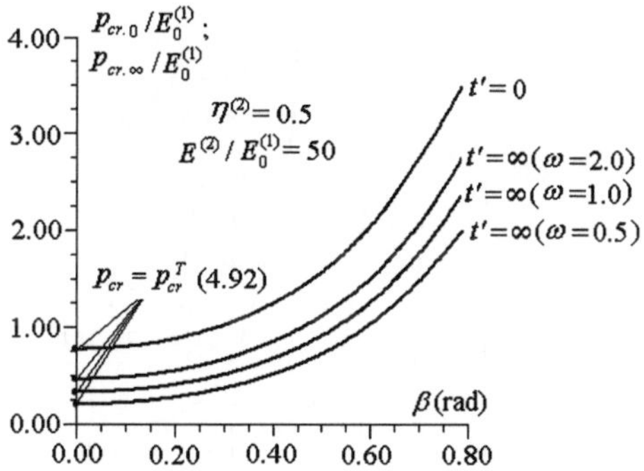


**Fig. 5.16** The influence of the values $\omega$ on the dependencies among $p_{cr.0}\big/E_0^{(1)}$, $p_{cr.\infty}\big/E_0^{(1)}$ and $\beta$ under $\eta^{(2)} = 0.5$

$G_{12}$ in (5.80) by replacing $\mu^{(1)}$ and $\nu^{(1)}$ with $\mu_0^{(1)}$ and $\nu_0^{(1)}$ ($\mu_\infty^{(1)}$ and $\nu_\infty^{(1)}$) respectively. At the same time, in the references (Guz 1969, 1990, 2008a, b) it was established that the fracture (failure) occurs on the plane which is almost perpendicular to the compression direction. Consequently, according the approach (Guz 1969, 1990), the absolute minimum of the dependencies among $p_{cr.0}\big/E_0^{(1)}$, $p_{cr.\infty}\big/E_0^{(1)}$ and $\beta$ must occur at $\beta \approx 0$. This conclusion is also confirmed by the numerical results given in Figs. 5.15 and 5.16. Moreover these results show that, as can be predicted, the influence of the angle $\beta$ on the values of $p_{cr.0}\big/E_0^{(1)}$ and $p_{cr.\infty}\big/E_0^{(1)}$ is insignificant for the low concentration of the filler layers, for example, in the case where $\eta^{(2)} = 0.02$. However, the values of $p_{cr.0}\big/E_0^{(1)}$ and $p_{cr.\infty}\big/E_0^{(1)}$ increase significantly with $\beta$ under high concentration of the filler layers.

We now consider the influence of the problem parameters on the values of $t'_{cr}$. In this case we assume that the external compressive force satisfies the inequality $p_{cr.\infty}\big/E_0^{(1)} \le p\big/E_0^{(1)} \le p_{cr.0}\big/E_0^{(1)}$. Figure 5.17 shows the graphs of the dependence between $t'_{cr}$ and $\xi\,(= |p - p_{cr.\infty}|/p_{cr.\infty})$ for the cases where $\omega = 0.5$ (Fig. 5.17a) and $\omega = 1.0$ (Fig. 5.17b) under $\alpha = -0.5$ and $\eta^{(2)} = 0.5$. Note that these graphs are constructed under $\beta = 0°, 20°$ and $45°$. Figure 5.18 shows the foregoing graphs constructed for the various values of $\alpha$ and $\beta$ under $\omega = 1.0$ and $\eta^{(2)} = 0.5$. Note that the results obtained for the case where $\beta = 0°$ coincide with the corresponding ones obtained in the previous section.

It follows from these graphs that for fixed $t'_{cr}$ the values of $\xi$ decrease with $\beta$. Also the graphs show that before the certain values of $\xi$(denoted by $\xi^*$) the values of $t'_{cr}$ increase with $|\alpha|$but, for $\xi > \xi^*$ the values of $t'_{cr}$ decrease with $\xi$. Note that $\xi^*$ corresponds the case where $t'_{cr} = 0.5$ and this conclusion agrees with that which has been made in Sect. 2.5.1.

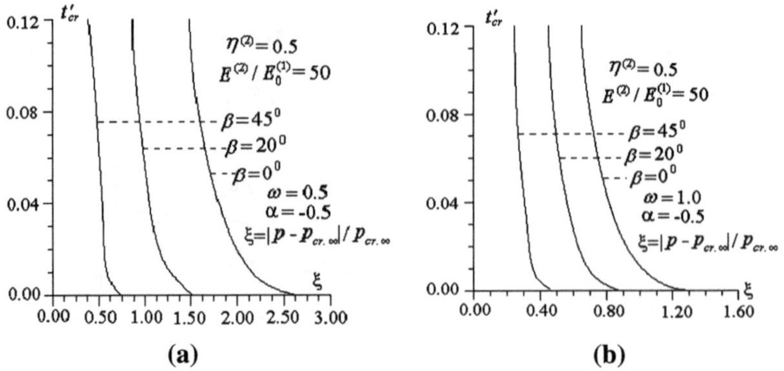


**Fig. 5.17** The graphs of the dependence between $t'_{cr}$ and $\xi$ for various values of the angle $\beta$ under $\alpha = -0.5$, $\eta^{(2)} = 0.5$, $\omega = 0.5$ (a) and $\omega = 1.0$(b)

## *5.4.3 Conclusions*

In both the present and previous sections within the framework of the piecewise homogeneous body model with the use of the exact geometrical non-linear field equations of the theory of elasticity and viscoelasticity, a method is developed for determination of the TSLC. We recall that under the TSLC, the $p_{cr}$ for the pure elastic composites and the $t'_{cr}$ for the viscoelastic composites are understood. The corresponding investigations are done on the layered composite consisting of the two layers of the homogeneous isotropic material. It was assumed that the reinforcing layers have initial local imperfections and these imperfections are moved the same distance from each other. Consequently, the aim of the investigations was to study the influence of this length on the values of the TSLC. In this case the length of this distance is expressed by the angle $\beta$ shown in Fig. 5.14 and based on the fracture criterion (i.e. the stability loss criterion according to which the TSLC is determined), the expression (5.104) for the elastic and viscoelastic composites is taken.

The numerical results regarding the influence of the aforementioned moving of the initial local imperfections on the values of the TSLC are presented and discussed. According to these results the following conclusions are established:

The fracture of the layered composite material considered in compression along the layers takes place over the plane which is perpendicular to the compression direction. This conclusion was first established in the work (Guz 1969) and later detailed in the monograph (Guz 1990). Moreover, in a qualitative sense this conclusion agrees with the experimental results obtained in the paper (Guz et al. 1969).

In (Guz 1969, 1990) corresponding investigations were made within the scope of the continuum approach. However, in the present and previous sections these results were replicated within the scope of the piecewise homogeneous body

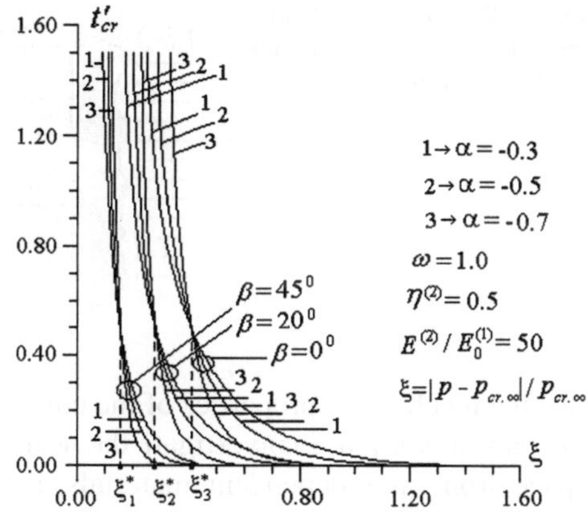


**Fig. 5.18** The graphs of the dependence between $t'_{cr}$ and $\xi$ for various values of $\beta$ and $\alpha$ under $\eta^{(2)} = 0.5$ and $\omega = 1.0$

model not only for elastic composite materials, but also for viscoelastic composites.

The values of the TSLC increase monotonically with $\beta$ (Fig. 5.14) under $0° \leq \beta \leq 45°$.

The dependence between the TSLC and $\beta$ is non-linear.

The TSLC decreases with a decrease in the rheological parameter $\omega$.

For the cases where $\xi > \xi^* \, (\xi < \xi^*)$ (in Fig. 5.18 the values of $\xi^*$ for $\beta = 45°$, $20°$ and $0°$ are indicated through $\xi_1^*$, $\xi_2^*$ and $\xi_3^*$ respectively) the values of $t'_{cr}$ decrease (increase) with $|\alpha|$.

## 5.5 Surface Undulation Instability of the Viscoelastic Half-Space Covered with a Stack of Layers in Bi-Axial Compression

So far, in the present chapter, the problems related to the inner stability loss in the structure of the composite materials have been studied. Consequently, in these studies it is assumed that the material occupies the 3D infinite space and the influence of the surfaces bonded to this material on the values of the critical parameters is not taken into account. However under this influence, in the cases where the abovementioned surfaces are free, the stability loss near these surfaces may appear in the earlier stage of the external loading than the internal stability loss stage. Therefore the necessity arises to investigate the near-surface stability loss. Corresponding investigations were made within the framework of the Euler approach and a review of these investigations was given in a paper (Babich et al. 2001). In these and other investigations (for example, (Guz et al. 2000)) the results regarding the viscoelastic materials were obtained in the framework of only the critical deformation criterion (Gerard and Gilbert 1958) which is very approximate. Consequently, in these investigations in a very strict sense, it was assumed

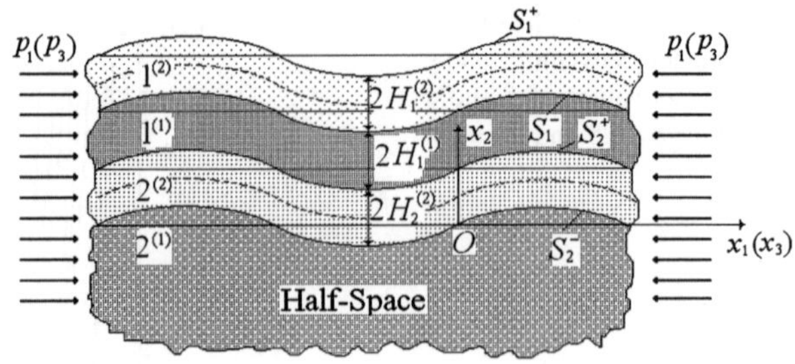


**Fig. 5.19** The geometry of the half-space covered by the stack consisting of three layers

that the mechanical properties of the considered materials are time-independent. In the present section, to investigate the near-surface stability loss of the layered half-space from time-dependent materials, an approach is developed and employed within the framework of the initial imperfection criterion. Below we follow papers (Akbarov and Tekercioglu 2006, 2007) and use the notation used in these papers.

## 5.5.1 Formulation of the Problem

We consider a semi-infinite half-space joined with the stack consisting of the finite number of the filler (reinforcing) and binder layers. Assume that the filler layers in this stack have an insignificant initial imperfection in the curving form. For simplicity of the consideration, we suppose that the stack consists of three layers as shown in Fig. 5.19.

Values related to the binder layers and half-space will be denoted by upper indices (1); values related to the filler layers by upper indices (2). Moreover, the values related with each selected layer and half-space will be denoted by the additional index indicated its sequence in the considered body. We associate the corresponding Lagrangian coordinates $O_m^{(k)}x_{1m}^{(k)}x_{2m}^{(k)}x_{3m}^{(k)}$ $(k = 1,2; m = 1,2)$ which in their natural state coincide with Cartesian coordinates and are obtained from $Ox_1x_2x_3$ by parallel transfer along the $Ox_2$ axis, with the middle surface of each layer of the filler and binder. The cross-section of the considered system at $x_1 = const\,(x_3 = const)$ is shown in Fig. 5.19.

The thickness of each filler layer will be assumed constant. It will also be assumed that the binder, filler and half-space materials are homogeneous, isotropic and non-aging (hereditary) linearly viscoelastic. Now we investigate the stress deformation state in the above body under compression at infinity by uniformly distributed normal forces with intensity $p_1\,(p_3)$ in the direction of the $Ox_1\,(Ox_3)$ axis.

For each layer and for the half-space we write the equilibrium equations, constitutive and geometrical relations given in (5.1). We assume that between the constituents of the system considered there is a complete cohesion

$$\left[\sigma_{jn}^{(2)1}\left(\delta_i^n+\frac{\partial u_i^{(2)1}}{\partial x_{n1}^{(2)}}\right)\right]\Bigg|_{S_1^-}n_j^{1,-}=\left[\sigma_{jn}^{(1)1}\left(\delta_i^n+\frac{\partial u_i^{(1)1}}{\partial x_{n1}^{(1)}}\right)\right]\Bigg|_{S_1^-}n_j^{1,-},\ u_i^{(2)1}\Big|_{S_1^-}$$
$$=u_i^{(1)1}\Big|_{S_1^-},\ \left[\sigma_{jn}^{(1)1}\left(\delta_i^n+\frac{\partial u_i^{(1)1}}{\partial x_{n1}^{(1)}}\right)\right]\Bigg|_{S_2^+}n_j^{2,+}=\left[\sigma_{jn}^{(2)2}\left(\delta_i^n+\frac{\partial u_i^{(2)2}}{\partial x_{n2}^{(2)}}\right)\right]\Bigg|_{S_2^+}n_j^{2,+},$$
$$u_i^{(1)1}\Big|_{S_2^+}=u_i^{(2)2}\Big|_{S_2^+},\ u_i^{(2)2}\Big|_{S_2^-}=u_i^{(1)2}\Big|_{S_2^-}$$
$$\left[\sigma_{jn}^{(2)2}\left(\delta_i^n+\frac{\partial u_i^{(2)2}}{\partial x_{n2}^{(2)}}\right)\right]\Bigg|_{S_2^-}n_j^{2,-}=\left[\sigma_{jn}^{(1)2}\left(\delta_i^n+\frac{\partial u_i^{(1)2}}{\partial x_{n2}^{(2)}}\right)\right]\Bigg|_{S_2^-}n_j^{2,-}. \tag{5.105}$$

In (5.105) the values indicated by the upper index (1)2 regard the half-space and these values satisfy also the following decay conditions

$$\sigma_{11}^{(1)2}\to p_1,\ \sigma_{33}^{(1)2}\to p_3,\ \sigma_{ij}^{(1)2}\to 0 \text{ for } ij\neq 11;33 \text{ as } x_{22}^{(2)}\to-\infty. \tag{5.106}$$

Moreover, on the upper surface of the $1^{(2)}$th layer the condition

$$\left[\sigma_{jn}^{(2)1}\left(\delta_i^n+\frac{\partial u_i^{(2)1}}{\partial x_{n1}^{(2)}}\right)\right]\Bigg|_{S_1^+}n_j^{1,+}=0 \tag{5.107}$$

is satisfied.

In (5.105) and (5.107), $n_j^{m,\pm}$ are the components of the unit normal vector to the surfaces $S_m^{\pm}$. The other notation used in the equations (5.105)–(5.107) is conventional.

The initial insignificant imperfection of the filler layers is given through the equations of the middle surface of those as

$$x_{2m}^{(2)}=F_m(x_{1m}^{(2)},x_{3m}^{(2)})=\varepsilon f_m(x_{1m}^{(2)},x_{3m}^{(2)}), \tag{5.108}$$

where $\varepsilon$ is a dimensionless small parameter $(0\leq\varepsilon\ll 1)$whose geometrical meaning is described by the specifically prescribed form of the Eq. (5.108).

Moreover, we suppose that the functions $F_m(x_{1m}^{(2)},x_{3m}^{(2)})$ and their first order derivatives are continuous and satisfy the following conditions:

$$\left(\frac{\partial F_m}{\partial x_{1m}^{(2)}}\right)^2+\left(\frac{\partial F_m}{\partial x_{3m}^{(2)}}\right)^2\ll 1. \tag{5.109}$$

This completes the formulation of the problem.

### 5.5.2 Method of Solution

We employ the approach developed in Sect. 5.2 for solution to the formulated problem above. Firstly, taking into account the condition of constant thickness of the filler layers and Eq. (5.108), the Eq. (5.13) are derived for interface surfaces $S_m^{\pm}$ and for the components of the unit normal vector of these surfaces (Fig. 5.19). Using the power series presentation (5.14) of the sought values with respect to the small parameter ε and doing almost the same operations described in Sect. 5.2, we obtain the corresponding field equations and contact conditions for the zeroth and the first approximations which are sufficient for investigation of the near-surface stability loss of the considered material. Note that under obtaining these equations and conditions, as in the previous sections we assume that $\partial u_i^{(k)m,0} \Big/ \partial x_{nm}^{(k)} \ll 1$ we take $\delta_i^n$ instead of $\delta_i^n + \partial u_i^{(k)m,0} \Big/ \partial x_{nm}^{(k)}$. Because of this assumption, we obtain field equations

$$\frac{\partial}{\partial x_{jm}^{(k)}} \sigma_{ji}^{(k)m,0} = 0,\ \varepsilon_{ij}^{(k),0} = \frac{1}{2}\left(\frac{\partial u_i^{(k)m,0}}{\partial x_{jm}^{(k)}} + \frac{\partial u_j^{(k)m,0}}{\partial x_{im}^{(k)}}\right),\ i;j = 1,2,3,\ k;m = 1,2,$$
$$\sigma_{ij}^{(k)m,0} = \lambda^{(k)m*} \theta^{(k)m,0} \delta_i^j + 2\mu^{(k)m*} \varepsilon_{ij}^{(k)m,0},$$
$$\theta^{(k)m,0} = \varepsilon_{11}^{(k)m,0} + \varepsilon_{22}^{(k)m,0} + \varepsilon_{33}^{(k)m,0}, \tag{5.110}$$

where

$$\left\{\begin{array}{c}\lambda^{(k)m*}\\ \mu^{(k)m*}\end{array}\right\} g(t) = \left\{\begin{array}{c}\lambda_0^{(k)m}\\ \mu_0^{(k)m}\end{array}\right\} g(t) + \int_0^t \left\{\begin{array}{c}\lambda^{(k)m}\\ \mu^{(k)m}\end{array}\right\}(t-\tau) g(\tau) d\tau, \tag{5.111}$$

contact conditions

$$\sigma_{2i}^{(2)1,0}\Big|_{(t_1,-H_1^{(2)},\ t_3)} = \sigma_{2i}^{(1)1,0}\Big|_{(t_1,+H_1^{(1)},\ t_3)},$$
$$u_i^{(2)1,0}\Big|_{(t_1,-H_1^{(2)},\ t_3)} = u_i^{(1)1,0}\Big|_{(t_1,+H_1^{(1)},\ t_3)},$$
$$\sigma_{2i}^{(1)1,0}\Big|_{(t_1,-H_1^{(1)},\ t_3)} = \sigma_{2i}^{(2)2,0}\Big|_{(t_1,+H_1^{(2)},\ t_3)},$$
$$u_i^{(1)1,0}\Big|_{(t_1,-H_1^{(1)},\ t_3)} = u_i^{(2)2,0}\Big|_{(t_1,+H_1^{(2)},\ t_3)},$$
$$\sigma_{2i}^{(2)2,0}\Big|_{(t_1,-H_1^{(2)},\ t_3)} = \sigma_{2i}^{(1)2,0}\Big|_{(t_1,-H_1^{(2)},\ t_3)},$$
$$u_i^{(2)2,0}\Big|_{(t_1,-H_1^{(2)},\ t_3)} = u_i^{(1)2,0}\Big|_{(t_1,-H_1^{(2)},\ t_3)}, \tag{5.112}$$

and boundary conditions

$$\sigma_{11}^{(1)2,0} \to p_1,\ \sigma_{33}^{(1)2,0} \to p_3,\ \sigma_{ij}^{(1)2,0} \to 0 \text{ for } ij \neq 11;33 \text{ as } x_{22}^{(2)} \to -\infty, \quad (5.113)$$

$$\sigma_{2i}^{(2)1,0}\Big|_{(t_1,+H_1^{(2)},\ t_3)} = 0, \quad (5.114)$$

for the zeroth approximation.

For solution to the problem formulated through the Eqs. (5.110)–(5.114) we employ the Laplace transformation (5.27) to these equations. Taking into account that $\sigma_{11}^{(1)2,0} = p_1$ and $\sigma_{33}^{(1)2,0} = p_3$, and doing some mathematical calculations we get

$$\begin{aligned}
\bar{\sigma}_{11}^{(k)m,0} &= \frac{\bar{E}^{(k)m*}(\bar{p}_1 + \bar{\nu}^{(k)m*}\bar{p}_3)}{\bar{E}^{(1)2*}(1-(\bar{\nu}^{(k)m*})^2)} - \frac{\bar{E}^{(k)m*}\bar{\nu}^{(1)2*}(\bar{p}_3 + \bar{\nu}^{(k)m*}\bar{p}_1)}{\bar{E}^{(1)2*}(1-(\bar{\nu}^{(k)m*})^2)}, \\
\bar{\sigma}_{33}^{(k)m,0} &= \frac{\bar{E}^{(k)m*}(\bar{p}_3 + \bar{\nu}^{(k)m*}\bar{p}_1)}{\bar{E}^{(1)2*}(1-(\bar{\nu}^{(k)m*})^2)} - \frac{\bar{E}^{(k)m*}\bar{\nu}^{(1)2*}(\bar{p}_1 + \bar{\nu}^{(k)m*}\bar{p}_3)}{\bar{E}^{(1)2*}(1-(\bar{\nu}^{(k)m*})^2)}, \\
\bar{\sigma}_{ij}^{(k)m,0} &= 0 \text{ for } ij \neq 11;33,
\end{aligned} \quad (5.115)$$

where

$$\begin{aligned}
\bar{\nu}^{(k)m*}(s) &= \frac{\bar{\lambda}^{(k)m*}(s)}{2(\bar{\lambda}^{(k)m*}(s) + \bar{\mu}^{(k)m*}(s))}, \\
\bar{E}^{(k)m*}(s) &= \frac{\bar{\mu}^{(k)m*}(s)(3\bar{\lambda}^{(k)m*}(s) + \bar{\mu}^{(k)m*}(s))}{\bar{\lambda}^{(k)m*}(s) + \bar{\mu}^{(k)m*}(s)}, \\
\bar{\lambda}^{(k)m*}(s) &= \lambda_0^{(k)m} + \bar{\lambda}^{(k)m}(s),\ \bar{\mu}^{(k)m*}(s) = \mu_0^{(k)m} + \bar{\mu}^{(k)m}(s).
\end{aligned} \quad (5.116)$$

The originals of the Laplace transformations of $\bar{\sigma}_{11}^{(k)m,0}$ and $\bar{\sigma}_{33}^{(k)m,0}$, i.e. the functions $\sigma_{11}^{(k)m,0}(t)$ and $\sigma_{33}^{(k)m,0}(t)$ are determine by applying the method (Schapery 1962, 1978) the detail of which was described in Sect. 5.2.

Thus, taking the foregoing results into account we get the following field equations

$$\begin{aligned}
&\frac{\partial \sigma_{ji}^{(k)m,1}}{\partial x_{jm}^{(k)}} + \sigma_{11}^{(k)m,0}\frac{\partial^2 u_i^{(k)m,1}}{\partial (x_{1m}^{(k)})^2} + \sigma_{33}^{(k)m,0}\frac{\partial^2 u_i^{(k)m,1}}{\partial (x_{3m}^{(k)})^2} = 0, \\
&\varepsilon_{ij}^{(k),1} = \frac{1}{2}\left(\frac{\partial u_i^{(k)m,1}}{\partial x_{jm}^{(k)}} + \frac{\partial u_j^{(k)m,1}}{\partial x_{im}^{(k)}}\right),\ \theta^{(k)m,1} = \varepsilon_{11}^{(k)m,1} + \varepsilon_{22}^{(k)m,1} + \varepsilon_{33}^{(k)m,1}, \\
&\sigma_{ij}^{(k)m,1} = \lambda^{(k)m*}\theta^{(k)m,1}\delta_i^j + 2\mu^{(k)m*}\varepsilon_{ij}^{(k)m,1},
\end{aligned} \quad (5.117)$$

contact conditions

$$\sigma_{2i}^{(2)1,1}\Big|_{x_{21}^{(2)}=-H_1^{(2)}} - \sigma_{2i}^{(1)1,1}\Big|_{x_{21}^{(1)}=+H_1^{(1)}} = \delta_i^1\left(\sigma_{11}^{(2)1,0} - \sigma_{11}^{(1)1,0}\right)\frac{\partial f_1}{\partial x_{11}^{(2)}}$$
$$+\delta_i^3\left(\sigma_{33}^{(2)1,0} - \sigma_{33}^{(1)1,0}\right)\frac{\partial f_1}{\partial x_{31}^{(2)}},$$
$$u_i^{(2)1,1}\Big|_{x_{21}^{(2)}=-H_1^{(2)}} - u_i^{(1)1,1}\Big|_{x_{21}^{(1)}=+H_1^{(1)}} = \delta_i^2 f_1\left(\frac{\partial u_2^{(2)1,0}}{\partial x_{21}^{(2)}} - \frac{\partial u_2^{(1)1,0}}{\partial x_{11}^{(2)}}\right),$$
$$\sigma_{2i}^{(1)1,1}\Big|_{x_{21}^{(1)}=-H_1^{(1)}} - \sigma_{2i}^{(2)2,1}\Big|_{x_{22}^{(2)}=+H_2^{(2)}} = \delta_i^1\left(\sigma_{11}^{(1)1,0} - \sigma_{11}^{(2)2,0}\right)\frac{\partial f_1}{\partial x_{12}^{(2)}}$$
$$+\delta_i^3\left(\sigma_{33}^{(1)1,0} - \sigma_{33}^{(2)2,0}\right)\frac{\partial f_1}{\partial x_{32}^{(2)}},$$
$$u_i^{(1)1,1}\Big|_{x_{21}^{(1)}=-H_1^{(1)}} - u_i^{(2)2,1}\Big|_{x_{22}^{(2)}=+H_2^{(2)}} = \delta_i^2\left(\frac{\partial u_2^{(1)1,0}}{\partial x_{21}^{(1)}} - \frac{\partial u_2^{(2)2,0}}{\partial x_{22}^{(2)}}\right),$$
$$\sigma_{2i}^{(2)2,1}\Big|_{x_{22}^{(2)}=-H_2^{(2)}} - \sigma_{2i}^{(1)2,1}\Big|_{x_{22}^{(1)}=-H_2^{(2)}} = \delta_i^1\left(\sigma_{11}^{(2)2,0} - p_1\right)\frac{\partial f_1}{\partial x_{12}^{(2)}}$$
$$+\delta_i^3\left(\sigma_{33}^{(2)2,0} - p_3\right)\frac{\partial f_1}{\partial x_{32}^{(2)}},$$
$$u_i^{(2)2,1}\Big|_{x_{22}^{(2)}=-H_2^{(2)}} - u_i^{(1)2,1}\Big|_{x_{22}^{(2)}=-H_2^{(2)}} = \delta_i^2\left(\frac{\partial u_2^{(2)2,0}}{\partial x_{22}^{(2)}} - \frac{\partial u_2^{(1)2,0}}{\partial x_{22}^{(2)}}\right), \tag{5.118}$$

and boundary conditions

$$\sigma_{21}^{(2)1,1}\Big|_{x_{21}^{(2)}=+H_1^{(2)}} = \sigma_{11}^{(2)1,0}\frac{\partial f_1}{\partial x_{11}^{(2)}},\ \sigma_{23}^{(2)1,1}\Big|_{x_{21}^{(2)}=+H_1^{(2)}} = \sigma_{33}^{(2)1,0}\frac{\partial f_1}{\partial x_{31}^{(2)}},$$
$$\sigma_{22}^{(2)1,1}\Big|_{x_{21}^{(2)}=+H_1^{(2)}} = 0,\ \sigma_{ij}^{(1)2,1} \to 0,\ u_i^{(1)2,1} \to 0 \text{ as } x_{22}^{(2)} \to -\infty. \tag{5.119}$$

Further development of the method of solution depends on the selection of the initial imperfection mode, i.e. on the selection of the function (5.108). In the present investigation, we assume that

$$F_1(x_{11}^{(2)}, x_{31}^{(2)}) = F_2(x_{12}^{(2)}, x_{32}^{(2)}) = F(x_1, x_3) = \frac{L}{\ell_1}\ell_1 \sin(\alpha_1 x_1)\cos(\gamma\alpha_1 x_3), \tag{5.120}$$

where

$$\alpha_1 = \pi/\ell_1,\ \gamma = \ell_1/\ell_3. \tag{5.121}$$

We suppose that $L \ll \ell_1$ and the dimensionless small parameter $\varepsilon$ is determined by $\varepsilon = L/\ell_1$. Consequently the expressions for $\partial f_1 / \partial x_{11}^{(2)}$, $\partial f_1 / \partial x_{31}^{(2)}$, $\partial f_2 / \partial x_{12}^{(2)}$ and $\partial f_2 / \partial x_{32}^{(2)}$ in (5.118) and (5.119) are defined as

$$\frac{\partial f_1}{\partial x_{11}^{(2)}} = \frac{\partial f_2}{\partial x_{12}^{(2)}} = \pi \cos(\alpha_1 x_1) \cos(\gamma \alpha_1 x_1),$$
$$\frac{\partial f_1}{\partial x_{31}^{(2)}} = \frac{\partial f_2}{\partial x_{32}^{(2)}} = -\pi\gamma \sin(\alpha_1 x_1) \sin(\gamma \alpha_1 x_3). \tag{5.122}$$

Thus, for considered case, the determination of the first approximation is reduced to the solution of the integro-differential Eqs. (5.117) and (5.111) with contact conditions (5.118) and (5.122) and boundary conditions (5.119) and (5.122). Note that the coefficient $\sigma_{11}^{(k)m,0}$ and $\sigma_{33}^{(k)m,0}$ in Eq. (5.117) are functions of time, i.e. $\sigma_{11}^{(k)m,0} = \sigma_{11}^{(k)m,0}(t)$ and $\sigma_{33}^{(k)m,0} = \sigma_{33}^{(k)m,0}(t)$. Just this statement violates the application of the principle of correspondence in the solution to the problems (5.111), (5.117)–(5.122). As in Sect. 5.2, we assume that

$$\int_0^\infty \sigma_{jj}^{(k)m,0}(t) \frac{\partial^2 u_i^{(k)m,1}}{\partial (x_{jm}^{(k)})^2} \exp(-st)dt = \sigma_{jj}^{(k)m,0}(t_*) \int_0^\infty \frac{\partial^2 u_i^{(k)m,1}}{\partial (x_{jm}^{(k)})^2} \exp(-st)dt$$
$$= \sigma_{jj}^{(k)m,0}(t_*) \frac{\partial^2 \bar{u}_i^{(k)m,1}}{\partial (x_{jm}^{(k)})^2}, j = 1, 3; \quad i = 1, 2, 3. \tag{5.123}$$

It is evident that finding a certain value of $t_*$ under which Eq. (5.123) is satisfied exactly is impossible or very difficult. However, it follows from the nature of the considered mechanical object that exact values of the critical parameters of the investigated stability loss problems must be limited with those obtained under $t_* = 0$ and $t_* = \infty$. Therefore under concrete numerical investigation we will select one of these values of $t_*$.

Thus, taking into account the above considerations, the Laplace transformation of the equilibrium equation in (5.117) can be written as follows:

$$\frac{\partial \bar{\sigma}_{ji}^{(k)m,1}}{\partial x_{jm}^{(k)}} + \sigma_{11}^{(k)m,0}(t_*) \frac{\partial^2 \bar{u}_i^{(k)m,1}}{\partial (x_{1m}^{(k)})^2} + \sigma_{33}^{(k)m,0}(t_*) \frac{\partial^2 \bar{u}_i^{(k)m,1}}{\partial (x_{3m}^{(k)})^2} = 0. \tag{5.124}$$

The remaining relations in the Eqs. (5.117)–(5.119) hold also for the Laplace transformation of the sought values with the changes

$$\left\{ \sigma_{ij}^{(k)m,1}; \varepsilon_{ij}^{(k)m,1}; u_i^{(k)m,1}; \lambda^{(k)m*}; \mu^{(k)m*} \right\} \Rightarrow$$
$$\left\{ \bar{\sigma}_{ij}^{(k)m,1}; \bar{\varepsilon}_{ij}^{(k)m,1}; \bar{u}_i^{(k)m,1}; \bar{\lambda}^{(k)m*}; \bar{\mu}^{(k)m*} \right\}. \tag{5.125}$$

Using (5.125) and the Laplace transformation of the mechanical and geometrical relations in (5.117), we obtain from (5.124) the equilibrium equation in displacement terms.

$$\nabla^2 \bar{u}_i^{(k)m,1} + (1+\frac{\bar{\lambda}^{(k)m}}{\bar{\mu}^{(k)m}})\frac{\partial \bar{\theta}^{(k)m}}{\partial x_{im}^{(k)}} + \frac{\sigma_{11}^{(k)m,0}(t_*)}{\bar{\mu}^{(k)m}}\frac{\partial^2 \bar{u}_i^{(k)m}}{\partial (x_{1m}^{(k)})^2} + \frac{\sigma_{33}^{(k)m,0}(t_*)}{\bar{\mu}^{(k)m}}\frac{\partial^2 \bar{u}_i^{(k)m}}{\partial (x_{3m}^{(k)})^2} = 0, \quad (5.126)$$

where

$$\nabla^2 = \frac{\partial^2}{\partial (x_{1m}^{(k)})^2} + \frac{\partial^2}{\partial (x_{2m}^{(k)})^2} + \frac{\partial^2}{\partial (x_{3m}^{(k)})^2}. \quad (5.127)$$

From Eq. (5.126) it is found that

$$(2+\frac{\bar{\lambda}^{(k)m}}{\bar{\mu}^{(k)m}})\nabla^2 \bar{\theta}^{(k)m} + \frac{\sigma_{11}^{(k)m,0}(t_*)}{\bar{\mu}^{(k)m}}\frac{\partial^2 \bar{\theta}^{(k)m}}{\partial (x_{1m}^{(k)})^2} + \frac{\sigma_{33}^{(k)m,0}(t_*)}{\bar{\mu}^{(k)m}}\frac{\partial^2 \bar{\theta}^{(k)m}}{\partial (x_{3m}^{(k)})^2} = 0,$$

$$\nabla^2 \bar{u}_i^{(k)m,1} + \frac{\sigma_{11}^{(k)m,0}(t_*)}{\bar{\mu}^{(k)m}}\frac{\partial^2 \bar{u}_i^{(k)m}}{\partial (x_{1m}^{(k)})^2} + \frac{\sigma_{33}^{(k)m,0}(t_*)}{\bar{\mu}^{(k)m}}\frac{\partial^2 \bar{u}_i^{(k)m}}{\partial (x_{3m}^{(k)})^2} = -(1+\frac{\bar{\lambda}^{(k)m}}{\bar{\mu}^{(k)m}})\frac{\partial \bar{\theta}^{(k)m}}{\partial x_{im}^{(k)}}. \quad (5.128)$$

According to the conditions (5.118) and (5.119) and the expressions in (5.120), we can write that

$$\begin{aligned}
\bar{\theta}^{(k)m,1} &= \bar{\theta}_1^{(k)m,1}(x_{2m}^{(k)}) \sin(\alpha_1 x_1)\cos(\gamma\alpha_1 x_3), \\
\bar{u}_1^{(k)m,1} &= \bar{\varphi}_1^{(k)m,1}(x_{2m}^{(k)}) \cos(\alpha_1 x_1)\cos(\gamma\alpha_1 x_3), \\
\bar{u}_2^{(k)m,1} &= \bar{\varphi}_2^{(k)m,1}(x_{2m}^{(k)}) \sin(\alpha_1 x_1)\cos(\gamma\alpha_1 x_3), \\
\bar{u}_3^{(k)m,1} &= \bar{\varphi}_3^{(k)m,1}(x_{2m}^{(k)}) \sin(\alpha_1 x_1)\sin(\gamma\alpha_1 x_3),
\end{aligned} \quad (5.129)$$

where $\alpha_1$ and $\gamma$ are determined through the expression (5.121).

Substituting (5.129) into Eq. (5.128) we get the following ordinary differential equations for the functions $\bar{\theta}_1^{(k)m,1}$ and $\bar{\varphi}_i^{(k)m,1}$ $(i = 1, 2, 3)$.

$$\begin{aligned}
&\frac{d^2 \bar{\theta}_1^{(k)m,1}}{d(\alpha_1 x_{2m}^{(k)})^2} - \left(1+\gamma^2+\frac{\sigma_{11}^{(k)m,0}(t_*)}{\bar{\lambda}^{(k)m}+2\bar{\mu}^{(k)m}}+\gamma^2\frac{\sigma_{33}^{(k)m,0}(t_*)}{\bar{\lambda}^{(k)m}+2\bar{\mu}^{(k)m}}\right)\bar{\theta}_1^{(k)m,1} = 0, \\
&\frac{d^2 \bar{\varphi}_i^{(k)m,1}}{d(\alpha_1 x_{2m}^{(k)})^2} - \left(1+\gamma^2+\frac{\sigma_{11}^{(k)m,0}(t_*)}{\bar{\mu}^{(k)m}}+\gamma^2\frac{\sigma_{33}^{(k)m,0}(t_*)}{\bar{\mu}^{(k)m}}\right)\bar{\varphi}_i^{(k)m,1} \\
&= -\left(1+\frac{\bar{\lambda}^{(k)m}}{\bar{\mu}^{(k)m}}\right)\left(\frac{d\bar{\theta}_1^{(k)m}}{dx_{2m}^{(k)}}\delta_i^2 + \bar{\theta}_1^{(k)m}\delta_i^1 - \gamma\bar{\theta}_1^{(k)m}\delta_i^3\right).
\end{aligned} \quad (5.130)$$

Below, for simplicity we will use $x_{2m}^{(k)}$ instead of $\alpha_1 x_{2m}^{(k)}$.

Thus, for the general solution to the first equation in (5.130) we find as follows:

$$\bar{\theta}_1^{(k)m,1} = A_1^{(k)m} e^{r_1^{(k)m} x_{2m}^{(k)}} + A_2^{(k)m} e^{-r_1^{(k)m} x_{2m}^{(k)}}, \tag{5.131}$$

where

$$r_1^{(k)m} = \sqrt{1 + \gamma^2 + \frac{\sigma_{11}^{(k)m,0}(t_*)}{\bar{\lambda}^{(k)m} + 2\bar{\mu}^{(k)m}} + \gamma^2 \frac{\sigma_{33}^{(k)m,0}(t_*)}{\bar{\lambda}^{(k)m} + 2\bar{\mu}^{(k)m}}}. \tag{5.132}$$

Substituting the expression (5.131) into the second equation in (5.130) we obtain

$$\begin{aligned}
&\frac{d^2\bar{\varphi}_i^{(k)m,1}}{d(\alpha_1 x_{2m}^{(k)})^2} - \left(1 + \gamma^2 + \frac{\sigma_{11}^{(k)m,0}(t_*)}{\bar{\mu}^{(k)m}} + \gamma^2 \frac{\sigma_{33}^{(k)m,0}(t_*)}{\bar{\mu}^{(k)m}}\right) \bar{\varphi}_i^{(k)m,1} \\
&= -\left(1 + \frac{\bar{\lambda}^{(k)m}}{\bar{\mu}^{(k)m}}\right)\left(A_1^{(k)m} r_1^{(k)m} e^{r_1^{(k)m} x_{2m}^{(k)}} - r_1^{(k)m} A_2^{(k)m} e^{-r_1^{(k)m} x_{2m}^{(k)}}\right)\delta_i^2 \\
&\quad - \left(1 + \frac{\bar{\lambda}^{(k)m}}{\bar{\mu}^{(k)m}}\right)\left(A_1^{(k)m} e^{r_1^{(k)m} x_{2m}^{(k)}} + A_2^{(k)m} e^{-r_1^{(k)m} x_{2m}^{(k)}}\right)\delta_i^1 \\
&\quad + \gamma\left(1 + \frac{\bar{\lambda}^{(k)m}}{\bar{\mu}^{(k)m}}\right)\left(A_1^{(k)m} e^{r_1^{(k)m} x_{2m}^{(k)}} + A_2^{(k)m} e^{-r_1^{(k)m} x_{2m}^{(k)}}\right)\delta_i^3.
\end{aligned} \tag{5.133}$$

In (5.130) and (5.133) $\delta_i^j (i;j = 1,2,3)$ is a Kronecker symbol.

By usual procedure, we determine the general solution to the Eq. (5.133). The expressions of these functions are

$$\begin{aligned}
\bar{\varphi}_1^{(k)m,1} &= C_{11}^{(k)m} e^{\kappa_1^{(k)m} x_{2m}^{(k)}} + C_{12}^{(k)m} e^{-\kappa_1^{(k)m} x_{2m}^{(k)}} \\
&+ A_1^{(k)m} a_{11}^{(k)m} e^{r_1^{(k)m} x_{2m}^{(k)}} + A_2^{(k)m} a_{11}^{(k)m} e^{-r_1^{(k)m} x_{2m}^{(k)}}, \\
\bar{\varphi}_2^{(k)m,1} &= C_{21}^{(k)m} e^{\kappa_1^{(k)m} x_{2m}^{(k)}} + C_{22}^{(k)m} e^{-\kappa_1^{(k)m} x_{2m}^{(k)}} \\
&+ A_1^{(k)m} r_1^{(k)m} a_{11}^{(k)m} e^{r_1^{(k)m} x_{2m}^{(k)}} - A_1^{(k)m} r_1^{(k)m} a_{11}^{(k)m} e^{-r_1^{(k)m} x_{2m}^{(k)}}, \\
\bar{\varphi}_3^{(k)m,1} &= C_{31}^{(k)m} e^{\kappa_1^{(k)m} x_{2m}^{(k)}} + C_{32}^{(k)m} e^{-\kappa_1^{(k)m} x_{2m}^{(k)}} \\
&- A_1^{(k)m} \gamma a_{11}^{(k)m} e^{r_1^{(k)m} x_{2m}^{(k)}} - A_1^{(k)m} \gamma a_{11}^{(k)m} e^{-r_1^{(k)m} x_{2m}^{(k)}},
\end{aligned} \tag{5.134}$$

where

$$\begin{aligned}
a_{11}^{(k)m} &= -\left(1 + \frac{\bar{\lambda}^{(k)m}}{\bar{\mu}^{(k)m}}\right)\left((r_1^{(k)m})^2 - \left(1 + \gamma^2 + \frac{\sigma_{11}^{(k)m,0}(t_*)}{\bar{\mu}^{(k)m}} + \gamma^2 \frac{\sigma_{33}^{(k)m,0}(t_*)}{\bar{\mu}^{(k)m}}\right)\right)^{-1} \\
&= \frac{\bar{\lambda}^{(k)m} + 2\bar{\mu}^{(k)m}}{\sigma_{11}^{(k)m,0}(t_*) + \gamma^2 \sigma_{33}^{(k)m,0}}
\end{aligned} \tag{5.135}$$

$$\kappa_1^{(k)m} = \sqrt{1 + \gamma^2 + \frac{\sigma_{11}^{(k)m,0}(t_*)}{\bar{\mu}^{(k)m}} + \gamma^2 \frac{\sigma_{33}^{(k)m,0}(t_*)}{\bar{\mu}^{(k)m}}}.$$

From the relations $\bar{\theta}^{(k)m,1} = \bar{\varepsilon}_{11}^{(k)m,1} + \bar{\varepsilon}_{22}^{(k)m,1} + \bar{\varepsilon}_{33}^{(k)m,1}$ and (5.129) we can write that

$$-\bar{\varphi}_1^{(k)m} + \frac{d\bar{\varphi}_2^{(k)m}}{dx_{2m}^{(k)}} + \gamma\bar{\varphi}_3^{(k)m} = A_1^{(k)m} e^{r_1^{(k)m} x_{2m}^{(k)}} + A_2^{(k)m} e^{-r_1^{(k)m} x_{2m}^{(k)}}. \tag{5.136}$$

Substituting the expressions in (5.134) into (5.136) we obtain

$$\left(-C_{11}^{(k)m}+\kappa_1^{(k)m}C_{21}^{(k)m}+\gamma C_{31}^{(k)m}\right)e^{\kappa_1^{(k)m}x_{2m}^{(k)}}$$
$$+\left(-C_{12}^{(k)m}-\kappa_1^{(k)m}C_{22}^{(k)m}+\gamma C_{32}^{(k)m}\right)e^{-\kappa_1^{(k)m}x_{2m}^{(k)}}$$
$$+A_1^{(k)m}\left(-1+\left(r_1^{(k)m}\right)^2-\gamma^2\right)a_{11}^{(k)m}e^{r_1^{(k)m}x_{2m}^{(k)}}$$
$$+A_2^{(k)m}\left(-1+\left(r_1^{(k)m}\right)^2-\gamma^2\right)a_{11}^{(k)m}e^{-r_1^{(k)m}x_{2m}^{(k)}}$$
$$=A_1^{(k)m}e^{r_1^{(k)m}x_{2m}^{(k)}}+A_2^{(k)m}e^{-r_1^{(k)m}x_{2m}^{(k)}}. \quad (5.137)$$

According to the expression of $a_{11}^{(k)m}$ in (5.135) and expression $r_1^{(k)m}$ in (5.132) we obtain that $(-1+\left(r_1^{(k)m}\right)^2-\gamma^2=1)$ and from this and from (5.137) equalities follows that

$$C_{11}^{(k)m}=\kappa_1^{(k)m}C_{21}^{(k)m}+\gamma C_{31}^{(k)m};\ C_{12}^{(k)m}=-\kappa_1^{(k)m}C_{22}^{(k)m}+\gamma C_{32}^{(k)m}. \quad (5.138)$$

Taking (5.138) into account the expression for the function $\bar{\varphi}_1^{(k)m,1}$ is obtained as follows:

$$\bar{\varphi}_1^{(k)m,1}=\left(\kappa_1^{(k)m}C_{21}^{(k)m}+\gamma C_{31}^{(k)m}\right)e^{\kappa_1^{(k)m}x_{2m}^{(k)}}$$
$$+\left(-\kappa_1^{(k)m}C_{22}^{(k)m}+\gamma C_{32}^{(k)m}\right)e^{-\kappa_1^{(k)m}x_{2m}^{(k)}} \quad (5.139)$$
$$+A_1^{(k)m}a_{11}^{(k)m}e^{r_1^{(k)m}x_{2m}^{(k)}}+A_2^{(k)m}a_{11}^{(k)m}e^{-r_1^{(k)m}x_{2m}^{(k)}}.$$

Thus, the sought functions contain six unknown constants $A_1^{(k)m}$, $A_2^{(k)m}$, $C_{21}^{(k)m}$, $C_{31}^{(k)m}$, $C_{22}^{(k)m}$ and $C_{32}^{(k)m}$, and these functions are determined for the $(k)m$th layer with constant thickness. Under the determination of the expressions of the mentioned functions related to the half-space the decay conditions in (5.119) must be taken into account, according to which, we get

$$\bar{\varphi}_1^{(1)2,1}=\left(\kappa_1^{(1)2}C_{21}^{(1)2}+\gamma C_{31}^{(1)2}\right)e^{\kappa_1^{(1)2}x_{22}^{(1)}}+A_1^{(1)2}a_{11}^{(1)2}e^{r_1^{(1)2}x_{22}^{(1)}},$$
$$\bar{\varphi}_2^{(1)2,1}=C_{21}^{(1)2}e^{\kappa_1^{(1)2}x_{22}^{(1)}}+A_1^{(1)2}r_1^{(1)2}a_{11}^{(1)2}e^{r_1^{(1)2}x_{22}^{(1)}},$$
$$\bar{\varphi}_3^{(1)2,1}=C_{31}^{(1)2}e^{\kappa_1^{(1)2}x_{22}^{(1)}}-A_1^{(1)2}\gamma a_{11}^{(1)2}e^{r_1^{(1)2}x_{22}^{(1)}}. \quad (5.140)$$

Thus, we determine completely the Laplace transformation of the displacements for each constituent of the system illustrated in Fig. 5.19. The unknown constants, which enter in these transformations, are determined from the contact (5.118) and boundary (5.119) conditions. Although, the solution procedure has been made for the system shown in Fig. 5.19, this procedure can be easily extended for the similar complicate systems. Moreover, taking $\gamma=0$ as a particular case we can describe the plane-strain state in the considered system through the functions $\bar{\varphi}_1^{(k)m,1}$ and $\bar{\varphi}_2^{(k)m,1}$. The originals of these functions is determined by

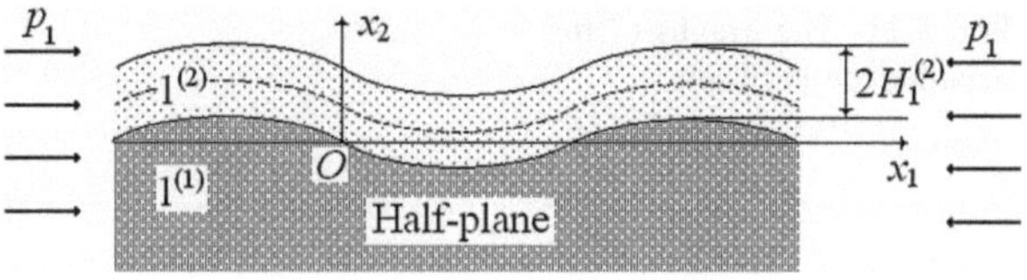


**Fig. 5.20** The geometry of the system consisting of the elastic layer and viscoelastic half-plane

employing the method (Schapery 1962, 1978) and the critical time is calculated from the criterion

$$\left|\varphi_2^{(2)1,1}(0)\right| \to \infty \text{ as } t \to t_{cr}. \tag{5.141}$$

In the case where $\gamma = \ell_1/\ell_3 = 0$ (i.e. under $\ell_3 \to \infty$) the results obtained above related to the corresponding two-dimensional problems in the $Ox_1x_2$ plane.

With the stated above, we restrict ourselves to consideration of the method of solution. This method with the corresponding changing can be applied for the investigation of the other similar problems. Moreover, the developed approach can also be applied for the time-independent materials. In this case by introducing the notation $p_3 = \rho p_1$ the criterion (5.141) must be replaced with the following one

$$\left|\varphi_2^{(2)1,1}(0)\right| \to \infty \text{ as } p_1 \to p_{1.cr}. \tag{5.142}$$

### 5.5.3 Numerical Results and Discussions

We assume that the operators (5.40)–(5.42), describe the viscoelasticity of the layers or half-space materials. Below the expressions in (5.40)–(5.42) will supplied by the corresponding additional indices. We analyze the results related to two- and three-dimensional problems separately.

#### 5.5.3.1 Two-Dimensional Problems

Consider the plane-strain state in the $Ox_1x_2$ plane. For this case we must write $\gamma = 0$ in the expressions obtained in the previous subsection. Analyze numerical results related to some particular cases.

The system consisting of the elastic covering layer and viscoelastic half-plane.

The structure of the mentioned system is shown in Fig. 5.20. Assume that the material of the covering layer is pure elastic with mechanical constants $E^{(2)1}$ (modulus of elasticity), $\nu^{(2)1}$ (Poisson's coefficient), but the material of the half-plane is viscoelastic with operators (5.40)–(5.42). We denote the instantaneous values of the modulus of elasticity and Poisson's coefficient through $E_0^{(1)1}$ and$\nu_0^{(1)1}$, respectively. Assume that $\nu^{(2)1} = \nu_0^{(1)1} = 0.3$.

First we consider the case where the half-plane material is also pure elastic with mechanical constants $E_0^{(1)1}$ and $\nu_0^{(1)1}$, and the critical values of the $p_1'$ $(= p_1/E_0^{(1)1})$

**Fig. 5.21** The graphs of the dependence between $|p_{1cr.0}|/E_0^{(1)1}$ and $\pi H_1^{(2)}/\ell_1$

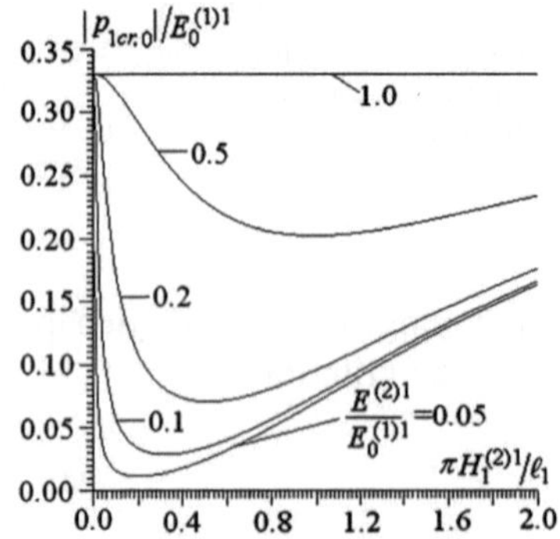


related to the mentioned case we denote by $p'_{1cr.0}\,(=p_{1cr.0}/E_0^{(1)1})$. Analyze the dependence between $p'_{1cr.0}$ and $\chi\,(=\pi H_1^{(2)}/\ell_1)$. The graphs of this dependence are given in Fig. 5.21 for various $E_0^{(1)1}/E^{(2)1}$.

Under construction of these graphs the values of $p'_{1cr.0}$ are calculated from the criterion (5.142) and these values of $p'_{1cr.0}$ coincide with the corresponding ones obtained within the scope of the Euler approach. However under employing the Euler approach the $\min|p'_{1cr.0}|$ and the $\chi_{cr.0}$ have a real meaning, where $\chi_{cr.0}$ is a value of $\chi$ at which the $\min|p'_{1cr.0}|$ arises. Note that such minimums in the dependence between $|p'_{1cr.0}|$ and $\chi$ arise only in the case where the material of the covering layer is stiffer than the material of the half-plane, i.e. only in the case where $E^{(2)1} > E_0^{(1)1}$. But under applying the approach based on the initial imperfection criterion on the investigation of the problem under consideration, all values of $|p'_{1cr.0}|$ calculated from the criterion (5.142) and corresponding $\chi$ may have a real meaning, if $|p'_{1cr.0}| < |p^{se}_{cr.}|$ and $\ell_1 < L_{se}$, where $|p^{se}_{cr.}|$ is the critical force corresponding the whole stability loss of the structural elements and $L_{se}$ is the characteristic (minimum) dimension of these elements.

It follows from the graphs given in Fig. 5.21 that the values of $|p'_{1cr.0}|$ decrease with increasing $E_0^{(1)1}/E^{(2)1}$ and under $E_0^{(1)1}/E^{(2)1}=1$the values of $|p'_{1cr.0}|$ do not depend on the stability loss mode, i.e. on the parameter $\chi$. Note that the value of $|p'_{1cr.0}|$obtained under $E_0^{(1)1}/E^{(2)1}=1$corresponds to the near surface stability loss of the half-plane without covering layer. Moreover, note that in general, $|p'_{1cr.0}| \to \min\{p^{*}_{cr.0}, p^{**}_{cr.0}\}$ as $\chi \to \infty$, where $p^{*}_{cr.0}$ is the critical force which corresponds the near surface stability loss of the interface plane between the half-plane and covering layer materials, $p^{**}_{cr.0}$ is the critical force which corresponds the near surface stability loss of the half-plane from the covering layer material. However, in the cases where $E_0^{(1)1}/E^{(2)1} < 1$, the relation $|p'_{1cr.0}| \to p^{*}_{cr.0}$ as $\chi \to \infty$takes place.

**Table 5.8** The values of $\min\left|p'_{1cr.\infty}\right|$ and $\chi_{cr.\infty}$ calculated for various $E_0^{(1)1}\big/E^{(2)1}$ and $\omega$

| $\frac{E_0^{(1)1}}{E^{(2)1}}$ | $\omega$ | $\min\left\|p'_{1cr.\infty}\right\| (\chi_{cr.\infty})$ | $\min\left\|p'_{1cr.0}\right\| (\chi_{cr.0})$ |
|---|---|---|---|
| 0.05 | 1 | 0.0046 (0.15) | 0.0114 (0.20) |
| | 2 | 0.0066 (0.15) | |
| | 3 | 0.0078 (0.15) | |
| 0.1 | 1 | 0.0113 (0.20) | 0.0288 (0.30) |
| | 2 | 0.0166 (0.25) | |
| | 3 | 0.0195 (0.25) | |
| 0.2 | 1 | 0.0285 (0.30) | 0.0706 (0.5) |
| | 2 | 0.0413 (0.40) | |
| | 3 | 0.0483 (0.4) | |
| 0.5 | 1 | 0.0903 (0.6) | 0.2022 (1.0) |
| | 2 | 0.1278 (0.75) | |
| | 3 | 0.1467 (0.80) | |

To determine the critical time for the case where the half-plane material is viscoelastic one, it is also necessary to know the critical values of the $p'_1$ calculated at $t' = \infty$. Let's remember that these values of the $p'_1$(denoted by $p'_{1cr.\infty}$) are calculated from the corresponding pure elastic stability loss problem by replacing $E_0^{(1)1}$ and $v_0^{(1)1}$ with $E_\infty^{(1)1}$ and $v_\infty^{(1)1}$respectively, where $E_\infty^{(1)1} = E_\infty^{(1)1}(1 - 1/\omega)$, $v_\infty^{(1)1} = v_0^{(1)1}(1 + (1 - 2v_0^{(1)1})\big/(2v_0^{(1)1}(1 + \omega)))$. Here $\omega$ is the dimensionless rheological parameter determined in Sect. 5.2.3. As an example, Table 5.8 shows the values of $\min\left|p'_{1cr.\infty}\right|$ calculated from the criterion (5.142) for various values of $E_0^{(1)1}\big/E^{(2)1}$ and of the parameter $\omega$. This table shows also (in parenthesis) the values of $\chi\,(= \chi_{cr.\infty})$ at which the values of $\left|p'_{1cr.\infty}\right|$ have its minimum.

It follows from Table 5.8 that $\min\left|p'_{1cr.\infty}\right| < \min\left|p'_{1cr.0}\right|$. The other numerical results also show that for each fixed $\chi$ the inequality $\left|p'_{1cr.\infty}\right| < \left|p'_{1cr.0}\right|$ occurs. Consequently, under investigation of the noted above viscoelastic stability loss problem the values of $p'_1$ must be selected according to the inequality $\left|p'_{1cr.\infty}\right| < \left|p'_1\right| < \left|p'_{1cr.0}\right|$.

Thus, within the scope of the foregoing assumptions, consider the influence of the rheological parameters $\omega$ and $\alpha$ on the values of $t'_{cr}$under $t'_* = 0$ in Eqs. (5.123) and (5.124). This influence is illustrated through the graphs given in Figs. 5.22 (for $\alpha = -0.5$) and 5.23 (for $\omega = 1.0$) under $\chi = 0.15$, $E_0^{(1)1}\big/E^{(2)1} = 0.05$ (Figs. 5.22a and 5.23a) and 0.10 (Figs. 5.22b and 5.23b).

The numerical results illustrated in Figs 5.21 and 5.22 in the qualitative sense agree well with the results obtained in Sect. 5.2.

Consider the influence of selection of the values for $t'_*$ in Eqs. (5.123) and (5.124) on the $t'_{cr}$. Figure 5.24 shows the graphs of the dependence between $t'_{cr}$ and $p'_1$ constructed in the cases where $t'_* = 0$ and $t'_* = \infty$ for various $\omega$ under $\alpha = -0.5$

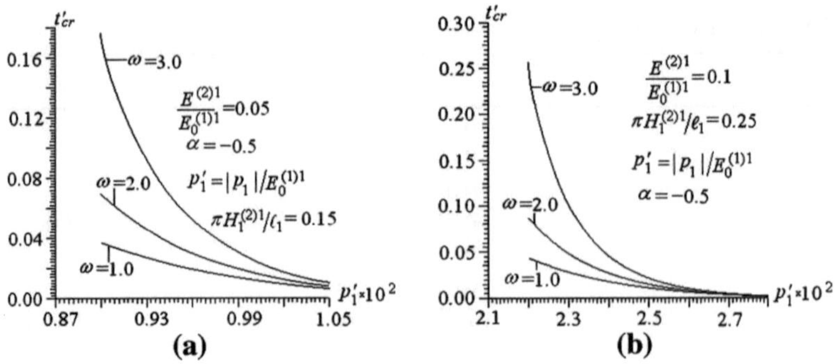


**Fig. 5.22** The influence of the rheological parameter $\omega$ on the values of $t'_{cr}$ under $\alpha = -0.5$, $E_0^{(1)1} \big/ E^{(2)1} = 0.05$ (**a**) and 0.1(**b**)

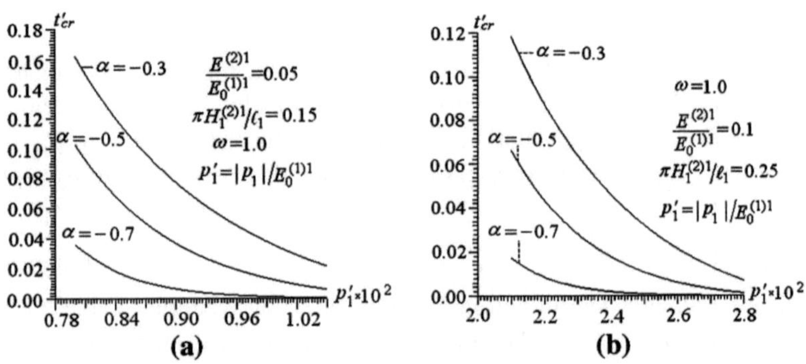


**Fig. 5.23** The influence of the rheological parameter $\alpha$ on the values of $t'_{cr}$ under $\omega = 1.0$, $E_0^{(1)1} \big/ E^{(2)1} = 0.05$, (**a**) and 0.1(**b**)

(Fig. 5.24a) and for various $\alpha$ under $\omega = 1.0$ (Fig. 5.24b) for $E_0^{(1)1} \big/ E^{(2)1} = 0.05$ and $\chi = 0.15$. Note that in this figure the value of $t'_{cr}$ obtained at $t'_* = 0$ (at $t'_* = \infty$) is denoted as $t'_{cr.0}$ (as $t'_{cr.\infty}$). Comparison the results obtained for $t'_{cr.0}$ with the corresponding results obtained for $t'_{cr.\infty}$ shows that $t'_{cr.0}$ and $t'_{cr.\infty}$ is very close to each other. Therefore, we will consider below only the case where $t'_* = 0$.

The system consisting of the elastic covering and viscoelastic bond layers, and elastic half-plane.

We assume that the stack of layers consists of two layers as shown in Fig. 5.25. Moreover, at first we assume that the materials of these layers and half-space are elastic and examine the case where $E^{(1)1}/E^{(2)1} = 0.125$, $E^{(2)2}/E^{(1)1} = 4.0$, $\nu^{(1)1} = \nu^{(2)1} = \nu^{(2)2} = 0.3$. Introduce a notation $H = H_1^{(2)} + H_1^{(1)}$ and consider the case where $H_1^{(2)} \big/ H_1^{(1)} = 1.0, 2/3, 3/7, 1/4 \text{and} 1/9$. The graphs of the dependencies between $|p_{1cr}|/E^{(2)2}$ and $\chi(= 2\pi H/\ell_1)$ for these cases are given in Fig. 5.26.

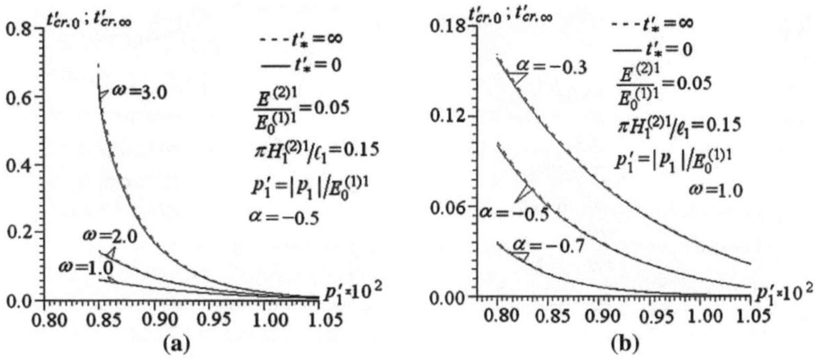


**Fig. 5.24** The comparison of $t'_{cr}$ obtained under $t'_* = 0$ with those obtained under $t'_* = \infty$

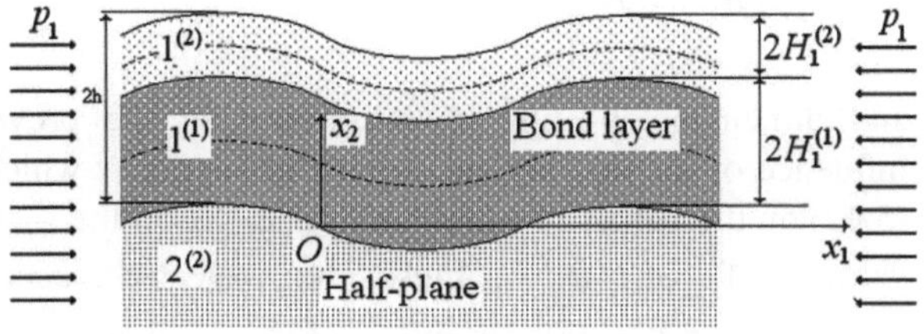


**Fig. 5.25** The geometry of the half-plane covered by the stack consisting of two layers

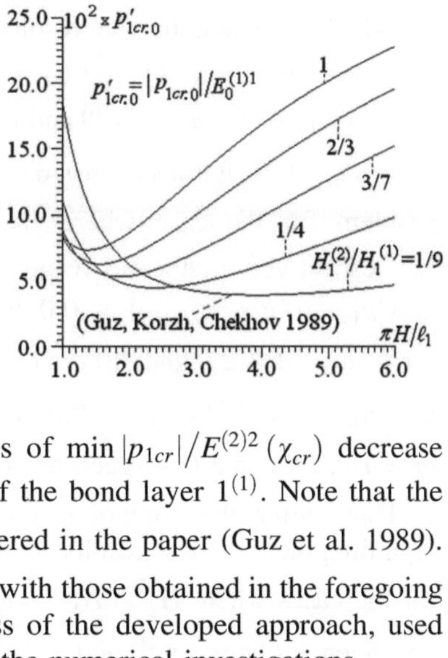


**Fig. 5.26** The graphs of the dependencies between $|p_{1cr}|/E^{(2)2}$ and $\pi H/\ell_1$ for the case shown in Fig. 5.25

It follows from the graphs that the values of $\min |p_{1cr}|/E^{(2)2}\ (\chi_{cr})$ decrease (increase) with an increase in the thickness of the bond layer $1^{(1)}$. Note that the case where $H_1^{(2)}/H_1^{(1)} = 1/9$ was also considered in the paper (Guz et al. 1989). The results obtained here coincide completely with those obtained in the foregoing paper. This situation holds for the correctness of the developed approach, used algorithm and PC programmes composed for the numerical investigations.

Now we assume that the material of the bond layer $1^{(1)}$ (Fig. 5.25) is viscoelastic one with operators (5.40)–(5.42) and $E_0^{(1)1}/E^{(2)1} = 0.125$, $E^{(2)2}/E_0^{(1)1} = 4.0$, $\nu_0^{(1)1} = \nu^{(2)1} = \nu^{(2)2} = 0.3$. We use the dimensionless rheological parameter $\omega$

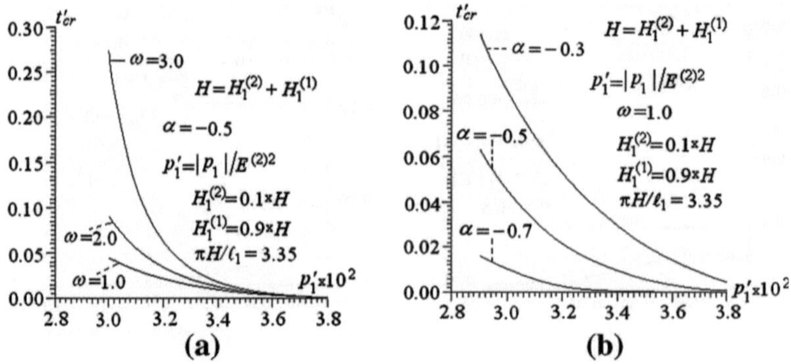


**Fig. 5.27** The influence of the rheological parameters $\omega$ (**a**) and $\alpha$ (**b**) on the values of $t'_{cr}$ in the case $H_1^{(2)}/H_1^{(1)}=1/9$

and dimensionless time $t'$ introduced in the previous sections. Consider the influence of the rheological parameters $\omega$ and $\alpha$ which enter the operators (5.40)–(5.42) on the values of the critical time, i.e. on the $t'_{cr}$. For calculation of the $t'_{cr}$, the values of $|p_1|/E^{(2)2}$ must be selected according to the inequality $|p_{1cr.\infty}|/E^{(2)2}<|p_1|E^{(2)2}<|p_{1cr.0}|/E^{(2)2}$. The meaning of the notation $p_{1cr.0}$ and $p_{1cr.\infty}$ is the same as in the previous sub-subsection and the values of $|p_{1cr.0}|/E^{(2)2}$ are already determined by the results given in Fig. 5.26, according to which, for example, $\min|p_{1cr.0}|/E^{(2)2}\,(\chi_{cr.0})=0.03899\ (3.90)$ and $0.07318\ (1.40)$ for the cases where $H_1^{(2)}/H_1^{(1)}=1/9$ and 1 respectively. For the determination of the values of $|p_{1cr.\infty}|/E^{(2)2}$ it is necessary to solve again the foregoing pure elastic stability loss problem for the case where $E_\infty^{(1)1}$ and $\nu_\infty^{(1)1}$ are taken instead of $E_0^{(1)1}$ and $\nu_0^{(1)1}$respectively. As a result of this solution it is determined that $\min|p_{1cr.\infty}|/E^{(2)2}\,(\chi_{cr.\infty})=\ 0.01686\ (2.80),\ 0.02347\ (3.20)$and $0.02707\ (3.35)$ for $\omega=\ 1.0, 2.0$ and $3.0$, respectively under $H_1^{(2)}/H_1^{(1)}=1/9$. But $\min|p_{1cr.\infty}|/E^{(2)2}\,(\chi_{cr.\infty})=0.04654(1.10), 0.05563(1.20)$ and $0.06004(1.25)$ for $\omega=1.0, 2.0$ and $3.0$, respectively occur under $H_1^{(2)}/H_1^{(1)}=1$.

Thus, after the foregoing preparation procedure we consider the results regarding the $t'_{cr}$. These results are given by the graphs given in Figs. 5.27 and 5.28 for the cases where $H_1^{(2)}/H_1^{(1)}=1/9$ and 1 respectively. These graphs show the dependencies between $t'_{cr}$ and $|p_1|/E^{(2)2}$ for various $\omega$ under $\alpha=-0.5$ (Figs. 5.27a and 5.28a) and for various $\alpha$ under $\omega=1.0$(Figs. 5.27b and 5.28b). It follows from these results that all conclusions made in the previous sub-subsection on the influence of the rheological parameters on the values of $t'_{cr}$occur also for the case under consideration.

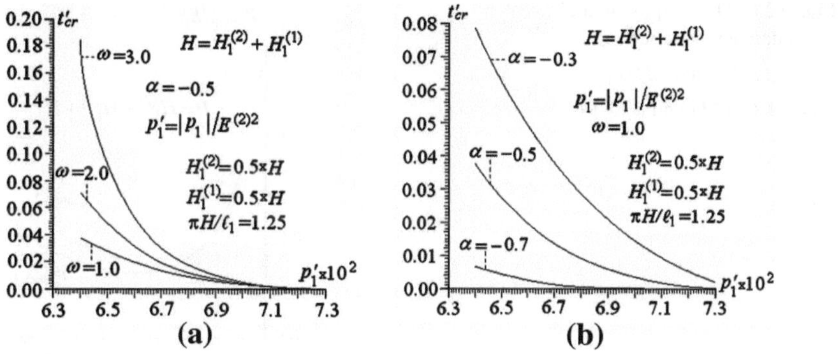


**Fig. 5.28** The influence of the rheological parameters $\omega$(**a**) and $\alpha$ (**b**) on the values of $t'_{cr}$ in the case $H_1^{(2)}/H_1^{(1)} = 1$

The system consisting of the elastic covering, viscoelastic and elastic bond layers and viscoelastic half-plane

We assume that the stack of layers consists of elastic covering, viscoelastic and again elastic bond layers as shown in Fig. 5.19. Also we assume that the material of the half-plane is viscoelastic. Consider two cases:

Case 1

$$\frac{E_0^{(1)2}}{E^{(2)1}} = 0.1, \frac{E_0^{(1)2}}{E_0^{(1)1}} = 1.0, \frac{E_0^{(1)2}}{E^{(2)2}} = 0.5, \tag{5.143}$$

Case 2

$$\frac{E_0^{(1)2}}{E^{(2)1}} = 0.5, \frac{E_0^{(1)2}}{E_0^{(1)1}} = 1.0, \frac{E_0^{(1)2}}{E^{(2)2}} = 0.1. \tag{5.144}$$

Assume that $\nu_0^{(1)2} = \nu_0^{(1)1} = \nu^{(2)1} = \nu^{(2)2} = 0.3$ and $H_1^{(2)} = H_1^{(1)} = H_2^{(2)}$, and introduce the notation $H = H_1^{(1)} + H_1^{(2)} + H_2^{(2)}$. Consider the graphs of the dependencies between $|p_{1cr.0}|/E_0^{(1)1}$ and parameter $\chi\,(= \pi H/\ell_1)$. These graphs are shown in Fig. 5.29. According to the foregoing assumptions and (5.143) and (5.144), for Cases 1 and 2 the materials of the bond viscoelastic layer $1^{(1)}$ and half-plane are the same. At the same time, in Case1 (in Case 2) the relation $E^{(2)1} > E^{(2)2} > E_0^{(1)1} = E_0^{(1)2}(E^{(2)2} > E^{(2)1} > E_0^{(1)1} = E_0^{(1)2})$ takes place. In the other words, in Case 1 the stiffer layer is the covering layer, but in Case 2 the stiffer layer is lower layer of the stack.

We denote the values of $\min|p_{1cr.0}|/E_0^{(1)1}(\chi_{cr})$ for Cases 1 and 2 by $\min|p'_{1cr.0}|/E_0^{(1)1}(\chi'_{cr})$ and $\min|p''_{1cr.0}|/E_0^{(1)1}(\chi''_{cr})$ respectively. It follows from

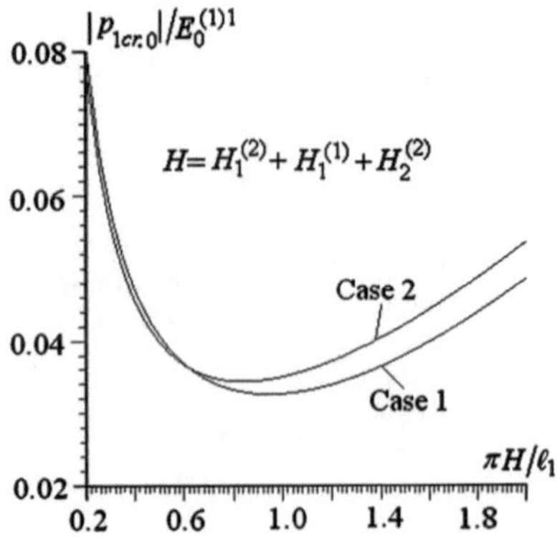


**Fig. 5.29** The graphs of the dependencies between $|p_{1cr.0}|\big/E_0^{(1)1}$ and $\pi H/\ell_1$ in Cases 1 (5.143) and 2 (5.144)

Fig. 5.29 that $\min|p'_{1cr.0}|\big/E_0^{(1)1} < \min|p''_{1cr.0}|\big/E_0^{(1)1}$ but $\chi'_{cr} > \chi''_{cr}$. These results can be used for controlling of the near-surface stability loss of the system under consideration with the location of the stiffer layer in the stack of the layers.

Consider numerical results related to the $t'_{cr}$ taking into account that in Case 1 $\min|p_{1cr.0}|\big/E_0^{(1)1}(\chi_{cr.0}) = 0.03265(0.95)$, and $\min|p_{1cr.\infty}|\big/E_0^{(1)1}(\chi_{cr.\infty}) =$ $0.01293\ (0.60), 0.01906(0.70)$and $0.02233(0.75)$, as well as, in Case 2 $\min|p_{1cr.0}|\big/E_0^{(1)1}(\chi_{cr.0}) = 0.03441(0.85)$, $\min|p_{1cr.\infty}|\big/E_0^{(1)1}(\chi_{cr.\infty}) = 0.01318(0.55)$, $0.01960(0.65)$and $0.02306(0.70)$ obtained under $\omega = 1.0, 2.0$ and $3.0$ respectively. These results are given through the graphs shown in Fig. 5.30. These graphs correspond the dependencies between $t'_{cr}$ and $|p_{1cr.0}|\big/E_0^{(1)1}$ for various $\omega$ under $\alpha = -0.5$(Fig. 5.30a), as well as, for various $\alpha$ under $\omega = 1.0$(Fig. 5.30b).

It follows from Fig. 5.30 that the values of $t'_{cr}$ obtained in Case 2 are greater than corresponding ones obtained in Case 1.

#### 5.5.3.2 Three-Dimensional Problems

We will use below all assumptions and notation used under analyzing of the two-dimensional stability loss problems above. The main attention in the numerical investigations below will be focused on the influence of the bi-axiality of the external loading (i.e. of the ratio $p_3/p_1$) and of the spatiality of the initial imperfection and stability loss modes (i.e. of the parameter $\gamma$ (5.121)) on the values of the critical parameters.

The system consisting of the elastic covering layer and viscoelastic half-plane.

The structure of the system under consideration is shown schematically in Fig. 5.20. Assume that the covering layer material is pure elastic but the half-space material is viscoelatic with operators (5.40)–(5.42). First we consider the numerical results related to the pure elastic stability loss problem in the case where the half-space material is also pure elastic one with modulus of elasticity $E_0^{(1)1}$ and

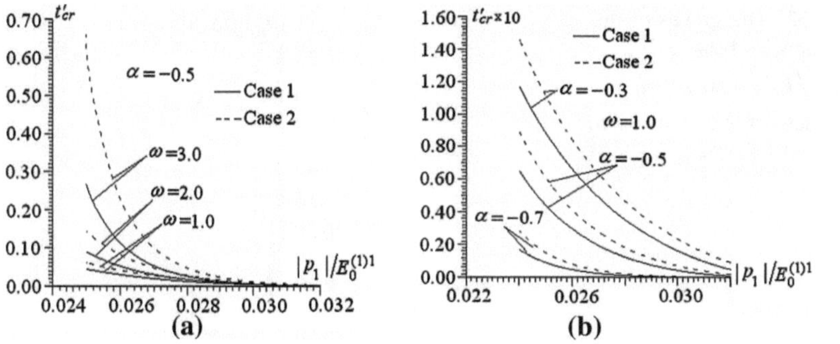


**Fig. 5.30** The influence of the rheological parameters $\omega$ (**a**) and $\alpha$ (**b**) on the values of $t'_{cr}$

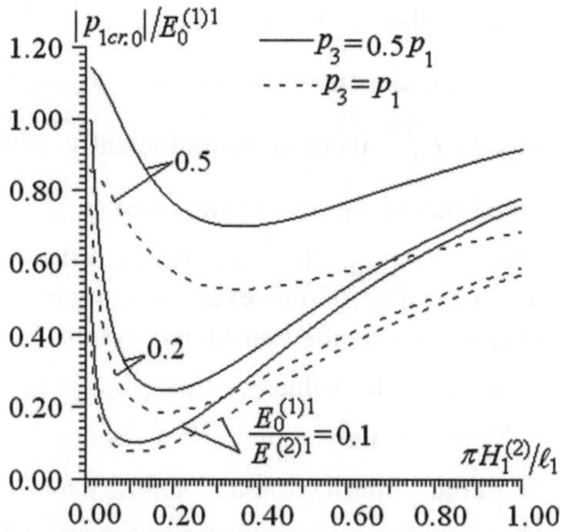


**Fig. 5.31** The graphs of the dependencies between $|p_{1cr.0}|\big/E_0^{(1)1}$ and $\chi$ constructed for various $E_0^{(1)1}\big/E^{(2)1}$ under $p_3 = p_1$ and $p_3 = 0.5p_1$

Poisson's coefficient $\nu_0^{(1)1}$ and assume that $\gamma = 1$ and $\nu_0^{(1)1} = \nu^{(2)1} = 0.3$. Analyze the graphs which show the dependencies between $|p_{1cr.0}|\big/E_0^{(1)1}$ and $\chi\,(= \pi H_1^{(2)}\big/\ell_1)$ for various values of the ratio $E_0^{(1)1}\big/E^{(2)1}$ under $p_3 = p_1$ and $p_3 = 0.5p_1$.

It follows from the graphs given in Fig. 5.31 that the values of $|p_{1cr.0}|\big/E_0^{(1)1}$ decrease with increasing of the $E_0^{(1)1}\big/E^{(2)1}$. Moreover, Fig. 5.31 shows that the values of $|p_{1cr.0}|\big/E_0^{(1)1}$ obtained under $p_3 = 0.5p_1$ are greater than the corresponding ones obtained under $p_3 = p_1$. This result alludes to the decreasing of the values $|p_{1cr.0}|\big/E_0^{(1)1}$ with the bi-axiality of the external compressive forces. For more detail analyses of the influence of the ratio $p_3/p_1$ on the values of $|p_{1cr.0}|\big/E_0^{(1)1}$ we consider the graphs shown in Fig. 5.32 which are constructed for

**Fig. 5.32** The graphs of the dependencies between $|p_{1cr.0}| / E_0^{(1)1}$ and $\chi$ constructed for various $p_3/p_1$ under $E_0^{(1)1} / E^{(2)1} = 0.1$

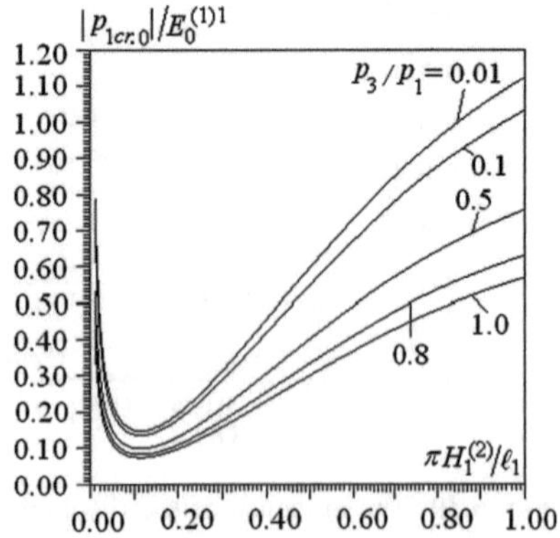


various values of this ratio under $E_0^{(1)1} / E^{(2)1} = 0.1$ and $\gamma = 1$. These graphs relate also to the dependencies between $|p_{1cr.0}| / E_0^{(1)1}$ and $\chi$ and show that the values of $|p_{1cr.0}| / E_0^{(1)1}$ increase monotonically with decreasing the ratio $p_3/p_1$. The mentioned increasing in the values of $|p_{1cr.0}| / E_0^{(1)1}$ is more considerable in the cases where $\chi > \chi_{cr.0}$. Thus, according to the foregoing results it can be concluded that the bi-axiality of the external compression, i.e. the existents of the additional compression of the considered system in the direction of the $Ox_3$ axis, causes to decrease of the values of $\min |p_{1cr.0}| / E_0^{(1)1}$ and $\chi_{cr.0}$.

Now we analyze numerical results illustrated the influence of the spatiality of the initial imperfection and stability loss mode on the values of $|p_{1cr.0}| / E_0^{(1)1}$. For this purpose we consider the graphs given in Fig. 5.33 which show this influence under $p_3/p_1 = 1.0$ (Fig. 5.33a) and $p_3/p_1 = 0.5$ (Fig. 5.33b). These graphs show that in the considered case for fixed $\ell_1$the influence of the change of the length of the period of the initial imperfection mode along the $Ox_3$ axis on the values of $\min |p_{1cr.0}| / E_0^{(1)1}$ is insignificant. However, because of this influence the values of $\chi_{cr.0}$ increase monotonically with decreasing $\gamma$. At the same time, the results show that after the certain value of $\chi$ (denoted by $\chi'$) the values of $|p_{1cr.0}| / E_0^{(1)1}$ increase with $\gamma$. Moreover, it should be noted that these results approach asymptotically to the corresponding ones obtained for the plane-strain state as $\gamma \to 0$.

Now we consider the results related to the case where the material of the half-space is the viscoelastic one and analyze the influence of the $|p_1| / E_0^{(1)1}$ and rheological parameters $\omega$ and $\alpha$ on the values of the critical time $t'_{cr}$. In this case the values of $|p_1| / E_0^{(1)1}$ is selected according to the relation $|p_{1cr.\infty}| / E_0^{(1)1} < |p_1| / E_0^{(1)1} < |p_{1cr.0}| / E_0^{(1)1}$ For the case where $E_0^{(1)1} / E^{(2)1} = 0.1$, $\alpha = -0.5$ for

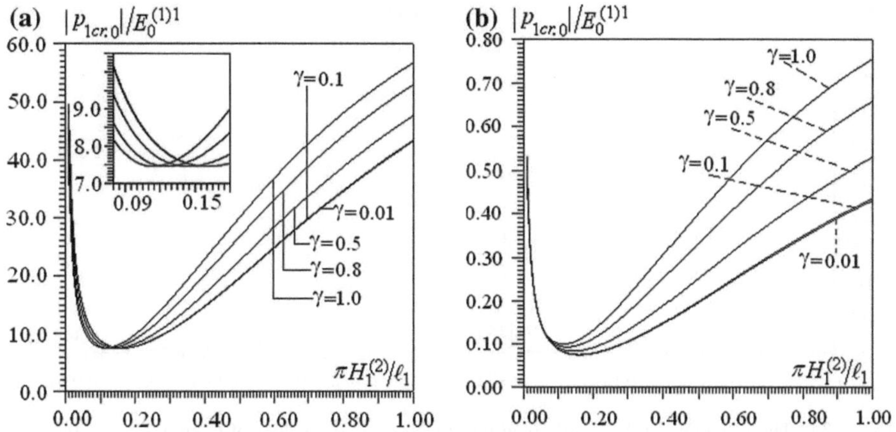


**Fig. 5.33** The influence of the spatiality of the stability loss and initial imperfection mode on the dependencies between $|p_{1cr.0}|/E_0^{(1)1}$ and $\chi$ in the cases where $p_3/p_1 = 1$(**a**) and 0.5 (**b**) under $E_0^{(1)1}/E^{(2)1} = 0.1$

**Table 5.9** The values of $\min|p'_{1cr.\infty}|(\chi_{cr.\infty})$ and $\min|p'_{1cr.0}|(\chi_{cr.0})$ obtained for various $\gamma$, $p_3/p_1$ and $\omega$

| $\gamma$ | $\frac{p_3}{p_1}$ | $\omega$ | $\min\|p'_{1cr.\infty}\|(\chi_{cr.\infty})$ | $\min\|p'_{1cr.0}\|(\chi_{cr.0})$ |
|---|---|---|---|---|
| 1.0 | 1.0 | 1.0 | 0.02956 (0.07) | 0.07471 (0.11) |
| | | 2.0 | 0.04338 (0.09) | |
| | | 3.0 | 0.05078 (0.09) | |
| | 0.5 | 1.0 | 0.03953 (0.07) | 0.09961 (0.11) |
| | | 2.0 | 0.05784 (0.09) | |
| | | 3.0 | 0.06768 (0.09) | |
| 0.8 | 1.0 | 1.0 | 0.02963 (0.08) | 0.07471 (0.13) |
| | | 2.0 | 0.04340 (0.10) | |
| | | 3.0 | 0.05074 (0.10) | |

the certain values of the problem parameters the values of $\min|p'_{1cr.\infty}| = \min|p_{1cr.\infty}|/E_0^{(1)1}(\chi_{cr.\infty})$ and $\min|p'_{1cr.0}| = \min|p_{1cr.0}|/E_0^{(1)1}(\chi_{cr.0})$ are given in Table 5.9 for various $\omega$.

Thus, we consider the graphs illustrated the influence of the rheological parameters $\omega$ and $\alpha$ on the dependence between $t'_{cr}$ and $|p_1|/E_0^{(1)1}$. These graphs are given in Figs. 5.34 and 5.35, which are constructed under $\alpha = -0.5$ (Fig.3.34) and $\omega = 1.0$ (Fig. 5.35) for $\chi = 0.09$ and $\gamma = 1.0$ in the cases where $p_3/p_1 = 1.0$(a) and 0.5(b). It follows from the graphs that in the qualitative sense these results agree well with the foregoing ones obtained for the two-dimensional

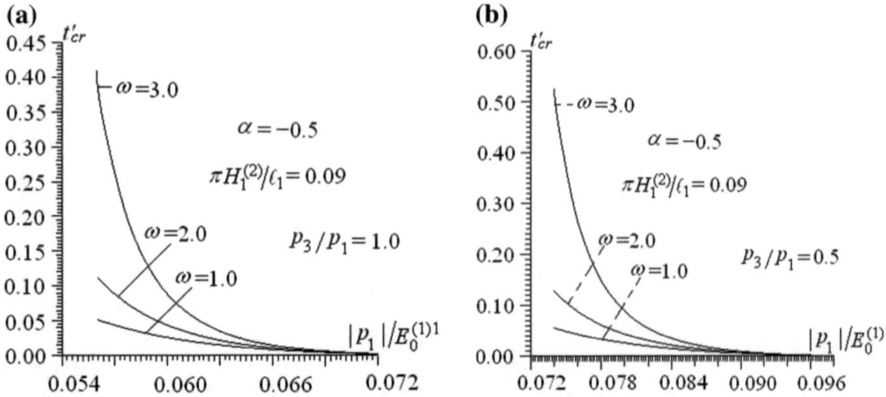


**Fig. 5.34** The influence of the rheological parameter $\omega$ on the values of $t'_{cr}$ under $\alpha = -0.5$, $\gamma = 1.0$ and $\chi = 0.09$ in the cases where $p_3/p_1 = 1.0$ (**a**) and 0.5 (**b**)

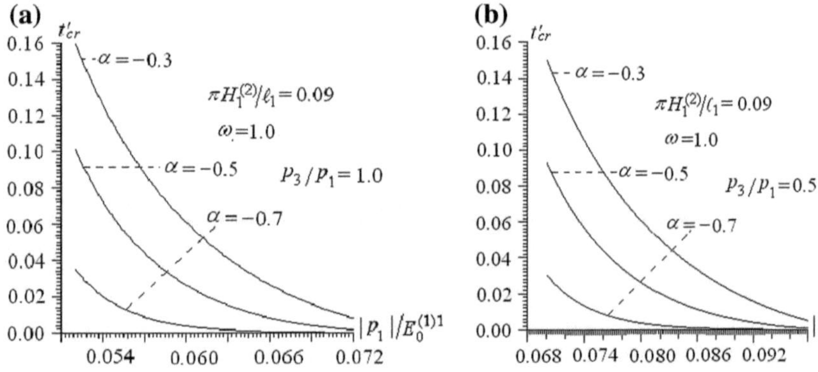


**Fig. 5.35** The influence of the rheological parameter $\alpha$ on the values of $t'_{cr}$ under $\omega = 1.0$, $\gamma = 1.0$ and $\chi = 0.09$ in the cases where $p_3/p_1 = 1.0$ (**a**) and 0.5 (**b**)

stability loss problems. Note that it is difficult to make any conclusion, based on the foregoing results, on the influence of the ratio $p_3/p_1$on the values of $t'_{cr}$. Because the intervals for the change of the values of $|p_1|/E_0^{(1)1}$, selected for the cases where $p_3/p_1 = 1.0$ and 0.5, do not have common parts.

Now we consider the results which show the influence of the parameter $\gamma$ on the values of $t'_{cr}$. Note that these results are given in Fig. 5.36 through the graphs of the dependencies between $t'_{cr}$ and $|p_1|/E_0^{(1)1}$ constructed at $\gamma = 1.0$ and 0.8 for various $\omega$ (Fig. 5.36a) under $\alpha = -0.5$ and for various $\alpha$ (Fig. 5.36b) under $\omega = 1.0$. Moreover note that under construction of these graphs it is assumed that $\chi = 0.09$, $p_3/p_1 = 1.0$ and $E_0^{(1)1}/E^{(2)1} = 0.1$. It follows from the graphs that the

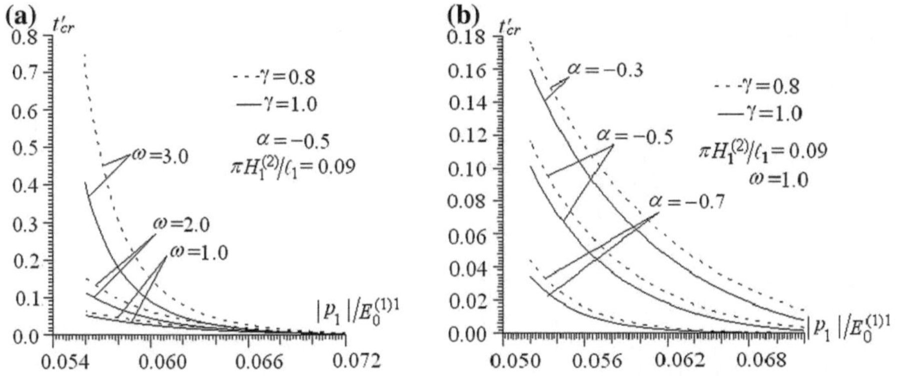


**Fig. 5.36** The influence of the parameter $\gamma$ on the values of $t'_{cr}$ obtained for various $\omega$ (**a**) for $\alpha = -0.5$ and for various $\alpha$ (**b**) for $\omega = 1.0$, in the cases where $\gamma = 1.0$ and 0.8 under $\chi = 0.09$ and $p_3/p_1 = 1.0$

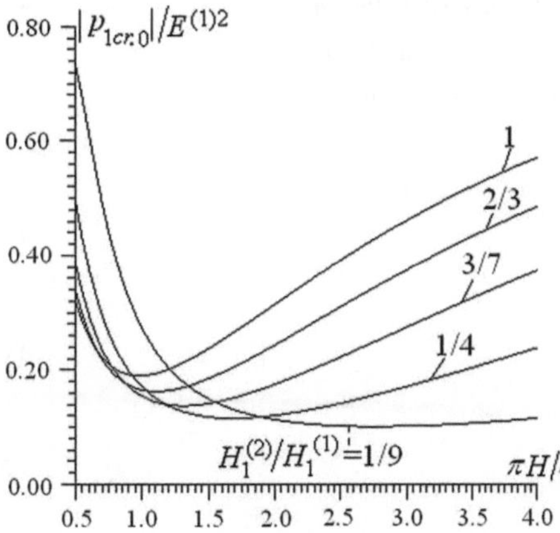


**Fig. 5.37** The graphs of the dependencies between $|p_{1cr.0}|/E^{(1)2}$ and $\chi$ obtained for various $H_1^{(2)}/H_1^{(1)}$

values of $t'_{cr}$ obtained for the case where $\gamma = 0.8$ are greater than the corresponding ones obtained in the case where $\gamma = 1.0$. Because the values of $p_{1cr.\infty}$ obtained under $\chi = 0.09$ for the case where $\gamma = 0.8$ are greater than those obtained in the case where $\gamma = 1.0$.

The system consisting of the elastic covering and viscoelastic bond layers, and elastic half-space

The structure of the considered system is shown in Fig. 5.25. Assume that $p_3 = p_1$ and $\gamma = 1.0$. Figure 5.37 shows the graphs of the dependencies between $|p_{1cr.0}|/E^{(1)2}$ and $\chi\,(= \pi H/\ell_1)$ (where $H = H_1^{(2)} + H_1^{(1)}$) for various values of $H_1^{(2)}/H_1^{(1)}$ in the case where the materials of the layers and half-space are elastic and $E_0^{(1)1}/E^{(2)1} = 0.125$, $E^{(1)2}/E^{(2)1} = 0.5$, $E^{(1)2}/E_0^{(1)1} = 4$ and $\nu_0^{(1)1} = \nu^{(1)2} = \nu^{(2)1} = 0.3$.

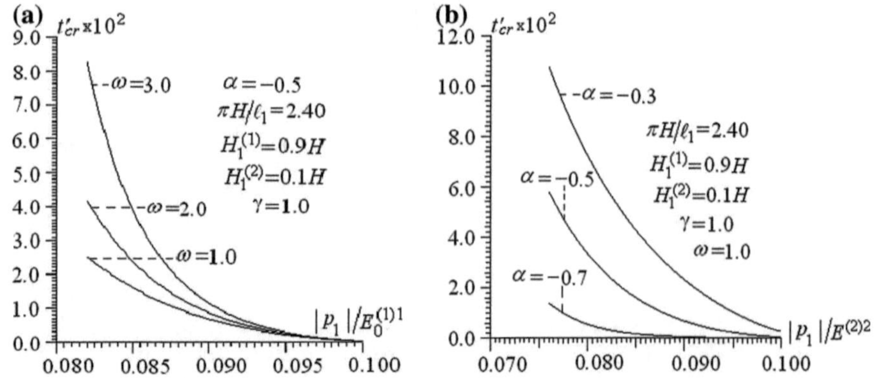


**Fig. 5.38** The graphs of the dependencies between $t'_{cr}$ and $|p_1|/E^{(1)2}$ for various values of $\omega$ under $\alpha = -0.5$ (**a**) and for various $\alpha$ under $\omega = 1.0$ (**b**)

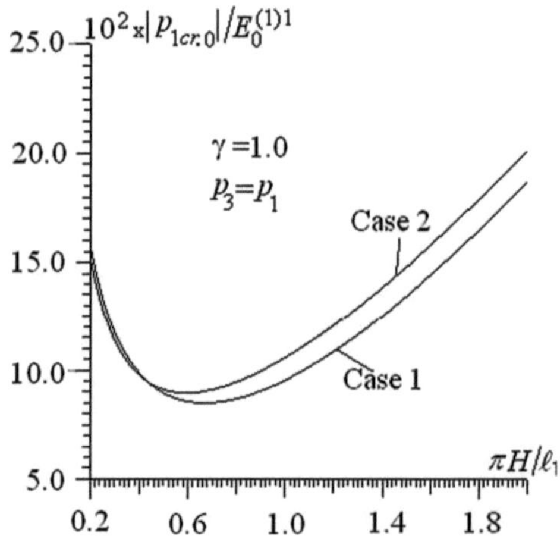


**Fig. 5.39** The graphs of the dependencies between $|p_{1cr.0}|/E_0^{(1)1}$ and $\pi H/\ell_1$ for the case shown in Fig. 5.19

It follows from these graphs that, as in the foregoing two-dimensional problem, the values of $\min|p_{1cr.0}|/E^{(1)2}(\chi_{cr.0})$ increase (decrease) monotonically with $H_1^{(2)}/H_1^{(1)}$. This situation agrees well with the mechanical considerations.

Now we assume that the material of the layer $1^{(1)}$ (Fig. 5.25) is viscoelastic. Investigate the influence of the rheological parameters of this material on the values of $t'_{cr}$. For this purpose we select the case where $H_1^{(2)}/H_1^{(1)} = 0.9$. It is obtained that for this case the values of $\min|p_{1cr.\infty}|/E^{(1)2}(\chi_{cr.\infty}) = 0.04388$ (2.00), 0.06108 (2.25) and 0.07044(2.40) for $\omega = 1, 2$ and 3 respectively, and $\min|p_{1cr.0}|/E^{(1)2}\,(\chi_{cr.0}) = 0.10138$ (2.75). The graphs of the dependencies between $t'_{cr}$ and $|p_1|/E^{(1)2}$ are shown in Fig 5.38 for various $\omega$ under $\alpha = -0.5$

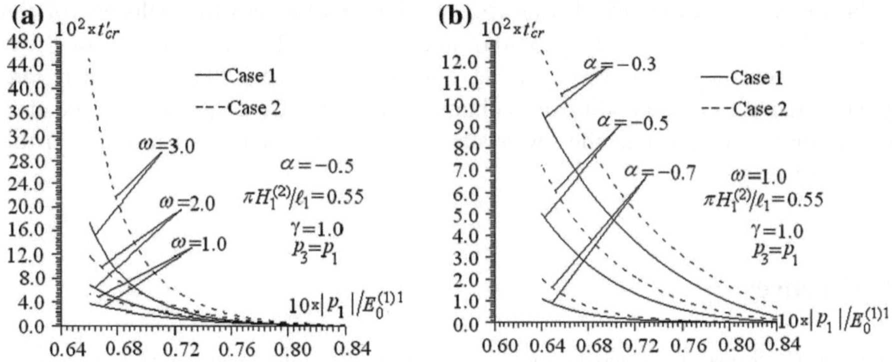


**Fig. 5.40** The graphs of the dependencies between $t'_{cr}$ and $|p_1|/E^{(1)2}$ for various values of $\omega$ under $\alpha = -0.5$(**a**) and for various $\alpha$ under $\omega = 1.0$(**b**) for the case shown in Fig. 5.19

(Fig. 5.38a) and for various $\alpha$ under $\omega = 1.0$ (5.38b). In the qualitative sense these results are similar to those obtained for the problems considered above.

The system consisting of the elastic covering, viscoelastic and elastic bond layers and viscoelastic half-space.

Finally we consider the case where the covering stack consists of three layers as shown in Fig. 5.19. At first we analyze the case where the materials of the layers and half-space are elastic and assume that $p_3 = p_1$ and $\gamma = 1.0$, $\nu^{(2)1} = \nu_0^{(1)1} = \nu^{(2)2} = \nu_0^{(1)2} = 0.3$. With respect to the ratio of the modulus of elasticity of the materials we consider two cases: Case 1 (5.143) and Case 2 (5.144). According to the Eqs. (5.143) and (5.144), the materials of the layer $1^{(1)}$ and half-space are the same. In Case 1 $E^{(2)1} > E^{(2)2} > E_0^{(1)1} = E_0^{(1)2}$, but in Case 2 $E^{(2)2} > E^{(2)1} > E_0^{(1)1} = E_0^{(1)2}$. In other words, in Case 1 the most rigid layer is the upper face layer of the stack, but in Case 2 the most rigid layer is the lower layer of the stack and contacted directly with the half-space. The graphs of the dependencies between $|p_{1cr.0}|/E_0^{(1)1}$ and $\chi\,(= \pi H/\ell_1)$, where $H = H_1^{(2)} + H_1^{(1)} + H_2^{(2)}$, $H_1^{(2)} = H_1^{(1)} = H_2^{(2)}$ are show in Fig. 5.39.

We denote the values of $\min |p_{1cr.0}|/E_0^{(1)1}$ and $\chi_{cr.0}$ for Cases 1 and 2 as $\min |p'_{1cr.0}|/E_0^{(1)1}$, $(\chi'_{cr.0})$ and $\min |p'_{1cr.0}|/E_0^{(1)1}$, $\chi'_{cr.0}$ respectively. It follows from the graphs given in Fig. 5.39 that $\min |p''_{1cr.0}|/E_0^{(1)1} > \min |p'_{1cr.0}|/E_0^{(1)1}$ but $\chi'_{cr.0} > \chi''_{cr.0}$. Consequently, according to these results, we can conclude that the resistance of the considered body with respect to the near-surface stability loss can be increased by the location of the most rigid layer in the lower level in the stack of the layers.

Now we assume that the material of the layer $1^{(1)}$(Fig. 5.19) and half-space are viscoelastic. Consider the influence of the rheological parameters $\omega$ and $\alpha$ on the values of $t'_{cr}$ for Cases (5.143) an (5.144). Fig. 5.40 shows the influence of the rheological parameters $\omega$ (Fig. 5.40a) and $\alpha$ (Fig. 5.40b) on the values of $t'_{cr}$. It follows from the graphs given in Fig. 5.40 that in Case 2 the values of $t'_{cr}$ are greater than the corresponding ones obtained in Case 1. Consequently, the location of the more stiff layer in the lower stage in the stack causes an increase in the values of $t'_{cr}$.

## References


Akbarov SD, Sisman T, Yahnioglu N (1997) On the fracture of the unidirectional composites in compression. Int J Eng Sci 35:1115–1135

Akbarov SD, Cilli A, Guz AN (1999) The theoretical strength limit in the compression of viscoelastic layered composite materials. Compos B Eng 30:465–472

Akbarov SD, Guz AN (2004) Mechanics of curved composites and some related problems for structural members. Mech Advan Mater Struct 11(6):445–515

Akbarov SD, Guz AN (2000) Mechanics of curved composites. Kluwer Academic Publishers, Boston

Akbarov SD, Tekercioglu R (2006) Near-surface buckling instability of a system consisting of a moderately rigid substrate, a viscoelastic bond layer, and an elastic covering layer. Mech Comp Mater 42(4):363–372

Akbarov SD, Tekercioglu R (2007) Surface undulation instability of the viscoelastic half-space covered with the stack of layers in bi-axial compression. Int J Mech Sci 49:778–789

Babich IYu, Guz AN, VikN Chekhov (2001) The three-dimensional theory of stability of fibrous and laminated materials. Int Appl Mech 37(9):1103–1141

Biot MA (1965) Mechanics of incremental deformations. Wiley, New York

Balint DS, Hutchinson JW (2003) Undulation instability of a compressed elastic film on a nonlinear creeping substrate. Acta Mater 51:3965–3983

Bowden N, Brittain S, Evans AG, Hutchinson JW, Whitesides GM (1998) Spontaneous formation of ordered structures in thin films of metals supported on an elastomeric polymer. Nature (London) 393:146–149

Budiansky B, Fleck NA (1994) Compressive king of fiber composites: a topical review. Appl Mech Rev Pt.2 47(6):246–270

Cilli A (2011) On the theoretical strength limit of the layered elastic and viscoelastic composites in compression. Appl Math Model 35:5470–5479

Dow NF, Gruntfest IJ (1960) Determination of most needed potentially possible improvements in materials for ballistic and space vehicles. General Electric Co., Space Sci. Lab., TISR 60 SD 389

Gerard G, Gilbert AC (1958) A critical strain approach to creep buckling of plates and shells. J. Aero/Space Sci 25(7):429–434

Greszczuk LB (1975) Microbuckling failure of circular fiber-reinforced composites. AIAAJ 13(10):1311–1318

Guz AN (1969) On determination of the theoretical strength limit in compression of reinforced materials. Dokl Akad Nauk USSR Ser A 3:236–238 (in Russian)

Guz AN (1990) Fracture mechanics of composites under compression. Naukova Dumka, Kiev (in Russian)

Guz AN (1999) Fundamentals of the three-dimensional theory of stability of deformable bodies. Springer, Berlin

Guz AN (2008a) Fundamentals of the compressive fracture mechanics of composites: fracture in structure of materials. Vol 1 Litera, Kiev (in Russian)

Guz AN (2008b) Fundamentals of the compressive fracture mechanics of composites: related mechanics of fracture. Vol 2 Litera, Kiev (in Russian)
Guz AN, Korzh VP, Chekhov VN (1989) Instability of layered bodies during compression taking into account the action of distributed surface loads. Int Appl Mech 25(5):435–443
Guz AN, Kritsuk AA, Emel'yanov RA (1969) Character of the failure of unidirectional glass-reinforced plastic in compression. Int Appl Mech 5(9):997–999
Guz AN, Tkachenko EA, VikN Chekhov (2000) Surface instability of laminated coatings upon inelastic deformation. Mech Comp Mater 36(6):475–480
Hoff NJ (1954) Buckling and stability. J Roy Aeron Soc 58:1
Huang R, Suo Z (2002) Wrinkling of a compressed elastic film on a viscous layer. J App Phys 91(3):1135–1142
Huck WTS, Bowden N, Onck P, Pardoen T, Hutchinson JW, Whitesides GM (2000) Ordering of spontaneously formed buckles on planar surfaces. Langmuir 16:3497–3501
Rabotnov YuN (1977). Elements of hereditary mechanics of solid bodies, Nauka, Moscow (in Russian)
Rosen BW (1965) Mechanics of composite strengthening. In: Fiber composite materials, American Society of Metals, Seminar, pp 37–75
Rosen BW, Dow NF (1972) The fracture mechanics of fiber composite materials. In: Liebowitz H (ed) Fracture of nonmetals and composites. Fracture. an advanced treatise, vol 7. Academic Press, New York
Schapery RA (1962) Approximate method of transform inversion for viscoelastic stress analysis. In: Proceedings of 4th US national congress of applied mechanics ASME, New York, pp 1075–1086
Schapery RA (1978) A viscoelastic behaviour of composite materials. In: Broutman LJ, Krock RH (eds) Composite materials, vol 1–7, Mir, Moscow (translated from English), vol 2.: Mechanics of composite materials, pp 102–195 (in Russian)

# Chapter 6
# Stability Loss in the Structure of Unidirected Fibrous Elastic and Viscoelastic Composites

Internal and near surface stability loss of unidirected fibrous elastic and viscoelastic composites are studied within the scope of the piecewise homogeneous body model using the approach based on the investigation of the evolution of the fibers' initial infinitesimal imperfections with compressed forces or with time. For this purpose, as in the previous chapters, 3D geometrically nonlinear field equations for viscoelastic bodies are employed. Concrete investigations are made for the internal stability loss of a single, two neighboring and periodically located row of fibers in an elastic and viscoelastic matrix. A solution method for the problems of the micromechanics of a periodically curved fiber, which is near a convex cylindrical surface, is proposed. Next, this method is employed for investigation of the near-surface stability loss of the fiber which is close to the convex cylindrical surface. The numerical results obtained for the critical forces and critical time are presented and discussed.

The results detailed in the present chapter are based on the investigations carried out in papers (Akbarov and Kosker 2001, 2004; Akbarov and Mamedov 2009, 2011).

## 6.1 Some General Remarks on the Field Equations, Problem Formulations and Method of Solution

### *6.1.1 General Remarks on the Field Equations and Problem Formulations*

We consider an infinite body containing any number of unidirected fibers each of which has an initial infinitesimal imperfection. The values related to the fibers are denoted by upper index $(2)m$, where $m$ shows the number of a fiber, and the values related to the matrix by upper index (1). The position of the points of the

S. D. Akbarov, *Stability Loss and Buckling Delamination*,
DOI: 10.1007/978-3-642-30290-9_6, © Springer-Verlag Berlin Heidelberg 2013

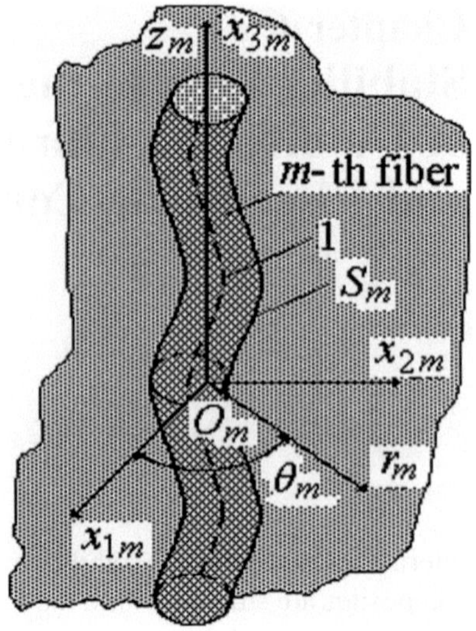


**Fig. 6.1** The schematic illustration of the structure of the composite around the *m*th fiber

constituents we determine through the Lagrange coordinates in the Cartesian coordinates $O_m x_{1m} x_{2m} x_{3m}$ and cylindrical system of coordinates $O_m r_m \theta_m z_m$ associated with *m*th fiber (Fig. 6.1). We will assume that the fibers lie along the $O_m x_{3m}$ axis and $x_{31} = x_{32} = \ldots = x_{3m} = x_3 = z = z_1 = z_2 = \ldots = z_m$.

We determine the initial imperfection of the *m*th fiber through the equation of the middle line (shown by the number 1 in Fig. 6.1) and examine the case where this line lies on the $O_m x_{2m} x_{3m}$ plane. At the same time, we assume that the cross sections of the fiber normal to the middle line of that are circles, with constant radius $R_m$ along the entire length of the *m*th fiber. Also, we assume that the materials of the fibers and matrix are homogeneous, isotropic and non-aging viscoelastic. Investigate the evolution of the foregoing imperfections of the fibers under compression of the body containing these fibers at infinity by uniformly distributed normal forces with intensity $p$ acting in the direction of the $Ox_3$ (or $Oz$) axis. We make this investigation within the scope of the piecewise homogeneous body model with the use of the 3D geometrically non-linear field equations of the theory of the elastic and hereditary-viscoelatic materials. The mentioned field equations are

$$\frac{\partial t_{rr}}{\partial r} + \frac{\partial t_{\theta r}}{r\partial \theta} + \frac{\partial t_{zr}}{\partial z} + \frac{1}{r}(t_{rr} - t_{\theta\theta}) = 0,$$
$$\frac{\partial t_{r\theta}}{\partial r} + \frac{\partial t_{\theta\theta}}{r\partial \theta} + \frac{1}{r}(t_{\theta r} + t_{r\theta}) + \frac{\partial t_{z\theta}}{\partial z} = 0,$$
$$\frac{\partial t_{rz}}{\partial r} + \frac{\partial t_{\theta z}}{r\partial \theta} + \frac{1}{r}t_{rz} + \frac{\partial t_{zz}}{\partial z} = 0, \quad (6.1)$$

$$
\begin{aligned}
t_{rr} &= \sigma_{rr}\left(1+\frac{\partial u_r}{\partial r}\right)+\sigma_{r\theta}\left(\frac{\partial u_r}{r\partial\theta}-\frac{u_\theta}{r}\right)+\sigma_{rz}\frac{\partial u_r}{\partial z},\\
t_{r\theta} &= \sigma_{rr}\frac{\partial u_\theta}{\partial r}+\sigma_{r\theta}\left(1+\frac{\partial u_\theta}{r\partial\theta}+\frac{u_r}{r}\right)+\sigma_{rz}\frac{\partial u_\theta}{\partial z},\\
t_{rz} &= \sigma_{rr}\frac{\partial u_z}{\partial r}+\sigma_{r\theta}\frac{\partial u_z}{r\partial\theta}+\sigma_{rz}\left(1+\frac{\partial u_z}{\partial z}\right),\\
t_{\theta r} &= \sigma_{\theta r}\left(1+\frac{\partial u_r}{\partial r}\right)+\sigma_{\theta\theta}\left(\frac{\partial u_r}{r\partial\theta}-\frac{u_\theta}{r}\right)+\sigma_{\theta z}\frac{\partial u_r}{\partial z},\\
t_{\theta\theta} &= \sigma_{\theta r}\frac{\partial u_\theta}{\partial r}+\sigma_{\theta\theta}\left(1+\frac{\partial u_\theta}{r\partial\theta}+\frac{u_r}{r}\right)+\sigma_{\theta z}\frac{\partial u_\theta}{\partial z},\\
t_{\theta z} &= \sigma_{\theta r}\frac{\partial u_z}{\partial r}+\sigma_{\theta\theta}\frac{\partial u_z}{r\partial\theta}+\sigma_{\theta z}\left(1+\frac{\partial u_z}{\partial z}\right),\\
t_{zr} &= \sigma_{zr}\left(1+\frac{\partial u_r}{\partial r}\right)+\sigma_{z\theta}\left(\frac{\partial u_r}{r\partial\theta}-\frac{u_\theta}{r}\right)+\sigma_{zz}\frac{\partial u_r}{\partial z},\\
t_{z\theta} &= \sigma_{zr}\frac{\partial u_\theta}{\partial r}+\sigma_{z\theta}\left(1+\frac{\partial u_\theta}{r\partial\theta}+\frac{u_r}{r}\right)+\sigma_{zz}\frac{\partial u_\theta}{\partial z},\\
t_{zz} &= \sigma_{zr}\frac{\partial u_z}{\partial r}+\sigma_{z\theta}\frac{\partial u_z}{r\partial\theta}+\sigma_{zz}\left(1+\frac{\partial u_z}{\partial z}\right)
\end{aligned}
\tag{6.2}
$$

$$
\begin{aligned}
\varepsilon_{rr} &= \frac{\partial u_r}{\partial r}+\frac{1}{2}\left\{\left(\frac{\partial u_r}{\partial r}\right)^2+\left(\frac{\partial u_\theta}{\partial r}\right)^2+\left(\frac{\partial u_z}{\partial r}\right)^2\right\},\\
\varepsilon_{r\theta} &= \frac{1}{2}\left(\frac{\partial u_\theta}{\partial r}+\frac{\partial u_r}{r\partial\theta}-\frac{u_\theta}{r}\right)\\
&\quad+\frac{1}{2}\left\{\frac{\partial u_r}{\partial r}\left(\frac{\partial u_r}{r\partial\theta}-\frac{u_\theta}{r}\right)+\frac{\partial u_\theta}{\partial r}\left(\frac{\partial u_\theta}{r\partial\theta}+\frac{u_r}{r}\right)+\frac{\partial u_z}{\partial r}\frac{\partial u_z}{r\partial\theta}\right\}\\
\varepsilon_{rz} &= \frac{1}{2}\left(\frac{\partial u_r}{\partial z}+\frac{\partial u_z}{\partial r}\right)+\frac{1}{2}\left\{\frac{\partial u_r}{\partial r}\frac{\partial u_r}{\partial z}+\frac{\partial u_\theta}{\partial r}\frac{\partial u_\theta}{\partial z}+\frac{\partial u_z}{\partial r}\frac{\partial u_z}{\partial z}\right\}\\
\varepsilon_{\theta\theta} &= \frac{\partial u_\theta}{r\partial\theta}+\frac{u_r}{r}+\frac{1}{2}\left\{\left(\frac{\partial u_r}{r\partial\theta}-\frac{u_\theta}{r}\right)^2+\left(\frac{\partial u_\theta}{r\partial\theta}+\frac{u_r}{r}\right)^2+\frac{1}{r^2}\left(\frac{\partial u_z}{\partial\theta}\right)^2\right\}\\
\varepsilon_{\theta z} &= \frac{1}{2}\left(\frac{\partial u_z}{r\partial\theta}+\frac{\partial u_\theta}{\partial z}\right)+\frac{1}{2}\left\{\frac{\partial u_r}{\partial z}\left(\frac{\partial u_r}{r\partial\theta}-\frac{u_\theta}{r}\right)+\frac{\partial u_\theta}{\partial z}\left(\frac{\partial u_\theta}{r\partial\theta}+\frac{u_r}{r}\right)+\frac{1}{r}\frac{\partial u_z}{\partial\theta}\frac{\partial u_z}{\partial z}\right\},\\
\varepsilon_{zz} &= \frac{\partial u_z}{\partial z}+\frac{1}{2}\left\{\left(\frac{\partial u_r}{\partial z}\right)^2+\left(\frac{\partial u_\theta}{\partial z}\right)^2+\left(\frac{\partial u_z}{\partial z}\right)^2\right\}
\end{aligned}
\tag{6.3}
$$

Here the Eq. (6.1) is equilibrium equation in terms of the physical components $t_{rr}$, $t_{\theta\theta}$, $t_{zz}$, $t_{r\theta}$, $t_{\theta r}$, $t_{z\theta}$, $t_{\theta z}$, $t_{zr}$ and $t_{rz}$ of the non-symmetric Kirchhoff stress tensor, the Eq. (6.2) shows the relation between the physical components of non-symmetric Kirchhoff stress tensor and the physical components of the ordinary stress tensor in the cylindrical system of coordinates, the non-linear relation between the physical components $\varepsilon_{rr}$, $\varepsilon_{zz}$, $\varepsilon_{\theta\theta}, \varepsilon_{r\theta}$, $\varepsilon_{\theta r}$, $\varepsilon_{z\theta}$, $\varepsilon_{\theta z}$, $\varepsilon_{zr}$ and $\varepsilon_{rz}$ of the Green strain tensor and the physical components $u_r$, $u_\theta$ and $u_z$ of the displacement vector is illustrated by the Eq. (6.3).

The constitutive relations for the fibers and matrix materials in the cylindrical system of coordinates are given as follows:

$$\sigma_{(ij)} = \lambda^* \theta \delta_i^j + 2\mu^* \varepsilon_{(ij)}, \ (ij) = rr; \theta\theta; zz; r\theta; rz; z\theta, \tag{6.4}$$

$\lambda^*$ and $\mu^*$ are following operators

$$\left\{ \begin{matrix} \lambda^* \\ \mu^* \end{matrix} \right\} \phi(t) = \left\{ \begin{matrix} \lambda_0 \\ \mu_0 \end{matrix} \right\} \phi(t) + \int_0^t \left\{ \begin{matrix} \lambda_1(t-\tau) \\ \mu_1(t-\tau) \end{matrix} \right\} \phi(\tau) d\tau. \tag{6.5}$$

Here, $\lambda_0$ and $\mu_0$ are the instantaneous values of elastic constants, $\lambda_1(t)$ and $\mu_1(t)$ are the given functions which determine the hereditary properties of the materials of the constituents.

We assume that there is complete cohesion conditions on the fiber-matrix interfaces $S_m$:

$$\begin{gathered}
\left( t_{rr}^{(2)m} n_r^m + t_{\theta r}^{(2)m} n_\theta^m + t_{zr}^{(2)m} n_z^m \right)\Big|_{S_m} = \left( t_{rr}^{(1)} n_r^m + t_{\theta r}^{(1)} n_\theta^m + t_{zr}^{(1)} n_z^m \right)\Big|_{S_m}, \\
\left( t_{r\theta}^{(2)m} n_r^m + t_{\theta\theta}^{(2)m} n_\theta^m + t_{z\theta}^{(2)m} n_z^m \right)\Big|_{S_m} = \left( t_{r\theta}^{(1)} n_r^m + t_{\theta\theta}^{(1)} n_\theta^m + t_{z\theta}^{(1)} n_z^m \right)\Big|_{S_m}, \\
\left( t_{rz}^{(2)m} n_r^m + t_{\theta z}^{(2)m} n_\theta^m + t_{zz}^{(2)m} n_z^m \right)\Big|_{S_m} = \left( t_{rz}^{(1)} n_r^m + t_{\theta z}^{(1)} n_\theta^m + t_{zz}^{(1)} n_z^m \right)\Big|_{S_m}, \\
u_r^{(2)m}\Big|_{S_m} = u_r^{(1)}\Big|_{S_m}, \ u_\theta^{(2)m}\Big|_{S_m} = u_\theta^{(1)}\Big|_{S_m}, \ u_z^{(2)m}\Big|_{S_m} = u_z^{(1)}\Big|_{S_m},
\end{gathered} \tag{6.6}$$

where $n_r^m$, $n_\theta^m$ and $n_z^m$ are the components of the unit normal vector to the surface $S_m$ in the cylindrical system of coordinates $O_m r_m \theta_m z_m$.

The equation of the middle line of the $m$th fiber is given as follows

$$x_{2m} = F_m(x_3) = \varepsilon f_m(x_3), \ x_{1m} = 0, \tag{6.7}$$

where $\varepsilon$ is the dimensionless small parameter and $0 \leq \varepsilon \ll 1$. The geometric significance of this parameter will be indicated with a specifically prescribed form of the function (6.7).

Thus, the investigation of the evolution of the infinitesimal initial imperfections of the fibers in a matrix is reduced to the solution to the non-linear integro-differential boundary value problem (6.1)–(6.7).

### 6.1.2 General Remarks on the Method of Solution

Consider the general aspects of the method of solution for the problems formulated in the previous subsection. Note that this method can be briefly summarized as follows:

i. By employing the boundary form perturbation techniques the considered boundary-value problems for the non-linear integro-differential Eqs. (6.1)–(6.5) are reduced to the series boundary-value problems for the corresponding system of the linear integro-differential equations;
ii. Owing to both the expressions of the operators (6.5) and the convolution theorem, by using the Laplace transformation with respect to time these series of problems of the system of the linear integro-differential equations are reduced to the corresponding series boundary-value problems for the linear systems of the differential equations in the Laplace transformation parameter space;
iii. For each value of the Laplace transformation parameter the linear problems are solved by employing the corresponding analytical solution methods for the system of partial differential equations;
iv. By applying the inverse transformation method (Schapery 1962, 1978) the originals of the sought values are determined;
v. Finally, with the use of the corresponding initial imperfection criterion the values of the critical parameters are found.

It should be noted that in the case where the materials of both the fibers and matrix are pure elastic simultaneously, the operators (6.5) are replaced by the corresponding mechanical constants and therefore instead of the noted above integro-differential equations we obtain differential equations and the corresponding problems for these equations are also investigated in the framework of the above procedure but without employing the Laplace transformation.

Since, according to the procedure summarized above and the problem statement, first we derive the equation for contact surface $S_m$, between the $m$th fiber and matrix, in the cylindrical system of coordinates $O_m r_m \theta_m z_m$. For this purpose we introduce the parameter $t_3 \in (-\infty, +\infty)$, and rewrite the Eq. (6.7) through this parameter:

$$x_{1m} = 0,\ x_3 = t_3,\ x_{2m} = F_m(t_3) = \epsilon f_m(t_3). \tag{6.8}$$

Moreover, we write the equation of the plane which is perpendicular to the tangent of the middle line (6.8).

$$\frac{dF_m(t_3)}{dt_3}(x_{2m} - F_m(t_3)) + x_3 - t_3 = 0. \tag{6.9}$$

We add the condition on the cross section of the $m$th fiber, which can be written as

$$x_{1m}^2 + (x_{2m} - F_m(t_3))^2 + (x_3 - t_3)^2 = R_m^2. \tag{6.10}$$

For a point of the surface $S_m$ the coordinates $x_{1m}, x_{2m}$ and $x_3 (= x_{3m})$ satisfy the Eqs. (6.9) and (6.10) simultaneously. Thus, making transformations $x_{1m} = r_m \cos\theta_m, x_{2m} = r_m \sin\theta_m$ and using (6.9) and (6.10), we obtain the following equation for the surface $S_m$:

$$\begin{aligned} r_m = & r_m(\theta_m, t_3, \varepsilon) = \varepsilon f_m(t_3) \sin\theta_m \frac{1 + \varepsilon^2 \left(f_m'(t_3)\right)^2}{1 + \varepsilon^2 \left(f_m'(t_3)\right)^2 \sin^2\theta_m} \\ & + \left\{ \frac{(R_m)^2 - \varepsilon^2 (f_m(t_3))^2 \left(1 + \varepsilon^2 \left(f_m'(t_3)\right)^2\right)}{1 + \varepsilon^2 \left(f_m'(t_3)\right)^2 \sin^2\theta_m} \right. \\ & \left. + \varepsilon^2 (f_m(t_3))^2 \sin^2\theta_m \frac{\left(1 + \varepsilon^2 \left(f_m'(t_3)\right)^2\right)}{\left(1 + \varepsilon^2 \left(f_m'(t_3)\right)^2 \sin^2\theta_m\right)^2} \right\}^{\frac{1}{2}}, \\ z_m = & t_3 - \varepsilon f_m'(t_3)(r_m(\theta_m, t_3, \varepsilon) - \varepsilon f_m(t_3)),\ f_m'(t_3) = \frac{df_m(t_3)}{dt_3}. \end{aligned} \tag{6.11}$$

Supposing that $(\varepsilon f_m'(t_3))^2 \ll 1$, after some mathematical manipulations, we obtain the following equations:

$$\begin{aligned} & r_m = R_m + \varepsilon f_m(t_3) \sin\theta_m + O(\varepsilon^2),\ z_m = t_3 - \varepsilon R_m f_m'(t_3) \sin\theta_m + O(\varepsilon^2), \\ & n_r^m = 1 + O(\varepsilon^2),\ n_\theta^m = \varepsilon \frac{f_m(t_3)}{R_m} \cos\theta_m + O(\varepsilon^2), \\ & n_z^m = -\varepsilon f_m'(t_3) \sin\theta_m + O(\varepsilon^2). \end{aligned} \tag{6.12}$$

As in the previous chapters we will seek quantities characterizing the stress–strain state of the matrix and fiber in the form of series in positive powers of the small parameter $\varepsilon$:

$$\left\{\sigma_{(ij)}^{(k)m};\ \varepsilon_{(ij)}^{(k)m};\ u_{(i)}^{(k)m}\right\} = \sum_{q=0}^{\infty} \varepsilon^q \left\{\sigma_{(ij)}^{(k)m,q};\ \varepsilon_{(ij)}^{(k)m,q};\ u_{(i)}^{(k)m,q}\right\}, \tag{6.13}$$

where $(ij) = rr, \theta\theta, zz, r\theta, rz, \theta z$ and $(i) = r, \theta, z$.

Substituting Eq. (6.13) into Eqs. (6.1)–(6.4), we obtain a set of equations for each approximation (6.13). Using Eq. (6.12) we expand the values of each approximation (6.13) in series form in the vicinity of the point $\{r_m = R_m, z_m = t_3\}$. Substituting these last expressions in the contact conditions (6.6) and using the expressions given in (6.12), after some mathematical transformations we obtain contact conditions which are satisfied at $\{r_m = R_m, z_m = t_3\}$ for each approximation in (6.13). As in the previous chapters, in the investigations which will be carried out below, we will restrict ourselves with the zeroth and the first approximations only. Therefore we write here the equations and relations obtained for these approximations.

*The zeroth approximation*. It is evident that for the zeroth approximation Eqs. (6.1)–(6.4) are valid and contact condition (6.6) is replaced by the same one satisfied at $\{r_m = R_m, z_m = t_3\}$. We assume that the non-linear parts of the strain tensor components are very small and can be neglected with respect to their linear parts. According to this assumption, for the zeroth approximation we obtain the following field equations of the classical linear theory of viscoelasticity.

$$\frac{\partial \sigma_{rr}^{(k)m,0}}{\partial r_m} + \frac{\partial \sigma_{\theta r}^{(k)m,0}}{r_m \partial \theta_m} + \frac{\partial \sigma_{zr}^{(k)m,0}}{\partial z_m} + \frac{1}{r_m}\left(\sigma_{rr}^{(k)m,0} - \sigma_{\theta\theta}^{(k)m,0}\right) = 0,$$

$$\frac{\partial \sigma_{r\theta}^{(k)m}}{\partial r_m} + \frac{\partial \sigma_{\theta\theta}^{(k)m,0}}{r_m \partial \theta_m} + \frac{2}{r_m}\sigma_{\theta r}^{(k)m,0} + \frac{\partial \sigma_{z\theta}^{(k)m,0}}{\partial z_m} = 0,$$

$$\frac{\partial \sigma_{rz}^{(k)m,0}}{\partial r_m} + \frac{\partial \sigma_{\theta z}^{(k)m,0}}{r_m \partial \theta_m} + \frac{1}{r_m}\sigma_{rz}^{(k)m,0} + \frac{\partial \sigma_{zz}^{(k)m,0}}{\partial z_m} = 0,$$

$$\varepsilon_{rr}^{(k)m,0} = \frac{\partial u_r^{(k)m,0}}{\partial r}, \quad \varepsilon_{r\theta}^{(k)m,0} = \frac{1}{2}\left(\frac{\partial u_\theta^{(k)m,0}}{\partial r_m} + \frac{\partial u_r^{(k)m,0}}{r_m \partial \theta_m} - \frac{u_\theta^{(k)m,0}}{r_m}\right),$$

$$\varepsilon_{rz}^{(k)m,0} = \frac{1}{2}\left(\frac{\partial u_r^{(k)m,0}}{\partial z_m} + \frac{\partial u_z^{(k)m,0}}{\partial r_m}\right), \quad \varepsilon_{\theta\theta}^{(k)m,0} = \frac{\partial u_\theta^{(k)m,0}}{r_m \partial \theta_m} + \frac{u_r^{(k)m,0}}{r_m},$$

$$\varepsilon_{\theta z}^{(k)m,0} = \frac{1}{2}\left(\frac{\partial u_z^{(k)m,0}}{r_m \partial \theta m} + \frac{\partial u_\theta^{(k)m,0}}{\partial z_m}\right), \quad \varepsilon_{zz}^{(k)m,0} = \frac{\partial u_z^{(k)m,0}}{\partial z_m}, \tag{6.14}$$

and contact conditions

$$\sigma_{(i)r}^{(k)m,0}\Big|_{\left(\begin{array}{c} r_m = R_m, \\ z_m = t_3 \end{array}\right)} = \sigma_{(i)r}^{(1),0}\Big|_{\left(\begin{array}{c} r_m = R_m, \\ z_m = t_3 \end{array}\right)},$$

$$u_{(i)}^{(k)m,0}\Big|_{\left(\begin{array}{c} r_m = R_m, \\ z_m = t_3 \end{array}\right)} = u_{(i)}^{(1),0}\Big|_{\left(\begin{array}{c} r_m = R_m, \\ z_m = t_3 \end{array}\right)}. \tag{6.15}$$

where $(i) = r, \theta, z$.

Moreover, the values of the zeroth approximation must satisfy the condition

$$\left\langle \sigma_{zz}^{(2)m,0} \right\rangle \eta^{(2)} + \left\langle \sigma_{zz}^{(1),0} \right\rangle \eta^{(1)} = p \tag{6.16}$$

where $\langle \sigma_{zz}^{(2)m,0} \rangle (\langle \sigma_{zz}^{(1),0} \rangle)$ is the stress averaged over the representative volume and acting on the $m$th fiber (matrix) fraction of this element. Note that the condition (6.16) is written formally and must be corrected for each concrete considered case separately.

*The first approximation*. We write the field equations for this approximation within the scope of the assumption

$$\sigma_{rr}^{(2)m,0} = \sigma_{rr}^{(1),0} = \sigma_{r\theta}^{(2)m,0} = \sigma_{r\theta}^{(1),0} = \sigma_{rz}^{(2)m,0} = \sigma_{rz}^{(1),0} = \sigma_{z\theta}^{(2)m,0} = \sigma_{z\theta}^{(1),0}$$
$$= \sigma_{\theta\theta}^{(2)m,0} = \sigma_{\theta\theta}^{(1),0} = 0, \ \sigma_{zz}^{(2)m,0} \neq 0, \ \sigma_{zz}^{(1),0} \neq 0, \quad (6.17)$$

and the stresses $\sigma_{zz}^{(2)m,0}$, $\sigma_{zz}^{(1),0}$ are homogeneous ones with respect to coordinates. Note that namely the case (6.17) will be considered below. Thus, in the case where the relations in (6.17) satisfy we obtain the following field equations

$$\frac{\partial \sigma_{rr}^{(k)m,1}}{\partial r_m} + \frac{\partial \sigma_{\theta r}^{(k)m,1}}{r_m \partial \theta_m} + \frac{\partial \sigma_{zr}^{(k)m,1}}{\partial z_m} +$$
$$\frac{1}{r_m}\left(\sigma_{rr}^{(k)m,1} - \sigma_{\theta\theta}^{(k)m,1}\right) + \sigma_{zz}^{(k)m,0} \frac{\partial^2 u_r^{(k)m,1}}{\partial z_m^2} = 0,$$
$$\frac{\partial \sigma_{r\theta}^{(k)m,1}}{\partial r_m} + \frac{\partial \sigma_{\theta\theta}^{(k)m,1}}{r_m \partial \theta_m} + \frac{2}{r_m}\sigma_{\theta r}^{(k)m,1} + \frac{\partial \sigma_{z\theta}^{(k)m,1}}{\partial z_m} + \sigma_{zz}^{(k)m,0} \frac{\partial^2 u_\theta^{(k)m,1}}{\partial z_m^2} = 0,$$
$$\frac{\partial \sigma_{rz}^{(k)m,1}}{\partial r_m} + \frac{\partial \sigma_{\theta z}^{(k)m,1}}{r_m \partial \theta_m} + \frac{1}{r_m}\sigma_{rz}^{(k)m,1} + \frac{\partial \sigma_{zz}^{(k)m,1}}{\partial z_m} + \sigma_{zz}^{(k)m,0} \frac{\partial^2 u_z^{(k)m,0}}{\partial z_m^2} = 0,$$
$$\varepsilon_{rr}^{(k)m,1} = \frac{\partial u_r^{(k)m,1}}{\partial r_m}, \ \varepsilon_{r\theta}^{(k)m,1} = \frac{1}{2}\left(\frac{\partial u_\theta^{(k)m,1}}{\partial r_m} + \frac{\partial u_r^{(k)m,1}}{r_m \partial \theta_m} - \frac{u_\theta^{(k)m,1}}{r_m}\right),$$
$$\varepsilon_{rz}^{(k)m,1} = \frac{1}{2}\left(\frac{\partial u_r^{(k)m,1}}{\partial z_m} + \frac{\partial u_z^{(k)m,1}}{\partial r_m}\right), \ \varepsilon_{\theta\theta}^{(k)m,1} = \frac{\partial u_\theta^{(k)m,1}}{r_m \partial \theta_m} + \frac{u_r^{(k)m,1}}{r_m},$$
$$\varepsilon_{\theta z}^{(k)m,1} = \frac{1}{2}\left(\frac{\partial u_z^{(k)m,1}}{r_m \partial \theta_m} + \frac{\partial u_\theta^{(k)m,1}}{\partial z_m}\right), \ \varepsilon_{zz}^{(k)m,1} = \frac{\partial u_z^{(k)m,1}}{\partial z_m}. \quad (6.18)$$

and contact conditions

$$\left(\sigma_{(i)r}^{(2)m,1} - \sigma_{(i)r}^{(1),1}\right)\Big|_{\left(\substack{r_m = R_m, \\ z_m = t_3}\right)} + \gamma_{zm}\left(\sigma_{(i)z}^{(k)m,0} - \sigma_{(i)z}^{(1),0}\right)\Big|_{\left(\substack{r_m = R_m, \\ z_m = t_3}\right)} = 0,$$
$$\left(u_{(i)}^{(2)m,1} - u_{(i)}^{(1),1}\right)\Big|_{\left(\substack{r_m = R_m, \\ z_m = t_3}\right)} = 0 \quad (6.19)$$

for the first approximation, where $(i) = r, \theta, z$ and $\gamma_{zm} = -f_m'(t_3)\cos\theta_m$.

If we replace the upper index $(k)m$ in the Eqs. (6.13), (6.14) and (6.18) with the upper index $(2)m$ (with the upper index $(1)$) then we obtain the equations for the $m$th fiber (for the matrix). It is also necessary to add to these equations the constitutive relations

$$\sigma_{(ij)}^{(k)m,q} = \lambda^{(k)m*}\theta^{(k)m,q}\delta_i^j + 2\mu^{(k)m*}\varepsilon_{(ij)}^{(k)m,q}, \ (ij) = rr; \theta\theta; zz; r\theta; rz; z\theta, \quad (6.20)$$

where

$$\left\{ \begin{array}{c} \lambda^{(k)m*} \\ \mu^{(k)m*} \end{array} \right\} \phi(t) = \left\{ \begin{array}{c} \lambda_0^{(k)m} \\ \mu_0^{(k)m} \end{array} \right\} \phi(t) + \int_0^t \left\{ \begin{array}{c} \lambda_1^{(k)m}(t-\tau) \\ \mu_1^{(k)m}(t-\tau) \end{array} \right\} \phi(\tau) d\tau. \quad (6.21)$$

By direct verification we convince that the field Eqs. (6.18)–(6.21) coincide with the field equations of Three-Dimensional Linearized Theory of Stability (TDLTS) (see, for example, (Guz 1999)).

For solution of the boundary value problems related to the zeroth and the first approximations we employ the Laplace transformation

$$\bar{\psi} = \int_0^\infty \psi(t) e^{-st} dt \quad (6.22)$$

with $s > 0$, to all the equations and relations of these approximations. After this employing procedure, Eqs. (6.14)–(6.19) are valid for the Laplace transformations of the corresponding sought-for quantities, whereas constitutive relations (9) are transformed to the following ones

$$\bar{\sigma}_{(ij)}^{(k)m,q} = \bar{\lambda}^{(k)m*} \bar{\theta}^{(k)m,q} \delta_i^j + 2\bar{\mu}^{(k)m*} \bar{\varepsilon}_{(ij)}^{(k)m,q}, \quad (ij) = rr; \theta\theta; zz; r\theta; rz; z\theta, \quad (6.23)$$

where

$$\left\{ \begin{array}{c} \bar{\lambda}^{(k)m*} \\ \bar{\mu}^{(k)m*} \end{array} \right\} = \left\{ \begin{array}{c} \lambda_0^{(k)m} \\ \mu_0^{(k)m} \end{array} \right\} + \left\{ \begin{array}{c} \bar{\lambda}_1^{(k)m}(s) \\ \bar{\mu}_1^{(k)m}(s) \end{array} \right\}. \quad (6.24)$$

According to the equations in (6.14), contact condition (6.15) and "homogeneousation" condition (6.16), the determination of the Laplace transformations of the zeroth approximation is reduced to the investigation of the generalized plane-strain state in the infinite plane containing finite or infinite number of the circular inclusions. This investigation must be made in the case where

$$\begin{gathered} \bar{u}_z^{(1),0} = \bar{u}_z^{(2)1,0} = \bar{u}_z^{(2)2,0} = \ldots = \bar{u}_z^{(2)m,0} = \bar{\varepsilon}_{zz}^{(1),0} z = \\ \bar{\varepsilon}_{zz}^{(2)1,0} z = \bar{\varepsilon}_{zz}^{(2)2,0} z = \ldots = \bar{\varepsilon}_{zz}^{(2)m,0} z = \bar{\varepsilon}_{zz} z \end{gathered} \quad (6.25)$$

where $\bar{E}^{(1)*}$ is the Laplace transformation of the operator

$$E^{(1)*}\varphi(t) = E_0^{(1)}\varphi(t) + \int_0^t E_1^{(1)}(t-\tau)\varphi(\tau)d\tau, \quad (6.26)$$

i.e. $\bar{E}^{(1)*} = E_0^{(1)} + \bar{E}_1^{(1)}(s)$.

It should be noted that the method for solution to such problems has been improved by many researches; the list of contributors can be found in (Akbarov and Babazade 1987; Schwartz 1992) and elsewhere. It follows from the results of these investigations that the stresses arising under (6.25) in the matrix (plane) and fibers (in each circular inclusion) can be represented as follows:

$$\left\{\bar{\sigma}_{rr}^{(2)m,0}, \bar{\sigma}_{\theta\theta}^{(2)m,0}, \bar{\sigma}_{r\theta}^{(2)m,0}, \bar{\sigma}_{rr}^{(1),0}, \bar{\sigma}_{\theta\theta}^{(1),0}, \bar{\sigma}_{r\theta}^{(1),0}\right\} =$$
$$\left(\bar{\nu}^{(2)m} - \bar{\nu}^{(1)}\right)\left\{\bar{S}_{rr}^{(2)m,0}, \bar{S}_{\theta\theta}^{(2)m,0}, \bar{S}_{r\theta}^{(2)m,0}, \bar{S}_{rr}^{(1),0}, \bar{S}_{\theta\theta}^{(1),0}, \bar{S}_{r\theta}^{(1),0}\right\}, \qquad (6.27)$$

where $\bar{\nu}^{(2)m}$ and $\bar{\nu}^{(1)}$ are the Laplace transformations of the operators

$$\nu^{(1)*}\varphi(t) = \nu_0^{(1)}\varphi(t) + \int_0^t \nu_1^{(1)}(t-\tau)\varphi(\tau)d\tau$$

$$\nu^{(2)2*}\varphi(t) = \nu_0^{(2)m}\varphi(t) + \int_0^t \nu_1^{(2)m}(t-\tau)\varphi(\tau)d\tau \qquad (6.28)$$

i.e. $\bar{\nu}^{(1)*} = \nu_0^{(1)} + \bar{\nu}_1^{(1)}(s)$, $\bar{\nu}^{(2)m*} = \nu_0^{(2)m} + \bar{\nu}_1^{(2)m}(s)$.

The relation (6.27) shows that in the cases where

$$\bar{\nu}^{(2)m} = \bar{\nu}^{(1)} \qquad (6.29)$$

for each $m$, the zeroth approximation corresponds to the homogeneous stress state and only the stresses $\bar{\sigma}_{zz}^{(2)m,0}$ and $\bar{\sigma}_{zz}^{(1),0}$ are differ from zero. In other words, in the case where the relation (6.29) satisfies, the Laplace transformations of the stresses related to the zeroth approximation are determined through the expressions (6.17). For simplicity for future discussions we assume that the relation (6.29) holds. In this case we can write from (6.16) that

$$\bar{\sigma}_{zz}^{(2)m,0} = \bar{E}^{(2)m}\bar{\varepsilon}_{zz}, \quad \bar{\sigma}_{zz}^{(1),0} = \bar{E}^{(1)}\bar{\varepsilon}_{zz}$$
$$\left\langle\bar{\sigma}_{zz}^{(2)m,0}\right\rangle\eta^{(2)} + \left\langle\bar{\sigma}_{zz}^{(1),0}\right\rangle\eta^{(1)} = \bar{\sigma}_{zz}^{(2)m,0}\eta^{(2)} + \bar{\sigma}_{zz}^{(1),0}\eta^{(1)} = \bar{p}, \qquad (6.30)$$

from which we obtain:

$$\bar{\varepsilon}_{zz} = \bar{p}\left(\bar{E}^{(2)m}\eta^{(2)} + \bar{E}^{(1)}\eta^{(1)}\right)^{-1}. \qquad (6.31)$$

The originals of the values of $\bar{\sigma}_{zz}^{(1),0}$ and $\bar{\sigma}_{zz}^{(2)m,0}$ can be determined by the use of the various analytical or approximate methods one of which is the method (Schapery 1962, 1978) and as a result of this transformation we determine the functions $\sigma_{zz}^{(1),0} = \sigma_{zz}^{(1),0}(t)$ and $\sigma_{zz}^{(2)m,0} = \sigma_{zz}^{(2)m,0}(t)$ with respect to time. Note that these functions enter the coefficient of the Eqs. (6.18) and violate the employing of the correspondence principle for solution of the boundary value problem related to the first approximation. For prevention this difficulty we employ the "own time" or "freezing time" principle of TDLTS for the viscoelastic bodies (see, for example, (Guz 1999)), according to which, we take $\sigma_{zz}^{(1),0}(0)$ and $\sigma_{zz}^{(2)m,0}(0)$ instead of functions $\sigma_{zz}^{(1),0}(t)$ and $\sigma_{zz}^{(2)m,0}(t)$ respectively, in Eq. (6.18). Consequently, after this replacing we obtain the Eqs. (6.18), (6.20), (6.21) and contact condition (6.19) rewritten for the Laplace transformation of the corresponding quantities. According to (Guz 1999), for solution of the problem formulated through the Laplace transformation of the Eqs. (6.18)–(6.21) we use the following representations:

$$\begin{aligned}
\bar{u}_r^{(k)m,1} &= \frac{1}{r_m}\frac{\partial}{\partial\theta_m}\Psi^{(2)m} - \frac{\partial^2}{\partial r_m \partial z_m}X^{(2)m},\\
\bar{u}_\theta^{(k)m,1} &= -\frac{\partial}{\partial r_m}\Psi^{(2)m} - \frac{\partial^2}{r_m\partial\theta_m\partial z_m}X^{(2)m},\\
\bar{u}_z^{(k)m,1} &= (\bar{\lambda}^{(k)m} + \bar{\mu}^{(k)m})^{-1}\left[(\bar{\lambda}^{(k)m} + 2\bar{\mu}^{(k)m})\Delta_1 + \bar{\mu}^{(k)m}\frac{\partial^2}{\partial z_m^2}\right]X^{(k)m}.
\end{aligned} \tag{6.32}$$

The functions $\Psi^{(2)m}$ and $X^{(2)m}$ are determined from the equations

$$\begin{aligned}
&\left(\Delta_1 + \left(\xi_1^{(2)m}\right)^2\frac{\partial^2}{\partial z^2}\right)\Psi^{(2)m} = 0,\\
&\left(\Delta_1 + \left(\xi_2^{(2)m}\right)^2\frac{\partial^2}{\partial z^2}\right)\left(\Delta_1 + \left(\xi_3^{(2)m}\right)^2\frac{\partial^2}{\partial z^2}\right)X^{(2)m} = 0
\end{aligned} \tag{6.33}$$

where

$$\begin{aligned}
\xi_1^{(2)m} &= \sqrt{\frac{\bar{\mu}^{(k)m} + \sigma_{zz}^{(k)m,0}(0)}{\bar{\mu}^{(k)m}}}, \quad \xi_2^{(2)m} = \sqrt{\frac{\bar{\mu}^{(k)m} + \sigma_{zz}^{(k)m,0}(0)}{\bar{\mu}^{(k)m}}},\\
\xi_2^{(2)m} &= \sqrt{\frac{\bar{\lambda}^{(k)m} + 2\bar{\mu}^{(k)m} + \sigma_{zz}^{(k)m,0}(0)}{\bar{\lambda}^{(k)m} + 2\bar{\mu}^{(k)m}}}, \quad \Delta_1 = \frac{\partial^2}{\partial r_m^2} + \frac{1}{r_m}\frac{\partial}{\partial r_m} + \frac{1}{r_m^2}\frac{\partial^2}{\partial\theta_m^2}.
\end{aligned} \tag{6.34}$$

Note that foregoing solution procedure holds also for the case where the materials of the fibers and matrix are pure elastic simultaneously. In this case in the foregoing relations and equations the Laplace transformations of the operators $\lambda^{(k)m*}$ and $\mu^{(k)m*}$(6.21) are replaced with the corresponding constants $\lambda^{(k)m}$ and $\mu^{(k)m}$ respectively, and the solution procedure discussed above are made without employing the Laplace transformation (6.22).

This completes the general concepts of the method of solution, according to which, all future investigations will be made in the present chapter. Note that this method is the generalization and development of the method presented in the monograph (Akbarov and Guz 2000) and used for investigations of mechanics of the composite materials with curved structures. The review of the corresponding investigations was also given in the paper (Akbarov and Guz 2004).

## 6.2 Micro Buckling of a Single Fiber in the Viscoelastic Matrix

### 6.2.1 Formulation of the Problem and Method of Solution

We concretize the foregoing formulation and method of solution for the case where an infinite viscoelastic matrix contains a single fiber with initial infinitesimal imperfection. This imperfection is determined by the Eq. (6.7) of the middle line of the fiber. Assume that the cross sections of the fiber normal to the middle line of the fiber are circles, with constant radius $R$ along the entire of the fiber. In all the equations and relations we replace the upper index $(2)m$ with upper index $(2)$. Moreover, for the case under consideration we must take $\eta^{(2)} = 0$ and $\eta^{(1)} = 1$ in the Eqs. (6.16), (6.30) and (6.31), after which, we obtain $\bar{\varepsilon}_{zz} = \bar{p}/\bar{E}^{(1)}$. Assume that the initial imperfection of the middle line of the fiber is periodic one and is given as

$$x_2 = F(x_3) = L\sin\left(2\pi\frac{x_3}{\ell}\right) = \varepsilon\ell\sin(\alpha_1 x_3),\ x_1 = 0,\ L \ll \ell,\ \varepsilon = \frac{L}{\ell}. \tag{6.35}$$

Taking the foregoing assumptions into account, we obtain the following contact condition for the first approximation from the Eq. (6.19):

$$\bar{\sigma}_{rr}^{(2),1}\Big|_{\left(\begin{smallmatrix} r=R,\\ z=t_3\end{smallmatrix}\right)} - \bar{\sigma}_{rr}^{(1),1}\Big|_{\left(\begin{smallmatrix} r=R,\\ z=t_3\end{smallmatrix}\right)} = 0,\ \bar{\sigma}_{r\theta}^{(2),1}\Big|_{\left(\begin{smallmatrix} r=R,\\ z=t_3\end{smallmatrix}\right)} - \bar{\sigma}_{r\theta}^{(1),1}\Big|_{\left(\begin{smallmatrix} r=R,\\ z=t_3\end{smallmatrix}\right)} = 0,$$

$$\bar{\sigma}_{rz}^{(2),1}\Big|_{\left(\begin{smallmatrix} r=R,\\ z=t_3\end{smallmatrix}\right)} - \bar{\sigma}_{rz}^{(1),1}\Big|_{\left(\begin{smallmatrix} r=R,\\ z=t_3\end{smallmatrix}\right)} = 2\pi\left(\bar{\sigma}_{zz}^{(2),0} - \bar{\sigma}_{zz}^{(1),0}\right)\cos(\alpha_1 t_3)\sin\theta,$$

$$\bar{u}_{r;\theta;z}^{(2),1}\Big|_{\left(\begin{smallmatrix} r=R,\\ z=t_3\end{smallmatrix}\right)} - \bar{u}_{r;\theta;z}^{(1),1}\Big|_{\left(\begin{smallmatrix} r=R,\\ z=t_3\end{smallmatrix}\right)} = 0. \tag{6.36}$$

Also, we must take into account the following limited conditions for the quantities related to the matrix and fiber materials.

$$\bar{\sigma}_{(ij)}^{(1),1} \to 0,\ u_{(i)}^{(1),1} \to 0 \text{ as } r \to \infty, (ij) = rr, r\theta, \theta\theta, rz, \theta z, zz,\ (i) = r, \theta, z,$$

$$\left|\bar{\sigma}_{(ij)}^{(2),1}\right| < const,\ \left|\bar{u}_{(i)}^{(2),1}\right| < const, \text{ as } r \to 0. \tag{6.37}$$

Thus, according to (6.36) and (6.37), the solution to the Eq. (6.33) is determined as follows:

$$\Psi^{(2)} = A_1^{(2)} I_1(\xi_1^{(2)} \alpha_1 r) \sin(\alpha_1 z) \cos\theta,$$
$$X^{(2)} = \left[A_2^{(2)} I_1(\xi_2^{(2)} \alpha_1 r) + A_3^{(2)} I_1(\xi_3^{(2)} \alpha_1 r)\right] \cos(\alpha_1 z) \sin\theta,$$
$$\Psi^{(1)} = A_1^{(1)} K_1(\xi_1^{(1)} \alpha_1 r) \sin(\alpha_1 z) \cos\theta,$$
$$X^{(1)} = \left[A_2^{(1)} K_1(\xi_2^{(1)} \alpha_1 r) + A_3^{(1)} K_1(\xi_3^{(1)} \alpha_1 r)\right] \cos(\alpha_1 z) \sin\theta. \quad (6.38)$$

where $\alpha_1 = 2\pi/\ell$, $I_1(x)$ is the Bessel function of a purely imaginary argument of the first order and $K_1(x)$ is the Macdonald function of the first order.

Substituting (6.38) into (6.32) and using, also, the constitutive relations (6.23) and (6.24), we determine the Laplace transformations of the first approximation in terms of the unknown constants $A_i^{(k)}$ $(i = 1, 2, 3;\ k = 1; 2)$. The system of the linear algebraic equations for these unknowns is obtained from the contact conditions (6.36). Finally, the originals of the sought values are determined by employing the method (Schapery 1962, 1978).

According to this method, if the external forces are either constants or slowly changing function of time, then

$$u_r^{(1),1}(t) \approx s\bar{u}_r^{(1),1}(s)\big|_{s=1/(2t)}. \quad (6.39)$$

Since we assume that the external compressive force $p$ is constant, we may determine $\sigma_{(ij)}^{(k),1}$ and $u_{(i)}^{(k),1}$ by the use of the relation (6.39) through the values of $s\bar{\sigma}_{(ij)}^{(k),1}$ and $s\bar{u}_{(i)}^{(k),1}$ respectively at $s = 1/(2t)$. After the foregoing determination the critical time is found from the criterion

$$\max_{\substack{z\in[0,\ell];\\ \theta\in[0,\pi]}} \left|u_r^{(1),1}\right|\Big|_{r=R} \to \infty \text{ as } t \to t_{cr}. \quad (6.40)$$

Note that the solution procedure detailed above holds also for the case where the materials of both the fiber and matrix are pure elastic. In this case the operators $\lambda^{(k)*}$ and $\mu^{(k)*}$ are replaced by the corresponding mechanical constants and the solution procedure is employed without using the Laplace transformation (6.22). For a pure elastic case, the values of the critical force are determined from the criterion

$$\max_{\substack{z\in[0,\ell];\\ \theta\in[0,\pi]}} \left|u_r^{(1),1}\right|\Big|_{r=R} \to \infty \text{ as } p \to p_{cr}. \quad (6.41)$$

### 6.2.2 Numerical Results and Discussions

Consider the case where the material of the fiber is pure elastic with mechanical constants $E^{(2)}$ (modulus of elasticity) and $\nu^{(2)}$ (Poisson ratio), but the material of the matrix is a viscoelastic one with operators

$$E^{(1)*} = E_0^{(1)}\left[1 - \omega_0\Pi_\alpha^*(-\omega_0 - \omega_\infty)\right],$$
$$\nu^{(1)*} = \nu_0^{(1)}\left[1 + \frac{1 - 2\nu_0^{(1)}}{2\nu_0^{(1)}}\omega_0\Pi_\alpha^*(-\omega_0 - \omega_\infty)\right], \tag{6.42}$$

where $E_0^{(1)}$ and $\nu_0^{(1)}$ are the instantaneous values of elasticity modulus and Poisson ratio, respectively, $\alpha$, $\omega_0$ and $\omega_\infty$ are the rheological parameters, and $\Pi_\alpha^*$ is the fractional-exponential operator (Rabotnov 1977).

$$\Pi_\alpha^*(x)\varphi = \int_0^t \Pi_\alpha(x, t - \tau)\varphi(\tau)d\tau,\ \Pi_\alpha(x,t) = t^\alpha \sum_{n=0}^{\infty} \frac{x^n t^{n(1+\alpha)}}{\Gamma((1+n)(1+\alpha))}, \tag{6.43}$$

where $-1 < \alpha < 0$ and $\Gamma(x)$ is the Gamma function.

We introduce a dimensionless rheological parameter $\omega(= \omega_\infty/\omega_0)$ and a dimensionless time $t'(= t\omega_0^{1/(1+\alpha)})$ and assume that $\nu^{(2)} = \nu_0^{(1)} = 0.3$. At the same time, we introduce a dimensionless parameter $\chi(= 2\pi R/\ell)$ which characterizes the initial imperfection, as well as, the stability loss modes. First we consider numerical results related to the pure elastic stability loss problem in the cases where the material of the matrix is also pure elastic with the mechanical constants $E_0^{(1)}$, $\nu_0^{(1)}$ and $E_\infty^{(1)}$, $\nu_\infty^{(1)}$, where

$$E_\infty^{(1)} = E_0^{(1)}\left[1 - \frac{1}{1+\omega}\right],\ \nu_\infty^{(1)} = \nu_0^{(1)}\left[1 + \frac{1 - 2\nu_0^{(1)}}{2\nu_0^{(1)}}\frac{1}{1+\omega}\right]. \tag{6.44}$$

In other words, we can write that

$$\left\{E_0^{(1)}; \nu_0^{(1)}\right\} = \left\{(E^{(1)*}1)\big|_{t=0}; (\nu^{(1)*}1)\big|_{t=0}\right\},$$
$$\left\{E_\infty^{(1)}; \nu_\infty^{(1)}\right\} = \left\{(E^{(1)*}1)\big|_{t=\infty}; (\nu^{(1)*}1)\big|_{t=\infty}\right\}.$$

We denote the values of the critical forces obtained at $t' = 0$ (at $t' = \infty$) by $p_{cr.0}$ (by $p_{cr.\infty}$) and consider numerical results given in Table 6.1 which shows the values of $p_{cr.0}\big/E_0^{(1)}$ and $p_{cr.\infty}\big/E_0^{(1)}$ calculated from the initial imperfection criterion (6.41) for various values of $E^{(2)}\big/E_0^{(1)}$ and $\omega$. Under tabulating the data given in Table 6.1 the values of the parameter $\chi(= 2\pi R/\ell)$ are selected so that the dependence between $|p_{cr.0}|$ and $\chi$ has the well defined minimum. Note that such

**Table 6.1** The values of $|p_{cr.0}|/E_0^{(1)}$ and $|p_{cr.\infty}|/E_0^{(1)}$ obtained for various $E^{(2)}/E_0^{(1)}$ and $\omega$

| $\frac{E^{(2)}}{E_0^{(1)}}$ | $\frac{\lvert p_{cr.0}\rvert}{E_0^{(1)}}$ | $\frac{\lvert p_{cr.\infty}\rvert}{E_0^{(1)}}$ | | | |
|---|---|---|---|---|---|
| | | $\omega = 3.0$ | $\omega = 2.0$ | $\omega = 1.0$ | $\omega = 0.5$ |
| 20 ($\chi = 0.6$) | 0.2002 | 0.1679 | 0.1570 | 0.1353 | 0.1133 |
| 50 ($\chi = 0.4$) | 0.1330 | 0.1083 | 0.1002 | 0.0840 | 0.0678 |
| 100 ($\chi = 0.4$) | 0.0863 | 0.0729 | 0.0686 | 0.0600 | 0.0516 |

"well defined minimums" (in other words, internal stability loss in the structure of the body under consideration) arise only in the cases where $E^{(2)}\big/E_0^{(1)} > 1$, i.e. in the cases where the material of the fiber is more rigid one than the material of the matrix. But in the cases where $E^{(2)}\big/E_0^{(1)} < 1$ the dependence between $|p_{cr.0}|$ and $\chi$ has monotonic character and minimum of this dependence arises under $\chi = 0$ (or under $\ell = \infty$) which has not real meaning within the scope of the initial imperfection approach. Because under employing the initial imperfection approach the values of $\chi$ (or the values of $\ell$) are given aforetime. Note that more detail explanation of the internal stability loss in the structure of the unidirected fibrous materials will be considered in the next section.

Let's remind that the corresponding results within the scope of the Euler approach were also obtained in the work (Babich 1973) and detailed in the monographs (Guz 1990, 2008a, b). Note that the results obtained for $|p_{cr.0}|$ and shown in Table 6.1 coincide with the corresponding one obtained in (Babich 1973). Moreover, note that according to the Euler approach, the arising of the minimum in the dependencies between $|p_{cr.0}|$ and $\chi$ under $\chi = 0$ (or under $\ell = \infty$) in the cases where $E^{(2)}\big/E_0^{(1)} < 1$ means that the stability loss of the whole structural elements takes place earlier than the internal stability loss in the material of this structural element. According to the foregoing discussion, this conclusion holds also for the initial imperfection criterion. However, in the cases where $E^{(2)}\big/E_0^{(1)} > 1$ within the scope of the Euler approach the realizable values of the critical force and stability loss mode are determined from the mentioned above well defined minimum of the dependence between $|p_{cr.0}|$ and $\chi$, and namely these values of $|p_{cr.0}|$ and $\chi$ have a real meaning. However, it is evident that within the scope of the initial imperfection approach all values of $|p_{cr.0}|$, which are less than the critical force for the stability loss of the corresponding structural element and which are obtained for $\chi \neq 0$, may have real meaning.

Thus, we turn to consideration of the results given in Table 6.1. It follows from these results that the values of the $|p_{cr.0}|\big/E_0^{(1)}$ and $|p_{cr.\infty}|\big/E_0^{(1)}$ decrease with $E^{(2)}\big/E_0^{(1)}$. At the same time these results show that the values of $p_{cr.\infty}\big/E_0^{(1)}$ increase with rheological parameter $\omega$.

**Table 6.2** The values of $t'_{cr}$ obtained for various values of the rheological parameters $\alpha$ and $\omega$, and various values of the compressive force $|p|/E_0^{(1)}$ under $E^{(2)}/E_0^{(1)} = 20$

| $\frac{\lvert p\rvert}{E_0^{(1)}}$ | $\alpha = -0.5$ | | | $\omega = 2$ | | |
|---|---|---|---|---|---|---|
| | $\omega = 0.5$ | $\omega = 1.0$ | $\omega = 3.0$ | $\alpha = -0.3$ | $\alpha = -0.5$ | $\alpha = -0.7$ |
| 0.1700 | 0.0649 | 0.0966 | 6.6184 | 0.3534 | 0.3076 | 0.2225 |
| 0.1750 | 0.0381 | 0.0513 | 0.3992 | 0.1710 | 0.1113 | 0.0409 |
| 0.1800 | 0.0209 | 0.0260 | 0.0882 | 0.0877 | 0.0437 | 0.0086 |
| 0.1850 | 0.0103 | 0.0119 | 0.0250 | 0.0441 | 0.0167 | 0.0017 |
| 0.1900 | 0.0040 | 0.0044 | 0.0067 | 0.0198 | 0.0054 | 0.0002 |
| 0.1950 | 0.0009 | 0.0009 | 0.0011 | 0.0062 | 0.0010 | 0.0000 |

**Table 6.3** The values of $t'_{cr}$ obtained for various values of the rheological parameters $\alpha$ and $\omega$, and various values of the compressive force $|p|/E_0^{(1)}$ under $E^{(2)}/E_0^{(1)} = 50$

| $\frac{\lvert p\rvert}{E_0^{(1)}}$ | $\alpha = -0.5$ | | | $\omega = 2$ | | |
|---|---|---|---|---|---|---|
| | $\omega = 0.5$ | $\omega = 1.0$ | $\omega = 3.0$ | $\alpha = -0.3$ | $\alpha = -0.5$ | $\alpha = -0.7$ |
| 0.1040 | 0.1397 | 0.2581 | $\infty$ | 1.9070 | 3.2578 | 11.3649 |
| 0.1080 | 0.0840 | 0.1329 | $\infty$ | 0.5468 | 0.5667 | 0.6161 |
| 0.1120 | 0.0489 | 0.0688 | 1.0337 | 0.2351 | 0.1739 | 0.0860 |
| 0.1160 | 0.0269 | 0.0344 | 0.1530 | 0.1144 | 0.0634 | 0.0160 |
| 0.1200 | 0.0134 | 0.0159 | 0.0385 | 0.0564 | 0.0236 | 0.0030 |
| 0.1240 | 0.0055 | 0.0061 | 0.0102 | 0.0256 | 0.0078 | 0.0004 |

**Table 6.4** The values of $t'_{cr}$ obtained for various values of the rheological parameters $\alpha$ and $\omega$, and various values of the compressive force $|p|/E_0^{(1)}$ under $E^{(2)}/E_0^{(1)} = 100$

| $\frac{\lvert p\rvert}{E_0^{(1)}}$ | $\alpha = -0.5$ | | | $\omega = 2$ | | |
|---|---|---|---|---|---|---|
| | $\omega = 0.5$ | $\omega = 1.0$ | $\omega = 3.0$ | $\alpha = -0.3$ | $\alpha = -0.5$ | $\alpha = -0.7$ |
| 0.0725 | 0.0897 | 0.1445 | $\infty$ | 0.6202 | 0.6761 | 0.8267 |
| 0.0740 | 0.0620 | 0.0913 | 4.3489 | 0.3294 | 0.2788 | 0.1889 |
| 0.0755 | 0.0419 | 0.0574 | 0.5527 | 0.1924 | 0.1313 | 0.0538 |
| 0.0770 | 0.0275 | 0.0353 | 0.1613 | 0.1172 | 0.0656 | 0.0169 |
| 0.0785 | 0.0172 | 0.0210 | 0.0603 | 0.0720 | 0.0332 | 0.0054 |
| 0.0800 | 0.0101 | 0.0117 | 0.0243 | 0.0434 | 0.0163 | 0.0016 |

Now we consider numerical results related to the critical time, i.e. to the $t'_{cr}$. Tables 6.2, 6.3 and 6.4 show these results which are obtained under $E^{(2)}\big/E_0^{(1)} = 20$, 50 and 100, respectively, for various values of $|p|\big/E_0^{(1)}$ and rheolgical parameters $\omega$ and $\alpha$. Note that the values of $|p|\big/E_0^{(1)}$ are selected within the scope of the relation

$$|p_{cr.\infty}|\big/E_0^{(1)} < |p|\big/E_0^{(1)} < |p_{cr.0}|\big/E_0^{(1)}. \tag{6.45}$$

It follows from the foregoing tables that, $t'_{cr} \to 0 (t'_{cr} \to \infty)$ as $|p|/E_0^{(1)} \to |p_{cr.0}|/E_0^{(1)} (|p|/E_0^{(1)} \to |p_{cr.\infty}|/E_0^{(1)})$. Moreover, these tables show that an increase in the values of the rheological parameters $\omega$ causes to increase in the values of the critical time. But the character of the influence of the rheological parameter $\alpha$ on the values of the critical time depends on the values of $|p|\Big/E_0^{(1)}$ ,i.e. there exists such value of the $|p|\Big/E_0^{(1)}$ (denoted $\left(|p|\Big/E_0^{(1)}\right)_*$) before (after) which an increase in the absolute values of the rheological parameter $\alpha$ causes to increase (to decrease) in the values of the critical time. Note that the value of $\left(|p|\Big/E_0^{(1)}\right)_*$ corresponds to the $t'_{cr} = 0.5$. Moreover, note that the character of the influence of the problem parameters on the values of the critical time in the quantitative sense agree well with the corresponding ones obtained in the previous chapters.

## 6.3 Internal Stability Loss of Two Neighboring Fibers in a Viscoelastic Matrix

### *6.3.1 Formulation of the Problem and Method of Solution*

We consider the infinite body (matrix) containing two neighboring fibers, which have initial infinitesimal imperfections in the periodical curving form. With the middle line of each fiber we associate Lagrangian rectilinear $O_m x_{1m} x_{2m} x_{3m}$ and cylindrical $O_m r_m \theta_m z_m$ coordinate systems shown in Fig. 6.2. Here $m = 1$ and $m = 2$ are related to the first and the second fibers (Fig. 6.2a) respectively. The location of the origins of these coordinate systems at $z_1 = 0$ and $z_2 = 0$, and geometrical parameters related to the location of the fibers are shown in Fig. 6.2b. As can be seen in Fig. 6.2b between these coordinates we have the following relations

$$x_{11} = x_{12},\ x_{31} = x_{32},\ z_1 = z_2,\ r_1 \exp(i\theta_1) = R_{12} + r_2 \exp(i\theta_2). \tag{6.46}$$

We assume that the length of the period of the initial infinitesimal curving of the fibers is the same and with respect to location of the fibers according to each other consider the following two cases:

Case 1. In this case we suppose that the middle lines of the fibers are in the same plane, i.e. in the plane for which $x_{11} = x_{12} = 0$ and the equations of these lines are given as

$$x_{21} = L \sin\left(\frac{2\pi}{\ell} x_{31}\right) \text{ (the 1th fiber)};\ x_{22} = L \sin\left(\frac{2\pi}{\ell} x_{32}\right) \text{ (the 2nd fiber)}. \tag{6.47}$$

This case will be called as the co-phase imperfections (curving) in plane.

Case 2. We suppose that the middle lines of the fibers are in the parallel planes, i.e. the middle line of the first fiber is in the plane $x_{21} = 0$, but the middle line of the second fiber is in the plane $x_{22} = 0$ and the equations of these lines are given as

$$x_{11} = L\sin\left(\frac{2\pi}{\ell}x_{31}\right) \text{ (the 1th fiber); } x_{12} = L\sin\left(\frac{2\pi}{\ell}x_{32}\right) \text{(the 2nd fiber).} \tag{6.48}$$

This case will be called as the co-phase initial imperfections (curving) out of plane.

We will consider below the two types of stability loss modes (the co-phase stability loss in plane (the first mode) and the co-phase stability loss out of plane (the second mode)) of the fibers induced by the foregoing initial imperfection modes of those. Note that the selection of these initial imperfections and stability loss modes of the fibers is caused, namely, by the results of the paper (Akbarov and Guz 1985), according to which, among the possible stability loss modes of two neighboring fibers in a pure elastic matrix these modes are realized only. Note that in the paper (Akbarov and Guz 1985) the corresponding investigations were made within the scope of the Euler (bifurcation) approach and it was established the critical deformation obtained for the foregoing two stability loss modes are less than the corresponding ones obtained for the other possible stability loss modes, such as anti-phase stability loss in plane and anti-phase stability loss mode out of plane.

Thus, we return to the formulation of the considered problem and assume that the cross section of each fiber, which is perpendicular to the middle line, is a circle of constant radius $R$ (Fig. 6.2b) and this is invariant along the entire length of the fibers. Assume that, $L$ (the initial curving amplitude of the fiber) is smaller than $\ell$ (the length of the period of the initial curving); we introduce a small parameter $\varepsilon = L/\ell$ where $0 \le \varepsilon \ll 1$.

Below, the values related to the first and the second fibers will be denoted by upper indices (2)1 and (2)2 respectively, but those related to the infinite body (matrix) contained these fibers denoted by upper index (1). Assume that the fibers and matrix materials are isotropic, homogeneous and viscoelastic. Moreover assume that the materials of the fibers are the same.

Thus, within the scope of the foregoing assumptions we remake all solution procedures detailed in Sect. 6.1 for the problem under consideration. As a result of these procedures we obtain that the Laplace transformations of the values related to the zeroth approximation are determined through the Eqs. (6.17), (6.25), (6.26), (6.30) and (6.31) under $\eta^{(2)} = 0$ and $\eta^{(1)} = 1$, i.e. we obtain that the non-zero stresses are $\sigma_{zz}^{(2)1,0} (= \sigma_{zz}^{(2)2,0})$ and $\sigma_{zz}^{(1),0}$ only. The Laplace transformations of those are determined as follows:

$$\bar{\sigma}_{zz}^{(1),0} = \bar{p},\ \bar{\sigma}_{zz}^{(2)1,0} = \bar{\sigma}_{zz}^{(2)2,0} = \frac{\bar{E}^{(2)*}}{\bar{E}^{(1)*}}\bar{p}. \tag{6.49}$$

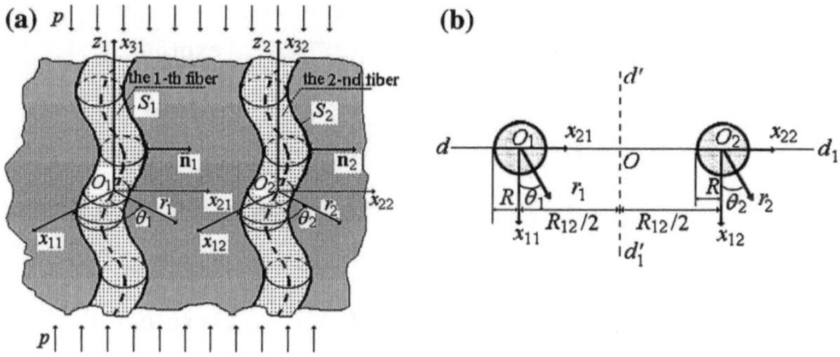


**Fig. 6.2** A co-phase initial imperfection modes of two neighboring fibers and the system of coordinates associated with them

Consider the determination of the Laplace transformation of the values related to the first approximation for which the following contact conditions are derived:

Case 1.

$$\begin{gathered}
\bar{\sigma}_{rr}^{(2)m,1}\Big|_{(r_m=R)} - \bar{\sigma}_{rr}^{(1),1}\Big|_{(r_m=R)} = 0,\ \bar{\sigma}_{r\theta}^{(2)m,1}\Big|_{(r_m=R)} - \bar{\sigma}_{r\theta}^{(1),1}\Big|_{(r_m=R)} = 0, \\
\bar{\sigma}_{rz}^{(2)m,1}\Big|_{(r_m=R)} - \bar{\sigma}_{rz}^{(1),1}\Big|_{(r_m=R)} = 2\pi\left(\bar{\sigma}_{zz}^{(2),0} - \bar{\sigma}_{zz}^{(1),0}\right)\cos(\alpha_1 t_3)\sin\theta_m, \\
\bar{u}_{r;\theta;z}^{(2)m,1}\Big|_{(r_m=R)} - \bar{u}_{r;\theta;z}^{(1),1}\Big|_{(r_m=R)} = 0,\ m = 1,2.
\end{gathered} \tag{6.50}$$

Case 2.

$$\begin{gathered}
\bar{\sigma}_{rr}^{(2)m,1}\Big|_{(r_m=R)} - \bar{\sigma}_{rr}^{(1),1}\Big|_{(r_m=R)} = 0,\ \bar{\sigma}_{r\theta}^{(2)m,1}\Big|_{(r_m=R)} - \bar{\sigma}_{r\theta}^{(1),1}\Big|_{(r_m=R)} = 0, \\
\bar{\sigma}_{rz}^{(2)m,1}\Big|_{(r_m=R)} - \bar{\sigma}_{rz}^{(1),1}\Big|_{(r_m=R)} = 2\pi\left(\bar{\sigma}_{zz}^{(2),0} - \bar{\sigma}_{zz}^{(1),0}\right)\cos(\alpha_1 t_3)\cos\theta_m \\
\bar{u}_{r;\theta;z}^{(2)m,1}\Big|_{(r_m=R)} - \bar{u}_{r;\theta;z}^{(1),1}\Big|_{(r_m=R)} = 0,\ m = 1,2.
\end{gathered} \tag{6.51}$$

For simplification of the consideration below here we will detail the determination of the Laplace transformation related to the first approximation in Case 1 only. Note that the solution procedure, which will be considered below, can also be applied for Case 2, by taking some obvious changing into account. However the numerical results which will be considered in the next section will be given both of the initial imperfections modes.

Thus, taking the expression of the right sides of the contact conditions in (6.50) into account the solution to the Eqs. (6.33) and (6.34) are found as follows:

$$\Psi^{(2)m} = \alpha_1 \sin(\alpha_1 z) \sum_{n=-\infty}^{+\infty} C_n^{(2)m} I_n(\xi_1^{(2)m} \alpha_1 r_m) \exp(in\theta_m),$$

$$X^{(2)m} = \cos(\alpha_1 z) \sum_{n=-\infty}^{+\infty} \Big[ A_n^{(2)m} I_n(\xi_2^{(2)m} \alpha_1 r_m) + B_n^{(2)m} I_n(\xi_3^{(2)m} \alpha_1 r_m) \Big] \exp(in\theta_m), \tag{6.52}$$

$$\Psi^{(1)} = \alpha_1 \sin(\alpha_1 z) \sum_{m=1}^{2} \sum_{n=-\infty}^{+\infty} C_n^{(1)m} K_n(\xi_1^{(1)} \alpha_1 r_m) \exp(in\theta_m),$$

$$X^{(1)} = \cos(\alpha_1 z) \sum_{m=1}^{2} \sum_{n=-\infty}^{+\infty} \Big[ A_n^{(1)m} K_n(\xi_2^{(1)} \alpha_1 r_m) + B_n^{(1)m} K_n(\xi_3^{(1)} \alpha_1 r_m) \Big] \exp(in\theta_m), \tag{6.53}$$

where $\alpha_1 = 2\pi/\ell$, $I_n(x)$ and $K_n(x)$ are the Bessel functions of a purely imaginary argument and Macdonald functions, in turn. Moreover the unknowns $C_n^{(2)m}$, $A_n^{(2)m}$, $B_n^{(2)m}$, $C_n^{(1)m}$, $A_n^{(1)m}$ and $B_n^{(1)m}$ are the complex numbers and satisfy the relations

$$A_n^{(k)m} = \overline{A_{-n}^{(k)m}},\ B_n^{(k)m} = \overline{B_{-n}^{(k)m}},\ C_n^{(k)m} = \overline{C_{-n}^{(k)m}},$$
$$\operatorname{Im} A_0^{(k)m} = \operatorname{Im} B_0^{(k)m} = \operatorname{Im} C_0^{(k)m} = 0. \tag{6.54}$$

Now we attempt to obtain the expressions of the values of the first approximation and satisfy the contact condition (6.50) or (6.51). For this purpose we must represent the expressions (6.52) and (6.53) in the $m$th ($m = 1, 2$) cylindrical coordinate system. The expressions in (6.52) are already presented in the $m$th cylindrical system of coordinates and for expressions (6.53) we use the summation theorem (Watson 1958) for the $K_n(x)$ function, which can be written for the case at hand as follows:

$$r_m \exp(i\theta_m) = R_{mk} \exp(i\varphi_{mk}) + r_k \exp(i\theta_k),$$

$$K_\nu(cr_k)\exp(i\nu\theta_k) = \sum_{q=-\infty}^{+\infty} (-1)^\nu I_q(cr_m) K_{\nu-q}(cR_{mk}) \times \exp(i(\nu-q)\varphi_{mk}) \exp(iq\theta_m); mk = 12; 21, m; k = 1, 2, r_m < R_{mk},$$

$$R_{12} = R_{21}, \varphi_{12} = 0, \varphi_{21} = \pi. \tag{6.55}$$

Thus, applying the summation theorem (6.55) to the solution (6.43), we obtain the following expressions for each cylindrical system of coordinates.

In the $O_1 r_1 \theta_1 z_1$ coordinate system:

$$\Psi^{(1)} = \alpha_1 \sin(\alpha_1 z) \sum_{n=-\infty}^{+\infty} \Bigg\{ C_n^{(1)1} K_n(\xi_1^{(1)} \alpha_1 r_1) + I_n(\xi_1^{(1)} \alpha_1 r_1)$$
$$\times \sum_{\nu=-\infty}^{+\infty} (-1)^\nu \exp(i(\nu - n)\frac{\pi}{2}) C_\nu^{(1)2} K_{\nu-n}(\xi_1^{(1)} \alpha_1 R_{12}) \Bigg\} \exp(in\theta_1),$$
$$X^{(1)} = \cos(\alpha_1 z) \sum_{n=-\infty}^{+\infty} \Bigg\{ A_n^{(1)1} K_n(\xi_2^{(1)} \alpha_1 r_1) + B_n^{(1)1} K_n(\xi_3^{(1)} \alpha_1 r_1) + I_n(\xi_2^{(1)} \alpha_1 r_1)$$
$$\times \sum_{\nu=-\infty}^{+\infty} (-1)^\nu \exp(i(\nu - n)\frac{\pi}{2}) A_\nu^{(1)2} K_{\nu-n}(\xi_2^{(1)} \alpha_1 R_{12}) + I_n(\xi_3^{(1)} \alpha_1 r_1)$$
$$\times \sum_{\nu=-\infty}^{+\infty} (-1)^\nu \exp(i(\nu - n)\frac{\pi}{2}) B_\nu^{(1)2} K_{\nu-n}(\xi_3^{(1)} \alpha_1 R_{12}) \Bigg\} \exp(in\theta_1). \quad (6.56)$$

In the $O_2\, r_2\, \theta_2\, z_2$ coordinate system:

$$\Psi^{(1)} = \alpha_1 \sin(\alpha_1 z) \sum_{n=-\infty}^{+\infty} \Bigg\{ C_n^{(1)2} K_n(\xi_1^{(1)} \alpha_1 r_2) + I_n(\xi_1^{(1)} \alpha_1 r_2)$$
$$\times \sum_{\nu=-\infty}^{+\infty} (-1)^\nu \exp(i(\nu - n)\frac{3\pi}{2}) C_\nu^{(1)1} K_{\nu-n}(\xi_1^{(1)} \alpha_1 R_{12}) \Bigg\} \exp(in\theta_2),$$
$$X^{(1)} = \cos(\alpha_1 z) \sum_{n=-\infty}^{+\infty} \Bigg\{ A_n^{(1)2} K_n(\xi_2^{(1)} \alpha_1 r_2) + B_n^{(1)2} K_n(\xi_3^{(1)} \alpha_1 r_2) + I_n(\xi_2^{(1)} \alpha_1 r_2)$$
$$\times \sum_{\nu=-\infty}^{+\infty} (-1)^\nu \exp(i(\nu - n)\frac{3\pi}{2}) A_\nu^{(1)1} K_{\nu-n}(\xi_2^{(1)} \alpha_1 R_{12}) + I_n(\xi_3^{(1)} \alpha_1 r_2)$$
$$\times \sum_{\nu=-\infty}^{+\infty} (-1)^\nu \exp(i(\nu - n)\frac{3\pi}{2}) B_\nu^{(1)1} K_{\nu-n}(\xi_3^{(1)} \alpha_1 R_{12}) \Bigg\} \exp(in\theta_2). \quad (6.57)$$

Using Eqs. (6.32), (6.34), (6.53), (6.55), (6.56), (6.57), (6.23) and (6.24) we obtain the expressions for Laplace transformations of the values of the first approximation related to the matrix. Since it is cumbersome, we do not present here all these expressions. As an example, in the $O_1 r_1 \theta_1 z_1$ coordinate system we present the expression of the Laplace transformation of the normal stress $\sigma_{rr}^{(1),1}$, which is

$$\bar{\sigma}_{rr}^{(1),1} = \alpha_1 \sin(\alpha_1 z_1) \sum_{n=-\infty}^{\infty} \Bigg\{ C_n^{(1)1} \Bigg[ -\frac{i\bar{\mu}^{(1)*} n}{r_1^2} K_n(\xi_1^{(1)} \alpha_1 r_1) - \frac{i\bar{\mu}^{(1)*} \xi_1^{(1)} \alpha_1}{r_1}$$
$$\times \left( K_{n-1}(\xi_1^{(1)} \alpha_1 r_1) + K_{n+1}(\xi_1^{(1)} \alpha_1 r_1) \right) \Bigg] + A_n^{(1)1} \Bigg[ -\frac{\bar{\lambda}^{(1)*} \xi_2^{(1)} \alpha_1}{2r_1} \times$$
$$\left( K_{n-1}(\xi_2^{(1)} \alpha_1 r_1) + K_{n+1}(\xi_2^{(1)} \alpha_1 r_1) \right) - \frac{\bar{\lambda}^{(1)*} n^2}{r_1^2} K_n(\xi_2^{(1)} \alpha_1 r_1) + \frac{(\xi_2^{(1)})^2 \alpha_1^2 (\bar{\lambda}^{(1)*} + 2\bar{\mu}^{(1)*})}{4}$$

$$\times \left(K_{n-2}(\xi_2^{(1)}\alpha_1 r_1) + 2K_n(\xi_2^{(1)}\alpha_1 r_1) + K_{n+2}(\xi_2^{(1)}\alpha_1 r_1)\right) - \bar{\lambda}^{(1)*}\alpha_1^2$$
$$\left.\times \frac{(\bar{\lambda}^{(1)*} + 2\bar{\mu}^{(1)*})\xi_2^{(1)} - \bar{\mu}^{(1)*} - \sigma_{zz}^{(1),0}}{\bar{\lambda}^{(1)*}2\bar{\mu}^{(1)*}} K_n(\xi_2^{(1)}\alpha_1 r_1)\right]$$
$$+ B_n^{(1)1}\left[-\frac{\bar{\lambda}^{(1)*}\xi_3^{(1)}\alpha_1}{2r_1}\left(K_{n-1}(\xi_3^{(1)}\alpha_1 r_1) + K_{n+1}(\xi_3^{(1)}\alpha_1 r_1)\right)\right.$$
$$- \frac{\bar{\lambda}^{(1)*}n^2}{r_1^2} K_n(\xi_3^{(1)}\alpha_1 r_1) + \frac{(\xi_3^{(1)})^2\alpha_1^2(\bar{\lambda}^{(1)*} + 2\bar{\mu}^{(1)*})}{4}\left(K_{n-2}(\xi_3^{(1)}\alpha_1 r_1)\right.$$
$$\left. + 2K_n(\xi_3^{(1)}\alpha_1 r_1) + K_{n+2}(\xi_3^{(1)}\alpha_1 r_1)\right) - \bar{\lambda}^{(1)*}\alpha_1^2$$
$$\left.\times \frac{(\bar{\lambda}^{(1)*} + 2\bar{\mu}^{(1)*})\xi_3^{(1)} - \bar{\mu}^{(1)*} - \sigma_{zz}^{(1),0}}{\bar{\lambda}^{(1)*} + 2\bar{\mu}^{(1)*}} K_n(\xi_3^{(1)}\alpha_1 r_1)\right]$$
$$+ \sum_{q=-\infty}^{+\infty} (-1)^q \left\{ C_q^{(1)2}\left[-\frac{2i\bar{\mu}^{(1)*}n}{r_1^2} I_n(\xi_1^{(1)}\alpha_1 r_1) + \frac{2i\bar{\mu}^{(1)*}\xi_1^{(1)}\alpha_1 n}{r_1} \times\right.\right.$$
$$\left.\left(I_{n-1}(\xi_1^{(1)}\alpha_1 r_1) + I_{n-1}(\xi_1^{(1)}\alpha_1 r_1)\right)\right] K_{q-n}(\xi_1^{(1)}\alpha_1 R_{12})$$
$$+ A_q^{(1)2}\left[\frac{\bar{\lambda}^{(1)*}\xi_2^{(1)}\alpha_1}{2r_1}\left(I_{n-1}(\xi_2^{(1)}\alpha_1 r_1) + I_{n+1}(\xi_2^{(1)}\alpha_1 r_1)\right)\right.$$
$$- \frac{\bar{\lambda}^{(1)*}n^2}{r_1^2} I_n(\xi_2^{(1)}\alpha_1 r_1) + \frac{(\xi_2^{(1)}\alpha_1)^2(\bar{\lambda}^{(1)*} + 2\bar{\mu}^{(1)*})}{4}\left(I_{n-2}(\xi_2^{(1)}\alpha_1 r_1)\right.$$
$$\left. + 2I_n(\xi_2^{(1)}\alpha_1 r_1) + I_{n+2}(\xi_2^{(1)}\alpha_1 r_1)\right) - \frac{(\bar{\lambda}^{(1)*} + 2\bar{\mu}^{(1)*})\xi_2^{(1)} - \bar{\mu}^{(1)*} - \sigma_{zz}^{(1),0}}{\bar{\lambda}^{(1)*} + \bar{\mu}^{(1)*}}$$
$$\left.\times \bar{\lambda}^{(1)*}\alpha_1^2 I_n(\xi_2^{(1)}\alpha_1 r_1)\right] K_{q-n}(\xi_2^{(1)}\alpha_1 R_{12})$$
$$+ B_q^{(1)2}\left[\frac{\bar{\lambda}^{(1)*}\xi_3^{(1)}\alpha_1}{2r_1}\left(I_{n-1}(\xi_3^{(1)}\alpha_1 r_1) + I_{n+1}(\xi_3^{(1)}\alpha_1 r_1)\right)\right.$$
$$- \frac{\bar{\lambda}^{(1)*}n^2}{r_1^2} I_n(\xi_3^{(1)}\alpha_1 r_1) + \frac{(\xi_3^{(1)}\alpha_1)^2(\bar{\lambda}^{(1)*} + 2\bar{\mu}^{(1)*})}{4}\left(I_{n-2}(\xi_3^{(1)}\alpha_1 r_1)\right.$$
$$\left. + 2I_n(\xi_3^{(1)}\alpha_1 r_1) + I_{n+2}(\xi_3^{(1)}\alpha_1 r_1)\right) - \frac{(\bar{\lambda}^{(1)*} + 2\bar{\mu}^{(1)*})\xi_3^{(1)} - \bar{\mu}^{(1)*} - \sigma_{zz}^{(1),0}}{\bar{\lambda}^{(1)*} + \bar{\mu}^{(1)*}}$$
$$\left.\left.\times \bar{\lambda}^{(1)*}\alpha_1^2 I_n(\xi_3^{(1)}\alpha_1 r_1)\right] K_{q-n}(\xi_3^{(1)}\alpha_1 R_{12})\right\} \exp(in\theta_1). \tag{6.58}$$

We write also the expression for the Laplace transformation of the normal stress $\bar{\sigma}_{rr}^{(2),1}$.

$$
\begin{aligned}
\bar{\sigma}_{rr}^{(2),1} =& \alpha_1 \sin(\alpha_1 z_1) \sum_{n=-\infty}^{\infty} \Bigg\{ C_n^{(2)1} \Bigg[ -\frac{i\bar{\mu}^{(2)*}n}{r_1^2} I_n(\xi_1^{(2)}\alpha_1 r_1) + \frac{i\bar{\mu}^{(2)*}\xi_1^{(2)}\alpha_1}{r_1} \\
& \times \Big( I_{n-1}(\xi_1^{(2)}\alpha_1 r_1) + I_{n+1}(\xi_1^{(2)}\alpha_1 r_1) \Big) \Bigg] + A_n^{(2)1} \Bigg[ \frac{\bar{\lambda}^{(2)*}\xi_2^{(2)}\alpha_1}{2r_1} \Big( I_{n-1}(\xi_2^{(2)}\alpha_1 r_1) \\
& + I_{n+1}(\xi_2^{(2)}\alpha_1 r_1) \Big) - \frac{\bar{\lambda}^{(2)*}n^2}{r_1^2} I_n(\xi_2^{(2)}\alpha_1 r_1) + \frac{(\xi_2^{(2)})^2 \alpha_1^2 (\bar{\lambda}^{(2)*} + 2\bar{\mu}^{(2)*})}{4} \\
& \times \Big( I_{n-2}(\xi_2^{(2)}\alpha_1 r_1) + 2I_n(\xi_2^{(2)}\alpha_1 r_1) + I_{n+2}(\xi_2^{(2)}\alpha_1 r_1) \Big) - \bar{\lambda}^{(2)*}\alpha_1^2 \\
& \times \frac{(\bar{\lambda}^{(2)*} + 2\bar{\mu}^{(2)*})\xi_2^{(2)} - \bar{\mu}^{(2)*} - \sigma_{zz}^{(2),0}}{\bar{\lambda}^{(2)*} + 2\bar{\mu}^{(2)*}} I_n(\xi_2^{(2)}\alpha_1 r_1) \Bigg] \\
& + B_n^{(2)1} \Bigg[ \frac{\bar{\lambda}^{(2)*}\xi_3^{(2)}\alpha_1}{2r_1} \Big( I_{n-1}(\xi_3^{(2)}\alpha_1 r_1) + I_{n+1}(\xi_3^{(2)}\alpha_1 r_1) \Big) \\
& - \frac{\bar{\lambda}^{(2)*}n^2}{r_1^2} I_n(\xi_3^{(2)}\alpha_1 r_1) + \frac{(\xi_3^{(2)})^2 \alpha_1^2 (\bar{\lambda}^{(2)*} + 2\bar{\mu}^{(2)*})}{4} \Big( I_{n-2}(\xi_3^{(2)}\alpha_1 r_1) \\
& + 2I_n(\xi_3^{(2)}\alpha_1 r_1) + I_{n+2}(\xi_3^{(2)}\alpha_1 r_1) \Big) - \bar{\lambda}^{(2)*}\alpha_1^2 \\
& \times \frac{(\bar{\lambda}^{(2)*} + 2\bar{\mu}^{(2)*})\xi_3^{(2)} - \bar{\mu}^{(2)*} - \sigma_{zz}^{(2),0}}{\bar{\lambda}^{(2)*} + 2\bar{\mu}^{(2)*}} I_n(\xi_3^{(2)}\alpha_1 r_1) \Bigg] \Bigg\} \exp(in\theta_1).
\end{aligned} \tag{6.59}
$$

Note that the expressions of the Laplace transformations of the other values of the considered approximation are similar to the expression (6.56) and (6.57). Substituting these expressions into Laplace transformation of the contact condition (6.50) from which we obtain an infinite system of linear algebraic equations for the unknowns $C_n^{(2)m}$, $A_n^{(2)m}$, $B_n^{(2)m}$, $C_n^{(1)m}$, $A_n^{(1)m}$ and $B_n^{(1)m}$. For numerical investigations we must replace this infinite system with the corresponding finite system of equations. We attempt to prove the validity of these replacements. For this purpose we introduce the following notation

$$
\begin{aligned}
& C_n^{(1)m} K_n(\xi_1^{(1)}\chi) = y_{n1}^{(1)m} + iz_{n1}^{(1)m}, \; A_n^{(1)m} K_n(\xi_2^{(1)}\chi) = z_{n2}^{(1)m} + iy_{n2}^{(1)m}, \\
& B_n^{(1)m} K_n(\xi_3^{(1)}\chi) = z_{n3}^{(1)m} + iy_{n3}^{(1)m}, \; C_n^{(2)m} I_n(\xi_1^{(2)}\chi) = y_{n1}^{(2)m} + iz_{n1}^{(2)m}, \\
& A_n^{(2)m} I_n(\xi_2^{(2)}\chi) = z_{n2}^{(2)m} + iy_{n2}^{(2)m}, \; B_n^{(2)m} I_n(\xi_3^{(2)}\chi) = z_{n3}^{(2)m} + iy_{n3}^{(2)m}, \\
& Z_n^{(k)q} = \left\| \begin{matrix} z_{n1}^{(k)q} \\ z_{n2}^{(k)q} \\ z_{n3}^{(k)q} \end{matrix} \right\|, \; Y_n^{(k)q} = \left\| \begin{matrix} y_{n1}^{(k)q} \\ y_{n2}^{(k)q} \\ y_{n3}^{(k)q} \end{matrix} \right\|, \; D_{nv}^{(1)q} = \left\| d_{rs}^{(1)q}(n,v) \right\|, \\
& D_n^{(2)q} = \left\| d_{rs}^{(2)q}(n) \right\|, \; F_{nv}^{(1)q} = \left\| f_{rs}^{(1)q}(n,v) \right\|, \; F_n^{(2)q} = \left\| f_{rs}^{(2)q}(n) \right\| \\
& k;q = 1,2, \; r;s = 1,2,3, \; \chi = \alpha_1 R.
\end{aligned} \tag{6.60}
$$

Using the notation (6.60) we can unite set equations in the following form:

$$
\begin{aligned}
&Z_n^{(1)1} + \sum_{v=0}^{\infty} D_{nv}^{(1)2} Z_v^{(1)2} + Z_n^{(2)1} D_n^{(2)1} = 0, \\
&Z_n^{(1)2} + \sum_{v=0}^{\infty} D_{nv}^{(1)1} Z_v^{(1)1} + Z_n^{(2)2} D_n^{(2)2} = 0, \\
&Y_n^{(1)1} + \sum_{v=0}^{\infty} F_{nv}^{(1)2} Y_v^{(1)2} + Y_n^{(2)1} F_n^{(2)1} = 2\pi\delta_n^3 (\sigma_{zz}^{(1),0} - \sigma_{zz}^{(2)1,0}), \\
&Y_n^{(1)2} + \sum_{v=0}^{\infty} F_{nv}^{(1)1} Y_v^{(1)1} + Y_n^{(2)2} F_n^{(2)2} = 2\pi\delta_n^3 (\sigma_{zz}^{(1),0} - \sigma_{zz}^{(2)2,0}),
\end{aligned} \quad (6.61)
$$

where $n = 0, 1, 2, \ldots, \infty$, $\delta_3^3 = 1$, $\delta_n^3 = 0$, if $n \neq 0$.

Note that the components of the matrices $D_{nv}^{(1)q}$, $F_{nv}^{(1)q}$, $D_n^{(2)q}$ and $F_n^{(2)q}$ are obtained from the (6.58) and (6.59) type expressions which are obtained from the Eqs. (6.32), (6.34), (6.53), (6.55), (6.56), (6.57), (6.23) and (6.24), and since it is unnecessary we omit here their detailed expressions.

It follows from the Eq. (6.61) and from the symmetry of the stress-strain state with respect to the $dd_1$ plane (Fig. 6.2b) that $Z_n^{(k)q} = 0$, $k = 1, 2$, $q = 1, 2$. Moreover, it follows from the mechanical consideration and from (6.61) that $Y_n^{(k)1} = Y_n^{(k)2}$ for the co-phase curving of the fibers. Taking these relations into account, we obtain from (6.61) that

$$
Y_n^{(1)1} + \sum_{v=0}^{\infty} F_{nv}^{(1)2} Y_v^{(1)1} + Y_n^{(2)1} F_n^{(2)1} = 2\pi\delta_n^3 (\sigma_{zz}^{(1),0} - \sigma_{zz}^{(2)1,0}). \quad (6.62)
$$

Note that in Case 2 the right sides of the last two Eq. (6.61) become zero, but the right sides of the first and second equations in (6.61) become nonzero as the right sides of the third and forth equations in (6.61). Consequently, according to the noted statement and according to the problem symmetry with respect to the $d' d_1'$ plane, in Case 2 we obtain that $Z_n^{(k)1} = Z_n^{(k)2}$, $Y_n^{(k)q} = 0$, $k = 1, 2$, $q = 1, 2$. Thus, taking these relations into consideration we obtain the equations for the unknowns $Z_n^{(1)1}$ and $Z_n^{(2)1}$ which can be written as equations in (6.62) by replacing $Y_n^{(1)1}$, $Y_n^{(2)1}$, $F_{nv}^{(1)2}$ and $F_n^{(2)1}$ with $Z_n^{(1)1}$, $Z_n^{(2)1}$, $-D_{nv}^{(1)2}$ and $D_n^{(2)1}$ respectively.

As it has been noted above, for the numerical investigations the infinite system of algebraic Eq. (6.62) must be approximated by a finite system. To validate such a replacement the determinant of this infinite system of equations must be a normal type (Kantarovich and Krilov 1962); this holds if we can prove the convergence of the series

$$M = \sum_{n=0}^{\infty} \sum_{\nu=0}^{\infty} \left| F_{n\nu}^{(1)2} \right|. \tag{6.63}$$

To investigate the series (6.63) we use the expressions of the first approximation, the examples for which are given by Eqs. (6.58) and (6.59), and the following asymptotic estimates of the functions $I_n(x)$ and $K_n(x)$:

$$I_n(x) < c_1 \frac{1}{n!} \left( \frac{|x|}{2} \right)^n, \; c_1 = const, \; K_n(x) \approx c_2 (n-1)! \left( \frac{2}{|x|} \right)^n, \; c_2 = const. \tag{6.64}$$

These hold for large $n$ and fixed $x$. We also use the following inequality:

$$\frac{R}{R_{12} - 2L} > \frac{R}{R_{12}}, \; \frac{R_{12}}{R} > 2, \tag{6.65}$$

which means that the fibers do not touch each other. Thus, taking (6.64) and (6.65) into account and analyzing the expressions for $F_{n\nu}^{(1)2}$ we obtain the following estimation for the series (6.63)

$$M < c_3 \sum_{n=0}^{\infty} n^{c_4} (\rho - 1)^{-n}, \; c_3; c_4 = const, \; \rho = R_{12}/R. \tag{6.66}$$

As the series on the right-hand side in (6.66) converges, so does (6.63).

Thus, the determinant of the infinite system of Eq. (6.62) is normal and this infinite system can be replaced by the corresponding finite system for numerical purposes. The required number of equations in this finite system must be determined from the convergence of the numerical results. In this way, we can determine the unknowns entering (6.62) for each selected values of the Laplace transformation parameter $s > 0$ (6.22). Consequently, according to the method (Schapery 1962, 1978), we can also determine the originals of the sought values for each value of a time $t$ and find the critical time from the criterion

$$\max_{\substack{\theta \in [0,\pi] \\ z \in [0,\ell]}} \left| u_r^{(2)1,1} \right|_{r_1 = R} = \max_{\substack{\theta \in [0,\pi] \\ z \in [0,\ell]}} \left| u_r^{(2)2,1} \right|_{r_1 = R} \to \infty. \tag{6.67}$$

This completes the consideration of the solution method. Note that this method can be also employed for determination of the self-balanced stresses arising as a result of the considered above initial imperfections of the fibers. Similar results for the fibrous composites are also discussed in the monograph (Akbarov and Guz 2000). But in the mentioned monograph the numerical results related to the single fiber in an infinite matrix were considered only. At the same time these results were obtained within the scope of the classical linear theory of elasticity. Consequently, the method developed in the present section allows us to extend the investigations carried out in the monograph (Akbarov and Guz 2000) to cases where the interactions between two neighboring fibers are taken into account. Moreover, the method developed in the present section allows us to take the effect of the geometrical non-

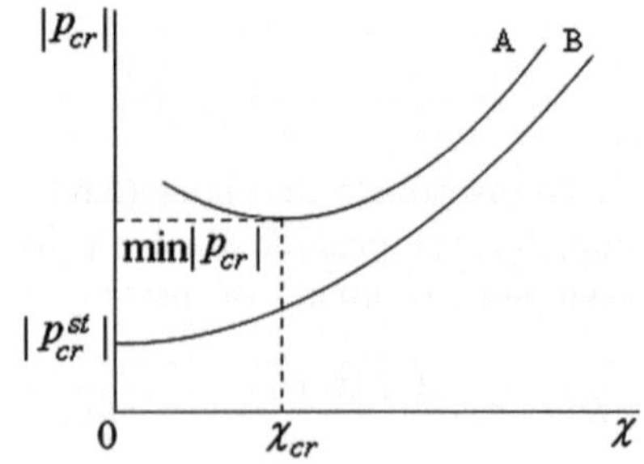


**Fig. 6.3** The types of the dependence between $|p_{cr}|$ and $\chi$

linearity on the noted above self-balanced stresses into account. Numerical results related to the distribution of the self-balanced stresses caused by the initial imperfections of these fibers will be considered in Supplement 2.

## *6.3.2 Numerical Results and Discussion*

We assume that the material of the fibers are pure elastic with mechanical constants $E^{(2)}$ (modulus of elasticity) and $\nu^{(2)}$ (Poisson ratio), but the material of the matrix is viscoelastic one with operators (6.42) and (6.43). We use the notation introduced in Sect. 6.2.2 and investigate the internal stability loss of two neighboring fibers in a viscoelastic matrix. We consider in detail the determination of the internal stability loss in the structure of the inhomogeneous materials. Note that this definition was first given in (Biot 1965) for incompressible material under plane-deformation state. The future development of the theory of the internal stability loss was made, mainly, in (Guz 1990). According to the mentioned references, under internal stability loss we understand the stability loss in the material structure, which arises for certain relations of the stiffness and geometrical parameters of the matrix and fibers and does not depend on boundary surfaces, sizes and forms of members of constructions. The conditions for realization of the internal stability loss for various members of constructions under pure elastic deformation can be presented as follows:

$$|p_{cr}| < |p_{cr}^{st}|, \ \ell_{cr} \ll L, \tag{6.68}$$

where $p_{cr}$ and $p_{cr}^{st}$ are the critical values which correspond to the internal stability loss and stability loss of the whole member of construction, $\ell_{cr}$ is a half-wave length of the internal stability loss mode, $L$ is a characteristic (minimum) size of the member of construction. Note that under investigation of the internal stability loss the values are given for parameter $\chi$ $(= 2\pi R/\ell)$ and the eigenvalues (i.e. the values of $p_{cr}$) are determined from the solution to the corresponding characteristic equations. After this procedure the graph of the dependence

$$p_{cr} = p_{cr}(\chi) \tag{6.69}$$

**Table 6.5** The values of $|p_{cr.0}|/E_0^{(1)}$ and $|p_{cr.\infty}|/E_0^{(1)}$ obtained for Case 1 with various $\rho$

| $\rho$ | $\lvert p_{cr.0}\rvert/E_0^{(1)}$ | $\lvert p_{cr.\infty}\rvert/E_0^{(1)}$ | | | |
|---|---|---|---|---|---|
| | | $\omega = 3.0$ | $\omega = 2.0$ | $\omega = 1.0$ | $\omega = 0.5$ |
| 2.1 | 0.1221 | 0.1007 | 0.0936 | 0.0792 | 0.0647 |
| 2.5 | 0.1141 | 0.0938 | 0.0872 | 0.0740 | 0.0609 |
| 3.0 | 0.1141 | 0.0938 | 0.0871 | 0.0739 | 0.0609 |
| 5.0 | 0.1206 | 0.0988 | 0.0916 | 0.0774 | 0.0632 |
| 6.0 | 0.1235 | 0.1010 | 0.0936 | 0.0789 | 0.0643 |
| 7.0 | 0.1258 | 0.1028 | 0.0952 | 0.0802 | 0.0652 |
| 9.0 | 0.1290 | 0.1053 | 0.0975 | 0.0819 | 0.0664 |
| 10.0 | 0.1301 | 0.1061 | 0.0982 | 0.0825 | 0.0668 |
| 20.0 | 0.1329 | 0.1082 | 0.1001 | 0.0840 | 0.0678 |
| $\infty$ | 0.1330 | 0.1083 | 0.1002 | 0.0840 | 0.0678 |

is constructed. Fig. 6.3 shows two types of graphs of the dependence (6.69). The curve $A$ has a clearly described minimum under $\chi \neq 0$. In this case it is said that the internal stability loss occurs and the realizable value of the compressed critical force (i.e. the value of $\min|p_{cr}|$) is determined by the minimization of the eigenvalues of $p_{cr}$, which corresponds to $\ell_{cr}$ (or $\chi_{cr}$) (Fig. 6.3). But in the second type dependence which is described by the curve $B$, the minimum of $|p_{cr}|$ arises at $\chi = 0$ (i.e. under $\ell = \infty$) and therefore we can not determine the length of $\ell$. Consequently, the second type dependence corresponds to the stability loss of the member of a construction. A lot of investigations detailed in the monograph (Guz 1990) show that the internal stability loss phenomenon for unidirecred fibrous composites occurs under $E^{(2)} > E_0^{(1)}$.

Note that the foregoing determination has been made within the scope of the Euler (bifurcation) approach. According to the mechanical consideration, the $B$ type dependence has not meaning within the scope of the initial imperfection approach also. But in the cases where there exist $A$ type dependencies between $|p_{cr}|$ and $\chi$ not only the values of $\min|p_{cr}|$ and corresponding value of $\chi$ ($= \chi_{cr}$) but also all values of $|p_{cr}|$ which is less than $|p_{cr}^{st}|$ and the corresponding values of $\chi$ have a real meaning within the scope of the initial imperfection approach. However, within the scope of the Euler approach the values of $\min|p_{cr}|$ and $\chi$ ($= \chi_{cr}$) have a the real meaning only.

Thus, taking the discussion above we analyze the numerical results obtained for the problem under consideration. First we consider numerical results related to the stability loss in the above sense for the pure elastic deformation state under $t' = 0$ and $t' = \infty$. Assume that $E^{(2)}/E_0^{(1)} = 50$, $\chi = 0.4$ and $\nu^{(2)} = \nu_0^{(1)} = 0.3$ and introduce the parameter $\rho = R_{12}/R$ through which we will characterize the interaction between the fibers under their stability loss. The corresponding results for $|p_{cr.0}|/E_0^{(1)}$(at $t' = 0$) and for $|p_{cr.\infty}|/E_0^{(1)}$(at $t' = \infty$) obtained for various

**Table 6.6** The values of $|p_{cr.0}|/E_0^{(1)}$ and $|p_{cr.\infty}|/E_0^{(1)}$ obtained for Case 2 with various $\rho$

| $\rho$ | $\lvert p_{cr.0}\rvert/E_0^{(1)}$ | $\lvert p_{cr.\infty}\rvert/E_0^{(1)}$ | | | |
|---|---|---|---|---|---|
| | | $\omega = 3.0$ | $\omega = 2.0$ | $\omega = 1.0$ | $\omega = 0.5$ |
| 2.1 | 0.1061 | 0.0880 | 0.0821 | 0.0703 | 0.0586 |
| 2.5 | 0.1094 | 0.0905 | 0.0843 | 0.0720 | 0.0598 |
| 3.0 | 0.1131 | 0.0934 | 0.0869 | 0.0740 | 0.0611 |
| 5.0 | 0.1233 | 0.1012 | 0.0939 | 0.0794 | 0.0648 |
| 6.0 | 0.1263 | 0.1035 | 0.0959 | 0.0809 | 0.0658 |
| 7.0 | 0.1283 | 0.1050 | 0.0973 | 0.0819 | 0.0665 |
| 9.0 | 0.1307 | 0.1067 | 0.0988 | 0.0830 | 0.0672 |
| 10.0 | 0.1314 | 0.1072 | 0.0993 | 0.0833 | 0.0674 |
| 20.0 | 0.1329 | 0.1083 | 0.1002 | 0.0840 | 0.0678 |
| $\infty$ | 0.1330 | 0.1083 | 0.1002 | 0.0840 | 0.0678 |

values of the parameter $\rho$ and $\omega$ for Cases 1 and 2 are given in Tables 6.5 and 6.6 respectively. It follows from these results that the values of $|p_{cr.0}|\big/E_0^{(1)}$ given in these tables coincide exactly with the corresponding results obtained in (Akbarov and Guz 1985). Moreover, these tables show that the results obtained for Cases 1 and 2 approaches to each other as $\rho \to \infty$ and the corresponding results obtained in the previous section. It follows from Tables 6.5 and 6.6 that in Case 1 the dependence among $|p_{cr.0}|\big/E_0^{(1)}$, $|p_{cr.\infty}|\big/E_0^{(1)}$ and $\rho$ has a non-monotonic character, but this dependence in Case 2 is monotonic one, i.e. the values of $|p_{cr.0}|\big/E_0^{(1)}$ (or the values of $|p_{cr.\infty}|\big/E_0^{(1)}$) increase monotonically with $\rho$ and approach asymptotically the corresponding values of that obtained for a single fiber (Babich 1973). Note that in Case 1 the mentioned non-monotonic character of the considered dependence arises in the cases where the fibers are very near to each other, i.e. in the cases where $2.1 \leq \rho \leq 2.5$.

After the above given preparing procedures, consider the numerical results related to $t'_{cr}$. In these cases the values selected for $|p|\big/E_0^{(1)}$ must satisfy the inequality (6.45).

In Figs. 6.4, 6.5 and 6.6 the graphs of the dependencies between $t'_{cr}$ and $\rho$ are given for Cases 1(Figs. 6.4a, 6.5a and 6.6a) and 2 (Figs. 6.4b, 6.5b and 6.6b) under $\omega = 0.5$, 1.0 and 2.0 respectively in the case where $\alpha = -0.5$. These graphs are constructed for various values of $|p|/E_0^{(1)}$ which are shown in the figures field. Note that the straight horizontal dashed lines in these figures indicate the corresponding values of $t'_{cr}$ obtained for the stability loss of a single fiber in a viscoelastic matrix.

In Tables 6.7 and 6.8 the values of $t'_{cr}$ are given for the Cases 1 and 2 respectively for various values of $\alpha$ which shows the order of the singularity of the operators (6.42) and (6.43). Note that these results are obtained for the case where

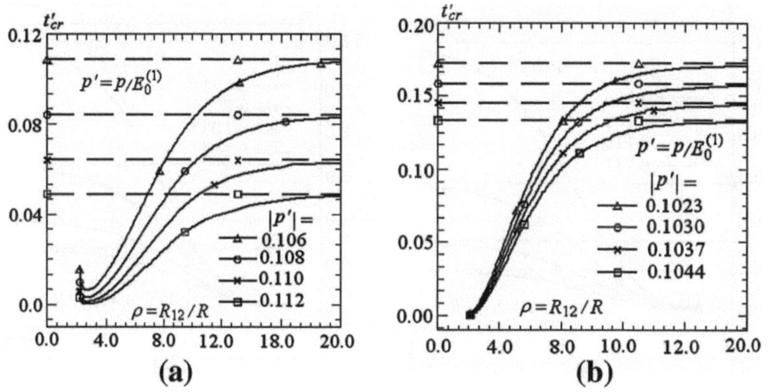


**Fig. 6.4** The graphs of the dependencies $t'_{cr}$ and $\rho$ for various values $|p|/E_0^{(1)}$ under $\alpha = -0.5$, $\omega = 0.5$: **a** Case 1, **b** Case 2

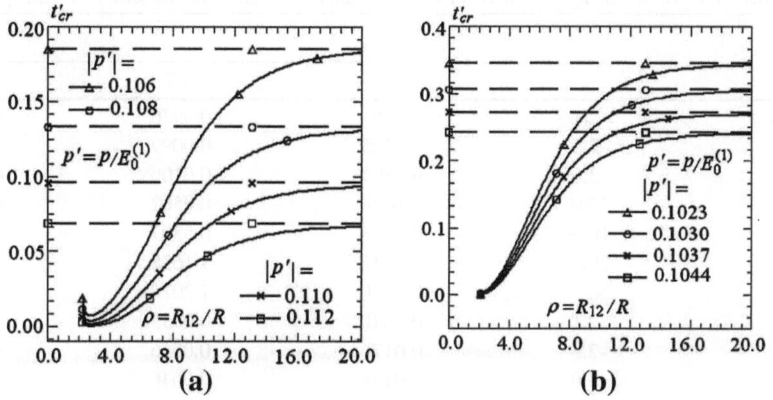


**Fig. 6.5** The graphs of the dependencies $t'_{cr}$ and $\rho$ for various values $|p|/E_0^{(1)}$ under $\alpha = -0.5$, $\omega = 1.0$: **a** Case 1, **b** Case 2

$\omega = 2.0$ with various $|p|/E_0^{(1)}$ and $\rho$. These tables show that under $t'_{cr} > 0.5$ ($t'_{cr} < 0.5$) the values of $t'_{cr}$ increase (decrease) monotonically with $|\alpha|$. Moreover, these tables show that an interaction between the fibers becomes more significantly with $|\alpha|$.

It follows from the foregoing figures that an interaction between the fibers causes to decrease the values of $t'_{cr}$ significantly. Moreover, it follows from the comparison of the values of $t'_{cr}$ obtained for various $\omega$ under the same $|p|/E_0^{(1)}$ that $t'_{cr}$ increase with $\omega$. Also, the foregoing graphs show that in Case 1 the dependencies between $t'_{cr}$ and $\rho$ are non-monotonic, but these dependencies for Case 2 are monotonic.

As has been noted above, the numerical results analyzed here are obtained within the framework of the first approximation. Under obtaining these results the

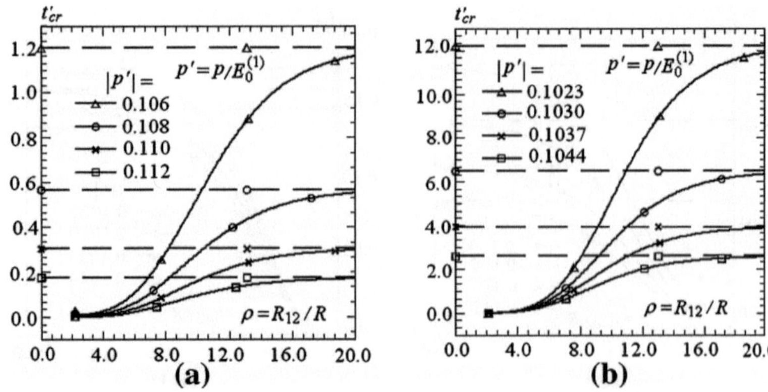


**Fig. 6.6** The graphs of the dependencies $t'_{cr}$ and $\rho$ for various values $|p|/E_0^{(1)}$ under $\alpha = -0.5$, $\omega = 2.0$: **a** Case 1, **b** Case 2

**Table 6.7** The values of $t'_{cr}$ obtained for Case 1 under $\omega = 2.0$ with various $\rho$, $\alpha$ and $|p|/E_0^{(1)}$

| $\lvert p\rvert/E_0^{(1)}$ | $\rho$ | $\alpha$ | | |
|---|---|---|---|---|
| | | −0.3 | −0.5 | −0.7 |
| 0.1060 | 2.1 | 0.1557 | 0.0976 | 0.0328 |
| | 2.5 | 0.0305 | 0.0099 | 0.0007 |
| | 3.0 | 0.0307 | 0.0100 | 0.0007 |
| | 5.0 | 0.1050 | 0.0562 | 0.0131 |
| | 10.0 | 0.5218 | 0.5307 | 0.5523 |
| | 20.0 | 0.9142 | 1.1638 | 2.0441 |
| | ∞ | 0.9350 | 1.2010 | 2.1541 |
| 0.1080 | 2.1 | 0.1040 | 0.0555 | 0.0128 |
| | 2.5 | 0.0176 | 0.0046 | 0.0002 |
| | 3.0 | 0.0178 | 0.0046 | 0.0002 |
| | 5.0 | 0.0706 | 0.0323 | 0.0052 |
| | 10.0 | 0.3316 | 0.2814 | 0.1918 |
| | 20.0 | 0.5367 | 0.5521 | 0.5898 |
| | ∞ | 0.5468 | 0.5667 | 0.6161 |
| 0.1120 | 2.1 | 0.0456 | 0.0175 | 0.0018 |
| | 2.5 | 0.0029 | 0.0003 | 0.0000 |
| | 3.0 | 0.0031 | 0.0004 | 0.0000 |
| | 5.0 | 0.0300 | 0.0097 | 0.0007 |
| | 10.0 | 0.1523 | 0.0947 | 0.0312 |
| | 20.0 | 0.2315 | 0.1701 | 0.0829 |
| | ∞ | 0.2351 | 0.1739 | 0.0860 |

infinite system of equations in (6.60) are replaced by the corresponding finite one. For illustration of the convergence of the numerical results with respect to the number of the equations in this finite system, the values of $t'_{cr}$ obtained for various number of equations are given in Tables 6.9 and 6.10 for Cases 1 and 2

**Table 6.8** The values of $t'_{cr}$ obtained for Case 2 under $\omega = 2.0$ with various $\rho$, $\alpha$ and $|p|/E_0^{(1)}$

| $\lvert p\rvert/E_0^{(1)}$ | $\rho$ | $\alpha$ | | |
|---|---|---|---|---|
| | | −0.3 | −0.5 | −0.7 |
| 0.1023 | 2.1 | 0.0095 | 0.0019 | 0.0000 |
| | 2.5 | 0.0270 | 0.0084 | 0.0005 |
| | 3.0 | 0.0624 | 0.0271 | 0.0038 |
| | 5.0 | 0.3883 | 0.3509 | 0.2771 |
| | 10.0 | 2.6512 | 5.1670 | 24.5142 |
| | 20.0 | 4.8106 | 11.8987 | 98.4435 |
| | $\infty$ | 4.9016 | 12.2149 | 102.8422 |
| 0.1030 | 2.1 | 0.0068 | 0.0012 | 0.0000 |
| | 2.5 | 0.0220 | 0.0063 | 0.0003 |
| | 3.0 | 0.0532 | 0.0217 | 0.0026 |
| | 5.0 | 0.3296 | 0.2790 | 0.1891 |
| | 10.0 | 1.8966 | 3.2329 | 11.2204 |
| | 20.0 | 3.0725 | 6.3518 | 34.5824 |
| | $\infty$ | 3.1175 | 6.4826 | 35.7774 |
| 0.1037 | 2.1 | 0.0045 | 0.0006 | 0.0000 |
| | 2.5 | 0.0177 | 0.0046 | 0.0002 |
| | 3.0 | 0.0452 | 0.0173 | 0.0018 |
| | 5.0 | 0.2817 | 0.2239 | 0.1311 |
| | 10.0 | 1.4283 | 2.1735 | 5.7896 |
| | 20.0 | 2.1539 | 3.8630 | 15.0977 |
| | $\infty$ | 2.1801 | 3.9290 | 15.5299 |
| 0.1044 | 2.1 | 0.0027 | 0.0003 | 0.0000 |
| | 2.5 | 0.0139 | 0.0033 | 0.0001 |
| | 3.0 | 0.0382 | 0.0136 | 0.0012 |
| | 5.0 | 0.2421 | 0.1812 | 0.0921 |
| | 10.0 | 1.1148 | 1.5364 | 3.2476 |
| | 20.0 | 1.6007 | 2.5494 | 7.5528 |
| | $\infty$ | 1.6175 | 2.5871 | 7.7398 |

**Table 6.9** The convergence of the numerical results with respect to number of equations in Case 1 obtained under $\alpha = -0.5$, $\rho = 2.1$ and $\omega = 3.0$

| $\lvert p\rvert/E_0^{(1)}$ | Number of equations | | | | | | |
|---|---|---|---|---|---|---|---|
| | 40 | 46 | 52 | 58 | 64 | 70 | 76 |
| 0.1060 | 0.2927 | 0.3035 | 0.3089 | 0.3115 | 0.3127 | 0.3134 | 0.3137 |
| 0.1080 | 0.1179 | 0.1215 | 0.1233 | 0.1242 | 0.1246 | 0.1248 | 0.1249 |
| 0.1010 | 0.0534 | 0.0549 | 0.0556 | 0.0560 | 0.0562 | 0.0562 | 0.0563 |
| 0.1120 | 0.0251 | 0.0259 | 0.0262 | 0.0264 | 0.0264 | 0.0265 | 0.0265 |

respectively under $\omega = 3.0$, $\rho = 2.1$ and $\alpha = -0.5$ for various $|p|/E_0^{(1)}$. It follows from these tables that the convergence of the solution method employed is highly effective.

**Table 6.10** The convergence of the numerical results with respect to number of equations in Case 2 obtained under $\alpha = -0.5$, $\rho = 2.1$ and $\omega = 3.0$

| $\|p\|/E_0^{(1)}$ | Number of equations | | | | |
|---|---|---|---|---|---|
| | 18 | 30 | 42 | 54 | 66 |
| 0.1023 | 0.0005 | 0.0017 | 0.0020 | 0.0021 | 0.0022 |
| 0.1030 | 0.0002 | 0.0010 | 0.0012 | 0.0013 | 0.0013 |
| 0.1037 | 0.0000 | 0.0005 | 0.0006 | 0.0007 | 0.0007 |
| 0.1044 | 0.0000 | 0.0002 | 0.0003 | 0.0003 | 0.0003 |

# 6.4 Internal Stability Loss of a Row of Unidirected Periodically Located Fibers in a Viscoelastic Matrix

## 6.4.1 Formulation of the Problem and Solution Method

We extended the problem formulation made in the previous section for the case where an infinite viscoelastic matrix contains periodically located row fibers (Fig. 6.7a). The position of the points is determined by the Lagrange coordinates in Cartesian $O_k x_{1k} x_{2k} x_{3k}$ and cylindrical $O_k r_k \theta_k z_k$ coordinates associated with the $k$th fiber. These coordinate systems are obtained from the system of coordinates $O_0 x_{10} x_{20} x_{30}$ and $O_0 r_0 \theta_0 z_0$ respectively, by parallel transfer along the $O_0 x_{20}$ axis (Fig. 6.7b).

The relations

$$x_{3j\pm} = x_{30} = x_3,\ x_{1j\pm} = x_{10} = x_1,\ x_{2j\pm} = x_{20} \mp jR_{12},\ x_2 = x_{20},\ j = 0, 1, 2, \ldots\infty,$$
$$x_{2j\pm} = x_{2k\pm} + (\mp j \mp k)R_{12},\ r_{j\pm}\exp(i\theta_{j\pm}) = r_0\exp(i\theta_0) \mp jR_{12}\exp(i\frac{\pi}{2}),$$
$$r_{j\pm}\exp(i\theta_{j\pm}) = r_{k\pm}\exp(i\theta_{\pm}) + (\mp j \mp k)R_{12}\exp(i\frac{\pi}{2}) \tag{6.70}$$

satisfy.

We select for consideration the case for which the following assumptions are satisfied:

i. The middle lines of the fibers are in the $x_{10} = 0$ plane.
ii. Fibers have the same initial infinitesimal imperfections in the periodically curving form.
iii. The location of these initial imperfections with respect to each other is "co-phase".

According to the foregoing assumptions, the equations of the middle lines of the fibers can be written as

$$x_{2k\pm} = \varepsilon f(x_3),\ x_1 = 0,\ f(x_3) = \ell\sin\alpha_1 x_3,\ \varepsilon = \frac{L}{\ell},\ \alpha_1 = \frac{2\pi}{\ell},\ L \ll \ell. \tag{6.71}$$

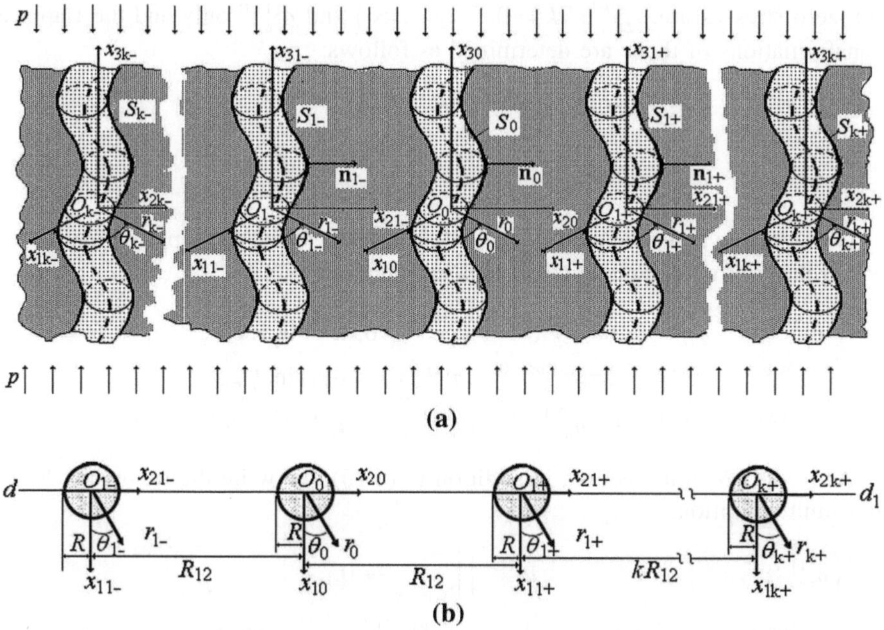

**Fig. 6.7** The geometry of periodically located row fibers with co-phase initial imperfections (**a**) and coordinate systems associated with these fibers (**b**)

The meaning of the parameters in (6.71) is detailed in the previous section. According to Fig. 6.7, the periodicity of the fibers location can be indicated as

$$\left\{\begin{matrix} u_i^{(1)} \\ \sigma_{ij}^{(1)} \end{matrix}\right\}(x_1, x_2 \pm jR_{12}, x_3) = \left\{\begin{matrix} u_i^{(1)} \\ \sigma_{ij}^{(1)} \end{matrix}\right\}(x_1, x_2, x_3). \tag{6.72}$$

Moreover, in the case considered the symmetry conditions

$$u_1^{(1)}(x_1, x_2, x_3) = -u_1^{(1)}(-x_1, x_2, x_3),\ u_2^{(1)}(x_1, x_2, x_3) = u_2^{(1)}(-x_1, x_2, x_3),$$
$$u_3^{(1)}(x_1, x_2, x_3) = u_3^{(1)}(-x_1, x_2, x_3), \tag{6.73}$$

takes place.

Thus, within the scope of the foregoing conditions we remake all solution procedures described in Sect. 6.1 for the problem considered. Note that under solution procedure the conditions (6.72) and (6.73) are satisfied for each approximation of the presentation (6.13) separately. In this case we obtain that the Laplace transformations of the values related to the zeroth approximation are determined through the Eqs. (6.17), (6.25), (6.26), (6.30) and (6.31) under $\eta^{(2)} = 0$ and $\eta^{(1)} = 1$. In other words, we obtain that in the zeroth approximation the non-zero stresses are $\sigma_{zz}^{(2)k\pm,0}(k=0,1,2,\ldots,\infty)$ and $\sigma_{zz}^{(1),0}$ only and the Laplace transformations of those are determined as follows:

$$\bar{\sigma}_{zz}^{(1),0}=\bar{p},\ \bar{\sigma}_{zz}^{(2)0+,0}=\bar{\sigma}_{zz}^{(2)1+,0}=\bar{\sigma}_{zz}^{(2)1-,0}=\ldots$$
$$=\bar{\sigma}_{zz}^{(2)k+,0}=\bar{\sigma}_{zz}^{(2)k-,0}=\ldots=\frac{\bar{E}_{(2)*}}{\bar{E}_{(1)*}}\bar{p}. \tag{6.74}$$

Consider the determination of the values related to the first approximation for which the following contact conditions are obtained.

$$\bar{\sigma}_{rr}^{(2)k\pm,1}-\bar{\sigma}_{rr}^{(1),1}=0,\ \bar{\sigma}_{r\theta}^{(2)k\pm,1}-\bar{\sigma}_{r\theta}^{(1),1}=0,\ k=0,1,2,\ldots,\infty,$$
$$\bar{\sigma}_{zz}^{(2)k\pm,1}-\bar{\sigma}_{zz}^{(1),1}=2\pi(\bar{\sigma}_{zz}^{(2)k\pm,0}-\bar{\sigma}_{zz}^{(1),0})\cos\alpha_1 t_3\sin\theta_{k\pm},$$
$$\bar{u}_{r}^{(2)k\pm,1}-\bar{u}_{r}^{(1),1}=0,\ \bar{u}_{\theta}^{(2)k\pm,1}-\bar{u}_{\theta}^{(1),1}=0,\ \bar{u}_{z}^{(2)k\pm,1}-\bar{u}_{z}^{(1),1}=0. \tag{6.75}$$

As in the previous section, in addition to (6.75), we write the following decay and limit conditions:

$$\left\{\bar{\sigma}_{rr}^{(1),1};\ldots;\bar{\sigma}_{rz}^{(1),1};\bar{u}_{r}^{(1),1};\ldots;\bar{u}_{z}^{(1),1}\right\}\Big|_{|x_1|\to\infty}\to 0,$$
$$\left\{\bar{\sigma}_{rr}^{(2)k\pm,1};\ldots;\bar{\sigma}_{rz}^{(2)k\pm,1};\bar{u}_{r}^{(2)k\pm,1};\ldots;\bar{u}_{z}^{(2)k\pm,1}\right\}\Big|_{|r_{k\pm}|\to 0}<M,\ M=const. \tag{6.76}$$

Moreover, according to the periodicity of the material structure we can write the relation:

$$\left\{\bar{\sigma}_{rr}^{(2)k\pm,1};\ldots;\bar{\sigma}_{rz}^{(2)k\pm,1};\bar{u}_{r}^{(2)k\pm,1};\ldots;\bar{u}_{z}^{(2)k\pm,1}\right\}$$
$$=\left\{\bar{\sigma}_{rr}^{(2)0,1};\ldots;\bar{\sigma}_{rz}^{(2)0,1};\bar{u}_{r}^{(2)0,1};\ldots;\bar{u}_{z}^{(2)0,1}\right\}. \tag{6.77}$$

Thus, the determination of the Laplace transformations related to the first approximation is reduced to the solution of the Eqs. (6.18), (6.20) and (6.21) in the framework of the conditions (6.75)–(6.77). For solution of this problem we use the presentations (6.32) and obtain the following solution to Eq. (6.33).

$$\Psi^{(2)k\pm}=\alpha_1\sin(\alpha_1 z)\sum_{n=-\infty}^{+\infty}C_n^{(2)k\pm}I_n(\xi_1^{(2)}\alpha_1 r_k)\exp(in\theta_{k\pm}),$$
$$\mathrm{X}^{(2)k\pm}=\cos(\alpha_1 z)\sum_{n=-\infty}^{+\infty}\Big[A_n^{(2)k\pm}I_n(\xi_2^{(2)}\alpha_1 r_k)$$
$$+B_n^{(2)k\pm}I_n(\xi_3^{(2)}\alpha_1 r_k)\Big]\exp(in\theta_{k\pm}), \tag{6.78}$$

$$\Psi^{(1)} = \alpha_1 \sin(\alpha_1 z)\left[\sum_{k=0}^{\infty}\sum_{n=-\infty}^{+\infty} C_n^{(1)k+} K_n(\xi_1^{(1)}\alpha_1 r_{k+})\exp(in\theta_{k+})\right.$$
$$\left. + \sum_{k=1}^{\infty}\sum_{n=-\infty}^{+\infty} C_n^{(1)k-} K_n(\xi_1^{(1)}\alpha_1 r_{k-})\exp(in\theta_{k-})\right],$$
$$X^{(1)} = \cos(\alpha_1 z)\left[\sum_{k=0}^{\infty}\sum_{n=-\infty}^{+\infty}\left[A_n^{(1)k+} K_n(\xi_2^{(1)}\alpha_1 r_{k+})\right.\right.$$
$$\left. + B_n^{(1)k+} K_n(\xi_3^{(1)}\alpha_1 r_{k+})\right]\exp(in\theta_{k+}) + \sum_{k=1}^{\infty}\sum_{n=-\infty}^{+\infty}\left[A_n^{(1)k-} K_n(\xi_2^{(1)}\alpha_1 r_{k-})\right.$$
$$\left.\left. + B_n^{(1)k-} K_n(\xi_3^{(1)}\alpha_1 r_{k-})\right]\exp(in\theta_{k-})\right], \tag{6.79}$$

where

$$A_{-n}^{(q)k\pm} = \overline{A_n^{(q)k\pm}},\ B_{-n}^{(q)k\pm} = \overline{B_n^{(q)k\pm}},\ C_{-n}^{(q)k\pm} = \overline{C_n^{(q)k\pm}},\ \mathrm{Im}\, A_0^{(q)k\pm} = 0,$$
$$\mathrm{Im}\, B_0^{(q)k\pm} = 0,\ \mathrm{Im} C_0^{(q)k\pm} = 0,\ q = 1,2,\ k = 0,1,2,\ldots,\infty. \tag{6.80}$$

According to the periodicity of the fibers location, we can write:

$$A_n^{(q)k\pm} = A_n^{(q)0},\ B_n^{(q)k\pm} = B_n^{(q)0},\ C_n^{(q)k\pm} = C_n^{(q)0}. \tag{6.81}$$

Taking (6.81) into account we can satisfy the contact conditions (6.75) on the $S_0$ surface only and find all unknowns. However, in this case we must rewrite the expressions (6.79) in the cylindrical coordinate system $O_0 r_0 \theta_0 z_0$. For this purpose we use the following summation theorem (Watson 1958) for the $K_n(x)$ functions:

$$K_n(cr_{k+})\exp(in\theta_{k+}) = \sum_{v=-\infty}^{+\infty}(-1)^n I_v(cr_0)K_{n-v}(ckR_{12})$$
$$\times \exp(i(n-v)\frac{3\pi}{2})\exp(iv\theta_0),$$
$$K_n(cr_{k-})\exp(in\theta_{k-}) = \sum_{v=-\infty}^{+\infty}(-1)^n I_v(cr_0)K_{n-v}(ckR_{12})$$
$$\times \exp(i(n-v)\frac{\pi}{2})\exp(iv\theta_0),$$
$$r_{k+}\exp(in\theta_{k+}) = r_0\exp(i\theta_0) + kR_{12}\exp(i\frac{3\pi}{2}),$$
$$r_{k-}\exp(in\theta_{k-}) = r_0\exp(i\theta_0) + kR_{12}\exp(i\frac{\pi}{2}),\ c = const,\ r_0 < R_{12}. \tag{6.82}$$

Substituting (6.82) in (6.79) and carrying out some transformations we obtain:

$$\Psi^{(1)}=\alpha_1\sin(\alpha_1 z)\sum_{n=-\infty}^{+\infty}\Big[C_n^{(1)0}K_n(\xi_1^{(1)}\alpha_1 r_0)+I_n(\xi_1^{(1)}\alpha_1 r_0)$$
$$\times\sum_{\nu=-\infty}^{\infty}(-1)^{\nu}C_\nu^{(1),0}\left(\exp(i(\nu-n)\frac{3\pi}{2})+\exp(i(\nu-n)\frac{3\pi}{2})\right)$$
$$\times\sum_{k=1}^{\infty}K_{\nu-n}(\xi_1^{(1)}\alpha_1 kR_{12})\Big]\exp(in\theta_0)$$
$$X^{(1)}=\cos(\alpha_1 z)\sum_{n=-\infty}^{+\infty}\Big[A_n^{(1)0}K_n(\xi_2^{(1)}\alpha_1 r_0)+I_n(\xi_2^{(1)}\alpha_1 r_0)$$
$$\times\sum_{\nu=-\infty}^{\infty}(-1)^{\nu}A_\nu^{(1),0}\left(\exp(i(\nu-n)\frac{3\pi}{2})+\exp(i(\nu-n)\frac{3\pi}{2})\right)\sum_{k=1}^{\infty}K_{\nu-n}(\xi_2^{(1)}\alpha_1 kR_{12})$$
$$+B_n^{(1)0}K_n(\xi_3^{(1)}\alpha_1 r_0)+I_n(\xi_3^{(1)}\alpha_1 r_0)\sum_{\nu=-\infty}^{\infty}(-1)^{\nu}B_\nu^{(1),0}\left(\exp(i(\nu-n)\frac{3\pi}{2})\right.$$
$$\left.+\exp(i(\nu-n)\frac{\pi}{2})\right)\sum_{k=1}^{\infty}K_{\nu-n}(\xi_3^{(1)}\alpha_1 kR_{12})\Big]\exp(in\theta_0). \tag{6.83}$$

Thus, substituting the expressions (6.78), (6.83) in (6.32) and using (6.23), (6.24) and (6.18) we determine the stresses and displacement for the matrix and for the *O*th fiber and introducing the notation for unknown constants as in (6.60) under $m = 0$, from the symmetry conditions (6.73) we obtain:

$$z_{n,1}^{(q)0}=z_{n,2}^{(q)0}=z_{n,3}^{(q)0}=0,\ q=1,2,\ n=-\infty,\ldots,0,\ldots,\infty. \tag{6.84}$$

Moreover, from the contact conditions (6.75) we obtain

$$Y_n^{(1)0}+\sum_{\nu=0}^{\infty}Y_\nu^{(1)0}\left(\sum_{k=1}^{\infty}F_{n\nu k}^{(1)0}\right)+Y_n^{(2)0}F_n^{(2)0}=2\pi\delta_n^3(\sigma_{zz}^{(1),0}-\sigma_{zz}^{(2),0}),\quad n=0,1,2,\ldots,\infty \tag{6.85}$$

where $\delta_n^3$ is determined as in (6.61). We omit the details of the expressions for $F_{n\nu k}^{(1)0}$.

As in the previous section, it can be proven that the type of the determinant of the infinite system of the linear algebraic Eqs. (6.85) is normal one (Kantarovich and Krilov 1962) and this infinite system can be replaced with the following finite system.

$$Y_n^{(1)0}+\sum_{\nu=0}^{N_\nu}Y_\nu^{(1)0}\left(\sum_{k=1}^{N_k}F_{n\nu k}^{(1)0}\right)+Y_n^{(2)0}F_n^{(2)0}=2\pi\delta_n^3(\sigma_{zz}^{(1),0}-\sigma_{zz}^{(2),0}),\quad n=0,1,2,\ldots,N_\nu. \tag{6.86}$$

**Table 6.11** The values of $|p_{cr.0}|/E_0^{(1)}$ and $|p_{cr.\infty}|/E_0^{(1)}$ obtained for various $\rho$ and $\omega$ under $E^{(2)}/E_0^{(1)} = 50$ and $\chi = 0.3$

| $\rho$ | $\lvert p_{cr.0}\rvert/E_0^{(1)}$ | $\lvert p_{cr.\infty}\rvert/E_0^{(1)}$ | | | |
|---|---|---|---|---|---|
| | | $\omega = 0.5$ | $\omega = 1.0$ | $\omega = 2.0$ | $\omega = 3.0$ |
| 2.1 | 0.1154 | 0.0527 | 0.0685 | 0.0842 | 0.0920 |
| 2.2 | 0.1017 | 0.0472 | 0.0606 | 0.0742 | 0.0810 |
| 2.5 | 0.0922 | 0.0437 | 0.0555 | 0.0675 | 0.0736 |
| 3.0 | 0.0931 | 0.0440 | 0.0559 | 0.0681 | 0.0742 |
| 4.0 | 0.1027 | 0.0474 | 0.0610 | 0.0747 | 0.0816 |
| 5.0 | 0.1128 | 0.0512 | 0.0664 | 0.0817 | 0.0895 |
| 10.0 | 0.1452 | 0.0639 | 0.0846 | 0.1050 | 0.1151 |
| $\infty$ | 0.1626 | 0.0704 | 0.0939 | 0.1170 | 0.1285 |

The values of $N_v$ and $N_k$ in Eq. (6.86) are determined from the convergence criterion of the numerical results.

Thus, in this way, we determine the Laplace transformations of the sought values related to the case under consideration. After this determination we can also find the originals of the stresses and displacements by employing the inverse transformation method (Schapery 1962, 1978).

Note that the foregoing solution method can be employed simultaneously both for the determination of the self-balanced stresses caused by the initial imperfections of the fibers and for the determination of the critical time or critical force from the criterion

$$\max_{\substack{\theta_0 \in [0,\pi] \\ z_0 \in [0,\ell]}} \left| u_r^{(2)0,1} \right| \to \infty. \tag{6.87}$$

This completes the solution method for the problem under consideration.

### 6.4.2 Numerical Results and Discussions

Within the scope of the notation and assumptions which are accepted in Sects. 6.2.2 and 6.3.2, first we consider numerical results related to $p_{cr.0}\big/E_0^{(1)}$ and $p_{cr.\infty}\big/E_0^{(1)}$ which are determined according to the criterion (6.87).

These results are given in Table 6.11 for various values of $\rho$ $(= R_{12}/R)$ and $\omega$ under $E^{(2)}\big/E_0^{(1)} = 50$, $\chi(= 2\pi R/\ell) = 0.3$. Note that these results coincide with corresponding ones obtained in the paper (Babaev et al. 1985).

It follows from Table 6.11 that the dependence among $p_{cr.0}\Big/E_0^{(1)}$, $p_{cr.\infty}\Big/E_0^{(1)}$ and $\rho$ is non-monotonic for the case under consideration. Nevertheless, the interaction between the fibers causes to decrease the values of $p_{cr.0}\Big/E_0^{(1)}$ and $p_{cr.\infty}\Big/E_0^{(1)}$. Note that the case where $\rho = \infty$ corresponds the stability loss of a single fiber in the infinite matrix.

Within the restriction (6.45) we consider numerical results obtained for $t'_{cr}$. These results are given in Table 6.12 and obtained for various $p\Big/E_0^{(1)}$, $\rho$ and $\omega$ under $E^{(2)}\Big/E_0^{(1)} = 50$, $\chi = 0.3$ and $\alpha = -0.5$.

Table 6.12 shows that an interaction between the fibers causes to decrease of the values of $t'_{cr}$. But dependence between $t'_{cr}$ and $\rho$ is non-monotonic for the in-plane co-phase stability loss mode of the fibers.

Note that the foregoing numerical results are obtained in the case where $N_v = 130$ and $N_k = 17$ in Eq. (6.86). The order of the difference between these and corresponding results obtained for $N_v = 124$ and $N_k = 16$ is not more than $10^{-4}$.

## 6.5 Stability Loss of a Micro-Fiber in an Elastic and a Viscoelastic Matrix Near the Free Convex Cylindrical Surface

This section basis on papers (Akbarov and Mamedov 2009, 2011). Therefore, we will follow the notation and the style of consideration used in these papers.

### *6.5.1 Formulation of the Problem*

We consider a cylinder with infinite length and assume that the cross section of this cylinder is a circle the radius of which is $R$ (Fig. 6.8). Assume that this cylindrical body contains a fiber which has an initial infinitesimal imperfection in the periodical curving form along its length. We suppose that the fiber's cross section, which is perpendicular to its middle line tangent vector, is a circle with the constant radius $R_0$. In the natural state, we associate Lagrangian cylindrical system of coordinates $Or\theta z(O_0 r_0 \theta_0 z_0)$ and Cartesian system of coordinates $Ox_1x_2x_3$ $(O_0x_{10}x_{20}x_{30})$ with cylinder (with the fiber). Between these coordinates the following relations are satisfied.

$$x_2 = x_{20},\ x_3 = x_{30},\ z = z_0,\ r\exp(i\theta) = R_{10} + r_0\exp(i\theta_0). \tag{6.88}$$

**Table 6.12** The values of $t'_{cr}$ obtained for various $|p|/E_0^{(1)}$ $\rho$ and $\omega$ under $E^{(2)}/E_0^{(1)} = 50$, $\chi = 0.3$ and $\alpha = -0.5$

| $\lvert p\rvert/E_0^{(1)}$ | $\rho$ | $\omega$ | | | |
|---|---|---|---|---|---|
| | | 0.5 | 1.0 | 2.0 | 3.0 |
| 0.0860 | 2.1 | 0.1782 | 0.3621 | 16.3100 | $\infty$ |
| | 2.2 | 0.0350 | 0.0465 | 0.0962 | 0.3055 |
| | 2.5 | 0.0045 | 0.0049 | 0.0061 | 0.0077 |
| | 3.0 | 0.0060 | 0.0067 | 0.0086 | 0.0114 |
| | 4.0 | 0.0369 | 0.0536 | 0.1187 | 0.4516 |
| | 5.0 | 0.1283 | 0.2301 | 2.2258 | $\infty$ |
| | $\infty$ | 5.6900 | $\infty$ | $\infty$ | $\infty$ |
| 0.0875 | 2.1 | 0.1469 | 0.2764 | 4.2000 | $\infty$ |
| | 2.2 | 0.0265 | 0.0339 | 0.0619 | 0.1476 |
| | 2.5 | 0.0024 | 0.0026 | 0.0030 | 0.0035 |
| | 3.0 | 0.0035 | 0.0038 | 0.0045 | 0.0056 |
| | 4.0 | 0.0304 | 0.0396 | 0.0766 | 0.2072 |
| | 5.0 | 0.1050 | 0.1769 | 1.0776 | $\infty$ |
| | $\infty$ | 4.5500 | $\infty$ | $\infty$ | $\infty$ |
| 0.0890 | 2.1 | 0.1208 | 0.2124 | 1.7536 | $\infty$ |
| | 2.2 | 0.0197 | 0.0243 | 0.0400 | 0.0778 |
| | 2.5 | 0.0010 | 0.0011 | 0.0012 | 0.0013 |
| | 3.0 | 0.0017 | 0.0018 | 0.0021 | 0.0024 |
| | 4.0 | 0.0229 | 0.0288 | 0.0499 | 0.1068 |
| | 5.0 | 0.0857 | 0.1364 | 0.5980 | $\infty$ |
| | $\infty$ | 3.6900 | $\infty$ | $\infty$ | $\infty$ |
| 0.0905 | 2.1 | 0.0991 | 0.1640 | 0.8987 | $\infty$ |
| | 2.2 | 0.0142 | 0.0170 | 0.0256 | 0.0429 |
| | 2.5 | 0.0002 | 0.0003 | 0.0003 | 0.0003 |
| | 3.0 | 0.0006 | 0.0007 | 0.0007 | 0.0008 |
| | 4.0 | 0.0170 | 0.0206 | 0.0324 | 0.0585 |
| | 5.0 | 0.0696 | 0.1053 | 0.3597 | 15.6000 |
| | $\infty$ | 3.0300 | $\infty$ | $\infty$ | $\infty$ |

The middle line of the fiber is given by the equation

$$x_{30} = t_3,\ x_{10} = L\sin(\alpha_1 t_3)\cos\beta,\ x_{20} = L\sin(\alpha_1 t_3)\sin\beta, \tag{6.89}$$

where $t_3$ is a parameter and $t_3 \in (-\infty, +\infty)$, $\alpha_1 = 2\pi/\ell$, $L$ is the amplitude of the periodic curving form, $\ell$ is the period of the curving form and $\beta$ is an angle between the plane $O_0x_{10}x_{30}$ and the plane on which the middle line of the fiber lies. Suppose that $L \ll \ell$, we introduce the dimensionless small parameter $\varepsilon = L/\ell$.

In the discussion below, the values related to the cylinder and the fiber will be denoted by upper indices (1) and (0) respectively.

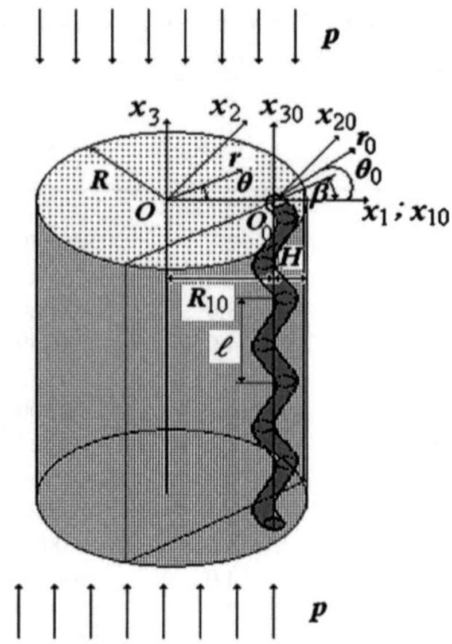


**Fig. 6.8** The geometry of the location of the fiber in a cylinder

We assume that the fiber and the surrounding cylinder material are isotropic, homogeneous and linearly hereditary. Within the scope of the piecewise homogeneous body model, through the use of the three-dimensional geometrically non-linear exact equations of the theory of viscoelasticity, we investigate the evolution of the infinitesimal initial waviness of the fiber with time in the case where the body is loaded at infinity by uniformly distributed normal compressed forces with an intensity $p$ acting in the direction of the $Oz$ axis. For this purpose we write the field Eqs. (6.1)–(6.5) which are satisfied within the fiber and surrounding cylinder separately. Assume that on the interface between the fiber and the surrounding medium (this surface is denoted by $S_0$) the complete contact conditions

$$\left(t_{rr}^{(1)} n_{0r} + t_{\theta r}^{(1)} n_{0\theta} + t_{zr}^{(1)} n_{0z}\right)\Big|_{S_0} = \left(t_{rr}^{(0)} n_{0r} + t_{\theta r}^{(0)} n_{0\theta} + t_{zr}^{(0)} n_{0z}\right)\Big|_{S_0},$$
$$\left(t_{r\theta}^{(1)} n_{0r} + t_{\theta\theta}^{(1)} n_{0\theta} + t_{z\theta}^{(1)} n_{0z}\right)\Big|_{S_0} = \left(t_{r\theta}^{(0)} n_{0r} + t_{\theta\theta}^{(0)} n_{0\theta} + t_{z\theta}^{(0)} n_{0z}\right)\Big|_{S_0},$$
$$\left(t_{rz}^{(1)} n_{0r} + t_{\theta z}^{(1)} n_{0\theta} + t_{zz}^{(1)} n_{0z}\right)\Big|_{S_0} = \left(t_{rz}^{(0)} n_{0r} + t_{\theta z}^{(0)} n_{0\theta} + t_{zz}^{(0)} n_{0z}\right)\Big|_{S_0},$$
$$u_r^{(1)}\big|_{S_0} = u_r^{(0)}\big|_{S_0},\ u_\theta^{(1)}\Big|_{S_0} = u_\theta^{(0)}\Big|_{S_0},\ u_z^{(1)}\big|_{S_0} = u_z^{(0)}\big|_{S_0} \quad (6.90)$$

are satisfied, where $n_{0r}$, $n_{0\theta}$ and $n_{0z}$ are components of the unit normal vector to the surface $S_0$.

Moreover on the cylindrical free surface, i.e. at $r = R$ the following conditions are also satisfied.

$$t_{rr}^{(1)}\big|_{r=R} = 0,\ t_{r\theta}^{(1)}\Big|_{r=R} = 0,\ t_{rz}^{(1)}\big|_{r=R} = 0. \quad (6.91)$$

Thus, with this the formulation of the considered problem has been exhausted and it follows from this formulation that the evolution of the infinitesimal initial periodical imperfection of the fiber with a time for the fixed value of the initial compressed force $p$ (for the case where only one of the constituents' materials are viscoelastic) or with an initial compressed force $p$ (for the case where the constituents' materials are pure elastic) will be considered within the framework of the field Eqs. (6.1)–(6.5), contact conditions (6.90) and the boundary condition (6.91).

### 6.5.2 Method of Solution

According to the general remarks on the method of solution detailed in Sect. 6.1.2, first we derive the equation for the interface surface $S_0$. Using the condition of the fiber's cross section we can conclude that the coordinates of this surface must simultaneously satisfy the following equations:

$$\begin{aligned}
&\varepsilon f'(t_3)(x_{10}-\varepsilon f(t_3)\cos\beta)\cos\beta+\varepsilon f'(t_3)\\
&\times(x_{20}-\varepsilon f(t_3)\sin\beta)\sin\beta+x_{30}-t_3=0,\\
&(-x_{10}\sin\beta+x_{20}\cos\beta)^2+(x_{30}-t_3)^2+(x_{10}\cos\beta\\
&+x_{20}\sin\beta-\varepsilon f(t_3))^2=R_0^2,
\end{aligned}\tag{6.92}$$

where $f(t_3)=\ell\sin(\alpha_1 t_3)$, $f'(t_3)=2\pi\cos(\alpha_1 t_3)$; $x_{10}$, $x_{20}$, $x_{30}$ are the coordinates of the surface $S_0$. Note that the first equation in (6.92) is an equation of the plane perpendicular to the vector which is the tangent vector to the middle line of the fiber at the point that corresponds to the fixed value of the parameter $t_3$; but the second equation in (6.92) is the equation of the circle which is counter to the cross section of the fiber which lies on the foregoing plane.

Using the relations $x_{10}=r_0\cos\theta_0$ and $x_{20}=r_0\sin\theta_0$, and the assumption $(\varepsilon f'(t_3))^2\ll 1$, we derive the following asymptotic expression for equation of the surface $S_0$ and for the components of the unit normal vector to this surface in the cylindrical system of coordinates $O_0r_0\theta_0z_0$:

$$\begin{aligned}
&r_0=R_0+\varepsilon f(t_3)\cos(\theta_0-\beta)+O(\varepsilon^2),\ z_0=t_3-\varepsilon R_0 f'(t_3)\cos(\theta_0-\beta)+O(\varepsilon^2),\\
&n_{0r}=1+O(\varepsilon^2),\ n_{0\theta}=-\varepsilon\frac{f(t_3)}{R_0}\sin(\theta_0-\beta)+O(\varepsilon^3),\\
&n_{0\theta}=-\varepsilon f'(t_3)\cos(\theta_0-\beta)+O(\varepsilon^3).
\end{aligned}\tag{6.93}$$

According to the series presentation (6.13), we obtain the field equations in (6.14) and (6.18) for the zeroth and the first approximations. The constitutive relations for these approximations are given through the Eqs. (6.20) and (6.21). In this case for the zeroth approximation we derive the following contact

$$\left(\sigma_{(ij)}^{(0),0}-\sigma_{(ij)}^{(1),0}\right)\Big|_{r_0=R_0}=0,\ \left(u_{(i)}^{(0),0}-u_{(i)}^{(1),0}\right)\Big|_{r_0=R_0}=0,\tag{6.94}$$

and boundary

$$\sigma_{(ij)}^{(1),0}\Big|_{r=R} = 0 \tag{6.95}$$

conditions, where $(ij) = rr; r\theta; rz$, $(i) = r; \theta; z$. As well as, for the first approximation we derive the contact

$$\left(\sigma_{rr}^{(0),1} - \sigma_{rr}^{(1),1}\right)\Big|_{r_0=R_0} = 0, \ \left(\sigma_{r\theta}^{(0),1} - \sigma_{r\theta}^{(1),1}\right)\Big|_{r_0=R_0} = 0,$$
$$\left(\sigma_{rz}^{(0),1} - \sigma_{rz}^{(1),1}\right)\Big|_{r_0=R_0} = f'(t_3)(\sigma_{zz}^{(0),0} - \sigma_{zz}^{(1),0})(\cos\theta_0\cos\beta + \sin\theta_0\sin\beta),$$
$$\left(u_{(i)}^{(0),1} - u_{(i)}^{(1),1}\right)\Big|_{r_0=R_0} = 0 \tag{6.96}$$

and the boundary

$$\sigma_{(ij)}^{(1),1}\Big|_{r=R} = 0 \tag{6.97}$$

conditions.

Now we determine the unknown values belonging to these approximations. For generality of the consideration we assume that the materials of both the cylinder and the fiber are viscoelastic ones with operators $E^{(1)*}$, $\nu^{(1)*}$ (for the cylinder's material) and $E^{(0)*}$, $\nu^{(0)*}$ (for the fiber's material) and these operators are determined through the expressions (6.26) and (6.28).

Thus, after accepting the foregoing assumptions we employ the Laplace transformation (6.22) to the all equations and relations related to the zeroth and the first approximations. As in the previous sections we suppose that $\bar{\nu}^{(1)*} = \bar{\nu}^{(0)*}$ and the Laplace transformations of the values of the zeroth approximation are determined as follows:

$$\bar{\sigma}_{zz}^{(1),0} = \bar{p}\left[1 + \frac{R_0^2}{R^2}\left(\frac{\bar{E}^{(0)*}}{\bar{E}^{(1)*}} - 1\right)\right], \ \bar{\sigma}_{zz}^{(0),0} = \bar{\sigma}_{zz}^{(1),0}\frac{\bar{E}^{(0)*}}{\bar{E}^{(1)*}},$$
$$\bar{\varepsilon}_{zz}^{(0),0} = \bar{\varepsilon}_{zz}^{(1),0} = \frac{\bar{\sigma}_{zz}^{(0),0}}{\bar{E}^{(0)*}}, \ \bar{u}_z^{(1),0} = \bar{u}_z^{(0),0} = \bar{\varepsilon}_{zz}^{(0),0} z. \tag{6.98}$$

Other stresses and strains of the zeroth approximation, which are not shown in (6.98), are equal to zero.

Consider the determination of the Laplace transformation of the first approximation. We use the presentations in (6.32) and taking the expression of the right side of the contact condition (6.96) into account the solution to the equations in (6.33) we find as follows:

$$\bar{\psi}^{(0)} = \alpha_1 \sin \alpha_1 z \sum_{n=-\infty}^{+\infty} C_{n1}^{(0)}(s) I_n\left(\xi_1^{(0)}(s)\alpha_1 r_0\right) e^{in\theta_0},$$

$$\bar{\chi}^{(0)} = \cos \alpha_1 z \sum_{n=-\infty}^{+\infty} \left[C_{n2}^{(0)}(s) I_n\left(\xi_2^{(0)}(s)\alpha_1 r_0\right) + C_{n3}^{(0)}(s) I_n\left(\xi_3^{(0)}(s)\alpha r_0\right)\right] e^{in\theta_0}, \tag{6.99}$$

$$\bar{\psi}^{(1)} = \alpha_1 \sin \alpha_1 z \sum_{n=-\infty}^{+\infty} \left[B_{n1}^{(1)}(s) I_n\left(\xi_1^{(1)}(s)\alpha_1 r\right) e^{in\theta} + C_{n1}^{(1)}(s) K_n\left(\xi_1^{(1)}(s)\alpha_1 r_0\right) e^{in\theta_0}\right],$$

$$\bar{\chi}^{(1)} = \cos \alpha_1 z \sum_{n=-\infty}^{+\infty} \left\{\left[B_{n2}^{(1)}(s) I_n\left(\xi_2^{(1)}(s)\alpha_1 r\right) + B_{n3}^{(1)}(s) I_n\left(\xi_3^{(1)}(s)\alpha_1 r\right)\right] e^{in\theta} + \left[C_{n2}^{(1)}(s) K_n\left(\xi_2^{(1)}(s)\alpha_1 r_0\right) + C_{n3}^{(1)}(s) K_n\left(\xi_3^{(1)}(s)\alpha_1 r_0\right)\right] e^{in\theta_0}\right\}, \tag{6.100}$$

where $\alpha_1 = 2\pi/\ell$, $I_n(x)$ and $K_n(x)$ are the Bessel function of a purely imaginary argument and Macdonald function, in turn. Moreover, the unknowns $C_{ni}^{(k)}$, $B_{ni}^{(k)}$ $(i = 1, 2, 3;\ k = 0, 1)$ are complex numbers and satisfy the relations:

$$C_{ni}^{(k)} = \overline{C_{-ni}^{(k)}},\ B_{ni}^{(k)} = \overline{B_{-ni}^{(k)}},\ \operatorname{Im} C_{01}^{(k)} = \operatorname{Im} B_{01}^{(k)} = 0. \tag{6.101}$$

Thus, using the solution in (6.99) and (6.100) we obtain the expressions for the sought values from (6.32), (6.18), (6.23) and (6.24). Note that these expressions contain the unknown constants (6.101). For the determination of these unknowns we must obtain the corresponding algebraic system of equations from the contact conditions (6.96) and boundary conditions (6.97). In this case, under the satisfaction of the contact conditions (6.96), all the expressions mentioned must be written in the coordinates $(r_0, \theta_0)$. In other words, the terms in these expressions attained from the terms $B_{nk}^{(1)} I_n(\xi_k^{(1)} \alpha_1 r) \exp(in\theta)$ must be written in the coordinates $(r_0, \theta_0)$. For this purpose, we use the summation theorem (Watson 1958) for $I_n(x)$ function, which can be written for the case at hand as follows.

$$I_\nu(cr) e^{i\nu\theta} = \sum_{m=-\infty}^{+\infty} I_{\nu-m}(cR_{10}) I_m(cr_0) e^{im\theta_0},\ c = const. \tag{6.102}$$

According to (Watson 1958), in order to employ the expression (6.102) to satisfy contact conditions (6.96) at $r_0 = R_0$, the inequality

$$R_0 < R_{10} \tag{6.103}$$

must occur.

Thus, as the inequality (6.103) satisfies for the problem considered (Fig. 6.8) using the summation theorem (6.102), after some mathematical manipulations the functions in (6.100) can be written as follows:

$$
\begin{aligned}
\bar{\psi}^{(1)} =& \alpha_1 \sin\alpha_1 z \left[ \sum_{n=0}^{+\infty} \varepsilon_n Y_{n1}^{(1)}(s) K_n\left(\xi_1^{(1)}(s)\alpha_1 r_0\right) \cos n\theta_0 \right. \\
&+ \sum_{n=1}^{+\infty} X_{n1}^{(1)}(s) K_n\left(\xi_1^{(1)}(s)\alpha_1 r_0\right) \sin n\theta_0 \\
&+ \sum_{n=0}^{+\infty} \left[ \sum_{m=0}^{+\infty} \varepsilon_n Z_{m1}^{(1)}(s) \lambda_{nm}^{+}\left(\xi_1^{(1)}(s)\alpha_1 R_{10}\right) I_m\left(\xi_1^{(1)}(s)\alpha_1 r_0\right) \right] \cos n\theta_0 \\
&+ \left. \sum_{n=1}^{+\infty} \left[ \sum_{m=1}^{+\infty} T_{m1}^{(1)}(s) \lambda_{nm}^{-}\left(\xi_1^{(1)}(s)\alpha_1 R_{10}\right) I_m\left(\xi_1^{(1)}(s)\alpha_1 r_0\right) \right] \sin n\theta_0 \right\}, \\
\bar{\chi}^{(1),1} =& \cos\alpha_1 z \left\{ \sum_{n=0}^{+\infty} \varepsilon_n Y_{n2}^{(1)}(s) K_n\left(\xi_2^{(1)}(s)\alpha_1 r_0\right) \cos n\theta_0 \right. \\
&+ \sum_{n=1}^{+\infty} X_{n2}^{(1)}(s) K_n\left(\xi_2^{(1)}(s)\alpha_1 r_0\right) \sin n\theta_0 \\
&+ \sum_{n=0}^{+\infty} \varepsilon_n Y_{n3}^{(1)}(s) K_n\left(\xi_3^{(1)}(s)\alpha_1 r_0\right) \cos n\theta_0 + \sum_{n=1}^{+\infty} X_{n3}^{(1)}(s) K_n\left(\xi_3^{(1)}(s)\alpha_1 r_0\right) \sin n\theta_0 \\
&+ \sum_{n=0}^{+\infty} \left[ \sum_{m=0}^{+\infty} \varepsilon_n Z_{m2}^{(1)}(s) \lambda_{nm}^{+}\left(\xi_2^{(1)}(s)\alpha_1 R_{10}\right) I_m\left(\xi_2^{(1)}(s)\alpha_1 r_0\right) \right] \cos n\theta_0 \\
&+ \sum_{n=0}^{+\infty} \left[ \sum_{m=0}^{+\infty} \varepsilon_n Z_{m3}^{(1)}(s) \lambda_{nm}^{+}\left(\xi_3^{(1)}(s)\alpha_1 R_{10}\right) I_m\left(\xi_3^{(1)}(s)\alpha_1 r_0\right) \right] \cos n\theta_0 \\
&+ \sum_{n=1}^{+\infty} \left[ \sum_{m=1}^{+\infty} T_{m2}^{(1)}(s) \lambda_{nm}^{-}\left(\xi_2^{(1)}(s)\alpha_1 R_{10}\right) I_m\left(\xi_2^{(1)}(s)\alpha_1 r_0\right) \right] \sin n\theta_0 \\
&+ \left. \sum_{n=1}^{+\infty} \left[ \sum_{m=1}^{+\infty} T_{m3}^{(1)}(s) \lambda_{nm}^{-}\left(\xi_3^{(1)}(s)\alpha_1 R_{10}\right) I_m\left(\xi_3^{(1)}(s)\alpha_1 r_0\right) \right] \sin n\theta_0 \right\}.
\end{aligned}
\tag{6.104}
$$

where

$$
\begin{gathered}
\varepsilon_0 = \frac{1}{2}, \ \varepsilon_n = 1.0 \ \text{ for } \ n \geq 1, \ \lambda_{nm}^{\pm}(x) = I_{n+m}(x) \pm I_{n-m}(x), \\
Y_{nj}^{(1)} = 2\,\mathrm{Re}\, C_{nj}^{(1)}, \ Z_{nj}^{(1)} = 2\,\mathrm{Re}\, B_{nj}^{(1)}, \ X_{nj}^{(1)} = -2\,\mathrm{Im}\, C_{nj}^{(1)}, \ T_{nj}^{(1)} = -2\,\mathrm{Im}\, B_{nj}^{(1)}.
\end{gathered}
\tag{6.105}
$$

Thus, from Eqs. (6.104), (6.32), (6.18), (6.23) and (6.24) we obtain the expressions for the Laplace transformation of the stresses and displacements of the surrounding cylinder and fiber. These expressions can be presented as follows.

$$\left\{\bar{\sigma}_{rr}^{(1),1};\bar{\sigma}_{r\theta}^{(1),1};\bar{\sigma}_{rz}^{(1),1};\bar{u}_{r}^{(1),1};\bar{u}_{\theta}^{(1),1};\bar{u}_{z}^{(1),1}\right\}^T$$
$$=\{\sin\alpha_1 z;\sin\alpha_1 z;\cos\alpha_1 z;\sin\alpha_1 z;\sin\alpha_1 z;\cos\alpha_1 z\}^T$$
$$\times\left\{\sum_{n=0}^{+\infty}\left\{\bar{\sigma}_{rrc}^{(1),1}(s,n);\bar{\sigma}_{r\theta c}^{(1),1}(s,n);\bar{\sigma}_{rzc}^{(1),1}(s,n);\right.\right.$$
$$\left.\bar{u}_{rc}^{(1),1}(s,n);\bar{u}_{\theta c}^{(1),1}(s,n);\bar{u}_{zc}^{(1),1}(s,n)\right\}^T\cos n\theta_0$$
$$+\sum_{n=1}^{+\infty}\left\{\bar{\sigma}_{rrs}^{(1),1}(s,n);\bar{\sigma}_{r\theta s}^{(1),1}(s,n);\bar{\sigma}_{rzs}^{(1),1}(s,n);\right.$$
$$\left.\left.\bar{u}_{rs}^{(1),1}(s,n);\bar{u}_{\theta s}^{(1),1}(s,n);\bar{u}_{zs}^{(1),1}(s,n)\right\}^T\sin n\theta_0\right\} \tag{6.106}$$

$$\left\{\bar{\sigma}_{rr}^{(0),1};\bar{\sigma}_{r\theta}^{(0),1};\bar{\sigma}_{rz}^{(0),1};\bar{u}_{r}^{(0),1};\bar{u}_{\theta}^{(0),1};\bar{u}_{z}^{(0),1}\right\}^T$$
$$=\{\sin\alpha_1 z;\sin\alpha_1 z;\cos\alpha_1 z;\sin\alpha_1 z;\sin\alpha_1 z;\cos\alpha_1 z\}^T$$
$$\times\left\{\sum_{n=0}^{+\infty}\left\{\bar{\sigma}_{rrc}^{(0),1}(n);\bar{\sigma}_{r\theta c}^{(0),1}(n);\bar{\sigma}_{rzc}^{(0),1}(n);\bar{u}_{rc}^{(0),1}(n);\bar{u}_{\theta c}^{(0),1}(n);\bar{u}_{zc}^{(0),1}(n)\right\}^T\cos n\theta_0\right.$$
$$\left.+\sum_{n=1}^{+\infty}\left\{\bar{\sigma}_{rrs}^{(0),1}(n);\bar{\sigma}_{r\theta s}^{(0),1}(n);\bar{\sigma}_{rzs}^{(0),1}(n);\bar{u}_{rs}^{(0),1}(n);\bar{u}_{\theta s}^{(0),1}(n);\bar{u}_{zs}^{(0),1}(n)\right\}^T\sin n\theta_0\right\} \tag{6.107}$$

Where

$$\bar{\sigma}_{rrc}^{(0),1}(n)=X_{n1}^{(0),1}a_{11n}^{(0)}\left(\xi_1^{(0)}\alpha_1 r_0\right)+Y_{n2}^{(0),1}a_{12n}^{(0)}\left(\xi_2^{(0)}\alpha_1 r_0\right)$$
$$+Y_{n3}^{(0),1}a_{13n}^{(0)}\left(\xi_3^{(0)}\alpha_1 r_0\right),$$
$$\bar{\sigma}_{rrs}^{(0),1}(n)=Y_{n1}^{(0),1}b_{11n}^{(0)}\left(\xi_1^{(0)}\alpha_1 r_0\right)+X_{n2}^{(0),1}b_{12n}^{(0)}\left(\xi_2^{(0)}\alpha_1 r_0\right)$$
$$+X_{n3}^{(0),1}b_{13n}^{(0)}\left(\xi_3^{(0)}\alpha_1 r_0\right),$$

$$\bar{\sigma}_{rrc}^{(1),1}(s,n)=X_{n1}^{(1)}a_{12n}^{(1)}\left(\xi_1^{(1)}\alpha_1 r_0\right)+\sum_{m=0}^{+\infty}T_{m1}^{(1)}\lambda_{nm}^{-}(\xi_1^{(1)}\alpha_1 R_{10})a_{11m}^{(1)}\left(\xi_1^{(1)}\alpha_1 r_0\right)$$
$$+Y_{n2}^{(1)}a_{14n}^{(1)}\left(\xi_2^{(1)}\alpha_1 r_0\right)+\sum_{m=0}^{+\infty}Z_{m2}^{(1)}\lambda_{nm}^{+}(\xi_2^{(1)}\alpha_1 R_{10})a_{13m}^{(1)}\left(\xi_2^{(1)}\alpha_1 r_0\right)$$
$$+Y_{n3}^{(1)}a_{16n}^{(1)}\left(\xi_3^{(1)}\alpha_1 r_0\right)+\sum_{m=0}^{+\infty}Z_{m3}^{(1)}\lambda_{nm}^{+}(\xi_3^{(1)}\alpha_1 R_{10})a_{15m}^{(1)}\left(\xi_3^{(1)}\alpha_1 r_0\right),$$

$$\bar{\sigma}_{rrs}^{(1),1}(s,n) = Y_{n1}^{(1)} b_{12n}^{(1)}\left(\xi_1^{(1)}\alpha_1 r_0\right) + \sum_{m=0}^{+\infty} Z_{m1}^{(1)} \lambda_{nm}^{+}(\xi_1^{(1)}\alpha_1 R_{10}) b_{11m}^{(1)}\left(\xi_1^{(1)}\alpha_1 r_0\right)$$
$$+X_{n2}^{(1)} b_{14n}^{(1)}\left(\xi_2^{(1)}\alpha_1 r_0\right) + \sum_{m=0}^{+\infty} T_{m2}^{(1)} \lambda_{nm}^{-}(\xi_2^{(1)}\alpha_1 R_{10}) b_{13m}^{(1)}\left(\xi_2^{(1)}\alpha_1 r_0\right)$$
$$+ X_{n3}^{(1)} b_{16n}^{(1)}\left(\xi_3^{(1)}\alpha_1 r_0\right) + \sum_{m=0}^{+\infty} T_{m3}^{(1)} \lambda_{nm}^{+}(\xi_3^{(1)}\alpha_1 R_{10}) b_{15m}^{(1)}\left(\xi_3^{(1)}\alpha_1 r_0\right). \quad (6.108)$$

The expressions for $\bar{\sigma}_{r\theta s}^{(1),1}(s,n)$, $\bar{\sigma}_{rzc}^{(1),1}(s,n)$, $\bar{u}_{rc}^{(1),1}(s,n)$, $\bar{u}_{\theta s}^{(1),1}(s,n)$ and $\bar{u}_{zc}^{(1),1}(s,n)$ (for $\bar{\sigma}_{r\theta c}^{(1),1}(s,n)$, $\bar{\sigma}_{rzs}^{(1),1}(s,n)$, $\bar{u}_{rs}^{(1),1}(s,n)$, $\bar{u}_{\theta c}^{(1),1}(s,n)$, $\bar{u}_{zs}^{(1),1}(s,n)$) are obtained from (6.104) by replacing $a_{1\bullet\bullet}^{(1)}(\bullet)$ $(b_{1\bullet\bullet}^{(1)}(\bullet))$ with $a_{2\bullet\bullet}^{(1)}(\bullet)$, $a_{3\bullet\bullet}^{(1)}(\bullet)$, $a_{4\bullet\bullet}^{(1)}(\bullet)$, $a_{5\bullet\bullet}^{(1)}(\bullet)$ and $a_{6\bullet\bullet}^{(1)}(\bullet)$ $(b_{2\bullet\bullet}^{(1)}(\bullet)$, $b_{3\bullet\bullet}^{(1)}(\bullet)$, $b_{4\bullet\bullet}^{(1)}(\bullet)$, $b_{5\bullet\bullet}^{(1)}(\bullet)$ and $b_{6\bullet\bullet}^{(1)}(\bullet))$, as well as the expressions for $\bar{\sigma}_{r\theta s}^{(0),1}(s,n)$, $\bar{\sigma}_{rzc}^{(0),1}(s,n)$, $\bar{u}_{rc}^{(0),1}(s,n)$, and $\bar{u}_{zc}^{(0),1}(s,n)$ (for $\bar{\sigma}_{r\theta c}^{(0),1}(s,n)$, $\bar{\sigma}_{rzs}^{(0),1}(s,n)$, $\bar{u}_{rs}^{(0),1}(s,n)$, $\bar{u}_{\theta c}^{(0),1}(s,n)$ and $\bar{u}_{zs}^{(0),1}(s,n)$) are obtained from (6.104) by replacing $a_{1\bullet\bullet}^{(0)}(\bullet)$ $(b_{1\bullet\bullet}^{(0)}(\bullet))$ with $a_{2\bullet\bullet}^{(0)}(\bullet)$, $a_{3\bullet\bullet}^{(0)}(\bullet)$, $a_{4\bullet\bullet}^{(0)}(\bullet)$, $a_{5\bullet\bullet}^{(0)}(\bullet)$ and $a_{6\bullet\bullet}^{(0)}(\bullet)$ $(b_{2\bullet\bullet}^{(0)}(\bullet)$, $b_{3\bullet\bullet}^{(0)}(\bullet)$, $b_{4\bullet\bullet}^{(0)}(\bullet)$, $b_{5\bullet\bullet}^{(0)}(\bullet)$ and $b_{6\bullet\bullet}^{(0)}(\bullet))$, respectively. The explicit expressions of the coefficients $a_{1\bullet\bullet}^{(k)}(\bullet)$ $a_{2\bullet\bullet}^{(k)}(\bullet)$, $a_{3\bullet\bullet}^{(k)}(\bullet)$, $a_{4\bullet\bullet}^{(k)}(\bullet)$, $a_{5\bullet\bullet}^{(k)}(\bullet)$, $a_{6\bullet\bullet}^{(k)}(\bullet)$, $b_{1\bullet\bullet}^{(k)}(\bullet)$, $b_{2\bullet\bullet}^{(k)}(\bullet)$, $b_{3\bullet\bullet}^{(k)}(\bullet)$, $b_{4\bullet\bullet}^{(k)}(\bullet)$, $b_{5\bullet\bullet}^{(k)}(\bullet)$ and $b_{6\bullet\bullet}^{(k)}(\bullet)$ are

$$a_{11n}^{(1)}\left(\xi_1^{(1)}x\right) = -b_{11n}^{(1)}\left(\xi_1^{(1)}x\right) = \left(\bar{\lambda}^{(1)*}2\bar{\mu}^{(1)*}\right)\left[(2n/x^2)I_n(\xi_1^{(1)}x)\right.$$
$$\left. -\left(n\xi_1^{(1)}/2x\right)\left(I_{n+1}(\xi_1^{(1)}x) + I_{n-1}(\xi_1^{(1)}x)\right)\right]$$
$$+ \bar{\lambda}^{(1)*}\left[n\xi_1^{(1)}\left(I_{n+1}(\xi_1^{(1)}x) + I_{n-1}(\xi_1^{(1)}x)\right) - (2n/x^2)I_n(\xi_1^{(1)}x)\right],$$
$$a_{12n}^{(1)}\left(\xi_1^{(1)}x\right) = -b_{12n}^{(1)}\left(\xi_1^{(1)}x\right) = \left(\bar{\lambda}^{(1)*} + 2\bar{\mu}^{(1)*}\right)$$
$$\times \left[(2n/x^2)K_n(\xi_1^{(1)}x) + \left(n\xi_1^{(1)}/2x\right)\left(K_{n+1}(\xi_1^{(1)}x) + K_{n-1}(\xi_1^{(1)}x)\right)\right]$$
$$+ \bar{\lambda}^{(1)*}\left[-n\xi_1^{(1)}\left(K_{n+1}(\xi_1^{(1)}x) + K_{n-1}(\xi_1^{(1)}x)\right) - (2n/x^2)K_n(\xi_1^{(1)}x)\right],$$

$$a_{13n}^{(1)}\left(\xi_2^{(1)}x\right) = b_{13n}^{(1)}\left(\xi_2^{(1)}x\right) = \left(\bar{\lambda}^{(1)*} + 2\bar{\mu}^{(1)*}\right)$$
$$\times \left[\varepsilon_n\left(\left(\xi_2^{(1)}\right)^2/2\right)\left(I_{n+2}(\xi_2^{(1)}x) + 2I_n(\xi_2^{(1)}x) + I_{n-2}(\xi_2^{(1)}x)\right)\right]$$
$$+ \bar{\lambda}^{(1)*}(2n^2/x)I_n(\xi_2^{(1)}x) + \bar{\lambda}^{(1)*}\left(\xi_2^{(1)}/x\right)\left(I_{n+1}(\xi_2^{(1)}x) + I_{n-1}(\xi_2^{(1)}x)\right)$$
$$- \bar{\lambda}^{(1)*}2c_2^{(1)}I_n(\xi_2^{(1)}x), a_{14n}^{(1)}\left(\xi_2^{(1)}x\right) = b_{14n}^{(1)}\left(\xi_2^{(1)}x\right) = \left(\bar{\lambda}^{(1)*} + 2\bar{\mu}^{(1)*}\right)$$

$$\times\left[\varepsilon_n\left(\xi_2^{(1)}\right)^2\left(K_{n+2}(\xi_2^{(1)}x)+2K_n(\xi_2^{(1)}x)+K_{n-2}(\xi_2^{(1)}x)\right)/2\right]$$
$$+\bar{\lambda}^{(1)*}(2n^2/x)K_n(\xi_2^{(1)}x)-\bar{\lambda}^{(1)*}\left(\xi_2^{(1)}/x\right)\left(K_{n+1}(\xi_2^{(1)}x)+K_{n-1}(\xi_2^{(1)}x)\right)$$
$$-\bar{\lambda}^{(1)*}2c_2^{(1)}K_n(\xi_2^{(1)}x),a_{15n}^{(1)}(\xi_3^{(1)}x)$$
$$=b_{15n}^{(1)}(\xi_3^{(1)}x)=a_{13n}^{(1)}(\xi_3^{(1)}x),a_{16n}^{(1)}(\xi_3^{(1)}x)=b_{16n}^{(1)}(\xi_3^{(1)}x)=a_{14n}^{(1)}(\xi_3^{(1)}x),$$

$$a_{11n}^{(0)}\left(\xi_1^{(0)}x\right)=-b_{11n}^{(0)}\left(\xi_1^{(0)}x\right)=\left(\bar{\lambda}^{(0)*}+2\bar{\mu}^{(0)*}\right)$$
$$\times\left[(2n/x^2)I_n(\xi_1^{(0)}x)-\left(n\xi_1^{(0)}/2x\right)\left(I_{n+1}(\xi_1^{(0)}x)+I_{n-1}(\xi_1^{(0)}x)\right)\right]$$
$$+\bar{\lambda}^{(0)*}\left[n\xi_1^{(0)}\left(I_{n+1}(\xi_1^{(0)}x)+I_{n-1}(\xi_1^{(0)}x)\right)-(2n/x^2)I_n(\xi_1^{(0)}x)\right],$$

$$a_{12n}^{(0)}\left(\xi_2^{(0)}x\right)=b_{12n}^{(0)}\left(\xi_2^{(0)}x\right)=\left(\bar{\lambda}^{(0)*}+2\bar{\mu}^{(0)*}\right)$$
$$\times\left[\varepsilon_n\left(\xi_2^{(0)}\right)^2\left(I_{n+2}(\xi_2^{(0)}x)+2I_n(\xi_2^{(0)}x)+I_{n-2}(\xi_2^{(0)}x)\right)/2\right]$$
$$+\bar{\lambda}^{(0)*}(2n^2/x)I_n(\xi_2^{(0)}x)$$
$$+\bar{\lambda}^{(0)*}\left(\xi_2^{(0)}/x\right)\left(I_{n+1}(\xi_2^{(0)}x)+I_{n-1}(\xi_2^{(0)}x)\right)$$
$$-\bar{\lambda}^{(0)*}2c_2^{(0)}\varepsilon_nI_n(\xi_2^{(0)}x);a_{13n}^{(0)}(\xi_3^{(0)}x)$$
$$=b_{13n}^{(0)}(\xi_3^{(0)}x)=a_{12n}^{(0)}(\xi_3^{(0)}x),$$

$$a_{21n}^{(1)}\left(\xi_1^{(1)}x\right)=b_{21n}^{(1)}\left(\xi_1^{(1)}x\right)=2\bar{\mu}^{(1)*}\left[\left(\xi_1^{(1)}\right)^2\left(I_{n+2}(\xi_1^{(1)}x)+2I_n(\xi_1^{(1)}x)+I_{n-2}(\xi_1^{(1)}x)\right)/2\right.$$
$$\left.+2(n^2/x^2)I_n(\xi_1^{(1)}x)-\left(\left(\xi_1^{(1)}\right)^2/x\right)\left(I_{n+1}(\xi_1^{(1)}x)+I_{n-1}(\xi_1^{(1)}x)\right)\right],$$

$$a_{22n}^{(1)}\left(\xi_1^{(1)}x\right)=b_{22n}^{(1)}\left(\xi_1^{(1)}x\right)=2\bar{\mu}^{(1)*}\left[\left(\xi_1^{(1)}\right)^2\left(K_{n+2}(\xi_1^{(1)}x)+2K_n(\xi_1^{(1)}x)+K_{n-2}(\xi_1^{(1)}x)\right)/2\right.$$
$$\left.+2(n^2/x^2)K_n(\xi_1^{(1)}x)+\left(\left(\xi_1^{(1)}\right)^2/x\right)\left(K_{n+1}(\xi_1^{(1)}x)+K_{n-1}(\xi_1^{(1)}x)\right)\right]$$

$$a_{23n}^{(1)}\left(\xi_2^{(1)}x\right)=-b_{23n}^{(1)}\left(\xi_2^{(1)}x\right)=2\bar{\mu}^{(1)*}\left[-(n/x)\xi_2^{(1)}\left(I_{n+1}(\xi_2^{(1)}x)+I_{n-1}(\xi_2^{(1)}x)\right)-(n/x)\xi_2^{(1)}\right.$$
$$\left.\times\left(I_{n+1}(\xi_2^{(1)}x)+I_{n-1}(\xi_2^{(1)}x)\right)-(2n/x^2)I_n(\xi_2^{(1)}x)\right],$$

$$a_{24n}^{(1)}\left(\xi_2^{(1)}x\right)=-b_{24n}^{(1)}\left(\xi_2^{(1)}x\right)=2\bar{\mu}^{(1)*}\left[(n/x)\xi_2^{(1)}\left(K_{n+1}(\xi_2^{(1)}x)+K_{n-1}(\xi_2^{(1)}x)\right)\right.$$
$$\left.+(n/x)\xi_2^{(1)}\left(K_{n+1}(\xi_2^{(1)}x)+K_{n-1}(\xi_2^{(1)}x)\right)-(2n/x^2)K_n(\xi_2^{(1)}x)\right],$$

$$a_{25n}^{(1)}(\xi_3^{(1)}x)=-b_{25n}^{(1)}(\xi_3^{(1)}x)=a_{23n}^{(1)}(\xi_3^{(1)}x);a_{26n}^{(1)}(\xi_3^{(1)}x)=-b_{26n}^{(1)}(\xi_3^{(1)}x)=a_{24n}^{(1)}(\xi_3^{(1)}x),$$

$$a_{21n}^{(0)}\left(\xi_1^{(0)}x\right) = b_{21n}^{(0)}\left(\xi_1^{(0)}x\right) = 2\bar{\mu}^{(0)*}\left[\left(\xi_1^{(0)}\right)^2\left(I_{n+2}(\xi_1^{(0)}x) + 2I_n(\xi_1^{(0)}x) + I_{n-2}(\xi_1^{(0)}x)\right)/2\right.$$

$$\left. + 2(n^2/x^2)I_n(\xi_1^{(0)}x) - \left(\left(\xi_1^{(0)}\right)^2/x\right)\left(I_{n+1}(\xi_1^{(0)}x) + I_{n-1}(\xi_1^{(0)}x)\right)\right],$$

$$a_{22n}^{(0)}\left(\xi_2^{(0)}x\right) = -b_{22n}^{(0)}\left(\xi_2^{(0)}x\right) = 2\bar{\mu}^{(0)*}\left[-(n/x)\xi_2^{(0)}\left(I_{n+1}(\xi_2^{(0)}x) + I_{n-1}(\xi_2^{(0)}x)\right) - (n/x)\xi_2^{(0)}\right.$$

$$\left.\times\left(I_{n+1}(\xi_2^{(0)}x) + I_{n-1}(\xi_2^{(0)}x)\right) - (2n/x^2)I_n(\xi_2^{(0)}x)\right],$$

$$a_{23n}^{(0)}(\xi_3^{(0)}x) = -b_{23n}^{(0)}(\xi_3^{(0)}x) = a_{22n}^{(0)}(\xi_3^{(0)}x),$$

$$a_{31n}^{(1)}\left(\xi_1^{(1)}x\right) = -b_{31n}^{(1)}\left(\xi_1^{(1)}x\right) = -2\bar{\mu}^{(1)*}(2n/x)I_n(\xi_1^{(1)}x),$$

$$a_{32n}^{(1)}\left(\xi_1^{(1)}x\right) = -b_{32n}^{(1)}\left(\xi_1^{(1)}x\right) = -2\bar{\mu}^{(1)*}(2n/x)K_n(\xi_1^{(1)}x),\ a_{33n}^{(1)}\left(\xi_2^{(1)}x\right) = b_{33n}^{(1)}\left(\xi_2^{(1)}x\right)$$

$$= 2\bar{\mu}^{(1)*}\left(1 + c_2^{(1)}\right)\xi_2^{(1)}\left(I_{n+1}(\xi_2^{(1)}x) + I_{n-1}(\xi_2^{(1)}x)\right),$$

$$a_{34n}^{(1)}\left(\xi_2^{(1)}x\right) = b_{34n}^{(1)}\left(\xi_2^{(1)}x\right) = -2\bar{\mu}^{(1)*}\left(1 + c_2^{(1)}\right)\xi_2^{(1)}\left(K_{n+1}(\xi_2^{(1)}x) + K_{n-1}(\xi_2^{(1)}x)\right),$$

$$a_{35n}^{(1)}(\xi_3^{(1)}x) = b_{35n}^{(1)}(\xi_3^{(1)}x) = a_{33n}^{(1)}(\xi_3^{(1)}x),$$

$$a_{36n}^{(1)}(\xi_3^{(1)}x) = b_{36n}^{(1)}(\xi_3^{(1)}x) = a_{34n}^{(1)}(\xi_3^{(1)}x),$$

$$a_{31n}^{(0)}\left(\xi_1^{(0)}x\right) = -b_{31n}^{(0)}\left(\xi_1^{(0)}x\right) = -2\bar{\mu}^{(0)*}(2n/x)I_n(\xi_1^{(0)}x),$$

$$a_{32n}^{(0)}\left(\xi_2^{(0)}x\right) = b_{32n}^{(0)}\left(\xi_2^{(0)}x\right) = \varepsilon_n 2\bar{\mu}^{(0)*}\xi_2^{(0)}\left(1 + c_2^{(0)}\right)\left(I_{n+1}(\xi_2^{(0)}x) + I_{n-1}(\xi_2^{(0)}x)\right),$$

$$a_{33n}^{(0)}(\xi_3^{(0)}x) = b_{33n}^{(0)}(\xi_3^{(0)}x) = a_{32n}^{(0)}(\xi_3^{(0)}x),$$

$$a_{41n}^{(1)}\left(\xi_1^{(1)}x\right) = -b_{41n}^{(1)}\left(\xi_1^{(1)}x\right) = -(2n/x)I_n(\xi_1^{(1)}x), a_{42n}^{(1)}\left(\xi_1^{(1)}x\right)$$

$$= -b_{42n}^{(1)}\left(\xi_1^{(1)}x\right) = -(2n/x)K_n(\xi_1^{(1)}x),$$

$$a_{43}^{(1)}\left(\xi_2^{(1)}x\right) = b_{43}^{(1)}\left(\xi_2^{(1)}x\right) = \xi_2^{(1)}\left(I_{n+1}(\xi_2^{(1)}x) + I_{n-1}(\xi_2^{(1)}x)\right),$$

$$a_{44n}^{(1)}\left(\xi_2^{(1)}x\right) = b_{44n}^{(1)}\left(\xi_2^{(1)}x\right) = -\xi_2^{(1)}\left(K_{n+1}(\xi_2^{(1)}x) + K_{n-1}(\xi_2^{(1)}x)\right),$$

$$a_{45n}^{(1)}\left(\xi_3^{(1)}x\right) = b_{45n}^{(1)}\left(\xi_3^{(1)}x\right) = a_{43n}^{(1)}\left(\xi_3^{(1)}x\right),$$

$$a_{46n}^{(1)}\left(\xi_3^{(1)}x\right) = b_{46n}^{(1)}\left(\xi_3^{(1)}x\right) = a_{44n}^{(1)}\left(\xi_3^{(1)}x\right),$$

$$a_{41n}^{(0)}\left(\xi_1^{(0)}x\right) = -b_{41n}^{(0)}\left(\xi_1^{(0)}x\right) = -(2n/x)I_n(\xi_1^{(0)}x),$$

$$a_{42n}^{(0)}\left(\xi_2^{(0)}x\right) = b_{42n}^{(0)}\left(\xi_2^{(0)}x\right) = \varepsilon_n\xi_2^{(0)}\left(I_{n+1}(\xi_2^{(0)}x) + I_{n-1}(\xi_2^{(0)}x)\right),$$

$$a_{43n}^{(0)}(\xi_3^{(0)}x) = b_{43n}^{(0)}(\xi_3^{(0)}x) = a_{42n}^{(0)}(\xi_3^{(0)}x),$$

$$a_{51n}^{(1)}\left(\xi_1^{(1)}x\right) = b_{51n}^{(1)}\left(\xi_1^{(1)}x\right) = \xi_1^{(1)}\left(I_{n+1}(\xi_1^{(1)}x) + I_{n-1}(\xi_1^{(1)}x)\right),$$

$$
\begin{aligned}
&a_{52n}^{(1)}\left(\xi_1^{(1)}x\right) = b_{52n}^{(1)}\left(\xi_1^{(1)}x\right) = -\xi_1^{(1)}\left(K_{n+1}(\xi_1^{(1)}x) + K_{n-1}(\xi_1^{(1)}x)\right),\\
&a_{53n}^{(1)}\left(\xi_2^{(1)}x\right) = -b_{53n}^{(1)}\left(\xi_2^{(1)}x\right) = -(2n/x)I_n(\xi_2^{(1)}x),\\
&a_{54n}^{(1)}\left(\xi_2^{(1)}x\right) = -b_{54n}^{(1)}\left(\xi_2^{(1)}x\right) = -(2n/x)K_n(\xi_2^{(1)}x),\\
&a_{55n}^{(1)}\left(\xi_3^{(1)}x\right) = -b_{55n}^{(1)}\left(\xi_3^{(1)}x\right) = a_{53n}^{(1)}\left(\xi_3^{(1)}x\right), a_{56n}^{(1)}\left(\xi_3^{(1)}x\right) = -b_{56n}^{(1)}\left(\xi_3^{(1)}x\right) = a_{54n}^{(1)}\left(\xi_3^{(1)}x\right),\\
&a_{51n}^{(0)}\left(\xi_1^{(0)}x\right) = b_{51n}^{(0)}\left(\xi_1^{(0)}x\right) = \xi_1^{(0)}\left(I_{n+1}(\xi_1^{(0)}x) + I_{n-1}(\xi_1^{(0)}x)\right),\\
&a_{52n}^{(0)}\left(\xi_2^{(0)}x\right) = -b_{52n}^{(0)}\left(\xi_2^{(0)}x\right) = -(2n/x)I_n(\xi_2^{(0)}x),\\
&a_{53n}^{(0)}(\xi_3^{(0)}x) = -b_{53n}^{(0)}(\xi_3^{(0)}x) = a_{52n}^{(0)}(\xi_3^{(0)}x), a_{61n}^{(1)}\left(\xi_1^{(1)}x\right) = b_{61n}^{(1)}\left(\xi_1^{(1)}x\right) = 0.0,\\
&a_{62n}^{(1)}\left(\xi_1^{(1)}x\right) = b_{62n}^{(1)}\left(\xi_1^{(1)}x\right) = 0.0, a_{63n}^{(1)}\left(\xi_2^{(1)}x\right) = b_{63n}^{(1)}\left(\xi_2^{(1)}x\right) = \varepsilon_n 2c_2^{(1)}I_n(\xi_2^{(1)}x),\\
&a_{64n}^{(1)}\left(\xi_2^{(1)}x\right) = b_{64n}^{(1)}\left(\xi_2^{(1)}x\right) = \varepsilon_n 2c_2^{(1)}K_n(\xi_2^{(1)}x),\\
&a_{65n}^{(1)}\left(\xi_3^{(1)}x\right) = b_{65n}^{(1)}\left(\xi_3^{(1)}x\right) = a_{63n}^{(1)}\left(\xi_3^{(1)}x\right), a_{66n}^{(1)}\left(\xi_3^{(1)}x\right) = b_{66n}^{(1)}\left(\xi_3^{(1)}x\right) = a_{64n}^{(1)}\left(\xi_3^{(1)}x\right),\\
&a_{61n}^{(0)}\left(\xi_1^{(0)}x\right) = b_{61n}^{(0)}\left(\xi_1^{(0)}x\right) = 0.0, a_{62n}^{(0)}\left(\xi_2^{(0)}x\right) = \varepsilon_n 2c_2^{(0)}I_n(\xi_2^{(0)}x),\\
&a_{63n}^{(0)}(\xi_3^{(0)}x) = a_{62n}^{(0)}(\xi_3^{(0)}x), \varepsilon_0 = 0.5, \varepsilon_n = 1.0 \mathit{for}\, n \geq 1, c_2^{(k)}\\
&\quad = \left[\left(\bar{\lambda}^{(k)*}/\bar{\mu}^{(k)*} + 2\right)\left(\xi_2^{(k)}\right)^2 - 1 - \bar{\sigma}_{33}^{(k),0}/\bar{\mu}^{(k)*}\right]\\
&\quad \times \bar{\mu}^{(k)*}/\left(\bar{\lambda}^{(k)*} + \bar{\mu}^{(k)*}\right), k = 0, 1.
\end{aligned}
\tag{6.109}
$$

Thus, we have prepared all the needed mathematical calculations to satisfy the contact conditions (6.96) at $r_0 = R_0$.

Now we consider the satisfaction of the boundary conditions in (6.97). For this purpose we must present the terms written in the coordinates $(r_0, \theta_0)$ which enter the expressions of $\sigma_{rr}^{(1),1}$, $\sigma_{r\theta}^{(1),1}$ and $\sigma_{rz}^{(1),1}$ in the system of coordinates $(r, \theta)$ . These expressions are presented in series form, the terms of which are $(un.\ const)_n \times K_{n+k}(cr_0)\cos n\theta_0$ or $(un.\ const)_n \times K_{n+k}(cr_0)\sin n\theta_0$ $(k = 0, 1, 2, 3)$. It should be noted that for the presentation of these terms in the system of coordinates $(r, \theta)$, we cannot employ the summation theorem for functions $K_{n+k}(cr_0)\cos n\theta_0$ and $K_{n+k}(cr_0)\sin n\theta_0$, because, in the considered case, the inequality $R > R_{10}$ occurs, i.e. the condition given as (6.103) does not occur under satisfaction of the boundary conditions at $r = R$ (6.97). Therefore, we propose here the following algorithm for this satisfaction. To simplify the consideration below we will make all the discussions for the stress $\sigma_{rr}^{(1),1}$ only. These discussions can be easily transformed for the stresses $\sigma_{r\theta}^{(1),1}$ and $\sigma_{rz}^{(1),1}$.

First we present the expressions of the stress $\sigma_{rr}^{(1),1}$ as follows:

$$\bar{\sigma}_{rr}^{(1),1} = \sin\alpha_1 z\left\{\sum_{n=0}^{+\infty} X_{n1}^{(1)} a_{12n}^{(1)}(\xi_1^{(1)}\alpha_1 r_0)\cos n\theta_0\right.$$
$$+\sum_{n=0}^{+\infty} T_{n1}^{(1)} a_{11n}^{(1)}(\xi_1^{(1)}\alpha_1 r)\cos n\theta + \sum_{n=0}^{+\infty} Y_{n2}^{(1)} a_{14n}^{(1)}(\xi_2^{(1)}\alpha_1 r_0)\cos n\theta_0$$
$$+\sum_{n=0}^{+\infty} Z_{n2}^{(1)} a_{13n}^{(1)}(\xi_2^{(1)}\alpha_1 r)\cos n\theta + \sum_{n=0}^{+\infty} Y_{n3}^{(1)} a_{16n}^{(1)}(\xi_3^{(1)}\alpha_1 r_0)\cos n\theta_0$$
$$\left.+\sum_{n=0}^{+\infty} Z_{n3}^{(1)} a_{15n}^{(1)}(\xi_3^{(1)}\alpha_1 r)\cos n\theta\right\} + \sin\alpha_1 z\left\{\sum_{n=1}^{+\infty} Y_{n1}^{(1)} b_{12n}^{(1)}(\xi_1^{(1)}\alpha_1 r_0)\sin n\theta_0\right.$$
$$+\sum_{n=1}^{+\infty} Z_{n1}^{(1)} b_{11n}^{(1)}(\xi_1^{(1)}\alpha_1 r)\sin n\theta + \sum_{n=1}^{+\infty} X_{n2}^{(1)} b_{14n}^{(1)}(\xi_2^{(1)}\alpha_1 r_0)\sin n\theta_0$$
$$+\sum_{n=1}^{+\infty} T_{n2}^{(1)} b_{13n}^{(1)}(\xi_2^{(1)}\alpha_1 r)\sin n\theta + \sum_{n=1}^{+\infty} X_{n3}^{(1)} b_{16n}^{(1)}(\xi_3^{(1)}\alpha_1 r_0)\sin n\theta_0$$
$$\left.+\sum_{n=1}^{+\infty} T_{n3}^{(1)} b_{15n}^{(1)}(\xi_3^{(1)}\alpha_1 r)\sin n\theta\right\}. \tag{6.110}$$

Under satisfying the boundary condition (6.97) at $r = R$ with the use of Eq. (6.110), instead of $r_0$ and $\theta_0$ we write the following expressions obtained from Eq. (6.88).

$$r_0 = r_0(r,\theta) = R\sqrt{1 - 2\frac{R_{10}}{R}\cos\theta + \left(\frac{R_{10}}{R}\right)^2},\ \theta_0 = \theta_0(r,\theta)$$
$$= \arg\cos\left(\left(\cos\theta - \frac{R_{10}}{R}\right)\Big/\sqrt{1 - 2\frac{R_{10}}{R}\cos\theta + \left(\frac{R_{10}}{R}\right)^2}\right). \tag{6.111}$$

Taking the Eq. (6.111) into account, we expand the terms $a_{1jn}^{(1)}(cr_0)\cos n\theta_0$ and $b_{4jn}^{(1)}(cr_0)\sin n\theta_0$ with the following cosine and sine Fourier series, respectively.

$$a_{1jn}^{(1)}(cr_0)\cos n\theta_0 = \sum_{k=0}^{+\infty}\alpha_{1jnk}^{(1)}(cR)\cos k\theta,$$
$$b_{1jn}^{(1)}(cr_0)\sin n\theta_0 = \sum_{k=1}^{+\infty}\beta_{1jnk}^{(1)}(cR)\sin k\theta \tag{6.112}$$

where

$$\alpha_{1jn0}^{(1)}(cR) = \frac{2}{\pi}\int_0^{\pi} a_{1jn}^{(1)}(cr_0)\cos n\theta_0 d\theta$$

$$\alpha_{1jnk}^{(1)}(cR) = \frac{1}{\pi}\int_0^{\pi} a_{1jn}^{(1)}(cr_0)\cos n\theta_0 \cos k\theta d\theta \text{ for } k \geq 1,$$

$$\beta_{1jnk}^{(1)}(cR) = \frac{1}{\pi}\int_0^{\pi} b_{1jn}^{(1)}(cr_0)\sin n\theta_0 \sin k\theta d\theta. \qquad (6.113)$$

Substituting expression (6.113) into Eq. (6.110) and doing some mathematical transformations, we obtain the following expression for $\sigma_{rr}^{(1)}$ at $r = R$.

$$\begin{aligned}
\bar{\sigma}_{rr}^{(1),1}\big|_{r=R} = \sin\alpha_1 z \Bigg\{ \sum_{n=0}^{+\infty} \Bigg[ & \sum_{k=0}^{+\infty} X_{k1}^{(1)}\alpha_{12nk}^{(1)}(\xi_1^{(1)}\alpha_1 R) \\
& + T_{n1}^{(1)} a_{11n}^{(1)}(\xi_1^{(1)}\alpha_1 R) + \sum_{k=0}^{+\infty} Y_{k2}^{(1)}\alpha_{14nk}^{(1)}(\xi_2^{(1)}\alpha_1 R) \\
& + Z_{n2}^{(1)} a_{13n}^{(1)}(\xi_2^{(1)}\alpha_1 R) + \\
& \sum_{k=0}^{+\infty} Y_{k3}^{(1)}\alpha_{16nk}^{(1)}(\xi_3^{(1)}\alpha_1 R) + Z_{n3}^{(1)} a_{15n}^{(1)}(\xi_3^{(1)}\alpha_1 R) \Bigg] \cos n\theta \\
& + \sum_{n=1}^{+\infty} \Bigg[ \sum_{k=1}^{+\infty} Y_{k1}^{(1)}\beta_{12nk}^{(1)}(\xi_1^{(1)}\alpha_1 R) + Z_{n1}^{(1)} b_{11n}^{(1)}(\xi_1^{(1)}\alpha_1 R) \\
& + \sum_{k=1}^{+\infty} X_{k2}^{(1)}\beta_{14nk}^{(1)}(\xi_2^{(1)}\alpha_1 R) + T_{n2}^{(1)} b_{13n}^{(1)}(\xi_2^{(1)}\alpha_1 R) + \\
& \sum_{k=1}^{+\infty} X_{k3}^{(1)}\beta_{16nk}^{(1)}(\xi_3^{(1)}\alpha_1 R) + T_{n3}^{(1)} b_{15n}^{(1)}(\xi_3^{(1)}\alpha_1 R) \Bigg] \sin n\theta \Bigg\}. \qquad (6.114)
\end{aligned}$$

In a similar manner we obtain the following expressions for the stresses $\bar{\sigma}_{r\theta}^{(1),1}$ and $\bar{\sigma}_{rz}^{(1),1}$ at $r = R$.

$$\begin{aligned}
\bar{\sigma}_{r\theta}^{(1),1}\Big|_{r=R} = \sin\alpha_1 z \Bigg\{ \sum_{n=1}^{+\infty} \Bigg[ \sum_{k=1}^{+\infty} X_{k1}^{(1)} \alpha_{22nk}^{(1)} (\xi_1^{(1)} \alpha_1 R) \\
+ T_{n1}^{(1)} a_{21n}^{(1)} (\xi_1^{(1)} \alpha_1 R) + \\
\sum_{k=1}^{+\infty} Y_{k2}^{(1)} \alpha_{24nk}^{(1)} (\xi_2^{(1)} \alpha_1 R) + Z_{n2}^{(1)} a_{23n}^{(1)} (\xi_2^{(1)} \alpha_1 R) \\
+ \sum_{k=1}^{+\infty} Y_{k3}^{(1)} \alpha_{26nk}^{(1)} (\xi_3^{(1)} \alpha_1 R) + Z_{n3}^{(1)} a_{25n}^{(1)} (\xi_3^{(1)} \alpha_1 R) \Bigg] \sin n\theta \\
+ \sum_{n=0}^{+\infty} \Bigg[ \sum_{k=0}^{+\infty} Y_{k1}^{(1)} \beta_{22nk}^{(1)} (\xi_1^{(1)} \alpha_1 R) + Z_{n1}^{(1)} b_{21n}^{(1)} (\xi_1^{(1)} \alpha_1 R) \\
+ \sum_{k=0}^{+\infty} X_{k2}^{(1)} \beta_{24nk}^{(1)} (\xi_2^{(1)} \alpha_1 R) + T_{n2}^{(1)} b_{23n}^{(1)} (\xi_2^{(1)} \alpha_1 R) + \\
\sum_{k=0}^{+\infty} X_{k3}^{(1)} \beta_{26nk}^{(1)} (\xi_3^{(1)} \alpha_1 R) + T_{n3}^{(1)} b_{25n}^{(1)} (\xi_3^{(1)} \alpha_1 R) \Bigg] \cos n\theta \Bigg\},
\end{aligned}$$

$$\begin{aligned}
\bar{\sigma}_{rz}^{(1),1}\Big|_{r=R} = \cos\alpha_1 z \Bigg\{ \sum_{n=0}^{+\infty} \Bigg[ \sum_{k=0}^{+\infty} X_{k1}^{(1)} \alpha_{32nk}^{(1)} (\xi_1^{(1)} \alpha_1 R) \\
+ T_{n1}^{(1)} a_{31n}^{(1)} (\xi_1^{(1)} \alpha_1 R) + \\
\sum_{k=0}^{+\infty} Y_{k2}^{(1)} \alpha_{34nk}^{(1)} (\xi_2^{(1)} \alpha_1 R) + Z_{n2}^{(1)} a_{33n}^{(1)} (\xi_2^{(1)} \alpha_1 R) + \\
\sum_{k=0}^{+\infty} Y_{k3}^{(1)} \alpha_{36nk}^{(1)} (\xi_3^{(1)} \alpha_1 R) + Z_{n3}^{(1)} a_{35n}^{(1)} (\xi_3^{(1)} \alpha_1 R) \Bigg] \cos n\theta + \\
\sum_{n=1}^{+\infty} \Bigg[ \sum_{k=1}^{+\infty} Y_{k1}^{(1)} \beta_{32nk}^{(1)} (\xi_1^{(1)} \alpha_1 R) + Z_{n1}^{(1)} b_{31n}^{(1)} (\xi_1^{(1)} \alpha_1 R) + \\
\sum_{k=1}^{+\infty} X_{k2}^{(1)} \beta_{34nk}^{(1)} (\xi_2^{(1)} \alpha_1 R) + T_{n2}^{(1)} b_{33n}^{(1)} (\xi_2^{(1)} \alpha_1 R) + \\
\sum_{k=1}^{+\infty} X_{k3}^{(1)} \beta_{36nk}^{(1)} (\xi_3^{(1)} \alpha_1 R) + T_{n3}^{(1)} b_{33n}^{(1)} (\xi_3^{(1)} \alpha_1 R) \Bigg] \sin n\theta \Bigg\} \quad (6.115)
\end{aligned}$$

where

$$\alpha_{2jn0}^{(1)}(cR) = \frac{2}{\pi}\int_0^{\pi} a_{2jn}^{(1)}(cr_0)\cos n\theta_0 d\theta,$$

$$\alpha_{2jnk}^{(1)}(cR) = \frac{1}{\pi}\int_0^{\pi} a_{2jn}^{(1)}(cr_0)\cos n\theta_0 \cos k\theta d\theta \text{ for } k \geq 1,$$

$$\beta_{2jnk}^{(1)}(cR) = \frac{1}{\pi}\int_0^{\pi} b_{2jn}^{(1)}(cr_0)\sin n\theta_0 \sin k\theta d\theta,$$

$$\beta_{2jn0}^{(1)}(cR) = \frac{2}{\pi}\int_0^{\pi} b_{2jn}^{(1)}(cr_0)\cos n\theta_0 d\theta$$

$$\beta_{2jnk}^{(1)}(cR) = \frac{1}{\pi}\int_0^{\pi} b_{2jn}^{(1)}(cr_0)\cos n\theta_0 \cos k\theta d\theta \text{ for } k \geq 1,$$

$$\alpha_{2jnk}^{(1)}(cR) = \frac{1}{\pi}\int_0^{\pi} a_{2jn}^{(1)}(cr_0)\sin n\theta_0 \sin k\theta d\theta. \tag{6.116}$$

Thus, using Eqs. (6.105)–(6.108) we can satisfy the contact conditions in (6.96), and using Eqs. (6.115) and (6.116) we can satisfy the boundary condition (6.97). In this way from this boundary and contact conditions we attain the infinite system of linear algebraic equations for the determination of the unknown constants in the foregoing expressions. The analyses of these expressions show that the mentioned unknowns can be divided into two uncrossing sets (groups): the first of them (denoted by $S_I$) is

$$S_I = \Big\{X_{n1}^{(1),1}, T_{n1}^{(1),1}, Y_{n2}^{(1),1}, Y_{n3}^{(1),1}, Z_{n2}^{(1),1}, Z_{n3}^{(1),1}, X_{n1}^{(0),1}, Y_{n2}^{(0),1}, Y_{n3}^{(0),1}\Big\}, \tag{6.117}$$

but the second of them (denoted by $S_{II}$) is

$$S_{II} = \Big\{Y_{n1}^{(1),1}, Z_{n1}^{(1),1}, X_{n2}^{(1),1}, X_{n3}^{(1),1}, T_{n2}^{(1),1}, T_{n3}^{(1),1}, Y_{n1}^{(0),1}, X_{n2}^{(0),1}, X_{n3}^{(0),1}\Big\}. \tag{6.118}$$

From the foregoing contact and boundary conditions we attain the system of equations for each group of unknowns separately.

The equations for the $S_I$ (6.117) group of unknowns are as follows:

$$\sum_{k=0}^{+\infty} X_{k1}^{(1)} \alpha_{i2nk}^{(1)}(\xi_1^{(1)}\alpha_1 R) + T_{n1}^{(1)} a_{i1n}^{(1)}(\xi_1^{(1)}\alpha_1 R)$$
$$+ \sum_{k=0}^{+\infty} Y_{k2}^{(1)} \alpha_{i4nk}^{(1)}(\xi_2^{(1)}\alpha_1 R) + Z_{n2}^{(1)} a_{i3n}^{(1)}(\xi_2^{(1)}\alpha_1 R)$$
$$+ \sum_{k=0}^{+\infty} Y_{k3}^{(1)} \alpha_{i6nk}^{(1)}(\xi_3^{(1)}\alpha_1 R) + Z_{n3}^{(1)} a_{i5n}^{(1)}(\xi_3^{(1)}\alpha_1 R) = 0,$$

$$X_{n1}^{(1)} a_{j2n}^{(1)}\left(\xi_1^{(1)}\alpha_1 R_0\right) + \sum_{m=1}^{+\infty} T_{m1}^{(1)} \lambda_{nm}^{-}(\xi_1^{(1)}\alpha_1 R_{10}) a_{j1m}^{(1)}\left(\xi_1^{(1)}\alpha_1 R_0\right)$$
$$+ Y_{n2}^{(1)} a_{j4n}^{(1)}\left(\xi_2^{(1)}\alpha_1 r_0\right) + \sum_{m=0}^{+\infty} Z_{m2}^{(1)} \lambda_{nm}^{+}(\xi_2^{(1)}\alpha_1 R_{10}) a_{j3m}^{(1)}\left(\xi_2^{(1)}\alpha_1 R_0\right)$$
$$+ Y_{n3}^{(1)} a_{j6n}^{(1)}\left(\xi_3^{(1)}\alpha_1 R_0\right) + \sum_{m=0}^{+\infty} Z_{m3}^{(1)} \lambda_{nm}^{+}(\xi_3^{(1)}\alpha_1 R_{10}) a_{j5m}^{(1)}\left(\xi_3^{(1)}\alpha_1 R_0\right) =$$
$$\delta_j^6 \delta_1^n 2\pi\left(\bar{\sigma}_{33}^{(1),0} - \bar{\sigma}_{33}^{(0),0}\right) \cos\beta,$$
$$n = 0, 1, 2, \ldots, \infty; \; i = 1, 2, 3; \; j = 1, 2, 3, 4, 5, 6. \tag{6.119}$$

The equations for the $S_{II}$ (6.119) group of unknowns are as follows:

$$\sum_{k=1}^{+\infty} Y_{k1}^{(1)} \beta_{i2nk}^{(1)}(\xi_1^{(1)}\alpha_1 R) + Z_{n1}^{(1)} b_{i1n}^{(1)}(\xi_1^{(1)}\alpha_1 R)$$
$$+ \sum_{k=1}^{+\infty} X_{k2}^{(1)} \beta_{i4nk}^{(1)}(\xi_2^{(1)}\alpha_1 R) + T_{n2}^{(1)} b_{i3n}^{(1)}(\xi_2^{(1)}\alpha_1 R)$$
$$+ \sum_{k=1}^{+\infty} X_{k3}^{(1)} \beta_{i6nk}^{(1)}(\xi_3^{(1)}\alpha_1 R) + T_{n3}^{(1)} b_{i3n}^{(1)}(\xi_3^{(1)}\alpha_1 R) = 0,$$

$$Y_{n1}^{(1)} b_{j2n}^{(1)}\left(\xi_1^{(1)}\alpha_1 R_0\right) + \sum_{m=0}^{+\infty} Z_{m1}^{(1)} \lambda_{nm}^{+}(\xi_1^{(1)}\alpha_1 R_{10}) b_{j1m}^{(1)}\left(\xi_1^{(1)}\alpha_1 R_0\right)$$
$$+ X_{n2}^{(1)} b_{j4n}^{(1)}\left(\xi_2^{(1)}\alpha_1 R_0\right) + \sum_{m=1}^{+\infty} T_{m2}^{(1)} \lambda_{nm}^{-}(\xi_2^{(1)}\alpha_1 R_{10}) b_{j3m}^{(1)}\left(\xi_2^{(1)}\alpha_1 R_0\right)$$
$$+ X_{n3}^{(1)} b_{j6n}^{(1)}\left(\xi_3^{(1)}\alpha_1 R_0\right) + \sum_{m=0}^{+\infty} T_{m3}^{(1)} \lambda_{nm}^{+}(\xi_3^{(1)}\alpha_1 R_{10}) b_{j5m}^{(1)}\left(\xi_3^{(1)}\alpha_1 R_0\right)$$
$$= \delta_j^6 \delta_1^n 2\pi\left(\bar{\sigma}_{33}^{(1),0} - \bar{\sigma}_{33}^{(0),0}\right) \sin\beta,$$
$$n = 0, 1, 2, \ldots, \infty; \; i = 1, 2, 3; \; j = 1, 2, 3, 4, 5, 6. \tag{6.120}$$

As $S_I \cap S_{II} = \varnothing$, the system of equations in (6.119) and (6.120) can be solved separately. But for this purpose the contact and boundary conditions must also be separated into two parts with respect to the presentation of the corresponding terms by $\sin n\theta$ and $\cos n\theta$.

Thus, we have exhausted the consideration of the solution method as well as the computer modeling algorithm for determination of the Laplace transformation of the values related to the first approximation. The original of the sought values, as in the previous sections, are determined by the method (Schapery 1962, 1978). The critical values for the stability loss of the fiber will be determined from the criterion

$$\max_{\substack{z\in(0,\ell)\\ \theta\in(0,\pi)}} \left|u_r^{(0),1}\right| \to \infty \text{ as } t \to t_{cr.} \text{ (or as } p \to p_{cr.} \text{ for the pure elastic case).} \quad (6.121)$$

It should be noted that the main difference in the method and algorithm developed above from the methods and algorithms developed in the previous sections of the present chapter and the from that developed in papers (Guz and Lapusta 1986, 1988, 1999; Lapusta 1988) is the expansion of the sought values in the cosine and sine Fourier series under the satisfaction of the boundary conditions at $r = R$, for which the summation theorem for the cylindrical functions is not applicable.

### *6.5.3 Numerical Results and Discussions*

Assume that the material of the fiber are pure elastic with mechanical constants $E^{(0)}$(modulus of elasticity) and $\nu^{(0)}$(Poisson ratio), but the material of the cylinder is viscoelastic one with the operators (6.42) and (6.43). Moreover, we use the notation introduced in Subsection 6.2.2 and introduce also a dimensionless parameter $\delta = \sigma_{zz}^{(1),0} \big/ \mu_0^{(1)} = \sigma_{zz}^{(0),0} / \mu^{(0)}$. At the same time, we assume that $\nu^{(0)} = \nu_0^{(1)} = 0.3$.

Under obtaining the numerical results, the infinite series in Eqs. (6.119) and (6.120) must be replaced by the corresponding finite series, for example,

$$\sum_{k=0}^{+\infty} X_{k1}^{(1)} a_{i2nk}^{(2)} \left(\xi_1^{(1)} \alpha_1 R\right) \approx \sum_{k=0}^{+M} X_{k1}^{(1)} a_{i2nk}^{(2)} \left(\xi_1^{(1)} \alpha_1 R\right). \quad (6.122)$$

From the comparison of the corresponding numerical results obtained for various $M$ in (6.122), the final value of $M$ is determined. For example, if

$$\left| |\delta||_M - |\delta||_{M-1} \right| / |\delta||_{M-1} \le 10^{-4}, \quad (6.123)$$

then an increase in the number $M$ or an increase in the numbers of terms in the series (6.119) and (6.120) is stopped. It should be noted that for the validity of the replacement of the infinite series with the finite ones, i.e. for the validity of the Eq. (6.122), it is necessary to prove that the determinant of coefficients of the unknowns in Eqs. (6.119) and (6.120) is a normal type determinant (Kantarovich and Krilov 1962). Similar proofs were also presented in Sect. 6.3, therefore we will not stop here to show the procedure for this proof; rather, we continue to analyze the numerical results attained for the critical compressing deformation under which the micro-buckling of the fiber occurs.

In the numerical investigation, the integrals (6.113) and (6.116) are calculated by the use of the Gauss integration algorithm. In this case the interval $[0, \pi]$ is divided into a certain number of short intervals. The number of these intervals is determined from the numerical convergence of the values of the integrals.

The numerical investigation in this subsection focuses on the near-surface stability loss of the fiber which is near the convex free cylindrical surface of the surrounding cylinder which contains this fiber. Note that under these investigations the values are given for the parameter $\chi (= 2\pi R_0/\ell)$ and the critical values of $\delta$ (i.e. $\delta_{cr.}$) are determined from the criterion (6.121). After this procedure, the graph of the dependence $\delta = \delta(\chi)$ is constructed. Figure 6.3 shows two types of graphs of the dependence $\delta = \delta(\chi)$ which may occur. One of them is similar to the curve *A* (Fig. 6.3), but the other one is similar to the curve *B* (Fig. 6.3). The curve *A* has a clearly described minimum under $\chi \neq 0$. This means that the near-surface stability loss of the fiber in the cylinder takes place in an earlier stage of the external loading than the stability loss of the whole cylinder.

Namely, such type dependencies have a real mechanical meaning under the investigation of the near-surface stability loss of the fiber. But, the curve *B* has a clearly described minimum under $\chi = 0$, i.e. under $\ell \to \infty$. Note that in this case the stability loss of the whole cylinder takes place earlier than the near-surface stability loss of the fiber. Consequently, such cases have not any meaning in the near-surface stability loss sense. Therefore, the subject of the numerical investigations in the present section relates to the dependence of $\delta = \delta(\chi)$, the graphs of which is similar to the curve *A*. At the same time, we will also consider some examples for which the graphs of the dependence of $\delta = \delta(\chi)$ are similar to the curve *B*.

First we consider numerical results related to the near-surface stability loss in the above sense for the pure elastic deformation state under $t' = 0$ and $t' = \infty$. The large number of the numerical results show that the near-surface stability loss of the fiber may occur in the cases where $E^{(0)} > E_0^{(1)}$, only. According to this statement, we assume that $E^{(0)} \Big/ E_0^{(1)} > > 1$ and we differ the micro-fiber from each other by the value of ratio $E^{(0)} \Big/ E_0^{(1)}$. So that, according to the papers (Qian et al. 2000; Guz et al. 2008, Zhuk and Guz 2007; Maligino et al. 2009) and others, within the scope of the continuum mechanics the cases where $50 \leq E^{(0)}/E^{(1)} \leq 1000$ are considered as the polymer matrix + micro-fiber. Taking this statement into account, we will consider the cases where $E^{(0)}/E^{(1)} = 50$ and $E^{(0)}/E^{(1)} = 500$.

The location of the fiber in the surrounding cylinder will be determined by the values of the ratios $R_{10}/R_0$, $H/R_0$ and of the angle $\beta$ the meaning of which follow obviously from Fig. 6.8. The values of $\delta_{cr.}$ obtained for the case where $t' = 0$ $(t' = \infty)$ will be denoted as $\delta_{cr.0}(\delta_{cr.\infty})$.

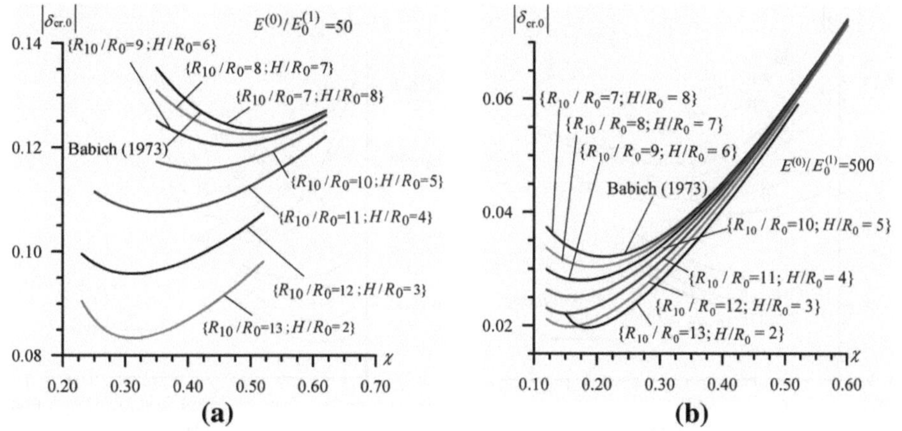


**Fig. 6.9** The influence of the location of the fiber on the dependencies between $\delta_{cr.0}$ and $\chi$ under $E^{(0)}/E_0^{(1)} = 50$ (**a**) and $E^{(0)}/E_0^{(1)} = 500$ (**b**)

Figure 6.9a (under $E^{(0)}\big/E_0^{(1)} = 50$) and 6.9b (under $E^{(0)}\big/E_0^{(1)} = 500$) show the influence of the distance of the fiber from the center of the surrounding circular cylinder on the dependence between $|\delta_{cr.0}|$ and $\chi$ for the case where $R/R_0 = 15$ and $\beta = 0$. It follows from the results that the dependence considered is similar to the $A$ type dependence shown in Fig. 6.3. Moreover, it follows from the results that the values of $\min|\delta_{cr.0}|$ increase with $H/R_0$ (i.e. with distance between the fiber and the free convex cylindrical surface) and approach to the corresponding values attained for the stability loss of a fiber in the infinite elastic matrix (see: Babich 1973).This statement agrees with the well-known mechanical consideration and ensures the trustiness of the solution method and calculation algorithm used in the present investigation.

According to the physical intuition, it can be predicted that for the fixed values of $H/R_0$ an increase in the ratio $R_{10}/R_0$ must cause the approach of the $|\delta_{cr.0}|$ to the corresponding results attained for the problem related to the fiber stability loss which is near the plane bounded with a half-space (see: Guz and Lapusta 1986, 1999). This prediction is proven with the results given in Figs. 6.10a, b, 6.11a and b. Note that the results given in Figs. 6.10a and b (Figs. 6.11a and b) are obtained for the case where $E^{(0)}/E_0^{(1)} = 50$ ($E^{(0)}/E_0^{(1)} = 500$). Moreover, in Figs. 6.10a and b the results relate to the cases where $\beta = 0$ and $\beta = \pi/2$ are illustrated simultaneously, but in Figs. 6.11a and b the case where $\beta = 0$ is considered only.

The aforementioned results show that the "limit" value of $\min|\delta_{cr.0}|$ for the case where $E^{(0)}\big/E_0^{(1)} = 500$ are attained under $R_{10}/R_0 \approx 10$(for $H/R_0 = 3$); $R_{10}/R_0 \approx 13$(for $H/R_0 = 2$), but for the case where $E^{(0)}\big/E_0^{(1)} = 50$ this "limit" value is attained under $R_{10}/R_0 \geq 100$.

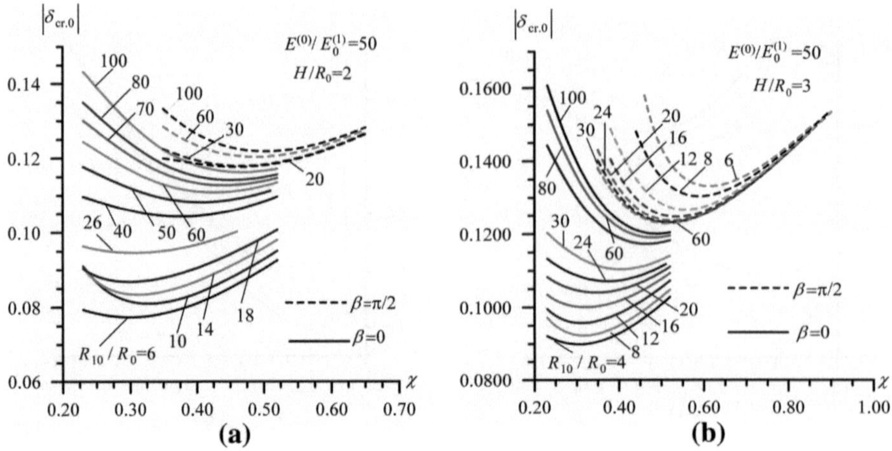


**Fig. 6.10** The influence of the change of distance between the fiber and center of the cylinder on the dependencies between $\delta_{cr.0}$ and $\chi$ under $E^{(0)}/E_0^{(1)} = 50$, $H/R_0 = 2$ (**a**) and $H/R_0 = 3$ (**b**)

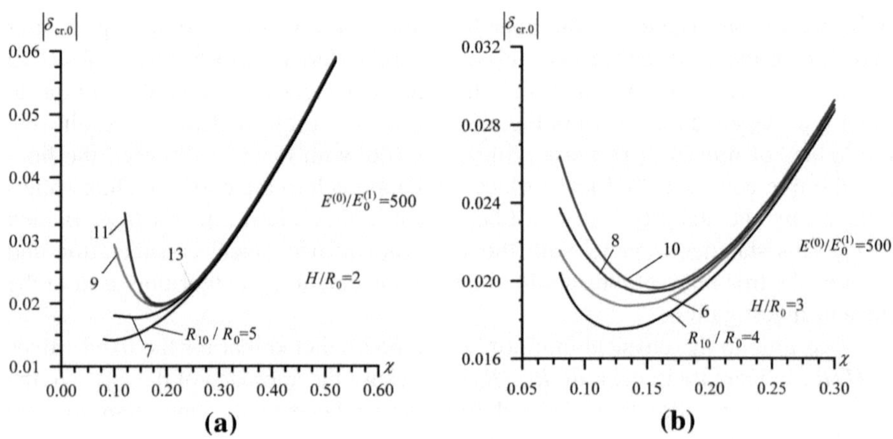


**Fig. 6.11** The influence of the distance between the fiber and the center of the cylinder on the dependencies between $\delta_{cr.0}$ and $\chi$ under $E^{(0)}/E_0^{(1)} = 500$, $H/R_0 = 2$ (**a**) and $H/R_0 = 3$ (**b**)

The comparison of the results shows that the values of $|\delta_{cr.0}||_{\beta=\pi/2}$ (i.e. the values of $|\delta_{cr.0}|$ obtained under $\beta = \pi/2$) for the considered change range of the parameter $\chi$ are greater significantly than the corresponding values of $|\delta_{cr.0}||_{\beta=0}$(i.e. the values of $|\delta_{cr.0}|$ obtained under $\beta = 0$). Moreover, the character of the influence of the ratio $R_{10}/R_0$ on the dependence between $|\delta_{cr.0}||_{\beta=\pi/2}$ and $\chi$ is more complicate. For example, under $E^{(0)}\Big/E_0^{(1)} = 50$ in the case where

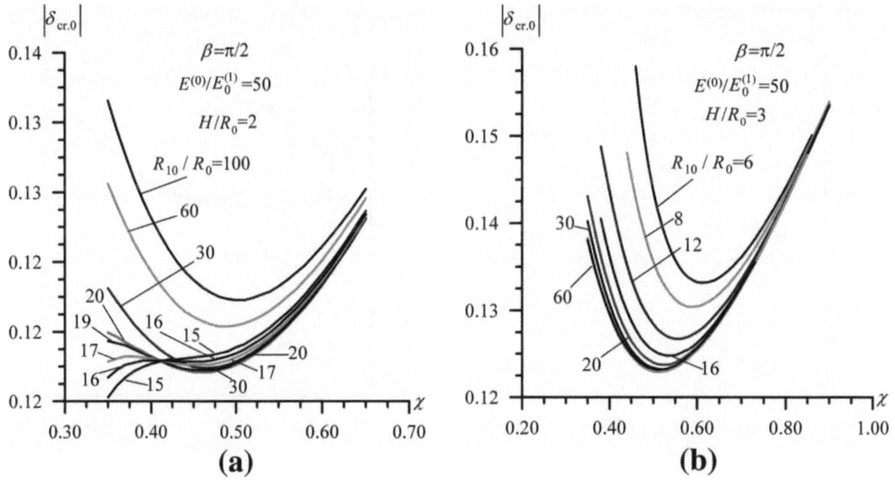


**Fig. 6.12** More detail illustration of the influence of the distance between the fiber and the center of the cylinder on the dependencies between $\delta_{cr.0}$ and $\chi$ under $\beta = \pi/\pi2$, $E^{(2)}/E_0^{(1)} = 50$, $H/R_0 = 2$ (**a**) and $H/R_0 = 3$ (**b**)

$H/R_0 = 2$ the values of $|\delta_{cr.0}||_{\beta=\pi/2}$ increase with $R_{10}/R_0$, but in the case where $H/R_0 = 3$, decrease with $R_{10}/R_0$. Figures 6.12a and b illustrate more clearly this conclusion. This statement can be explained with the non-monotonic character of the dependence between $|\delta_{cr.0}||_{\beta=\pi/2}$ and $R_{10}/R_0$ in the case where $H/R_0 = 2$. In other words, the character of the dependence between $|\delta_{cr.0}||_{\beta=\pi/2}$ and $R_{10}/R_0$ depends on the values of $H/R_0$. Moreover, Fig. 6.12a and b show that under $\beta = \pi/2$ and $E^{(0)}\Big/E_0^{(1)} = 50$ the character of the graphs between $|\delta_{cr.0}|$ and $\chi$ also depends on the ratio $H/R_0$. For example, under $H/R_0 = 2$ and $R_{10}/R_0 \geq 19$ the mentioned dependence consists of two types of curves, the first of which occurs for relatively small values of $\chi$ (for instance, under $\chi < \chi'$) is similar to the $B$ type curve in Fig. 6.3, but the second one which occurs for $\chi > \chi'$ is similar to the A type curve in Fig. 6.3.

Also consider the results regarding the dependence between $|\delta_{cr.0}||_{\beta=\pi/2}$ and $\chi$ attained for the case where $E^{(0)}\Big/E_0^{(1)} = 500$ for various values of $R_{10}/R_0$ and $H/R_0$. These results are illustrated in Figs. 6.13a (for $H/R_0 = 2$) and 6.13b (for $H/R_0 = 3$). It follows from these results that the foregoing conclusions on the type of the graphs hold also for the stability loss of the fiber for which $E^{(0)}\Big/E_0^{(1)} = 500$. However, for this fiber there exist only the case where the type of the graphs of the considered dependencies is the $B$ type graph (see Fig. 6.13a).

It should be noted that, in the scope of the bifurcation approach the $B$ type and the mixed $(A + B)$ type results have not any real meaning on the fiber stability loss in the structure of the surrounding cylinder's material, because in these cases the

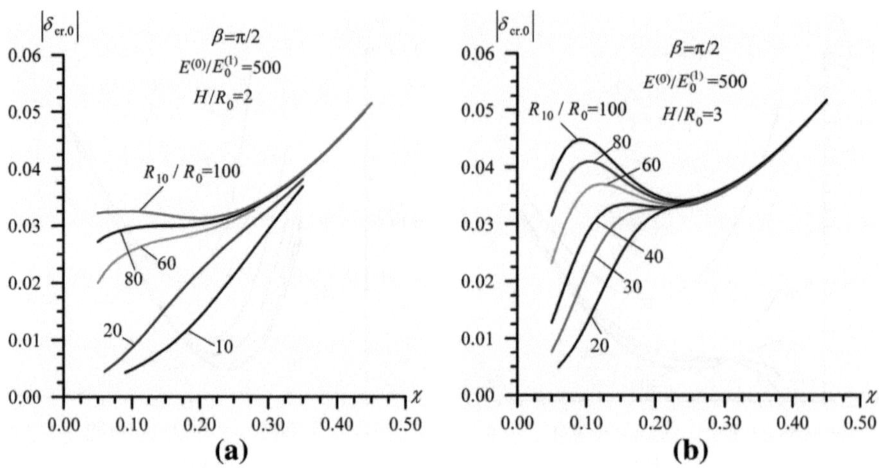


**Fig. 6.13** The influence of the change of the distance between the fiber and the center of the cylinder on the dependencies between $\delta_{cr.0}$ and $\chi$ under $\beta = \pi/22$, $E^{(0)}/E_0^{(1)} = 500$, $H/R_0 = 2$ (**a**) and $H/R_0 = 3$ (**b**)

$\min|\delta_{cr.0}|$ is obtained under $\ell \to \infty$ and this means that the stability loss of the member of construction will be earlier than the stability loss of the fiber in the structure of the material of this construction. However, in the scope of the initial imperfection approach the $B$ type and the mixed $(A + B)$ type results have also the real mechanical meaning, because in the scope of the initial imperfection approach the values of the critical parameters are determined not only with the mechanical relations of the fiber and surrounding materials, but also with the initial imperfection mode. In the foregoing sense, the application fields of the results attained within the scope of the initial imperfection approach are more widely than those attained within the scope of the bifurcation approach.

Consider the dependence between $\min|\delta_{cr.0}|$ and $R_{10}/R_0$ under $\beta = 0$. As has been noted above, the graph of this dependence must have an asymptote as $R_{10}/R_0 \to \infty$ and this asymptote must coincide with the results obtained in (Guz and Lapusta 1986). This consideration is proven by the graphs given in Figs. 6.14a and b. It follows from the graphs given in Fig. 6.14b that for the case where $E^{(0)}\Big/E_0^{(1)} = 500$ the dependence between $\min|\delta_{cr.0}|$ and $R_{10}/R_0$ has non-monotonic character in the case where $H/R_0 = 2$.

Thus, we restrict ourselves to the consideration of the numerical results related to the pure elastic stability loss. Now we consider the numerical results related to the viscoelastic stability loss. It is known that for the occurrence of the viscoelastic stability loss the values of the external compressive force must satisfy the inequality

$$|\delta_{cr.\infty}| < |\delta| < |\delta_{cr.0}| \quad (6.124)$$

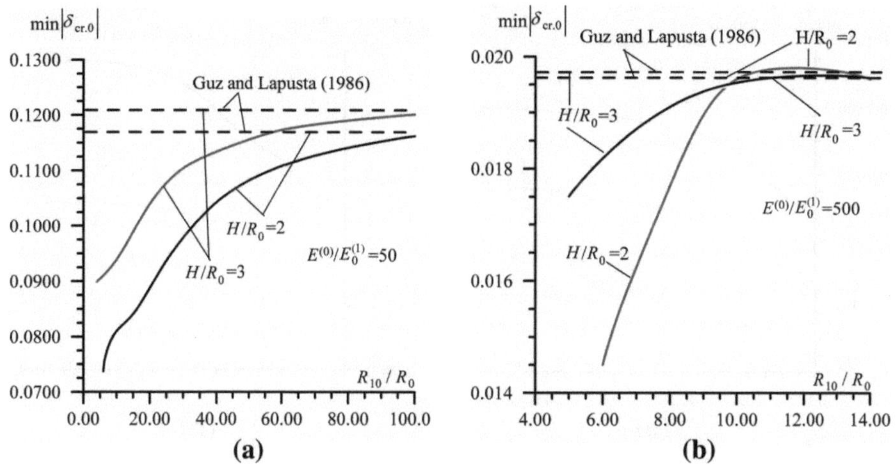


**Fig. 6.14** The approaching of the values of $\min|\delta_{cr.0}|$ to their asymptotic values with increasing of the radius of the surrounding cylinder under $\beta = 0$, $E^{(0)}/E_0^{(1)} = 50$ (**a**) and $E^{(2)}/E_0^{(1)} = 500$ (**b**)

where $\delta_{cr.\infty}$ indicates the critical value of the parameter $\beta$ obtained in the case where $t' = \infty$. In other words, the values of $\delta_{cr.\infty}$ are also determined as a result of the solution of the pure elastic stability loss problem; however, in this case the values of the modulus of elasticity and Poisson's coefficient of the surrounding cylinder material are taken as $E_\infty^{(1)}$ and $\nu_\infty^{(1)}$ which are determined by expression (6.43).

It follows from the expression (6.43) that the values of $\delta_{cr.\infty}$ also depend on the rheological parameter $\omega$. The graphs of the dependence between $\min|\delta_{cr.\infty}|$ and $R_{10}/R_0$ for various values of $\omega$ are given in Fig. 6.15a (for the case where $E^{(0)}/E_0^{(1)} = 50$) and 6.15b (for the case where $E^{(0)}/E_0^{(1)} = 500$) under $H/R_0 = 3$ and $\beta = 0$. According to these results, the values of $\min|\delta_{cr.\infty}|$ decrease with the decreasing $\omega$. Moreover, these results allow us to select the values for the parameter $\delta$ under which the viscoelastic stability loss takes place

Thus, we consider the numerical results regarding the $t'_{cr}$ which are given in Fig. 6.16a (for $\omega = 0.5$), b (for $\omega = 1.0$). Note that these results are attained for the case where $E^{(0)}/E_0^{(1)} = 50$ under $H/R_0 = 3$ and $R_{10}/R_0 = 60$. The similar results attained for the case where $E^{(0)}/E_0^{(1)} = 500$ under $H/R_0 = 3$ and $R_{10}/R_0 = 10$ are given in Fig. 6.17a (for $\omega = 0.5$) and b (for $\omega = 1.0$). It follows from these results that, as it can be predicted, the following asymptotic relations $t'_{cr.} \to 0$ as $|\delta| \to \min|\delta_{cr.0}|$, $t'_{cr.} \to \infty$ as $|\delta| \to \min|\delta_{cr.\infty}|$ occur. Moreover it follows from these results that there exists a certain value of $\delta$ (denote it by $\delta'$), before which (i.e. under $\delta < \delta'$) the values of $t'_{cr}$ increase with $|\alpha'|$; but, for$\delta > \delta'$, the values of $t'_{cr}$ decrease with $|\alpha'|$; the value of $\delta'$ corresponds to the $t'_{cr} = 0.5$.

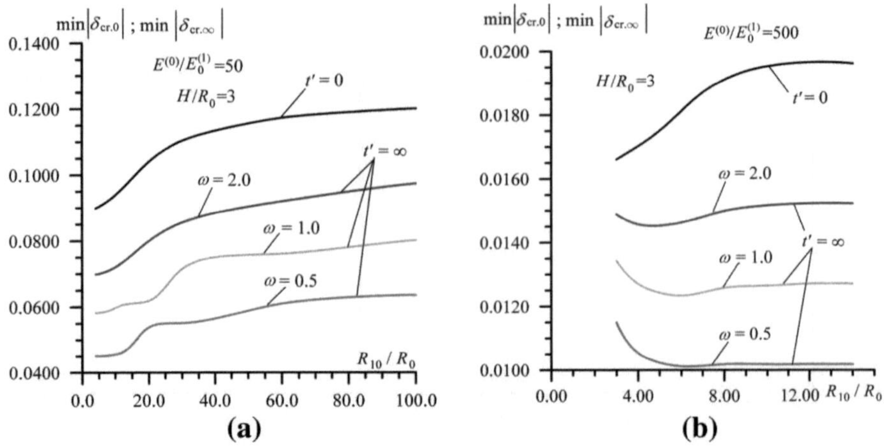


**Fig. 6.15** The influence of the rheological parameter $\omega$ on the critical values of the parameter $\delta$ attained in the case where $E^{(0)}/E_0^{(1)} = 50$ (**a**) and $E^{(0)}/E_0^{(1)} = 500$ (**b**) under $\beta = 0$, $H/R_0 = 3$

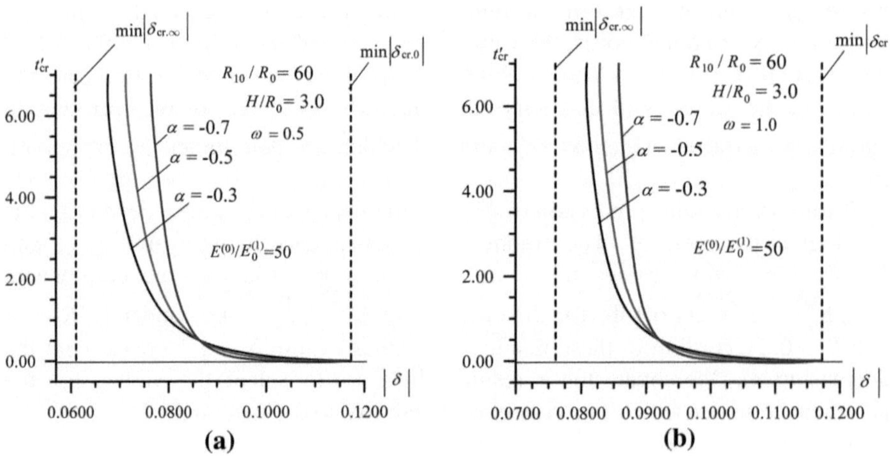


**Fig. 6.16** The graphs illustrating the influence of the rheological parameter $\alpha$ on the values of the critical time $t'_{cr.}$ attained under $E^{(0)}/E_0^{(1)} = 50$ in the cases where $\omega = 0.5$ (**a**) and $\omega = 1.0$ (**b**)

Note that under obtaining the foregoing numerical results Laplace transformation of the term $\sigma_{zz}^{(1),0}\partial^2 u_{(i)}^{(1),1}/\partial z^2$ in Eq. (6.18) is taken as $\sigma_{zz}^{(1),0}(0)\partial^2 \bar{u}_{(i)}^{(1),1}/\partial z^2$.

For the illustration of the convergence of the numerical results obtained for various $M$ in (6.122), under $H/R_0 = 3$, $R_{10}/R_0 = 13$ and $E^{(0)}/E_0^{(1)} = 500$, the values of $|\delta_{cr.0}|$ calculated for various $\chi$ are given in Table 6.13. It follows from

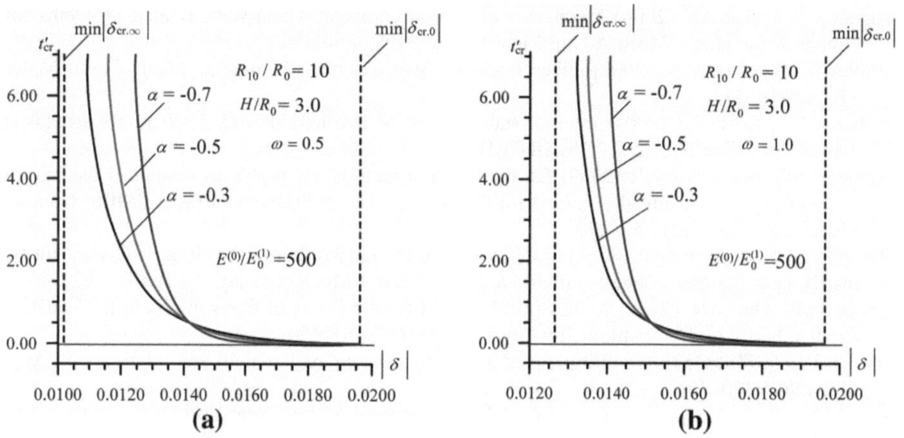


**Fig. 6.17** The graphs illustrating the influence of the rheological parameter $\alpha$ on the values of the critical time $t'_{cr}$ attained under $E^{(0)}/E_0^{(1)} = 500$ in the cases where $\omega = 0.5$ (**a**) and $\omega = 1.0$ (**b**)

**Table 6.13** The convergence of the numerical results attained for various values of $M$ in (6.128) and related stability loss of the fiber in the case where $H/R_0 = 3$, $R_{10}/R_0 = 13$, $E^{(0)}/E_0^{(1)} = 500$ and $\beta = 0$

| M | $\lvert\delta_{cr.0}\rvert \times 10^2$ | | | | | | | |
|---|---|---|---|---|---|---|---|---|
| | $\chi$ | | | | | | | |
| | 0.15 | 0.17 | 0.19 | 0.21 | 0.23 | 0.25 | 0.27 | 0.29 |
| 15 | 2.257 | 2.305 | 2.384 | 2.480 | 2.606 | 2.750 | 2.912 | 3.087 |
| 24 | 2.077 | 2.143 | 2.239 | 2.353 | 2.492 | 2.648 | 2.822 | 3.009 |
| 33 | 1.993 | 2.047 | 2.137 | 2.257 | 2.402 | 2.564 | 2.744 | 2.942 |
| 42 | 1.975 | 2.005 | 2.089 | 2.203 | 2.341 | 2.510 | 2.690 | 2.888 |
| 51 | 1.987 | 2.005 | 2.071 | 2.179 | 2.317 | 2.480 | 2.660 | 2.858 |

these results that the case where $M = 51$ is enough for obtaining numerical results with accuracy determined by the expression (6.123).

## References


Akbarov SD, Babazade MB (1987) On the methods of problem solving in mechanics of fibrous composite materials with curved fibers. Deposited in VINITI, CI, ONT. No 4993-B87, 67p (in Russian)

Akbarov SD, Guz AN (1985) Stability of two fibers in an elastic matrix with small strains. Int Appl Mech 21:1–7

Akbarov SD, Guz AN (2000) Mechanics of curved composites. Kluwer Academic Publishers, Dortrecht

Akbarov SD, Guz AN (2004) Mechanics of curved composites and some related problems for structural members. Mech Advan Mater Struct 11(6):445–515
Akbarov SD, Kosker R (2001) Fiber buckling in a viscoelastic matrix. Mech Comp Mater 37(4):299–306
Akbarov SD, Kosker R (2004) Internal stability loss of two neighboring fibers in a viscoelastic matrix. Int J Eng Sci 42 (17/18):1847–1873
Akbarov SD, Mamedov AR (2009) On the solution method for problems related to the micromechanics of a periodically curved fiber near a convex cylindrical surface. CMES: Comput Model Eng Sci 42(3):257–296
Akbarov SD, Mamedov AR (2011) Stability loss of the micro-fiber in the elastic and viscoelastic matrix near the free convex cylindrical surface. Eur J Mech A/Solids 22(2):167–182
Babaev MS, Guz AN, Cherevko MA (1985) Stability of a series of fibers in an elastic matrix at small subcritical deformations. Int Appl Mech 21(5):443-450
Babich IYu (1973) On the stability loss of a fiber in a matrix under small deformations. Int Appl Mech 9(4):370–373
Biot MA. (1965) Mechanics of incremental deformations. Wiley, New York
Guz AN (1990) Fracture mechanics of composites under compression. Naukova Dumka, Kiev (in Russian)
Guz AN (1999) Fundamentals of the Three-dimensional theory of stability of deformable bodies. Springer, Berlin
Guz AN (2008a) Fundamentals of the compressive fracture mechanics of composites: fracture in structure of materials. vol 1 Litera, Kiev (in Russian)
Guz AN (2008b) Fundamentals of the compressive fracture mechanics of composites: related mechanics of fracture. vol 2 Litera, Kiev (in Russian)
Guz AN, Lapusta YuN (1986) Stability of a fiber near a free surface. Int Appl Mech 22 (8):711–719
Guz AN, Lapusta YuN (1988) Stability of fibers near a free cylindrical surface. Int Appl Mech 24(10):939–944
Guz AN, Lapusta YuN (1999) Three-dimensional problems of the near-surface instability of fiber composites in compression (Model of a piecewise- uniform medium) (survey). Int Appl Mech 35(7):641–670
Guz AN, Rushchitsky JJ, Guz IA (2008) Comparative computer modeling of carbon-polimer composites with carbon or graphite microfibers or carbon nanofibers. CMES: Comput Model Eng Sci 26(3):139–156
Kantarovich LV, Krilov VI (1962) Approximate methods in advanced calculus. Moscow, Fizmatgiz, (in Russian).
Lapusta YuN (1988) Stability of fibers near the free surface of a cavity during finite precritical strains. Int Appl Mech 24(5):453–458
Maligino AR, Warrior NA, Long AC (2009) Effect of inter-fibre spacing on damage evolution in unidirectional (UD) fibre-reinforced composites. Eur J Mech A/Solids 28:768–776
Qian D, Dickey EC, Andrews R, Rantell T (2000) Load transfer and deformation mechanisms of carbon nanotube-plytyrene composites. Appl Phys Lett 76(20):2868–2870
Rabotnov YuN (1977) Elements of hereditary mechanics of solid bodies, Nauka, Moscow (in Russian)
Schapery RA (1962) Approximate method of transform inversion for viscoelastic stress analysis. In: Proceeding 4th US National Congress Applied Mechanics ASME : 1075–1085
Schapery RA. (1978) A viscoelastic behaviour of composite materials, in Composite materials, Vol. 1-7, In: Broutman LJ, Krock RH (Eds.) Mir Moscow, translated from English), vol. 2. Mechanics of composite materials, 102–195(in Ruission, translated from English)
Schwartz MM (Editor in Chief) (1992) Composite materials handbook, Second edition, Mc Graw-Hill, Inc. New York
Watson GM (1958) Theory of Bessel functions. Cambridge at the University Press
Zhuk YuA, Guz IA (2007) Features of plane wave propagation along the layers of a pre-strained nanocomposites. Int Appl Mech 43(4):361–379

# Supplement 1: Applications of the Approach Developed in Chap. 4 on the Problems Related to the Stress Concentration in Initially Stressed Bodies

In the present supplement we attempt, after some obvious changes, to employ the approach developed in Chap. 4 for determination of the influence of the initial stresses on the stress concentration which appears around cracks and holes. Namely, the influence of the homogeneous initial stresses on the Energy Release Rate (ERR) or Stress Intensity Factor (SIF) at crack tips contained by a plate-strip is investigated. For this purpose a plate-strip fabricated from orthotropic material and containing a crack whose edges are parallel to the face planes of the plate is considered. Moreover detailed results are provided of investigations related to the study of the influence of the initial tension of a plate-strip containing a rectangular hole on the stress concentration around this hole caused by additional loading with the uniformly distributed normal forces acting on the upper face plane of the plate-strip.

## S1.1 The Influence of the Initial Stresses on the SIF and ERR at Crack Tips in a Plate-Strip from Orthotropic Material

A review of the investigations on the mechanics of brittle fractures of pre-stressed materials is given in the papers (Guz 2000; Guz et al. 2004). However, the results reviewed in these papers relate to the case where the crack is in the infinite body. Consequently, studies of the effect of the initial tension or compression along cracks located in the finite region on the SIF are almost completely absent. The importance of such an inquiry is evident and deserves to be a topic of separate investigations.

The investigation carried out in this section can be briefly summarized as follows. It is assumed that the strip is stretched (or compressed) initially along the crack edges by uniformly distributed external normal forces acting on the simply supported ends of the plate-strip. After this initial stretching (or compressing) the crack edges are loaded by the additional uniformly distributed normal (opening)

S. D. Akbarov, *Stability Loss and Buckling Delamination*,
DOI: 10.1007/978-3-642-30290-9, © Springer-Verlag Berlin Heidelberg 2013

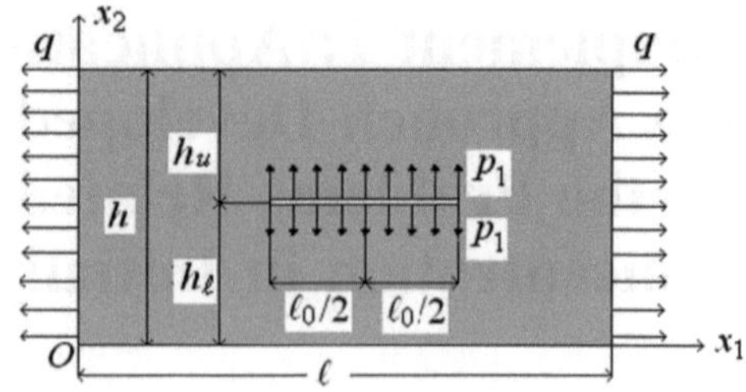


**Fig. S1.1** The geometry of the considered plate-strip and the crack

forces. As a result of the action of these additional forces the stress concentration characterized by the SIF of mode I or by the ERR of mixed mode arises at the crack tips. The influence of the initial stresses on the SIF or ERR is modeled mathematically within the scope of the equations and relations of the first approximation detailed in Chap. 4.

This section is based on a paper (Akbarov and Turan 2009).

## *S1.1.1 Formulation of the Problem*

We consider a plate-strip which occupies a region specified by $\{0 \le x_1 \ell, 0 \le x_2 \le h, -\infty < x_3 < +\infty\}$ and associate it with Lagrange coordinates which, in the natural state, coincide with Cartesian coordinates $Ox_1x_2x_3$ (Fig. S1.1). Let us assume that the plate-strip contains a crack located at $\{x_2 = h_\ell, \ell/2 - \ell_0/2$ (Fig. S1.1). Moreover, let us assume that at first this plate-strip is stretched (or compressed) along the crack edges (i.e. along the $Ox_1$ axis) by the uniformly distributed normal forces with intensity $q$ which exerted at the ends of the strip. After this initial stretching (or compression) the ends of the strip are simply supported and additional normal forces of intensity $p_1 = \varepsilon p$ are exerted on the edges of the crack. Here $\varepsilon$ is a small parameter and $\varepsilon < 1$; $\varepsilon = O(p_1/q)$. We assume that $p_1 < q$.

It follows from the noted above statements that in the initial state the existence of the crack does not cause any stress concentration in the plate-strip. However, the additional normal forces with intensity $p_1 = \varepsilon p$ which act on the crack edges causes the stress concentration at the crack tips. Assume that the material of the strip is an orthotropic one with principal axes $Ox_1$, $Ox_2$ and $Ox_3$.

We employ the approach developed in Sect. 4.2 (without the boundary form perturbation technique) for solution to this problem within the scope of the zeroth and first approximations and consider the plane-strain state in $Ox_1x_2$ plane. Below, the notation "the initial state" and "the perturbed state" will be also used instead of the zeroth and first approximations respectively. In this case, as in Sect. 4.2, the stresses of the zeroth approximation are determined as follows:

$$\sigma_{11}^{(0)} = q, \; \sigma_{ij}^{(0)} = 0 \text{ for } ij \neq 11 \tag{S1.1}$$

As well as, the determination of the values of the first approximation is reduced to the solution of the following boundary-value problem.

Equilibrium equations:

$$\frac{\partial\sigma_{11}^{(1)}}{\partial x_1}+\frac{\partial\sigma_{12}^{(1)}}{\partial x_2}+\sigma_{11}^{(0)}\frac{\partial^2 u_1^{(1)}}{\partial x_1^2}=0,\ \frac{\partial\sigma_{12}^{(1)}}{\partial x_1}+\frac{\partial\sigma_{22}^{(1)}}{\partial x_2}+\sigma_{11}^{(0)}\frac{\partial^2 u_2^{(1)}}{\partial x_1^2}=0. \tag{S1.2}$$

Mechanical relations:

$$\sigma_{11}^{(1)}=A_{11}\varepsilon_{11}^{(1)}+A_{12}\varepsilon_{22}^{(1)},\ \sigma_{22}^{(1)}=A_{12}\varepsilon_{11}^{(1)}+A_{22}\varepsilon_{22}^{(1)},\ \sigma_{12}^{(1)}=G_{12}\varepsilon_{12}^{(1)}. \tag{S1.3}$$

Geometrical relations:

$$\varepsilon_{11}^{(1)}=\frac{\partial u_1^{(1)}}{\partial x_1},\ \varepsilon_{22}^{(1)}=\frac{\partial u_2^{(1)}}{\partial x_2},\ \varepsilon_{12}^{(1)}=\frac{1}{2}\left(\frac{\partial u_1^{(1)}}{\partial x_2}+\frac{\partial u_2^{(1)}}{\partial x_1}\right). \tag{S1.4}$$

According to Lekhnitski (1963), we write the expressions for $A_{11}, A_{12}, A_{22}$ and $G_{12}$ in (S1.3) through the elasticity modulus and Poisson's ratios of the orthotropic body

$$A_{11}=\frac{a_{22}}{a_{11}a_{22}-a_{12}^2},\ A_{12}=-\frac{a_{12}}{a_{11}a_{22}-a_{12}^2},\ A_{22}=\frac{a_{11}}{a_{11}a_{22}-a_{12}^2},$$

$$a_{11}=\frac{1}{E_1}\left(1-\frac{E_3}{E_1}\nu_{13}^2\right),\ a_{12}=\frac{1}{E_1}\left(-\nu_{12}-\frac{E_3}{E_2}\nu_{13}\nu_{23}\right),$$

$$a_{22}=\frac{1}{E_2}\left(1-\frac{E_3}{E_2}\nu_{23}^2\right), \tag{S1.5}$$

where $E_1$, $E_2$ and $E_3$ are the modules of elasticity of the strip material in the direction of the $Ox_1$, $Ox_2$ and $Ox_3$ axes respectively, $\nu_{12}$, $\nu_{23}$, $\nu_{13}$ are the Poisson's ratios of this material. $G_{12}$ in the relations (S1.3) denotes the shear modulus of the strip material in the $Ox_1x_2$ plane.

The boundary conditions for the first approximation (or for the perturbed state), i.e. for the stress–strain state caused by the additional normal forces with intensity $p_1=\varepsilon p$ are

$$u_2^{(1)}\Big|_{\substack{x_1=0;\ell\\ x_2=[0,h]}}=0,\ \sigma_{11}^{(1)}\Big|_{\substack{x_1=0;\ell\\ x_2=[0,h]}}=0,\ \sigma_{12}^{(1)}\Big|_{\substack{x_2=0;h\\ x_1\in(0,\ell)}}=\sigma_{22}^{(1)}\Big|_{\substack{x_2=0;h\\ x_1\in(0,\ell)}}=0,$$

$$\sigma_{12}^{(1)}\Big|_{\substack{x_2=h_\ell\pm0\\ x_1\in(\ell/2-\ell_0/2,\ell/2+\ell_0/2)}}=0,\ \sigma_{22}^{(1)}\Big|_{\substack{x_2=h_\ell\pm0\\ x_1\in(\ell/2-\ell_0/2,\ell/2+\ell_0/2)}}=-p. \tag{S1.6}$$

Thus, within the framework of the above-stated we investigate the influence of the initial tension (or compression) (S1.1) of the plate-strip on the stress concentration at the crack tips arising as a result of the action of the additional forces by the intensity $p_1$.

Thus, with the above-stated the mathematical formulation of the considered problem is exhausted.

### S1.1.2 FEM Modeling

According to investigations carried out in Chap. 4, for FEM modeling of the problem formulated above, we introduce the functional

$$\Pi = \underline{\frac{1}{2}\iint\limits_{\Omega-\Omega'} \left[ T_{11}^{(1)}\frac{\partial u_1^{(1)}}{\partial x_1} + T_{12}^{(1)}\frac{\partial u_2^{(1)}}{\partial x_1} + T_{21}^{(1)}\frac{\partial u_1^{(1)}}{\partial x_2} + T_{22}^{(1)}\frac{\partial u_2^{(1)}}{\partial x_2} \right] dx_1 dx_2} \\ - \int\limits_{(\ell-\ell_0)/2}^{(\ell+\ell_0)/2} p u_2^{(1)}\Big|_{x_2=h_\ell-0} dx_1 + \int\limits_{(\ell-\ell_0)/2}^{(\ell+\ell_0)/2} p u_2^{(1)}\Big|_{x_2=h_\ell+0} dx_1, \tag{S1.7}$$

where

$$T_{11}^{(1)} = \sigma_{11}^{(1)} + \sigma_{11}^{(0)}\frac{\partial u_1^{(1)}}{\partial x_1} = \omega_{1111}\frac{\partial u_1^{(1)}}{\partial x_1} + \omega_{1122}\frac{\partial u_2^{(1)}}{\partial x_2},$$

$$T_{12}^{(1)} = \sigma_{12} + \sigma_{11}^{(0)}\frac{\partial u_2^{(1)}}{\partial x_1} = \omega_{1212}\frac{\partial u_1^{(1)}}{\partial x_2} + \omega_{1221}\frac{\partial u_2^{(1)}}{\partial x_1},$$

$$T_{21}^{(1)} = \sigma_{21}^{(1)} = \omega_{2112}\frac{\partial u_1^{(1)}}{\partial x_2} + \omega_{2121}\frac{\partial u_2^{(1)}}{\partial x_1},$$

$$T_{22}^{(1)} = \sigma_{22}^{(1)} + \sigma_{22}^{(0)}\frac{\partial u_2^{(1)}}{\partial x_2} = \omega_{2222}\frac{\partial u_2^{(1)}}{\partial x_2} + \omega_{2211}\frac{\partial u_1^{(1)}}{\partial x_1},$$

$$\begin{aligned} \Omega &= \{0 \le x_2 \le h;\ 0 \le x_1 \le \ell\}, \\ \Omega' &= \{x_2 = h_\ell - 0,\ (\ell-\ell_0)/2 < x_1 < (\ell+\ell_0)/2\} \\ &\quad \cup \{x_2 = h_\ell + 0,\ (\ell-\ell_0)/2 < x_1 < (\ell+\ell_0)/2\}. \end{aligned} \tag{S1.8}$$

According to (S1.3), we obtain that

$$\omega_{1111} = A_{11} + q,\ \omega_{1122} = A_{12},\ \omega_{1212} = G_{12},\ \omega_{1221} = G_{12} + q,\ \omega_{2112} = G_{12}, \\ \omega_{2121} = G_{12},\ \omega_{2211} = A_{12},\ \omega_{2222} = A_{22}. \tag{S1.9}$$

Using the index notation and the relation (S1.8) the underlined part of the functional (S1.7) can be rewritten as follows

$$\frac{1}{2}\iint\limits_{\Omega-\Omega'} T_{ij}^{(1)}\frac{\partial u_j^{(1)}}{\partial x_i} dx_1 dx_2 = \iint\limits_{\Omega-\Omega'} \frac{1}{2}\omega_{ijmn}\frac{\partial u_n^{(1)}}{\partial x_m}\frac{\partial u_j^{(1)}}{\partial x_i} dx_1 dx_2. \tag{S1.10}$$

In Eq. (S1.10) and below by the repeated indices summation is implied over their change within the range. Note that the components $\omega_{ijmn}$ which do not enter the relations (S1.9) are equal to zero.

Taking the relations (S1.8) and (S1.9) into account, by direct verification it is proven that

$$\omega_{ijmn} = \omega_{nmji}. \tag{S1.11}$$

To calculate the variation of the functional (S1.10) we will employ Gauss' theorem and symmetry conditions (S1.11), so that

$$\begin{aligned}
&\delta\left(\iint\limits_{\Omega-\Omega'} \frac{1}{2}\omega_{ijmn}\frac{\partial u_n^{(1)}}{\partial x_m}\frac{\partial u_j^{(1)}}{\partial x_i}dx_1dx_2\right) \\
&= \iint\limits_{\Omega-\Omega'} \frac{1}{2}\omega_{ijmn}\left[\frac{\partial u_n^{(1)}}{\partial x_m}\frac{\partial \delta u_j^{(1)}}{\partial x_i} + \frac{\partial \delta u_n^{(1)}}{\partial x_m}\frac{\partial u_j^{(1)}}{\partial x_i}\right]dx_1dx_2 \\
&= \iint\limits_{\Omega-\Omega'} \frac{1}{2}\left(\omega_{ijnm}+\omega_{mnji}\right)\frac{\partial u_n^{(1)}}{\partial x_m}\frac{\partial \delta u_j^{(1)}}{\partial x_i}dx_1dx_2 \\
&= \iint\limits_{\Omega-\Omega'} \omega_{ijnm}\frac{\partial u_n^{(1)}}{\partial x_m}\frac{\partial \delta u_j^{(1)}}{\partial x_i}dx_1dx_2 \\
&= \iint\limits_{\Omega-\Omega'} \left[\frac{\partial}{\partial x_i}\left(\omega_{ijnm}\frac{\partial u_n^{(1)}}{\partial x_m}\delta u_j^{(1)}\right) - \delta u_j^{(1)}\frac{\partial}{\partial x_i}\left(\omega_{ijnm}\frac{\partial u_n^{(1)}}{\partial x_m}\right)\right]dx_1dx_2 \\
&= -\iint\limits_{\Omega-\Omega'} \delta u_j^{(1)}\frac{\partial}{\partial x_i}\left(\omega_{ijnm}\frac{\partial u_n^{(1)}}{\partial x_m}\right)dx_1dx_2 + \int\limits_{\Im} N_i\omega_{ijnm}\frac{\partial u_n^{(1)}}{\partial x_m}\delta u_j^{(1)}d\xi,
\end{aligned} \tag{S1.12}$$

where

$$\begin{aligned}
\Im = &\{x_1=0,\ 0\le x_2\le h\}\cup\{x_1=\ell,\ 0\le x_2\le h\}\cup\{x_2=0,\ 0\le x_1\le \ell\}\cup \\
&\{x_2=h,\ 0\le x_1\le \ell\}\cup\{x_2=h_\ell-0,\ (\ell-\ell_0)/2<x_1<(\ell+\ell_0)/2\}\cup \\
&\{x_2=h_\ell+0,\ (\ell-\ell_0)/2<x_1<(\ell+\ell_0)/2\},
\end{aligned} \tag{S1.13}$$

and $N_i$ is a component of the unit normal vector of the strip's contour in the $Ox_1x_2$ plane.

Thus, using (S1.12), (S1.13) and (S1.7), from

$$\delta\Pi = 0 \tag{S1.14}$$

we obtain Eq. (S1.2) and the boundary conditions for stresses given in (S1.6). In this way, we prove the validity of the functional (S1.7) for the FEM modeling of the considered problem.

Since the problem is symmetric with respect to $x_1 = \ell/2$ (Fig. S1.1), under FEM modeling we consider only half of the region $\Omega$. In this case, the part around the crack tip is modeled by singular triangular finite elements shown in Fig. 4.14b. For the remaining part of the region, the standard quadratic Lagrange-family rectangular finite elements (Fig. 4.14a) are used.

Thus, the investigation of the stress-strain state in the considered plate-strip is reduced to the solution to the system of the linear algebraic equations obtained from Eq. (S1.14) with respect to the displacement of the nodal points. After determination of the stress-strain state we can calculate ERR (denoted by $\gamma$) by the use of the expression

$$\gamma = -\frac{1}{2}\frac{\delta U(\ell_0)}{\delta \ell_0}, \tag{S1.15}$$

where

$$U(\ell_0) = \frac{1}{2}\iint\limits_{\Omega-\Omega'} \left( T_{11}^{(1)}\frac{\partial u_1^{(1)}}{\partial x_1} + T_{12}^{(1)}\frac{\partial u_2^{(1)}}{\partial x_1} + T_{21}^{(1)}\frac{\partial u_1^{(1)}}{\partial x_2} + T_{22}^{(1)}\frac{\partial u_2^{(1)}}{\partial x_2} \right) dx_1 dx_2. \tag{S1.16}$$

### *S1.1.3 Numerical Results and Discussion*

According to (S1.15), under calculation of the ERR we use the approximate expression

$$\gamma \approx -\frac{1}{2}\frac{U(\ell_0 + \Delta\ell_0) - U(\ell_0)}{\Delta\ell_0}. \tag{S1.17}$$

The values of the increment $\Delta\ell_0$ are determined from the convergence requirement of the values $\gamma$ for each combination of the problem parameters $h_\ell/\ell$, $\ell_0/\ell$, $h/\ell, G_{12}/E_1, E_2/E_1, E_3/E_1, \nu_{12}, \nu_{13}, \nu_{23}$ and $q/E_1$. The influence of the initial stress in the strip on the values of $\gamma$ will be estimated through the parameter $\tilde{q} = q/E_1$.

Under $h_\ell = h/2$ (Fig. S1.1) for the considered problem mode I takes place and in this case we can calculate the values of SIF ($K_I$) through the values of $\gamma$. But in the cases for which $h_\ell \neq h/2$ the mixed mode arises at the crack tips and in such cases we will analyze the influence of the initial stress in the strip on the values of $\gamma$ (ERR).

According to Guz (1983), to calculate $K_I$ through the values of $\gamma$ we use the expression

$$K_I^2(\gamma_2^{(1)}\mu_2\gamma_{21}^{(2)} - \gamma_2^{(2)}\mu_1\gamma_{21}^{(1)}) = 4\gamma i(\mu_2\gamma_{21}^{(2)} - \mu_1\gamma_{21}^{(1)}),\ i = \sqrt{-1}. \tag{S1.18}$$

In (S1.18) the following notation is used.

$$\mu_{1,2} = i\sqrt{A \pm \sqrt{A^2 - A_1}},\ \gamma_{21}^{(j)} = \mu_j^{-2}\left(\omega_{2112}\omega_{1122}\mu_j^2 - \omega_{1111}\omega_{1212}\right)B_j^{-1}$$

$$\gamma_2^{(j)} = \mu_j^{-1}\left(\omega_{2112}\mu_j^2 + \omega_{1111}\right)B_j^{-1},\ j = 1,2, \tag{S1.19}$$

**Table S1.1** The values of $K_I$ obtained for isotropic plate-strip for various values of $\ell_0/\ell$ and $\ell_0/h$ under $h_\ell = h/2$

| $\ell_0/\ell$ | $\ell_0/h$ | $K_I^{(s)}/K_{I\infty}$ | $K_I^{(f)}/K_{I\infty}$ | $K_I^{(E)}/K_{I\infty}$ |
|---|---|---|---|---|
| 0.080 | 0.80 | 1.2406 | 1.2406 | 1.1930 |
| 0.075 | 0.75 | 1.2009 | 1.2009 | 1.1716 |
| 0.060 | 0.60 | 1.1444 | 1.1444 | 1.1108 |
| 0.050 | 0.50 | 1.0936 | 1.0931 | 1.0729 |
| 0.040 | 0.40 | 1.0473 | 1.0473 | 1.0409 |

where

$$A = \frac{1}{2\omega_{2222}\omega_{2112}}\left[\omega_{1111}\omega_{2222} + \omega_{2112}\omega_{1221} - (\omega_{2211} + \omega_{1212})^2\right],$$

$$A_1 = \omega_{1111}\omega_{1221}(\omega_{2222}\omega_{2112})^{-1},$$

$$B_j = \omega_{2222}\omega_{2112}\mu_j^2 + \omega_{1111}\omega_{2222} - \omega_{1122}(\omega_{1122} + \omega_{1212}),\ j = 1, 2. \quad \text{(S1.20)}$$

Now we consider the validity of the algorithm and programmes which are composed by the author and his students and are used in the present numerical investigations. For this purpose consider the case where the initial stresses are absent in the plate-strip, i.e. $\tilde{q} = 0$ and the material of the plate-strip is isotropic, i.e. $E_1 = E_2 = E_3 = E,\ \nu_{12} = \nu_{13} = \nu_{23} = \nu,\ G_{12} = E/2(1+\nu)$. Assume that $\nu = 0.3$. It is evident that in this case the obtained results must coincide with the corresponding ones obtained within the framework of the classical linear theory of elasticity.

Assume that $h_\ell = h/2$ and analyze the values of $K_I$ obtained for various values of $\ell_0/\ell$ and $\ell_0/h$. Table S1.1 shows the values of $K_I^{(s)}/K_{I\infty}$, $K_I^{(f)}/K_{I\infty}$ and $K_I^{(E)}/K_{I\infty}$ where $K_{I\infty} = p\sqrt{\pi\ell_0}$, $K_I^{(s)}$ and $K_I^{(f)}$ are the values of the SIF for mode I calculated by using the exact solution for an infinite plate, by the approximate series given in the handbook (Sih 1973) and by the present approach, from the values of the nodal displacements of the singular triangular finite elements shown in Fig. 4.14b, respectively. Moreover, in Table S1.1, $K_I^{(E)}$ shows the values of SIF for mode I calculated from the values of $\gamma$ by the use of the relations (S1.9), (S1.10).

The agreements of the corresponding results given in Table S1.1 provide support for the numerical approach used and proposed. Moreover, the results given in Table S1.1 agree with the mechanical consideration, according to which, the values of $K_I^{(s)}/K_{I\infty}$, $K_I^{(f)}/K_{I\infty}$ and $K_I^{(E)}/K_{I\infty}$ must simultaneously tend to unity with decreasing $\ell_0/\ell$ and $\ell_0/h$.

Now we turn to consideration of the results which characterize the influence of the initial stresses in the plate-strip on the values of $\gamma$ (ERR) and $K_I$. Assume that $h/\ell = 0.20$; $\nu_{12} = \nu_{13} = \nu_{23} = 0.3$ and analyze the effect of the parameters $G_{12}/E_1$, $E_2/E_1$, $E_3/E_1$, $h_\ell/\ell$ and $\ell_0/\ell$ on this influence. Before the analyses of the numerical results we must note the following statement.

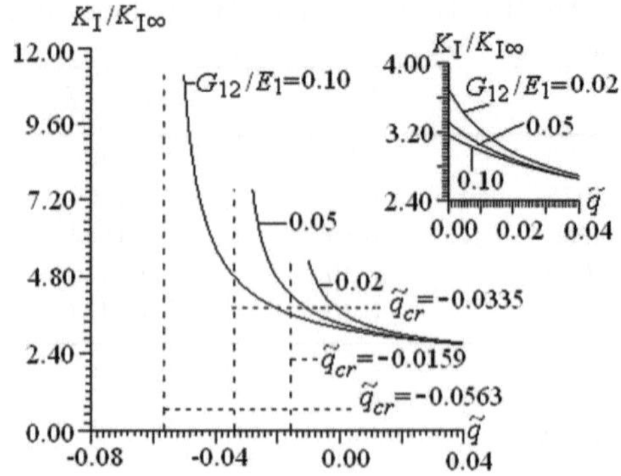


**Fig. S1.2** The influence of the initial stress, i.e. of the $\tilde{q}$ on the values of the $K_I/K_{I\infty}$ for the various values of $G_{12}/E_1$ under $\ell_0/2\ell = 0.15$, $E_2/E_1 = E_3/E_1 = 0.5$

Under initial compression of the strip for certain (critical) values of $\tilde{q}\,(= q/E_1)$(denoted by $\tilde{q}_{cr}$) the stability loss of that takes place. As the values of $\tilde{q}_{cr}$ depend on the problem parameters, therefore in the case where the initial stress in the plate-strip is compressive one for each selected combination of these parameters it is necessary to define these values. Note that the values of $\tilde{q}_{cr}$ are determined by employing of the approach developed in Sect. 4.2.

Thus, we analyze the results regarding the influence of the initial stress $\tilde{q}$ on the fracture parameters of the strip. First, we assume that $h_\ell = h/2$, according to which, there exists symmetry with respect to the $x_2 = h/2$ plane. Therefore, as it has been noted above, in this case we can calculate the values of $K_I$ by the use of the expression (S1.8)–(S1.10).

Figures S1.2, S1.3 and S1.4 show the graphs of the dependencies between $K_I/K_{I\infty}$ (where $K_{I\infty} = p\sqrt{\pi\ell_0}$) and $\tilde{q}$ for various $G_{12}/E_1$ (for the case where $E_2/E_1 = E_3/E_1 = 0.5$, Fig. S1.2), $E_3/E_1 (= E_2/E_1)$ (for the case where $G_{12}/E_1 = 0.02$, Fig. S1.3), $E_2/E_1$ (for the case where $E_3/E_1 = 0.5, G_{12}/E_1 = 0.01$, Fig. S1.4). Under construction of these graphs it was assumed that $\ell_0/2\ell = 0.2$.

It follows from the foregoing graphs that, for the all cases the existence of the initial stretching (compressing) of the strip causes to decrease (to increase) of the values of $K_I$. Moreover, it follows from these graphs that the values of $K_I$ increase with decreasing $G_{12}/E_1$. However, $K_I$ decreases with decreasing $E_2/E_1$ and $E_3/E_1$.

The results show that $K_I \to \infty$ as $\tilde{q} \to \tilde{q}_{cr}$. Consequently, if the initial stress is compressive and the values of that approach the $\tilde{q}_{cr}$ then the "resonance" type phenomenon takes place. In all the considered cases the influence of $G_{12}/E_1$, on the values of $\tilde{q}_{cr}$ is more significant than the influence of $E_2/E_1$ and $E_3/E_1$ on that.

Note that the values of the $\tilde{q}_{cr}$ in the figures are indicated by the vertical dashed lines. In Fig. S1.2 these values are shown for each selected value of $G_{12}/E_1$, however in Figs. S1.3 and S1.4 the value of $\tilde{q}_{cr}$ is indicated only for the case where $E_2/E_1 = E_3/E_1 = 0.05$ and $E_2/E_1 = 0.1$, respectively. Because, the influence of the $E_2/E_1$ and $E_3/E_1$ on the values of $\tilde{q}_{cr}$ is insignificant.

Consider the influence of the crack length on values of $K_I/K_{I\infty}$ and $\tilde{q}_{cr}$. The graphs given in Fig. S1.5 show this influence under $G_{12}/E_1 = 0.02$, $E_2/E_1 = E_3/E_1 = 0.5$. According to the known mechanical considerations, the

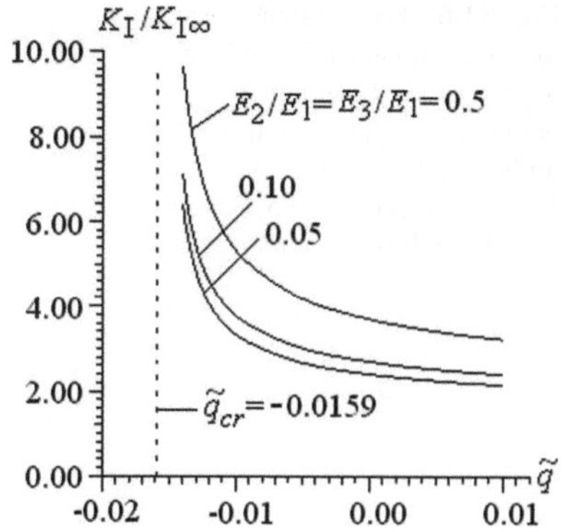


**Fig. S1.3** The graphs of the dependencies between $K_I/K_{I\infty}$ and $\tilde{q}$ for the various values of $E_2/E_1(=E_3/E_1)$ under $\ell_0/2\ell = 0.15$, $G_{12}/E_1 = 0.02$

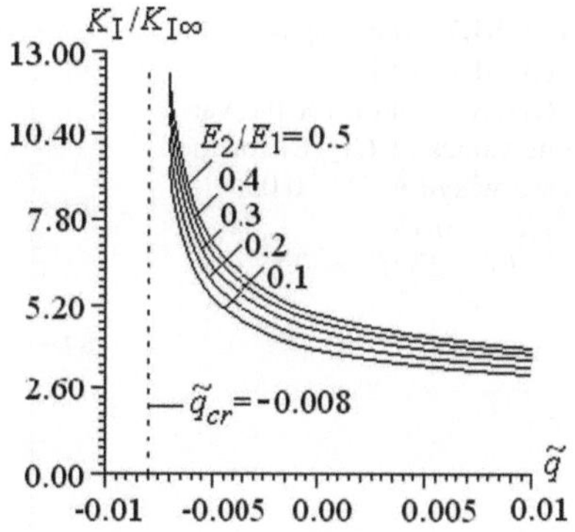


**Fig. S1.4** The dependencies between $K_I/K_{I\infty}$ and $\tilde{q}$ for the various values of $E_2/E_1$ for the case where $E_3/E_1 = 0.5$, $\ell_0/2\ell = 0.20$, $G_{12}/E_1 = 0.01$

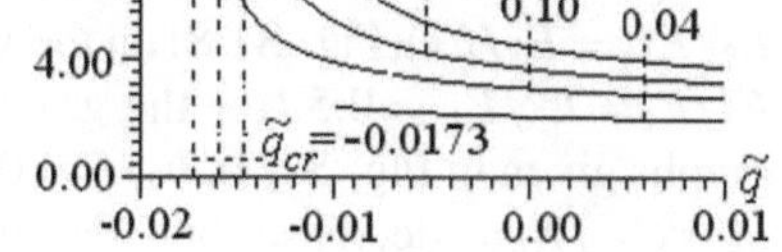


**Fig. S1.5** The influence of the crack length, i.e. of the $\ell_0/2\ell$ on the dependence between $K_I/K_{I\infty}$ and $\tilde{q}$ under $E_2/E_1 = E_3/E_1 = 0.5$, $G_{12}/E_1 = 0.02$

absolute values of $\tilde{q}_{cr}$ must decrease, but the values of $K_I/K_{I\infty}$ must increase with $\ell_0/2\ell$. This prediction is also confirmed by the graphs.

Now we assume that $h_u < h_\ell$ and consider the dependence between $\gamma G_{12}/K_{I\infty}$ and $\tilde{q}$ for various values of $h_u/\ell$ under $G_{12}/E_1 = 0.02$, $E_2/E_1 = E_3/E_1 = 0.5$ and $\ell_0/2\ell = 0.15$. The graphs of this dependence and corresponding values of $\tilde{q}_{cr}$ are given in Fig. S1.6, according to which, the values of $\gamma G_{12}/K_{I\infty}$ increase as the crack location approaches to the free face plane of the plate-strip.

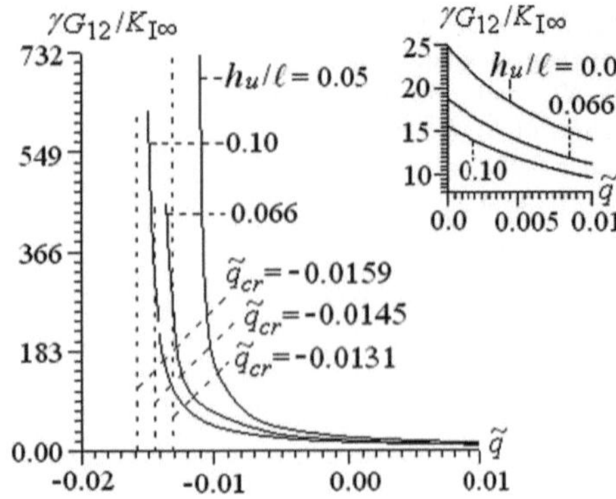


**Fig. S1.6** The influence of the crack location i.e. of the $h_u/\ell$ on the dependence between $\gamma G_{12}/K_{I\infty}$ and $\tilde{q}$ under $\ell_0/2\ell = 0.15$, $E_2/E_1 = E_3/E_1 = 0.5$, $G_{12}/E_1 = 0.02$

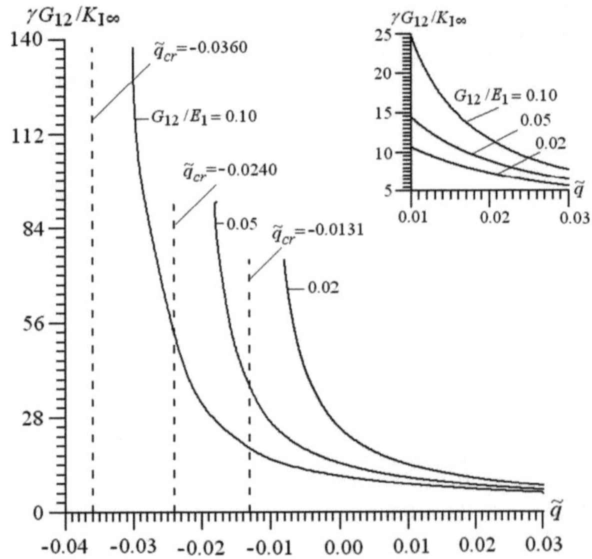


**Fig. S1.7** The graphs of the dependencies between $\gamma G_{12}/K_{I\infty}$ and $\tilde{q}$ for the various values of $G_{12}/E_1$ for the case where $h_u/\ell = 0.05$, $\ell_0/2\ell = 0.15$, $E_2/E_1 = E_3/E_1 = 0.5$

Figures S1.7 and S1.8 show the influence of the $G_{12}/E_1$ (Fig. S1.7) and $E_2/E_1(= E_3/E_1)$(Fig. S1.8) on the values of $\gamma G_{12}/K_{I\infty}$ and $\tilde{q}_{cr}$ under $\ell_0/2\ell = 0.15$, $E_2/E_1 = E_3/E_1 = 0.5$ (for the graphs given in Fig. S1.7), $G_{12}/E_1 = 0.02$ (for the graphs given in Fig. S1.8), $h_u/\ell = 0.05$. It follows from these results that the values of $\gamma G_{12}/K_{I\infty}$ decrease but the absolute values of $\tilde{q}_{cr}$ increase significantly with $G_{12}/E_1$. At the same time, the values $\gamma G_{12}/K_{I\infty}$ decrease also with decreasing $E_2/E_1(= E_3/E_1)$. These results also agree with the known mechanical considerations.

Figures S1.9 shows the dependence between $\gamma G_{12}/K_{I\infty}$ and $\tilde{q}$ for various $\ell_0/2\ell$ under $G_{12}/E_1 = 0.02$, $E_2/E_1 = E_3/E_1 = 0.5$ and $h_u/\ell = 0.05$. According to these results, we can conclude that the values of $\gamma G_{12}/K_{I\infty}$ increase, but the absolute values of $\tilde{q}_{cr}$ decrease with $\ell_0/2\ell$.

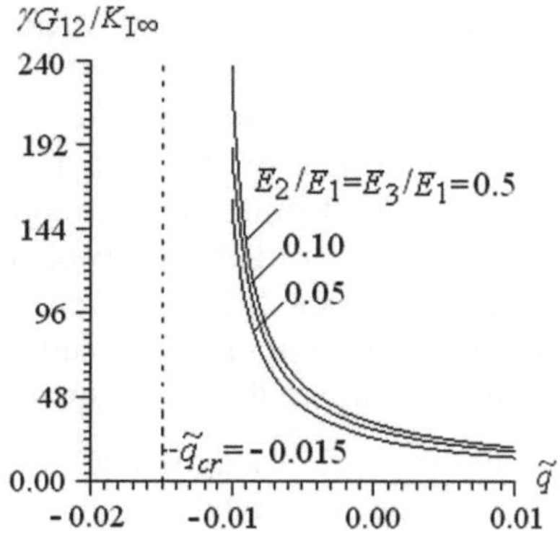


**Fig. S1.8** The dependencies between $\gamma G_{12}/K_{I\infty}$ and $\tilde{q}$ for the various values of $E_2/E_1 = E_3/E_1$ under $h_u/\ell = 0.05$, $\ell_0/2\ell = 0.15$, $G_{12}/E_1 = 0.02$

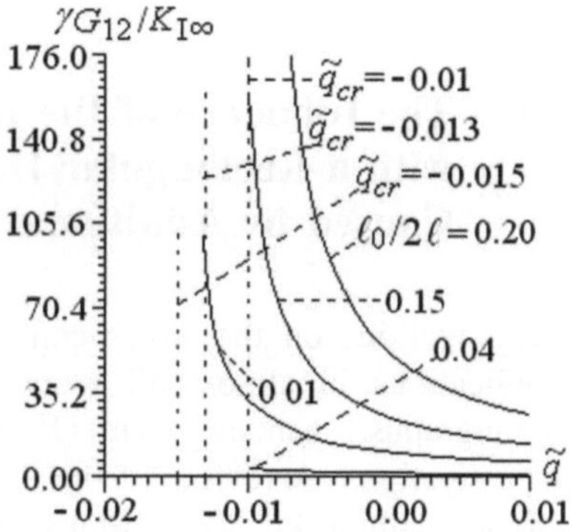


**Fig. S1.9** The influence of the crack length, i.e. of the $\ell_0/2\ell$ on the dependence between $\gamma G_{12}/K_{I\infty}$ and $\tilde{q}$ under $h_u/\ell = 0.05$, $E_2/E_1 = E_3/E_1 = 0.5$, $G_{12}/E_1 = 0.02$

## *Conclusions*

Thus, in the present section, within the scope of the approach developed in Chap. 4, the influence of the initial tension and compression of a simply supported strip containing a crack on the values of SIF of mode I and on the values of ERR of mixed mode has been investigated. It was assumed that the initial stresses operated along the crack whose edges are parallel to the free face planes of the strip. Moreover, it was assumed that the material of the strip was orthotropic. The investigations were carried out by employing FEM. From the numerical results obtained the following inferences can be drawn:

- The values of SIF and ERR decrease (increase) monotonically with initial tension (compression);
- The influence of the initial stresses on the SIF and ERR increases monotonically with crack length and with decreasing $G_{12}/E_1$, $E_2/E_1$, $E_3/E_1$ where $G_{12}$ is the shear modulus of the strip material in the strip plane i.e. in the plane $Ox_1x_2$; $E_1$, $E_2$ and $E_3$ are the moduli of elasticity in the directions of the $Ox_1$, $Ox_2$ and $Ox_3$ axes respectively.

- The values of SIF and ERR increase (decrease) with decreasing $G_{12}/E_1$ ($E_2/E_1 = E_3/E_1$).
- The absolute values of $\tilde{q}_{cr}$ decrease with decreasing $G_{12}/E_1$.
- The influence of the ratio $G_{12}/E_1$ on the values of SIF and ERR is more significant than that of the ratio $E_2/E_1$ and $E_3/E_1$.
- The values of ERR (the absolute values of $\tilde{q}_{cr}$) increase (decrease) as the crack approaches the free face plane of the strip.
- The obtained numerical results are also important in the classical sense, i.e. in the case where the initial stresses are absent ($\tilde{q} = 0$).

In the paper (Akbarov and Turan 2011) the problem considered in the present section was extended for the finitely pre-strained plate strip made from high elastic material.

## S1.2 The Influence of the Initial Tension of a Strip with a Rectangular Hole on the Stress Concentration Caused by Additional Loading

Investigations on the stress concentration around holes have a wide range of applications in almost all branches of modern industry. There are a lot of monographs, such as Savin (1951, 1968), which contain the results of these investigations. A review of the aforementioned study is given in various papers (Kosmodamianskii 2002; Maksimyuk et al. 2003; Lei et al. 2001) and many others. It should be noted that many researchers are continuing to study these investigations at present. It follows from analyses of these investigations that there are none among them regarding the study of the influence of the initial stresses arising because of an initial stretching or another type of initial loading, on the stress concentration caused by an additional loading for the case in which the superposition principle is not applicable. Here under non-applicability of the superposition principle it is understood that the stress field caused by the additional loading depends significantly on the initial loading. A theoretical investigation of the phenomenon corresponding to these cases requires the use of complicated geometrical nonlinear equations of mechanics of the deformable body. However, according to well-known mechanical considerations, for the cases where the magnitude of the initial loading is greater than that of the additional loading, these investigations can be carried out within the framework of the zeroth and first approximations of the approach developed and employed in the previous sections. In the present section, by applying the approach developed in Chap. 4 an attempt is made in this field and the influence of the initial tension of a simply supported strip containing a rectangular hole on the stress concentration around the hole caused by bending the strip under the action of the uniformly distributed normal forces on the plane of its upper face is presented and discussed.

This section is based on a paper (Akbarov et al. 2004a).

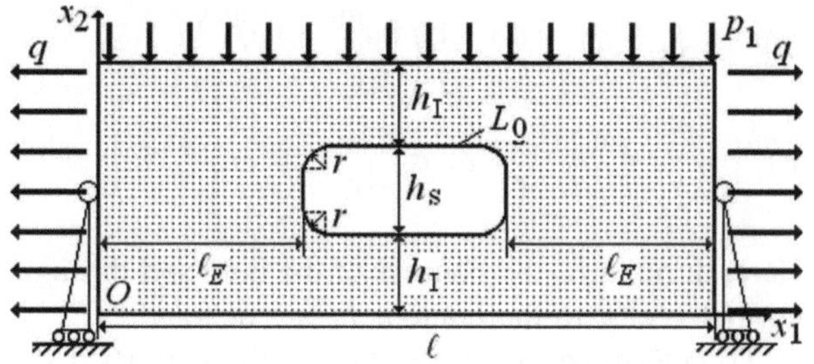


**Fig. S1.10** The geometry of the plate-strip and hole

## *S1.2.1 Formulation of the Problem*

Consider a plate-strip containing a rectangular hole or slot with rounded-off corners. The geometry of the plate-strip is shown in Fig. S1.10. We distinguish three states of the strip the first of which shall be regarded as a natural un-deformed state, the second and the third ones as initial and perturbed state respectively. The mentioned initial and perturbed states can be considered as the zeroth and first approximations respectively within the scope of the approach developed in Chap. 4. Shall henceforth label all the quantities referred to the unperturbed and perturbed states by the upper indices (0) and (1) respectively.

With the strip the Cartesian coordinate system $Ox_1x_2x_3$ so as to be the Lagrange coordinates in the initial state is associated. Assume that the plate-strip occupies the region $\{0 \le x_1 \le \ell, 0 \le x_2 \le h, -\infty < x_3 < +\infty\}$ and the axis $Ox_3$ is directed in a direction which is perpendicular to the one in Fig. S1.10, therefore the axis $Ox_3$ is not shown in this figure. Investigate a plane-strain state (in the $Ox_1x_2$ plane) in the plate-strip in the perturbed state within the framework of the equations and relations of the first approximation of the approach proposed in Sect. 4.2.

Suppose that the material of the strip is orthotropic with principal axes $Ox_1$, $Ox_2$ and $Ox_3$. Moreover, assume that the plate-strip is simply supported at the ends and in the initial state the uniformly distributed normal stretching forces with intensity $q$ act on these ends (Fig. S1.10). However, in the perturbed state on the upper face plane of the strip the additional uniformly distributed normal forces with intensity $p_1 = \varepsilon p$ act. As in previous section, we assume that $\varepsilon$ is a small parameter and $\varepsilon < 1$. We assume that $p_1 < q$ and $\varepsilon = O(p_1/q)$.

We investigate the influence of the initial stretching of the strip on the stress concentration around the hole caused by the additional forces with intensity $p_1$. For this purpose we use the approach developed in Sect. 4.2 (without the boundary form perturbation technique). In this case the determination of the values of the zeroth approximation (or the values related to the initial state) is reduced to the solution to the following boundary value problem.

$$\frac{\partial \sigma_{ij}^{(0)}}{\partial x_j} = 0,\ \sigma_{11}^{(0)} = A_{11}\varepsilon_{11}^{(0)} + A_{12}\varepsilon_{22}^{(0)},\ \sigma_{22}^{(0)} = A_{12}\varepsilon_{11}^{(0)} + A_{22}\varepsilon_{22}^{(0)},$$

$$\sigma_{12}^{(0)} = 2A_{66}\varepsilon_{12}^{(0)},\ \varepsilon_{ij}^{(0)} = \frac{1}{2}\left(\frac{\partial u_i^{(0)}}{\partial x_j} + \frac{\partial u_j^{(0)}}{\partial x_i}\right),\ u_2^{(0)}\Big|_{x_1=0;\ell} = 0,$$

$$\sigma_{11}^{(0)}\Big|_{x_1=0} = \sigma_{11}^{(0)}\Big|_{x_1=\ell} = q,\ \sigma_{i1}^{(0)}\Big|_{x_2=0;h} = 0,\ \sigma_{ji}^{(0)} n_j\Big|_{L_0} = 0,\ i;j = 1,2. \tag{S1.21}$$

Here, through $L_0$ the contour of the hole and through $n_j$ the component of the unit normal vector to the contour $L_0$ are denoted.

To determine the stress state in the perturbed state, i.e. the values of the first approximation the following boundary value problem must be solved.

$$\frac{\partial}{\partial x_j}\left[\sigma_{ji}^{(1)} + \sigma_{in}^{(0)}\frac{\partial u_i^{(1)}}{\partial x_n}\right] = 0,\ \sigma_{11}^{(1)} = A_{11}\varepsilon_{11}^{(1)} + A_{12}\varepsilon_{22}^{(1)},\ \sigma_{22}^{(1)} = A_{12}\varepsilon_{11}^{(1)} + A_{22}\varepsilon_{22}^{(1)},$$

$$\sigma_{12}^{(1)} = 2A_{66}\varepsilon_{12}^{(1)},\ \varepsilon_{ij}^{(1)} = \frac{1}{2}\left(\frac{\partial u_i^{(1)}}{\partial x_j} + \frac{\partial u_j^{(1)}}{\partial x_i}\right),\ u_2^{(1)}\Big|_{x_1=0;\ell} = 0,$$

$$\left[\sigma_{11}^{(1)} + \sigma_{1n}^{(0)}\frac{\partial u_1^{(1)}}{\partial x_n}\right]\Bigg|_{x_1=0;\ell} = 0,\ \sigma_{i2}^{(1)}\Big|_{x_2=h} = p\delta_i^2,\ \sigma_{i2}^{(1)}\Big|_{x_2=0} = 0,$$

$$\delta_1^2 = 0;\ \delta_2^2 = 1,\ \left(\sigma_{ji}^{(1)} + \sigma_{in}^{(0)}\frac{\partial u_i^{(1)}}{\partial x_n}\right)\Bigg|_{L_0} n_j = 0,\ i;j;n = 1,2. \tag{S1.22}$$

It follows from the problem formulation that $\sigma_{ij}^{(0)} \sim O(q)$ and $\sigma_{ij}^{(1)} \sim O(p)$, and the error of the numerical results obtained within the framework of the used approach has the order $o(\varepsilon)$. Consequently, for the given values of the initial loading $q$ and for the required error order it can be limited the values of $p$ for which the numerical results obtained below will be applicable.

### *S1.2.2 FEM Modeling*

For the FEM modeling of the boundary value problem (S1.21) the functional

$$\Pi^{(0)} = \frac{1}{2}\iint\limits_{\Omega-\Omega_{L_0}} \sigma_{ij}^{(0)}\varepsilon_{ij}^{(0)} dx_1 dx_2 - \int_0^h q u_1^{(0)}\Big|_{x_1=0} dx_2 - \int_0^h q u_1^{(0)}\Big|_{x_1=\ell} dx_2 \tag{S1.23}$$

is used, where

$$\Omega = \{0 \le x_1 \le \ell,\ 0 \le x_2 \le h\} \tag{S1.24}$$

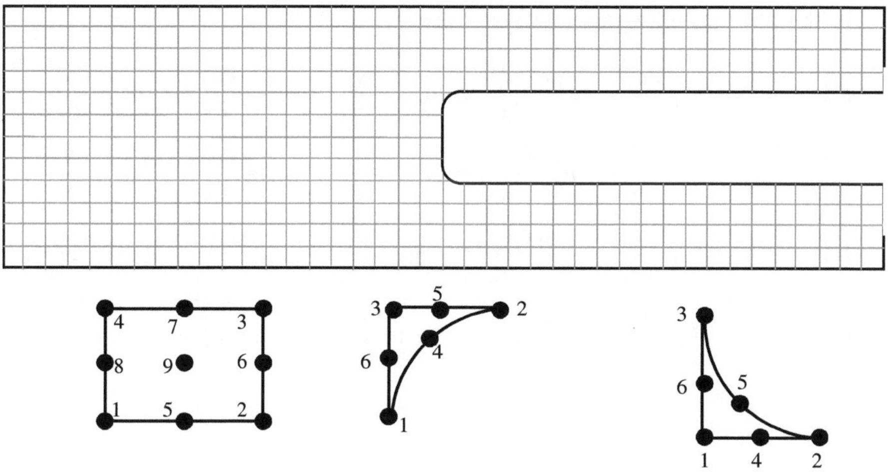


**Fig. S1.11** Finite element mesh and used finite elements with nodes

and $\Omega_{L_0}$ is the region occupied by the hole. Using the virtual work principle and employing the well known Ritz technique FEM modeling of the problem (S1.21) is obtained from the equation $\delta\Pi^{(0)} = 0$. In this case the region $\Omega - \Omega_{L_0}$ is divided into 804 rectangular Lagrange family quadratic elements with 6898 NDOF. Around the corners of the hole, the curvilinear triangular finite elements with six nodes are used. The used finite element mesh of the plate-strip and the geometry of the finite elements are shown in Fig. S1.11. The selection of the NDOF values follows from the requirements that the boundary conditions should be satisfied with very high accuracy and the numerical results obtained for various NDOF should converge.

For the FEM modeling of the problem (S1.22) the following functional is introduced.

$$\Pi^{(1)} = \frac{1}{2} \iint\limits_{\Omega-\Omega_{L_0}} \left( T_{11}^{(1)} \frac{\partial u_1^{(1)}}{\partial x_1} + T_{12}^{(1)} \frac{\partial u_2^{(1)}}{\partial x_1} + T_{21}^{(1)} \frac{\partial u_1^{(1)}}{\partial x_2} + T_{22}^{(1)} \frac{\partial u_2^{(1)}}{\partial x_2} \right) dx_1 dx_2$$

$$- \int\limits_0^{\ell} p u_2^{(1)} \Big|_{x_2=h} dx_1. \tag{S1.25}$$

In (S1.25) the following notation is used.

$$
\begin{aligned}
T_{11}^{(1)} &= \sigma_{11}^{(1)} + \sigma_{11}^{(0)}\frac{\partial u_1^{(1)}}{\partial x_1} + \sigma_{12}^{(0)}\frac{\partial u_1^{(1)}}{\partial x_2} = \omega_{1111}\frac{\partial u_1^{(1)}}{\partial x_1} + \omega_{1122}\frac{\partial u_2^{(1)}}{\partial x_2} + \omega_{1112}\frac{\partial u_1^{(1)}}{\partial x_2},\\
T_{12}^{(1)} &= \sigma_{12}^{(1)} + \sigma_{11}^{(0)}\frac{\partial u_2^{(1)}}{\partial x_1} + \sigma_{12}^{(0)}\frac{\partial u_2^{(1)}}{\partial x_2} = \omega_{1212}\frac{\partial u_1^{(1)}}{\partial x_2} + \omega_{1222}\frac{\partial u_2^{(1)}}{\partial x_2} + \omega_{1221}\frac{\partial u_2^{(1)}}{\partial x_1},\\
T_{21}^{(1)} &= \sigma_{12}^{(1)} + \sigma_{12}^{(0)}\frac{\partial u_1^{(1)}}{\partial x_1} + \sigma_{22}^{(0)}\frac{\partial u_1^{(1)}}{\partial x_2} = \omega_{2112}\frac{\partial u_1^{(1)}}{\partial x_2} + \omega_{2121}\frac{\partial u_2^{(1)}}{\partial x_1} + \omega_{2111}\frac{\partial u_1^{(1)}}{\partial x_1},\\
T_{22}^{(1)} &= \sigma_{22}^{(1)} + \sigma_{12}^{(0)}\frac{\partial u_2^{(1)}}{\partial x_1} + \sigma_{22}^{(0)}\frac{\partial u_2^{(1)}}{\partial x_2} = \omega_{2211}\frac{\partial u_1^{(1)}}{\partial x_2} + \omega_{2222}\frac{\partial u_2^{(1)}}{\partial x_2} + \omega_{2221}\frac{\partial u_2^{(1)}}{\partial x_1},
\end{aligned}
\tag{S1.26}
$$

where

$$
\begin{aligned}
&\omega_{1111} = A_{11} + \sigma_{11}^{(0)},\ \omega_{1122} = A_{12},\ \omega_{1112} = \sigma_{12}^{(0)},\ \omega_{1212} = A_{66},\ \omega_{1222} = \sigma_{12}^{(0)},\\
&\omega_{1221} = A_{66} + \sigma_{11}^{(0)},\ \omega_{2112} = A_{66} + \sigma_{22}^{(0)},\ \omega_{2121} = A_{66},\ \omega_{2111} = \sigma_{12}^{(0)},\\
&\omega_{2211} = A_{12},\ \omega_{2222} = A_{22} + \sigma_{22}^{(0)},\ \omega_{2221} = \sigma_{12}^{(0)}.
\end{aligned}
\tag{S1.27}
$$

Doing some mathematical manipulations the expression

$$
\begin{aligned}
\delta\Pi^{(1)} = &\iint\limits_{\Omega-\Omega_{L_0}}
\left\{\left[\omega_{1111}\frac{\partial u_1^{(1)}}{\partial x_1} + \frac{1}{2}\omega_{2111}\frac{\partial u_1^{(1)}}{\partial x_2} + \frac{1}{2}\omega_{1122}\frac{\partial u_2^{(1)}}{\partial x_2} + \frac{1}{2}\omega_{1112}\frac{\partial u_1^{(1)}}{\partial x_2} + \frac{1}{2}\omega_{2211}\frac{\partial u_2^{(1)}}{\partial x_2}\right]\delta\left(\frac{\partial u_1^{(1)}}{\partial x_1}\right)\right.\\
&+\left[\omega_{2222}\frac{\partial u_2^{(1)}}{\partial x_2} + \frac{1}{2}\omega_{2221}\frac{\partial u_2^{(1)}}{\partial x_1} + \frac{1}{2}\omega_{2211}\frac{\partial u_1^{(1)}}{\partial x_1} + \frac{1}{2}\omega_{1222}\frac{\partial u_2^{(1)}}{\partial x_1} + \frac{1}{2}\omega_{1122}\frac{\partial u_1^{(1)}}{\partial x_1}\right]\delta\left(\frac{\partial u_2^{(1)}}{\partial x_2}\right)\\
&+\left[\frac{1}{2}\omega_{1112}\frac{\partial u_1^{(1)}}{\partial x_1} + \frac{1}{2}\omega_{1212}\frac{\partial u_2^{(1)}}{\partial x_1} + \omega_{2112}\frac{\partial u_1^{(1)}}{\partial x_2} + \frac{1}{2}\omega_{2121}\frac{\partial u_2^{(1)}}{\partial x_1} + \frac{1}{2}\omega_{2111}\frac{\partial u_1^{(1)}}{\partial x_1}\right]\delta\left(\frac{\partial u_1^{(1)}}{\partial x_2}\right)\\
&\left.+\left[\frac{1}{2}\omega_{1212}\frac{\partial u_1^{(1)}}{\partial x_2} + \frac{1}{2}\omega_{1221}\frac{\partial u_2^{(1)}}{\partial x_1} + \omega_{1222}\frac{\partial u_2^{(1)}}{\partial x_2} + \frac{1}{2}\omega_{2121}\frac{\partial u_1^{(1)}}{\partial x_2} + \frac{1}{2}\omega_{2221}\frac{\partial u_2^{(1)}}{\partial x_2}\right]\delta\left(\frac{\partial u_2^{(1)}}{\partial x_1}\right)\right\}dx_1dx_2\\
&-\int\limits_0^{\ell} p\delta u_2^{(1)}\Big|_{x_2=h}dx_1
\end{aligned}
\tag{S1.28}
$$

is obtained from (S1.25) and (S1.26).

Using the expressions (S1.27) by direct verification it is proven that $\omega_{ijnm} = \omega_{mnji}$. Taking these equalities into account the following expression for $\delta\Pi^{(1)}$ is obtained from (S1.28).

$$\delta\Pi^{(1)} = \iint\limits_{\Omega-\Omega_{L_0}} \left\{ \left[ \omega_{1111}\frac{\partial u_1^{(1)}}{\partial x_1} + \omega_{2111}\frac{\partial u_1^{(1)}}{\partial x_2} + \omega_{1122}\frac{\partial u_2^{(1)}}{\partial x_2} \right] \delta\left(\frac{\partial u_1^{(1)}}{\partial x_1}\right) \right.$$
$$+ \left[ \omega_{2222}\frac{\partial u_2^{(1)}}{\partial x_2} + \omega_{2221}\frac{\partial u_2^{(1)}}{\partial x_1} + \omega_{2211}\frac{\partial u_1^{(1)}}{\partial x_1} \right] \delta\left(\frac{\partial u_2^{(1)}}{\partial x_2}\right)$$
$$+ \left[ \omega_{2111}\frac{\partial u_1^{(1)}}{\partial x_1} + \omega_{2121}\frac{\partial u_2^{(1)}}{\partial x_1} + \omega_{2112}\frac{\partial u_1^{(1)}}{\partial x_2} \right] \delta\left(\frac{\partial u_1^{(1)}}{\partial x_2}\right)$$
$$\left. + \left[ \omega_{1212}\frac{\partial u_1^{(1)}}{\partial x_2} + \omega_{1221}\frac{\partial u_2^{(1)}}{\partial x_1} + \omega_{1222}\frac{\partial u_2^{(1)}}{\partial x_2} \right] \delta\left(\frac{\partial u_2^{(1)}}{\partial x_1}\right) \right\} dx_1 dx_2$$
$$- \int_0^{\ell} p\delta u_2^{(1)}\Big|_{x_2=h} dx_1. \tag{S1.29}$$

From the equation $\delta\Pi^{(1)} = 0$ with the use of the expressions (S1.29) and (S1.26) after doing some well-known transformations the equations

$$\partial\left[\sigma_{ji}^{(1)} + \sigma_{in}^{(0)}\partial u_i^{(1)} \big/ \partial x_n\right] \big/ \partial x_j = 0$$

and boundary conditions

$$\left[\sigma_{ji}^{(1)} + \sigma_{in}^{(0)}\partial u_i^{(1)} \big/ \partial x_n\right]\Big|_{L_0} \times n_j = 0,\ \sigma_{i2}^{(1)}\Big|_{x_2=h} = p\delta_i^2,\ \sigma_{i2}^{(1)}\Big|_{x_2=0} = 0$$

are attained.

In this way, it is proven the validity of the functional (S1.25) for the problem (S1.22). Note that, with the same finite elements and with the same arrangement of those which are used for the FEM modeling of the problem (S1.21), the FEM modeling for the problem (S1.22) is formulated from the equation $\delta\Pi^{(1)} = 0$.

### *S1.2.3 Numerical Results and Discussions*

It is assumed that the material of the strip is a composite consisting of a large number of alternating layers of two materials. Suppose that the material of each layer is isotropic and these layers are located on the planes $x_2 = const$. A pack consisting of these two layers is a representative element of the considered composite. According to the continuum approach, the elasticity relations between the averaged values of stresses and strains are determined through the effective mechanical constants. Note that for the considered composite the averaging procedure is made over the thickness of the representative pack and the values of these effective mechanical constants are determined through the mechanical constants of the components of the pack by the use the well known formulae.

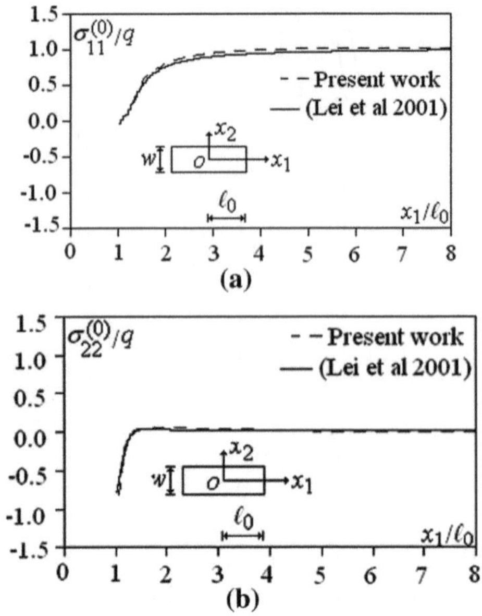


**Fig. S1.12** The comparison of the present results obtained by present algorithm with corresponding results obtained in Lei et al. (2001)

As usually, the components of the representative element (pack) are divided into reinforcing and matrix elements (layers) and as reinforcing element a comparatively rigid element (layer) is taken.

As in Sect. 3.3.3, below the values related to the matrix and to the reinforcing material will be tabulated by 1 and 2 respectively: $\lambda^{(k)}$, $\mu^{(k)}$ are Lamé constants; $E^{(k)}$ are Young's modulus; $\nu^{(k)}$ are Poisson's ratios and $\eta^{(k)}$ are the concentration of the components in the representative pack. It is known that in the considered case the material of the representative pack (or the material of the plate-strip) within the framework of the continuum approach can be modeled as homogeneous transversely isotropic one the effective mechanical constants $A_{ij}$ of which are determined by expressions given in (3.83). Assume that $\nu^{(1)} = \nu^{(2)} = 0.3$, $\eta^{(1)} = \eta^{(2)} = 0.5$, $h/\ell = 0.15$, $r/\ell = 1/80$.

It is evident that for efficient applicability of the numerical results that will be discussed below the thickness of the representative pack must be significantly smaller than the size of the considered hole. Note that the above described solution procedure can be also applied without concrete definition of the type of the composite material from which the strip is fabricated. In this case, the values of the mechanical constants $A_{ij}$ must be given directly.

Thus, return to the case for which the constants $A_{ij}$ are determined through the formulae (3.83) and first compare some numerical results obtained by the present algorithm and programmes with those obtained in the paper (Lei et al. 2001). This comparison is given in Fig. S1.12 for the distribution of the stresses $\sigma_{11}^{(0)} \big/ q$ (Fig. S1.12a) and $\sigma_{22}^{(0)} \big/ q$ (Fig. S1.12b) with respect to $x_1$ for $x_1 \geq \ell_0$, $x_2 = 0$, in the case where $E^{(2)}/E^{(1)} = 1$, $\nu^{(1)} = \nu^{(2)} = 0.25$, $2\ell_0/w = 3$ (Fig. S1.12), $h/\ell = 0.15$, $h_S/(\ell - 2\ell_E) = w/(2\ell_0)$, $h_S/\ell = 0.025$, $\ell_E/\ell = 0.4625$ (Fig. S1.10).

It should be noted that, in paper Lei et al. (2001) the results presented in Fig. S1.12 are obtained for an infinite plate by using the well-known complex variable theory and the confirm transformation technique developed in (Mushelishvili 1954). Hence, the validity of the algorithm and used programmes follows from these comparisons. Moreover, these results show that for the cases where $\ell_E/\ell \geq 0.4625$, $h/\ell \geq 0.15$, $h_S/\ell \leq 0.025$ the considered stress distribution around the hole (i.e. finite element modeling used above) can also be taken for that in the infinite plane.

Figure S1.13 shows the distribution of the stress $\sigma_{11}^{(1)}$ with respect to $x_1/\ell$ in the cross-sections determined as $x_2 = h$ (Fig. S1.13a), $5h/6$ (Fig. S1.13b), $2h/3$ (Fig. S1.13c), $h/3$ (Fig. S1.13d), $h/6$ (Fig. S1.13e), 0 (Fig. S1.13f) for $h/\ell = 0.15$, $\ell_E/\ell = 0.25$, $E^{(2)}/E^{(1)} = 10$ and $h_S/\ell = h_I/\ell = 0.05$. The results given in this figure illustrate the influence of the initial stretching (i.e. the influence of the parameter $q/E^{(1)}$) on the values of the stress $\sigma_{11}^{(1)}$. First, it must be noted that for the considered problem parameters the influence of the initial stretching of the strip without hole on the same stress distribution is insignificant and can be neglected. The graphs of the distribution of the stress $\sigma_{11}^{(1)}$ in the plate-strip without hole show that for all considered cross-section (i.e. for all layer determined as $x_2 = const$) the absolute values of $\sigma_{11}^{(1)}/p$ increase with $x_1/\ell$ for $x_1 \in [0, \ell/2]$. It follows from the results given in Fig. S1.13 that the existence of the hole significantly changes this distribution. For example, according to these results it can be concluded that around of the hole corner (i.e. around of the $x_1 = \ell_E + r$) the character of the distribution of the stress $\sigma_{11}^{(1)}$ changes significantly. In the part of the strip which is in the upper of the hole the mentioned changing is illustrated by the graphs given in Fig. S1.13a–c, according to which, $\sigma_{11}^{(1)}$ becomes tensional stress in the near vicinity of the point $x_1 = \ell_E$ in the upper face plane of the plate-strip. However, by approaching from upper boundary of the strip to the upper boundary of the hole this tensional character of the $\sigma_{11}^{(1)}$ near $x_1 = \ell_E$ disappears and the compressed stress appears instead of that. The latter has its peak at the point $x_1 = \ell_E + r$ in the upper boundary of the hole. However for $x_2 = h$ and $5h/6$ this peak appears at the point $x_1 = \ell/2$.

The graphs given in Fig. S1.13d–f show that in the part of the strip which is in the lower of the hole the local peaks of the absolute values of $\sigma_{11}^{(1)}$ appear at $x_1 = \ell_E + r$ for the lower boundary of the hole and for the lower face plane of the strip, i.e. for $x_2 = h/3$ and 0.

The observed character of the distribution of the stress $\sigma_{11}^{(1)}$ agrees with the well known engineering considerations. But, in all the considered cases the existence of the initial stretching causes the decreasing of the absolute values of the stress $\sigma_{11}^{(1)}$. This influence is more considerable at $x_1 = \ell/2$ ($x_1 = \ell_E + r$) for the section,

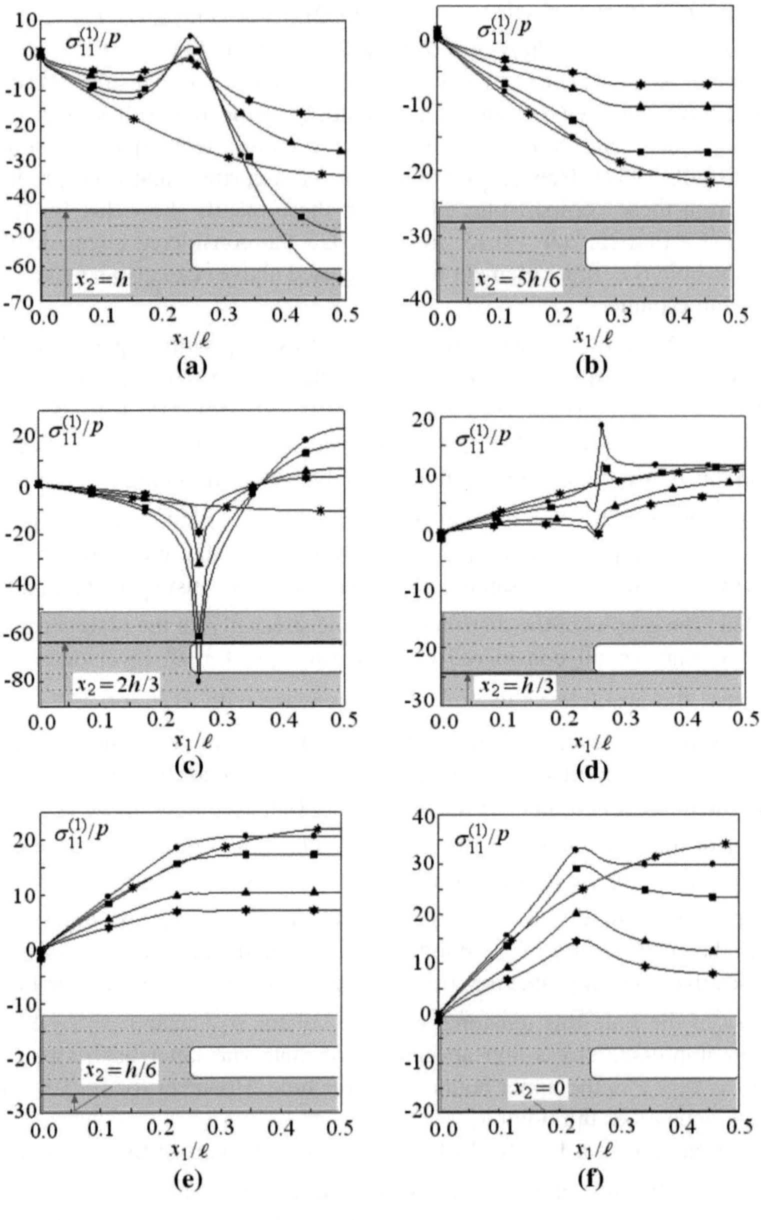


**Fig. S1.13** The distribution of $\sigma_{11}^{(1)}\big/p$ with respect to $x_1/\ell$ under various initial stretching at the cross-section: **a** $x_2/\ell = h/\ell$, **b** $x_2/\ell = 10h/12\ell$, **c** $x_2/\ell = 8h/12\ell$, **d** $x_2/\ell = 4h/12\ell$, **e** $x_2/\ell = 2h/12\ell$, **f** $x_2/\ell = 0$ in the case where •—$q/E^{(1)} = 0.0$; ■—$q/E^{(1)} = 0.0001$; ▲—$q/E^{(1)} = 0.0005$; ✱—$q/E^{(1)} = 0.001$; ✳—strip without hole

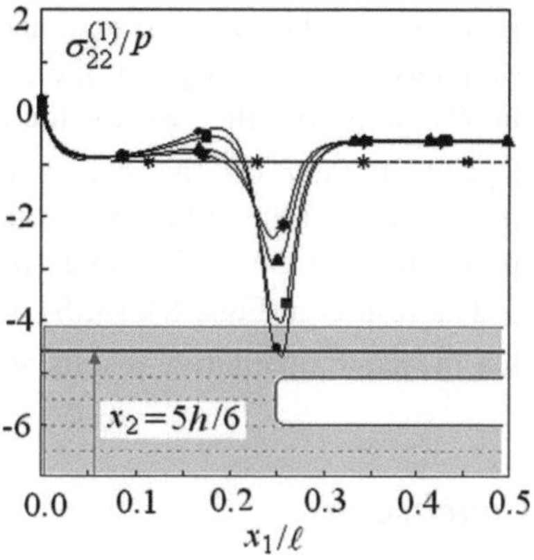


**Fig. S1.14** The distribution of $\sigma_{22}^{(1)}/p$ with respect to $x_1/\ell$ for various initial stretching, in the case where
•—$q/E^{(1)} = 0.0$;
■—$q/E^{(1)} = 0.0001$;
▲—$q/E^{(1)} = 0.0005$;
★—$q/E^{(1)} = 0.001$;
✳—strip without hole

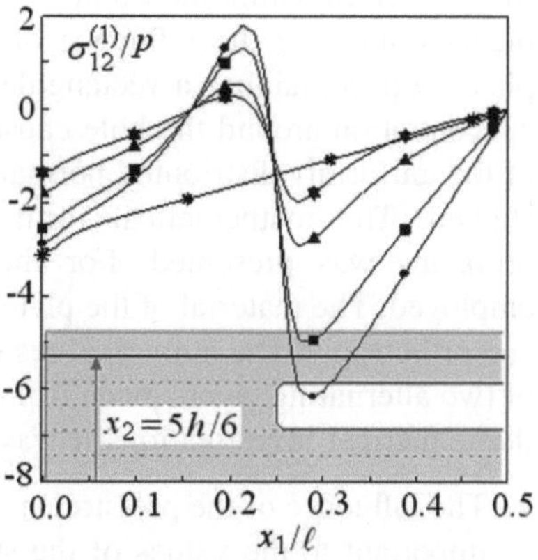


**Fig. S1.15** The distribution of $\sigma_{12}^{(1)}/p$ with respect to $x_1/\ell$ for various initial stretching, in the case where
•—$q/E^{(1)} = 0.0$;
■—$q/E^{(1)} = 0.0001$;
▲—$q/E^{(1)} = 0.0005$;
★—$q/E^{(1)} = 0.001$;
✳—strip without hole

i.e. for the selected values of $x_2$, which is far from of (near) the upper or lower boundaries of the hole. In the quantitative sense the degree of the initial stretching influence to the distribution of the $\sigma_{11}^{(1)}$ increases monotonically with $q/E^{(1)}$.

Figures S1.14 and S1.15 show the distribution of the stresses $\sigma_{22}^{(1)}$ and $\sigma_{12}^{(1)}$ in a characteristic cross-section of the plate-strip and the influence of the initial stretching of the strip, i.e. the influence of the parameter $q/E^{(1)}$ to this distribution for $h/\ell = 0.15$, $\ell_E/\ell = 0.25$, $E^{(2)}/E^{(1)} = 10$ and $h_S/\ell = h_I/\ell = 0.05$. Note that the influence of the initial stretching to the distribution of these stresses in the plate-strip without hole is insignificant and can be neglected. According to these graphs, it can be concluded that the peak of the absolute values of the stresses $\sigma_{22}^{(1)}$ and $\sigma_{12}^{(1)}$ arise at $x_1 = \ell_E + r$ and in the near vicinity of $x_1 = \ell_E + r$ respectively. In this case the absolute values of these peaks decrease monotonically with $q/E^{(1)}$.

It follows from the Figs. S1.13, S1.14, S1.15 that the values of $\sigma_{22}^{(1)}/p$ and $\sigma_{12}^{(1)}/p$ are significantly smaller than the corresponding values of $\sigma_{11}^{(1)}/p$.

Nevertheless the stresses $\sigma_{22}^{(1)}$ and $\sigma_{12}^{(1)}$ have also a great significance for the estimation of the strength of the considered composite material in the direction of the $Ox_2$ axis. In other words, the adhesion strength of the considered composite depends mainly on the values of the stresses $\sigma_{22}^{(1)}$ and $\sigma_{12}^{(1)}$. Therefore, the investigation of the influence of the pre-stretching to these stresses has also a great importance for the engineering application.

The results in Figs. S1.13, S1.14, S1.15 are obtained for $E^{(2)}/E^{(1)} = 10$. Note that the same qualitative results are obtained for other values of $E^{(2)}/E^{(1)}$.

## Conclusions

In this section within the framework of the approach developed in Chap. 4 under a plane-strain state the influence of the initial stretching of the simply supported plate strip containing a rectangular hole with rounded-off corners on the stress concentration around the hole caused by the bending of the strip under the action of the uniformly distributed normal forces on the plane of its upper face, has been studied. The mathematical formulation of the corresponding boundary-value problems was presented. For the solution to these problems the FEM was employed. The material of the plate-strip is selected as linear elastic, homogeneous and orthotropic. The numerical results were presented for the composite consisting of two alternating layers made from isotropic linear elastic materials. As a result of the numerical investigations it was established that:

1. The influence of the pre-stretching of the plate-strip becomes significantly more important to the values of the stresses caused by the additional loading in the case where the considered hole exists.
2. In the quantitative sense the peak of the influence of the pre-stretching arises in the near vicinity of the corner of the hole and in the middle of the plate-strip, i.e. at $x_1 = \ell/2$.
3. In all the considered cases the absolute values of the stresses $\sigma_{ij}^{(1)}$ decrease monotonically with an increase in the pre-stretching, i.e. with an increase in the values of $q/E^{(1)}$.
4. The results obtained here relate to the composite material consisting of alternating layers of two isotropic, homogeneous materials. Nevertheless, these results have significance in the qualitative sense for other materials. However, in the quantitative sense, for the other materials these results must be recalculated with the use of the approach developed in the present section.
5. The results obtained here can be applied to an engineering design which has a hole under loading, which is similar to that considered here. In this case the results can be applied directly and can also be taken as fundamentally true.
6. The results given in the Figs. S1.13, S1.14, S1.15 and obtained in the case where $q/E^{(1)} = 0.0$ are the corresponding ones obtained within the framework

of the classical linear theory of elasticity. In this case, from the equations and relations (S1.21) it is obtained that $\sigma_{ij}^{(0)} \equiv 0$ and the equations and relations (S1.28) of the linearized theory of elasticity coincide with the corresponding ones of the classical linear theory of elasticity. Consequently, the numerical results presented in this section also have significance in the application and theoretical sense.

In papers (Akbarov et al. 2008, 2010a) the investigations carried out in the present section are continued for other stress concentration problems in pre-stressed plates.

# Supplement 2: Self-Balanced Stresses Caused by Periodical Curving of Two Neighboring and Periodically Located Row of Fibers in an Infinite Matrix

In Supplement 2, numerical results related to the self-balanced normal and shear stresses caused by the periodical curving of the fibers and acting on the interface surface of the fibers and matrix are presented and detailed. These results are obtained by employing the approach developed in Chap. 6. First, consideration is made for the case where an infinite elastic matrix contains two neighboring fibers and then this consideration is continued for the case where the mentioned elastic matrix contains periodically located row of fibers. The main attention is focused on the influence of the interaction between the fibers on the values of the studied stresses. The influence of the geometrical nonlinearity on these stresses is also considered in detail.

## S2.1 Numerical Results Related to the Case Where an Infinite Matrix Contains Two Neighboring Fibers

The results which are detailed in the present section were obtained in papers (Akbarov and Kosker 2003a, b; Kosker and Akbarov 2003). The corresponding investigations are carried out within the scope of the assumption and notation accepted in the corresponding section of Chap. 6.

Thus, assume that the middle lines of the fibers are in the same plane, i.e. in the $x_{11} = x_{12} = 0$ plane. With respect to the reciprocal location of the fibers we select for consideration the cases where the initial imperfections of the fibers are co-phase (Fig. S2.1a) and anti-phase (Fig. S2.1b). We will call these cases briefly as "co-phase" and "anti-phase" ones below. The equations of the middle lines of the fibers in "co-phase" case are given already by equations in (6.47). But these equations for "anti-phase" case are written as follows

$$x_{21} = L\sin\left(\frac{2\pi}{\ell}x_{31}\right), \quad x_{22} = -L\sin\left(\frac{2\pi}{\ell}x_{32}\right). \tag{S2.1}$$

S. D. Akbarov, *Stability Loss and Buckling Delamination*,
DOI: 10.1007/978-3-642-30290-9, © Springer-Verlag Berlin Heidelberg 2013

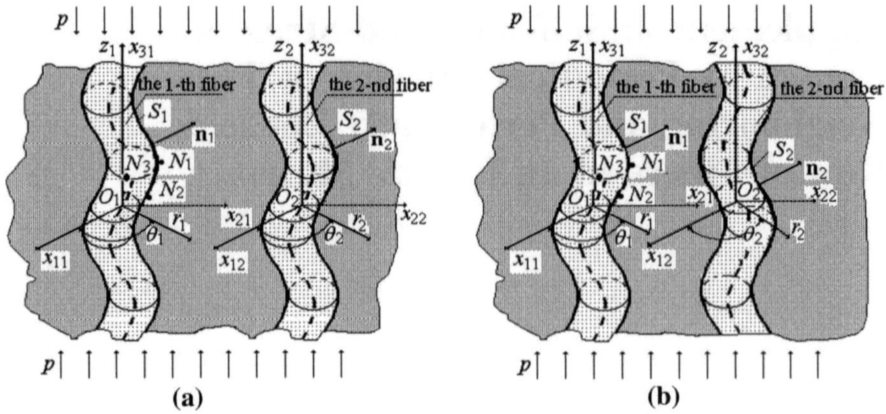


**Fig. S2.1** Co-phase (**a**) and anti-phase (**b**) periodical curving of two neighboring fibers and characteristic points on the interface surface at which the self-balanced stresses are calculated

Note that in the quantitative sense in the selected above cases for investigation of the self balanced stresses which will be discussed below become more considerable than those in the other possible reciprocal location cases of the initial imperfections of the fibers.

Thus, we consider the numerical results related to the distribution of the self-balanced stresses

$$\sigma_{nn} = \varepsilon\sigma_{rr}^{(1)}\Big|_{r_1=R}, \quad \sigma_{n\tau} = \varepsilon\sigma_{rz}^{(1)}\Big|_{r_1=R}, \quad \sigma_{ne} = \varepsilon\sigma_{r\theta}^{(1)}\Big|_{r_1=R}. \tag{S2.2}$$

Note that these stresses arise as a result of the initial imperfections of the fibers and act on the interface surfaces between the fibers and matrix. Therefore the adhesion strength of the fibrous composites depends mainly on the values of these stresses (see Corten 1967). Consequently, before the stability loss of the fibers which has been discussed in Chap. 6, the separation of these fibers from the matrix may be take place. Therefore, the investigations of these stresses have a great significance.

We use the notation introduced in Sect. 6.3.2 and taking the problem symmetry with respect to the $d'd'_1$ (Fig. 6.2) plane into account the values of the stresses (S2.2) on the points of the $S_1$ surface (Fig. S2.1) only will be considered. According to the monograph (Akbarov and Guz 2000), absolute maximum values of the stresses $\sigma_{nn}$, $\sigma_{n\tau}$ and $\sigma_{ne}$ are appear around the points $N_1$, $N_2$ and $N_3$ respectively. Note that coordinates of these points are determined from Eq. (6.12) written for the surface $S_1$ under $\{\alpha_1 t_3 = \pi/2,\ \theta = \pi/2\}$, $\{\alpha_1 t_3 = 0,\ \theta = \pi/2\}$ and $\{\alpha_1 t_3 = \pi/2,\ \theta = 0\}$ respectively.

For estimation the influence of the geometrical non-linearity of the problem statement on the values of the foregoing stresses we use the parameter $p/E^{(1)}$ where $E^{(1)}$ is a modulus of elasticity of the matrix material. Assume that $\varepsilon = 0.015$, $E^{(2)}/E^{(1)} = 50$, $\nu^{(2)} = \nu^{(1)} = 0.3$ where $\nu^{(1)}$ is a Poisson ratio of the

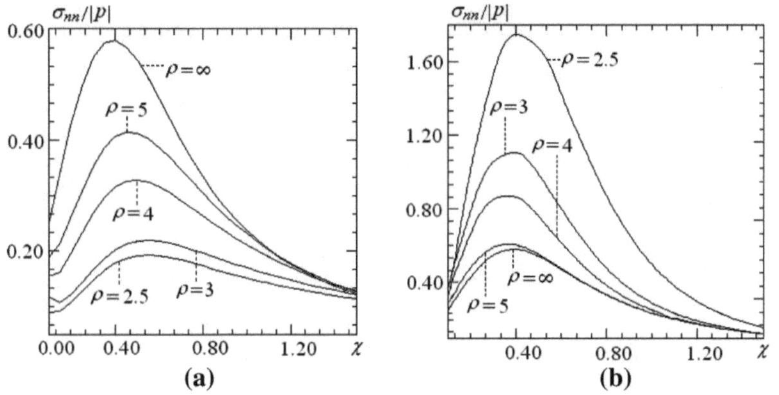


**Fig. S2.2** The graphs of the dependencies between $\sigma_{nn}/|p|$ and $\chi$ constructed for various $\rho$: **a** "co-phase" case; **b** "anti-phase" case

matrix material. Note that the mentioned effect increase with $p/E^{(1)}$ and in the cases where $|p|/E^{(1)} \sim O(10^{-5})$ that can be neglected with very high accuracy.

It is evident that as a result of the geometrical non-linearity the absolute values of the stresses (S2.2) obtained for cases where $p > 0$ and $p < 0$ become differ from each other, while under neglecting the effect of the geometrical non-linearity the mentioned values of the stresses coincide.

Taking the foregoing discussion into account, for illustration of the influence of the parameters $\rho$ $(= R_{12}/R)$ and $\chi$ $(= 2\pi R/\ell)$ (Fig. 6.2) on the values of the stresses (S2.2) we select the case where $|p|/E^{(1)} = 5 \times 10^{-5}$ and consider graphs given in Figs. S2.2, S2.3 and S2.4. These graphs illustrate dependencies among $\sigma_{nn}/|p|$, $\sigma_{n\tau}/|p|$, $\sigma_{ne}/|p|$ and $\chi$ obtained for various values of the parameter $\rho$ through which the influence of the interaction between the fibers on the stresses considered is estimated. Note that the graphs shown in Figs. S2.2a, S2.3a and S2.4a (in Figs. S2.2b, S2.3b and S2.4b) relate to "co-phase" (to "anti-phase") case.

Figure S2.2 shows that in "co-phase" case (in "anti-phase" case) the values of $\sigma_{nn}/|p|$ decrease (increase) monotonically with approaching of the fibers to each other. However, it follows from Figs. S2.3 and S2.4 that in "co-phase" case (in "anti-phase" case) the absolute values of $\sigma_{n\tau}/|p|$ and $\sigma_{ne}/|p|$ increase (decrease) monotonically with decreasing $\rho$. In this case the considerable increasing is observed in the absolute values of the shear stress $\sigma_{n\tau}/|p|$. Moreover, it follows from the obtained numerical results that with increasing $\rho$ the values of the stresses considered approach to the corresponding ones obtained for a single fiber (Akbarov and Guz 2000).

Consider the influence of the geometrical non-linearity, i.e. the influence of the $p/E^{(1)}$ on the values of the foregoing stresses. We investigate this influence for the shear stress $\sigma_{n\tau}/|p|$ in "co-phase" case and for the normal stress $\sigma_{nn}/|p|$ in "anti-phase" case.

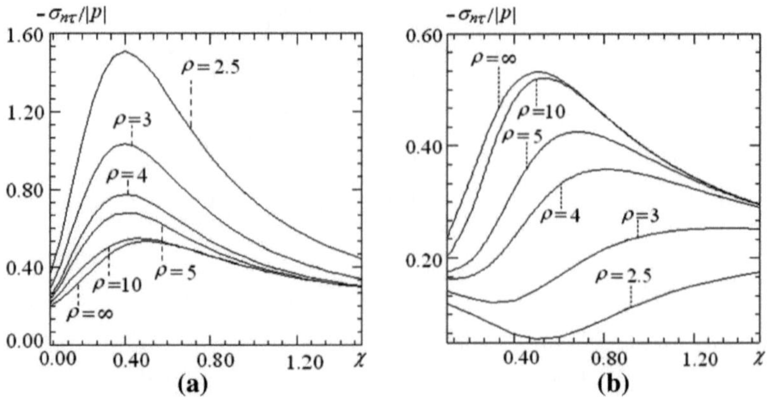


**Fig. S2.3** The graphs of the dependencies between $\sigma_{n\tau}/|p|$ and $\chi$ constructed for various $\rho$: **a** "co-phase" case; **b** "anti-phase" case

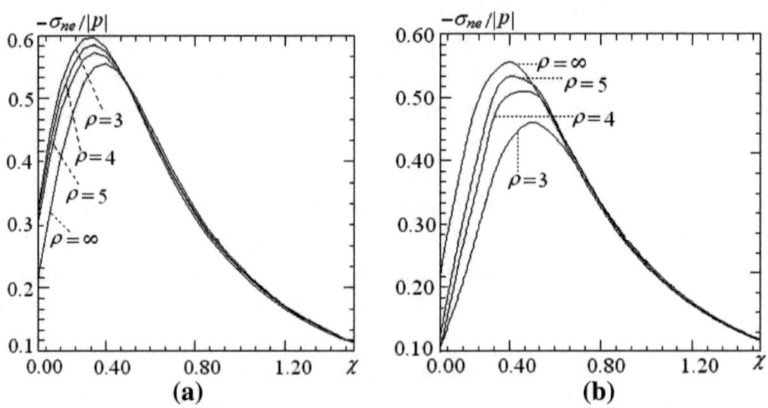


**Fig. S2.4** The graphs of the dependencies between $\sigma_{ne}/|p|$ and $\chi$ constructed for various $\rho$: **a** "co-phase" case; **b** "anti-phase" case

Data given in Tables S2.1 and S2.2 illustrate the mentioned influence on the shear stress $\sigma_{n\tau}/|p|$ in "co-phase" case for various $\chi$ under $\rho = 2.5$ (Table S2.1) and for various $\rho$ under $\chi = 0.4$ (Table S2.2). As well as, data given in Tables S2.3 and S2.4 show the influence considered on the normal stress $\sigma_{nn}/|p|$ in "anti-phase" case for various $\chi$ under $\rho = 2.5$ (Table S2.3) and for various $\rho$ under $\chi = 0.4$ (Table S2.4). It follows from these tables that as a result of the geometrical non-linearity the absolute values of the self-balanced stresses $\sigma_{nn}/|p|$ and $\sigma_{n\tau}/|p|$ increase in compression, but decrease in tension. For illustration the convergence of the numerical results, as an example, in Table S2.5 the values of $\sigma_{nn}/|p|$ obtained for "anti-phase" case under $\rho = 2.5$ and $\chi = 0.3$ in various number of the linear equations are given. It follows from this table that the used solution method is highly effective in the convergence sense.

**Table S2.1** The influence of the geometrical non-linearity, i.e. the parameter $p/E^{(1)}$ on the values of $\sigma_{n\tau}/|p|$ in "co-phase" case for various $\chi$ under $\rho = 2.5$

| $\chi$ | $p/E^{(1)}$ | | | | | | | |
|---|---|---|---|---|---|---|---|---|
| | Tension | | | | Compression | | | |
| | $5\times10^{-5}$ | $5\times10^{-4}$ | $5\times10^{-3}$ | $5\times10^{-2}$ | $-5\times10^{-5}$ | $-5\times10^{-4}$ | $-5\times10^{-3}$ | $-5\times10^{-2}$ |
| 0.3 | −1.4017 | −1.3983 | −1.3659 | −1.1318 | 1.4024 | 1.4058 | 1.4411 | 2.0152 |
| 0.4 | −1.5075 | −1.5032 | −1.4622 | −1.1757 | 1.5085 | 1.5128 | 1.5579 | 2.3568 |
| 0.5 | −1.4348 | −1.4307 | −1.3914 | −1.1168 | 1.4357 | 1.4399 | 1.4831 | 2.2522 |
| 0.6 | −1.2847 | −1.2812 | −1.2484 | −1.0151 | 1.2854 | 1.2889 | 1.3249 | 1.9400 |
| 0.7 | −1.1211 | −1.1184 | −1.0925 | −0.9048 | 1.1217 | 1.1244 | 1.1525 | 1.6091 |
| 0.8 | −0.9719 | −0.9698 | −0.9499 | −0.8017 | 0.9723 | 0.9744 | 0.9960 | 1.3289 |
| 0.9 | −0.8451 | −0.8435 | −0.8281 | −0.7114 | 0.8455 | 0.8471 | 0.8636 | 1.1081 |
| 1.0 | −0.7405 | −0.7393 | −0.7273 | −0.6347 | 0.7408 | 0.7420 | 0.7548 | 0.9376 |

**Table S2.2** The influence of the geometrical non-linearity, i.e. the parameter $p/E^{(1)}$ on the values of $\sigma_{n\tau}/|p|$ in "co-phase" case for various $\rho$ under $\chi = 0.4$

| $p/E^{(1)}$ | $\rho$ | | | | | |
|---|---|---|---|---|---|---|
| | $\infty$ | 10 | 5 | 4 | 3 | 2.5 |
| $5\times10^{-5}$ | −0.5097 | −0.5366 | −0.6775 | −0.7738 | −1.0328 | −1.5075 |
| $5\times10^{-4}$ | −0.5087 | −0.5355 | −0.6757 | −0.7717 | −1.0299 | −1.5032 |
| $5\times10^{-3}$ | −0.4987 | −0.5242 | −0.6589 | −0.7515 | −1.0018 | −1.4622 |
| $5\times10^{-2}$ | −0.4262 | −0.4428 | −0.5394 | −0.6093 | −0.08053 | −1.1757 |
| $-5\times10^{-2}$ | 0.6952 | 0.7496 | 1.0119 | 1.1833 | 1.6150 | 2.3568 |
| $-5\times10^{-3}$ | 0.5218 | 0.5503 | 0.6981 | 0.7986 | 1.0674 | 1.5579 |
| $-5\times10^{-4}$ | 0.5110 | 0.5381 | 0.6796 | 0.7764 | 1.0364 | 1.5128 |
| $-5\times10^{-5}$ | 0.5100 | 0.5369 | 0.6779 | 0.7743 | 1.0335 | 1.5085 |

In the paper (Akbarov et al. 2005) the foregoing method, i.e. the method detailed in Chap. 6, was also employed for investigation of the stress distribution caused by a locally curved single fiber in an infinite elastic matrix. Moreover, in other papers (Akbarov et al. 2011; Cinar et al. 2010; Kosker and Cinar 2009) the foregoing method was employed for investigation of self-balanced stresses in an infinite elastic matrix containing two neighboring locally curved fibers. Note that these investigations were made by using the exponential Fourier transformation with respect to the coordinate $z$.

**Table S2.3** The influence of the geometrical non-linearity, i.e. the parameter $p/E^{(1)}$ on the values of $\sigma_{nn}/|p|$ in "anti-phase" case for various $\chi$ under $\rho = 2.5$

| $\chi$ | $p/E^{(1)}$ | | | | | | | |
|---|---|---|---|---|---|---|---|---|
| | Tension | | | | Compression | | | |
| | $5 \times 10^{-5}$ | $5 \times 10^{-4}$ | $5 \times 10^{-3}$ | $5 \times 10^{-2}$ | $-5 \times 10^{-5}$ | $-5 \times 10^{-4}$ | $-5 \times 10^{-3}$ | $-5 \times 10^{-2}$ |
| 0.3 | 2.1135 | 2.1117 | 2.0934 | 1.9259 | −2.1140 | −2.1158 | −2.1345 | −2.3404 |
| 0.4 | 2.4453 | 2.4418 | 2.4071 | 2.1063 | −2.4461 | −2.4496 | −2.4854 | −2.9074 |
| 0.5 | 2.4791 | 2.4742 | 2.4261 | 2.0275 | −2.4802 | −2.4851 | −2.5355 | −3.1707 |
| 0.6 | 2.2656 | 2.2602 | 2.2072 | 1.7823 | −2.2668 | −2.2723 | −2.3283 | −3.0743 |
| 0.7 | 1.9218 | 1.9167 | 1.8670 | 1.4761 | −1.9229 | −1.9281 | −1.9809 | −2.7064 |
| 0.8 | 1.5555 | 1.5512 | 1.5091 | 1.1804 | −1.5565 | −1.5608 | −1.6056 | −2.2278 |
| 0.9 | 1.2277 | 1.2242 | 1.1906 | 0.9274 | −1.2284 | −1.2319 | −1.2677 | −1.7641 |
| 1.0 | 0.9588 | 0.9561 | 0.9299 | 0.7238 | −0.9594 | −0.9621 | −0.9899 | −1.3725 |

**Table S2.4** The influence of the geometrical non-linearity, i.e. the parameter $p/E^{(1)}$ on the values of $\sigma_{nn}/|p|$ in "anti-phase" case for various $\rho$ under $\chi = 0.4$

| $p/E^{(1)}$ | $\rho$ | | | | | |
|---|---|---|---|---|---|---|
| | $\infty$ | 10 | 5 | 4 | 3 | 2.5 |
| $5 \times 10^{-5}$ | 0.5774 | 0.6069 | 0.8585 | 1.0784 | 1.6388 | 2.4453 |
| $5 \times 10^{-4}$ | 0.5757 | 0.6050 | 0.8561 | 1.0757 | 1.6356 | 2.4418 |
| $5 \times 10^{-3}$ | 0.5590 | 0.5874 | 0.8330 | 1.0493 | 1.6042 | 2.4071 |
| $5 \times 10^{-2}$ | 0.4363 | 0.4570 | 0.6560 | 0.8421 | 1.3444 | 2.1063 |
| $-5 \times 10^{-2}$ | −0.8856 | −0.9288 | −1.2452 | −1.4955 | −2.0911 | −2.9074 |
| $-5 \times 10^{-3}$ | −0.5977 | −0.6283 | −0.8862 | −1.1097 | −1.6756 | −2.4854 |
| $-5 \times 10^{-4}$ | −0.5796 | −0.6091 | −0.8614 | −1.0817 | −1.6427 | −2.4496 |
| $-5 \times 10^{-5}$ | −0.5778 | −0.6073 | −0.8590 | −1.0790 | −1.6395 | −2.4461 |

## S2.2 Numerical Results Related to the Case Where an Infinite Matrix Contains Periodically Located Row of Fibers

The results which will be detailed in present section are obtained in papers (Akbarov et al. 2004b, 2006, 2010b). The corresponding investigations were carried out within the scope of the assumption and notation accepted in corresponding section of Chap. 6.

As in Sect. S2.1, we select for consideration the cases where the initial imperfections of the fibers are co-phase (Fig. S2.5a) and anti-phase (Fig. S2.5b) and we call these cases briefly "co-phase" and "anti-phase" ones respectively below. Note that the schema shown in Fig. S2.5a is a part of that shown in Fig. 6.11.

In the selected cases it is assumed that the middle lines of the fibers are in the $x_{1k} = 0$ plane and equations of these lines in "co-phase" case are given by

**Table S2.5** Convergence of the numerical results related to $\sigma_{nn}/|p|$ in "anti-phase" case via the number of the equations under $\rho = 2.5$ and $\chi = 0.3$

| $p/E^{(1)}$ | Number of equations | | | | | | |
|---|---|---|---|---|---|---|---|
| | 16 | 22 | 28 | 34 | 40 | 46 | 52 |
| $5 \times 10^{-5}$ | 1.4391 | 1.7569 | 1.9337 | 2.0276 | 2.0761 | 2.1009 | 2.1135 |
| $5 \times 10^{-4}$ | 1.4378 | 1.7554 | 1.9320 | 2.0258 | 2.0743 | 2.0991 | 2.1117 |
| $5 \times 10^{-3}$ | 1.4246 | 1.7400 | 1.9153 | 2.0082 | 2.0563 | 2.0809 | 2.0934 |
| $5 \times 10^{-2}$ | 1.3045 | 1.5992 | 1.7617 | 1.8476 | 1.8918 | 1.9144 | 1.9259 |
| $-5 \times 10^{-5}$ | −1.4394 | −1.7573 | −1.9341 | −2.0280 | −2.0765 | −2.1013 | −2.1140 |
| $-5 \times 10^{-4}$ | −1.4407 | −1.7588 | −1.9358 | −2.0297 | −2.0783 | −2.1032 | −2.1158 |
| $-5 \times 10^{-3}$ | −1.4542 | −1.7746 | −1.9529 | −2.0477 | −2.0967 | −2.1217 | −2.1345 |
| $-5 \times 10^{-2}$ | −1.6040 | −1.9482 | −2.1418 | −2.2452 | −2.2989 | −2.3264 | −2.3404 |

expression (6.71). But in "anti-phase" case the mentioned middle lines are determined by equations

$$x_{2k} = \varepsilon(-1)^k f(x_3), \; x_{1k} = 0, \; f(x_3) = \ell \sin(\alpha_1 x_3),$$
$$\varepsilon = \frac{L}{\ell}, \; \alpha_1 = \frac{2\pi}{\ell}, \; L \ll \ell. \tag{S2.3}$$

The solution procedure of the corresponding boundary value problem for "co-phase" case was already described in Sect. 6.4.1. Almost the same procedures are also employed for solution to the problem related to "anti-phase" case and the equation

$$Y_n^{(1)0} + \sum_{\nu=0}^{N_\nu} (-1)^\nu Y_\nu^{(1)0} \left( \sum_{k=1}^{N_k} F_{n\nu k}^{(1)0} \right) + Y_n^{(2)0} F_n^{(2)0} = 2\pi\delta_n^3 (\sigma_{zz}^{(1),0} - \sigma_{zz}^{(2),0}),$$
$$n = 0, 1, 2, \ldots, N_\nu \tag{S2.4}$$

is obtained instead of Eq. (6.86). Note that under obtaining Eq. (S2.4) the periodicity condition

$$\left\{ \begin{matrix} u_i^{(1)} \\ \sigma_{ij}^{(1)} \end{matrix} \right\} (x_1, x_2, x_3) = (-1)^j \left\{ \begin{matrix} u_i^{(1)} \\ \sigma_{ij}^{(1)} \end{matrix} \right\} (x_1, x_2 \pm jR_{12}, x_3) \tag{S2.5}$$

is used instead of (6.72).

We consider numerical results related to the distribution of the self-balanced stresses

$$\sigma_{nn} = \varepsilon \sigma_{rr}^{(1),1} \Big|_{r_0 = R}, \; \sigma_{n\tau} = \varepsilon \sigma_{r\theta}^{(1),1} \Big|_{r_0 = R}. \tag{S2.6}$$

Taking the problem symmetry into account the values of the stresses (S2.6) at points $N_1$ (for $\sigma_{nn}$) and $N_2$ (for $\sigma_{n\tau}$) on the $S_0$ surface (Fig. S2.5) only will be considered. Because absolute maximum values of these stresses appear at these points coordinates of which are determined from Eq. (6.12) rewritten for the

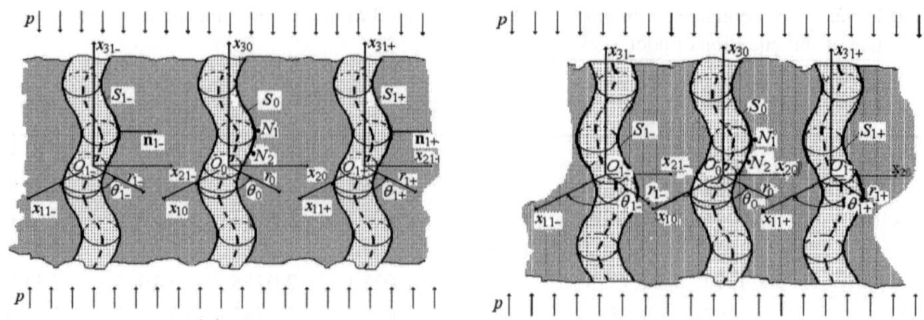


**Fig. S2.5** The geometry of material structure and chosen points on surface $S_0$: **a** "co-phase" and **b** "anti-phase" cases

surface $S_0$ under $\{\alpha_1 t_3 = \pi/2, \theta = \pi/2\}$ (for the point $N_1$) and $\{\alpha_1 t_3 = 0, \theta = \pi/2\}$ (for the point $N_2$).

Assume that the materials of the fibers and matrix are pure elastic and through $E^{(1)}$ and $\nu^{(1)}$ we denote modulus of elasticity and Poisson ratio of the matrix material. Moreover, we use the notation accepted in the previous section and assume that $\nu^{(2)} = \nu^{(1)} = 0.3$ and $\varepsilon = 0.015$. According to the results detailed in Sect. S2.1, in "co-phase" case ("anti-phase" case) location of the two fibers the shear (the normal) stress $\sigma_{n\tau}(\sigma_{nn})$ has dominating values. Note that similar results take place also for the problem under consideration. Taking this statement into account we consider the influence of the problem parameters on the values of $\sigma_{n\tau}$ only in "co-phase" case. But in "anti-phase" case we investigate this influence for the normal stress $\sigma_{nn}$ only.

Thus, consider the graphs given in Figs. S2.6, S2.7 and S2.8 which show the dependencies between $\sigma_{n\tau}/p$ and $\chi$ for $\rho$ =2.1, 2.5 and 5, respectively, for various suitable values of the parameter $p/E^{(1)}$ under $E^{(2)}/E^{(1)} = 50$. Note that the graphs of the dependencies between $\sigma_{nn}/p$ and $\chi$ with the same sequences of the problem parameters are given in Figs. S2.9, S2.10 and S2.11. In these figures the graphs denoted by letters (a) and (b) correspond to the tension and compression of the considered body, respectively. Note that under consideration of compression we assume that $|p|/E^{(1)} < \min|p_{cr}|/E^{(1)}$. The values of $|p_{cr}|/E^{(1)}$ are determined in Chap. 6.

The numerical results given in the foregoing figures show that the dependencies among $\sigma_{n\tau}/p$, $\sigma_{nn}/p$ and $\chi$, as the corresponding ones obtained in the previous section for two neighboring fibers, have non-monotonic character, i.e. there is such value of the parameter $\chi$ (denote it by $\chi*$) under which the absolute values of the considered stresses have its absolute maximum. It follows from the numerical results, the values of $\chi*$ decrease with increasing $\rho$, i.e. with increasing the distance between the two neighboring fibers. Absolute maximum values of $\sigma_{n\tau}/p$ in "co-phase" case, and absolute maximum values of $\sigma_{nn}/p$ in "anti-phase" case increase with decreasing $\rho$. In this case, as a result of the geometrical non-linearity

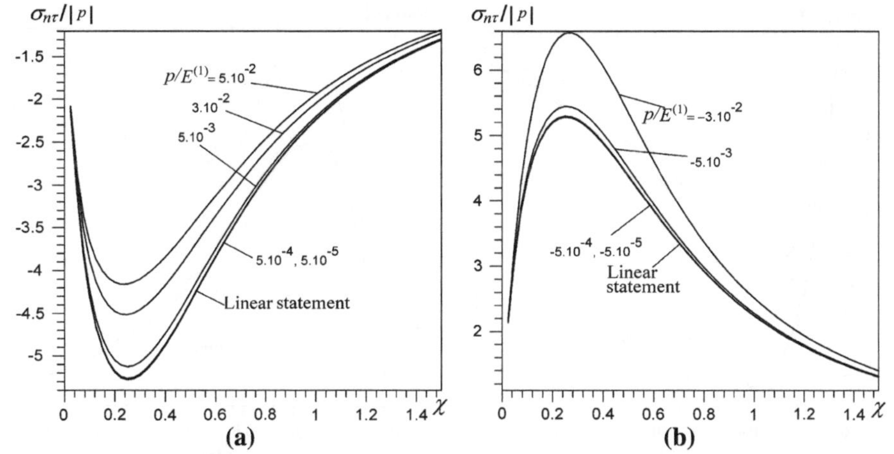


**Fig. S2.6** The graphs of the dependencies between $\sigma_{n\tau}/p$ and parameter $\chi$ for various values of $p/E^{(1)}$ under $\rho = 2.1$: **a** tension; **b** compression

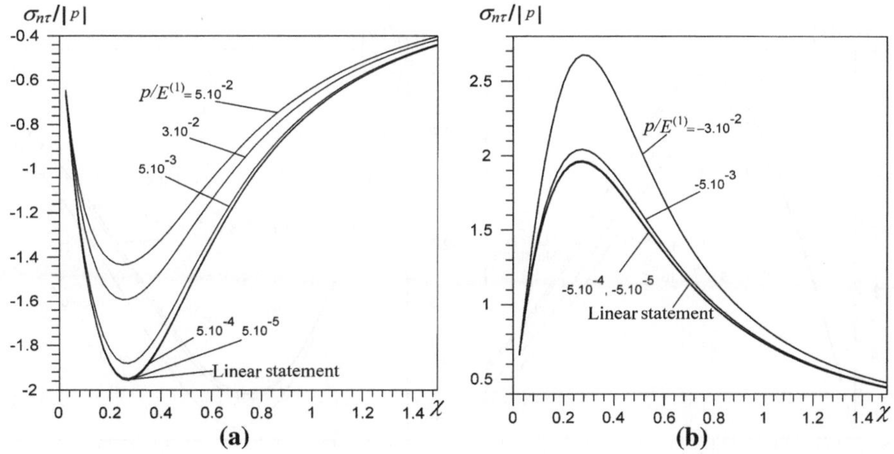


**Fig. S2.7** The graphs of the dependencies between $\sigma_{n\tau}/p$ and parameter $\chi$ for various values of $p/E^{(1)}$ under $\rho = 2.5$: **a** tension; **b** compression

the absolute values of $\sigma_{n\tau}/p$ and $\sigma_{nn}/p$ decrease (increase) under tension (compression) with the parameter $p/E^{(1)}$. It follows clearly from the foregoing numerical results that the maximum effect of the influence of the geometrical non-linearity arise for the cases where $\chi = \chi*$.

Moreover, this effect increases with increasing $\rho$. Note that the numerical results obtained under compression and tension in the case where $p/E^{(1)} \sim O(10^{-5})$ coincide with each other and with the corresponding ones obtained in the framework of the classical linear theory of elasticity. Moreover,

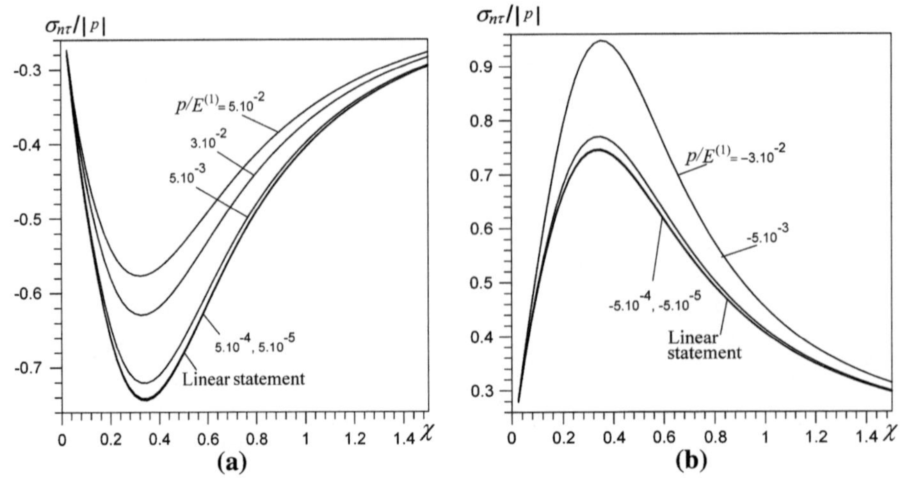


**Fig. S2.8** The graphs of the dependencies between $\sigma_{n\tau}/p$ and parameter $\chi$ for various values of $p/E^{(1)}$ under $\rho = 5.0$: **a** tension; **b** compression

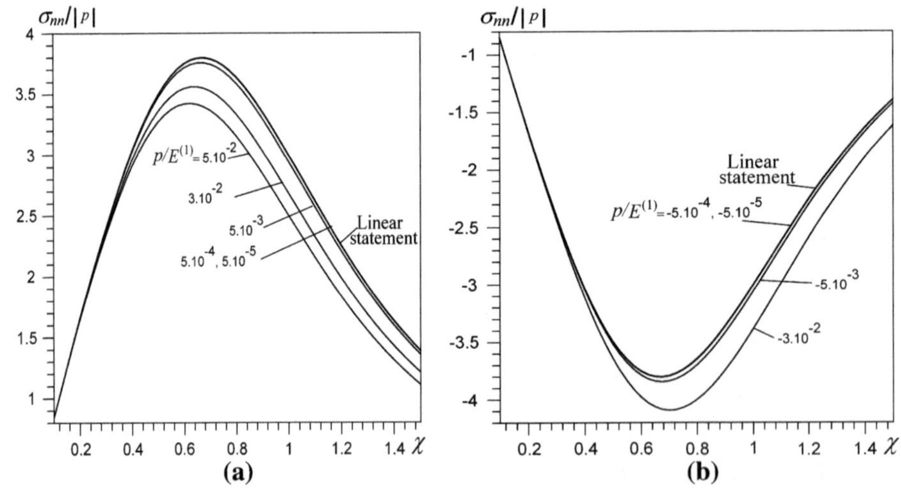


**Fig. S2.9** The graphs of the dependencies between $\sigma_{nn}/p$ and parameter $\chi$ for various values of $p/E^{(1)}$ under $\rho = 2.1$: **a** tension; **b** compression

with increasing $\rho$ and decreasing $p/E^{(1)}$ these results approach to the corresponding ones obtained for a single periodically curved fiber and treated in the monograph (Akbarov and Guz 2000). This situation also agrees well with the mechanical consideration and confirms the trustiness of the algorithm and programs used in the present numerical investigations.

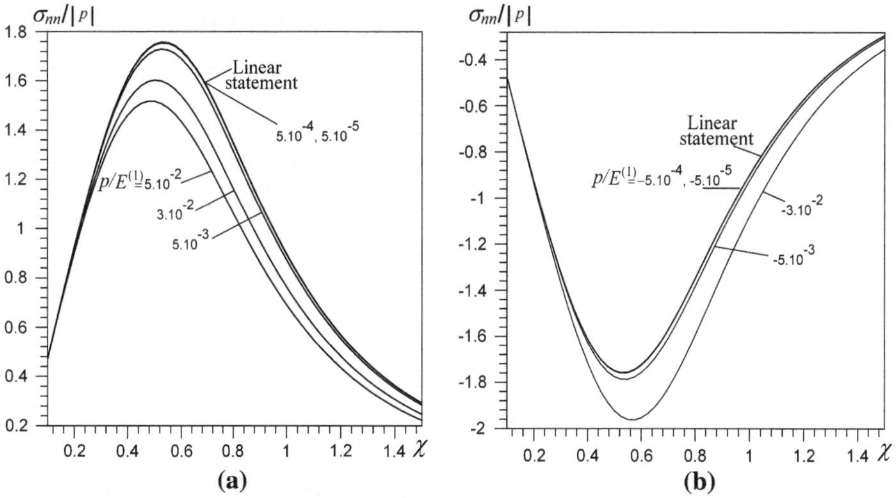


**Fig. S2.10** The graphs of the dependencies between $\sigma_{nn}/p$ and parameter $\chi$ for various values of $p/E^{(1)}$ under $\rho = 2.5$: **a** tension; **b** compression

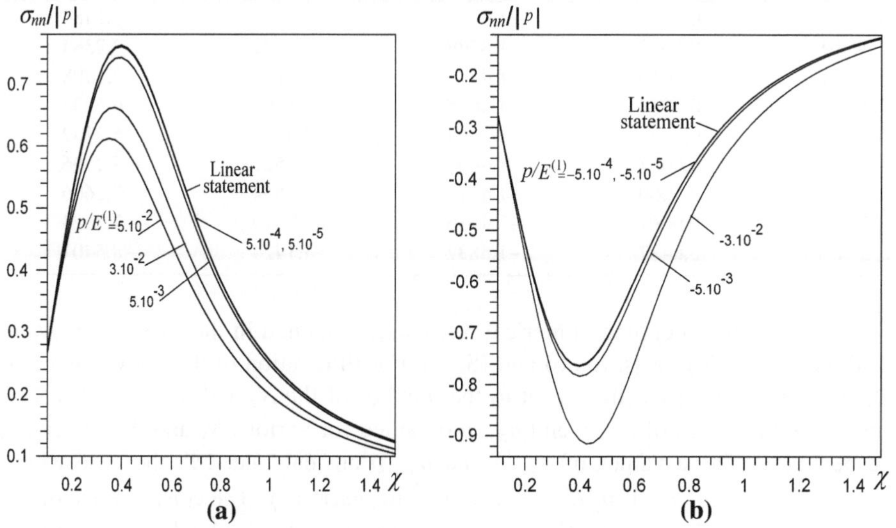


**Fig. S2.11** The graphs of the dependencies between $\sigma_{nn}/p$ and parameter $\chi$ for various values of $p/E^{(1)}$ under $\rho = 5$: **a** tension; **b** compression

Consider Tables S2.6 and S2.7 which show the values of $\sigma_{n\tau}/p$ (for "co-phase" case) and $\sigma_{nn}/p$ (for "anti-phase" case) respectively obtained for various $p/E^{(1)}$ and $E^{(2)}/E^{(1)}$ under $\rho = 2.1$ and $\chi = \chi*$. It follows from these tables that the influence of the geometrical non-linearity on the values of the considered stresses increase with $E^{(2)}/E^{(1)}$.

**Table S2.6** The values of $\sigma_{n\tau}/|p|$ obtained in "co-phase" case for various values of $E^{(2)}/E^{(1)}$ and $p/E^{(1)}$ under $\rho = 2.1$

| $p/E^{(1)}$ | $E^{(2)}/E^{(1)}$ | | | |
|---|---|---|---|---|
| | 10 ($\chi = 0.4$) | 20 ($\chi = 0.4$) | 50 ($\chi = 0.3$) | 100 ($\chi = 0.2$) |
| $5 \times 10^{-5}$ | −1.1545 | −2.2484 | −5.2264 | −5.1913 |
| $5 \times 10^{-4}$ | −1.1540 | −2.2459 | −5.2119 | −5.1780 |
| $5 \times 10^{-3}$ | −1.1485 | −2.2214 | −5.0731 | −5.0501 |
| $3 \times 10^{-2}$ | −1.1208 | −2.1020 | −4.4581 | −4.4773 |
| $5 \times 10^{-2}$ | −1.1015 | −2.0232 | −4.0995 | −4.1386 |
| $-5 \times 10^{-5}$ | 1.1546 | 2.2490 | 5.2296 | 5.1942 |
| $-5 \times 10^{-4}$ | 1.1552 | 2.2515 | 5.2442 | 5.2077 |
| $-5 \times 10^{-3}$ | 1.1609 | 2.2774 | 5.3969 | 5.3476 |
| $-3 \times 10^{-2}$ | 1.2287 | 2.6183 | 8.0753 | 7.7063 |

**Table S2.7** The values of $\sigma_{nn}/|p|$ obtained in "co-phase" case for various values of $E^{(2)}/E^{(1)}$ and $p/E^{(1)}$ under $\rho = 2.1$

| $p/E^{(1)}$ | $E^{(2)}/E^{(1)}$ | | | |
|---|---|---|---|---|
| | 10 ($\chi* = 0.8$) | 20 ($\chi* = 0.8$) | 50 ($\chi* = 0.6$) | 100 ($\chi* = 0.6$) |
| $5 \times 10^{-5}$ | 0.6056 | 1.4401 | 3.7514 | 7.2484 |
| $5 \times 10^{-4}$ | 0.6052 | 1.4389 | 3.7482 | 7.2384 |
| $5 \times 10^{-3}$ | 0.6015 | 1.4270 | 3.7162 | 7.1398 |
| $3 \times 10^{-2}$ | 0.5818 | 1.3638 | 3.5467 | 6.6336 |
| $5 \times 10^{-2}$ | 0.5667 | 1.3164 | 3.4204 | 6.2732 |
| $-5 \times 10^{-5}$ | −0.6057 | −1.4404 | −3.7521 | −7.2506 |
| $-5 \times 10^{-4}$ | −0.6060 | −1.4416 | −3.7554 | −7.2606 |
| $-5 \times 10^{-3}$ | −0.6097 | −1.4536 | −3.7879 | −7.3623 |
| $-3 \times 10^{-2}$ | −0.6487 | −1.5837 | −4.1424 | −8.5407 |

Note that the foregoing numerical results are obtained in the case where $N_\nu = 130$ and $N_q = 17$ in Eqs. (6.86) and (S2.4). For illustration of the convergence of the numerical results with respect to the number of the $N_\nu$ and $N_q$ in Tables S2.8 and S2.9 the values of $\sigma_{n\tau}/p$ and $\sigma_{nn}/p$ obtained for various $N_\nu$ and $N_q$ are given, respectively. These values of $\sigma_{n\tau}/p$ and $\sigma_{nn}/p$ are calculated in the case where $E^{(2)}/E^{(1)} = 50$, $\rho = 2.1, p/E^{(1)} = 5 \times 10^{-2}$ for various $\chi$. The convergence of the results obtained for various $N_\nu$ and $N_q$ confirms that the used solution method is also highly effective in the convergence sense.

Thus, from the analyses of the foregoing results the following conclusions are derived:

- in "anti-phase" ("co-phase") case as a result of the interaction of the fibers the normal stress (shear stress) acting on the interface surface between the fiber and matrix increases with a decrease in the distance between the fibers from one another;
- as a result of the geometrical non-linearity, the absolute values of the considered stresses increase in compression but decrease in tension;

**Table S2.8** The values of $\sigma_{n\tau}/|p|$ obtained in the co-phase curving case for various values of $N_q$ and $N_v$ in Eq. (6.88) in the case where $E^{(2)}/E^{(1)} = 50$, $\rho = 2.1$, $p/E^{(1)} = 5 \times 10^{-2}$

| $\chi$ | $N_q$ ($N_v = 130$) | | | | |
|---|---|---|---|---|---|
| | 13 | 14 | 15 | 16 | 17 |
| 0.1 | −3.6196 | −3.6366 | −3.6496 | −3.6595 | −3.6671 |
| 0.2 | −4.1381 | −4.1384 | −4.1386 | −4.1386 | −4.1386 |
| 0.3 | −4.0997 | −4.0996 | −4.0996 | −4.0995 | −4.0995 |
| 0.4 | −3.8435 | −3.8435 | −3.8435 | −3.8435 | −3.8435 |
| | $N_v$ ($N_q = 17$) | | | | |
| | 82 | 94 | 106 | 118 | 130 |
| 0.1 | −3.6280 | −3.6482 | −3.6587 | −3.6643 | −3.6671 |
| 0.2 | −4.0944 | −4.1173 | −4.1291 | −4.1354 | −4.1386 |
| 0.3 | −4.0558 | −4.0784 | −4.0901 | −4.0963 | −4.0995 |
| 0.4 | −3.8026 | −3.8237 | −3.8347 | −3.8405 | −3.8435 |

**Table S2.9** The values of $\sigma_{nn}/|p|$ obtained in "anti-phase" case for various values of $N_q$ and $N_v$ in Eq. (6.91) in the case where $E^{(2)}/E^{(1)} = 50$, $\rho = 2.1$, $p/E^{(1)} = 5 \times 10^{-2}$

| $\chi$ | $N_q$ ($N_v = 136$) | | | | |
|---|---|---|---|---|---|
| | 16 | 17 | 18 | 19 | 20 |
| 0.1 | 0.8421 | 0.8455 | 0.8427 | 0.8449 | 0.8432 |
| 0.2 | 1.6430 | 1.6437 | 1.6432 | 1.6436 | 1.6433 |
| 0.3 | 2.3488 | 2.3489 | 2.3488 | 2.3489 | 2.3488 |
| 0.4 | 2.9069 | 2.9069 | 2.9069 | 2.9069 | 2.9069 |
| | $N_v$ ($N_q = 20$) | | | | |
| | 88 | 100 | 112 | 124 | 136 |
| 0.1 | 0.8324 | 0.8381 | 0.8410 | 0.8424 | 0.8432 |
| 0.2 | 1.6224 | 1.6335 | 1.6391 | 1.6419 | 1.6433 |
| 0.3 | 2.3189 | 2.3347 | 2.3427 | 2.3468 | 2.3488 |
| 0.4 | 2.8699 | 2.8895 | 2.8994 | 2.9044 | 2.9069 |

- the maximum effect of the geometrical non-linearity to the stresses arises under certain values of $\chi$ $(= 2\pi R/\ell)$(where $R$ is the radius of the fiber cross section and $\ell$ is the length of the imperfection mode);
- the effect of the geometrical non-linearity on the considered stresses increases with $E^{(2)}/E^{(1)}$, where $E^{(2)}(E^{(1)})$ is Young's modulus of the fiber's (matrix's) material;
- the aforementioned effect of the geometrical non-linearity also increases with an increase in the distance between the neighboring fibers;
- the numerical results obtained agree well with well-known mechanical considerations and in particular cases coincide with the corresponding known results.

## References

Akbarov SD, Guz AN (2000) Mechanics of curved composites. Kluwer Academic Pubishers, Dortrecht
Akbarov SD, Kosker R (2003a) Stress distribution caused by antiphase periodical curving of two neighboring fibers in a composite materials. Eur J Mech A Solids 22:243–256
Akbarov SD, Kosker R (2003b) On the stress analyses in the infinite elastic body with two neighbouring curved fibers. Compos B Eng 34(2):143–150
Akbarov SD, Turan A (2009) Mathematical modeling and the study on influence of initial stresses on the SIF and ERR at the crack tips in a plate-strip of orthotropic material. Appl Math Mod 33(9):3682–3692
Akbarov SD, Turan A (2011) On the energy release rate at the crack tips in a finite pre-strained strip. CMC Comput Mater Continua 24(3):257–270
Akbarov SD, Yahnioglu N, Yucel AM (2004a) On the influence of the initial tension of a strip with a rectangular hole on the stress concentration caused by additional loading. J Strain Anal 39(6):615–624
Akbarov SD, Kosker R, Ucan Y (2004b) Stress distribution in an elastic body with a periodically curved row of fibers. Mech Compos Mater 40(3):191–202
Akbarov SD, Kosker R, Simsek K (2005) Stress distribution in an infinite elastic body with a locally curved fiber in a geometrically nonlinear statement. Mech Compos Mater 41(4):291–302
Akbarov SD, Kosker R, Ucan Y (2006) Stress distribution in a composite material with the row of anti-phase periodically curved fibers. Int Appl Mech 42(4):486–493
Akbarov SD, Yahnioglu N, Babuşcu Yesil U (2008) Interaction between two neighboring circular holes in a pre-stretched simply supported orthotropic strip under bending. Mech Compos Mater 44(6):827–838
Akbarov SD, Yahnioglu N, Babuscu Yesil U (2010a) Forced vibration of an initially stressed thick rectangular plate made of an orthotropic material with a cylindrical hole. Mech Compos Mater 46(3):287–298
Akbarov SD, Kosker R, Ucan Y (2010b) The effect of the geometrical non-linearity on the stress distribution in the infinite elastic body with a periodically curved row of fibers. CMC: Comput Mater Continua 17(2):77–102
Akbarov SD, Kosker R, Cinar NT (2011) Stress distribution in an infinite body containing two neighboring locally curved nanofibers. CMC: Comput Mater Continua 21(2):119–146
Cinar NT, Kosker R, Akbarov SD, Akat E (2010) Stress distribution caused by two neighboring out-of-plane locally cophasally curved fibers in a composite material. Mech Comp Mat 46(5):555–572
Corten HT (1967) Fracture of reinforcing plastics. In: Broutman LJ, Krock RH (eds) Modern composite materials. Addison-Wesley, Reading
Guz AN (1983) Brittle fracture mechanics of pre-stressed materials. Naukova Dumka, Kiev
Guz AN (2000) Description and study of some nonclassical problems of fracture mechanics and related mechanisms. Int Appl Mech 36(12):1537–1564
Guz AN, Dyshel MS, Nazarenko VM (2004) Fracture and stability of materials and structural members with cracks: approaches and results. Int Appl Mech 40(12):1323–1359
Kosker R, Akbarov SD (2003) Influence of the interaction between periodically curved fibers on the stress distribution in a composite material. Mech Compos Mater 39(2):165–176
Kosker R, Cinar NT (2009) Stress distribution in an infinite elastic body containing two neighboring locally curved fibers. Mech Compos Mater 45(3):315–330
Kosmodamianskii AS (2002) Accumulation of internal energy in multiply connected bodies. Int Appl Mech 38(4):399–422
Lei GH, Charles WWNg, Rigby DB (2001) Stress and displacement around an elastic artificial rectangular hole. J Eng Mech 127(9):880–890
Lekhnitski SG (1963) Theory of elasticity of an anisotropic body. Holden Day, San Francisko

Maksimyuk V, Mulyar VP, Chernyshenko I (2003) Stress state of thin spherical shells with an off-centre elliptic hole. Int Appl Mech 39(5):595–598

Muskhelishvili NI (1954) Some basic problems of the mathematical theory of elasticity, 4th edn (trans: Radok JRM). Noordhoff International Publishing, Leyden

Savin GN (1951) Stress concentration around holes. Pergamon Press, Oxford

Savin GN (1968) Stress distribution around holes. Naukova Dumka, Kiev (in Russian)

Sih G (1973) Handbook of stress intensity factors. Lehigh University, Bethlehem

# Index



S. D. Akbarov, *Stability Loss and Buckling Delamination*,
DOI: 10.1007/978-3-642-30290-9, © Springer-Verlag Berlin Heidelberg 2013

## D

The EU authorised representative is:
Springer Nature Customer Service Center GmbH
Europaplatz 3, 69115 Heidelberg, Germany
Printed by Libri Plureos GmbH
in Hamburg, Germany

MIX
Papier aus verantwortungsvollen Quellen
Paper from responsible sources
FSC® C105338

If you have any concerns about our products,
you can contact us on
**ProductSafety@springernature.com**

In case Publisher is established outside the EU,
the EU authorized representative is:
**Springer Nature Customer Service Center GmbH**
**Europaplatz 3, 69115 Heidelberg, Germany**

Printed by Libri Plureos GmbH
in Hamburg, Germany